D1189915

TECHNIQUES OF CHEMISTRY

ARNOLD WEISSBERGER, *Editor*

VOLUME I

PHYSICAL METHODS OF CHEMISTRY

PART IIID

X-Ray, Nuclear, Molecular Beam, and Radioactivity Methods

TECHNIQUES OF CHEMISTRY

ARNOLD WEISSBERGER, *Editor*

VOLUME I

PHYSICAL METHODS OF CHEMISTRY, in Five Parts
(INCORPORATING FOURTH COMPLETELY REVISED AND AUGMENTED EDITION OF PHYSICAL METHODS OF ORGANIC CHEMISTRY)
Edited by Arnold Weissberger and Bryant W. Rossiter

VOLUME II

ORGANIC SOLVENTS, Third Edition
John A. Riddick and William S. Bunger

VOLUME III

PHOTOCHROMISM
Edited by Glenn H. Brown

TECHNIQUES OF CHEMISTRY

VOLUME I

PHYSICAL METHODS OF CHEMISTRY

INCORPORATING FOURTH COMPLETELY REVISED AND AUGMENTED EDITION OF TECHNIQUE OF ORGANIC CHEMISTRY, VOLUME I, PHYSICAL METHODS OF ORGANIC CHEMISTRY

Edited by

ARNOLD WEISSBERGER

AND

BRYANT W. ROSSITER

Research Laboratories
Eastman Kodak Company
Rochester, New York

PART III

Optical, Spectroscopic, and Radioactivity Methods

PART IIID

X-Ray, Nuclear, Molecular Beam, and Radioactivity Methods

WILEY-INTERSCIENCE

A DIVISION OF JOHN WILEY & SONS, INC.

New York · London · Sydney · Toronto

Library of Congress Catalog Card Number: 45-8533

ISBN 0-471-92733-3

Printed in the United States of America.

10 9 8 7 6 5 4 3 2 1

PLAN FOR

PHYSICAL METHODS OF CHEMISTRY

PART I

Components of Scientific Instruments, Automatic Recording and Control, Computers in Chemical Research

PART II

Electrochemical Methods

PART III

Optical, Spectroscopic, and Radioactivity Methods

PART IV

Determination of Mass, Transport, and Electrical-Magnetic Properties

PART V

Determination of Thermodynamic and Surface Properties

AUTHORS OF PART IIID

ROYAL G. ALBRIDGE

Department of Physics and Astronomy, Vanderbilt University, Nashville, Tennessee

ISADOR AMDUR (*deceased*)

LAWRENCE S. BARTELL

Department of Chemistry, University of Michigan, Ann Arbor, Michigan

HERBERT M. CLARK

Department of Physical and Nuclear Chemistry, Rensselaer Polytechnic Institute, Troy, New York

JAMES H. GREEN

New England Institute, Ridgefield, Connecticut

VINCENT P. GUINN

Activation Analysis Program, Gulf General Atomic, Inc., San Diego, California

WALTER C. HAMILTON

Department of Chemistry, Brookhaven National Laboratory, Associated Universities, Inc., Upton, New York

JULIUS M. HASTINGS
Department of Chemistry, Brookhaven National Laboratory, Associated Universities, Inc., Upton, New York

YEHONATHAN HAZONY
School of Engineering and Applied Science, Princeton University, Princeton, New Jersey

ROLFE H. HERBER
Department of Chemistry, Rutgers University, New Brunswick, New Jersey

ROBERT A. JACOBSON
Institute for Atomic Research, Iowa State University, Ames, Iowa

JOHN E. JORDAN
Department of Chemistry, Massachusetts Institute of Technology, Cambridge, Massachusetts

WILLIAM N. LIPSCOMB
Department of Chemistry, Harvard University, Cambridge, Massachusetts

EDWARD A. MASON
Department of Chemistry, Brown University, Providence, Rhode Island

JOSEPH A. MERRIGAN
Research Laboratories, Eastman Kodak Company, Rochester, New York

JULIAN M. NIELSEN
Radiological Sciences Department, Battelle Memorial Institute, Pacific Northwest Laboratories, Richland, Washington

SHU-JEN TAO
New England Institute, Ridgefield, Connecticut

NEW BOOKS AND NEW EDITIONS OF BOOKS OF THE TECHNIQUE OF ORGANIC CHEMISTRY SERIES WILL NOW APPEAR IN TECHNIQUES OF CHEMISTRY. A LIST OF PRESENTLY PUBLISHED VOLUMES IS GIVEN BELOW.

TECHNIQUE OF ORGANIC CHEMISTRY
ARNOLD WEISSBERGER, *Editor*

Volume I: Physical Methods of Organic Chemistry
Third Edition—in Four Parts

Volume II: Catalytic, Photochemical, and Electrolytic Reactions
Second Edition

Volume III: Part I. Separation and Purification
Part II. Laboratory Engineering
Second Edition

Volume IV: Distillation
Second Edition

Volume V: Adsorption and Chromatography

Volume VI: Micro and Semimicro Methods

Volume VII: Organic Solvents
Second Edition

Volume VIII: Investigation of Rates and Mechanisms of Reactions
Second Edition—in Two Parts

Volume IX: Chemical Applications of Spectroscopy
Second Edition—in Two Parts

Volume X: Fundamentals of Chromatography

Volume XI: Elucidation of Structures by Physical and Chemical Methods
In Two Parts

Volume XII: Thin-Layer Chromatography

Volume XIII: Gas Chromatography

Volume XIV: Energy Transfer and Organic Photochemistry

INTRODUCTION TO THE SERIES

Techniques of Chemistry is the successor to the Technique of Organic Chemistry Series and its companion—Technique of Inorganic Chemistry. Because many of the methods are employed in all branches of chemical science, the division into techniques for organic and inorganic chemistry has become increasingly artificial. Accordingly, the new series reflects the wider application of techniques, and the component volumes for the most part provide complete treatments of the methods covered. Volumes in which limited areas of application are discussed can be easily recognized by their titles.

Like its predecessors, the series is devoted to a comprehensive presentation of the respective techniques. The authors give the theoretical background for an understanding of the various methods and operations and describe the techniques and tools, their modifications, their merits and limitations, and their handling. It is hoped that the series will contribute to a better understanding and a more rational and effective application of the respective techniques.

Authors and editors hope that readers will find the volumes in this series useful and will communicate to them any criticisms and suggestions for improvements.

Research Laboratories
Eastman Kodak Company
Rochester, New York

ARNOLD WEISSBERGER

PREFACE

Physical Methods of Chemistry succeeds, and incorporates the material of, three editions of *Physical Methods of Organic Chemistry* (1945, 1949, and 1959). It has been broadened in scope to include physical methods important in the study of all varieties of chemical compounds. Accordingly, it is published as Volume I of the new Techniques of Chemistry series.

Some of the methods described in *Physical Methods of Chemistry* are relatively simple laboratory procedures, such as weighing and the measurement of temperature or refractive index and determination of melting and boiling points. Other techniques require very sophisticated apparatus and specialists to make the measurements and to interpret the data; x-ray diffraction, mass spectrometry, and nuclear magnetic resonance are examples of this class. Authors of chapters describing the first class of methods aim to provide all information that is necessary for the successful handling of the respective techniques. Alternatively, the aim of authors treating the more sophisticated methods is to provide the reader with a clear understanding of the basic theory and apparatus involved, together with an appreciation for the value, potential, and limitations of the respective techniques. Representative applications are included to illustrate these points, and liberal references to monographs and other scientific literature providing greater detail are given for readers who want to apply the techniques. Still other methods that are successfully used to solve chemical problems range between these examples in complexity and sophistication and are treated accordingly. All chapters are written by specialists. In many cases authors have acquired a profound knowledge of the respective methods by their own pioneering work in the use of these techniques.

In the earlier editions of *Physical Methods* an attempt was made to arrange the chapters in a logical sequence. In order to make the organization of the treatise lucid and helpful to the reader, a further step has been taken in the new edition—the treatise has been subdivided into technical families and parts:

Part I Components of Scientific Instruments, Automatic Recording and Control, Computers in Chemical Research
Part II Electrochemical Methods

Part III Optical, Spectroscopic, and Radioactivity Methods
Part IV Determination of Mass, Transport, and Electrical-Magnetic Properties
Part V Determination of Thermodynamic and Surface Properties

This organization into technical families provides more consistent volumes and should make it easier for the reader to obtain from a library or purchase at minimum cost those parts of the treatise in which he is most interested.

The more systematic organization has caused additional labor for the editors and the publisher. We hope that it is worth the effort. We thank the many authors who made it possible by adhering closely to the agreed dates of delivery of their manuscripts and who promptly returned their proofs. To those authors who were meticulous in meeting deadlines we offer our apologies for delays caused by late arrival of other manuscripts, in some cases necessitating rewriting and additions.

The changes in subject matter from the Third Edition are too numerous to list in detail. We thank previous authors for their continuing cooperation and welcome new authors to the series. New authors of Part III-D are Dr. R. G. Albridge, Dr. I. Amdur, Dr. L. S. Bartell, Dr. H. M. Clark, Dr. J. H. Green, Dr. V. C. Guinn, Dr. W. C. Hamilton, Dr. Y. Hazony, Dr. R. H. Herber, Dr. R. A. Jacobson, Dr. J. E. Jordan, Dr. E. A. Mason, Dr. J. A. Merrigan, Dr. J. M. Nielsen, and Dr. S. Tao.

The Chapter on "Mass Spectrometry" by Dr. Guy P. Arsenault was not received by the publication deadline. This chapter will appear in a supplemental volume, since it would have been unfair to the authors of the other chapters to delay publication of their work any longer.

We are grateful to the many colleagues who advised us in the selection of authors and helped in the evaluation of the manuscripts. These are, for Part III-D, Dr. Thomas Carlson, Mr. John A. Hamilton, Mrs. Ardelle Kocher, Dr. Edwin P. Przybylowicz, and Dr. Douglas L. Smith. Also, the help of Mrs. Donna S. Roets is gratefully acknowledged.

The editors express their deep sorrow upon learning of the untimely death of Professor Isador Amdur.

The senior editor expresses his gratitude to Bryant W. Rossiter for joining him in the work and taking on the very heavy burden with exceptional devotion and ability.

September 1971
Research Laboratories
Eastman Kodak Company
Rochester, New York

ARNOLD WEISSBERGER
BRYANT W. ROSSITER

CONTENTS

X-RAY, NUCLEAR, MOLECULAR BEAM, AND RADIOACTIVITY METHODS

Chapter 1

X-RAY CRYSTAL STRUCTURE ANALYSIS

William N. Lipscomb and Robert A. Jacobson

In Section 1 the reader is introduced to the general principles of a complete structure determination, and two specific methods of structure determination, symbolic addition and Patterson superposition, are chosen to illustrate in a simple way some of the methods that practicing crystallographers now use to solve complex structures. Many other methods are available, and a few of these are also described briefly here. Equations for the computation of amplitudes of scattering, the electron density, and the vector map are derived in Section 2. No mathematics beyond elementary calculus is assumed, and all the necessary properties of vectors, complex numbers, and Fourier series are presented in detail. Section 3 is a geometrical description of symmetry in crystals with emphasis on the symmetries most frequently encountered in complex crystals. This section owes very much to Professor J. H. Sturdivant, who kindly gave his permission for the use of his notes on symmetry which formed the basis from which Section 3 was written. In Section 4 the relation between the experimental methods and the preceding sections are outlined, and one complete structure determination is described in detail.

These sections are independent of one another and can be read separately. Section 1 is designed primarily for the reader who wishes to obtain a general idea of how complex structures are determined. The later sections elucidate different aspects of these methods of structure investigation, and they have been used during the past several years for notes in a course on X-ray crystallography.

I INTRODUCTION AND ILLUSTRATION OF SOME TYPICAL METHODS FOR THE DETERMINATION OF COMPLEX STRUCTURES*

Introduction

Some of the finest recent achievements of X-ray crystallography are undoubtedly the determinations of the structures of myogloblin [1], hemoglobin [2, 3], lysozyme [4], carboxypeptidase A [5], ribonuclease [6, 7], and other large biological molecules [8]. These structures are among the most complex elucidated at present and represent, of course, many man-years of effort. However, the routine determination of almost any complex structure containing say 40 or more atoms can be expected from any one of many crystallographic laboratories at the present time. The purpose of this chapter is to introduce the general principles of X-ray crystallography and to illustrate a few of the frequently used methods for the solution of such structures.

The complexity of structures solved by X-ray diffraction techniques has increased greatly since about 1950. New developments in the methods by

* For more detailed discussion of the topics in this section, the reader should consult C. W. Bunn, *Chemical Crystallography*, 2d ed., Oxford University Press, New York, 1961; K. Lonsdale, *Crystals and X-rays*, Van Nostrand, New York, 1949; J. M. Robertson, *Organic Crystals and Molecules*, Cornell University Press, Ithaca, N.Y., 1953; G. H. Stout and L. H. Jenson, *X-ray Structure Determination*, Macmillan, New York, 1968.

which the first approximately correct atomic arrangement (trial structure) in the unit cell of a previously unsolved structure is obtained have greatly increased the power of this technique. At present it is reasonable to expect a complete structure determination of nearly any molecule containing up to forty atoms (excluding hydrogens) with the expenditure of a week to 6 months of effort.

A structure that has been established by careful X-ray diffraction techniques is unlikely to be incorrect [9]. The crystallographer is in a very favorable position with respect to the ratio of observations to unknowns, which usually ranges from 6 up to 10 to 1 when complete data are obtained. For example, in the structure determination of di-μ-diphenylphosphinatoacetylacetonatochromium(III) [10] some 4,400 independent X-ray diffraction maxima were observed. The unknowns in this case are the x, y, z coordinates of the chromium, oxygen, phosphorus, and carbon atoms of which there are a total of 60 atoms, a scale parameter, six anisotropic temperature parameters for each of the chromium, oxygen, and phosphorus atoms, and single isotropic temperature parameters for each of the carbon atoms, giving 140 temperature parameters in all. The total number of parameters is thus 321.

Unfortunately the equations relating these 321 unknowns to the 4400 observations are not simple. For example, the 180 distance parameters are related to the observations by complicated trigonometric expressions. Hence it is usually necessary to combine mathematical methods with methods into which chemical reasonableness can be introduced. We describe in this introduction a few of the common methods of structure determination based either on the direct use of the intensities or on the analysis of the Patterson function, a three-dimensional map of all the vectors that can be drawn between pairs of atoms in the structure and then plotted on a common origin. This map can be computed directly from the observed data with no chemical assumptions. These methods are not the only ones available but in general have proved to be the ones of most generality, hence most widely used.

The final test of correctness of a structure is the agreement between the observed and the calculated intensities. How good this agreement is varies according to the structure, the methods used to acquire the data, and other things, but it should be comparable with that obtained in similar structural studies, which are in general readily available. The trial structure may be obtained by mathematical techniques involving few or no chemical assumptions or by other techniques involving intuition, guesses, and the background available in the "scientific method," but the test of obtaining good agreement with large numbers of observations with relatively few parameters is so severe that the results can usually be considered equally as good whatever the method.

The beginning stages of a structure determination are similar in all methods of attack; hence we start with a brief description of what the chemist can learn from the preliminary stages. Briefly, these stages are the identification from powder photographs or from a determination of the unit cell and molecular weight, the unit cell symmetry, and the complete structure determination.

Identification by Powder Photography

X-rays are very short wavelength electromagnetic radiations produced when L electrons fall into the K shell of elements of moderately high or high atomic number, when M electrons fall into the L shell of the elements of high atomic number, or when electrons are decelerated rapidly. These three types of radiation are called characteristic K, characteristic L, or white radiation, respectively. The vacancies necessary for these transitions to occur are produced when electrons are accelerated through a potential of about 50,000 V toward a water-cooled target of a heavy metal such as copper (1.54-Å radiation) or molybdenum (0.711-Å radiation). X-rays are not appreciably deviated by an ordinary lens system; hence pinholes are used to produce a collimated beam, or a self-focusing arrangement is used in which the source, a large area of sample, and the film or counter detector all lie on the circumference of a circle. If a flat sample is used in this way, absorption problems are usually negligible. However, if a cylindrical specimen of powder or a single crystal is used, the X-rays must pass through part of the sample, and excessive absorption of X-rays can occur if the sample is too large. For a sample of a given size the amount of reduction of intensity is proportional to the intensity and to the thickness of the sample, that is, $dI = -\mu I\, dx$, where μ is called the linear absorption coefficient. It may be calculated from $\mu = \rho[f_A\mu_A^{(m)} + f_B\mu_B^{(m)} + f_C\mu_C^{(m)} + \cdots]$, where ρ is the density of the sample and f_A is the weight fraction of element A having mass absorption coefficient $\mu_A^{(m)}$. Tables of $\mu_A^{(m)}$ can be found in *International Tables for X-ray Crystallography* (The Kynoch Press, Birmingham, England, 1952). Integration of this expression gives $I = I_0e^{-\mu x}$, where I is the resultant intensity for a sample of thickness x on which an initial intensity of I_0 falls. The intensity of diffracted radiation depends on I and on the total size of the specimen, which is proportional to x^2 for a cylindrical specimen. Hence $I_{\text{diff}} = KIx^2$. The maximum value of I_{diff} is obtained by setting $dI_{\text{diff}}/dx = KI_0\, d(x^2e^{-\mu x})/dx = 0$, from which the optimum size of sample is found to be $x = 2/\mu$. Typical values of $1/\mu$ are 0.5 mm for Cu$K\alpha$ radiation and 4 mm for Mo$K\alpha$ radiation for diamond, and 0.006 mm for Cu$K\alpha$ and 0.05 mm for Mo$K\alpha$ radiation for I_2. In order to avoid appreciable corrections for absorption it is necessary to keep the sample thickness to about $\frac{1}{4}$ or less of $1/\mu$. More often, however, one does not wish to sacrifice intensities of weak reflections; hence absorption corrections are made.

When K radiation is produced from a target, the $K\alpha$ doublet (resolved only at high diffraction angles) is the desired line. However an additional line ($K\beta$) is produced which can effectively be reduced by the use of special absorption filters. For example, a 10-μ ($1\ \mu = 10^{-4}$ cm) nickel filter will reduce the Cu$K\alpha$ line to about $\frac{1}{2}$ of its original intensity, while at the same time reducing the Cu$K\beta$ line to 1/85 of its original intensity or about 1/300 of the intensity of the reduced Cu$K\alpha$ intensity. For molybdenum radiation a zirconium filter of 80 to 100 μ is employed to obtain similar results.

A given diffraction maximum from a small crystal is quite sharply defined. It is intrinsically small because of the large number of unit cells, for reasons discussed below, but for practical purposes most of its width comes from divergence of the X-ray beam and from crystal imperfections. Its angle of deviation (2θ) from the direct beam of X-rays is given by the Bragg law, which arises from the condition that the path difference for a set of equivalent planes into which the structure is resolved shall be some integral number of wavelengths. If the spacing of these planes is called d'_{hkl} and if the integer is n (the order of the reflection), the relation is

$$n\lambda = 2d'_{hkl} \sin \theta.$$

The derivation of this equation and the definition of hkl, the three numbers describing the reflecting plane, are discussed in detail below. In a finely powdered sample all orientations of crystals can occur; thus a given plane diffracting in a given order, say $n = 1$, would give a diffraction maximum whenever it is oriented so that the Bragg law is satisfied for specular reflection. Thus all reflections from a given plane in a powdered sample would lie on a cone making an angle of 2θ with respect to its axis (Fig. 1.1). Because 2θ can have values from 0 to 180°, a cylindrical strip of film is ordinarily used instead of a flat plate (Fig. 1.2).

A powder sample that is not sufficiently randomly oriented or that is too coarse will give discrete spots instead of continuous powder lines. Hence it is usual to grind the sample until it passes through about a 300-mesh screen, which will give a crystallite size of about 40 μ or less. In addition, the sample

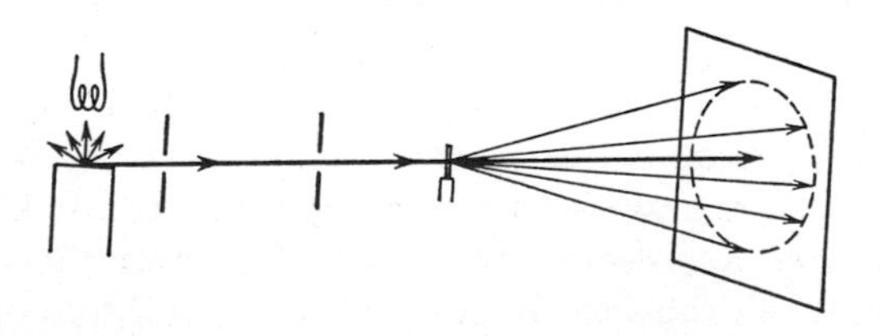

Fig. 1.1 Experimental arrangement showing a single powder line at an angle of deviation 2θ from the X-ray beam.

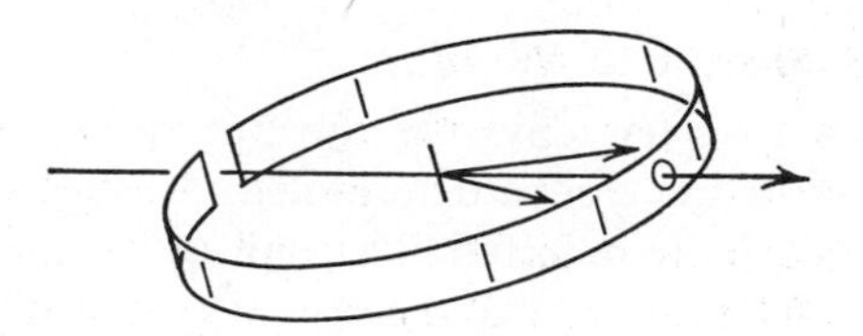

Fig. 1.2 Use of a cylindrical strip of film to record a powder pattern.

is usually rotated during the photography to increase the randomness of the orientation. Several acceptable techniques for mounting the sample include introduction into a 0.1 to 1.0-mm Pyrex tube, similar to a melting-point tube, or adhering the material to a Pyrex tube or rod.

A satisfactory film exposure or counter trace can usually be obtained in $\frac{1}{2}$ to 10 hours. A pattern of discrete sharp lines indicates that the material is crystalline; broadened lines indicate small particle size as discussed below; and diffuse halos indicate amorphous material, glassy material, or liquid crystals.

The method of X-ray powder photography is at its best when the identity of two samples is being tested. If the patterns are identical and if this identity is supported by other physical and chemical techniques, this method is one of the most powerful. For the usual crystalline compound both the positions and intensities of about 100 distinct lines are required to be identical; hence the test of identity is a severe one. If, on the other hand, the two samples give obviously different diffraction patterns, no positive statements can be made. Probably the two compounds are different, but there always remains the possibility that the same compound has merely crystallized in two different crystal structures (polymorphism). Samples as small as 10 to 100 μg of material will give satisfactory photographs, but the quality will usually be a bit better if approximately 1 mg of material is available. The sample is nearly always left unchanged; therefore recovery of the material is usually complete.

The solution of a complex structure from powder photographs alone is practically impossible, except in rare cases in which the investigator can guess the complete solution on structural chemical grounds. However, for the simpler inorganic compounds some investigators have been remarkably successful. Pauling's early work on the structures of minerals and Zachariasen's studies on the compounds of the thoride and actinide series are among the most extensive and impressive series of complete structure determination from powder photographs. Usually though the chemist must look to studies of single crystals if he wants more than a method of characterization or a comparison that tests the identity of two samples.

The Unit Cell and Molecular Weights

Within a few hours to a few days the dimensions of the unit cell of a previously unknown structure can be determined. A single crystal of dimensions 0.1 to 1.0 mm is usually required. The unit cell is the smallest unit of volume that shows the full symmetry of the crystal system and from which the complete three-dimensional arrangement of molecules is generated by translations along the axes. Excluding the cases, very rare among complex crystals, of translational disorder, the unit cell necessarily contains an integral number (n) of molecules of molecular weight M. From the unit-cell dimensions and angles the volume V can be obtained, as described below. An accurate determination of the density ρ of the crystal then yields nM, an integer times the molecular weight, that is, $\rho = nM/(NV)$, where N is Avogadro's number. If a rough measurement of the molecular weight is known, say from colligative property measurements, then the value of n can be found, and an accurate molecular weight can be determined from the equation above. The X-ray measurements of the unit-cell parameters can be made so precisely that the final accuracy of the molecular weight is limited only by the accuracy of the density determination, which in favorable cases is a few tenths of a percent. Actually, a few of the present atomic weights have been determined by this method.

For many compounds the internal symmetry of the unit cell frequently gives a lower limit to n, often 2 or 4, and the known molecular volumes of similar compounds can be used to give either the value of n or of an upper limit. Frequently these limits are the same, and then the molecular weight can be determined without aid of the cryoscopic method.

The Symmetry of the Unit Cell

The determination of the unit-cell size and its symmetry is the first stage of a complete structure determination. However, such a study often yields valuable information; hence it is frequently carried out even when a complete structural study is not contemplated. It usually takes a few days or less, and yields (*a*) the molecular weight as described above, (*b*) a method of characterizing the compound, which is more compact and more nearly unique than X-ray powder patterns, and (*c*) frequently, some information about molecular symmetry. A detailed discussion of symmetry of crystals is given below; hence only a few general conclusions are stated here. Molecular symmetry information from the symmetry of the unit cell is most often effective when the molecule has a center of symmetry. This molecular center of symmetry is very frequently used also as a center of symmetry in the crystal structure and can very often be established uniquely from the unit-cell information only. For example, many centrosymmetric molecules crystallize in the ubiquitous space group $P2_1/c$, described below, with two molecules in the unit cell. Establishment of this crystallographic information, in about 2 days from a single crystal 0.3 mm in size, then proves that the molecule has a center of symmetry, providing no disorder is present. (Orientational disorder is rare, and when it occurs it is often apparent from the abnormally rapid decline in intensities with increasing angle of scattering and from the presence of abnormal streaks of radiation associated with only partial order of neighboring molecules.) When four molecules are present in the unit cell of symmetry $P2_1/c$, no conclusion can be drawn about the molecular symmetry. However, when six molecules are present, at least two, hence presumably all, have centrosymmetric molecular symmetry. A very interesting example of this is biphenylene [11], for which a possible structure involving one six-membered and two five-membered rings was eliminated immediately by these very simple symmetry arguments. This study was also carried to the final stages and proved the correctness of the presently accepted structure with benzene rings fused to a four-membered ring in the centrosymmetric arrangement.

Unfortunately, with some exceptions, a molecule does not make use of other symmetry elements, such as a plane or an axis of symmetry, in its crystal structure. Also, centrosymmetric molecules do not always make use of the centers of symmetry in the crystal structure. Hence proof that a center of symmetry is absent in the unit cell means nothing about the molecular symmetry. However, noncentrosymmetric molecules frequently crystallize in the orthorhombic symmetry group $P2_12_12_1$ with four molecules, or the monoclinic symmetry group $P2_1$ with two molecules, particularly if they are optically active organic molecules. A detailed description of the meaning of these symbols is given in a later section.

General Principles of a Complete Structure Determination

In order to determine a complex molecular structure by the X-ray diffraction method the complete three-dimensional arrangement of all atoms in the unit cell must be found. Normally this type of study is carried out under the direction of an experienced crystallographer, with a considerable amount of computational aid. If the heavy-atom method described below is to be used, then a single crystal of dimensions 0.1 to 1.0 mm is required of a single heavy-atom derivative, such as a chloride, a bromide, a sodium or potassium salt, or a molecular crystal containing some other molecule with heavy atoms. It usually makes little difference to the crystallographer whether the heavy atom in the unit cell is bound to the molecule or not, and it is often helpful to make several different types of heavy-atom derivatives. A few days' survey may show that one of these derivatives can be solved with far less work than the others. Even though these derivatives may seem little different to the chemist, they may present very different problems to the crystallographer.

The accumulation of a few thousand intensities by counter or film methods usually takes a few days to a month. Computation of the series described below and the complete solution of the structure requires less than a week to a few months or so, depending on the complexity of the structure, even with the aid of high-speed computers. The improvements in computers and the advances being made in the automatic recording of intensities promise further reductions in this total time, but even now the number of man-years of effort required is often much less than that expended by, for instance, the organic chemist using the classical methods on a new natural product.

Scattering by Electrons

The periodically varying field of an electromagnetic wave is capable of accelerating an electron. But an electron so accelerated emits an electromagnetic wave, which has a frequency essentially identical to that of the incident wave and a definite phase relation that we shall take as identical to that of the incident wave. Thus the maximum of the incident wave corresponds to the maximum of the scattered wave. The energy of X-rays is usually large in comparison with all the binding energies, except for the K electrons of the elements of high atomic number and the K and L electrons of the very heavy elements. Hence to the approximation usually employed all electrons are said to scatter X-rays independently of all other electrons, with the same amplitude and phase. However, some interference can occur in an atom between scattered waves from different electrons because of their different positions in space.

The use of the anomalous behavior of the K and L electrons in heavy elements is becoming increasingly important as a basis for solving certain complex structures as the accuracy of measurement of X-ray intensities increases and as a wide variety of target materials for generating X-rays of different wavelength becomes available.

Scattering by an Atom

Consider an atom at the origin of a coordinate system. If all of the electrons of this atom were at the origin, the amplitudes of scattering from them would then be additive. Then the atomic form factor, the ratio of the amplitude of scattering of the atom to that of a single electron at the origin, would be Z, the atomic number.

The acceleration produced by the electric vector of the light wave has a strong inverse dependence on the mass of the particle. Hence the nucleus produces a negligible scattering of X-rays in spite of its larger charge.

However, the electron density of an atom is spread out in space to an extent consistent with the coulombic binding energies. Except for the relatively small effects of chemical binding and the usually more important

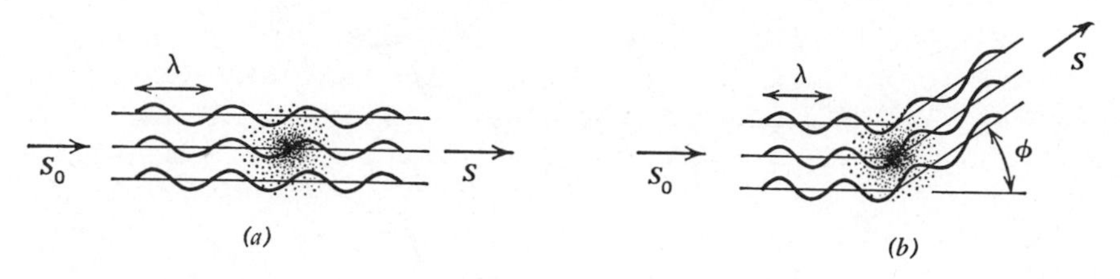

Fig. 1.3 (*a*) Scattering from an atom at zero angle, showing no path difference; (*b*) scattering at angle ϕ, showing path difference.

effects of anisotropy of thermal vibrations, atoms may be regarded as spherically symmetrical. They have an approximately exponential decrease of electron density from the center and an effective radius of about 1 to 3 Å. Hence interference can take place between different parts of the electron cloud about the atom. As illustrated in Fig. 1.3*a*, at zero angle of scattering there is no path difference between the incident and the scattered waves; hence the atomic form factor is just Z, the total number of electrons. As the scattering angle increases (Fig. 1.3*b*) some interference occurs, and the form factor is less than Z. When one takes into account the manner in which the electon density decreases as distance from the nucleus increases, a resulting form factor is obtained that decreases as angle of scattering increases, as shown in Fig. 1.4. In addition, temperature vibrations and other disorders further increase the diffuseness of the electron distribution. These effects together with the form factor account for the general decline of intensities as the angle of scattering increases, a feature that is characteristic of all X-ray diffraction films. The small size of the spots is due to the periodic nature of the crystal, and the streak associated with each reflection is due to the white radiation produced when electrons from the filament are decelerated upon impact with the target.

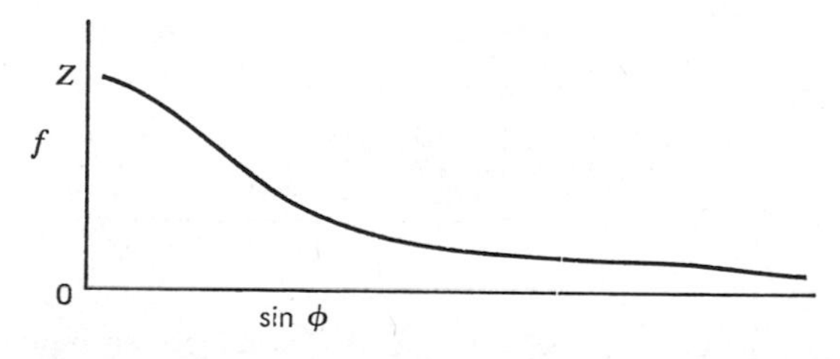

Fig. 1.4 Decrease of amplitude of scattering from an atom of atomic number, Z, as a function of angle of scattering, ϕ.

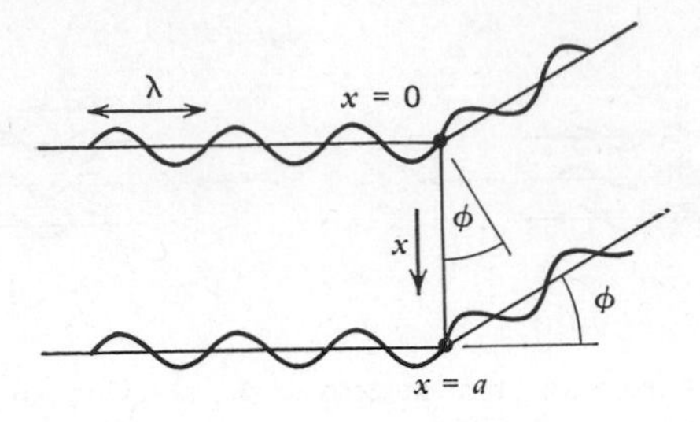

Fig. 1.5 The condition for a maximum when two atoms scatter X-rays is that the path difference is $h\lambda$. In this example $h = 1$. In general, the condition for a maximum is that $a \sin \phi = h\lambda$.

Scattering by Two or More Atoms

Waves scattered from an atom at $x = 0$ and an atom at $x = a$ reinforce one another when the path difference is $h\lambda$, where h is an integer and λ is the wavelength of the X-rays (Fig. 1.5). If we now add a third atom at $x = 2a$, the maximum for all three atoms will still occur at the same angle ϕ, but it will be sharper because slight deviations from the correct value of ϕ will produce some interference between atoms at $x = 0$ and $x = 2a$. As this distance between end atoms is increased by the addition of more and more atoms to the row at intervals a distance a apart, the maximum becomes sharper and sharper. This result is proved in a later section, where it is generalized to directions of incidence at angles other than 90° to the row of atoms. Further generalization to three dimensions, also discussed below, then shows that when the crystal structure is resolved into planes (called equivalent planes) that have identical atomic environment, the Bragg law $n\lambda = 2d'_{hkl} \sin \theta$ gives the condition for a maximum (Fig. 1.6).

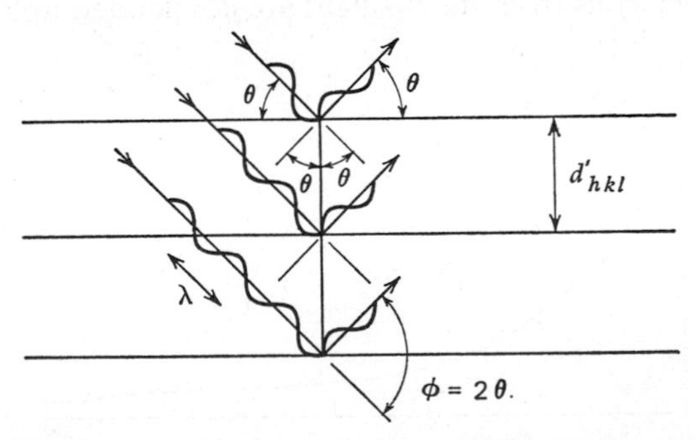

Fig. 1.6 The Bragg law for the case $n = 2$. The condition for a maximum is $n\lambda = 2d'_{hkl} \sin \theta$, where $2d'_{hkl} \sin \theta$ is the path difference associated with equivalent planes which are spaced by a distance d'_{hkl}. The letters hkl identify the planes.

A large number of identical atoms spaced a distance a apart as in Fig. 1.5 would simply show successive maxima for increasing values of h. These maxima would occur at increasing values of θ and would show the normal decline in intensity because of atomic shape and thermal vibrations. Suppose, now, that we add another atom of the same or different kind at a point X, a certain distance between 0 and a, and similarly at $X + a$, in accord with the repeat unit a of the one-dimensional row (Fig. 1.7). If, for example, the two larger atoms were at $X = a/2$ and $3a/2$, they would send out waves exactly out of phase with respect to the waves scattered with a path difference of λ from the atoms at 0 and a. Or, if the path difference, between any two atoms at 0

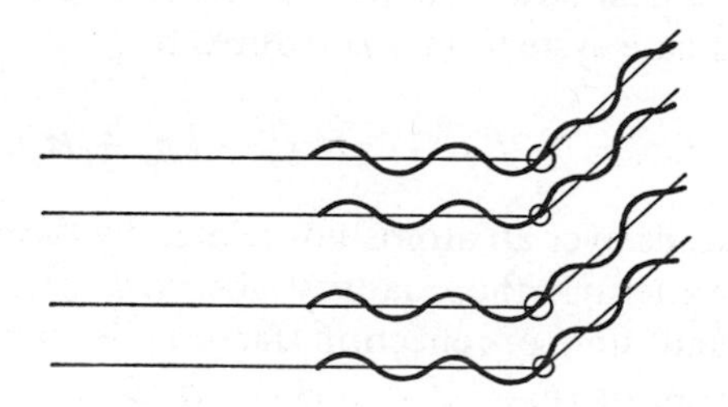

Fig. 1.7 The values of x coordinate for these lines are 0, X, a, and $X + a$, reading from the bottom. The larger atoms at X and $X + a$ send out waves that interfere with those from the atoms at 0 and a.

and a is 2λ, then the waves from the atoms at $a/2$ and $3a/2$ will be in phase with respect to the atoms at 0 and a. Thus for odd orders interference occurs, and for even orders reinforcement occurs. When the larger atoms are not greatly different from the smaller ones (Fig. 1.7), an alternation of intensities occurs as one examines successive orders of reflection, with the odd orders weak and the even orders strong. If all the atoms were identical, the odd orders would be missing completely for this choice of atomic positions.

Another simple example occurs when the atoms are identical and the atoms designated by the larger circles are not quite at $X = a/2$ and $3a/2$. Then interference is not quite complete and the value of the intensities can be used to locate the positions of these atoms relative to those at $X = 0$ and a. The quantitative relation for the amplitude of scattering is derived in a later section, but for this simple example it is $F_h = 2f_1 + 2f_2e^{-2\pi ihx}$, where f_1 and f_2 are the amplitudes of scattering from the two kinds of atoms, h is the order of the reflection, $x = X/a$ is the atomic position expressed as fraction of the unit-cell dimension, and $i = \sqrt{-1}$. The use of the complex exponential notation instead of the trigonometric function is explained below, but for the simple case in which all atoms are identical ($f_1 = f_2 = f$) and a new origin is chosen between the two types of atoms, F_h reduces to the trigonometric function $4f \cos (2\pi hx/2)$. It should be reasonable to expect the amplitude of scattering to be proportional to the amplitudes f of the individual atoms and to be a trigonometric function of both the position and the order of the reflection, since path differences vary with both x and h.

The generalization of the above argument to three dimensions and any number N of atoms in the unit cell leads to the expression

$$F_{hkl} = \sum_{j=1}^{N} f_j \exp [-2\pi i(hx_j + ky_j + lz_j)]. \tag{1.1a}$$

For the case of a crystal structure with a center of symmetry taken as the origin of the coordinate system, (1.1a) reduces to

$$F_{hkl} = \sum_{j=1}^{N/2} 2f_j \cos 2\pi(hx_j + ky_j + lz_j), \tag{1.1b}$$

where this sum extends over all atoms not related by the center of symmetry. The intensity of scattering, the quantity observed after routine correction for polarization and time-of-reflection factors, is simply the square of F_{hkl} when F_{hkl} is real, or $|F_{hkl}|^2$ if F_{hkl} is complex. Both of these correction factors and the method of their computation are described below. For our present purpose we assert that the amplitudes or intensities of scattering can be computed from a set of atoms at known positions x_j, y_j, z_j in the unit cell. Comparison with the corresponding observed quantities constitutes a test of the correctness of these atomic positions.

Fourier Series for the Electron Density

What the investigator really wishes to do in an X-ray diffraction study is to start with the intensities and derive the atomic positions x_j, y_j, z_j. Unfortunately this procedure is much more difficult and not at all a routine matter or merely a computational problem. The problem is most clearly illustrated by the crystal structure with a center of symmetry chosen as the origin. Because the crystal structure is a periodic function it should be no surprise that it is possible and convenient to express the electron density as a three-dimensional Fourier series. For the centrosymmetric case the expression is

$$\rho(x, y, z) = \frac{1}{V} \sum F_{hkl} \cos 2\pi(hx + ky + lz), \tag{1.2}$$

where the sum extends over all planes and orders of scattering, that is, all observed reflections, and V is the volume of the unit cell. The proof that the coefficients in this expansion are F_{hkl}/V and the derivation of the general expression for the noncentrosymmetric case are both given in a later section. The maxima in the function $\rho(x, y, z)$ are at or very near the atomic positions.

If the amplitudes of scattering, F_{hkl}, were known, the complete electron density map of the unit cell could be computed from (1.2). But one observes F_{hkl}^2 in this centrosymmetric example. Whereas the magnitude of F_{hkl} can be easily found by taking the square root, the sign of F_{hkl} is unknown. A random selection of signs or phrases followed by an electron-density calculation is impractical because of the large number of combinations possible. Direct methods for selection of the appropriate signs have had considerable, but not universal, success, and we discuss one such method below, as well as a few other methods that are somewhat more indirect. If the signs of most of the large terms in (1.2) are determined and attached to the observed F_{hkl}, the series "refines"; that is, the peaks in $\rho(x, y, z)$ occur at positions somewhat shifted from those atomic positions used to calculate the signs from (1.1*b*). If the majority of signs are correct and because the observed F_{hkl} are used with these computed signs, the atomic shifts will be ultimately toward the correct final atomic positions. Unfortunately one must have nearly all the atoms within a few tenths of an angstrom of their correct positions before this refinement process converges to the correct structure; that is, a structure that produces a set of F_{hkl} from (1.1*b*), the magnitudes of which agree with the observed F_{hkl}. We shall call a structure that is sufficiently close to the correct one to refine "a trial structure," and now we describe a few methods of obtaining a correct trial structure for complex molecules.

The Heavy-Atom Method

The heavy-atom method is useful for both centrosymmetric and noncentrosymmetric molecules and crystals, but because the description of the method is simpler for centrosymmetric crystals, this case is discussed first.

We have seen from (1.1) that the total amplitude of scattering depends on numbers f_j, proportional to the number of electrons in each atom j, as well as on their relative positions. If one atom is of high atomic number compared with the rest, say iodine compared with carbon, nitrogen, or oxygen, it will largely determine both the signs and the magnitudes of the amplitudes of scattering, F_{hkl}. Suppose that by trial-and-error methods the positions of the few heavy atoms in the unit cell were found. Then these heavy atoms would give the correct signs to nearly all of the larger F_{hkl}'s, and when these signs are used with the corresponding observed F_{hkl}'s in the computation of the electron density map (1.2), the positions of the light atoms as well as slightly better positions of the heavy atoms can be found. If one repeats this computation with revised heavy-atom positions as well as with the positions of such light atoms as are clearly indicated, this series also refines in the same sense, and after a few cycles the positions of all atoms are known.

When the structure is not centrosymmetric, the principles are the same, but the statements are not as simple because the mathematical expressions are more complex. Since these expressions are derived in detail in a later section, no further discussion is given here.

The limitation of the heavy-atom method as described occurs when the number of light atoms becomes so great that the heavy atoms do not determine enough signs for the electron-density map to give a recognizable molecular structure. It is not necessary that the number of the electrons in the heavy atom be greater than the total in all the light atoms because these light atoms are distributed throughout the unit cell. Thus their contributions to the total amplitudes of scattering are smaller, on the average, than the contribution of an equal number of electrons in a few heavy atoms. The use of statistical arguments leads to the conclusion that the heavy-atom method as described above should succeed when, roughly,

$$2N^2 \geq \sum_j n_j^2,$$

where N is the atomic number of the heavy atom and n_j is the atomic number of the jth light atom. The vitamin B_{12} structure [12], determined partly by the use of the heavy-atom method, is sufficiently complex that this limit is approached.

The Vector Map

The location of heavy atoms and the solution of complex structures is greatly facilitated by the computation of the "vector map," developed by Patterson [13].

$$P(u, v, w) = \frac{1}{V} \sum |F_{hkl}|^2 \cos 2\pi(hu + kv + lw). \tag{1.3}$$

This function can be computed directly from the observed $|F_{hkl}|^2$ without any knowledge or assumptions about the signs or phases of the amplitudes F_{hkl}. The vector map (1.3) is obtained from the product $\rho(x, y, z)\rho(x + u, y + v, z + w)$ of the electron densities at x, y, z and $x + u, y + v, z + w$. When the expression in (1.2) is used for the electron density and all contributions at different x, y, z are added, (1.3) results. These mathematical details are described in a later section.

Physically the function $P(u, v, w)$ is large whenever both $\rho(x, y, z)$ and $\rho(x + u, y + v, z + w)$ are large. Hence $P(u, v, w)$ has maxima whenever the difference u, v, w between the coordinates at these two points corresponds to a vector between a pair of atoms. Moreover the maxima in $P(u, v, w)$ are proportional to the product of the scattering amplitudes (approximately the atomic numbers) of the pair of atoms. Hence $P(u, v, w)$ is a map of all interatomic vectors redrawn from a single origin. Its great advantage is that it can be computed directly from observed data. Its disadvantage is that it is often poorly resolved, particularly for complex structures. If the unit cell contains N atoms, the vector map contains $N(N - 1)$ peaks, plus the origin peak.

Consider $Ni(L)Br_2$, where L is the tridentate ligand, 6,6′-dimethyl-di(2-pyridylmethyl)amine, as an example. There are four formula weights in the unit cell. If we ignore hydrogen atoms, there are 4 nickel atoms, 8 bromine atoms, and $4 \times 17 = 68$ light atoms in the unit cell of dimensions $a = 11.23$, $b = 16.19$, $c = 8.75$ Å, $\beta = 92.67°$, and $\alpha = \gamma = 90°$. Thus there will be $80 \times 79 = 6320$ peaks in $P(u, v, w)$, but $12 \times 11 = 132$ will be prominent Br–Br, Ni–Ni, or Br–Ni vectors, and the others will be sufficiently small and randomly distributed for the most part, so that to a first approximation they can be ignored. Hence a reasonable procedure for solving this structure is to analyze these 132 prominent peaks to determine the nickel and bromine atom positions. These positions and the known amplitudes of scattering can then be substituted into (1.1*a*) to yield the amplitudes of scattering for the structure containing only bromine and nickel atoms. The magnitudes of these amplitudes will not agree well with the observed amplitudes because the contributions of all the light atoms are omitted, but most of the phases will be correct, and these phases along with the observed amplitudes are used to compute an electron-density map that should show a number of the light atoms in approximately the correct positions along with better positions for the bromine and nickel atoms.

Thus once the heavier atoms have been located, the positions of the lighter atoms can be readily determined. However, with 132 peaks in the Patterson function, the determination of the heavier-atom positions is not trivial. These positions can be found though by a systematic method of Patterson analysis called the Patterson superposition technique. We consider this method in the following section, along with examples of vector maps for both an artificial and a real structure.

The Patterson Superposition Method

Illustration of the Principle

Some of the most complex structures have been solved by this method. As a very simple example we consider the hypothetical two-dimensional structure in Fig. 1.8*a*. If one draws all the vectors between pairs of atoms in this structure and plots these vectors from a common origin, the resulting vector map is that shown in Fig. 1.8*b*. For a given pair of atoms A and B one must remember that there is a vector from A to B, and also in the opposite direction from B to A. Hence the vector map always has a center of symmetry at the origin whether or not the original structure is centrosymmetric. A vector map always has a very large peak at the origin that represents the interactions

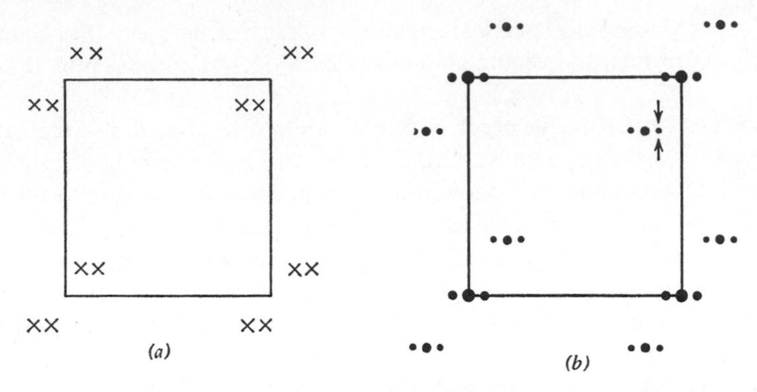

Fig. 1.8 (*a*) Unit cell containing four atoms, related in pairs by center of symmetry at origin. (*b*) Vector map of the structure of (*a*). The single interaction to be used for superposition is indicated by arrows. Next largest peaks are double interactions, and the largest peak is the origin peak that indicates interaction within atoms.

between different parts of the electron density within each of the atoms. Suppose in this example that we take atomic scattering amplitudes proportional to the number of electrons and assume that each of the atoms has five electrons. Aside from the factor V in (1.3), the origin peak has the weight $4 \times 5^2 = 100$; the double interactions have weight $2 \times 5^2 = 50$; and the single interactions have weight $5^2 = 25$. For this centrosymmetric structure there are four single interactions plus four double interactions in the unit cell of the vector map, exclusive of the origin peak. In general there are $n(n - 1)$ vectors if there are n atoms in the unit cell; this is so because we can choose any one of n atoms for the specific one. If the structure is centrosymmetric, there are n single interactions, and the remaining interactions are all double, giving rise to $n(n - 2)/2$ double peaks. Often one knows from the absence of certain reflections (see a later section) that either the three-dimensional structure or one of its projections onto two dimensions has a center of symmetry. If so, the analysis given above can be helpful in starting to analyze the vector map. In particular, from the knowledge of which atoms are present, a single interaction can often be located which will form a good starting point for the analysis of the vector map.

In the Patterson superposition or vector convergence method a single interaction in the vector map is located, and a second vector map of the same crystal is superimposed with its origin on this single interaction, as illustrated in Fig. 1.9. When these two maps are added, the highest peaks, indicated by the arrows, give the structure [14]. In the following sections we give an illustration of this method for a real crystal, the Ni–Br complex mentioned above.

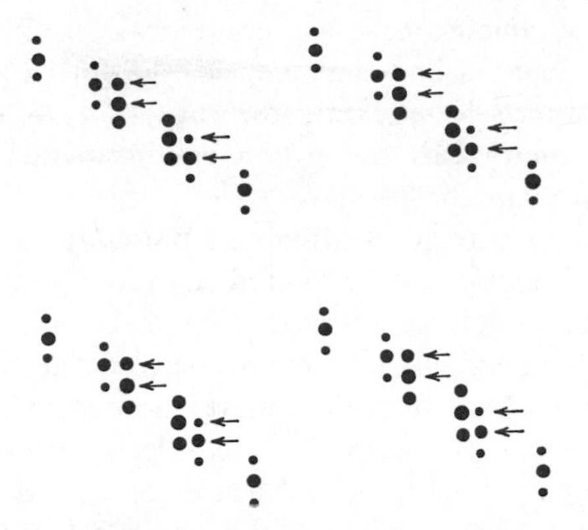

Fig. 1.9 Result of slightly displaced superposition of vector map on single interaction of the same vector map. Heaviest peaks of heights 125 to 100 are indicated by arrows; next highest peaks, 50 or lower, are ignored. The heavy peaks represent the structure (Fig. 1.8).

Dibromo-6,6′-Dimethyl-Di(2-Pyridylmethyl)Amine Nickel(II)

The Patterson superposition method as described above would appear to be restricted to a centrosymmetric crystal. However this restriction can be removed if we choose a noncentrosymmetric arrangement of atoms for the superposition. The structure of $Ni(L)Br_2$ is a striking example of the power of the method in a noncentrosymmetric application.

A brief algebraic analysis may be helpful in describing this method. Assume that N atoms are arranged in a unit cell at positions $A_1, A_2, \ldots, A_N$. Then the Patterson function can be described by the vector set $(A_j - A_i)$, where both i and j can take on any values from 1 to N. Assume that we have selected a well-resolved single peak that represents the vector $A_\beta - A_\alpha$. Now shift the Patterson by this vector and select those vectors that result which are common to the unshifted vector set. The origin is taken as that for the unshifted map. Thus for $(A_j - A_i) + (A_\beta - A_\alpha)$, the subsets $(A_j - A_\alpha)$, $j = 1, \ldots, N$, and $(A_\beta - A_i)$, $i = 1, \ldots, N$ will be in common as well as points resulting from any other cases where vector triangles can be formed using as one side $(A_\beta - A_\alpha)$. If point atoms and a single interacton are assumed, then there will be essentially no points of the latter type. These two subsets represent, respectively, an image of the atomic arrangement in the unit cell with atom α at the origin, and an inverted image of the atom arrangement with atom β at the origin. It can be readily shown that these two subsets are related by a center of symmetry one-half way along the vector $A_\beta - A_\alpha$. Thus if $A_\beta - A_\alpha$ is a single vector in a centrosymmetric structure (β and α are thus related by the center of symmetry), the two images coincide, and one superposition is sufficient to determine the atomic arrangement in the unit cell.

In the noncentrosymmetric case a second peak in either of the subsets listed above can be selected and the superposition process repeated. If, for example, $A_\gamma - A_\alpha$ were selected, only the subset $(A_j - A_\alpha)$ remains. Thus in the noncentrosymmetric case the atomic arrangement in the unit cell can be determined by two superpositions.

In practice however we are not dealing with point atoms, and the Patterson peaks then have appreciable width. Therefore even if sharpened Pattersons, described in a later section, are employed, appreciable numbers of extraneous peaks remain, and great care must be taken to avoid selection of one of these peaks after the first superposition. Of course, more than two superpositions can be carried out as long as one consistently selects vectors belonging to the same subset. Recognizable structural features or use of space-group symmetry are valuable aids in ensuring that vectors in other subsets are not chosen.

To illustrate this superposition method, let us consider the $Ni(L)Br_2$ complex, the unit cell of which is monoclinic, acentric, and contains four molecules [15]. In addition, the absence of certain reflections ($h0l$, $h + l$ odd) indicates the presence of a symmetry element called an n-glide, which means that for every atom at x, y, z in the unit cell, an equivalent atom must be located at $\frac{1}{2} + x, -y, \frac{1}{2} + z$. We consider this effect of symmetry in a later section in more detail, but for the moment it means that the coordinates of the atoms of only two of the four molecules must be specified.

A projection of the sharpened Patterson along the c-axis for this Ni–Br complex is shown in Fig. 1.10. Since there must be 132 heavy-atom vectors in this map, it is obvious that many peaks must be overlapping or partially overlapping in this projection. It must also be remembered that if a multiple peak of multiplicity m is chosen for the first superposition, then $2m$ images will result. Therefore it is best to select a peak of as low a multiplicity as possible. The peak marked with an A was selected since this peak is small and well-resolved. The result of the superposition with this peak is shown in Fig. 1.11. This map was calculated on a computer by selecting at each point the minimum value of the Patterson and the shifted Patterson. The n-glide symmetry condition requires that related to atoms at the head and the tail of the shift vector there must be other atoms related by this symmetry operation, hence at positions displaced by one-half the x-direction. The y-coordinate will be determined by the position of the plane relative to the superposition map origin (which in general is not the same as the origin in the electron-density space), but once this position has been selected all atoms must obey this symmetry operation.

The two peaks marked B and C were selected as those related to the head and the tail of the first shift vector, and a further superposition was carried out using the peak marked B. This result is shown in Fig. 1.12a. As can be

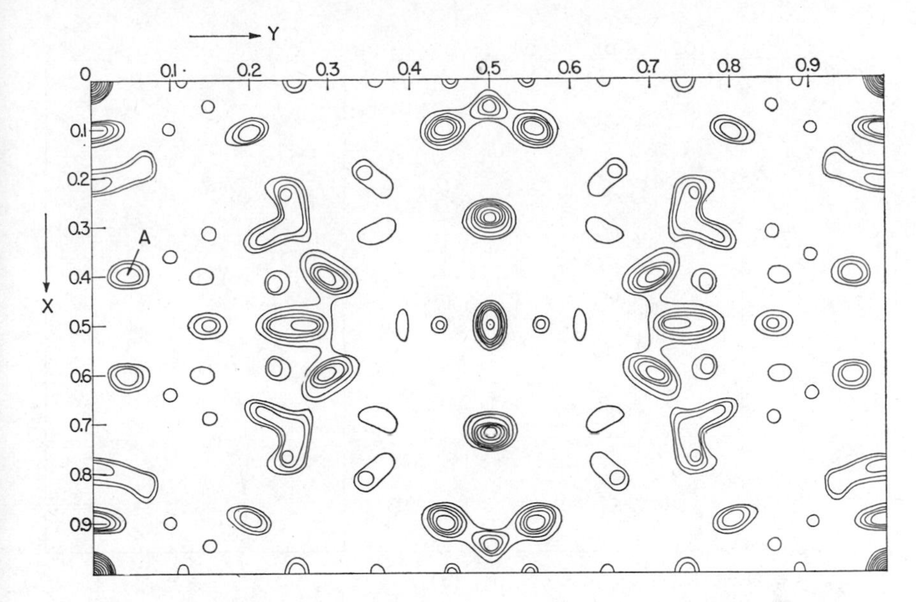

Fig. 1.10 A sharpened Patterson from the observed data for $Ni(L)Br_2$, where L is the tridentate ligand 6,6′-dimethyl di(2-pyridylmethyl)amine, and projected along the *c*-axis. The peak selected for the first superposition is indicated by an *A*.

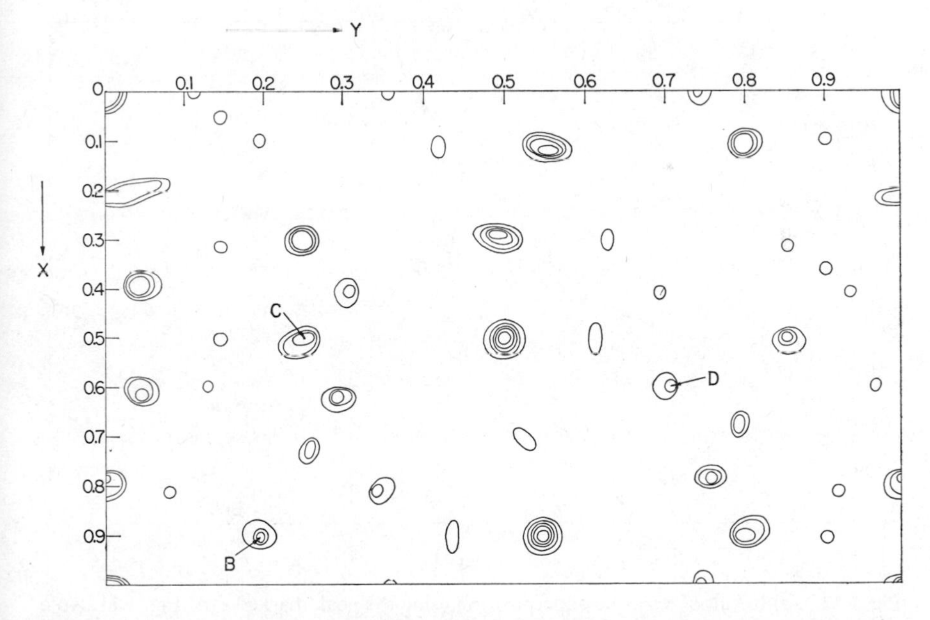

Fig. 1.11 Result of the first superposition using peak *A* of Fig. 1.10. A minimum function was used to compute this map. Peaks marked *B* and *C* were selected as those related to the origin and to *A* by the *n*-glide operation.

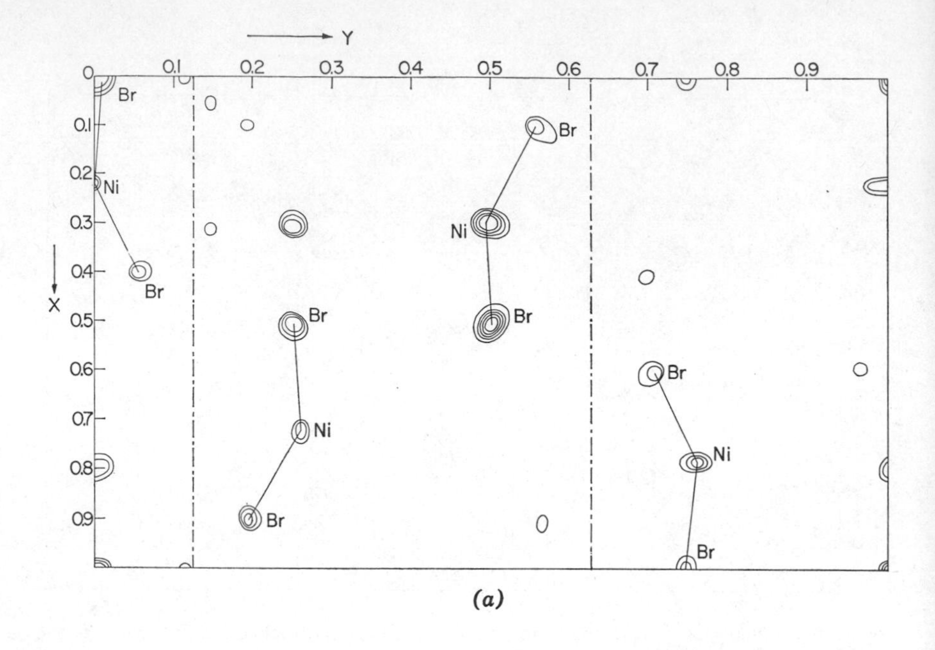

(a)

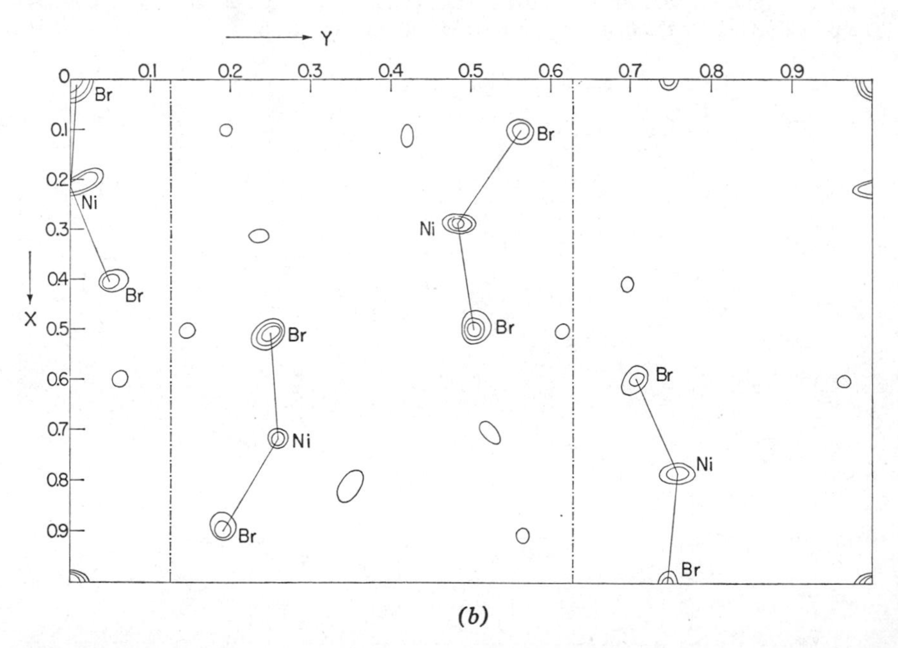

(b)

Fig. 1.12 (*a*) Result of a second superposition using the peak marked *B* of Fig. 1.11. Note that of those peaks higher than one contour, one extraneous peak remains per asymmetric unit. (*b*) Result of a second superposition using the peak marked *D* in Fig. 1.11. The image of the structure remains with no extraneous peaks of any reasonable height.

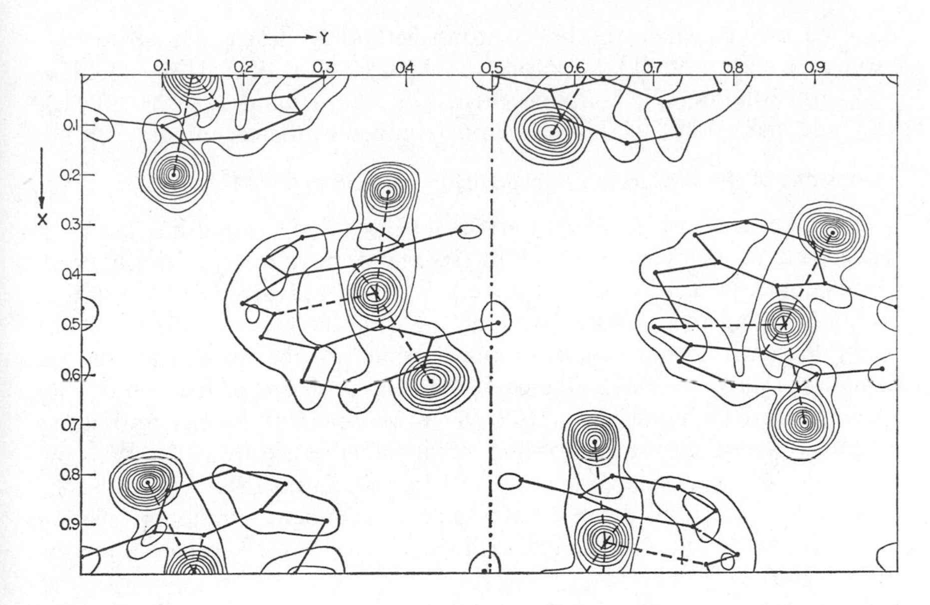

Fig. 1.13 A *c*-axis projection of the electron density map of the unit cell of $Ni(L)Br_2$. The origin of the unit cell has been shifted to the *n*-glide position. The ligand atom positions were determined from a three-dimensional electron-density calculation.

seen from this figure, the nickel and bromine atom positions can for the most part be readily determined since these atoms should correspond to peaks with at least two contours. However it should also be noted that there are two extraneous peaks in this map that appear to be related by the *n*-glide. Such an effect is common, and thus often more than two superpositions are required to eliminate enough accidental coincidences to obtain a good trial structure. Any number of superpositions can be carried out as long as one stays in the correct subset, that is, the image that has been selected by the first two superpositions. Such accidental coincidences are, of course, less frequent in three dimensions than in two. Other peaks could also be selected for the second superposition, which would give fewer extraneous peaks; for example, if peak *D* (Fig. 1.11) were selected for the second superposition, the map shown in Fig. 1.12*b* would result. A projection of the final electron-density map using the nickel and bromine atom positions so determined is shown in Fig. 1.13. The light-atom positions could also have been determined from a superposition carried out in three dimensions where overlap would not so hinder search for this type of vector.

In fact the most extensive use of the Patterson superposition approach is in those cases in which many atoms of about equal atomic number are involved and in which the heavy-atom method is clearly not applicable. Structures such as glycl-L-trypotophan dihydrate [16], B_8Cl_8 [17], cellobiose [18], α-L-sorbose [19], L-aspartic acid [20], β-D-mannitol [21], $(NPCl_2)_5$ [22], and many others [23] have been determined by this technique.

Extensions of the Patterson Superposition Method

One manner in which the Patterson superposition method has been extended is in the complete use of all the space group symmetry of the unit cell. For example, in the problem we have just considered, since for every atom at x, y there must be a related atom at $\frac{1}{2} + x, \bar{y}$, the peaks on the $(\frac{1}{2}, v)$ line in the Patterson give evidence for the possible presence or at least for the definite absence of atoms at the corresponding positions in electron density space. If more than one symmetry element were present, further restrictions could be placed on the probability of appreciable electron density being located at any particular point in the unit cell. From these kinds of considerations a map called a symmetry map or symmetry minimum function can be produced and can be used with the Patterson in a superposition type of procedure [24].

It has also been shown that significant improvement is obtained in the resultant superposition map if, when the chosen shift vector is between atoms of different atomic numbers, one Patterson is weighted differently from the other [25]. In general one Patterson is weighted by the ratio of the atomic numbers involved, relative to the other kept at unit weight. Such a procedure in general results in fewer ambiguities in the resultant map, and partially destroys the center of symmetry usually introduced by the superposition technique.

Known geometry can also be used, and this approach is discussed briefly on p. 35.

Another direction in which the method can be extended is based on the discovery of a method [26] of obtaining a noncentrosymmetric vector map in some acentric cases, based on the anomalous scattering amplitudes of *K* and *L* electrons near their absorption edges.

This method requires careful experimental work and is not described in detail here, but it should be clear to the reader how useful a noncentrosymmetric vector map would be in the solution of complex structures. For example, for crystals of some symmetry types the vector convergence method is ambiguous. This ambiguity can be removed by the use of this noncentrosymmetric vector map. In addition, there are in general many fewer images of the unit cell in this map, and the superposition method properly applied leads to an image of the structure much more directly [27]. These and future developments should permit successful attacks on the structure determinations of increasingly complex molecules.

Direct Methods

The Symbolic Addition Method–Centrosymmetric Case

As we have indicated, the Patterson superposition method can be used to attempt to obtain a trial structure with either a centrosymmetric or a non-centrosymmetric structure. Another general method has enjoyed a good deal of success, primarily in the case of centrosymmetric structures, and we briefly discuss in this section the method and its application. This is the symbolic addition method, which involves the direct use of the intensities rather than their indirect use, as in a Patterson function. Since, as we have seen, it is possible to deconvolute the Patterson function to obtain an image of the structure, it should be possible to manipulate the magnitudes of the structure amplitudes directly to obtain the same result.

The symbolic addition method starts from an equation suggested by Sayre [28], and was developed into an addition procedure and extended by Karle and Hauptman [29]. Consider the structure factor written in the form

$$F_{\mathbf{k}} = \sum_j f_j e^{-2\pi i(\mathbf{k}\cdot\mathbf{r}_j)},$$

where f_j contains the temperature factor, and anticipating a later discussion, $\mathbf{k}\cdot\mathbf{r}_j = hx_j + ky_j + lz_j$. Now define

$$|E_{\mathbf{k}}|^2 = \frac{|F_{\mathbf{k}}|^2}{\epsilon \sum f_j^2}, \tag{1.4}$$

where ϵ is a factor to account for degeneracy, but which for most reflections can be assumed to be unity [29]. Taking the case of N equal atoms,

$$E_{\mathbf{k}} = \frac{1}{N^{1/2}} \sum_{j=1}^{N} \exp(-2\pi i\mathbf{k}\cdot\mathbf{r}_j). \tag{1.5}$$

Then

$$E_{\mathbf{k}}E_{\mathbf{h}-\mathbf{k}} = \frac{1}{N}\left[\sum_j \exp(-2\pi i\mathbf{h}\cdot\mathbf{r}_j) + \sum_i \sum_{\substack{j \\ i\neq j}} \exp\{-2\pi i[\mathbf{k}\cdot\mathbf{r}_j + (\mathbf{h}-\mathbf{k})\cdot\mathbf{r}_i]\}\right],$$

where the vector quantities $\mathbf{h}$ and $\mathbf{k}$ refer in general to different sets of indices. Depending on the value of $\mathbf{k}$, the second term can assume either positive or negative values at random, and therefore averaging over all values of $\mathbf{k}$, we obtain

$$E_{\mathbf{h}} = N^{1/2}\langle E_{\mathbf{k}}E_{\mathbf{h}-\mathbf{k}}\rangle_{\mathbf{k}}. \tag{1.6}$$

In the unequal atom case it can be shown that

$$E_{\mathbf{h}} \approx \sigma_2^{3/2}\sigma_3^{-1}\langle E_{\mathbf{k}}E_{\mathbf{h}-\mathbf{k}}\rangle, \tag{1.7}$$

where

$$\sigma_n = \sum_{j=1}^{N} Z_j^{\,n}.$$

However, it is not necessary to average over all values of **k** to determine the sign of $|E_{\mathbf{h}}|$. In fact, for $|E|$'s sufficiently large, $S_{\mathbf{h}} = S_{\mathbf{k}}S_{\mathbf{h-k}}$, where $S_{\mathbf{h}}$ is the sign of the corresponding $E_{\mathbf{h}}$. This can be seen by considering the restriction placed on the individual terms making up $E_{\mathbf{k}}$, if $|E_{\mathbf{k}}|$ is to be very large:

$$E_{\mathbf{k}} = \frac{1}{N^{1/2}}[\exp(-2\pi i\mathbf{k}\cdot\mathbf{r}_1) + \exp(-2\pi i\mathbf{k}\cdot\mathbf{r}_2)\ \cdots + \exp(-2\pi i\mathbf{k}\cdot\mathbf{r}_N)].$$

For $|E_{\mathbf{k}}|$ to be near maximum value, all terms in the summation must be nearly equal so that a common phase factor can be factored out; that is,

$$E_{\mathbf{k}} \approx N^{1/2}\exp(-2\pi i\mathbf{k}\cdot\mathbf{r}).$$

The exponential term represents the phase of $|E_{\mathbf{k}}|$. (Remember that $|e^{-i\theta}|^2 = 1$.) Thus if $|E_{\mathbf{h-k}}|$, $|E_{\mathbf{k}}|$, and $|E_{\mathbf{h}}|$ are all very large,

$$\begin{aligned} E_{\mathbf{k}}E_{\mathbf{h-k}} &\approx [N^{1/2}\exp(-2\pi i\mathbf{k}\cdot\mathbf{r})][N^{1/2}\exp(-2\pi i(\mathbf{h}-\mathbf{k})\cdot\mathbf{r})] \\ &\approx N\exp(-2\pi i\mathbf{h}\cdot\mathbf{r}) \\ &\approx N^{1/2}E_{\mathbf{h}}. \end{aligned}$$

In other words, the larger the $|E|$ values, the greater the probability that the relation $S_{\mathbf{h}} = S_{\mathbf{k}}S_{\mathbf{h-k}}$ would be expected to hold. Cochran and Woolfson [30] have shown that this is indeed the case and have derived the following probability expression for $|E_{\mathbf{h}}|$ being positive:

$$P_+(\mathbf{h}) \approx \tfrac{1}{2} + \tfrac{1}{2}\tanh \sigma_3\sigma_2^{-3/2}\,|E_{\mathbf{h}}|\sum_{\mathbf{k}} E_{\mathbf{k}}E_{\mathbf{h-k}}.$$

To illustrate the usefulness of this method and to describe it in somewhat more detail, we consider its application to the structure determination of tetrahydrofurandione. This compound crystallizes [31] in the centrosymmetric orthorhombic space group *Pnma* or in the noncentrosymmetric space group $Pn2_1a$ with four molecules per unit cell. Ordinarily when such a space group ambiguity occurs and when the expected molecular symmetry can be made compatible with the centrosymmetric space group, one proceeds to use this more symmetrical space group. If one fails, then he turns to the non-symmetric space group. Here we begin by assuming that the molecular plane of symmetry coincides with the plane of symmetry in the space group *Pnma*. Statistical tests on the intensities also indicated the centrosymmetric space group. The $|E_{\mathbf{h}}|$ values are listed in Table 1.1. Listings of the $|E_{\mathbf{k}}|\,|E_{\mathbf{h-k}}|$ were made for each $E_{\mathbf{h}}$ value (only the larger E's being considered), and after we referred to this listing, the signs of the following three reflections were chosen to fix an origin: $S_{211} = -$, $S_{441} = +$, and $S_{174} = -$. In addition, the signs of three reflections were represented by the algebraic symbols a through c.

Table 1.1 List of Reflections with $|E| \geq 2.30$

h	k	l	$\|E\|$	*Sign*
2	1	1	3.95	−1
4	0	2	3.23	−1
8	4	4	2.87	a
0	8	2	2.87	c
5	0	3	2.86	c
9	0	4	2.86	ab
4	4	1	2.82	1
4	1	6	2.80	$-a$
1	5	8	2.72	$-ac$
2	0	5	2.68	a
2	4	4	2.62	a
6	8	5	2.61	$-ab$
2	7	3	2.60	b
2	4	2	2.53	$-a$
4	10	5	2.52	−1
8	12	1	2.51	$-b$
5	13	4	2.50	$-b$
3	14	3	2.48	$-b$
0	5	5	2.46	$-a$
1	7	4	2.45	−1
1	12	6	2.44	—
3	1	7	2.43	ac
4	13	5	2.42	ab
5	4	3	2.39	ab
8	0	6	2.34	—
10	8	1	2.34	$-ab$
6	0	5	2.33	a
5	0	5	2.32	$-c$
1	0	7	2.32	c
3	0	2	2.31	$-ac$
0	0	2	2.31	−1
2	4	6	2.30	$-a$
6	0	3	2.30	$-a$

Taking only those probabilities greater than 97%, the signs of as many as possible of the largest E's were determined from $|E_{\mathbf{k}}||E_{\mathbf{h}-\mathbf{k}}|$ products. (Of course one must keep in mind that the signs of all symmetry-related reflections can be referred to one selected in the set.) Using this approach, the signs indicated in the last column of Table 1.1 were obtained. Relations are also usually obtained between some or all of the algebraic symbols, and thus these symbols can be reduced in number or in some cases they can be completely determined. In this particular example it was possible to determine signs for two of the three symbols used, and the signs for 171 of the largest E's were thus obtained. A typical listing for the 244 reflection is shown in Table 1.2. From this table it is apparent that the sign corresponding to the letter b is probably $-$, and that the internal consistency of the sign indications for this 244 reflection is excellent. If p unknown symbols were left, then 2^p electron-density type maps could be computed and investigated for structural features characteristic of the compound under investigation. Other

Table 1.2 Some Products Determining S_{244}

$\mathbf{k}$	$\mathbf{h} - \mathbf{k}$	$S_{\mathbf{k}}$	$S_{\mathbf{h}-\mathbf{k}}$	$S_{\mathbf{h}} = S_{\mathbf{k}} S_{\mathbf{h}-\mathbf{k}}$	$\beta \lvert E_{\mathbf{h}} \rvert \lvert E_{\mathbf{h}-\mathbf{k}} \rvert \lvert E_{\mathbf{k}} \rvert$ [a]
$44\bar{1}$	$\bar{2}05$	-1	$-a$	a	3.79
$\bar{4}4\bar{1}$	605	$+1$	$+a$	a	3.30
$\bar{4}\bar{4}\bar{1}$	685	$+1$	$-ab$	$-ab$	3.69
211	033	-1	ab	$-ab$	4.10
$40\bar{2}$	$\bar{2}46$	-1	$-a$	a	3.74
082	$2\bar{4}2$	b	$-a$	$-ab$	3.64
$08\bar{2}$	$2\bar{4}6$	b	$-a$	$-ab$	3.31

[a] $\beta = \sigma_3 \sigma_2^{-3/2}$.

tests involving internal sign consistencies can be employed that will also help reduce the number of possibilities. In this example the sign corresponding to the letter a was still in doubt but the calculation of electron-density maps clearly indicated that the negative sign was the most appropriate choice.

In Fig. 1.14 is a composite Fourier map computed from the large E's and projected down the c-axis. The four tetrahydrofuran molecules can be readily seen in this map. This example is one involving a fairly simple structure in which the symbolic addition method works quite well. In more complicated structures one would expect many more extraneous peaks in one of these E-maps, and more difficulty would be encountered in finding the correct atomic positions. In general, those structures that have planes of symmetry or other features that will tend to give rise to larger E values are more susceptible to solution by the symbolic addition method.

Noncentrosymmetric Algebraic Methods and the Tangent Formula

There have been recent developments of powerful algebraic–statistical methods for noncentrosymmetric structures similar to and extended beyond the methods just described for centrosymmetric structures [28, 32, 33]. The problem is the determination of the phase angle $\varphi_{\mathbf{h}}$ for each reflection $\mathbf{h}$,

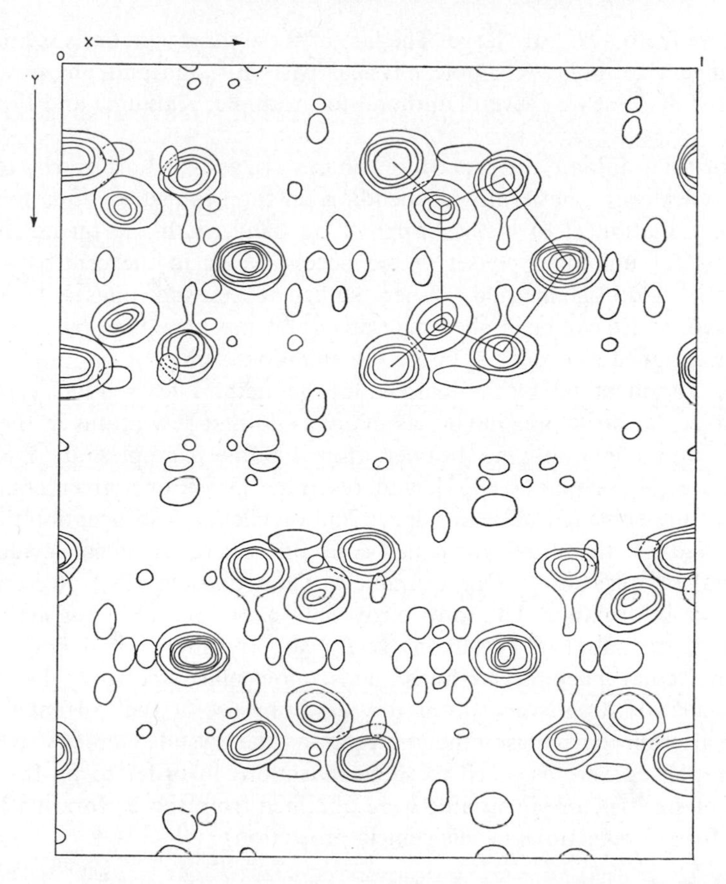

Fig. 1.14 A Fourier map computed using the large E's for tetrahydro-3,4-furandione. Signs for these E's were determined by the symbolic addition method.

where **h** is some hkl and the phase angle is in the range from 0 to 2π radians (or $-\pi$ to π). The main formulas and procedure for the most widely used such method are summarized by Karle and Karle [29, 34] as follows:

$$\varphi \approx \langle \varphi_{\mathbf{k}} + \varphi_{\mathbf{h-k}} \rangle_{\mathbf{k}r} \tag{1.8}$$

$$\varphi \approx \frac{\sum_{\mathbf{k}r} |E_{\mathbf{k}} E_{\mathbf{h-k}}| (\varphi_{\mathbf{k}} + \varphi_{\mathbf{h-k}})}{\sum_{\mathbf{k}r} |E_{\mathbf{k}} E_{\mathbf{h-k}}|} \tag{1.9}$$

$$\tan \varphi_{\mathbf{h}} \approx \frac{\sum_{\mathbf{k}} |E_{\mathbf{k}} E_{\mathbf{h-k}}| \sin (\varphi_{\mathbf{k}} + \varphi_{\mathbf{h-k}})}{\sum_{\mathbf{k}} |E_{\mathbf{k}} E_{\mathbf{h-k}}| \cos (\varphi_{\mathbf{k}} + \varphi_{\mathbf{h-k}})}, \tag{1.10}$$

where **k** refers to hkl other than **h**, and $\mathbf{k}_r$ refers to those **k** for which the unitary structure factors $|E_{\mathbf{k}}|$ are large. The last of these three equations is known as the tangent formula. A simple physical basis for sign and phase relations has been discussed by several authors, for example, Kainuma and Lipscomb [35].

In order to initiate the procedure, phases are assigned arbitrarily to up to three reflections (the number depending on the symmetry), thus fixing the origin. Equation (1.8) is then applied by hand, with the introduction of symbols for unknown phases where necessary, as in the centrosymmetric symbolic addition method. When sufficient starting phases have been obtained, (1.10) can be applied reiteratively by machine. One arbitrary phase can be assigned in order to choose the enantiomorph.

In the applications by the Karles [36] the method has produced solutions for molecular structures having as many as 46 first-row atoms in the asymmetric unit containing no heavier atoms. Some examples are L-arginine dihydrate [37], panamine [34], and reserpine [38]. For a given complexity of structure this method is less dependent on chemical information than any other and is therefore particularly useful where chemical evidence is minimal or misleading. One example is veratrobasine [39], in which the chemical information had previously led to an incorrect ring system and therefore to failure of Patterson, vector superposition, rigid body search, and molecular packing methods. This molecular structure, having the formula $C_{27}H_{41}O_3N$, was solved very easily by use of the tangent formula. There are four molecules in the unit cell, and the crystal symmetry is $P2_12_12_1$. Hence phases were assigned to three reflections in order to fix the origin. Signs of six structure invariants were obtained from the Σ_1 formula [40, 32] (e.g., for $h0l$ reflections in this centric projection)

$$SE_{2h,0,2l} = S\sum_{k}(-1)^{k+l}(E^2_{hkl} - 1), \tag{1.11}$$

and two phases were assigned symbols of unknown values, one of which determined the enantiomorph. In (1.11) S denotes "sign of" what follows. Fifty-two phases were then established by the symbolic addition procedure, with the use of (1.8), and then these phases were used as input to the tangent formula, which gave phases for 636 reflections, each having a value of $|E_{\mathbf{h}}| > 1.1$. This procedure converged in only some 14 minutes on an IBM 7094 and led quickly to the full structure based upon 3105 unique reflections.

It is especially important that the initial set of $|E_{\mathbf{h}}|$'s be among the largest and that they produce many relations (to other $|E_{\mathbf{h}}|$'s) of the types required in these formulas. Also, when symbols are used for the unknown phases of some reflections, great care must be exercised in accepting relationships among these symbols. Usually one should delay accepting these relationships as long as possible in the procedure. Further developments can be expected

[41], and solutions of complex structures somewhat larger than those solved to date will occur in many laboratories, now that the appropriate computer programs are available from several laboratories. One such development is the use of the tangent formula for refining phases obtained by other methods for a trial structure or part of a structure.

Protein Structure Determination

Following the crystallization of urease by Summer in 1926, Bernal and Crowfoot [42] (Mrs. D. C. Hodgkin) showed in 1934 that high-quality diffraction maxima could be obtained from a crystal of pepsin maintained in a small amount of its mother liquor. Perutz began a systematic study of hemoglobin crystals in 1938, and shortly thereafter was joined by Kendrew who began a parallel investigation of myoglobin. Heavy-atom techniques became powerful enough for scientists to attempt to determine protein structure only when Bokhoven, Schoone, and Bijvoet [43] proposed the multiple isomorphous replacement method, which is described below. An initial important achievement by Perutz and his co-workers [44] was the attachment of heavy atoms to hemoglobin and the demonstration that the derivatives were isomorphous with the protein. The structure of myoglobin (mol. wt. 17,000) reached 6 Å resolution [45] in 1958 and atomic resolution (2.0 Å) [46] in 1960. Hemoglobin (mol. wt. 67,000) was solved at 5.5 Å [2] in 1960, and at atomic [3] resolution (2.8 Å) in 1968. The first enzyme brought to atomic resolution was lysozyme [4] (mol. wt. 14,600) at 2.0 Å in 1965. Carboxypeptidase A [47] (mol. wt. 34,600) was at atomic resolution (2.8 Å) in 1966, but the absence of a chemical sequence has delayed the full atomic interpretation, except for the probable mechanisms of action [48]. Ribonuclease [6] (mol. wt. 13,700), ribonuclease-S [49] (mol. wt. 11,500), chymotrypsin [50] (mol. wt. 25,000), papaine [51] (mol. wt. 23,000) and subtilisin [52] (mol. wt. 27,600) complete the list of published structures at atomic resolution as of June 1969, but a number of additional structures will soon appear: elastase, insulin, α-lactalbumin, glucagon, trypsin, pepsin, and others. At a resolution of about 3 Å or less it normally becomes possible to obtain atomic coordinates with the combination of the chemical sequence and the electron-density map obtained by isomorphous replacement methods. Thus even though individual atoms are not resolved, the solution of the structure can usually be obtained in the range of 3.5 to 2.0 Å resolution. We now consider some of the simpler aspects of the isomorphous replacement method.

It will be shown in (1.36) that the electron density $\rho(x, y, z)$ at every point x, y, z can be computed from (*a*) the indices h, k, l, which are known from the geometry of the experiment, (*b*) the amplitudes of X-ray scattering $|F_{hkl}|$, which are directly obtainable from the measured diffraction maxima, and (*c*) the phase angle α_{hkl} for each maximum. Thus the determination of α_{hkl} completes the solution to the problem.

Suppose, for the moment, that the structure is known and that we also have located the heavy atoms in two isomorphous derivatives. Then the amplitude (a complex number) is

$$F_P = \sum f_j \exp[-2\pi i(hx_j + ky_j + lz_j)]$$

$$F_H = \sum f_j \exp[-2\pi i(hx_j + ky_j + lz_j)] + \sum f_{j'} \exp[-2\pi i(hx_{j'} + ky_{j'} + lz_{j'})]$$

and

$$F_H = \sum f_i \exp[-2\pi i(hx_j + ky_j + lz_j)] + \sum f_{j''} \exp[-2\pi i(hx_{j''} + ky_{j''} + lz_{j''})]$$

for the protein, heavy-atom derivatives 1 and 2, respectively. We rewrite these last two equations as

$$F_P = -f_{H_1} + F_{H1}$$

$$F_P = -f_{H_2} + F_{H2},$$

where $f_{H_1} = \sum f_{j'} \exp[-2\pi i(hx_{j'} + ky_{j'} + lz_{j'})]$ can be computed for each *hkl* if we locate the positions of heavy atoms in the first derivative. A similar statement applies to the second heavy-atom derivative. This situation is illustrated in Fig. 1.15*a*.

Now let us assume the actual situation: we have located the heavy atoms in each of these two derivatives, but we know only the magnitudes, $|F_P|$, $|F_{H_1}|$, and $|F_{H_2}|$ from the measured intensities of maxima from the protein and the two heavy-atom derivatives. Thus the situation as it actually exists in practice is illustrated in Fig. 1.15*b*. However, the known complex numbers $-f_{H_1}$ and $-f_{H_2}$, when added to $|F_{H_1}|$ and $|F_{H_2}|$, respectively, must yield $|F_P|$, but they can only do so at certain phase angles α, as shown in Fig 1.15*b*. There are normally two solutions for each heavy-atom derivative, but usually only one solution in common between two heavy-atom derivatives. The common solution is the correct one for each maximum *hkl*. Inasmuch as the phase repeats after 2π, the information in Fig. 1.15 can be plotted as shown in Fig. 1.16. Because the phases of both F_P and F_{H_1} are initially unknown but the magnitudes $|F_P|$ and $|F_{H_1}|$ are known, a complete circle having a line width corresponding to the error in $|F_P|$ (or $|F_{H_1}|$) represents our initial knowledge. But remember that both magnitude and phase of f_{H_1} are known from the heavy-atom positions, and so we must find where to place $-f_{H_1}$ (in Fig. 1.15 or Fig. 1.16) in order to solve the equation $F_P = -f_{H_1} + F_{H_1}$. As before, there are usually two solutions. Again there are usually two

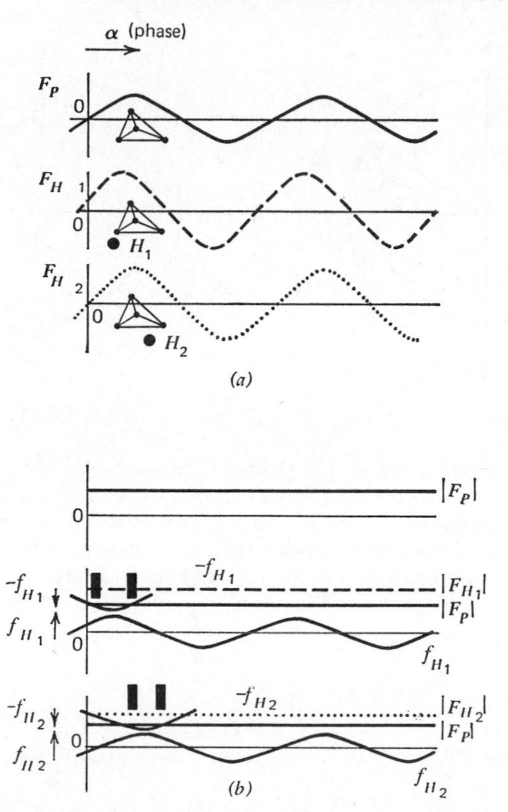

Fig. 1.15 (*a*) The wave scattered by the four small atoms, representing the protein, has a maximum coinciding with most of the scattering matter. If a heavy atom H_1 (or H_2) is added, the maximum shifts in a direction that reflects its contribution. (*b*) The X-ray experiment yields knowledge only of the magnitudes of the scattering amplitude (F_P) for the protein (solid lines), (F_{H_1}) for the first derivative (dashed lines), and (F_{H_2}) for the second derivative (dotted lines). Location of the heavy atoms yields both their amplitudes and phases (sine curves) with maxima as indicated from their average contribution at the phase position indicated by large dots above. If these sine waves are redrawn as $-f_{H_1}$ and $-f_{H_2}$ and measured from the lines (F_{H_1}) and (F_{H_2}) as indicated, then their intersections with the (F_P) line satisfy the equations $(F_P) = -f_{H_1} + (F_{H_1})$ and $(F_P) = -f_{H_2} + (F_{H_2})$ in the approximation that F_{H_1}, F_{H_2} and F_P have the same phase angle α.

solutions for the other derivative, but only one solution for the phase angle α in common for these two derivatives for each reflection *hkl*. Finally, we show in Fig. 1.17 the more usual method of replotting the phase circles and of obtaining a graphical solution. In practice, careful treatment of errors and least-squares procedures are required in order to find the best, properly weighted, α_{hkl} values for a given set of heavy-atom derivatives. A computer

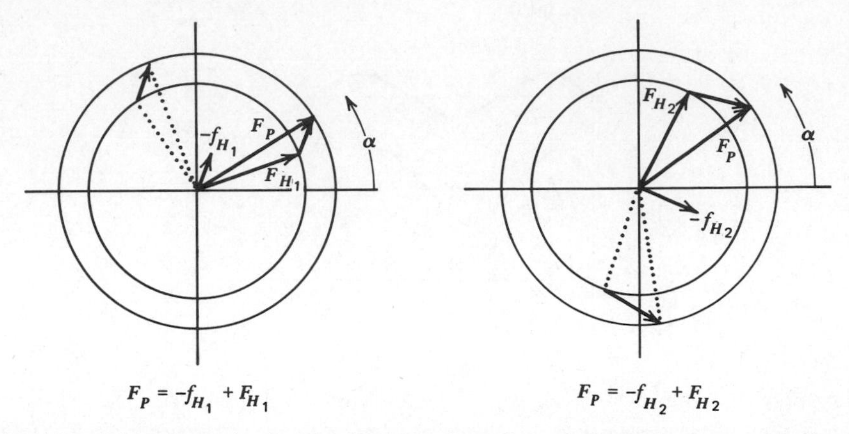

Fig. 1.16 The information in Fig. 15*b* plotted as phase circles. Only the magnitudes of F_P, F_{H_1}, and F_{H_2} are known from the X-ray experiment. Location of the heavy atoms yields both the magnitude and the phase of f_{H_1} and f_{H_2}, each of which can be placed only in the two indicated positions in order to satisfy the vector equations $F_P = -f_{H_1} + F_{H_1}$ and $F_P = -f_{H_2} + F_{H_2}$. The phase α that is common to both is correct.

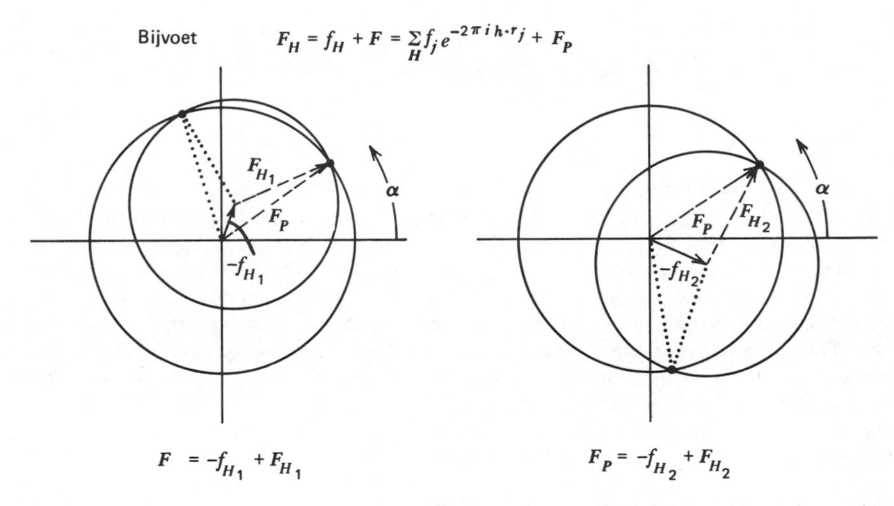

Fig. 1.17 Phase circles in the complex plane representing the structure factor (scattering amplitude) with its real component to the right and imaginary component vertical. Location of the heavy-atom positions and occupancies gives the vectors f_{H_1} for derivative 1 and f_{H_2} for derivative 2. The completely unknown phase, but known magnitude leads to the circles for F_P for the protein and F_{H_1} and F_{H_2} for the total scattering of each reflection for the derivatives. The intersections of circles are the only points that satisfy the vector equation $F_{H_1} = F_P + f_{H_1}$ and $F_{H_2} = F_P + f_{H_2}$. The intersection common to both derivatives is the correct phase angle α.

program for this procedure written at Harvard by H. M. Muirhead is in general use.

Heavy-atom derivatives can be prepared by immersing single crystals into a solution of a heavy-atom salt, but it is frequently desirable to use a dialysis procedure (Fig. 1.18) in order to make gradual changes in concentration of salt or *p*H. For a fairly large protein molecule, such as carboxypeptidase A, the intensity changes produced by the heavy atoms are not always large, but are frequently quite observable (Fig. 1.19). Some 20,600 independent diffraction maxima were obtained for the protein, but only the 6000 largest reflections of the protein were measured in the 2.0 to 3.0 Å range (14,000 reflections possible) on the three best heavy-atom derivatives. A region of the electron-density map at 2.0-Å resolution is shown in Fig. 1.20. Even the waterlike structure can be seen in the region between four neighboring molecules (Fig. 1.21). The large amount of water in protein crystals is important in that poor substrates or inhibitors can be diffused into the crystals, and the binding can be deduced from X-ray diffraction data on the complex with the use of phases α_{hkl} of the parent protein. One example of the excess density associated with binding of glycyl-L-tyrosine to carboxypeptidase A is given in Fig. 1.22, and the mechanistic deductions from this study are summarized in Fig. 1.23. Finally, a drawing of the polypeptide backbone is shown in Fig. 1.24.

Search Methods

Other types of methods that have been receiving increasing attention in the past few years are those that can in general be classified as search methods. Such methods work best when the approximate molecular structure, or at least a reasonable fragment of the molecular structure, can be assumed. Either of two general approaches can be followed. One is to use known space group and unit-cell information and to try to devise "reasonable" packing arrangements that would be consistent with such a unit cell and symmetry. By "reasonable" is meant either that intermolecular contacts must be larger than some prescribed minimum [53], or that the configurations found lead to a minimum in the intermolecular energy [54].

A second approach makes use of the fact that a Patterson function contains all interatomic vectors and therefore would contain the vectors between atoms in the fragment in question. By appropriate methods the resultant vectors can be fitted to the Patterson, and thus the orientation of the fragment can be determined [55–57].

Interpretive Problems

It would be a mistake to leave the reader with the impression that structure determination is a routine procedure. Not infrequently the heavy atoms occur

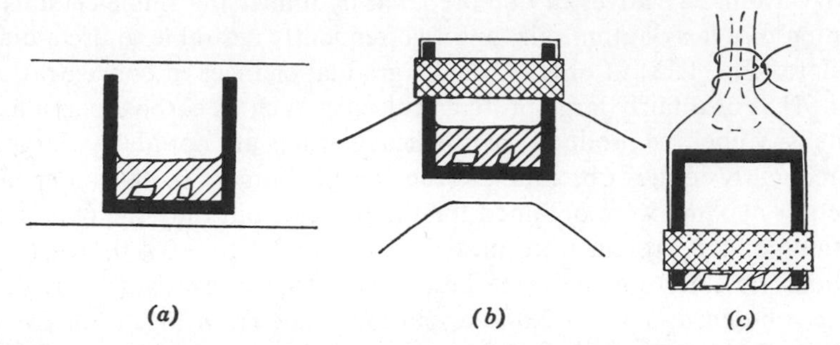

Fig. 1.18 Dialysis cell, 0.8 to 1.0 cm in diameter by 1.5 to 2.0 cm, (*a*) placed in a length of dialysis tubing which, in (*b*), is pulled tightly over the open end of the cell, and then secured by means of a polyethylene ring (cross-hatched). In (*c*) the cell is inverted, and the dialysis tubing tied. [W. N. Lipscomb, J. C. Coppola, J. A. Hartsuck, M. L. Ludwig, H. Muirhead, J. Searl and T. A. Steitz, *J. Mol. Biol.*, **19,** 423 (1966).]

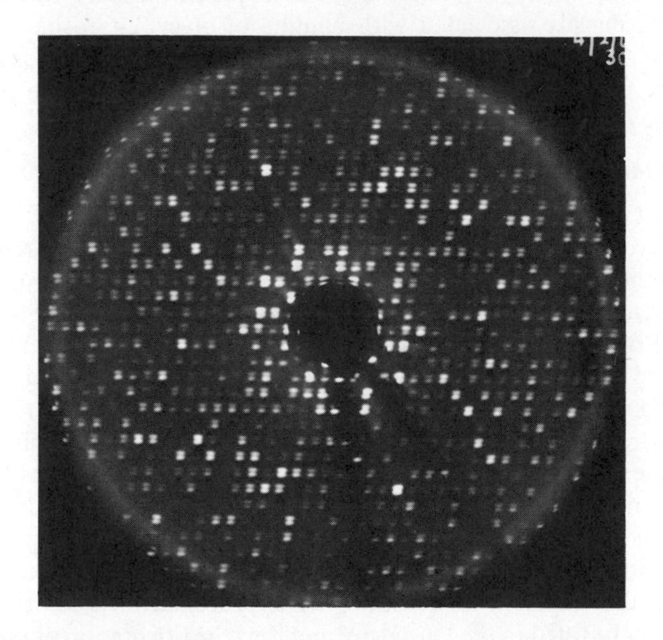

Fig. 1.19 The *h*0*l* X-ray reflections of carboxypeptidase A are the lower one of each pair of spots, while the corresponding reflections from a derivative having two lead atoms per molecule are shown as the upper one of each pair.

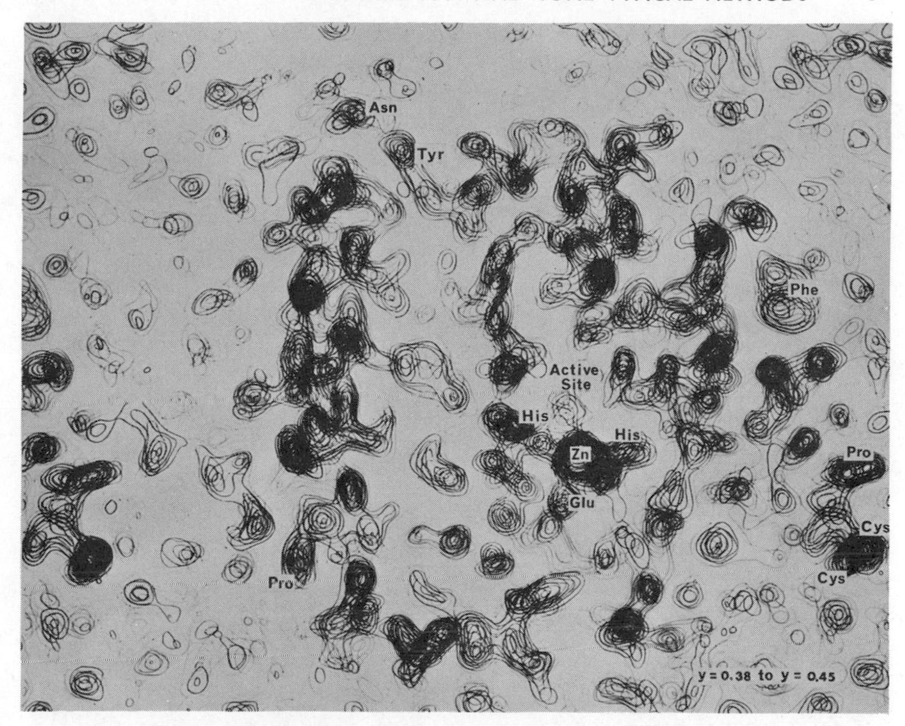

Fig. 1.20 Composite of sections $y = 0.38$ to 0.45 of the electron density at 2.0-Å resolution of carboxypeptidase A. A helix extends along the left-hand side of the map from Pro 288 up to the C-terminal Asn 307, which is hydrogen bonded to residue Tyr 265. The enzyme contains zinc, which has three ligands to the protein (His 69, Glu 72, and His 196), and one more ligand to H_2O, which is displaced by peptide substrates and by some inhibitors. An extended chain, part of an extensive β-pleated sheet structure, is parallel to the helix, and at the lower right a disulfide bond can be seen.

at or near special positions in the unit cell of a crystal in such a way that they do not contribute greatly to a reasonably large fraction of the intensities. Usually the resulting electron-density maps computed either with the contributions of the heavy atoms only or by use of any of the other methods described above present a further problem of interpretation. Sometimes this problem is one of resolution of two molecules superimposed on each other but with different orientations. Sometimes the problem is one of recognition of the light-atom part of the molecule when it is only very badly resolved. Lack of great care in the recognition of atoms not put into the structure can cause one to arrive at a false structure unless he places high demands on the detailed agreement of observed and calculated intensities, on equally good

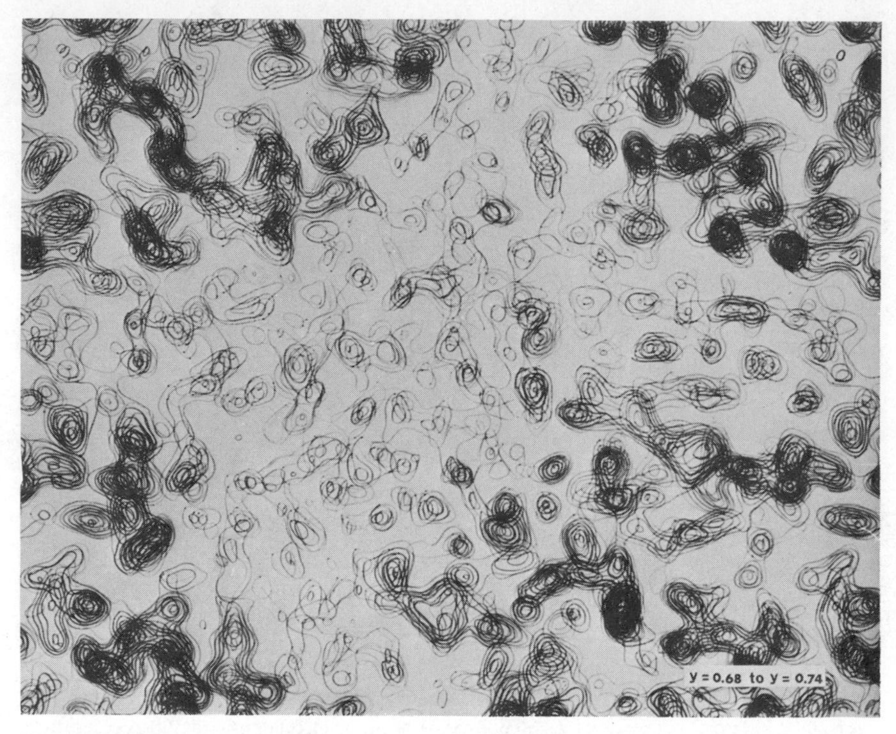

Fig. 1.21 The four molecules are shown only in part as the denser regions in the four corners. The major central region is the solvent (mostly 0.2 *M* LiCl solution) in which the protein is present in the crystalline phase.

resolution of atoms in the map, on equally normal behavior of all coordinate and temperature parameters, and on good chemical sense.

We illustrate only one of the kinds of interpretative problems here. The solution of the iresin *p*-bromobenzoate diester structure by X-ray diffraction methods [58] was completed in a total time of about 6 months, the first three of which were occupied with the photography and measurement* of 1432 diffraction maxima. The chemical structure is shown in Fig. 1.25. The final electron density map is shown in Fig. 1.26, in which each atom is taken from the particular section of electron density corresponding to its z-coordinate, and a composite projection is then made from these various sections. The three stages immediately preceding the final stage of Fig. 1.26 are shown as F-2, F-3, and F-4 in Fig. 1.27.

* The time for this data collection has by now been reduced to a few days.

Successive refinements of the structure operate, as is clearly shown by the penultimate stage F-4. The model from which F-4 was computed included the contributions of all atoms except O-5 and C-12. Phase and signs from this model were combined with the observed data, which of course contain contributions from all atoms, to give map F-4. This map shows especially clearly that atoms that have not yet been included tend to appear with only about one-third of their proper electron density.

It is small wonder, then, that at earlier stages one can easily fall into the trap of assigning atomic positions to false detail. This happened in F-2, which was computed at an earlier stage of the investigation when the positions were known of the 2 bromine atoms and 10 carbon or oxygen atoms,

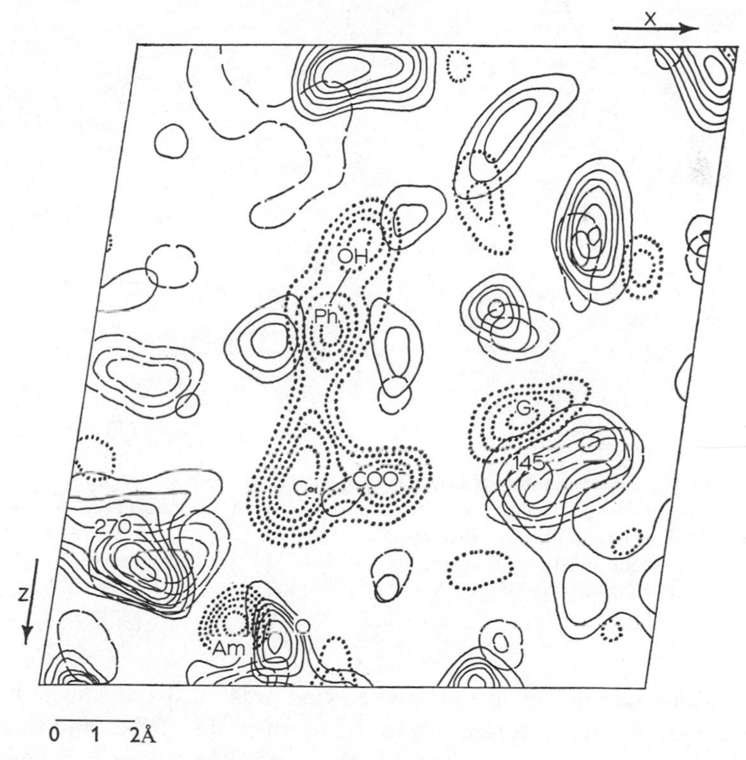

Fig. 1.22 The solid contours are those of carboxypeptidase A at 2.0-Å resolution. The dotted contours indicate the excess electron density when glycyl-L-tyrosine is added to the active site, and negative contours are shown as dashed lines. In this section at $y = 0.50$ the guanidinium group of Arg 145 moves about 2 Å from the negative contours to the dotted contours. Also the carboxyl group of Glu 270 moves out of this section by about 2 Å, as indicated by the negative contours. Features of the substrate Gly–Tyr are part of the phenyl ring (Ph), the OH group, the CO_2^- group, the C_α position and the probable position of the amino group (Am).

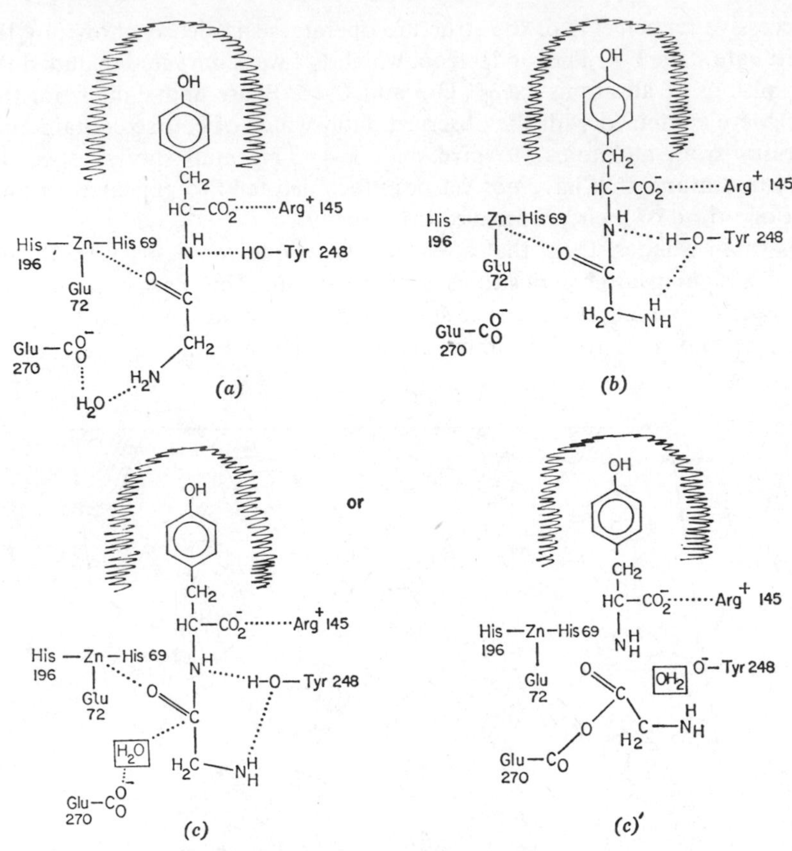

Fig. 1.23 (*a*) Details of binding of Gly–Tyr. (*b*) Probable mode of binding of Gly–Tyr when carboxypeptidase A cleaves the peptide bond. (*c*) General base mode of attack by H_2O, promoted by Glu 270. (*c*′) Nucleophilic attack by Glu 270 on the carbon of the carbonyl group of the substrate, followed by attack of H_2O. It is not certain whether the amino terminus of Gly–Tyr is NH_2 or NH_3^+.

none of which was in this five-membered-ring area. A five-membered ring, shown in F-3, was then "forced" onto the set of peaks of F-2, and then the electron-density map was recomputed with the result shown in F-3. Only the lack of resolution of atoms O-5 and C-12, and especially the nonspherical nature of C-12, prevented an erroneous interpretation. Note that inclusion of the contribution of oxygen at the position of O-5 in F-3, in a totally erroneous position, actually produces a peak more than two-thirds of its proper height. Thus in the computation of the next stage, F-4, the atoms labeled O-5 and C-12 were omitted.

Fig. 1.24 Backbone model, showing each peptide unit as a rod, and each C_α as a dot. The zinc atom is a shaded circle near the center, and arrows indicate the three ligands from the protein to zinc.

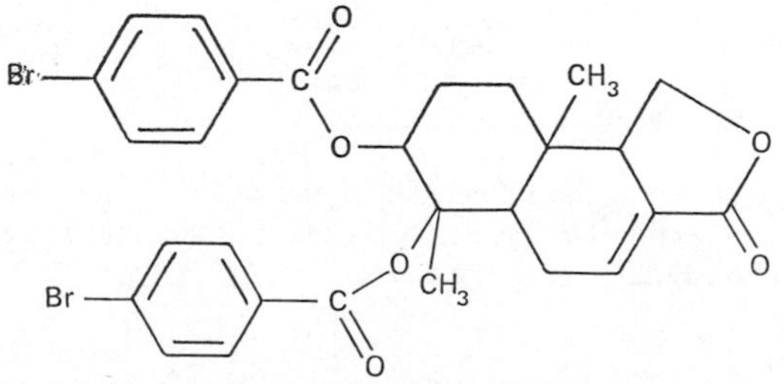

Fig. 1.25 The *p*-bromobenzoate diester of the sesquiterpenoid iresin.

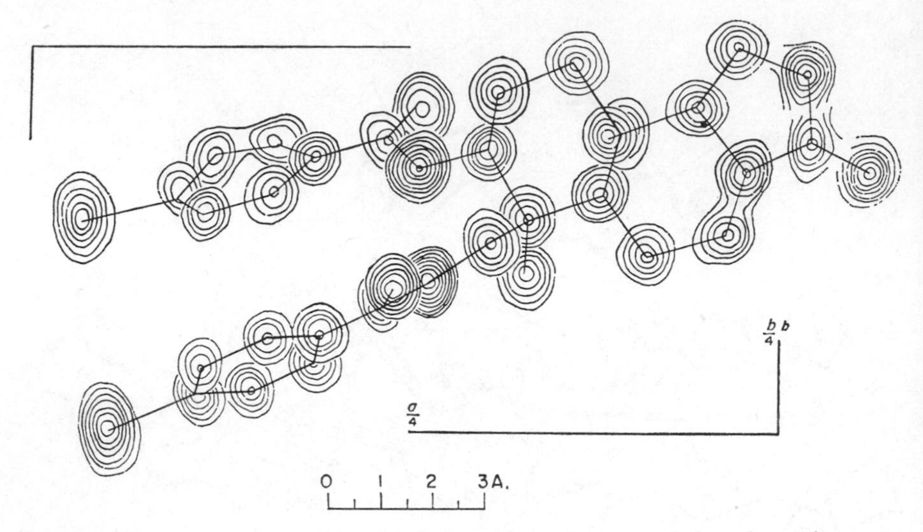

Fig. 1.26 Composite three-dimensional electron density may of the *p*-bromobenzoate diester of iresin. Electron-density contours are at intervals of 1 electron/Å^3 starting at 2 electron/Å^3, except for bromine, for which intervals are 5 electron/Å^3.

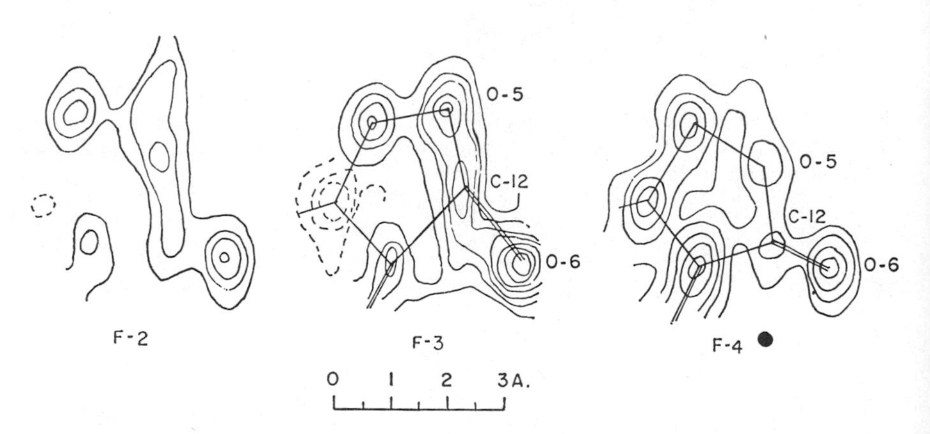

Fig. 1.27 Successive stages in the refinement of the region of the five-membered ring. Solid contours are in the section 14/20 *c*, nearly the plane of the ring, while dashed contours are in the nearby section 13/20 *c* along the *z*-axis.

It is thus common, but not general, experience that in the latter stages of an investigation atoms entered into false positions appear at about two-thirds of their proper height and atoms which have not yet been entered appear at only about one-third of their proper height. If the background of these maps is not sufficiently smooth for these deviations to be recognized, the problem of interpretation of these maps is more difficult. Even at best the procedure is dangerous. Fortunately, there are other guideposts, as mentioned in the first paragraph of this section, and the method is powerful to the point of being very nearly certain in the hands of an experienced and careful investigator.

2 KINEMATICAL THEORY OF X-RAY SCATTERING FROM CRYSTALS*

The assumptions of the kinematical theory—that X-rays suffer no change in wavelength, that boundary effects and refractive index effects are negligible, and that multiple scattering effects are negligible—have generally proved to be quite adequate in the determination of crystal and molecular structures. When difficulties arise their effects usually are easy to circumvent experimentally or to take into account, within the accuracy that one usually desires. Hence in the following sections we outline only the simpler aspects of the theory of scattering of X-rays from crystals. When mathematics beyond ordinary calculus is required we outline it in some detail, and we often appeal to physical grounds in preference to mathematical rigor.

Mathematical Introduction

Vectors

A physical quantity that has both magnitude and direction can be represented by a vector. The length is taken as proportional to the magnitude. Addition of vectors **A** and **B** to give a resultant **C** is carried out by placing the tail of **B** on the head of **A**, or vice versa:

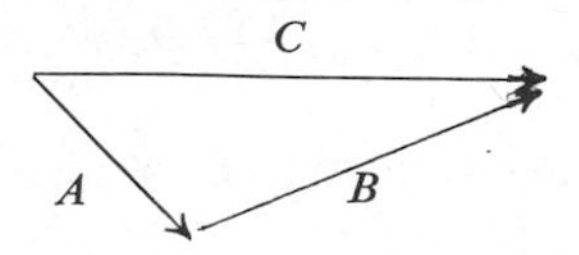

Multiplication of **A** by a scalar g gives a new vector $g\mathbf{A}$, which has the same direction as **A** and has length $g\,|\mathbf{A}|$ or gA, where $|\mathbf{A}|$, or A represents the

* For more detailed discussion of the topics in this section see L. V. Aza'roff, *Elements of X-Ray Crystallography*. McGraw-Hill, New York, 1968; M. J. Buerger, *Crystal Structure Analysis*, Wiley, New York, 1960; W. H. Zachariasen, *Theory of X-Ray Diffraction in Crystals*, Wiley, New York, 1945.

length of **A**. A unit vector may be defined by dividing **A** by its magnitude A. It is then possible to define three mutually orthogonal vectors: **i** along x, **j** along y, and **k** along z. Let the components of **A** be A_x, A_y, and A_z; then $\mathbf{A} = \mathbf{i}A_x + \mathbf{j}A_y + \mathbf{k}A_z$. From the Pythagorean theorem, $|\mathbf{A}| = (A_x^2 + A_y^2 + A_z^2)^{1/2}$ is the length of **A**.

The addition of $\mathbf{A} + \mathbf{B} = \mathbf{C}$ then becomes $A_x + B_x = C_x$, $A_y + B_y = C_y$, and $A_z + B_z = C_z$ in components. Subtraction is now a special case of addition, if one defines $-\mathbf{A}$ as a vector with the same magnitude as **A** but with opposite direction. The associative law, $(\mathbf{A} + \mathbf{B}) + \mathbf{D} = \mathbf{A} + (\mathbf{B} + \mathbf{D})$; the commutative law, $\mathbf{A} + \mathbf{B} = \mathbf{B} + \mathbf{A}$; and the distributive law for scalar multiplication, $g(\mathbf{A} + \mathbf{B}) = g\mathbf{A} + g\mathbf{B}$ are easily proved.

The angle between **A** and **B** is defined as positive if measured from **A** to **B**, when **A** and **B** are drawn from the same point. If θ is the angle $\angle \mathbf{A}, \mathbf{B}$, $-\theta$ is the angle $\angle \mathbf{B}, \mathbf{A}$.

The Scalar (Dot) Product

The scalar, or dot, product is represented by $\mathbf{A} \cdot \mathbf{B}$ and is by definition equal to $AB \cos(\mathbf{A}, \mathbf{B})$. Geometrically this product is A times the projection $B \cos(\mathbf{A}, \mathbf{B})$, or **B** on **A**; or B times the projection of **A** on **B**. The reverse product $\mathbf{B} \cdot \mathbf{A} = BA \cos(\mathbf{B}, \mathbf{A})$; but the angle $(\mathbf{B}, \mathbf{A})$ is $-(\mathbf{A}, \mathbf{B})$ and $\cos(-\theta) = \cos(\theta)$, so that $\mathbf{A} \cdot \mathbf{B} = \mathbf{B} \cdot \mathbf{A}$. The reader can show geometrically that $\mathbf{A} \cdot (\mathbf{B} + \mathbf{C}) = \mathbf{A} \cdot \mathbf{B} + \mathbf{A} \cdot \mathbf{C}$, where **A**, **B**, and **C** are any three vectors. This is the distributive law.

Note that for our **i**, **j**, **k** set of orthogonal vectors $\mathbf{i} \cdot \mathbf{i} = \mathbf{j} \cdot \mathbf{j} = \mathbf{k} \cdot \mathbf{k} = 1$ and $\mathbf{i} \cdot \mathbf{j} = \mathbf{j} \cdot \mathbf{k} = \mathbf{k} \cdot \mathbf{i} = 0$. The distributive law then gives $\mathbf{A} \cdot \mathbf{B}$ in terms of components as $\mathbf{A} \cdot \mathbf{B} = (\mathbf{i}A_x + \mathbf{j}A_y + \mathbf{k}A_z) \cdot (\mathbf{i}B_x + \mathbf{j}B_y + \mathbf{k}B_z) = A_xB_x + A_yB_y + A_zB_z$. A special case is $A^2 = \mathbf{A} \cdot \mathbf{A} = A_x^2 + A_y^2 + A_z^2$, from which the length of **A** may be obtained if the components are known. A relation useful for calculating bond angles in an orthogonal system is

$$\cos(\mathbf{A}, \mathbf{B}) = \frac{\mathbf{A} \cdot \mathbf{B}}{AB} = \frac{A_xB_x + A_yB_y + A_zB_z}{(A_x^2 + A_y^2 + A_z^2)^{1/2}(B_x^2 + B_y^2 + B_z^2)^{1/2}}.$$

The equation $\cos(\mathbf{A}, \mathbf{B}) = (\mathbf{A} \cdot \mathbf{B})/\mathrm{AB}$ may also be expanded in nonorthogonal systems.

The Vector (Cross) Product

The vector, or cross, product is designated by $\mathbf{A} \times \mathbf{B}$ and is defined as $\mathbf{p}AB \sin(\mathbf{A}, \mathbf{B})$, where **p** is a unit vector perpendicular to the plane of **A** and **B**, and **A**, **B**, and **p** form a right-handed system. First note that $AB \sin(\mathbf{A}, \mathbf{B})$ is the area of the parallelogram defined by **A** and **B**. Note also since $\sin(\mathbf{A}, \mathbf{B}) = -\sin(\mathbf{B}, \mathbf{A})$, the cross product is not commutative; that is,

$\mathbf{A} \times \mathbf{B} = -\mathbf{B} \times \mathbf{A}$. However, the cross product is distributive: $\mathbf{A} \times (\mathbf{B} + \mathbf{C}) = \mathbf{A} \times \mathbf{B} + \mathbf{A} \times \mathbf{C}$. An interesting way to restate these equations is in terms of a Cartesian system of unit vectors **i**, **j**, and **k**. The definition then gives $\mathbf{i} \times \mathbf{i} = \mathbf{j} \times \mathbf{j} = \mathbf{k} \times \mathbf{k} = 0$, and $\mathbf{i} \times \mathbf{j} = \mathbf{k} = -\mathbf{j} \times \mathbf{i}$, $\mathbf{k} \times \mathbf{i} = \mathbf{j} = -\mathbf{i} \times \mathbf{k}$, $\mathbf{j} \times \mathbf{k} = \mathbf{i} = -\mathbf{k} \times \mathbf{j}$. The product $\mathbf{A} \times \mathbf{B}$ then becomes, for example,

$$\begin{aligned}\mathbf{A} \times \mathbf{B} &= (\mathbf{i}A_x + \mathbf{j}A_y + \mathbf{k}A_z) \times (\mathbf{i}B_x + \mathbf{j}B_y + \mathbf{k}B_z) \\ &= \mathbf{i}(A_yB_z - A_zB_y) + \mathbf{j}(A_zB_x - A_xB_z) + \mathbf{k}(A_xB_y - A_yB_x) \\ &= \begin{vmatrix} \mathbf{i} & \mathbf{j} & \mathbf{k} \\ A_x & A_y & A_z \\ B_x & B_y & B_z \end{vmatrix}.\end{aligned}$$

The use of the unit vector **p** will become obvious in later discussions. One important application is that if we have three noncoplanar vectors, the scalar triple product

$$\mathbf{A} \cdot \mathbf{B} \times \mathbf{C} = \begin{vmatrix} A_x & A_y & A_z \\ B_x & B_y & B_z \\ C_x & C_y & C_z \end{vmatrix}$$

is the volume of the parallelepiped defined by **A**, **B**, and **C** (Fig. 1.28). This follows because $|\mathbf{B} \times \mathbf{C}|$ is the area of the base and $\mathbf{A} \cdot \mathbf{p}$ is the altitude measured along a line perpendicular to the base. An expression for the volume of the unit cell in terms of the measured values of a, b, c, α, β, and γ

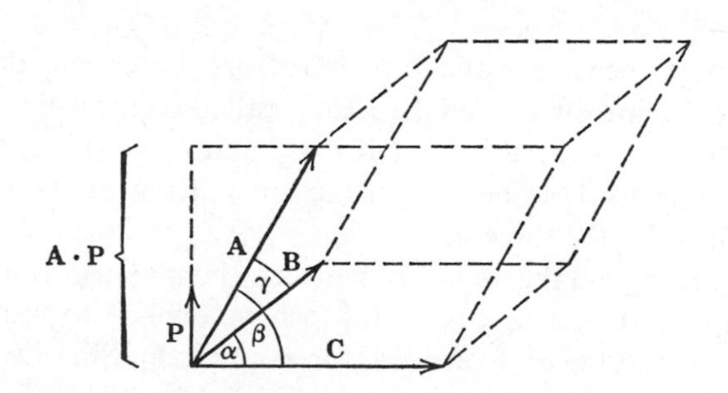

Fig. 1.28 The volume of the parallelepiped defined by **A**, **B**, and **C** is $\mathbf{A} \cdot \mathbf{B} \times \mathbf{C}$.

can now readily be found as follows,

$$V^2 = (\mathbf{a} \cdot \mathbf{b} \times \mathbf{c})^2 = \begin{vmatrix} a_x & a_y & a_z \\ b_x & b_y & b_z \\ c_x & c_y & c_z \end{vmatrix} \begin{vmatrix} a_x & b_x & c_x \\ a_y & b_y & c_y \\ a_z & b_z & c_z \end{vmatrix}$$

$$= \begin{vmatrix} a_x^2 + a_y^2 + a_z^2 & a_xb_x + a_yb_y + a_zb_z & a_xc_x + a_yc_y + a_zc_z \\ a_xb_x + a_yb_y + a_zb_z & b_x^2 + b_y^2 + b_z^2 & b_xc_x + b_yc_y + b_zc_z \\ a_xc_x + a_yc_y + a_zc_z & b_xc_x + b_yc_y + b_zc_z & c_x^2 + c_y^2 + c_z^2 \end{vmatrix}$$

$$= \begin{vmatrix} \mathbf{a} \cdot \mathbf{a} & \mathbf{a} \cdot \mathbf{b} & \mathbf{a} \cdot \mathbf{c} \\ \mathbf{a} \cdot \mathbf{b} & \mathbf{b} \cdot \mathbf{b} & \mathbf{b} \cdot \mathbf{c} \\ \mathbf{c} \cdot \mathbf{a} & \mathbf{c} \cdot \mathbf{b} & \mathbf{c} \cdot \mathbf{c} \end{vmatrix}$$

$$= (\mathbf{a} \cdot \mathbf{a})(\mathbf{b} \cdot \mathbf{b})(\mathbf{c} \cdot \mathbf{c}) + (\mathbf{a} \cdot \mathbf{b})(\mathbf{b} \cdot \mathbf{c})(\mathbf{c} \cdot \mathbf{a}) + (\mathbf{a} \cdot \mathbf{c})(\mathbf{b} \cdot \mathbf{a})(\mathbf{c} \cdot \mathbf{b})$$

$$- (\mathbf{b} \cdot \mathbf{b})(\mathbf{a} \cdot \mathbf{c})^2 - (\mathbf{c} \cdot \mathbf{c})(\mathbf{a} \cdot \mathbf{b})^2 - (\mathbf{a} \cdot \mathbf{a})(\mathbf{c} \cdot \mathbf{b})^2$$

$$= a^2b^2c^2(1 + 2 \cos \alpha \cos \beta \cos \gamma - \cos^2 \alpha - \cos^2 \beta - \cos^2 \gamma).$$

The two properties of determinants used here are that rows and columns may be interchanged without altering the value and that they multiply as indicated. Note that although a Cartesian system was used for the components of **a**, **b**, and **c**, the resulting dot products are independent of the choice of coordinate system and the result has general validity.

Reciprocal Vectors

Let $\mathbf{a}_1$, $\mathbf{a}_2$, and $\mathbf{a}_3$ be three noncoplanar vectors. Later they define the unit cell of a crystal. Define three related vectors, called reciprocal vectors, by the equations $\mathbf{a}_i \cdot \mathbf{b}_j = \delta_{ij}$, where $\delta_{ij} = 1$ if $i = j$ and $\delta_{ij} = 0$ if $i \neq j$. Later the vectors $\mathbf{b}_j$ are used to describe the diffraction pattern of the crystal whose unit cell is defined by the three $\mathbf{a}_i$.

We measure the $\mathbf{a}_i$ and the angles α_i between them (α_1 is the angle between $\mathbf{a}_2$ and $\mathbf{a}_3$, and so on). Hence it is useful to have explicit formulas for the $\mathbf{b}_j$ in terms of the $\mathbf{a}_i$ and the α_i. Consider, for example $\mathbf{b}_1$. Since $\mathbf{a}_2 \cdot \mathbf{b}_1 = 0$ and $\mathbf{a}_3 \cdot \mathbf{b}_1 = 0$, $\mathbf{b}_1$ must be perpendicular to both $\mathbf{a}_2$ and $\mathbf{a}_3$. Therefore $\mathbf{b}_1$ must be parallel to $\mathbf{a}_2 \times \mathbf{a}_3$. If ϵ is some scalar to be determined, then $\mathbf{b}_1 = \epsilon(\mathbf{a}_2 \times \mathbf{a}_3)$.

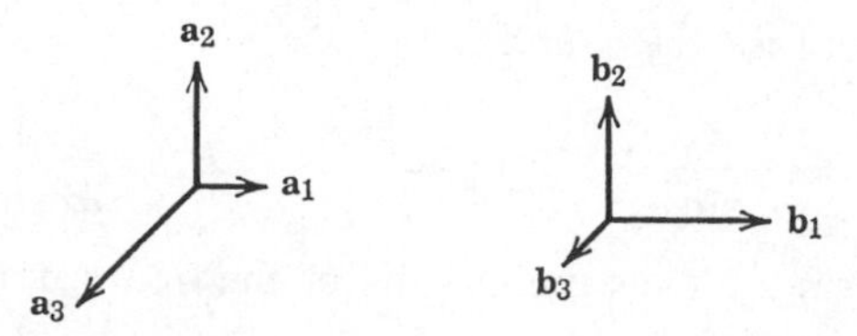

Fig. 1.29 Relations between crystal vectors, **a**, and reciprocal vectors, **b**, in the orthorhombic system. Although the origins are separated here, it is sometimes useful to regard them as common. If the $\mathbf{a}_i$ are measured in angstroms, the $\mathbf{b}_j$ are measured in Å^{-1}.

But on taking the dot product on both sides with $\mathbf{a}_1$, we obtain $1 = \epsilon(\mathbf{a}_1 \cdot \mathbf{a}_2 \times \mathbf{a}_3) = \epsilon v_a$, where v_a is the cell volume. Hence

$$\mathbf{b}_1 = \frac{\mathbf{a}_2 \times \mathbf{a}_3}{v_a}, \qquad \mathbf{b}_2 = \frac{\mathbf{a}_3 \times \mathbf{a}_1}{v_a}, \qquad \mathbf{b}_3 = \frac{\mathbf{a}_1 \times \mathbf{a}_2}{v_a} \tag{1.12}$$

$$\mathbf{a}_1 = \frac{\mathbf{b}_2 \times \mathbf{b}_3}{v_b}, \qquad \mathbf{a}_2 = \frac{\mathbf{b}_3 \times \mathbf{b}_1}{v_b}, \qquad \mathbf{a}_3 = \frac{\mathbf{b}_1 \times \mathbf{b}_2}{v_b}, \tag{1.13}$$

which follow because the relations are symmetrical to a cyclic permutation, and the definition is symmetrical in $\mathbf{a}_i$ and $\mathbf{b}_j$. The volumes v_a and v_b are reciprocally related ($v_a = 1/v_b$):

$$v_a v_b = \begin{vmatrix} a_{1x}a_{1y}a_{1z} \\ a_{2x}a_{2y}a_{2z} \\ a_{3x}a_{3y}a_{3z} \end{vmatrix} \begin{vmatrix} b_{1x}b_{2x}b_{3x} \\ b_{1y}b_{2y}b_{3y} \\ b_{1z}b_{2z}b_{3z} \end{vmatrix} = \begin{vmatrix} \mathbf{a}_1 \cdot \mathbf{b}_1 & \mathbf{a}_1 \cdot \mathbf{b}_2 & \mathbf{a}_1 \cdot \mathbf{b}_3 \\ \mathbf{a}_2 \cdot \mathbf{b}_1 & \mathbf{a}_2 \cdot \mathbf{b}_2 & \mathbf{a}_2 \cdot \mathbf{b}_3 \\ \mathbf{a}_3 \cdot \mathbf{b}_1 & \mathbf{a}_3 \cdot \mathbf{b}_2 & \mathbf{a}_3 \cdot \mathbf{b}_3 \end{vmatrix} = \begin{vmatrix} 100 \\ 010 \\ 001 \end{vmatrix} = 1.$$

A geometrical interpretation is very helpful. Let $\alpha_1 = \alpha_2 = \alpha_3 = 90°$ and $a_1 < a_2 < a_3$. These relations are valid in the orthorhombic system. Then $b_1 = 1/a_1$, $b_2 = 1/a_2$, $b_3 = 1/a_3$, $\mathbf{b}_1 \parallel \mathbf{a}_1$, $\mathbf{b}_2 \parallel \mathbf{a}_2$, $\mathbf{b}_3 \parallel \mathbf{a}_3$, $v_a = a_1 a_2 a_3$, and $v_b = 1/a_1 a_2 a_3$ (Fig. 1.29). Now consider the monoclinic system, where $\alpha_1 = \alpha_3 = 90°$, and choose $\alpha_2 = 110°$. Again let $a_1 < a_2 < a_3$ (Fig. 1.30).

Fig. 1.30 Relations between crystal vectors $\mathbf{a}_i$ and reciprocal vectors $\mathbf{b}_j$ in the monoclinic system. $\beta_2 = 180° - \alpha_2$.

The equations between the a_i and the b_j are

$$b_1 = \frac{1}{a_1 \sin \alpha_2}, \qquad b_2 = \frac{1}{a_2}, \qquad b_3 = \frac{1}{a_3 \sin \alpha_2},$$

and $v_a = a_1 a_2 a_3 \sin \alpha_2$. Thus the lengths of the $\mathbf{b}_j$ are not always the simple reciprocals of the a_i.

We make use of these geometrical interpretations later. Meanwhile let us consider another application of reciprocal vectors, the evaluation of the product $\mathbf{a}_1 \times (\mathbf{a}_2 \times \mathbf{a}_3)$. The result of this operation, by the definitions of the cross product, must be a vector in $\mathbf{a}_1$, $\mathbf{a}_2$, and $\mathbf{a}_3$ space. Therefore

$$\mathbf{a}_1 \times (\mathbf{a}_2 \times \mathbf{a}_3) = l\mathbf{a}_1 + m\mathbf{a}_2 + n\mathbf{a}_3.$$

Now taking the dot product of both sides with $\mathbf{b}_1$, $\mathbf{b}_2$, and $\mathbf{b}_3$, we obtain the three equations

$$l = \mathbf{b}_1 \cdot \mathbf{a}_1 \times (\mathbf{a}_2 \times \mathbf{a}_3)$$

$$m = \mathbf{b}_2 \cdot \mathbf{a}_1 \times (\mathbf{a}_2 \times \mathbf{a}_3)$$

$$n = \mathbf{b}_3 \cdot \mathbf{a}_1 \times (\mathbf{a}_2 \times \mathbf{a}_3).$$

Recognizing that $\mathbf{a}_2 \times \mathbf{a}_3 = v_a \mathbf{b}_1$,

$$l = (\mathbf{b}_1 \cdot \mathbf{a}_1 \times \mathbf{b}_1) v_a$$

$$m = (\mathbf{b}_2 \cdot \mathbf{a}_1 \times \mathbf{b}_1) v_a$$

$$n = (\mathbf{b}_3 \cdot \mathbf{a}_1 \times \mathbf{b}_1) v_a.$$

Rearranging,

$$l = (\mathbf{a}_1 \cdot \mathbf{b}_1 \times \mathbf{b}_1) v_a = 0$$

$$m = (\mathbf{a}_1 \cdot \mathbf{b}_1 \times \mathbf{b}_2) v_a = (\mathbf{a}_1 \cdot \mathbf{a}_3) v_a v_b = (\mathbf{a}_1 \cdot \mathbf{a}_3)$$

$$n = (\mathbf{a}_1 \cdot \mathbf{b}_1 \times \mathbf{b}_3) v_a = -(\mathbf{a}_1 \cdot \mathbf{a}_2).$$

Therefore

$$\mathbf{a}_1 \times (\mathbf{a}_2 \times \mathbf{a}_3) = (\mathbf{a}_1 \cdot \mathbf{a}_3)\mathbf{a}_2 - (\mathbf{a}_1 \cdot \mathbf{a}_2)\mathbf{a}_3. \tag{1.14}$$

A number of further relations are now easily proved. For example, $(\mathbf{A} \times \mathbf{B}) \cdot (\mathbf{C} \times \mathbf{D}) = (\mathbf{C} \cdot \mathbf{A})(\mathbf{D} \cdot \mathbf{B}) - (\mathbf{C} \cdot \mathbf{B})(\mathbf{A} \cdot \mathbf{D})$, and $(\mathbf{A} \times \mathbf{B}) \times (\mathbf{C} \times \mathbf{D}) = \mathbf{C}(\mathbf{D} \cdot \mathbf{A} \times \mathbf{B}) - \mathbf{D}(\mathbf{C} \cdot \mathbf{A} \times \mathbf{B})$. Interchange of the dot and the cross is possible: $\mathbf{A} \cdot \mathbf{B} \times \mathbf{C} = \mathbf{A} \times \mathbf{B} \cdot \mathbf{C}$. The useful relation

$$\cos \beta_3 = \frac{\cos \alpha_1 \cos \alpha_2 - \cos \alpha_3}{\sin \alpha_1 \sin \alpha_2}$$

may be proved by expressing all the $\mathbf{b}_j$ in $\mathbf{b}_1 \cdot \mathbf{b}_2 = (\mathbf{b}_1 \cdot \mathbf{b}_1)^{1/2}(\mathbf{b}_2 \cdot \mathbf{b}_2)^{1/2} \cos \beta_3$ in terms of the $\mathbf{a}_i$ and then making use of some of the preceding equations. Finally, the expression of a vector $\mathbf{V}$ in terms of either the $\mathbf{a}_i$ or the $\mathbf{b}_j$ is sometimes useful. First write

$\mathbf{V} = r\mathbf{a}_1 + s\mathbf{a}_2 + t\mathbf{a}_3$; then form the products $\mathbf{b}_1 \cdot \mathbf{V} = r$, and so on. The result is

$$\begin{aligned} \mathbf{V} &= (\mathbf{b}_1 \cdot \mathbf{V})\mathbf{a}_1 + (\mathbf{b}_2 \cdot \mathbf{V})\mathbf{a}_2 + (\mathbf{b}_3 \cdot \mathbf{V})\mathbf{a}_3 \\ &= (\mathbf{a}_1 \cdot \mathbf{V})\mathbf{b}_1 + (\mathbf{a}_2 \cdot \mathbf{V})\mathbf{b}_2 + (\mathbf{a}_3 \cdot \mathbf{V})\mathbf{b}_3. \end{aligned} \tag{1.15}$$

Complex Numbers

A complex number $Z = A + iB$, where $i = \sqrt{-1}$, is like a vector in two dimensions. The components A and B can be plotted in an orthogonal system, in which the unit vectors are 1 along the horizontal axis and i along the vertical axis (Fig. 1.31). The real and the imaginary parts are treated like components of a vector in addition: $Z_1 + Z_2 = A_1 + A_2 + i(B_1 + B_2)$. The magnitude of a complex number is the length

$$|Z| = (ZZ^*)^{1/2} = [(A + iB)(A - iB)]^{1/2} = (A^2 + B^2)^{1/2}.$$

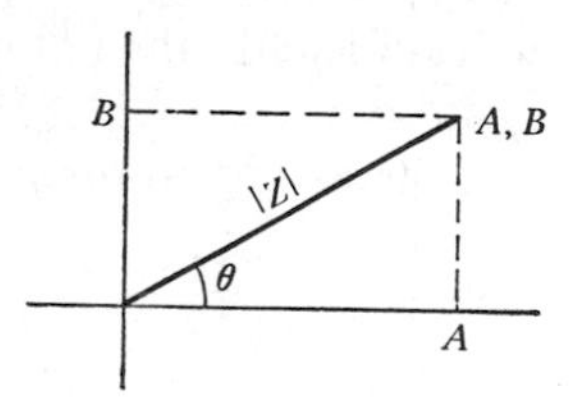

Fig. 1.31 The complex number $Z = A + iB$.

The polar representation is also useful. Define $R = |Z|$, and θ by the equations $\sin\theta = B/R$ and $\cos\theta = A/R$. Then $Z = A + iB = R(\cos\theta + i\sin\theta)$, which can be written more conveniently as $Re^{i\theta}$ from Euler's formula. To prove this, define the function $f(\theta)$ to be determined as $f(\theta) = \cos\theta + i\sin\theta$. Differentiation then gives $df(\theta)/d\theta = -\sin\theta + i\cos\theta$. Comparison with $f(\theta)$ then shows that $df(\theta)/d\theta = if(\theta)$, or $df(\theta)/f(\theta) = i\,d\theta$, which on integration gives $\ln f(\theta) = i\theta + \ln K$, where K is a constant. But, when $\theta = 0$, $f(\theta) = 1$; hence $K = 1$, and $f(\theta) = e^{i\theta}$, which is $\cos\theta + i\sin\theta$. The complex conjugate is $e^{-i\theta} = \cos\theta - i\sin\theta$. Note that the magnitude of $e^{i\theta}$ is $e^{i\theta}e^{-i\theta} = 1 = \cos^2\theta + \sin^2\theta$, so that the points on the plane of Fig. 1.31 representing $e^{i\theta}$ all lie on a circle of radius 1. Later it will be useful to remember this circle when expressions such as $e^{2\pi ih/4}$ are encountered. When h is an integer, 0, 1, 2, . . . , this expression has the values 1, i, -1, $-i$, and so on.

Fourier Series

A function $f(x)$, reasonably well-behaved (discontinuous at a finite number of points) and periodic with interval p, can be expressed as

$$f(x) = \sum_{n=-\infty}^{\infty} a_n \exp\left(\frac{2\pi inx}{p}\right), \tag{1.16}$$

where

$$a_n = \frac{1}{p}\int_{-p/2}^{p/2} f(x) \exp\left(\frac{-2\pi i n x}{p}\right) dx.$$

Extending p to infinity and considering functions that are small for numerically large values of x, we obtain

$$f(x) = \int_{-\infty}^{\infty} F(t) \exp(2\pi i t x)\, dt$$

$$F(t) = \int_{-\infty}^{-\infty} f(x) \exp(-2\pi i t x)\, dx.$$

The function $F(t)$ is called the Fourier transform of $f(x)$. If x is measured in units of length, then t must have units of 1/length, since tx must be a unitless quantity.

Generalizing to three dimensions,

$$f(\mathbf{r}) = \int F(\mathbf{h}) \exp(2\pi i \mathbf{h} \cdot \mathbf{r})\, dv_b \qquad (1.17a)$$

$$F(\mathbf{h}) = \int f(\mathbf{r}) \exp(-2\pi i \mathbf{h} \cdot \mathbf{r})\, dv_a, \qquad (1.17b)$$

where $\mathbf{r} = x\mathbf{a}_1 + y\mathbf{a}_2 + z\mathbf{a}_3$, expressed in terms of three noncoplanar vectors, $\mathbf{a}_1$, $\mathbf{a}_2$, and $\mathbf{a}_3$. Since $\mathbf{h}$ must be a vector of dimensions $(\text{length})^{-1}$, it can then be expressed in terms of components of the $\mathbf{b}_1$, $\mathbf{b}_2$, and $\mathbf{b}_3$ vectors as follows:

$$\mathbf{h} = h\mathbf{b}_1 + k\mathbf{b}_2 + l\mathbf{b}_3.$$

Thus

$$f(\mathbf{r}) = \iiint F(\mathbf{h}) \exp[2\pi i(hx + ky + lz]v_b\, dh\, dk\, dl \qquad (1.18a)$$

$$F(\mathbf{h}) = \iiint f(\mathbf{r}) \exp[-2\pi i(hx + ky + lz)]v_a\, dx\, dy\, dz, \qquad (1.18b)$$

where v_a and v_b are the volumes in real and reciprocal space, respectively.

Superposition of Waves

Consider the function $E = E_o \cos 2\pi(\alpha x - \beta t + \delta)$, where E_o, α, β, and δ are constants. At $t = 0$, $E_{t=0} = E_o \cos 2\pi(\alpha x + \delta)$, which is a function that repeats itself in space when the argument $(\alpha x + \delta) = 0, 1, 2$, and so on; that is, when $x = -\delta/\alpha$, $-\delta/\alpha + 1/\alpha$, $-\delta/\alpha + 2/\alpha$, and so forth. The distance between successive crests (Fig. 1.32) is thus $1/\alpha$, and is defined as the wavelength $\lambda = 1/\alpha$. At a later time t_1 the wave looks just the same except that it has moved along the x-axis by an amount $\beta t_1/\alpha$ (Fig. 1.32*b*). In other

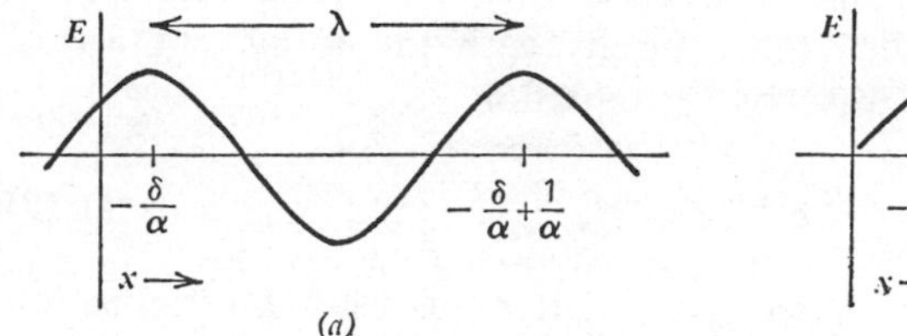

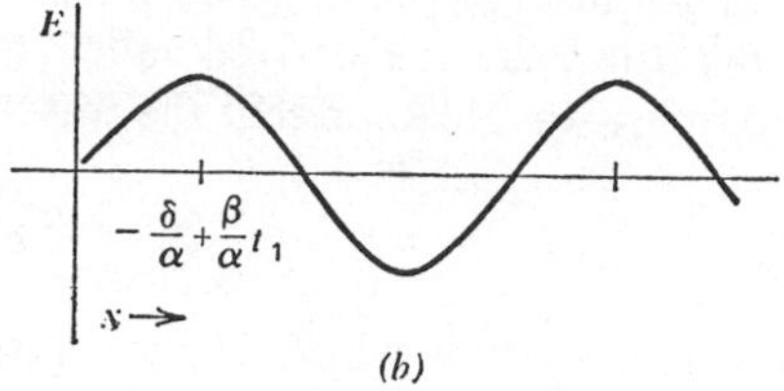

Fig. 1.32 (*a*) The wave $E_{t=0} = E_0 \cos 2\pi(\alpha x + \delta)$ at $t = 0$. (*b*) The wave $E_{t=t_1} = E_0 \cos 2\pi(\alpha x - \beta t_1 + \delta)$ at a slightly later time t_1.

words, maxima occur for $E_{t=t_1}$ when $\alpha x - \beta t_1 + \delta = 0, 1, 2, \ldots$, that is, when $x = -\delta/\alpha + \beta t_1/\alpha, -\delta/\alpha + 1/\alpha + \beta t_1/\alpha, -\delta/\alpha + 2/\alpha + \beta t_1/\alpha$, and so on. The wave velocity u is the distance $\beta t_1/\alpha$ that any given crest has moved divided by the time t_1; hence $u = \beta/\alpha$, and since $\lambda = 1/\alpha$, we find $\beta = u/\lambda$. But this is the frequency, $\nu = u/\lambda$, the number of waves per unit of time. We can then write the wave in the form

$$E = E_o \cos 2\pi(x/\lambda - \nu t + \delta),$$

where the wave velocity $u = \lambda\nu$. This idealization will do for the usual crystallographic studies in which refractive indices can be assumed to be unity within 1 part in 10^5, boundary effects are neglected, coherent attenuation of the wave in the crystal is neglected, and a strictly monochromatic wave is assumed.

The scattering between two different electrons at different points in space (different δ's) is then obtained by adding the amplitudes of scattering from each. The result is a simple trigonometric function. First expand

$$\begin{aligned} E = E_1 + E_2 &= E_{o_1} \cos 2\pi(x/\lambda - \nu t + \delta_1) + E_{o_2} \cos 2\pi(x/\lambda - \nu t + \delta_2) \\ &= (E_{o_1} \cos 2\pi\delta_1 + E_{o_2} \cos 2\pi\delta_2) \cos 2\pi(x/\lambda - \nu t) \\ &\qquad - (E_{o_1} \sin 2\pi\delta_1 + E_{o_2} \sin 2\pi\delta_2) \sin 2\pi(x/\lambda - \nu t). \end{aligned}$$

If we define $E_o' \cos 2\pi\delta_3 = E_{o_1} \cos 2\pi\delta_1 + E_{o_2} \cos 2\pi\delta_2$, and $E_o' \sin 2\pi\delta_3 = E_{o_1} \sin 2\pi\delta_1 + E_{o_2} \sin 2\pi\delta_2$, we find $E = E_o' \cos 2\pi(x/\lambda - \nu t + \delta_3)$. We can also find E_o', and δ_3 from the defining equations,

$$(E_o')^2 = E_{o_1}^2 + E_{o_2}^2 + 2E_{o_1}E_{o_2} \cos 2\pi(\delta_1 - \delta_2).$$

Thus the amplitudes are additive, but the intensities that are proportional to the square of the amplitudes [59] are not additive in general and depend in a sensitive and important way on the constants δ_1 and δ_2, which we later relate to the positions of the scattering centers in space.

This last equation, which demonstrates the effect of interference on the intensities, can be derived much more briefly with the equivalent use of complex waves. Represent the wave by the equation

$$E = E_o e^{2\pi i(x/\lambda - vt + \delta)}, \tag{1.19}$$

the real part of which is the same as we had previously. Multiplication by the complex conjugate gives E_o^2, the square of the amplitude constant. The two superimposed waves are

$$E = E_1 + E_2 = E_{o_1} e^{2\pi i(x/\lambda - vt + \delta_1)} + E_{o_2} e^{2\pi i(x/\lambda - vt + \delta_2)}.$$

The intensity is then, aside from a proportionality constant,

$$EE^* = E_{o_1}^2 + E_{o_2}^2 + 2E_{o_1}E_{o_2} \cos 2\pi(\delta_1 - \delta_2),$$

where we have used the relations $2 \cos \theta = e^{i\theta} + e^{-i\theta}$, and $\cos \theta = \cos(-\theta)$.

Scattering of X-Rays by a Single Electron

The periodically varying electric field of a light wave accelerates the electron into a periodic motion, and hence it emits an electromagnetic wave in all directions, except precisely along the direction of its acceleration. The necessity that the electric vector of a light wave shall be perpendicular to the direction of propagation causes the scattered wave to be polarized (Fig. 1.33). Consider an initially unpolarized wave incident along $\mathbf{s}_0$, and resolve all electric vectors into components along $E_\perp{}^i$ or $E_\parallel{}^i$. Since these two directions are equally probable in an unpolarized wave, $(E_\perp{}^i)^2 = (E_\parallel{}^i)^2$. In the scattering process the amplitude in any given direction is reduced. At zero angle of scattering the two directions ($\perp$ and $\parallel$) for E, the amplitude of the scattered wave, are equally probable, and so $(E_\perp)^2 = (E_\parallel)^2$. But when scattering at an angle ψ takes place, $E_\parallel$ is reduced to $E_\parallel \cos \psi$, the component of $E_\parallel$ which is perpendicular to the direction of

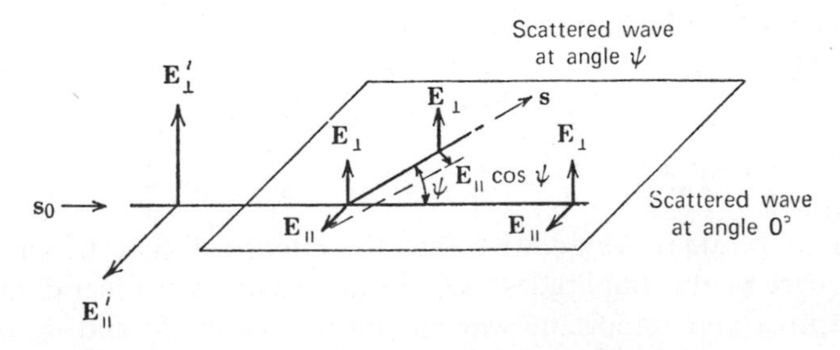

Fig. 1.33 Polarization of wave scattered at angle ψ.

propagation. Hence

$$P = \frac{\text{intensity of scattering at angle } \psi}{\text{intensity of scattering at angle } 0°} = \frac{E_\perp{}^2 + (E_\parallel \cos \psi)^2}{E_\perp{}^2 + E_\parallel{}^2}.$$

But $E_\perp{}^2 = E_\parallel{}^2$, and hence

$$P = \frac{1 + \cos^2 \psi}{2} \tag{1.20}$$

is the polarization factor, the factor by which the density of scattering is reduced by polarization of part of the scattered radiation. Later we show that $\psi = 2\theta$, where θ is the Bragg angle.

In the complete expression for the intensity I_e of a wave I scattered by a single electron at the origin,

$$I_e = I_o \frac{e^4}{r^2 m^2 c^4} \left(\frac{1 + \cos^2 \psi}{2}\right),$$

only the polarization factor is of use to us later. The expected inverse square dependence does not appear since we shall integrate the total intensity over the area of a reflection in a later section. The physical constants e, m, and c, the charge and mass of the electron and the velocity of light, are not usually needed because the scale of the diffracted maxima is ordinarily obtained from a comparison of observed and calculated amplitudes of scattering from the crystal, as is described later. In summary, we shall show that the integrated intensity of scattering I_{hkl} in a maximum characterized by the numbers h, k, and l is

$$I_{hkl} = kLPT\,|F_{hkl}|^2, \tag{1.21}$$

where k is the scale factor, P is the polarization factor, L is the Lorentz factor, which arises because of the integration of the intensity over the maximum, T is the temperature factor, $e^{-B(\sin^2\theta/\lambda^2)}$, where B is determined empirically and $|F_{hkl}|^2$ is calculated from the expression

$$F_{hkl} = \sum_{\substack{\text{unit}\\\text{cell}}} f_j e^{-2\pi i(hx_j + ky_j + lz_j)}. \tag{1.22}$$

These expressions will now be derived, and the symbols are defined in the following sections.

The Atomic Scattering Factor, *f*, and the Structure Factor, F_{hkl}

Define F_{hkl} as the ratio of the amplitude of scattering by the contents of the unit cell to the amplitude of scattering by a single electron at the origin of the unit cell.

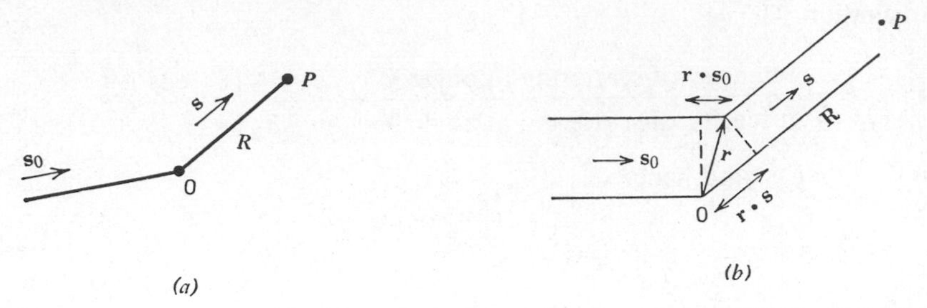

Fig. 1.34 (*a*) The unit vectors $\mathbf{s}_0$ and $\mathbf{s}$ represent directions of the incident and scattered waves, respectively. (*b*) Scattering from a point not at the origin. The point of observation P is very far away compared with the distance $\mathbf{r}$.

Consider a wave (Fig. 1.34*a*) incident upon an electron at the origin. The scattered wave has amplitude $E = E_0 e^{2\pi i(|\mathbf{R}|/\lambda - vt + \delta)}$, where $|\mathbf{R}| = \mathbf{R} \cdot \mathbf{s}$ is the distance from the origin O to the point of observation P. If there are n electrons, then

$$E = E_o \exp\,[2\pi i(|\mathbf{R}|/\lambda - vt)][e^{2\pi i\delta_1} + e^{2\pi i\delta_2} + e^{2\pi i\delta_3} + \cdots + e^{2\pi i\delta_n}],$$

where $\delta_1, \delta_2, \delta_3, \ldots$ are the phases of the n-component waves. If $\mathbf{r}_j$ is drawn from electron 1 at the origin to electron j, then the difference in phase, $\delta_j - \delta_1$, is given by $(\mathbf{r}_j \cdot \mathbf{s}_0 - \mathbf{r}_j \cdot \mathbf{s})/\lambda$, as is shown in Fig. 1.34*b*. Thus

$$E = E_o \exp\,[2\pi i(|\mathbf{R}|/\lambda - vt + \delta_1)]F',$$

where

$$F' = 1 + \exp\,[-2\pi i\mathbf{r}_2 \cdot (\mathbf{s} - \mathbf{s}_0)/\lambda] + \cdots + \exp\,[-2\pi i\mathbf{r}_n \cdot (\mathbf{s} - \mathbf{s}_0)/\lambda].$$

But this expression is just a special form of a Fourier series. More generally,

$$F' = \int_V \rho(\mathbf{r}) \exp\,[-2\pi i\mathbf{r} \cdot (\mathbf{s} - \mathbf{s}_0)/\lambda]\, dV, \tag{1.23}$$

where $\rho(\mathbf{r})$ is the electron density.

Let θ be any angle between $\mathbf{r}$ and $\mathbf{s} - \mathbf{s}_0$. Then the volume can be expressed in spherical coordinates as

$$dV = r^2 \sin\theta\, dr\, d\theta\, d\varphi.$$

Also,

$$(\mathbf{s} - \mathbf{s}_0)^2 = (\mathbf{s}^2 + \mathbf{s}_0^{\ 2} - 2|\mathbf{s}| \cdot |\mathbf{s}_0| \cos\psi) = 4\left(\frac{\sin\psi}{2}\right)^2$$

since both $\mathbf{s}$ and $\mathbf{s}_0$ are unit vectors, where ψ is the angle between $\mathbf{s}$ and $\mathbf{s}_0$.

Therefore

$$F'(\psi) = \int_V \rho(\mathbf{r}) \exp(ikr\cos\theta) r^2 \sin\theta \, dr \, d\theta \, d\varphi, \tag{1.24}$$

where

$$k = \frac{4\pi \sin \psi/2}{\lambda}.$$

We have made no assumptions yet as to the state of the material giving rise to the diffraction; hence this expression is perfectly general.

Now if $\rho(\mathbf{r})$ is a function of r only,

$$F'(\psi) = 4\pi \int_{-\infty}^{-\infty} \rho(r) r^2 \frac{\sin kr}{kr} \, dr \tag{1.25}$$

and would be the appropriate expression to use for diffraction of gases or most liquids.

In the case of a crystal

$$\mathbf{r} = (x + p)\mathbf{a}_1 + (y + m)\mathbf{a}_2 + (z + n)\mathbf{a}_3,$$

where p, m, and n are integers. Now since the dot product of $\mathbf{r}$ and $(\mathbf{s} - \mathbf{s}_0)/\lambda$ must be a dimensionless quantity, the latter must be expressible in terms of any three basis vectors in the reciprocal space, hence in terms of $\mathbf{b}_1$, $\mathbf{b}_2$, and $\mathbf{b}_3$:

$$(\mathbf{s} - \mathbf{s}_0)/\lambda = h\mathbf{b}_1 + k\mathbf{b}_2 + l\mathbf{b}_3 = \mathbf{h}. \tag{1.26}$$

Therefore

$$\mathbf{r} \cdot (\mathbf{s} - \mathbf{s}_0)/\lambda = hx + ky + lz + hp + km + ln$$

and

$$F'(\psi) = \sum_{p,m,n} \int_{\substack{\text{unit}\\\text{cell}}} \rho(\mathbf{r}) \exp[-2\pi i(hx + ky + lz + hp + km + ln)] \, dV$$

$$= \sum_{p,m,n} \exp[-2\pi i(hp + km + ln)] \int_V \rho(\mathbf{r}) \exp[-2\pi i(hx + ky + lz)] \, dV,$$

where the volume is now that of the volume of the unit cell. If the special case of one electron at the origin of each unit cell is considered, the equation reduces to

$$g(\psi) = \sum_p \sum_m \sum_n e^{-2\pi i(hp+km+ln)}$$

$$= \left(\sum_{p=0}^{N_1} e^{-2\pi i hp}\right)\left(\sum_{m=0}^{N_2} e^{-2\pi i km}\right)\left(\sum_{n=0}^{N_3} e^{-2\pi i ln}\right),$$

where N_1, N_2, and N_3 are the number of unit cells in each of the three directions, respectively.

Diffraction maxima occur whenever the phase difference is equal to an integer. Therefore

$$hp + km + ln = \text{integer}$$

for all values of p, m, and n, but since p, m, and n must be integers, it follows that h, k, and l must also be integers.

We have shown that the function g can have maxima only at integer values of h, k, and l. Next let us consider the width of these diffraction maxima. The sum of the series can be calculated as

$$\sum_{p=0}^{N_1} e^{-2\pi i h p} = \frac{1 - e^{-2\pi i N_1 h}}{1 - e^{-2\pi i h}},$$

when $h \neq$ integer. However, when $h =$ integer, the series sum has the value

$$= N_1 + 1.$$

Expressing nonintegral values of h as $h' + \Delta h$, where h' is integer, the ratio of the intensity for a general value of h to that at diffraction maxima is

$$\frac{I(h)}{I(h')} = \frac{2}{(N_1 + 1)^2(1 - \cos 2\pi\,\Delta h)}.$$

Thus for $N_1 = 1000$, the ratio of the intensity at $\Delta h = 0.01$ is approximately 0.001 for the value at the maximum of the crystal peak. It is apparent that the diffraction maxima for crystals are very sharp indeed. Because of the mosaic character of crystals and because of experimental conditions, maxima broaden slightly, and consequently integrated intensities are ordinarily used.

The general expression for $F'(\psi)$ can now be written as

$$F'(\psi) = g\left(\int_V \rho(\mathbf{r}) e^{-2\pi i(hx+ky+lz)}\, dV\right),$$

and $F'(\psi)$ can have nonzero values only where h, k, and l are all integers, since g has nonzero values only at these points. Consequently, if we divide $F'(\psi)$ by g, the scattering caused by one electron at the origin of each cell, we obtain

$$F_{hkl} = \frac{F'(\psi)}{g} = \int_V \rho(\mathbf{r}) e^{-2\pi i(hx+ky+lz)}\, dV, \tag{1.22a}$$

where F_{hkl} is called the structure factor. Note that the intensities at these hkl points will be proportional to $|F_{hkl}|^2$ and thus depend upon the arrangement of electrons, and hence atoms in the unit cell, but the positions of the maxima will be determined by $\mathbf{h}$ and therefore by the reciprocal vectors $\mathbf{b}_1$, $\mathbf{b}_2$, and $\mathbf{b}_3$.

The expression for the structure factor can be modified somewhat by substituting

$$\rho(\mathbf{r}) = \sum_{j=1}^{N} \rho_j(\mathbf{r}' + \mathbf{r}_j),$$

where ρ_j is the electron density associated with atom j, $\mathbf{r}'$ is measured from the atom center, and $\mathbf{r}_j$ denotes the position of atom j. Thus

$$F_{hkl} = \sum_{j=1}^{N} \int_{\text{atom}} \rho_j(\mathbf{r}') e^{-2\pi i[h(x_j+x')+k(y_j+y')+l(z_j+z')]} \, dV$$

$$= \sum_{j=1}^{N} e^{-2\pi i(hx_j+ky_j+lz_j)} \int_{\text{atom}} \rho_j(\mathbf{r}') e^{-2\pi i(hx'+ky'+lz')} \, dV'.$$

Furthermore if the assumption is made that in the molecule $\rho_j(\mathbf{r}')$ is essentially the same as for the free atom, then this integral can be evaluated once an approximate wave function, ψ, is known from quantum mechanics, since

$$\rho(\mathbf{r}') = \psi\psi^*.$$

Defining f, the atomic scattering factor, as

$$f = \int_{\text{atom}} \rho(\mathbf{r}') e^{-2\pi i(hx'+ky'+lz')} \, dV'$$

$$= \int_{\text{atom}} \rho(\mathbf{r}') e^{ikr' \cos\theta} \, dr' \, d\theta \, d\varphi \, r'^2 \sin\theta, \tag{1.27}$$

the structure factor may now be written as

$$F_{hkl} = \sum_{j=1}^{N} f_j e^{-2\pi i(hx_j+ky_j+lz_j)}, \tag{1.22}$$

where now x_j, y_j, and z_j are the coordinates of the atom centers, and the summation is over the N atoms in the unit cell.

Let us assume that $\rho(x, y, z)$ is known from solutions of the Schroedinger wave equation for many electron atoms and that it is a function of r only; in other words, that the atoms are spherically symmetrical. Then $\rho(x, y, z) = \rho(r)$, and we can integrate over θ and φ in polar coordinates in (1.27), obtaining

$$f = 4\pi \int_0^{\infty} r^2 \rho(r) \frac{\sin kr}{kr} \, dr. \tag{1.28}$$

As an example consider the **evaluation** of the scattering factor for the hydrogen atom. From the solution of the Schroedinger wave equation, $\psi = (\pi a^3)^{-1/2} e^{-r/a}$ for the lowest electronic state, where $a = 0.53$ Å is the Bohr radius. The value of $\rho(r)$ is then $(\pi a^3)^{-1} e^{-2r/a}$ which, when substituted into (1.28), integrates to

$$f = 1 + \left(\frac{k^2a^2}{4}\right)^{-2}.$$

A graph of f as a function of $k = (4\pi/\lambda) \sin \psi/2$ is shown in Fig. 1.35. The shape of this curve is representative of atomic scattering factors in general. Note that at zero angle of scattering all parts of the electronic cloud scatter in phase and hence for all atoms the value of f at $k = 0$ is just the number of

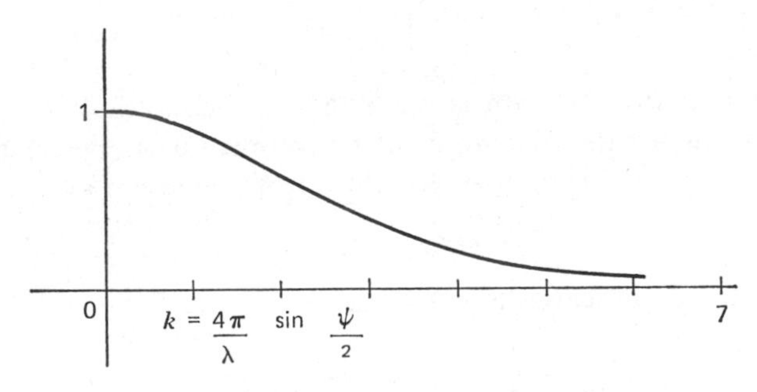

Fig. 1.35 Atomic scattering factor for hydrogen.

electrons. As the angle of scattering increases, interference begins to occur between different parts of the electron cloud, and hence the atomic scattering factor falls off with angle. Values of f for atoms and many ions are listed in the International Tables [60] and in other sources [61].

The Reciprocal Lattice Vector h

We have stated that the vector **h** is defined by $\mathbf{h} = h\mathbf{b}_1 + k\mathbf{b}_2 + l\mathbf{b}_3$ and have shown that $\mathbf{h} = (\mathbf{s} - \mathbf{s}_0)/\lambda$. The latter equation is called the Laue equation and contains all of the geometrical information of X-ray diffraction photographs. So far all that we know about h, k, and l is that they are the orders of the X-ray diffraction maxima. Their relation to the planes into which a crystal can be resolved will now be proved.

Assume that we have a plane passing through the unit cell with intercepts along the three axes given by

$$\mathbf{X}_1 = \mathbf{a}_1/h, \qquad \mathbf{X}_2 = \mathbf{a}_2/k, \qquad \mathbf{X}_3 = \mathbf{a}_3/l.$$

We now prove that the vector **h** is perpendicular to this plane. The vectors $\mathbf{X}_3 - \mathbf{X}_1$ and $\mathbf{X}_3 - \mathbf{X}_2$ must be parallel to this plane; also the vector

$(\mathbf{X}_3 - \mathbf{X}_1) \times (\mathbf{X}_3 - \mathbf{X}_2)$ must by definition of the cross product be perpendicular to the plane:

$$\begin{aligned}(\mathbf{X}_3 - \mathbf{X}_1) \times (\mathbf{X}_3 - \mathbf{X}_2) &= (\mathbf{a}_3/l - \mathbf{a}_1/h) \times (\mathbf{a}_3/l - \mathbf{a}_2/k) \\ &= -\mathbf{a}_3 \times \mathbf{a}_2/kl - \mathbf{a}_1 \times \mathbf{a}_3/hl + \mathbf{a}_1 \times \mathbf{a}_2/hk \\ &= (v/hkl)(h\mathbf{b}_1 + k\mathbf{b}_2 + l\mathbf{b}_3) \\ &= (v_a/hkl)(\mathbf{h}).\end{aligned}$$

Hence $\mathbf{h}$ is perpendicular to the plane with indices hkl obtained in the manner described above. The indices so used are usually termed crystallographic indices, and they differ slightly from Miller indices in that Miller indices are defined with the auxiliary condition that the indices be three relatively prime numbers. Thus it is proper to refer to a 224 plane if crystallographic indices are being used, but the same plane in Miller index notation would be the second order of the 112 plane.

Now we shall also prove that $|\mathbf{h}| = (d)^{-1}$, where d is the perpendicular distance between planes of the type hkl. Select any one of the vectors describing the plane intercept along one of the axes, say $\mathbf{X}_1$. The component of this vector along a unit vector perpendicular to the plane of interest is just equal to the distance between planes, d. However, we have just shown that $\mathbf{h}$ is a vector perpendicular to the plane; hence

$$\begin{aligned}d &= \mathbf{X}_1 \cdot \mathbf{h}/|\mathbf{h}| \\ &= (\mathbf{a}_1/h) \cdot (h\mathbf{b}_1 + k\mathbf{b}_2 + l\mathbf{b}_3)/|\mathbf{h}| \\ &= 1/|\mathbf{h}|.\end{aligned}$$

We have said earlier that

$$|(\mathbf{s} - \mathbf{s}_0)|/\lambda = 2(\sin \theta)/\lambda,$$

where $\theta = \psi/2$. Now by the Laue equation,

$$\mathbf{h} = (\mathbf{s} - \mathbf{s}_0)/\lambda,$$

and hence on taking magnitudes on both sides, this vector equation reduces to the scalar equation

$$(d)^{-1} = 2(\sin \theta)/\lambda,$$

or

$$\lambda = 2d \sin \theta, \tag{1.29}$$

which is just Bragg's law. An alternate form of this equation would involve Miller indices, in which case

$$n\lambda = 2d' \sin \theta,$$

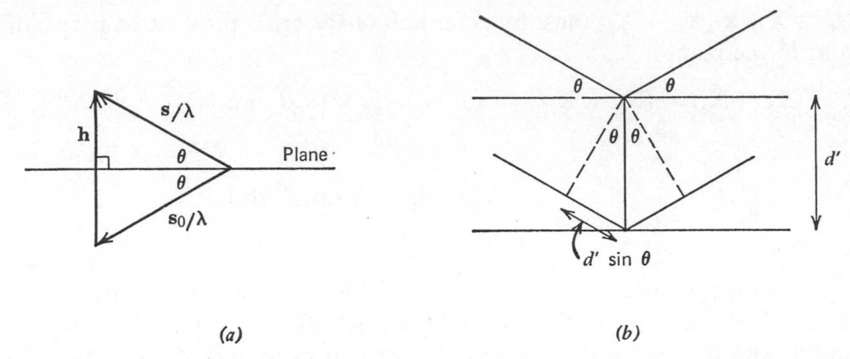

Fig. 1.36 (*a*) Graphical representation of $\mathbf{s}/\lambda - \mathbf{s}_0/\lambda = \mathbf{h}$. (*b*) Path difference between successive planes a distance d' apart is $2d' \sin \theta$, which for a maximum is set equal to $n\lambda$.

where $1/d = n/d'$. Figs. 1.36*a* and *b* illustrate these two equivalent approaches to consideration of diffraction.

The reciprocal lattice vector **h** is quite helpful for proof of certain useful relations. For example the interplanar spacing $d = 1/|\mathbf{h}|$ in any crystallographic set of coordinates can be calculated by forming the product $|\mathbf{h}|^2 = \mathbf{h} \cdot \mathbf{h}$, substituting $\mathbf{h} = h\mathbf{b}_1 + k\mathbf{b}_2 + l\mathbf{b}_3$, writing the **b**'s in terms of the **a**'s and then using the equations among the vectors discussed earlier. For the monoclinic system the resulting equation is

$$d = \left[\frac{1}{\sin^2 \beta}\left(\frac{h^2}{a^2} + \frac{l^2}{c^2} - \frac{2hl}{ac}\cos \beta\right) + \frac{k^2}{b^2}\right]^{-1/2}.$$

Angles between plane normals are obtainable from such equations as $\cos(\mathbf{h}_p, \mathbf{h}_q) = \mathbf{h}_p \cdot \mathbf{h}_q/(|\mathbf{h}_p|\,|\mathbf{h}_q|)$. If the normals **h** for a set of planes are all perpendicular to an axis $\mathbf{T} = u_1\mathbf{a}_1 + u_2\mathbf{a}_2 + u_3\mathbf{a}_3$ of the crystal, then the equation $\mathbf{h} \cdot \mathbf{T} = 0$ holds. Therefore the indices *hkl* of these planes satisfy the equation $hu_1 + ku_2 + lu_3 = 0$. For example if the axis in question is the monoclinic **b**-axis, [010], then all planes with indices ($h0l$) satisfy this equation for all values of h and l. It is then said that the ($h0l$) planes lie in the zone of the **b**-axis.

The vectors $\mathbf{h} = h\mathbf{b}_1 + k\mathbf{b}_2 + l\mathbf{b}_3$ for the various *hkl* describe a three-dimensional lattice. With each point of this lattice we associate the three numbers *hkl* and the value of F_{hkl}. This weighted reciprocal lattice is a complete description of the information ordinarily obtainable in X-ray diffraction studies of single crystals. If we combine the reciprocal lattice with the graphical representation of the Laue equation, shown in Fig. 1.36*a*, the result is the Ewald construction (Fig. 1.37). The origin of the reciprocal lattice can be thought of as the point at which the crystal vectors and the reciprocal lattice vectors have a common origin and fixed relative orientation. Thus as the crystal rotates the reciprocal lattice also rotates. The vector $\mathbf{s}_0/\lambda$ is defined by the direction of the incident X-ray beam. The length of this vector, $1/\lambda$, is measured in the same units as the lengths of the reciprocal lattice vectors. In order to satisfy the Laue equation we draw $\mathbf{s}_0/\lambda$ so that its head is at the common origin of all reciprocal lattice vectors. A sphere of radius $1/\lambda$ about its tail describes the locus of all possible vectors $\mathbf{s}/\lambda$ that could represent

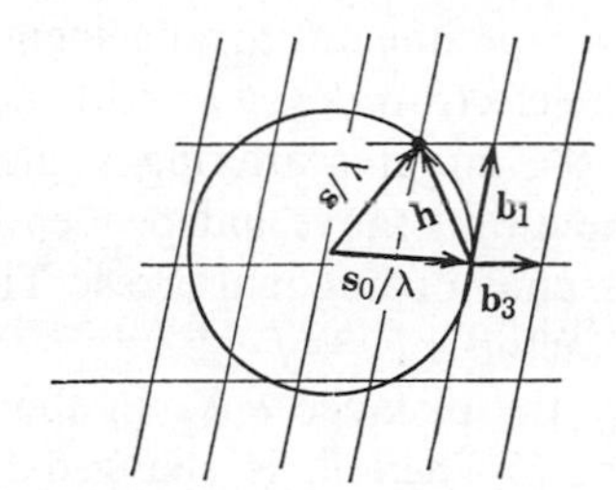

Fig. 1.37 The Ewald construction. Let the reciprocal lattice net represent the $h0l$ zone of a monoclinic crystal. Then the $\mathbf{a}_2$ axis [010] is perpendicular to the plane of the paper. As the crystal is rotated about this axis, various points of the reciprocal lattice cross the circle of reflection; as they do so, the equation $\mathbf{s}/\lambda - \mathbf{s}_0/\lambda = \mathbf{h}$ is satisfied, and a diffracted beam occurs. Diffraction by the [101] plane is illustrated; $\mathbf{h}$ is normal to this plane.

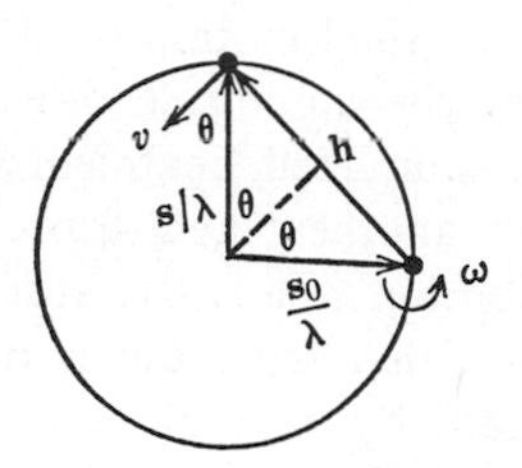

Fig. 1.38 The Lorentz factor is proportional to $v_\perp$ along a radius of the sphere.

diffraction directions. When a point in reciprocal space touches the sphere, the Laue equation is satisfied for the particular set of parallel planes and reflection results represented by the numbers hkl that identify the reciprocal lattice point. Then, if $F_{hkl} \neq 0$, a diffraction maxima will occur in the direction of $\mathbf{s}$.

The Lorentz and Temperature Factors

The Lorentz factor arises because different reflections are in a position to reflect for different lengths of time. If, for example, in Fig. 1.37 the crystal is rotated about the **b**-axis with uniform angular velocity ω, not all of the points in reciprocal space cross the circle of reflection with the same velocity. As an illustration, the Lorentz factor is derived for this case. Define the Lorentz factor as a function inversely proportional to the linear velocity normal to the sphere, $v_\perp$, of a reciprocal lattice point (Fig. 1.38).

Now the linear velocity of the end of the vector **h** is $\omega|\mathbf{h}|$, and its component normal to the surface of the sphere is $\omega|\mathbf{h}| \cos \theta = v_{\perp}$. But $\lambda = 2d \sin \theta$ and $d = 1/|\mathbf{h}|$, so that $v_{\perp} = (\omega/\lambda)\, 2 \sin \theta \cos \theta = (\omega/\lambda) \sin 2\theta$. Hence aside from a proportionality constant,

$$L = \lambda/(\omega \sin 2\theta)$$

is the Lorentz factor for a zero-level oscillation, rotation, or Weissenberg photograph. Graphical charts for Lorentz or combined Lorentz and polarization factors are available for other experimental methods. The principle of the calculation is the same for all experimental arrangements.

The thermal motion of atoms spreads out the electron density and hence reduces the amplitude of scattering more at the larger scattering angles. In this simplified treatment we assume that the atoms move independently of one another, and for convenience we assume orthogonal axes. The contribution of a single atom to the structure factor is $F_1 = f e^{-2\pi i(hx+ky+lz)}$. Suppose that the atom is displaced (in Å) by the distance $a\,\Delta x$ in the x direction, by $b\,\Delta y$ along y, and by $c\,\Delta z$ along z. Then F_1 is changed by ΔF_1, so that

$$\begin{aligned} F_1 + \Delta F_1 &= f e^{-2\pi i[h(x+\Delta x)+k(y+\Delta y)+l(z+\Delta z)]} \\ &= F_1 e^{-2\pi i(h\,\Delta x+k\,\Delta y+l\,\Delta z)}. \end{aligned}$$

Now assume that x, y, and z are independent variables, and average over all displacements.

$$\overline{F_1} = \overline{F_1 + \Delta F_1} = F_1 \overline{e^{-2\pi i h\,\Delta x}}\,\overline{e^{-2\pi i k\,\Delta y}}\,\overline{e^{-2\pi i l\,\Delta z}}$$

A representative term is $e^{i\alpha}$, where $\alpha = 2\pi h\,\Delta x$ is small. But the mean value of

$$e^{-i\alpha} = 1 - \frac{i\alpha}{1!} + \frac{(i\alpha)^2}{2!} - \frac{(i\alpha)^3}{3!} + \frac{(i\alpha)^4}{4!} + \cdots$$

over positive and negative values of α is

$$\overline{e^{-i\alpha}} = 1 - \frac{\overline{\alpha^2}}{2} + \frac{\overline{\alpha^4}}{24} - \cdots,$$

which is the same as the mean value of

$$e^{-\alpha^2/2} = 1 - \frac{\alpha^2}{2} + \frac{\alpha^4}{8}$$

within a negligible error. Hence

$$\overline{F_1} = F_1 e^{-2\pi^2[h^2\overline{(\Delta x)^2}+k^2\overline{(\Delta y)^2}+l^2\overline{(\Delta z)^2}]}.$$

This is the form of an anisotropic temperature factor in an orthogonal or nearly orthogonal system, when the largest or smallest amplitudes are primarily along x, y, or z. The constants $\overline{(\Delta x)^2}$, $\overline{(\Delta y)^2}$, and $\overline{(\Delta z)^2}$ are determined empirically when the atomic arrangement is known, but they can also be obtained approximately at the start of an X-ray diffraction study from mean values of F_{hkl}. Generally, however, the cross terms do not completely average to zero, and the more general form

$$\overline{F_1} = F_1 e^{-(\beta_{11}h^2+\beta_{22}k^2+\beta_{33}l^2+\beta_{12}hk+\beta_{13}hl+\beta_{23}kl)}$$

is therefore required, where the β_{ij} are constants to be determined empirically.

Often an isotropic temperature factor is satisfactory if the crystal is not bound by forces of widely different strength in various directions. If so, we define the mean square amplitude of vibration as $\mu^2 = a^2\,\overline{(\Delta x)^2} = b^2\,\overline{(\Delta y)^2} = c^2\,\overline{(\Delta z)^2}$, and remember that $\lambda = 2d\sin\theta$ and that $d = (h^2/a^2 + k^2/b^2 + l^2/c^2)^{-1/2}$ in the orthorhombic system. Then

$$\overline{F_1} = F_1 e^{-8\pi^2\mu^2\sin^2\theta/\lambda^2} = F_1 e^{-B\sin^2\theta/\lambda^2},$$

where B is the temperature factor constant, to be determined empirically.

Electron Density and Patterson Functions

The Electron Density Function

As we have said earlier, (1.22*a*) is a Fourier integral representation of the structure factor. Therefore the Fourier transform of this function should exist and, with proper scaling, should give the electron density function, $\rho(\mathbf{r})$:

$$\rho = \sum_{\substack{\text{all}\\ hkl}} C_{hkl} e^{2\pi i(hx+ky+lz)}. \tag{1.30}$$

The coefficients are as yet unknown but can be determined by (1.22*a*), which we can rewrite as

$$F_{HKL} = \int_0^1\int_0^1\int_0^1 \rho e^{-2\pi i(Hx+Ky+Lz)}\,\mathbf{a}\,dx\cdot\mathbf{b}\,dy\times\mathbf{c}\,dz$$

in order to distinguish the particular HKL from those in (1.30). Now substitute (1.30) into (1.22*a*). In each term of the sum the integrals over x, y, and z can be expressed as a product of three integrals, one of which is

$$\int_0^1 e^{2\pi i(h-H)x}\,dx = \left.\frac{e^{2\pi i(h-H)x}}{2\pi i(h-H)}\right|_0^1 = 0 \qquad \text{if } h \neq H$$

$$= \int_0^1 dx = 1 \qquad \text{if } h = H.$$

There are similar integrals over y and z. Hence all terms vanish except the term for which $h = H, k = K$, and $l = L$. The result is $F_{hkl} = \mathbf{a} \cdot \mathbf{b} \times \mathbf{c}\, C_{hkl}$, or $C_{hkl} = (1/V)F_{hkl}$. The three-dimensional electron-density function is thus

$$\rho(x, y, z) = \frac{1}{V} \sum_{h=-\infty}^{\infty} \sum_{k=-\infty}^{\infty} \sum_{l=-\infty}^{\infty} F_{hkl} e^{2\pi i(hx+ky+lz)}. \tag{1.31}$$

The electron density can be projected along any axis, say the y-axis, as follows:

$$\rho(x, z) = \int_0^1 \rho(x, y, z) b\, dy.$$

Again we can separate the exponential into products so that we need only consider the factor

$$\int_0^1 e^{2\pi iky}\, dy$$

which $= 0$ if $k \neq 0$ and $= 1$ if $k = 0$. Hence

$$\rho(x, z) = \frac{b}{V} \sum_{h=-\infty}^{\infty} \sum_{l=-\infty}^{\infty} F_{h0l} e^{2\pi i(hx+lz)}. \tag{1.32}$$

In general the structure factor $F_{hkl} = A_{hkl} - i\, B_{hkl}$, has both real and imaginary parts. We know that $\rho(x, y, z)$ is an observable quantity and hence must be real, and thus equal to its complex conjugate

$$\rho(x, y, z) = \frac{1}{V} \sum_h \sum_k \sum_l F_{hkl}^* e^{-2\pi i(hx+ky+lz)}.$$

If this equation is now added to (1.31) and the result is divided by 2, we obtain

$$\rho(x, y, z) = \frac{1}{V} \sum_h \sum_k \sum_l [A_{hkl} \cos 2\pi(hx + ky + lz) + B_{hkl} \sin 2\pi(hx + ky + lz)]. \tag{1.33}$$

For a centrosymmetric crystal in which the origin is chosen at a center of symmetry, for every atom at x_j, y_j, z_j, there is another atom at $-x_j, -y_j, -z_j$. Hence F_{hkl} becomes

$$\begin{aligned} F_{hkl} &= \sum_{j=1}^{N} f_j e^{-2\pi i(hx_j+ky_j+lz_j)} \\ &= \sum_{j=1}^{N/2} f_j [e^{-2\pi i(hx_j+ky_j+lz_j)} + e^{2\pi i(hx_j+ky_j+lz_j)}] \\ &= \sum_{j=1}^{N/2} 2f_j \cos 2\pi(hx_j + ky_j + lz_j), \end{aligned} \tag{1.34}$$

where the sum extends over the $N/2$ atoms not related by the center of symmetry. The electron-density function is then given by

$$\rho(x, y, z) = \frac{1}{V} \sum_h \sum_k \sum_l F_{hkl} \cos 2\pi(hx + ky + lz), \qquad (1.35)$$

since $A_{hkl} = F_{hkl}$ and $B_{hkl} = 0$. The result for the acentric case can be expressed as

$$\rho(x, y, z) = \frac{1}{V} \sum_h \sum_k \sum_l \{|F_{hkl}| \cos [2\pi(hx + ky + lz) + \alpha_{hkl}]\}, \qquad (1.36)$$

where $|F_{hkl}| = (A_{hkl}^2 + B_{hkl}^2)^{1/2}$, and α_{hkl} (the so-called phase angle) is determined by the relations $\cos \alpha_{hkl} = A_{hkl}/|F_{hkl}|$ and $\sin \alpha_{hkl} = -B_{hkl}/|F_{hkl}|$. The reader can verify that the substitution of these relations into (1.36) gives (1.33).

The crystallographer must obtain the correct signs or phases of the F_{hkl} in order to calculate the complete three-dimensional map. The observed data give only $|F_{hkl}|^2$, and even in the centric case there are far too many intensities to attempt all possible sign combinations, even with the fastest computers. (Note that 20 terms would involve approximately one million electron density maps if all possible sign combinations were tried.)

However, (1.22) and (1.36) are usually employed in the refinement of structures in the following process:

1. Calculate $|F_{hkl}|$ by taking the positive square root of the observed $|F_{hkl}|^2$.
2. Calculate A_{hkl} and B_{hkl} from a trial structure obtained by methods described above.
3. If centric, assign the same sign to $|F_{hkl}|$ as found for the calculated $A_{hkl}(B_{hkl} = 0)$, or if acentric, compute α_{hkl}.
4. From the resultant electron-density map, obtain new atomic coordinates x_j, y_j, z_j. Then repeat steps 2, 3, and 4, until the coordinates no longer change.

The Patterson Function

THE FUNCTION

After correction for the Lorentz and the polarization factors, each observed intensity yields $|F_{hkl}|^2$, which here includes the temperature factor. A Fourier series could be readily computed with these terms as coefficients, since no phase ambiguity would be involved, but one might well ask what is the physical meaning of such a function.

The electron density values $\rho(x, y, z)$ at x, y, z, and $\rho(x + u, y + v, z + w)$ at $x + u$, $y + v$, $z + w$ are both especially large if these coordinates are chosen at atomic positions. The product of these two functions is therefore large whenever the differences between these coordinates corresponds to an interatomic distance; that is, when u, v, w represents the components of a vector between these two atoms. The totality of all such vectors, all drawn from a common origin, is obtained by summing this product over all space, here over the unit cell.

$$P(u, v, w) = \int_0^1 \int_0^1 \int_0^1 \rho(x, y, z)\rho(x + u, y + v, z + w)\mathbf{a}\, dx \cdot \mathbf{b}\, dy \times \mathbf{c}\, dz$$

Remembering that ρ is real and therefore equal to its complex conjugate, we substitute (1.31) for $\rho(x, y, z)$ and

$$\rho(x + u, y + v, z + w) = \frac{1}{V} \sum_H \sum_K \sum_L F^*_{HKL} e^{-2\pi i[H(x+u)+K(y+v)+L(z+w)]}$$

into $P(u, v, w)$. A representative integral is

$$\int_0^1 e^{2\pi i h x}\, e^{-2\pi i H(x+u)}\, dx = e^{-2\pi i H u} \int_0^1 e^{2\pi i(h-H)x}\, dx = 0$$

unless $H = h$, in which case the value is $e^{-2\pi i h u}$. Hence only those terms remain for which $H = h$, $K = k$, $L = l$, and

$$P(u, v, w) = \frac{1}{V} \sum_h \sum_k \sum_l |F_{hkl}|^2\, e^{-2\pi i(hu+kv+lw)}$$

$$= \frac{1}{V} \sum_{h=-\infty}^{\infty} \sum_{k=-\infty}^{\infty} \sum_{l=-\infty}^{\infty} |F_{hkl}|^2 \cos 2\pi(hx + ky + lz). \qquad (1.37)$$

This last result follows because $P(u, v, w)$ and $|F_{hkl}|^2 = F_{hkl}F^*_{hkl}$ are both real, and hence the imaginary part of $P(u, v, w)$ vanishes. The Patterson function is always centrosymmetric whether or not the structure is; for if we have a vector from atom one to atom two we also have the vector in the reverse direction. Projections can be made and are analogous to these for the electron-density map.

At the present time, methods based upon the three-dimensional Patterson function are among the most powerful methods for the solution of crystal structures. Its power is greatly increased by establishing an approximate absolute scale for the observed $|F_{hkl}|$ and by sharpening the peaks. These two modifications are described.

SCALE FACTOR AND TEMPERATURE FACTOR ESTIMATES

Let $|F_o|$ be the observed $|F^{\text{obs}}_{hkl}|$, all of which are assumed to be on the same relative scale; they all need to be multiplied by a scale factor k to bring them into agreement with $|F_c|$, where $|F_c| = |F^{\text{calc}}_{hkl}|$. We shall suppose that the

$|F_c|$ do not contain the temperature factor; hence

$$k\,|F_o| = |F_c| \exp(-B \sin^2 \theta/\lambda^2), \tag{1.38}$$

where

$$F_c = \sum_{j=1}^{N} f_j \exp(-2\pi i \mathbf{h} \cdot \mathbf{r}_j).$$

At the beginning of the structure investigation the atomic coordinates r_j are unknown, but the number of atoms N and their scattering factors are known. Now

$$|F_c|^2 = \left(\sum_{j=1}^{N} f_j \exp(-2\pi i \mathbf{h} \cdot \mathbf{r}_j)\right)\left(\sum_{k=1}^{N} f_k \exp(2\pi i \mathbf{h} \cdot \mathbf{r}_k)\right)$$

$$= \sum_{j=1}^{N} f_j^2 + \sum_{j}\sum_{\substack{k \\ j \neq k}} f_j f_k \exp[-2\pi i \mathbf{h} \cdot (\mathbf{r}_k - \mathbf{r}_j).]$$

Suppose that the structure is sufficiently complex that the $\mathbf{r}_k - \mathbf{r}_j$ are roughly random and that a large fraction of the spacings d are not $\gg |\mathbf{r}_k - \mathbf{r}_j|$; these qualifications require modification in the discussion for simple structures, highly disordered structures, and proteins.

Aside from these cases, the average value of $|F_c|^2$ over many reflections is $\sum_{j=1}^{N} f_j^2$. Since both f_j and the temperature factor term vary with $\sin\theta$, it is most reasonable to average over regions about 0.1 Å^{-1} in $\sin\theta/\lambda$ and plot

$$\log \frac{\overline{|F_o|^2}}{\overline{|F_c|^2}} = \log\left(\frac{\overline{|F_o|^2}}{\sum_{j=1}^{N} f_j^2}\right) = -\log k^2 - \left(\frac{2B}{2.303}\right) \sin^2\theta/\lambda^2; \tag{1.39}$$

that is, plot $\log(\overline{|F_o|^2})/(\sum f_j^2)$ as ordinate against $\sin^2\theta/\lambda^2$ as abscissa. The slope, $-2B/2.3$, will give the value of B, and the intercept on the vertical axis, $-\log k^2$, will give the value of k. Usually the values of k are accurate to about 10%, which is sufficient to establish the identity of single interactions in the Patterson function if they are resolved.

THE SHARPENED PATTERSON FUNCTION

As we have seen, peaks in the Patterson function result from the products of electron density peaks. The electron density peaks are appreciably wide, and thus the peaks in Patterson space are even wider. Since in the Patterson function there are $N(N-1)$ peaks in general, resolution becomes a real problem. It is advantageous therefore to sharpen these peaks; that is, to reduce the half-width as much as possible. To do so we might ask how we could modify F_{hkl} such that the scattering would correspond to that of a point atom. Such modification could be made as follows: replace the atomic scattering factor f_j by the corresponding atomic number Z_j, and then multiply by a term of the type $\exp(B \sin^2\theta/\lambda^2)$ to counteract the smearing out of the atom due to thermal motion effects. The shape of all scattering factors is similar (excluding hydrogen), and therefore dividing $|F_{hkl}|$ by $\hat{f} = f/Z$ effectively replaces f by Z. If more than one type of atom is involved, then $\hat{f}$ can be computed by $\sum f_j / \sum Z_j$. Note that a Patterson function involving truly point atoms would require an infinite number of terms in the Fourier series. Therefore a factor of the type $e^{-B' \sin^2 \theta/\lambda^2}$ is included to reduce the series termination effects. Thus the new coefficients for the Patterson function calculation are $(|F_{hkl}|^2/\hat{f}^2) \exp [(2B - B') \sin^2 \theta/\lambda^2$.

It is also possible to obtain a sharpened Patterson-like function by considering the product of two derivative electron density functions in place of the electron-density functions in the usual formulation of the Patterson function [18]. Since a product is involved, the result is a function having large and sharper peaks at those positions corresponding to interatomic distances and small ripples around the peak, but these ripples can be essentially completely removed by a slight modification of the coefficients. In general, both types of sharpening approaches can be used together. The effect of sharpening is illustrated in Figs. 1.39*a* and *b*, which show, respectively, an unsharpened and a sharpened Patterson projection of the $Ni(L)Br_2$ complex discussed earlier.

SYMMETRY IN THE PATTERSON FUNCTION

The effect of symmetry on the Patterson function can be illustrated by a twofold screw axis, 2_1, parallel to **b**. We shall consider two pairs of atoms, 1 and 2, with coordinates x_1, y_1, z_1; $\bar{x}_1$, $y_1 + \frac{1}{2}$, $\bar{z}_1$, and x_2, y_2, z_2 and $\bar{x}_2$, $y_2 + \frac{1}{2}$, $\bar{z}_2$. First (Fig. 1.40*a*) we shall assume that $y_1 \neq y_2$, in which case the vectors with component $\frac{1}{2}$ along **b** are $2x_1$, $\frac{1}{2}$, $2z_1$, and $2x_2$, $\frac{1}{2}$, $2z_2$. These vectors occur as single interactions in the section $v = \frac{1}{2}$ in Patterson space. The difficulty that often occurs in this interpretation is that when two atoms have by accident the same y-coordinate, additional vectors occur in the $v = \frac{1}{2}$ section that have twice the weight of the single interactions (Fig. 1.40*b*). There are similar effects due to other symmetry elements, but all of these form part of the general analysis of the Patterson function. The original paper on this subject by Patterson [62] is very clearly written, and is recommended to the reader.

The Least-Squares Method of Refinement

Let us assume that approximate positions have been derived from the Patterson function or from some other method. We wish now to refine these trial coordinates to the best possible values consistent with the observed X-ray data. The most satisfactory method of refinement is the least-squares method [63]. Its advantages over methods based upon Fourier series are that corrections for nonconvergence need not be made and weighting factors for the

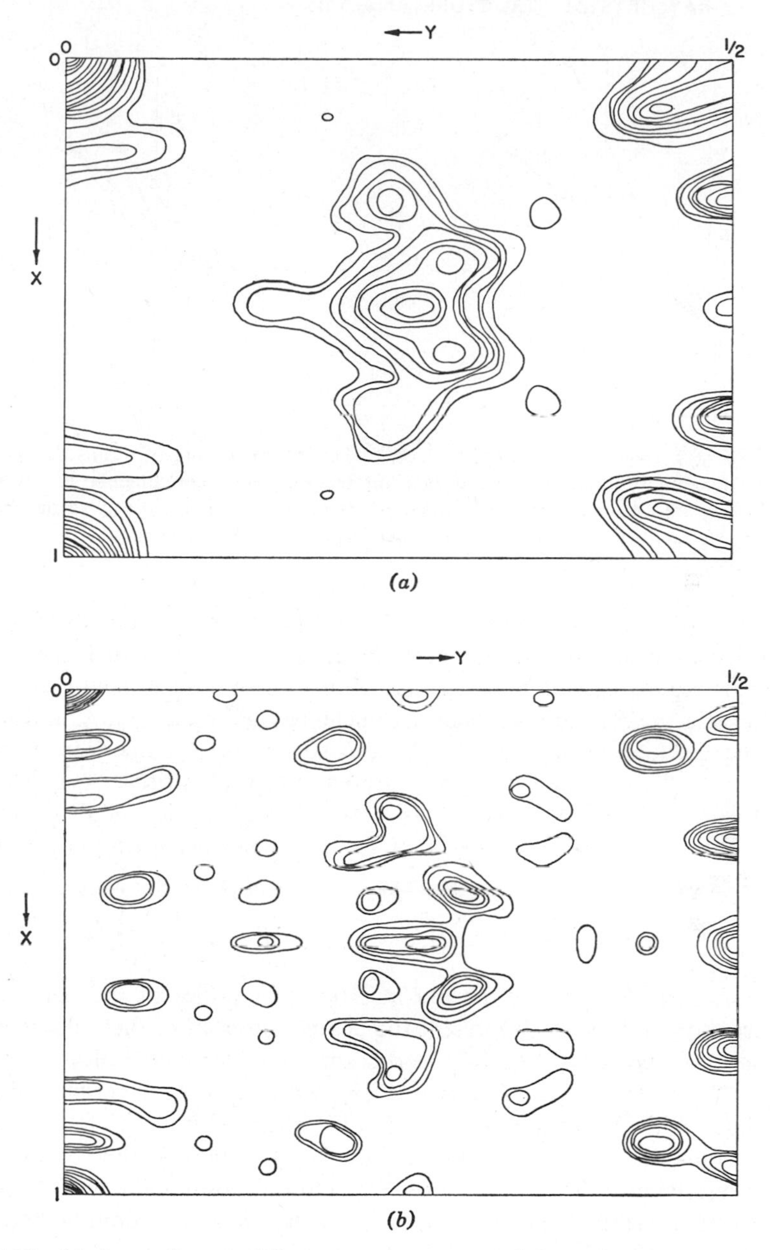

Fig. 1.39 (*a*) An unsharpened Patterson map computed from the observed data for $Ni(L)Br_2$, where L is the tridentate ligand 6,6′-dimethyl di(2-pyridylmethyl) amine, and projected along the *c*-axis. (*b*) A sharpened Patterson map for $Ni(L)Br_2$ projected along the *c*-axis. A B' value of 3.0 $Å^2$ was used and the derivative sharpening function was also employed.

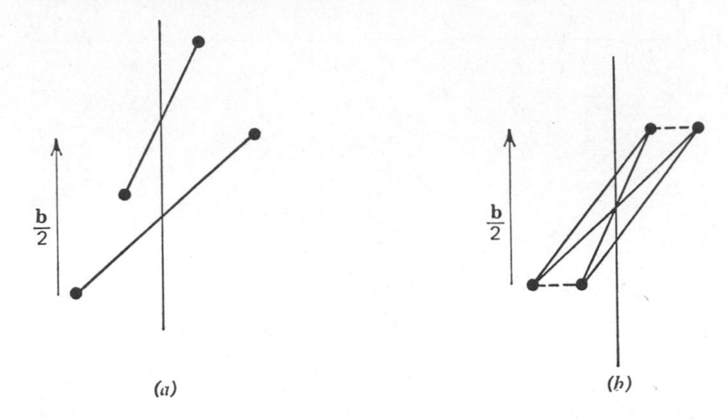

Fig. 1.40 (*a*) Vectors with components $\mathbf{b}/2$ when the two pairs of atoms do not have coincident y-coordinates. (*b*) Additional interactions with components $\mathbf{b}/2$ when y-coordinates are coincident. The two interactions that occur in the $y = 0$ section can be used to help identify these additional interactions in the $y = \frac{1}{2}$ section.

observations can be related to the experimental errors in these observations. Also, errors in the final positional parameters can be readily calculated [64].

Let us regard the intensity I_{hkl} as a function of a scale factor $k' = 1/k$, multiplying the F_c, the positional parameters $x_1, y_1, z_1, x_2, y_2, z_2$, and so on, and the thermal parameters B_1, B_2, and so on, or β_{ij}, the corresponding anisotropic β's. Let the symbol x_i represent a typical variable. Let I_j^c represent for plane j the calculated intensity, which is a function of all these variables. Now I_j^c can be developed in a Taylor's series neglecting terms higher than 2 as

$$I_j^c(x_1', x_2', \ldots, x_n') = I_j^c(x_1, x_2, \ldots, x_n) + \sum_i \left(\frac{\partial I_j^c}{\partial x_i}\right) \Delta x_i, \tag{1.40}$$

where x_i and x_i' represent the old and the new values of the parameters, respectively, and the Δx_i represent the changes in each of these parameters; that is, $\Delta k'$, Δx_1, Δy_1, Δz_1, ΔB_1, and so on. Now for each reflection we set $\Delta I_j^c = I_j^o - I_j^c - \epsilon_j = I_j^c(x') - I_j^c(x)$, where ϵ_j is the experimental error associated with the reflection j and I_j^o is the observed intensity for plane j.

Assuming that the errors follow a Gaussian distribution, the "best values" of the parameter shifts, Δx_i, are given by a minimization of the sum of squares of the errors. Weighting factors, to be introduced later, are omitted here. Let

$$\begin{aligned} S &= \sum_j \epsilon_j^2 \\ &= \sum_j \left[(I_j^o - I_j^c) - \sum_i c_{ij} \Delta x_i\right]^2, \end{aligned}$$

where $c_{ij} = (\partial I_j)/(\partial x_i)$. The derivative of S with respect to each parameter change is set equal to 0,

$$\frac{\partial S}{\partial \Delta x_k} = 2 \sum_j \left[(I_j^o - I_j^c) - \sum_i c_{ij} \Delta x_i \right] c_{kj} = 0, \tag{1.41}$$

since this is the condition for a minimum in S relative to some particular Δx_k. This procedure reduces the large number of equations to the same number of equations as unknowns,

$$\begin{aligned}
\sum_j c_{1j}^2 \Delta x_1 + \sum_j c_{1j} c_{2j} \Delta x_2 + \cdots + \sum_j c_{1j} c_{nj} \Delta x_n &= \sum_j c_{1j}(I_j^o - I_j^c) \\
\sum_j c_{2j} c_{1j} \Delta x_1 + \sum_j c_{2j}^2 \Delta x_2 + \cdots + \sum_j c_{2j} c_{nj} \Delta x_n &= \sum_j c_{2j}(I_j^o - I_j^c) \\
\sum_j c_{nj} c_{1j} \Delta x_1 + \sum_j c_{nj} c_{2j} \Delta x_2 + \cdots + \sum_j c_{nj}^2 \Delta x_n &= \sum_j c_{nj}(I_j^o - I_j^c)
\end{aligned} \tag{1.42}$$

which can be solved simultaneously. The c_{ij}'s have been assumed to be constant in this discussion, but their values are slightly different after the new parameters x_k' are obtained from the above equations. Hence the whole process is repeated with these new vaules of x_k as starting values, until the changes in x_k are well within their probable errors.

Not all of the observations should contribute equally in the final set of equations. Because some observations may be more accurate than others, they should be weighted more heavily than the less accurate. Hence we minimize the weighted sum of errors, $S' = \sum_j w_j \epsilon_j^2$. This process has the net effect of multiplying the observational equations by $\sqrt{w_j}$:

$$\sqrt{w_j}\, \epsilon_j = \sqrt{w_j}\,(I_j^o - I_j^c) - \sqrt{w_j} \sum_i c_{ij} \Delta x_i.$$

The best choice of the $\sqrt{w_j}$ is that they be inversely proportional to σ_j, the standard deviation associated with the observation of the intensity I_j.

However a large number of observations are required for the determination of σ_j. Therefore, in practice, one of two schemes is usually followed, depending on the method used to take the data. For film data the usual choice is

$$w_j = \left(\frac{4I_{\min}^o}{I_j^o} \right)^2 \quad \text{for} \quad I_j^o \geq 4I_{\min}^o \tag{1.43}$$

and

$$= 1 \quad \text{for} \quad I_j^o \leq 4I_{\min}^o,$$

where $I_{\min}^o$ is the minimum observable intensity for the appropriate region of the film. These weighting factors arise because the larger intensities have a constant percentage error, whereas the smaller intensities have a constant error.

For data recorded by counter techniques it is common to evaluate $w_j = \sigma(I)^{-2}$ based on errors in counting statistics plus an attempt to include effects due to systematic errors that have not been completely eliminated. Therefore an expression similar to

$$\sigma(I) = \left[T_c + \left(\frac{t_c}{2t_b}\right)^2(B_1 + B_2) + (pI)^2\right]^{1/2} \tag{1.44}$$

is used for the standard deviation in the net intensity, where T_c is the total count in a scan of time t_c, B_1 and B_2 are the background counts each obtained in time t_b, and $I = T_c - (t_c/2t_b)(B_1 + B_2)$. The last term in (1.44) is used to take into account effects due to systematic errors, and a value of p of 0.03 to 0.05 is usually selected.

These least-squares equations are in a form readily handled on a digital computer. The weighting factors are quite different from those normally used in Fourier refinement, and hence it is reasonable to favor the results of least squares over results obtained from Fourier methods. In particular, the weighting factors inherent in the usual Fourier method give incorrect standard deviations as compared with the proper weighting factors.

An example illustrating refinement techniques is given in the Appendix, along with a discussion of the plotting techniques used to display the results.

Similar least-squares equations based on F_{hkl} instead of I_{hkl} could also be used.

3 SYMMETRY OF CRYSTALS*

Symmetry Operations and Symmetry Elements

It is almost always unnecessary to locate all of the atoms in the unit cell because some are generated by symmetry elements from the others. For a discussion of internal symmetry we assume that the structure extends indefinitely in three dimensions. The smallest unit that shows the full symmetry (aside from purely translational symmetry) and that is repeated by translational elements only is generally taken as the unit cell.

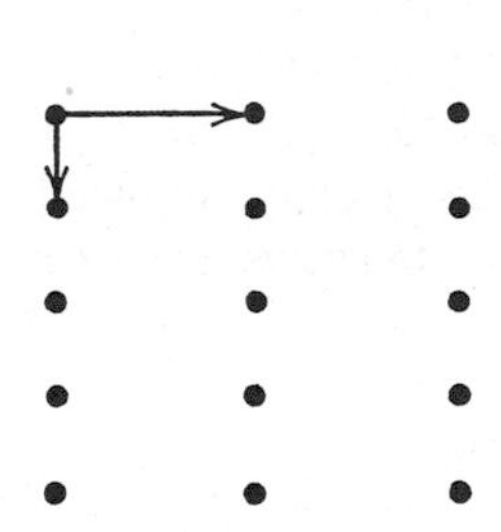

Fig. 1.41 A two-dimensional lattice generated by the orthogonal vectors $\mathbf{a}_1$ and $\mathbf{a}_2$.

Consider an infinite two-dimensional array of points (a two-dimensional lattice), as shown in Fig. 1.41, in which the pair of vectors $\mathbf{a}_1$ and $\mathbf{a}_2$ are perpendicular. The lattice is generated from this pair of vectors by the set of translations $\mathbf{T} = l_1\mathbf{a}_1 + l_2\mathbf{a}_2$, where l_1 and l_2 are integers, positive, negative, or zero. A translation of this

* For further discussion of the topics in this section, the reader should consult L. V. Aza'roff, *Elements of X-ray Crystallography*, McGraw-Hill, New York, 1968; M. J. Buerger, *Elementary Crystallography*, rev. ed., Wiley, New York, 1963.

two-dimensional net by, for example, $\mathbf{a}_1$, $l_1\mathbf{a}_1$, or $\mathbf{T}$ will leave this infinite net unchanged to an observer. This operation is called a symmetry operation, and the vectors $\mathbf{a}_1$, $l_1\mathbf{a}_1$, or $\mathbf{T}$ are symmetry elements. Similarly, this two-dimensional lattice is left unchanged by a rotation of π, 2π, 3π and so on, or by reflection across a line along $\mathbf{a}_1$, or a line along $\mathbf{a}_2$, or by a center of inversion (replace $+x$ and $+y$ by $-x$ and $-y$, respectively) at any lattice point. These are symmetry operations. The axis of rotation, the line of symmetry, and the center of inversion are symmetry elements. The symmetry operations considered here preserve distances and angles, and involve no distortions. If a symmetry element, such as a translation or a rotation, transforms an arbitrary figure into a congruent figure it is called a symmetry element of the first kind. On the other hand, if a symmetry element, such as a line or plane of symmetry or a center of inversion, transforms an arbitrary figure into the mirror image of a congruent figure it is called a symmetry element of the second kind. This is an important classification, and it forms the basis, for example, of a discussion of optical activity. No molecule having a symmetry element of the second kind can show optical activity.

It can be shown that the general symmetry operation of the first kind is a screw rotation. The axis of rotation and a translation are special cases. Any figure S_1 can be superimposed on a congruent figure S_2 by a translation, which superimposes the equivalent points of S_1 and S_2, and two rotations. Suppose that we have carried out this translation, and now draw a sphere about the point P that is common to S_1 and S_2 (Fig. 1.42). A careful inspection of this drawing, in which solid lines are drawn on the surface of the sphere and dotted lines toward the center at P, indicates that the superposition of three noncollinear points has now been achieved by the operation $\mathbf{T}C(\alpha)$. But this is not yet a simple screw rotation because the directions of $\mathbf{T}$ and $C(\alpha)$ are as yet unrelated. Resolve the vector $\mathbf{T}$ into components $\mathbf{T}_{\parallel}$ parallel to C and $\mathbf{T}_{\perp}$ normal to C, where $\mathbf{T} = \mathbf{T}_{\parallel} + \mathbf{T}_{\perp}$. The general operation is now $\mathbf{T}C(\alpha) = \mathbf{T}_{\parallel}\mathbf{T}_{\perp}C(\alpha)$. We can reduce $\mathbf{T}_{\perp}C(\alpha)$ to a single rotation $A(\alpha)$ as follows. Let $C(\alpha)$ be normal to the plane of the paper (Fig. 1.43). The operation $\mathbf{T}_{\perp}$ carries the point A into A', while the rotation $C(\alpha)$ carries A' back to A. A is so located along the perpendicular bisector of $\mathbf{T}_{\perp}$ that $\mathbf{T}_{\perp}$ subtends the angle α at A. Then A is invariant, and a rotation $A(\alpha)$ is equivalent to $\mathbf{T}_{\perp}C(\alpha)$. Thus the general operation of the first kind,

$$\mathbf{T}B_1(\beta)L_2(\lambda) = \mathbf{T}C(\alpha) = \mathbf{T}_{\parallel}\mathbf{T}_{\perp}C(\alpha) = \mathbf{T}_{\parallel}A(\alpha),$$

is a screw rotation.

The general operation of the second kind, which superimposes a figure S onto S', which is the mirror image of S, can also be simplified. First we

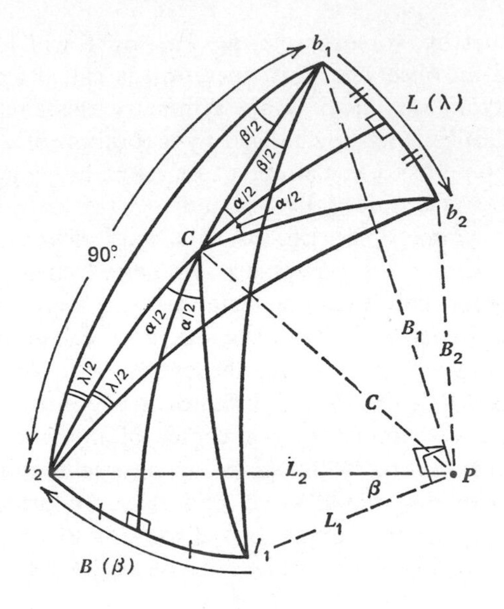

Fig. 1.42 A sphere is drawn around point P that is common to S_1 and S_2. The general operation which superimposes l_1 on l_2 and then b_1 on b_2 is $B_1(\beta)$, a rotation of β about B_1 as an axis, followed by $L_2(\lambda)$, a rotation of λ about L_2 as an axis, where L_2 is $\perp B_1$. The axis $C(\alpha) = B_1(\beta)L_2(\lambda)$ is located from the intersection of the bisectors of β and λ. The equalities marked in the drawing then indicate the equivalence of $C(\alpha)$ and $B_1(\beta)L_2(\lambda)$ for superposing l_1 on l_2 and b_1 on b_2 in a single rotation by the angle α about the axis C.

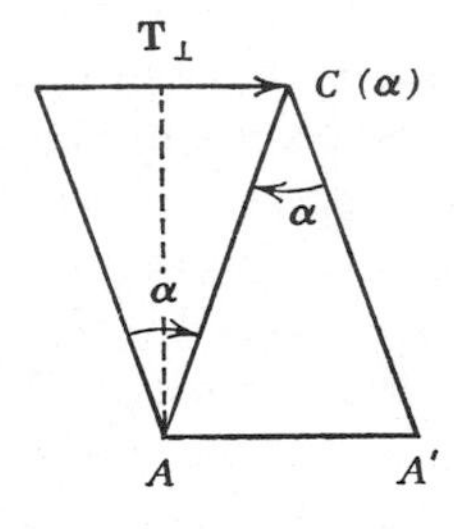

Fig. 1.43 T is followed by the rotation α about the axis C extending perpendicular to the plane of the paper. We locate the point A on the perpendicular bisector of $\mathbf{T}_\perp$ so that the angle α occurs as indicated at A. Then the rotation $A(\alpha) = \mathbf{T}_\perp C(\alpha)$, and the point A is invariant.

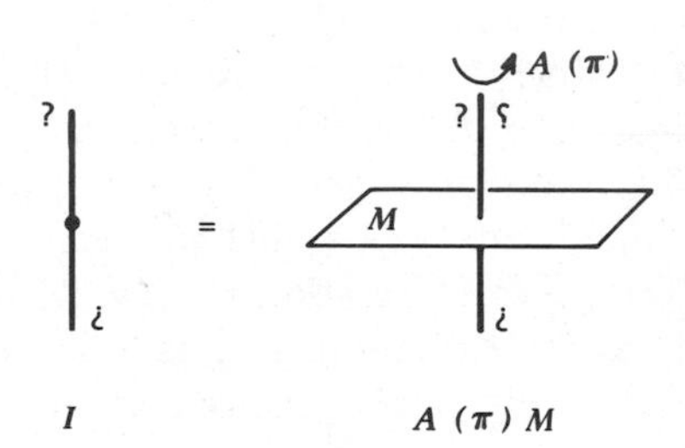

Fig. 1.44 Proof of the equivalence $I = A(\pi)M$. This is easily generalized to $A(\alpha)I = A(\alpha + \pi)M$. The order of the operations, which is immaterial in this case, is not always so. Note that $M \perp A$ and that the center of inversion lies on A.

translate S so that point P is brought into coincidence with the equivalent point P' of S'. Then we carry out an inversion with this common point as a center, thus converting S into a figure congruent with S'. This figure can then be superimposed by the single rotation $A(\alpha)$. Thus the general operation of the second kind is $\mathbf{T}IA(\alpha) = \mathbf{T}A(\alpha)I$. The order of these last two operations is immaterial, and the center of inversion lies on the axis of rotation. The operation $A(\alpha)I$ can be seen (Fig. 1.44) to be the same as $A(\alpha + \pi)M$, where M is a mirror plane perpendicular to the axis A. The general operation of the second kind is then $\mathbf{T}A(\alpha)I = \mathbf{T}A(\alpha')M$, where $\alpha' = \alpha + \pi$. A further simplification is possible, however. If S and S', the object and its mirror image, are parallel, no rotation is necessary, and the general operation reduces to $\mathbf{T}M$. But $\mathbf{T}$ can be resolved into $\mathbf{T}_{\parallel}$ parallel to the plane M, and $\mathbf{T}_{\perp}$ perpendicular to the plane M. Then $\mathbf{T}_{\perp}M$ can be shown to be equivalent to M', a mirror plane parallel to M and halfway along $\mathbf{T}$ (Fig. 1.45). Hence the general operation reduces (when S and S' are parallel) to

$$\mathbf{T}M = \mathbf{T}_{\parallel}\mathbf{T}_{\perp}M = \mathbf{T}_{\parallel}M',$$

which is a gliding reflection, with the glide along a line parallel to the plane. If a rotation is necessary, we may resolve the translation $\mathbf{T}$ into $\mathbf{T}_{\parallel}$ parallel to the axis and $\mathbf{T}_{\perp}$ perpendicular to this axis, and then the general operation becomes (when S and S' are not parallel)

$$\begin{aligned}\mathbf{T}A(\alpha)I = \mathbf{T}A(\alpha')M = \mathbf{T}MA(\alpha') = \mathbf{T}_{\perp}\mathbf{T}_{\parallel}MA(\alpha')\\ = \mathbf{T}_{\perp}M'A(\alpha') = M'A'(\alpha') = A'(\alpha)M.\end{aligned}$$

We have used our previous relations, remembering that $\mathbf{T}_{\parallel}$ parallel to the axis is perpendicular to the mirror plane. Thus the general operation of the second kind is a rotatory reflection with the mirror plane perpendicular to the axis, or a rotatory inversion with the center of inversion on the axis. It is convenient to keep the gliding reflection as a special case.

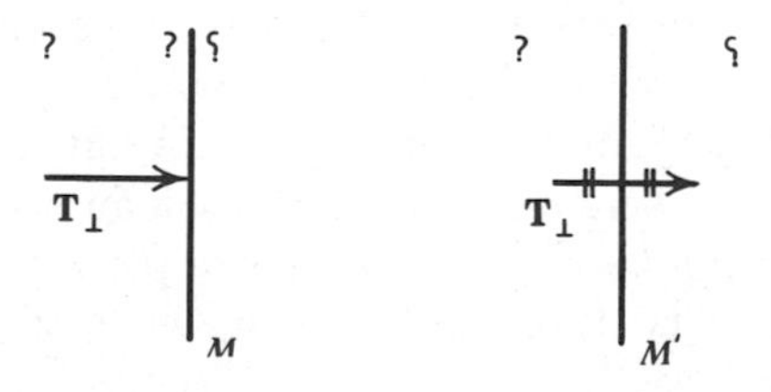

Fig. 1.45 Proof of the equivalence of $\mathbf{T}_{\perp}M = M'$, where M' is parallel to M and bisects $\mathbf{T}_{\perp}$.

Thus all symmetry operations are included in the *screw rotation*, the *gliding reflection*, or the *rotatory inversion*. The first is an operation of the first kind, the second two are operations of the second kind (p. 73). A figure left invariant after a symmetry operation is applied is said to be symmetric, and the symmetry elements are the vectors along which it is translated, planes across which it is reflected, centers through which it is inverted, or axes about which it is rotated.

The rotational axes of a crystal structure are restricted to one-, two-, three-, four-, and six-fold axes, which represent rotations by 0°, 180°, 120°, 90°, or 60°, respectively. This restriction arises because repeated application of the symmetry element must eventually restore the original position and hence the rotation is by $360°/n = 2\pi/n$, where n is an integer, and because only these types of axes will combine with the lattice vectors. Consider a rotation axis $A(\alpha)$, which is as yet arbitrary. We later include the axis 1, the rotation by 0°, which it is useful to include for completeness. Let $\mathbf{T}_1$ be a vector of the lattice not parallel to $A(\alpha)$. The operation $A(\alpha)$ then generates the vectors $\mathbf{T}_2$, $\mathbf{T}_3, \ldots$ from $\mathbf{T}_1$. The vectors $\mathbf{T}_1 - \mathbf{T}_2, \mathbf{T}_2 - \mathbf{T}_3$, and so on are perpendicular to $A(\alpha)$, and these vectors or other similarly formed vectors can be chosen to describe a two-dimensional lattice perpendicular to $A(\alpha)$. It is a property of this lattice (p. 72) that if a vector **a** occurs, the vector $-\mathbf{a}$ also occurs (Fig. 1.46), and if a rotation $A(\alpha)$ is possible, a rotation $A(-\alpha)$ is also possible. Since these rotations are symmetry operations the two points at the ends of the vector **b** are lattice points, and hence $b = ma$, where m is an integer. But $b = 2a \cos \alpha$, and we find from these two equations that $\cos \alpha = m/2$. Now $\cos \alpha$ lies within the range -1 to $+1$, so that $m = -2, -1, 0, 1, 2$ are the only permissible values. The resulting value of $\cos \alpha = -1, -\frac{1}{2}, 0, \frac{1}{2}, 1$ correspond, respectively, to rotations of 180°, 120° and 240°, 90° and 270°, 60° and 300°, and 360°. Hence only rotations by $2\pi/n$, where $n = 1, 2, 3, 4$, or 6 are permitted. In particular, a fivefold axis cannot occur, and axes where $n > 6$ cannot occur in crystal structures.

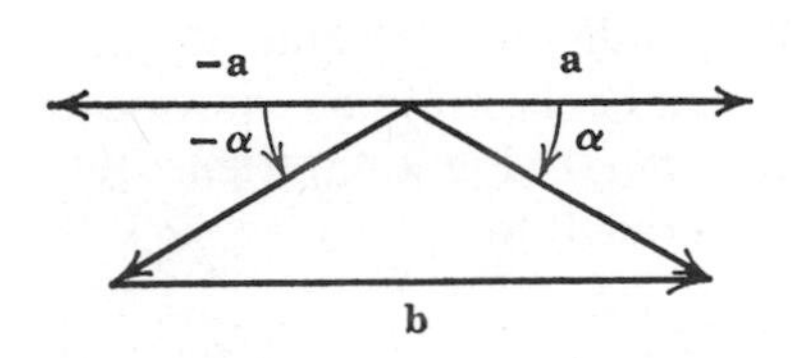

Fig. 1.46 The rotation axis $A(\alpha)$ at a lattice point. Note that the vector **b** is parallel to **a**. The reader should examine the situation when $A(\alpha)$ is halfway along the vector **a** and when $A(\alpha)$ is at some other position relative to **a**.

Similarly, the inversion axes $A(\alpha)I$, which are permissible in crystal structures, can be shown to be $\bar{1}$, $\bar{2}$, $\bar{3}$, $\bar{4}$, and $\bar{6}$, where the bar indicates the inversion.

Symmetry operations are classified into point groups and space groups, examples of both of which are discussed below. The symmetry operations form a group if (*a*) the successive operation R_1R_2, or operation R_1 followed by operation R_2 is equivalent to some other operation R_3 of the group; (*b*) the associative law holds, $R_1R_2R_3 = R_1(R_2R_3) = (R_1R_2)R_3$, where those operations in parentheses are carried out and the result is combined in the order indicated with the remaining operation; (*c*) the identity operation E is a member of the group; and (*d*) the inverse R^{-1} of every operation R is also a member of the group. These group properties determine whether one has finished discovering the number of operations in a group or the number of atoms equivalent by symmetry in any symmetric arrangement.

A space group occurs whenever a translation is involved. No single point is left invariant upon application of a symmetry element of which a translation is a part, such as a lattice translation, a screw axis, or a glide plane. Moreover, by virtue of the group properties, all translation groups are infinite, that is, property (*a*) leads to an infinite number of translational operations and to infinite lattices of points upon which structures are based. Of course, no structure is infinite, but this idealization forms the basis of classification of three-dimensional periodic structures in Euclidean space and serves as a basis for discussing the symmetries of crystals.

A point group occurs when the set of symmetry operations, such as a rotation axis, a center of inversion, a mirror plane, or an inversion axis, leaves at least one point in space fixed in position. Point groups form the basis for discussion of the external forms of crystals, and they occur as subgroups in space groups. A subgroup is any set of symmetry operations which among themselves satisfy the group properties listed above but which are part of a larger group. Because of this subgroup relation we find that a discussion of point groups, followed by their extension by translation groups, forms a simple geometrical basis for discovering the symmetry properties of crystals having no disorders.

Point Groups

There are 32 point groups of importance to crystallography. Not all of these are developed here, but the methods employed for the first few illustrate the principle.

Cyclic groups of the first kind with a single axis are simply the axial groups 1, 2, 3, 4, and 6, or, in alternative notation, C_1, C_2, C_3, C_4, and C_6. They can be represented in stereographic projections, the principle of which is illustrated in Fig. 1.47. Stereographic projections of these five cyclic groups are shown in Fig. 1.48. The order of the group is the total number of equivalent points (or objects) generated by the group when the first object does not lie on a symmetry element. The orders of C_1, C_2, C_3, C_4, and C_6 are 1, 2, 3, 4, and 6, respectively.

Rotation groups that have one n-fold axis but more than one twofold

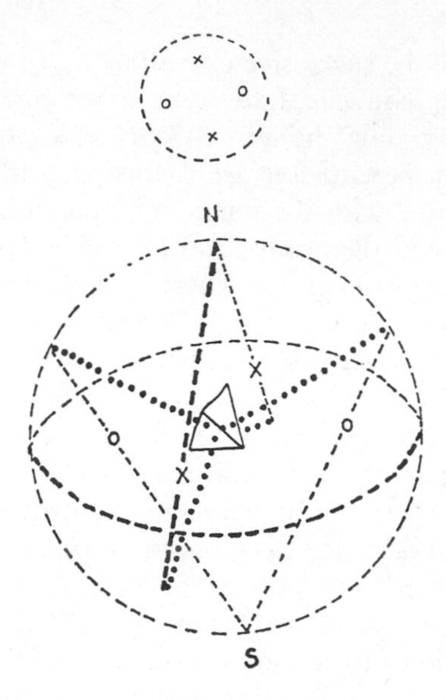

Fig. 1.47 Stereographic projection of the faces of a tetrahedron. A sphere is drawn about the crystal, in this case a tetrahedron. The normals to the faces are drawn out to where they intersect the sphere. Now define an equatorial plane, and a north pole, N, and south pole, S. Connect intersections of face normals in the northern hemisphere by straight lines to the S pole, and face normals in the southern hemisphere with the north pole. The intersections of these connecting lines with the equator give the stereographic projection. Points above the plane are represented by *o*, and those below the plane by x.

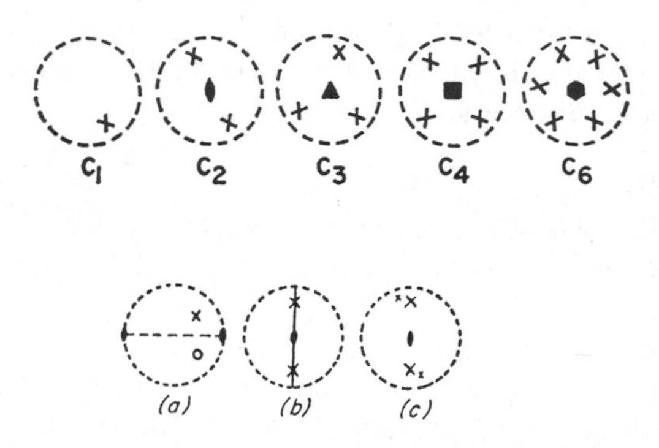

Fig. 1.48 The five cyclic groups of the first kind. Note that C_2 could also be represented as shown in (*a*), where the twofold axis is horizontal. The points x and *o* should be regarded as asymmetric so that we should not be tempted to introduce a mirror plane into C_2 as shown in (*b*); if any doubt remains concerning the symmetry, add another independent point as shown in (*c*), which clearly shows that (*b*) is incorrect.

axis can be found by extension of the cyclic groups of the first kind with a twofold axis. If we are to have only one n-fold axis, the twofold axis must be perpendicular to the n-fold axis; otherwise an additional n-fold axis will appear. Let us extend each of the cyclic groups by one twofold axis, as indicated in Fig. 1.49, and examine the symmetry of the resulting equivalent

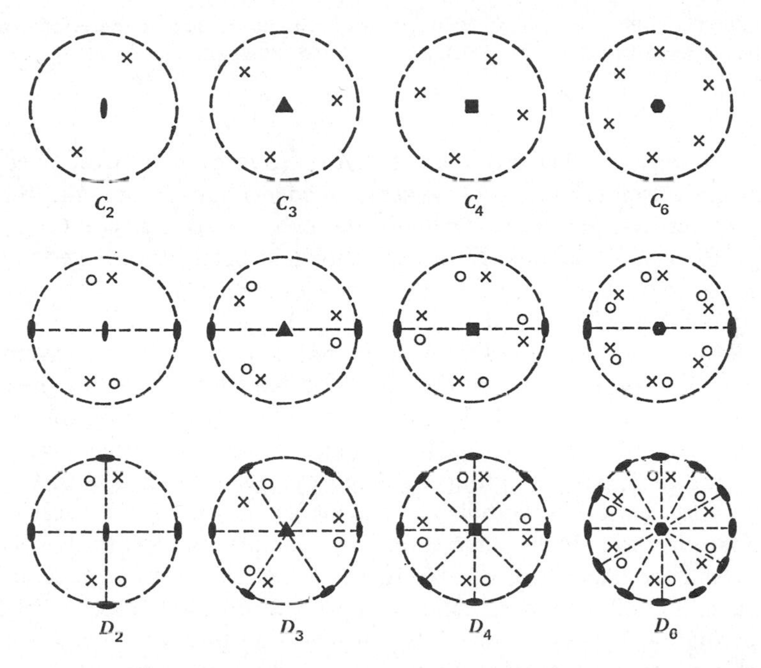

Fig. 1.49 The cyclic groups in the first line are extended by a single twofold axis in the second line, and the additional twofold axes so generated are indicated in the third line. The order of the group is the number of equivalent points generated when the points do not lie on a symmetry element.

objects, remembering that the objects themselves may be asymmetric. The additional symmetry elements, all twofold axes, so generated are shown in the last line along with their symbols.

Rotation groups with more than one axis of order greater than two form axial systems characteristic of regular polyhedra, as may be shown by starting with two intersecting axes of order greater than two and carrying out the rotations. A sphere drawn about the point of intersection of these axes is then covered by regular polyhedra when the symmetry operations of these axes are carried out. We shall not derive this result here, but the five regular polyhedra are the tetrahedron (4 faces, 4 vertices, 6 edges), the cube

Fig. 1.50 The cubic axial groups. Point groups having more than one threefold axis are described in terms of cubic axes, which are equal and orthogonal.

(6 faces, 8 vertices, 12 edges), the octahedron (8 faces, 6 vertices, 12 edges), the dodecahedron (12 faces, 20 vertices, 30 edges), and the icosahedron (20 faces, 12 vertices, 30 edges). Of these the last two have fivefold axes and hence are noncrystallographic, and the cube and octahedron have the same symmetry group. Thus there are only two new purely axial symmetry groups, T-23 and O-432, which are shown in Fig. 1.50.

We have now derived all of the 11 axial point groups. All of the symmetry elements of these groups are of the first kind. Now let us examine the cyclic groups of the second kind, $\bar{1}$, $\bar{2}$, $\bar{3}$, $\bar{4}$, and $\bar{6}$. Their stereographic projections are illustrated in Fig. 1.51. The principle of extension is illustrated here by the equivalence of $\bar{3}$, with the group obtained by extending C_3-3 with a center of inversion; hence the symbol C_{3i}-$\bar{3}$. Symbols such as S_4 are occasionally used for other axes in the literature. For example, S_6, identical with C_{3i}, occurs, and the reader will be able to show that the group generated by a rotation of 60° followed by reflection in a plane normal to this axis is identical with a rotation by 120° followed by an inversion through a center on the axis of rotation. The only group that could not have been obtained from extension of the cyclic groups of the first kind by i or m is $S_4 - \bar{4}$.

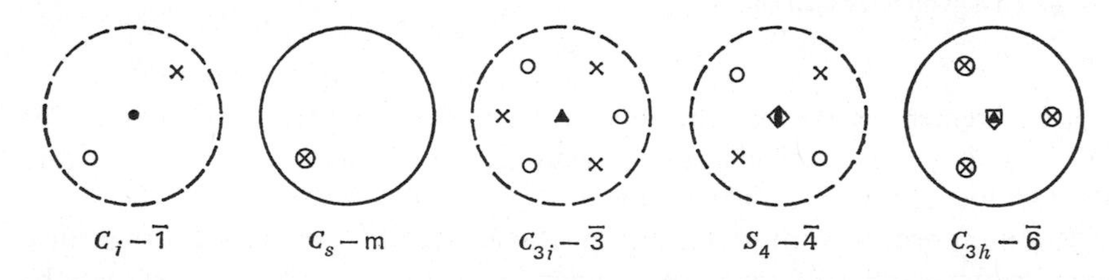

Fig. 1.51 Cyclic groups of the second kind. The operation $\bar{N}$ is a rotation by $(2\pi/N)$ followed by an inversion through a center of symmetry. The reader should check the equivalence of $\bar{1}$ to i, of $\bar{2}$ to m, of $\bar{3}$ to C_3 extended by i, and of $\bar{6}$ to C_3 extended by m_h. A mirror plane is called m_h if $\perp$ and m_v if parallel to the principal axis. A mirror plane in the plane of the stereographic projection is represented by a solid circle. Note that $C_s - m$, which is the same as $\bar{2}$, can also be represented, as shown in Fig. 1.52.

We have now derived 16 of the 32 point groups. The remainder are obtained by extending these 16 by mirror planes or centers of inversion added in such a way that the axial system remains invariant, for we have derived all of the axial groups. It is a simple exercise to finish the list. In doing so, be careful to exclude the noncrystallographic groups; that is, those that have axes of the first or second kind of order other than 1, 2, 3, 4, or 6. The remaining 16 are included in the complete list, in which all 32 are classified into reference systems of axes that prove convenient for their description (Fig. 1.52).

These point groups are extended, by the use of translation groups and the introduction of screw axes and glide planes, to the space groups in the following section. In addition, the centrosymmetric point groups, which are C_i, C_{2h}, D_{2h}, C_{4h}, D_{4h}, T_h, O_h, C_{3i}, D_{3d}, C_{6h}, and D_{6h}, are the possibilities shown by the diffraction pattern of any crystal [65]. The identification of this symmetry is usually the first step in a structure determination.

The knowledge that point groups must obey the mathematical rules for a group is useful in determining the presence of all necessary symmetry elements, that is, the completeness of the group. Consider for example the group C_{2h}. From the notation it is evident that the group must consist of a twofold axis and a mirror plane whose normal is in the direction of the twofold axis. The identity operation, E, must also be a member of the group. Since the inverse of the twofold axis and the mirror plane are just the same twofold axis and mirror plane, respectively, there are three symmetry elements thus far. To test for completeness, a multiplication table, called the Cayley table, is constructed. It is a square table in which the rows and columns are

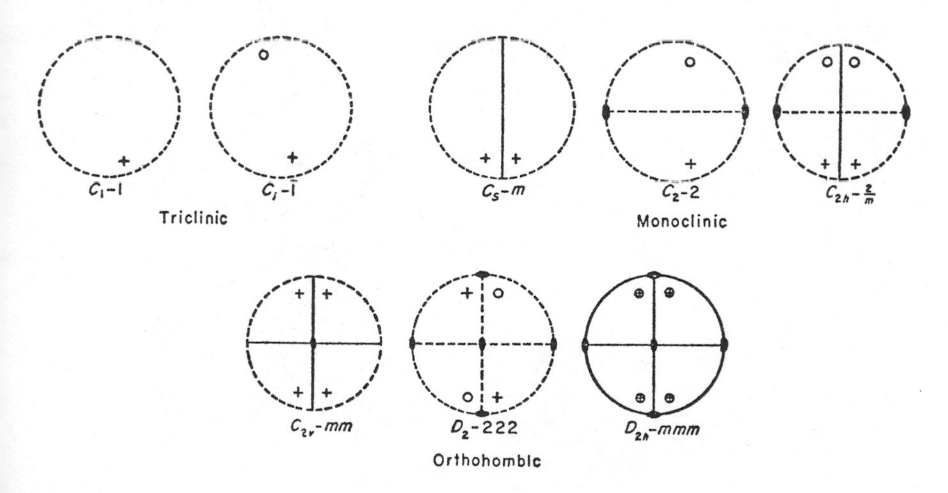

Fig. 1.52 The 32 point groups.

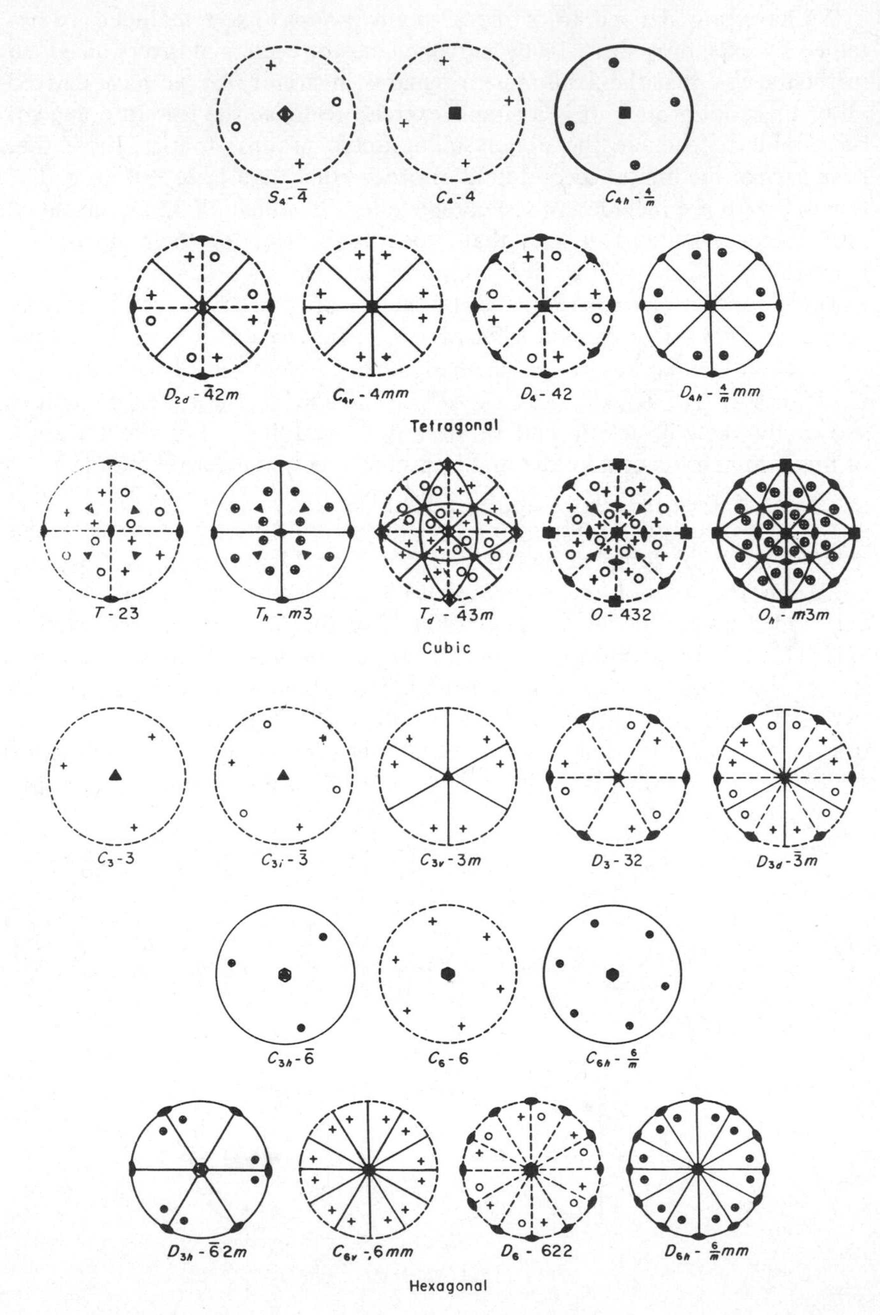

Fig. 1.52 (*contd.*)

Table 1.3 Cayley Table for the Point Group C_{2h}

	E	2	m	I
E	E	2	m	I
2	2	E	I	m
m	m	I	E	2
I	I	m	2	E

labeled with the operations of the group. In each space within the array is written the operation of the group that is equivalent to a row operation followed by a column operation. For the group C_{2h} the Cayley table is given in Table 1.3; the direction of the twofold axis is assumed to be the y-direction, and this is also then the direction of the normal to the mirror plane. Thus to complete the table an inversion operation must be added.

Space Groups [66]

Translation Groups in Two and Three Dimensions

The symmetry groups obtained by combining the screw rotation, the rotatory inversion, the gliding reflection, or some special case of these, with an infinite three-dimensional translation group are the space groups. Hence our first consideration is of the translation groups only, and we begin by listing the two-dimensional translation groups (Fig. 1.53). Any two-dimensional net can be expressed as a two-dimensional set of vectors $\mathbf{T} = l_1\mathbf{a}_1 + l_2\mathbf{a}_2$, where l_1 and l_2 are integers. A net so defined is called a primitive net. The obvious extension of this statement applies in three dimensions to lattices. But such a choice of a primitive pair of vectors for the C-centered orthorhombic net does not define a unit cell that exhibits the D_{2h} symmetry of the net, and hence a nonprimitive description is preferred. The nonprimitive set for this C-centered cell is $\mathbf{T} = l_1\mathbf{a}_1 + l_2\mathbf{a}_2 + (m/2)(\mathbf{a}_1 + \mathbf{a}_2)$, where $m = 1$ or 0, and l_1 and l_2 are integers. Although other-centered cells are possible, such as a centered cell of D_{4h} symmetry, the reader will be able to show that such a cell can also be described as a primitive cell by a new choice of axes less by $\sqrt{2}$ than the axes of the centered cell and oriented at 45° relative to them. Finally, recall that the only axial symmetries possible perpendicular to a net are 1, 2, 3, 4, and 6. Since the minimum symmetry at a point of a two-dimensional net is C_{2h}, such nets with axes 1 and 3 are not possible, and hence we have listed all of the possible two-dimensional nets in Fig. 1.53.

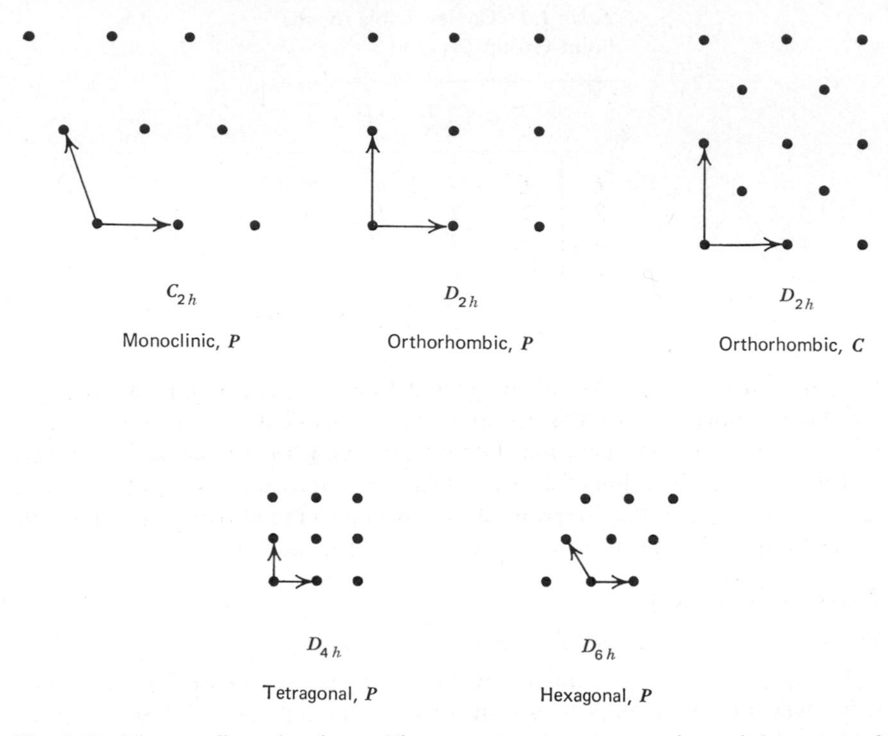

Fig. 1.53 The two-dimensional nets. The symmetry at every net point and the system of axes into which these nets most naturally fall is indicated. All of these nets indicated by *P* are primitive; the *C*-centered net is indicated by *C*.

The three-dimensional translation groups can be found in an analogous way. There are 14 such lattices, called the Bravais lattices, and they are described in Table 1.4. The coordinate systems that are most natural for descriptions of these lattices are indicated in the last column. The axes x, y, z are measured along the translations $\mathbf{a}_1$, $\mathbf{a}_2$, $\mathbf{a}_3$, respectively, of the lattices. Restrictions on the lengths of these axes and on the angles between them (α_1 is the angle between $\mathbf{a}_2$ and $\mathbf{a}_3$, and so on) required by the symmetry of each lattice are indicated in this last column. It must be emphasized that the symmetry, not some accidental coincidences of axial lengths or angles, determines the choice of axes and the crystal system.

To acquire additional familiarity with these lattices the reader may show that an orthorhombic cell centered on two pairs of opposite faces only is not possible, since such a set of vectors will generate equivalent points on the remaining pair of faces by virtue of the group properties, or prove the equivalence of *C* and *P* tetragonal cells or of *F* and *I* tetragonal cells. Which

Table 1.4 The Fourteen Bravias Lattices

Symmetry at a Lattice Point	*Name*	*Points per unit Cell*	*Elementary Translations*
$C_i - 1$	Triclinic	1	
$C_{2h} - \frac{2}{m}$	Simple monoclinic	1	$\alpha_1 = \alpha_3 = 90°$
	Side-centered monoclinic	2	
$D_{2h} - mmm$	Simple orthorhombic	1	$\alpha_1 = \alpha_2 = \alpha_3 = 90°$
	End-centered	2	
	Face-centered	4	
	Body-centered	2	
$D_{4h} - \frac{4}{m}mm$	Simple tetragonal (or	1	$\alpha_1 = \alpha_2 = \alpha_3 = 90°$
	end-centered)	2	$a_1 = a_2$
	Body-centered (or face-	2	
	centered)	4	
$D_{3d} - 3m$	Rhombohedral	1	$\alpha_1 = \alpha_2 = \alpha_3$
			$a_1 = a_2 = a_3$
$D_{6h} - \frac{6}{m}mm$	Hexagonal	1	$\alpha_1 = \alpha_2 = 90°$
			$\alpha_3 = 120°$
			$a_1 = a_2$
$O_h - m3m$	Simple cubic	1	$\alpha_1 = \alpha_2 = \alpha_3 = 90°$
	Face-centered cubic	4	$a_1 = a_2 = a_3$
	Body-centered cubic	2	

axis is labeled a, b, or c (a_1, a_2, or a_3) in the orthorhombic system is not determined by symmetry, and hence relabeling of axes converts an orthorhombic A- or B-centered cell into a C-centered cell. A similar statement holds for the equivalence of monoclinic A- and C-centered cells. A rhombohedral lattice can be indexed on hexagonal axes, and a hexagonal lattice can be indexed on rhombohedral axes [67], but the resulting transformations convert primitive cells to nonprimitive cells.

Space Operations

Screw axes permissible in three-dimensional crystallographic groups are limited to those involving rotations by 60°, 90°, 120°, 180°, and 360°, by the same argument as that for ordinary rotation axes. The operation for an n-fold screw axis A, taken always as a right-handed screw, is $A(2\pi/n, \mathbf{t})$ and its multiples. The operation for various fourfold screw axes is illustrated in Fig. 1.54. Diagrams for the twofold screw axis 2_1, the threefold screw axes

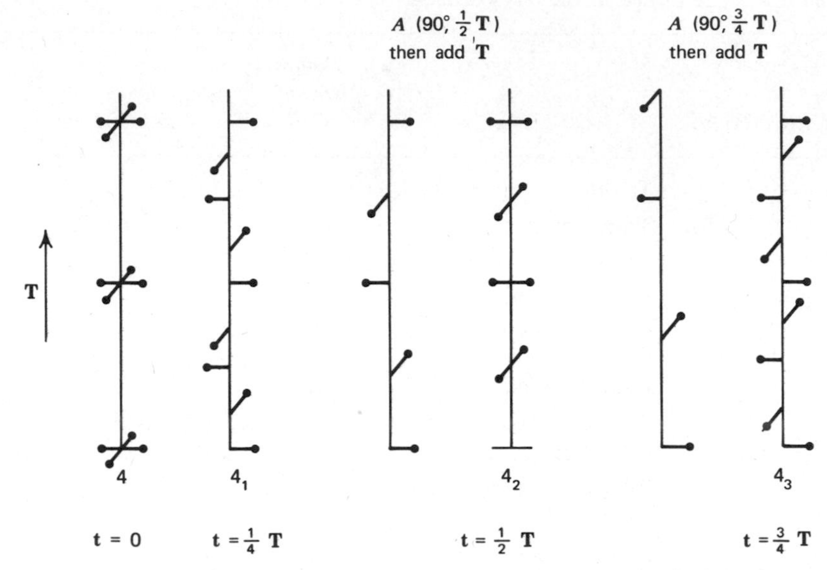

Fig. 1.54 Operations of fourfold screw axes. Although the operation is defined as right-handed, the result of 4_3 is a left-handed rotation after the effect of the primitive translation **T** is taken into account.

3_1 and 3_2, and the sixfold screw axes 6_1, 6_2, 6_3, 6_4, and 6_5 are analogous. A onefold screw axis is simply a lattice translation. Note that four successive applications of operation $A(2\pi/4, \mathbf{t})$, of the screw axis 4_1, result in a lattice translation **T**; and hence like any rotation axis the screw axis is always parallel to a row of the lattice and perpendicular to a net plane of the lattice.

The kinds of glide planes are also very limited in number. Let us represent a glide plane as $M(\mathbf{t})$, a reflection across a mirror plane m combined with a translation **t** parallel to the plane m (Fig. 1.55). If the operation is repeated the result is a translation of the original object by 2**t**, which must be a lattice translation **T**. The plane must therefore be parallel to a net plane of the lattice, and if this net plane is described by a primitive pair $\mathbf{a}_1$ and $\mathbf{a}_2$ the only possibilities are glides of $\frac{1}{2}\mathbf{a}_1$, $\frac{1}{2}\mathbf{a}_2$, or $\frac{1}{2}(\mathbf{a}_1 + \mathbf{a}_2)$. If the net is centered then a glide by $\frac{1}{4}(\mathbf{a}_1 + \mathbf{a}_2)$ is also possible. The glide planes corresponding to translations of $\frac{1}{2}\mathbf{a}_1$, $\frac{1}{2}\mathbf{a}_2$, $\frac{1}{2}\mathbf{a}_3$, $\frac{1}{2}(\mathbf{a}_1 + \mathbf{a}_2)$, and $\frac{1}{4}(\mathbf{a}_1 + \mathbf{a}_2)$ are called a, b, c, n, and d glide planes, respectively.

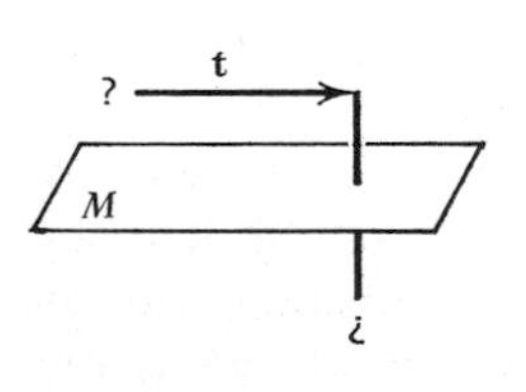

Fig. 1.55 The glide plane.

By combining these symmetry elements together the 230 space groups can be obtained.

The Hermann-Mauguin symbol is usually used in naming a space group. The first letter, a capital, indicates the type of cell, that is, primitive, body-centered, and so on, whereas the remaining symbols indicate the symmetry elements present in the space group, written in an order that refers to the direction of each symmetry element relative to the crystallographic axes.

In the triclinic system the only possible symmetry elements are 1 or $\bar{1}$. Thus only one symbol need be given after the cell-type designation.

In the monoclinic system there is one unique direction (usually the *b*-direction is chosen), and therefore in the Hermann-Mauguin symbol it is necessary to give only the symmetry element in that direction. If two symmetry elements have the same direction they are written in fractional form, for example, $2/m$.

In the orthorhombic system, however, symmetry elements lie along any of three directions. Therefore the symmetry element in the *x*-direction is given first, followed by the one in the *y*-direction, and then the one in the *z*-direction.

In the tetragonal and hexagonal systems the axis of order higher than two is assumed to be in the *c*-direction, and this symbol is listed first. The next symbol indicates the symmetry in the *a*-direction, and the third symbol indicates the symmetry along the *ab*-diagonal.

In the trigonal system the threefold or threefold inversion axis is along the body diagonal, and this symbol is listed following the cell type. The second symbol indicates the symmetry along a projection of *a* on the *abc* plane. The third symbol gives the symmetry 30° to this line in the *ab* plane.

In the cubic system the first symbol indicates the symmetry in the *a*-direction, the second along the body diagonal, and the third along the face diagonal.

The orthorhombic space group $Pna2_1$ indicates that the cell is primitive with a glide plane that reflects in the *x*-direction and translates in the *b* and *c* directions, a glide plane that reflects in the *y*-direction and translates in the *a*-direction, and a twofold screw axis in the *z*-direction.

Monoclinic Space Groups with Axes Only

As first examples of three-dimensional space groups, consider the possible combinations of the only lattices permitted in the monoclinic system, *P* and *C*, with the only axes permissible, 2 and 2_1. The origin of the translation group can be placed anywhere as long as the axis is parallel to a row of the lattice and perpendicular to a net plane. We lose no generality if we choose the origin (upper left corner) at the axis, which is perpendicular to the plane of the paper. The combination of a primitive translation group with a twofold axis is shown in Fig. 1.56. This unit cell may contain any number of molecules or atoms; for example, if it contained only one molecule this molecule would have to lie on a twofold axis and would itself have to have a

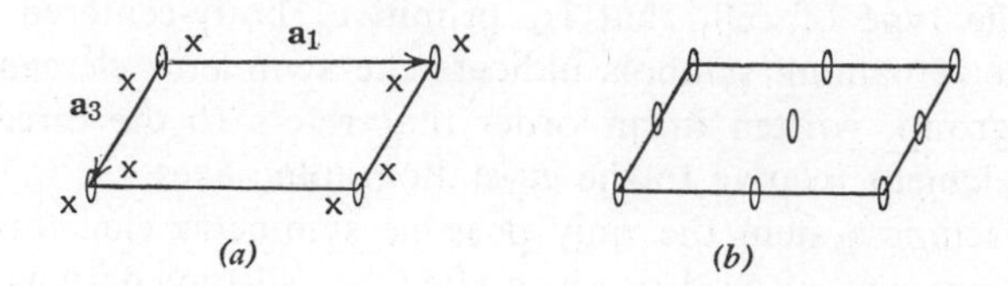

Fig. 1.56 The array of points generated by a twofold axis at the origin and the primitive translation group. (*a*) Incomplete. (*b*) Complete. Note that additional twofold axes at $\frac{1}{2}\mathbf{a}_1$, $\frac{1}{2}\mathbf{a}_3$, and $\frac{1}{2}(\mathbf{a}_1 + \mathbf{a}_3)$ have been generated as shown in part (*b*) of the figure, which is the complete representation of the space group $C_2{}^1 - P2$.

twofold axis of symmetry (assuming no disorder). If there are two molecules in the unit cell they could lie either on twofold axes or in the general position indicated in Fig. 1.56. Of course, more equivalent pairs of molecules in general positions may also occur. To illustrate one value of a knowledge of the space group: one needs to determine the coordinates of atoms in only one quadrant of this unit cell (Fig. 1.56), since the remainder of the cell is determined by symmetry.

If we combine a twofold screw axis with a primitive translation group, the result is shown in Fig. 1.57. This space group, $P2_1$, frequently occurs in crystals of optically active molecules, but perhaps less frequently than the orthorhombic space group $P2_12_12_1$ described below.

Now combine a *C*-centered lattice with a twofold axis. The set of equivalent points is shown in Fig. 1.58. The coordinates of the set of equivalent points produced by this space group are x, y, z; $\frac{1}{2} + x, \frac{1}{2} + y, z$; $\bar{x}, y, \bar{z}$; $\frac{1}{2} - x, \frac{1}{2} + y, -z$, where $\bar{x}$ means $-x$. If an atom or molecule lay on the twofold axis, the coordinates of its center would be $0, y, 0$; $\frac{1}{2}, \frac{1}{2} + y, 0$. On the other hand, if this center lay on the screw axis the coordinates would be $\frac{1}{4}, y, 0$; $\frac{3}{4}, \frac{1}{2} + y, 0$; $\frac{3}{4}, y, 0$; $\frac{1}{4}, \frac{1}{2} + y, 0$; thus the translational element

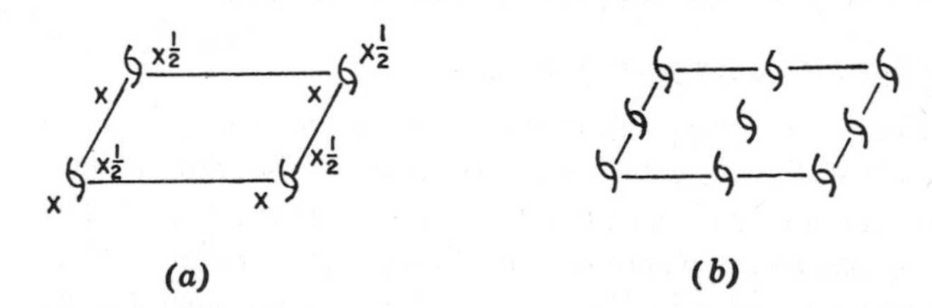

Fig. 1.57 Combination of a *P* lattice with 2_1. (*a*) Incomplete. (*b*) Complete. The twofold screw axis carries the point x halfway along the translation out of the plane of the paper and rotates it 180° around the screw axis, as indicated by the symbol $x\frac{1}{2}$. Note that the set of equivalent points so generated also requires the presence of screw axes halfway along the axes and at the center of the cell, as shown in the complete drawing (*b*) of the space group $C_2{}^2 - P2_1$.

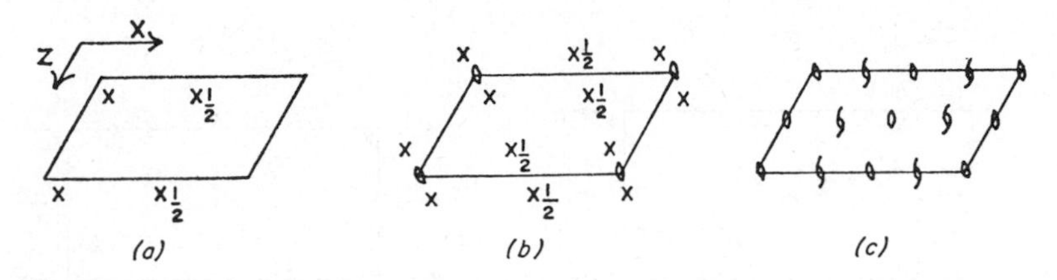

Fig. 1.58 The space group $C_2^3 - C2$ is shown in part (*c*). The *C*-centering is shown in (*a*); an atom at x, y, z has an equivalent atom at $\frac{1}{2} + x, \frac{1}{2} + y, z$. In (*b*) a twofold axis introduced at the origin produces the set of equivalent atoms shown, which gives the complete symmetry group shown in (*c*).

present in the screw axis would prevent the reduction in number of equivalent points possible for a twofold axis.

Finally, the combination of a *C* lattice with a 2_1 axis yields the same result as we have just obtained from *C* and 2. The choice of origin is arbitrary, and either the 2 or the 2_1 axis is a satisfactory choice.

Monoclinic Space Groups Related to $C_{2h} - 2/m$

Consider the extension of the space groups derived above by the possible kinds of symmetry planes, *m*, *a*, *c*, *n*, and *d*, $\perp$ the monoclinic axis 2 or 2_1. If we extend with a *d*-glide plane the cell becomes *B*-centered and can be reduced to a *P* cell. Hence the *d*-glide plane cannot occur in either a *P* or a *C* cell. The choice of axes is not determined by symmetry in the *P* cell, and hence the z-axis of the unit cell can be chosen along the glide plane. If this is done we need only consider extension of $P2$ and $P2_1$ by *m* or *c*. The extension of $C2$ is quite analogous, and is not considered here because the resulting space groups occur less frequently. These four possibilities are distinct and are summarized in Fig. 1.59. The reader will be able to show in detail how they arise from the extension of $P2$ and $P2_1$ by *m* or *c* planes.

A few details regarding the space group $P2_1/c$ may be helpful because this space group occurs very frequently, particularly for centrosymmetric organic molecules. Let the coordinates of the point x in $P2_1$ (Fig. 1.59) be x, y, z. Then the point o is at $-x, y + \frac{1}{2}, -z$. If the glide plane is introduced at $\frac{1}{4}b$, as indicated in $P2_1/c$, the additional points at $-x, \frac{1}{2} - y, \frac{1}{2} - z$ and $x, -y, \frac{1}{2} + z$ are generated. Let us refer these coordinates to the center of symmetry at $0\frac{1}{4}\frac{1}{4}$; that is, subtract $0\frac{1}{4}\frac{1}{4}$ from each set of coordinates. The result is

$$x, y - \tfrac{1}{4}, z - \tfrac{1}{4}; \qquad -x, y + \tfrac{1}{4}, -z - \tfrac{1}{4}$$

$$-x, -y + \tfrac{1}{4}, -z + \tfrac{1}{4}; \qquad x, -y - \tfrac{1}{4}, z + \tfrac{1}{4}.$$

Suppose we define $y_1 = y - \frac{1}{4}$ and also $z_1 = z - \frac{1}{4}$. Then the four equivalent points are x_1, y_1, z_1; $-x_1, -y_1, -z_1$; $x_1, \frac{1}{2} - y_1, \frac{1}{2} + z_1$; $-x_1, \frac{1}{2} + y_1, \frac{1}{2} - z_1$,

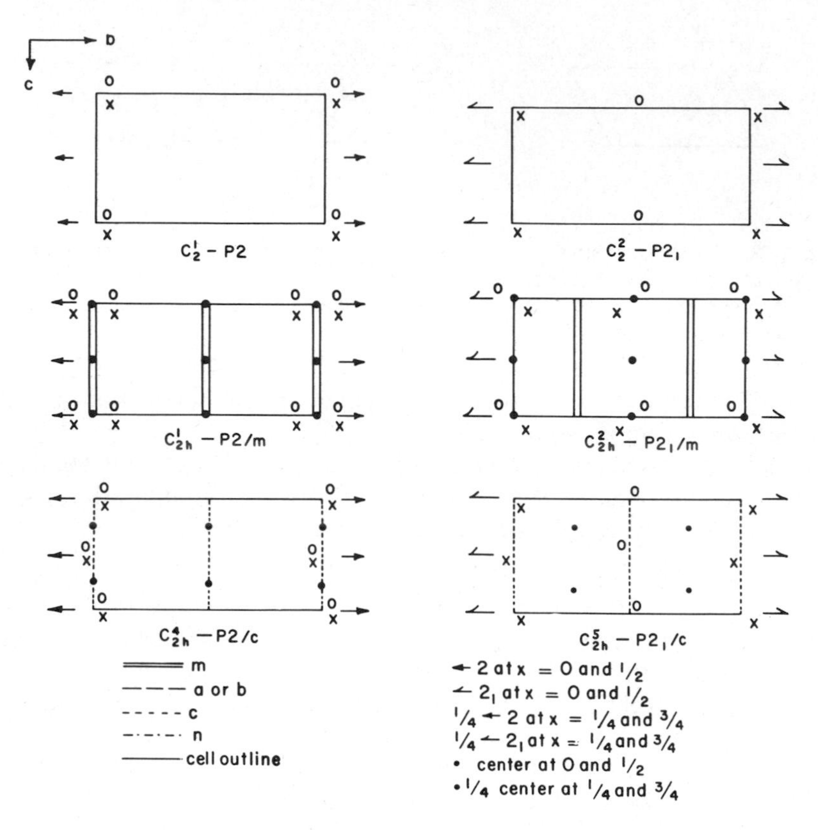

Fig. 1.59 Monoclinic space groups *P*2 and *P*2$_1$, and groups obtained upon extending these by *m* or *c*. A center of symmetry is usually chosen as origin.

where we remember that $+\frac{1}{2}$ is the same as $-\frac{1}{2}$ because of the lattice translation.

This result can also be obtained using a somewhat different approach. Define a general symmetry operator by the quantity $[\mathbf{S} + \mathbf{t}]$, where $\mathbf{S}$ is a point group type of operator and $\mathbf{t}$ is the associated translation. Then if the point x, y, z is transformed into the point $\frac{1}{2} - x$, $\frac{1}{2} + y$, $-z$ by a symmetry operation, the associated $\mathbf{S}$ and $\mathbf{t}$ operators are

$$\mathbf{S} = \begin{pmatrix} -1 & 0 & 0 \\ 0 & 1 & 0 \\ 0 & 0 & -1 \end{pmatrix} \qquad \mathbf{t} = \begin{pmatrix} \frac{1}{2} \\ \frac{1}{2} \\ 0 \end{pmatrix}.$$

Therefore the operation $[\mathbf{S}_1 + \mathbf{t}_1]$ followed by $[\mathbf{S}_2 + \mathbf{t}_2]$ would be equivalent

Table 1.5 Cayley Table for Space Groups Isomorphous with C_{2h}

	$[E/t_1]$	$[2/t_2]$	$[m/t_3]$	$[I/t_4]$
$[E/t_1]$	$[E/2t_1]$	$[2/t_2 + t_1]$	$[m/t_3 + t_1]$	$[I/t_4 + t_1]$
$[2/t_2]$	$[2/2t_1 + t_2]$	$[E/2t_2 + t_2]$	$[I/2t_3 + t_2]$	$[m/2t_4 + t_2]$
$[m/t_3]$	$[m/mt_1 + t_3]$	$[I/mt_2 + t_3]$	$[E/mt_3 + t_3]$	$[2/mt_4 + t_3]$
$[I/t_4]$	$[I/It_1 + t_4]$	$[m/It_2 + t_4]$	$[2/It_3 + t_4]$	$[E/It_4 + t_4]$

to the operation given by $[\mathbf{S}_2\mathbf{S}_1 + \mathbf{S}_2\mathbf{t}_1 + \mathbf{t}_2]$. Taking the space groups related to the point group C_{2h} we can write the Cayley table as shown in Table 1.5, where $[\mathbf{S}/\mathbf{t}] = [\mathbf{S} + \mathbf{t}]$. Consider the space group $P2_1/c$. Only the translations modulo the unit-cell translation are of interest. Since the cell is primitive, $\mathbf{t}_1 = 0$. Also from the space group symbol it is evident that $t_{2y} = t_{3z} = \frac{1}{2}$. Since this group possesses a center of symmetry it is most convenient to place the center of symmetry at the origin, that is, $\mathbf{t}_4 = 0$ and $2\mathbf{t}_3 + \mathbf{t}_2 = 0$. Solving for the individual components of these column vectors one obtains $t_{2x} = t_{3x} = 0$, $t_{2z} = \frac{1}{2}$, and $t_{3y} = \frac{1}{2}$.

Note that when the equivalent points lie on the screw axes or glide planes there are still four equivalent points in the unit cell, but if they lie on the symmetry centers they reduce to only two: 000 and $0\frac{1}{2}0$. Very frequently the determination of the unit cell and the space group leads to the unique conclusion that there are two molecules in a unit cell of symmetry $P2_1/c$ (or the equivalent $P2_1/a$ or $P2_1/n$). Except for the rare case of orientational disorder one can then be sure that the molecule has a center of symmetry. This particular case arises so frequently that it is worth testing if there is any possibility that a molecule has a center of symmetry, for the amount of time involved in such a determination is only a day or two.

The Space Group $P2_12_12_1$

In the orthorhombic system symmetry elements occur along the **a**, **b**, and **c** directions, and in particular the space group $P2_12_12_1$ is very frequently chosen by noncentrosymmetric organic molecules when they crystallize. Usually four molecules occur in the unit cell.

Consider the extension of $P2_1$ by a screw axis along another direction perpendicular to the first. Introduction of the second screw axis so that it intersects the first leads to the perfectly good space group $P2_12_12$, but we introduce the second screw axis so that it does not intersect the first one (Fig. 1.60). The points x, y, z and $-x, y + \frac{1}{2}, -z$ of $P2_1$ then create the additional points $\frac{1}{2} - x, -y, \frac{1}{2} + z$ and $\frac{1}{2} + x, \frac{1}{2} - y, \frac{1}{2} - z$, and these four points are equivalent in $P2_12_12_1$. If we list these points in a column, subtract

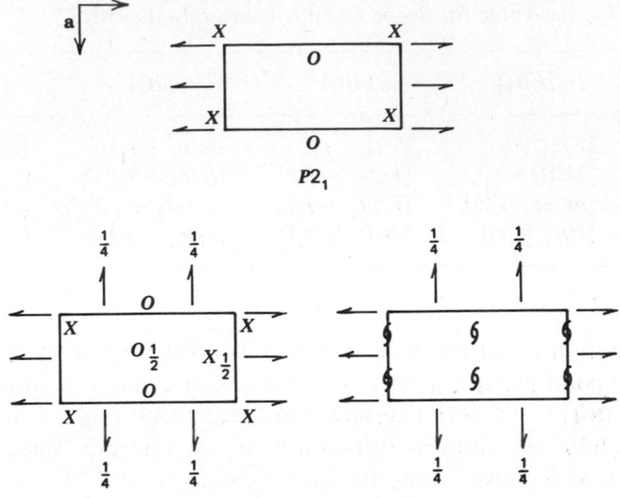

Fig. 1.60 The space group $P2_12_12_1$, obtained by extension of $P2_1$ with another 2_1 not intersecting the first.

$00\frac{1}{4}$ to change the origin by $\frac{1}{4}\mathbf{c}$ in the second column, and define $z_1 = z - \frac{1}{4}$, we find

x, y, z	$x, y, z - \frac{1}{4}$	x, y, z_1
$-x, y + \frac{1}{2}, -z$	$-x, y + \frac{1}{2}, -\frac{1}{4} - z$	$-x, y + \frac{1}{2}, \frac{1}{2} - z_1$
$\frac{1}{2} - x, -y, \frac{1}{2} + z$	$\frac{1}{2} - x, -y, \frac{1}{4} + z$	$\frac{1}{2} - x, -y, \frac{1}{2} + z_1$
$\frac{1}{2} + x, \frac{1}{2} - y, \frac{1}{2} - z$	$\frac{1}{2} + x, \frac{1}{2} - y, \frac{1}{4} - z$	$\frac{1}{2} + x, \frac{1}{2} - y, -z_1$

It is worth noting that in the space group $P2_12_12_1$ there are no special positions that lead to less than four equivalent atoms.

Effect of Symmetry Elements on X-Ray Intensities

Extinctions

Any symmetry element involving a translation produces systematic absences of certain types of X-ray diffraction maxima. These elements include centering operations in translation groups, glide planes, and screw axes. Three illustrations are given.

EXTINCTIONS DUE TO A *B*-CENTERED LATTICE

If a lattice is *B*-centered, then for every atom or molecule at x, y, z there exists an equivalent atom or molecule at $\frac{1}{2} + x, y, \frac{1}{2} + z$. The structure factor expression then becomes, for these two atoms only,

$$F_{hkl} = \sum_{j=1}^{N=2} f_j e^{-2\pi i(hx_j+ky_j+lz_j)}$$
$$= f[e^{-2\pi i(hx+ky+lz)} + e^{-2\pi i[h(1/2+x)+ky+l(1/2+z)]}]$$
$$= fe^{-2\pi i(hx+ky+lz)}[1 + e^{-2\pi i[(h+l)/2]}].$$

Remember that h and l are always integers and that $e^{i\theta} = \cos\theta + i\sin\theta$. Then the exponential, $e^{2\pi i[(h+l)/2]}$, is -1 when $h + l$ is odd and is $+1$ when $h + l$ is even. Hence $F_{hkl} = 0$ when $h + l$ is odd.

In general, if there are more than two atoms in the unit cell the reduction is

$$F_{hkl} = \sum_{j=1}^{N} f_j e^{-2\pi i(hx_j+ky_j+lz_j)} = \sum_{j=1}^{N/2} f_j e^{-2\pi i(hx_j+ky_j+lz_j)}(1 + e^{-2\pi i[(h+l)/2]}),$$

where the sum from $j = 1$ to $N/2$ extends over all atoms *not* related by the B centering.

EXTINCTION DUE TO A *b*-GLIDE PLANE PERPENDICULAR TO THE *c*-AXIS

A *b*-glide $\perp$ **c**, with z coordinate $z = 0$, means that if there is an atom at xyz, then there is an equivalent atom at $x, \frac{1}{2} + y, \bar{z}$. The structure factor then becomes

$$F_{hkl} = \sum_{j=1}^{N/2} f_j[e^{-2\pi i(hx_j+ky_j+lz_j)} + e^{-2\pi i[hx_j+k(1/2+y_j)-lz_j]}],$$

where the sum extends over the $N/2$ atoms not related by this *b*-glide. For the general *hkl* reflections, no systematic extinction results, but, for the *hk*0 reflections,

$$F_{hk0} = \sum_{j=1}^{N/2} f_j e^{-2\pi i(hx_j+ky_j)}[1 + e^{-2\pi i(k/2)}].$$

Hence all reflections of the type *hk*0 vanish when k is odd.

EXTINCTION DUE TO A TWOFOLD SCREW AXIS ALONG **b**

The symmetry operation of a 2_1 axis along **b**, with x and z coordinates zero, transforms an atom at x, y, z into an equivalent atom at $\bar{x}, y + \frac{1}{2}, \bar{z}$. The structure factor expression becomes

$$F_{hkl} = \sum_{j=1}^{N/2} f_j[e^{-2\pi i(hx_j+ky_j+lz_j)} + e^{-2\pi i[-hx_j+k(1/2+y_j)-lz_j]}].$$

Systematic absences occur for the 0*k*0 reflections. The expression reduces to

$$F_{0k0} = \sum_{j=1}^{N/2} f_j e^{-2\pi i(ky_j)}[1 + e^{-2\pi i(k/2)}],$$

where the sum extends over the $N/2$ atoms not related by the 2_1 axis. Thus the $0k0$ reflections are absent when k is odd.

EXTINCTIONS DUE TO OTHER SYMMETRY ELEMENTS

A complete list of extinctions due to symmetry elements are given in Table 1.6.

Statistical Relations

While symmetry elements not involving translations do not produce systematic extinctions they do influence the intensity distributions. For example, Wilson [69] has shown that for a crystal containing a reasonably large number of atoms of about the same atomic number in the unit cell, the ratio of the square $\langle F\rangle^2$ of the mean value $\langle F\rangle$ to the mean value $\langle I\rangle$, where $I = |F|^2$, is significantly different for centrosymmetric unit cells (0.785) than for noncentrosymmetric unit cells (0.637). A more detailed examination of this and related more powerful criteria [70] and a comparison with experimental data do indicate that this method is useful in deciding between centrosymmetric and noncentrosymmetric space groups for complex structures. However, some caution is necessary, for pseudo-symmetry may exist; that is, the structure may closely resemble a centrosymmetric arrangement without actually having this symmetry, and the statistical method may then give an incorrect result. Similar statistical relations exist for other symmetry elements, and are usually most apparent from examination of the limited amount of data most strongly influenced by these symmetry elements.

4 PRINCIPLES OF EXPERIMENTAL METHODS*

Introduction

In Section 1 only a few of the many methods available for the solution of structures were discussed in any detail. We follow the same philosophy in this, the experimental section. Hence some of the details of a relatively simple compound, hydrogen fluoride, are discussed here. The major reason for the selection of the compound is the relatively small amount of data involved. We only consider the location of the fluorine atoms and do not consider the hydrogen atom positions, which are however discussed in the original publication [71].

* For more detailed discussion of the experimental methods of this section consult U. W. Arndt and B. T. M. Willis, *Single Crystal Diffractometry*, Cambridge University Press, Cambridge, 1966; L. V. Aza'roff, *Elements of X-ray Crystallogrphy*, McGraw-Hill, New York, 1968; M. J. Buerger, *X-ray Crystallography*, Wiley, New York, 1942; and E. W. Nuffield, *X-ray Diffraction Methods*, Wiley, New York, 1966.

Table 1.6 Extinctions Due to Symmetry Elements

Class of Reflection	*Condition for Nonextinction (n an integer)*	*Interpretation of Extinction*	*Symbol for Symmetry Element*
hkl	$h + k + l = 2n$	Body-centered lattice	I
	$h + k = 2n$	C-centered lattice	C
	$h + l = 2n$	B-centered lattice	B
	$k + l = 2n$	A-centered lattice	A
	$h + k, h + l, h + l = 2n$	Face-centered lattice	F
	$-h + k + l = 3n$	Rhombohedral lattice indexed on hexagonal lattice	R
	$h + k + l = 3n$	Hexagonal lattice indexed on rhombohedral lattice	H
$0kl$	$k = 2n$	Glide plane $\perp\mathbf{a}$, translation $\mathbf{b}/2$	b
	$l = 2n$	Glide plane $\perp\mathbf{a}$, translation $\mathbf{c}/2$	c
	$k + l = 2n$	Glide plane $\perp\mathbf{a}$, translation $(\mathbf{b} + \mathbf{c})/2$	n
	$k + l = 4n$	Glide plane $\perp\mathbf{a}$, translation $(\mathbf{b} + \mathbf{c})/4$	d
$h0l$	$h = 2n$	Glide plane $\perp\mathbf{b}$, translation $\mathbf{a}/2$	a
	$l = 2n$	Glide plane $\perp\mathbf{b}$, translation $\mathbf{c}/2$	c
	$h + l = 2n$	Glide plane $\perp\mathbf{b}$, translation $(\mathbf{a} + \mathbf{c})/2$	n
	$h + l = 4n$	Glide plane $\perp\mathbf{b}$, $(\mathbf{a} + \mathbf{c})/4$	d
$hk0$	$h = 2n$	Glide plane $\perp\mathbf{c}$, translation $\mathbf{a}/2$	a
	$k = 2n$	Glide plane $\perp\mathbf{c}$, translation $\mathbf{b}/2$	b
	$h + k = 2n$	Glide plane $\perp\mathbf{c}$, translation $(\mathbf{a} + \mathbf{b})/2$	n
	$h + k = 4n$	Glide plane $\perp\mathbf{c}$, translation $(\mathbf{a} + \mathbf{b})/4$	d
hhl	$l = 2n$	Glide plane $\perp(\mathbf{a} - \mathbf{b})$, translation $\mathbf{c}/2$	c
	$2h + l = 2n$	Glide plane $\perp(\mathbf{a} - \mathbf{b})$, translation $(\mathbf{a} + \mathbf{b} + \mathbf{c})/2$	n
	$2h + l = 4n$	Glide plane $\perp(\mathbf{a} - \mathbf{b})$, translation $(\mathbf{a} + \mathbf{b} + \mathbf{c})/4$	d
$h00$	$h = 2n$	Screw axis $\parallel\mathbf{a}$, translation $\mathbf{a}/2$	2_1, 4_2
	$h = 4n$	Screw axis $\parallel\mathbf{a}$, translation $\mathbf{a}/4$	4_1, 4_3
$00l$	$l = 2n$	Screw axis $\parallel\mathbf{c}$, translation $\mathbf{c}/2$	2_1, 4_2, 6_3
	$l = 4n$	Screw axis $\parallel\mathbf{c}$, translation $\mathbf{c}/4$	4_1, 4_3
	$l = 3n$	Screw axis $\parallel\mathbf{c}$, translation $\mathbf{c}/3$	3_1, 3_2, 6_2 6_4
	$l = 6n$	Screw axis $\parallel\mathbf{c}$, translation $\mathbf{c}/6$	6_1, 6_5
$hh0$	$h = 2n$	Screw axis $\parallel(\mathbf{a} + \mathbf{b})$, translation $(\mathbf{a} + \mathbf{b})/2$	2_1

Some of the difficulties in this study somewhat resemble difficulties in studies of very complex structures. Low-temperature techniques [72] must be used here because of the low melting point of the crystal, and in proteins in order to decrease the rate of destruction of the crystal by the products of X-radiation. Also, the material must be sealed into capillaries, here in fluorothene (polytrifluorochloroethylene) capillaries, in order to place the liquid in an inert container, and in glass capillaries for compounds unstable in air, for proteins, or for other crystals containing solvent of crystallization, in order to prevent destruction of the single crystal by evaporation of the solvent.

A number of experimental methods are available from which to obtain space-group and unit-cell information and from which to collect intensity data, but the simplest of the film techniques is the precession method [73], developed by M. J. Buerger, and now widely in use. The precession method gives an undistorted picture of the reciprocal lattice. In fact it is useful to think of all other methods for the study of single crystals as giving photographs of distorted reciprocal space.

The Principles of the Precession Method

Undistorted photographs of layers of the reciprocal lattice may be obtained in experiments in which the crystal and X-ray film are always maintained in the same relative orientation as they are moved (Fig. 1.61). Then, as described in Section 2, as a reciprocal lattice point crosses the sphere of reflection of radius $1/\lambda$, a diffraction maximum occurs in the direction of **s**, the scattered ray. Since in this method the reciprocal lattice plane shown in Fig. 1.61 is always maintained parallel to the film, the resulting scattered rays merely project the reciprocal lattice plane onto the film.

In the precession method the normal N to the reciprocal lattice plane and the normal N' to the X-ray film are maintained parallel as they are rotated slowly, about once a minute, about the line OO' (Fig. 1.62). Remember from Section 2 that when a reciprocal lattice plane is being photographed, the direction of N coincides with an axis of the crystal. Usually Mo$K\alpha$ radiation ($\lambda = 0.711$ Å) is employed with which one can obtain a number of reflections at least comparable with other methods now in use. A usual crystal-to-film distance is 6.00 cm for zero-level photographs. If we recall from Section 2 that $d_{hkl} = 1/|\mathbf{h}|$ and that $\mathbf{s}$ and $\mathbf{s}_0$ are unit vectors, the interplanar spacing is, from similar triangles, $d_{hkl} = \lambda F/x$, as derived in the legend of Fig. 1.61. During the complete precession motion of N and N' about OO', a circle of radius OP is projected onto the X-ray film in the zero-level photograph illustrated in Fig. 1.62.

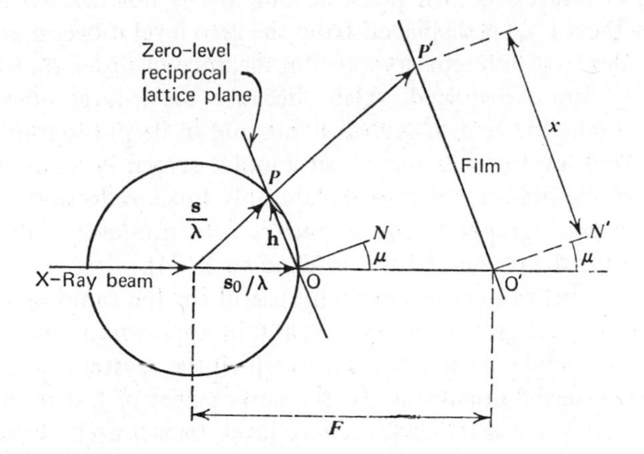

Fig. 1.61 Geometry of the de Jong and Bouman and the precession methods. The plane in reciprocal space extends above and below the plane of this figure. A reciprocal lattice point at the end of the vector **h** satisfies the conditions $\mathbf{s}/\lambda - \mathbf{s}_0/\lambda = \mathbf{h}$ for reflection. The position x at which the reflection occurs on the film is, from similar triangles, $x/F = |\mathbf{h}|/|\mathbf{s}_0/\lambda| = \lambda/d_{hkl}$. Hence $d_{hkl} = \lambda F/x$. In this construction, note that the crystal-film distance F is measured from the point at which the diffraction vector **s** originates; it is helpful to think of the conversion of the crystal into a reciprocal lattice in a volume which is comparable with the size of the crystal.

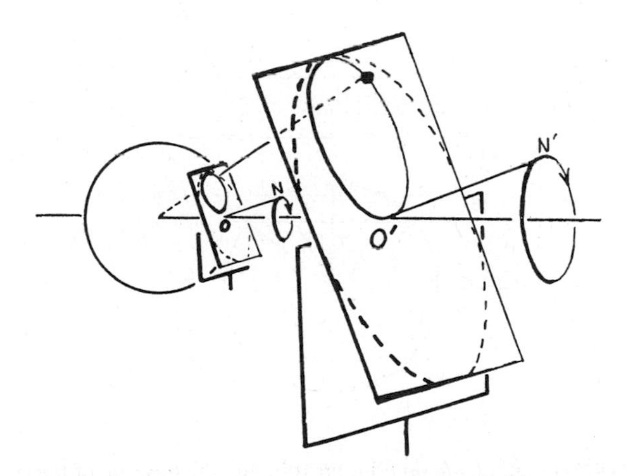

Fig. 1.62 As N and N' rotate about the line OO', the portion of the reciprocal lattice in the dotted circle about O is projected undistorted onto the film in the dotted circle about O' as a center.

Recording of *n*-levels is also possible but this is not described in detail here. Because the *n*-level is displaced from the zero level it becomes necessary to move the film to a different crystal-film distance in order that the *n*-level photograph is also undistorted. Also, because the *n*-level does not pass through the origin, the central region is missing in its photograph. Finally, in either the zero level or the *n*-level an annular screen is required between the crystal and the film in order to isolate only those reflections that are in the level being photographed. The geometry for an *n*-level is illustrated in Fig. 1.63, and the details can be found elsewhere [74].

The precession instrument is especially useful for the rapid determination of unit cells and space groups of crystals and in applications such as low- or high-temperature single-crystal studies, in which the crystal has to be grown or placed near external apparatus. In the early stages of a determination of of crystal symmetry one searches for a zero level, lines it up by trial, and then determines the symmetry of the *n*-levels parallel to this zero level. It is not necessary to take *n*-level photographs for this purpose, for the symmetry of the *n*-levels is observed on the zero-level photograph if the layer-line screen is left out of the camera. After the symmetry elements of the reciprocal lattice are located and the unit cell is chosen, the reflections can then be identified by their indices *hkl*.

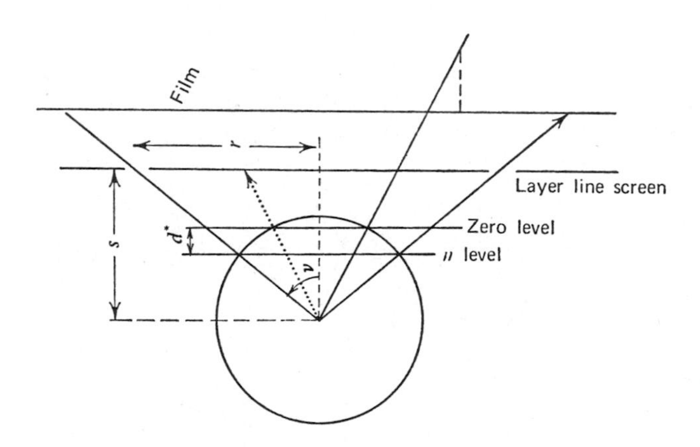

Fig. 1.63 The geometry of an *n* level photograph, and the position of the layer line screen, which excludes reflections from other levels, such as the zero-level reflection shown by a dotted line. The relation $\tan \nu = r/s$, which determines the setting of the layer line screen, may be expressed in terms of d^* and the precession angle.

Example: The Structure of Crystalline HF

The X-ray Diffraction Photographs

A large number of diffraction photographs is normally required for a structure determination; it is usually advisable to obtain as many non-zero reflections as possible in order to obtain certainty of the structure and to keep standard deviations of parameters as low as possible. Of the 20 or so diffraction photographs taken of a single crystal of HF, only three are shown here (Fig. 1.64).

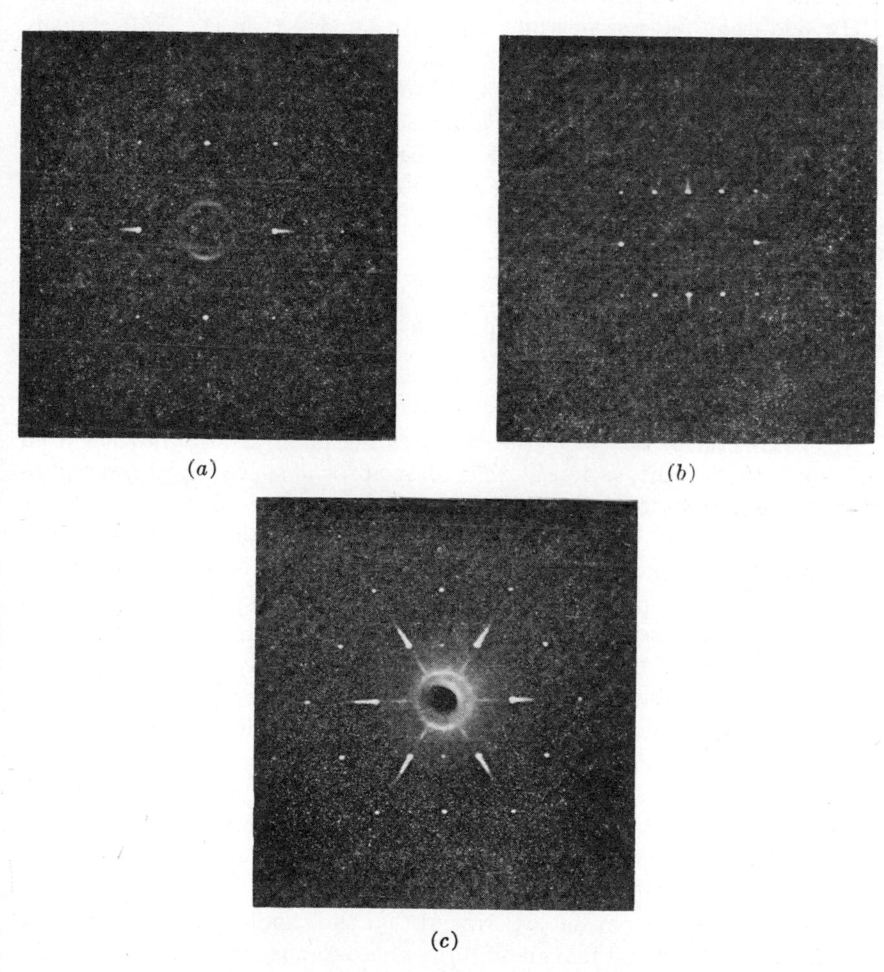

(*a*) (*b*) (*c*)

Fig. 1.64 Diffraction photographs of a single crystal of hydrogen fluoride.

Some of the features of these photographs are common to all X-ray diffraction photographs of single crystals. The finite size of the maxima is due mostly to crystal size and beam divergence and is far greater than the usual intrinsic width of a diffraction maximum as discussed in Section 2. The average overall decline of intensities with increasing angle of scattering is due mostly to the finite size of atoms and in particular to the approximately Gaussian decrease of electron density as one moves away from the nucleus; but there are other factors contributing to this decline of intensities, such as the thermal motions of the atoms, the polarization of radiation, and the time each reflection is in a position to reflect (Lorentz factor). All of these effects have been given a preliminary quantitative description in Section 2. In addition, the radial white radiation streaks (wheaks [75]) actually appear on both sides of the most intense reflections, but are much less just inside the spots because of absorption by the zirconium filter that was introduced to remove the radiation. The black area in the center of the photograph has been shielded from the direct X-ray beam by a circular piece of lead. A white ring around this dark area and the much less intense outer halos are due to the fluorothene polymer tubing. For the structure determination we shall be interested only in the symmetry, positions, and intensities of the diffraction maxima, which occur as white spots on these prints and as dark spots on the original negatives, which had been exposed to the X-ray scattering from the crystal.

Symmetry of the Reciprocal Lattice

The symmetry of the reciprocal lattice is the point group obtained from the symmetry of the various levels of reciprocal space on these photographs. Since the substance under study does not show appreciable X-ray absorption near the wavelength of Mo$K\alpha$ radiation, the reciprocal lattice necessarily has a center of symmetry. All three of these photographs are zero levels of reciprocal space, and hence contain the origin, so that a center of symmetry is therefore necessarily present. The additional lines of symmetry, 90° apart in each of these three photographs, then imply either a single line of symmetry plus the center, or two lines of symmetry 90° apart. Which of these possibilities is correct can be decided by additional photographs, which include the higher-level reflections on these zero levels, in experiments in which the annular layer line screen is omitted. These additional photographs, not shown here, indicate that there are two lines of symmetry 90° apart on photographs (*a*) and (*c*), but only one line of symmetry perpendicular to the horizontal line on photograph (*b*).

Now it is known from the experiment that the capillary axis is along the horizontal direction and that these three photographs are obtained with the crystal reoriented at 33 and 90° about the horizontal axis for photographs (*b*) and (*c*), respectively, relative to photograph (*a*). Thus these three photographs are to be regarded as three sections through the origin of reciprocal space all containing the horizontal direction in common and oriented at 0, 33, and 90° relative to one another. The symmetry of photographs (*a*) and (*c*) implies pairs of mirror planes 90° apart, and these, together with the center of symmetry of reciprocal space, require (Section 3) that the symmetry of reciprocal space is at least as high as D_{2h}-*mmm*. The single-mirror plane of photograph (*b*) is consistent with these observations. The directions from the origin to the first reflection along the vertical directions of photographs (*a*) and (*c*) and along the horizontal direction of all three photographs represent the directions of special symmetry, and hence will be chosen as the axial directions. That the distances to these first reflections in these three directions are not related by an integer or some multiple of $\sqrt{2}$ or $\sqrt{3}$ is a strong indication, later confirmed by many additional photographs, that the symmetry of reciprocal space is not higher than D_{2h}-*mmm*. We therefore conclude that the crystal structure is orthorhombic. Note that this conclusion is based upon symmetry requirements and not on any special axial lengths or angles between axes over and above those required by these symmetry requirements. Finally the symmetry of reciprocal space, that is, the diffraction symmetry, is not usually the same as the symmetry of the crystal itself. All orthorhombic crystals show D_{2h}-*mmm* diffraction symmetry, but there are many different orthorhombic crystal symmetries.

The Unit Cell

The simplest approach to the determination of the unit cell is to choose axes according to symmetry directions, find the unit cell in reciprocal space and convert this unit cell by the methods of Section 2 to the unit cell in crystal space. In the orthorhombic system the directions of all axes are determined by symmetry, and the directions of the reciprocal axes and crystal axes coincide. In reciprocal space these directions are perpendicular to the mirror planes; that is, along the twofold axes of the D_{2h}-*mmm* symmetry group. Hence we shall be required to choose one axis, say b_1, along the vertical direction of photograph (*a*); the second axis, say b_2, along the horizontal direction of photograph (*a*), (*b*), or (*c*); and the third axis, b_3, along the vertical direction of photograph (*c*). The distances of the closest reflections along b_1 and b_3, the relative orientations of the three photographs about b_2, and the smallest reciprocal cell in the b_1 and b_3 directions are summarized in Fig. 1.65. We must now choose a cell in three-dimensional reciprocal space in such a way that we account for all observed reflections in a three-dimensional array. Now the distances of the closest reflections to the center along b_1, b_2, and b_3 are, respectively, 2.49, 1.97, and 1.58 cm. But a unit cell of this size would not include the first reflection along the vertical direction

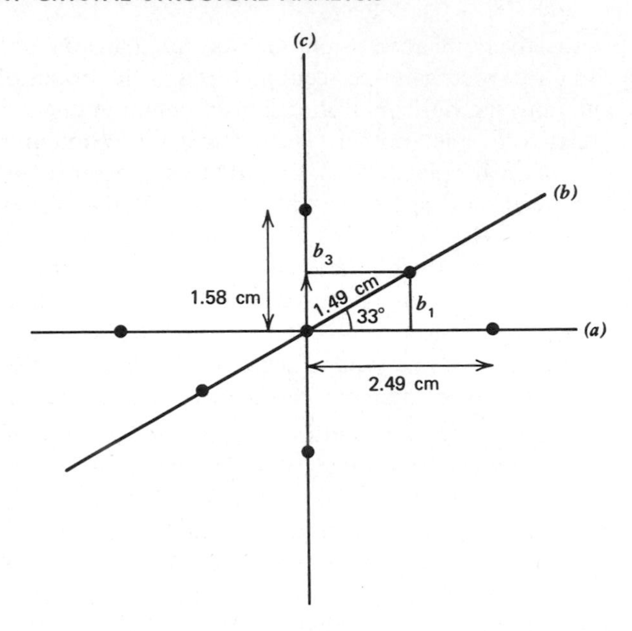

Fig. 1.65 Relative orientations of photographs (*a*), (*b*), and (*c*) about the horizontal axis common to all three. Measurements are those from the center to the first reflection in the vertical direction on each photograph. The smallest unit cell of reciprocal space that will account for all reflections has dimensions $b_1 = 1.245$ cm, $b_2 = 0.985$ cm ($\perp$ b_1 and b_3), and $b_3 = 0.79$ cm.

of photograph (*b*), and hence we are required to choose a cell half as big along b_1 and b_3, as illustrated in Fig. 1.65. Similarly we are required to halve the 1.97 direction along b_2, the horizontal direction on the photographs, if we are to avoid omitting about half of the reflections, which have components along b_2 of ($\frac{1}{2}$) 1.97 cm. Thus the reciprocal unit cell has dimensions of $b_1 = 1.245$ cm, $b_2 = 0.985$ cm, and $b_3 = 0.79$ cm. Note that this cell accounts for all observed reflections and that there are a number of absent reflections which are discussed below.

Having chosen a suitable cell in reciprocal space, we can now "index" the photographs, that is, assign three integers to each reflection. Along the horizontal direction of the photographs we find the 020 and 040 reflections. Along the three vertical directions of the photographs we find 200 on (*a*), 101 on (*b*), and a weak 002 and strong 004 both on (*c*). The strongest reflection on (*c*) is 012. As was proved in Section 2, reflections on (*a*) should be of the type *hk*0, those on (*b*) of the type *hkh*, and those on (*c*) of the type 0*kl*. The axes about which these photographs were taken are, respectively, [001] for (*a*), [$\bar{1}$01] for (*b*), and [100] for (*c*), and reflections (*hkl*) lying in the zero level about an axis $[uvw]$ satisfy the relation $hu + kv + lw = 0$, as was also proved in Section 2.

We have identified the vectors $\mathbf{h}_1 = \mathbf{b}_1 + \mathbf{b}_3$ and $\mathbf{h}_2 = \mathbf{b}_2$ in photograph (*b*) from our choice of unit cell. Let $\mathbf{T} = l_1\mathbf{a}_1 + l_2\mathbf{a}_2 + l_3\mathbf{a}_3$ be any vector in crystal space. Then, if $\mathbf{h}_1$ and $\mathbf{h}_2$ are perpendicular to $\mathbf{T}$, $\mathbf{T} \cdot \mathbf{h}_1 = 0 = \mathbf{T} \cdot \mathbf{h}_2$, so that $l_1 + l_3 = 0$ and $l_2 = 0$. Hence $l_1 = -l_3$ and $\mathbf{T} = l_1\mathbf{a}_1 - l_1\mathbf{a}_3 = l_1(\mathbf{a}_1 - \mathbf{a}_3)$, and the axis $[10\bar{1}]$ is therefore $\perp$ $\mathbf{h}_1$ and $\mathbf{h}_2$. Finally the general reciprocal lattice vector is $\mathbf{h} = h_1\mathbf{b}_1 + h_2\mathbf{b}_2 + h_3\mathbf{b}_3$. Those $\mathbf{h} \perp \mathbf{T} = \mathbf{a}_1 - \mathbf{a}_3$ must satisfy the relation $\mathbf{h} \cdot \mathbf{T} = 0 = h_1 - h_3$. Hence $h_1 = h_3$, but no restriction occurs for h_2, and planes with indices $h_1h_2h_1 = hkh$ lie in the zone of the $[10\bar{1}]$ axis. This general procedure is valuable for indexing photographs.

In general, the safest procedure to obtain the unit cell is first to calculate the dimensions of the reciprocal axes in Å^{-1} and the angles among them, and then to convert to the real cell with the use of the equations of Section 2. The geometry of the precession method was shown above to give the relation $h = 1/d = x/(\lambda F)$. Hence, with the knowledge that the crystal to film distance is 6.00 cm and the wavelength of the X-rays is 0.711 Å, we can then convert these film measurements into reciprocal axes in Å^{-1} units. Thus $b_1 = |h_{100}| = 1/d_{100} = 1.245\ \text{cm}/[(0.711\ \text{Å})\ (6.00\ \text{cm})] = 0.292\ \text{Å}^{-1}$. Then, in the orthorhombic system, $a_1 = 1/b_1 = 1/(0.292\ \text{Å}^{-1}) = 3.42\ \text{Å}$. Similarly, we find $a_2 = 4.32\ \text{Å}$ and $a_3 = 5.41\ \text{Å}$. The volume V of the unit cell is $80\ \text{Å}^3$.

Number of Molecules in the Unit Cell

The measured density of solid HF is $1.658\ \text{g/cm}^3$. If in the unit cell there are Z molecules of HF and if each molecule weighs $M/N = 20.008/6.023 \times 10^{23} = 3.32 \times 10^{-23}$ g, then the density is $D = (ZM)/(NV)$. Therefore there are

$$Z = \frac{DV}{(M/N)} = \frac{(1.658)(80 \times 10^{-24})}{3.32 \times 10^{-23}} = 4.0$$

molecules in the unit cell.

Sometimes an experimental value of the density of a substance is not available, but even so it is usually possible to obtain an unambiguous answer for the number of molecules in the unit cell. Usually one estimates either the density from densities of known related compounds or the volume per molecule from related molecules. Moreover, symmetry elements are usually present that may require multiples of two, three, four, and more, for the number of molecules in the unit cell. We shall see that the symmetry of HF requires that there be multiplies of four, that is, the unit cell of HF may only have 4, 8, 12, . . . molecules from symmetry arguments only. Thus the experimental value of the density need not be known with accuracy, and hence rough estimates often suffice.

Symmetry Elements from Extinctions

If we use the symbols s, m, w, v for strong, medium, weak, and very, we then find on photograph (*a*) the reflections 020 (s), 040 (m), 200 (s), 400 (vvw), 220 (m), 240 (vw); on photograph (*b*) the reflections 020 (s), 040 (m), 101 (vs), 111 (s), 121 (m), 131 (w), 141 (vw), 212 (w), 232 (vw); and on photograph (*c*) the reflections 020 (s), 040 (w), 002 (w), 012 (vs), 032 (m), 052 (w), 004 (m), 024 (w), 044 (vw), 016 (vw), 036 (w). In all, 42 reflections, listed below, were observed on all the films. Quantitative intensity measurements are of course required for the structure determination. Measurements of intensities with the use of scintillation, proportional counters, or Geiger counters are the most accurate, and are commonly used. However, in some cases experimental factors dictate the use of intensities estimated from films. One technique involves the use of a microdensitometer to evaluate the amount of film blackening. Another employs a standard scale of intensities and measurements by visual comparison of intensities. A standard scale (Fig. 1.66) is usually prepared by exposure of a single reflection at different places on the film for different periods of time. A 15% increase in times for successive exposures gives a very satisfactory scale. Films of a given zone are then correlated with the use of this standard scale, and these relative intensities are then corrected for Lorentz and polarization factors, as described in Section 2, usually on a computer. A graphical representation of the Lorentz and polarization factors [76] for precession photographs is illustrated in Fig. 1.67. Correlations are then made among photographs about different axes with the use of reflections common to pairs of photographs, and one thus obtains a complete list of all observed $|F_{obs}|^2$ on the same relative scale. The usual $|F_{obs}|$ is obtained by taking the positive square root of each $|F_{obs}|^2$. Systematic absences were observed for all reflections hkl when $h + l$ was odd, and a further systematic absence of $hk0$ reflections was noted when k was odd. Many well-exposed photographs with low background are required for one to be reasonably sure of these systematic absences, and erroneous structure interpretations have occurred because of lack of care at this stage of the investigation.

These absences as shown in Section 3 indicate a *B*-centered cell and a *b*-glide plane perpendicular to the *c*-axis. Thus the space group symbol is *B*— —*b*, where a symmetry element such as *m*, 2, or 2_1 is present in each of the two blank spaces. Note that the extinction of $0k0$, when k is odd,

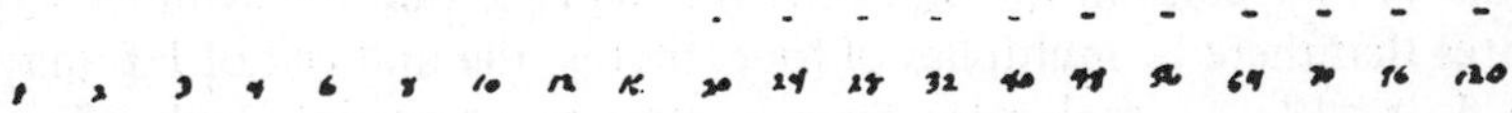

Fig. 1.66 Example of an intensity scale prepared by successive exposures of a reflection from a single crystal. Sometimes a scale turns out this way and needs to be done again.

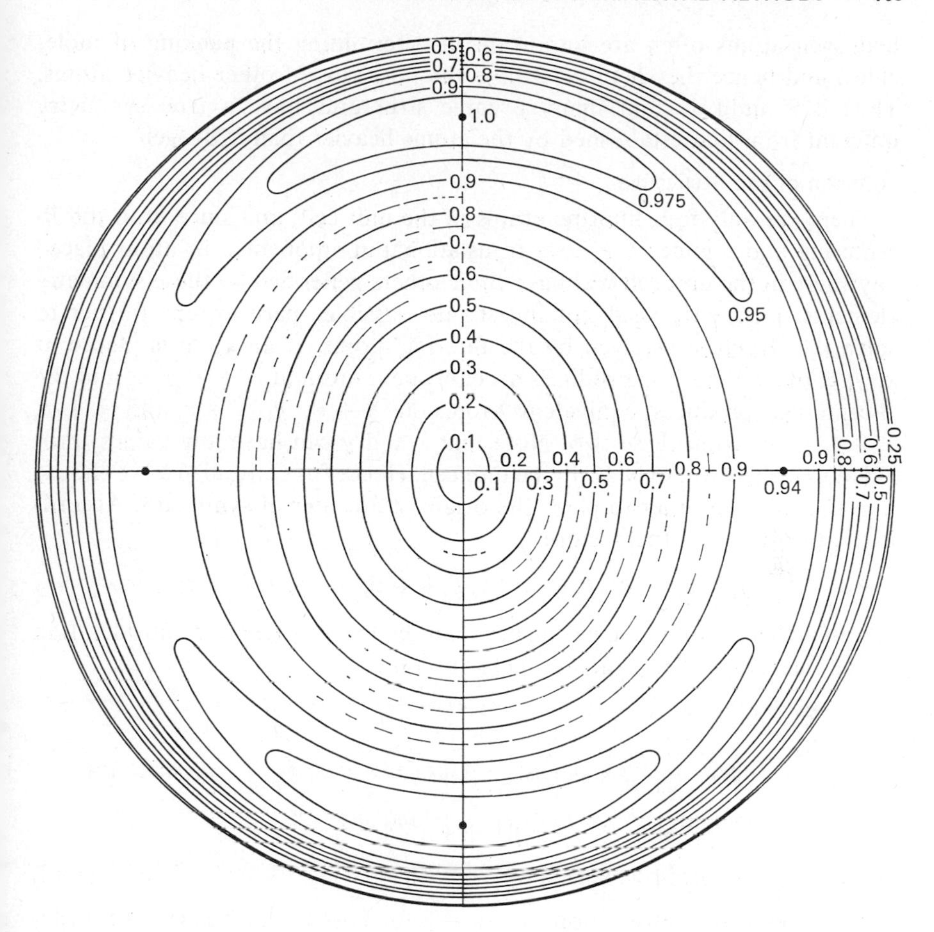

Fig. 1.67 Graphical representation of $10^3/(L \cdot P)$ factors for precession photographs taken at $\mu = 30°$. This template may be superimposed on the photographs of Fig. 1.64 and values of $10^3/(L \cdot P)$ read off where the reflections occur. Each estimated intensity is then multiplied by its $10^3/(L \cdot P)$ value, and then the positive square root is taken to find F_{obs}. For a definition of μ, see Fig. 1.61.

is already contained in the glide plane extinction, so the possible presence of 2_1 along b is obscured.

A word of caution about the space group may be added. The hydrogen atom scatters so little compared to fluorine that the true space group could be of different symmetry. This point is discussed more fully in the published paper, and in what follows we shall discuss the fluorine positions only. While this is a general comment that applies to organic crystals too, the hydrogen atoms often are important in determining the packing of molecules, and hence they help determine the positions of other heavier atoms. Thus it is unlikely that many organic structures have a true symmetry different from that established by the atoms heavier than hydrogen.

Solution of the Structure

There are only four fluorine atoms in the unit cell, and since both the ***B***-centering and *b*-glide $\perp$ *c* have translational components, an atom placed anywhere in the unit cell will have three others generated by these symmetry elements. Hence we need not investigate possible space groups further to solve the structure as given by the fluorine atoms. If an atom is placed at x, y, z, the *b*-glide $\perp$ *c* produces an equivalent atom at $x, \frac{1}{2} + y, \bar{z}$; and the *B*-centering produces equivalent atoms at $\frac{1}{2} + x,\ y,\ \frac{1}{2} + z$ and $\frac{1}{2} + x,\ \frac{1}{2} + y,\ \frac{1}{2} - z$ from these two. Note that x and y can have any values, provided that there are only 4F in the unit cell. Hence for convenience we choose $x = 0$, $y = \frac{1}{4}$, in order to place the origin at a center of symmetry. Accordingly, the 4F are at the positions

$$0, \tfrac{1}{4}, z; \qquad 0, \tfrac{3}{4}, \bar{z}; \qquad \tfrac{1}{2}, \tfrac{1}{4}, \tfrac{1}{2} + z; \qquad \tfrac{1}{2}, \tfrac{3}{4}, \tfrac{1}{2} - z;$$

and thus there is only one parameter, z, to be fixed by detailed consideration of the intensities. The structure factor becomes

$$\begin{aligned}
F_{hkl} &= \sum_{j=1}^{4} f_j e^{2\pi i h \cdot r_j} \\
&= f_F\{e^{2\pi i(k/4+lz)} + e^{-2\pi i(k/4+lz)} + e^{2\pi i[h/2+k/4+l(\frac{1}{2}+z)]} + e^{-2\pi i[h/2+k/4+l(\frac{1}{2}+z)]}\} \\
&= f_F[e^{2\pi i(k/4+lz)} + e^{-2\pi i(k/4+lz)}][1 + e^{2\pi i[(h+l)/2]}] \\
&= f_F 2\cos 2\pi(k/4 + lz)[1 + e^{2\pi i[(h+l)/2]}] \qquad (1.45)
\end{aligned}$$

We have used the substitution of $-\frac{1}{2} - z$ for $\frac{1}{2} - z$, which is a cell translation. Now the final bracket is 0 if $h + l$ is odd and 2 if $h + l$ is even, and hence

$$F_{hkl}(h + l \text{ even}) = f_F\, 4\cos 2\pi(k/4 + lz).$$

An approximate value for z can be found from the weak 002 reflection. If 002 were zero, then $z = \frac{1}{8}, \frac{3}{8}, \frac{5}{8}$, or $\frac{7}{8}$. These choices all lead to identical structures (Fig. 1.68) except for the choice of origin, and hence we discuss values of z near $\frac{1}{8}$. Now we consider a slight displacement of the fluorine atom from $z = \frac{1}{8}$ to $z = \frac{1}{8} \pm \delta$. Then

$$\begin{aligned}
F_{hkl}(h + l \text{ even}) = 4f_F\Big[&\cos 2\pi\frac{k}{4}\cos 2\pi l(\tfrac{1}{8} \pm \delta) \\
&- \sin 2\pi\frac{k}{4}\sin 2\pi l(\tfrac{1}{8} \pm \delta)\Big],
\end{aligned}$$

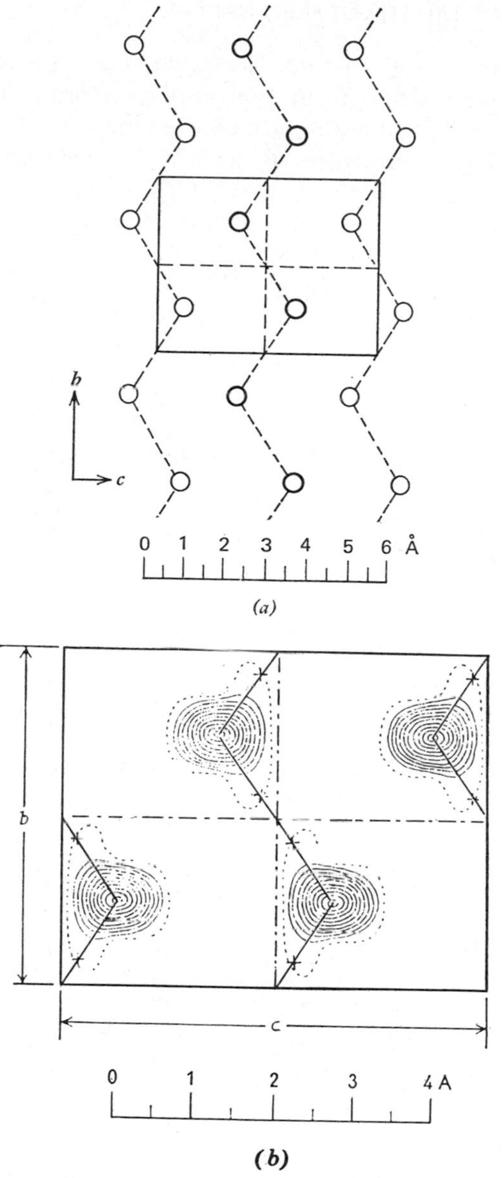

Fig. 1.68 (*a*) Arrangements of the fluorine atoms in the *a*-axis projection. Atoms represented by heavy and light circles are at $x = \frac{1}{2}$ and 0, respectively. Hydrogen bonds are indicated by broken lines. (*b*) Electron-density projection along the *a*-axis. Each contour line represents a density increment of $1e/\text{Å}^2$, the two-electron line being broken. The cross symbols, showing the possible positions of the hydrogen atoms, are based upon the HF distance of 0.92 Å in the gas phase and suggest some randomness with a half-hydrogen at each cross.

which for 002 becomes $\pm \sin 4\pi\delta$. Hence the magnitude of F_{002} is the same whether the displacement δ from $\frac{1}{8}$ is positive or negative. Only those reflections for which l is odd distinguish between these two possible structures; but only one of these structures is correct. The detailed calculations are carried out by substituting the values for f_F at the approximate value of $\sin\theta/\lambda$ from the *International Tables* into (1.45) and computing all F_{hkl}. The results are summarized in Table 1.7.

Table 1.7 Comparison of Observed and Calculated Values of F_{hkl}

hkl	F_{obs}	$F_{0.115}$	$F_{0.135}$	hkl	F_{obs}	$F_{0.115}$	$F_{0.135}$
200	14.6	16.2	16.2	303	2.4	−2.0	−3.0
400	1.8	3.2	3.2	204	4.4	−4.6	−4.6
020	20.8	−20.9	−20.9	103	6.6	−6.6	−9.7
040	8.5	8.6	8.6	206	<1.1	−0.7	0.7
002	3.4	2.8	−2.8	111	14.3	−14.0	−15.9
004	8.9	−8.4	−8.4	121	11.0	−11.5	−10.1
006	1.0	−1.1	1.1	131	7.4	7.0	7.9
220	10.0	−9.7	−9.7	141	5.0	5.4	4.8
240	4.4	5.2	5.2	151	3.8	−3.2	−3.6
420	1.5	−2.1	−2.1	212	9.0	−9.1	−9.1
012	18.5	−20.1	−20.1	222	<1.3	−1.0	1.0
022	1.2	−1.9	1.9	232	5.4	5.8	5.8
032	9.0	10.1	10.1	242	<1.3	0.5	−0.5
042	<1.0	0.9	−0.9	313	2.6	−2.9	−2.0
052	4.9	−4.6	−4.6	323	1.8	1.7	2.5
014	2.1	−2.0	2.0	333	1.9	2.1	1.4
024	7.0	6.6	6.6	214	<1.4	−1.1	1.1
034	1.0	1.3	−1.3	224	3.7	3.5	3.5
044	4.3	−3.8	−3.8	234	<1.4	0.7	−0.7
016	3.4	2.6	2.6	113	8.9	−9.0	−6.1
026	1.0	0.9	−0.9	123	5.3	5.0	7.3
036	3.8	−2.0	−2.0	133	4.9	5.5	3.8
101	18.4	17.8	15.7	143	2.8	−2.7	−4.0
202	1.2	1.2	−1.2	216	2.1	1.6	1.6

Values of $\dfrac{\sum \||F_o| - |F_c\||}{\sum |F_o|}$ are

	All observed F's	F's for $l = 1, 3$	F's for $l = 3$
for $z = 0.115$	0.08	0.06	0.05
for $z = 0.135$	0.13	0.21	0.34

Thus $z = 0.115$.

It is unusual that an incorrect structure can give such good agreement, but the error in the fluorine position is only 0.02b, about 0.1 Å. These results indicate that caution is necessary in deriving structures and that one must demand equally good agreement among all classes of reflections. Usually poor agreement for one projection, for example, as compared with the agreement for other projections, or for all reflections for $h + l$ odd as compared with $h + l$ even, or reflections for small sin θ as compared with those for large sin θ indicates that the structure is either incorrect or insufficiently refined.

The Weissenberg Method

The Weissenberg method [77] is another commonly used film method for the investigation of the reciprocal lattice; we discuss it only briefly here. In this method the crystal is rotated about an axis through a 180°+ range. Referring to the Laue diagram shown in Fig. 1.37, if the rotation axis is assumed to be perpendicular to the plane of the paper, then all the points in that zero-layer reciprocal lattice plane within the radius of $2/\lambda$ will pass through the sphere and hence give rise to diffraction maxima. The crystal is surrounded by a cylindrical film, the axis of the cylinder being coincident with the rotation axis. A cylindrical screen is employed to allow only reflections from the desired layer to reach the film. In order to resolve and index the resulting pattern of intensities the film carriage is translated while the crystal is being rotated (2°/mm). Because of the translation motion of the film, any radial line in reciprocal space is represented by a line making an angle of $\tan^{-1}(2) = 63°26'$ with the direct beam trace (the horizontal line of Fig. 1.69). Copper radiation is usually employed in the Weissenberg technique, and this method gives similar information to that obtained in the precession technique, that is, space group, unit cell, and intensity information. Note, however, that for a particular orientation of the crystal these two techniques give complimentary information because of the different way reciprocal space is sampled. More details on the Weissenberg method, including a discussion of upper-level photography, can be found in Buerger's *X-Ray Crystallography* [77].

The Four-Circle Single-Crystal Diffractometer

Single-crystal diffractometers are now commonly used to obtain more accurate intensities than can usually be obtained by film techniques, unless crystal instability in the X-ray beam or other experimental effects dictate the use of film methods. One of the most common diffractometers of this type is the four-circle diffractometer, illustrated in Fig. 1.70.

The principles of operation of this type of diffractometer are as follows. Using a Bragg picture the crystal can be thought of as consisting of numerous sets of planes. To observe a particular diffraction maximum, the appropriate

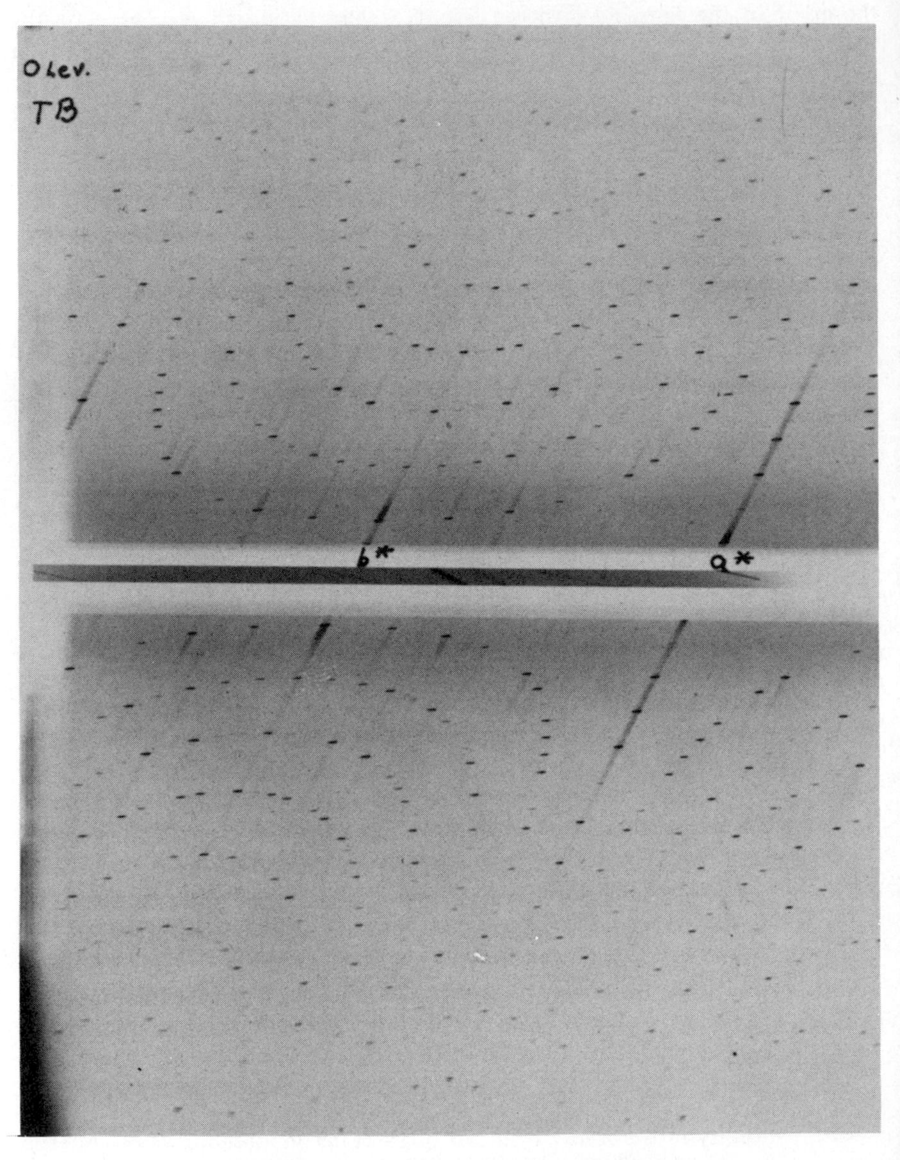

Fig. 1.69 Zero-level Weissenberg photograph.

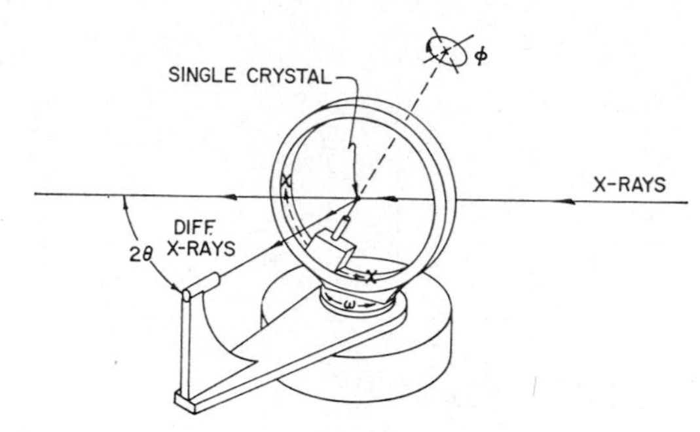

Fig. 1.70 Schematic of a four-circle diffractometer.

set of planes must be oriented in space such that the angle between the incident beam, which is fixed in space, and the set of planes in the crystal is made equal to the angle θ given by the Bragg equation, $\theta = \sin^{-1} (\lambda/2d)$. Since the counter (usually of the scintillation or proportional type) is constrained to move in the horizontal plane in the usual system, the crystal is rotated about the incident beam direction until the normal to the diffraction plane lies in the horizontal plane. All that then remains is to position the counter at an angle θ to the plane, and hence at an angle 2θ to the incident beam.

The crystal orientation axes are shown in Fig. 1.70: ω (also θ and 2θ), rotation about the vertical axes of the diffractometer; χ, angle of inclination about a horizontal axis through the center of the circle; and φ, rotation about the axis of the goniometer head.

Mathematically, these angles can be determined for a particular set of planes corresponding to the lattice vector $\mathbf{h}$ as follows. From Bragg's law θ can be readily determined, as we have already indicated, since $|\mathbf{h}| = d^{-1}$. The usual solution for ω is to set it equal to θ. The angles χ and φ are somewhat more difficult to determine. Three unit vectors can be thought of as superimposed on the four-circle diffractometer such that the **k**-direction is along the φ-axis, at $\chi = 0$; that is, the position at which the φ and the ω axes are collinear and the **i** vector is along the $\varphi = 0$ direction (Fig. 1.71). The three principal lattice vectors $\mathbf{b}_1$, $\mathbf{b}_2$, and $\mathbf{b}_3$ can be resolved into components along the three unit vectors, obtaining b_{1x}, b_{1y}, b_{1z}, b_{2x}, and so on. Then

$$\cos (90^\circ - \chi) = \frac{\mathbf{k} \cdot \mathbf{h}}{|\mathbf{h}|}$$

$$= \frac{\mathbf{k} \cdot (h\mathbf{b}_1 + k\mathbf{b}_2 + l\mathbf{b}_3)}{|\mathbf{h}|}$$

$$= \frac{hb_{1z} + kb_{2z} + lb_{3z}}{|\mathbf{h}|},$$

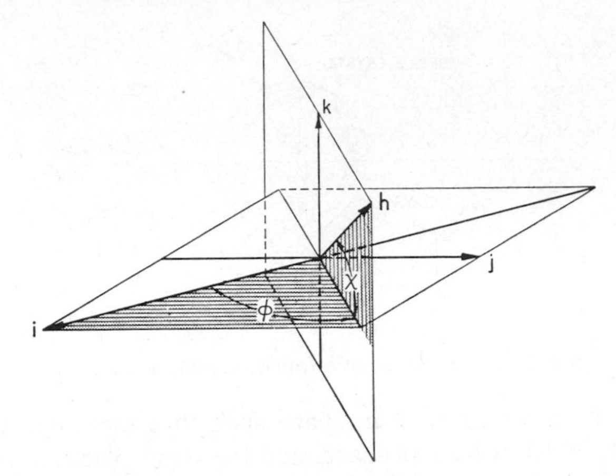

Fig. 1.71 Function of the ϕ and χ circles in bringing a reciprocal lattice point into the equatorial plane.

where h, k, and l are the appropriate indices for the reflection, **h**. Similarly,

$$\cos \varphi = \frac{\mathbf{i} \cdot \mathbf{h}_p}{|\mathbf{h}_p|},$$

where $\mathbf{h}_p$ is the projection of **h** on the **ij** plane. Now

$$\mathbf{h}_p = h(b_{1x}\mathbf{i} + b_{1y}\mathbf{j}) + k(b_{2x}\mathbf{i} + b_{2y}\mathbf{j}) + l(b_{3x}\mathbf{i} + b_{3y}\mathbf{j}).$$

Therefore

$$\cos \varphi = \frac{hb_{1x} + kb_{2x} + lb_{3x}}{|\mathbf{h}_p|}.$$

To resolve the ambiguity regarding the appropriate quadrant for the angle φ, the sign of the quantity

$$\mathbf{j} \cdot \mathbf{h}_p = hb_{1y} + kb_{2y} + lb_{3y}$$

can be used. A general description then is given by the following:

$$\begin{pmatrix} b_{1x} & b_{2x} & b_{3x} \\ b_{1y} & b_{2y} & b_{3y} \\ b_{1z} & b_{2z} & b_{3z} \end{pmatrix} \begin{pmatrix} h \\ k \\ l \end{pmatrix} = \begin{pmatrix} |\mathbf{h}_p| \cos \varphi \\ |\mathbf{h}_p| \sin \varphi \\ |\mathbf{h}| \sin \chi \end{pmatrix} = \begin{pmatrix} X_1 \\ X_2 \\ X_3 \end{pmatrix},$$

and $\tan \varphi = (X_2)/(X_1)$, $\tan \chi = X_3/(X_1^2 + X_2^2)^{1/2}$.

APPENDIX

Structure Determination and Refinement of D-glucono-(1-5)-lactone

A brief description of the structure determination and the refinement techniques used in the structure determination of D-glucono-(1-5)-lactone* is presented in this appendix as an illustration of some of these procedures and the methods used to display the results of a structure investigation.

Experimental Crystal Data

$C_6H_{10}O_6$, M = 178.14 g/m, Orthorhombic $P2_12_12_1$

$a = 7.838 \pm 0.001$, $b = 12.332 \pm 0.002$, and $c = 7.544 \pm 0.001$ Å

$V = 729.2$ Å^3, $D_c = 1.62$ g/cc, $Z = 4$, $F(000) = 376$,

Mo$K\alpha$ ($\lambda = 0.7107$ Å), $\mu = 1.60$ cm^{-1}.

Suitable crystals were obtained by recrystallizing commercially available glucono-δ-lactone from a saturated DMF (dimethylformamide) solution allowed to evaporate slowly. The colorless, tabular crystals grew, with b perpendicular to a and c along the diagonals of the broad face. Precession and Weissenberg photographs exhibited *mmm* Laue symmetry, with alternate extinctions along the axes indicating the orthorhombic space group $P2_12_12_1$. The unit-cell parameters and their standard deviations were obtained by a least-squares fit to 14 independent reflection angles whose centers were determined by left-right, top-bottom beam splitting on a previously aligned Hilger–Watts four-circle diffractometer (Mo$K\alpha$ radiation). Any error in the instrumental zero was eliminated by centering the reflection at both $+2\theta$ and -2θ.

A crystal of dimensions 0.24 × 0.20 × 0.12 mm was mounted on a glass fiber, with b along the spindle (χ) axis for data collection. Intensity data were taken at room temperature (24°C) using a fully automated Hilger–Watts four-circle diffractometer equipped with scintillation counter and interfaced with an SDS 910 computer in a real-time mode. Two equivalent octants of data were taken using zirconium-filtered Mo$K\alpha$ radiation within a θ sphere of 35° ($\sin\theta/\lambda = 0.8071$). The θ-2θ, step-scan technique, 0.01°/step counting for 0.4096 sec/step, was employed. To improve the efficiency of the data collection process, variable-step symmetric-scan ranges were used. The number of steps used for a particular reflection were determined by number of steps = $(50 + 2/°\theta)$. Individual backgrounds were obtained from stationary counter measurements for one-half the total scan time at each end of the scan.

* A more detailed discussion of the structure is given in M. L. Hackert and R. A. Jacobson, *Acta Cryst.* **B27**, 203 (1971).

As a general check on the electronic and crystal stability, the intensities of three standard reflections were measured periodically during the data collection. Monitoring options based on these standard counts were employed to maintain crystal alignment and to stop the data collection process if the standard counts fell below statistically allowed fluctuations. A total of 3762 reflections were recorded in this manner.

The intensity data were corrected for Lorentz-polarization effects. Owing to the small linear absorption coefficient (1.60 cm^{-1}, Mo$K\alpha$ radiation), no absorption correction was made. The minimum and maximum transmission factors were 0.96 and 0.98, respectively. Because absorption was negligible the consistency of equivalent data was easily checked. Those equivalent reflections differing by more than $5\sigma = 5\sqrt{TC}$ were retaken. This affected some 150 reflections. The individual values of $|F_o|^2$ from the equivalent octants were then averaged to yield 1851 independent $|F_o|^2$ values. Standard deviations were estimated from the average intensity and background values by

$$(\sigma_I)^2 = TC_{av} + BK_{av} + (0.05 * TC_{av})^2 + (0.05 * BK_{av})^2.$$

Of the 1851 independent reflections 974 had $|F_o|^2 > 2.5 * \sigma_I$. These were considered observed and were used in the refinement.

Solution and Refinement of the Structure

The observed data were used to compute an unsharpened Patterson map. The resulting map contained many broad, overlapping peaks, which made it unsuitable for superposition techniques. To reduce the peak width, sharpened coefficients were computed by

$$|F^s_{hkl}| = \left[\frac{|F^o_{hkl}|}{(\hat{f})}\right]^2 \exp\,[(2B - B')\sin^2\theta/\lambda^2],$$

where $\hat{f} = \sum f_j / \sum Z_j$, B is the overall isotropic temperature factor, and B' is a variable used to minimize rippling resulting from sharpening. Estimates of the overall temperature and scale factors were obtained from a Wilson plot. A sharpened Patterson map of good resolution was obtained using $2B - B' = 2.0$ Å^2.

Examination of the sharpened Patterson map and some initial superposition results gave promising starting fragments, but we were unable to develop these further. A symmetry map [24] was generated from the three Harker (symmetry) planes. Peaks only occur on the symmetry map at those electron density positions that are consistent with the Harker peaks of the Patterson map. To compute the symmetry map the value assigned to the point (x, y, z) is obtained by taking the minimum of the values of the associated points on the three Harker planes. Since a sharpened Patterson was used and since every other point in the Patterson would be skipped due to the form of the vector expressions for symmetry-related atoms in this space group, the maximum value of the point and the four other points immediately surrounding it in the Harker plane was taken before carrying out the minimum procedure. Since no origin has been specified, not 1 but 64 images of the unit cell appear on this map, because of the orthorhombic symmetry. However, it is possible to select a peak from the symmetry map and superimpose the sharpened Patterson on this and its symmetry-related points. This procedure is just one of the superposition techniques possible using a symmetry map. After a couple of unpromising starts, a single peak was selected and a set of four symmetry map–Patterson map superpositions were carried out. Analysis of the resulting map showed that there were only 32 consistent, independent peaks remaining. A second peak was chosen from what appeared to be the same image, and another set of four such superpositions were carried out. Analyzing the resultant map and comparing it with the first reduced the number of possible peak positions to 18, with a fragment of the molecule now readily observable. Although it was likely at this point that careful evaluation of these peaks on chemical grounds would have yielded the entire molecule, a third peak was chosen from the visible fragment and another set of superpositions was made. Comparison of the three maps easily resolved the previous ambiguities, locating all carbon and oxygen positions.

Three cycles of full-matrix least-squares refinement based on $|F|$ of these heavy-atom positional and isotropic thermal parameters gave a conventional discrepency factor $R = \sum ||F_o| - |F_c||/\sum |F_o| = 0.109$ and a weighted R-factor $\omega R = [\sum \omega(|F_o| - |F_c|)^2/\sum \omega|F_o|^2]^{1/2} = 0.133$ for the 974 observed reflections. The scattering factors used for carbon and oxygen were those of Doyle and Turner [78]. A difference electron-density map at this stage showed that all the nonhydrogen atoms had been accounted for, but that some anisotropic motion was evident. Anisotropic refinement of all heavy-atom positions for two additional cycles lowered the discrepancy factor to $R = 0.071$ and $\omega R = 0.090$. The resulting difference electron-density map clearly indicated the positions of all hydrogens bound to carbon. The scattering factor of Stewart and others [79] was used for the hydrogens, and the refinement proceeded smoothly, with all remaining hydrogen atom positions being obtained from subsequent difference electron density maps. Final values of R and ωR of 0.046 and 0.051, respectively, were obtained for the 974 observed reflections. At this point a structure factor calculation was carried out using all 1851 independent reflections recorded, and it gave $R = 0.095$ and $\omega R = 0.049$. A final electron density difference map showed no peaks greater than 0.3 $e/Å^3$. A statistical analysis of $\omega\Delta^2$ (where $\Delta^2 = (|F_o| - |F_c|)^2$) as a function of scattering angle and magnitude of F_o yielded a nearly straight line, indicating that the relative weighting scheme used was reasonable.

Table 1.8 Final Heavy-Atom Atomic Coordinates and Thermal Parameters[a,b]

Atom	x	y	z	β_{11}	β_{22}	β_{33}	β_{12}	β_{13}	β_{23}
C(1)	0.20707	0.22144	0.49678	990	438	735	−35	4	29
	35	20	32	42	17	38	24	36	22
C(2)	0.19107	0.12941	0.63116	823	297	855	29	24	19
	34	19	32	40	14	39	21	35	20
C(3)	0.10670	0.16315	0.80219	809	357	692	59	79	94
	34	19	32	40	14	39	21	35	20
C(4)	0.18678	0.26778	0.86527	958	301	668	61	14	37
	36	20	31	39	15	37	21	33	20
C(5)	0.15860	0.35808	0.73035	1003	365	676	71	−3	36
	35	20	33	44	14	36	22	35	19
C(6)	0.25498	0.45986	0.76700	1113	326	781	−19	−13	−1
	34	21	35	45	15	38	23	34	18
O(1)	0.22381	0.20231	0.34070	2196	557	716	567	156	−1
	30	16	23	48	15	29	23	34	18
O(2)	0.10411	0.04146	0.55196	1223	348	1185	−30	71	−115
	27	14	25	36	12	34	19	31	18
O(3)	0.13045	0.07824	0.92668	1351	375	975	152	242	215
	28	15	26	38	11	33	18	31	17
O(4)	0.10967	0.30610	1.02478	1920	470	708	83	189	5
	32	16	25	47	13	30	22	34	18
O(5)	0.21002	0.32330	0.55211	2194	369	603	−133	80	32
	30	13	23	48	12	26	20	33	15
O(6)	0.43311	0.44290	0.78525	1032	687	1019	−89	−64	92
	27	19	29	35	17	36	20	30	21

[a] Standard errors of the coordinates and the β_{ij} and their standard errors are $\times 10^5$.

[b] The β_{ij} are defined by:

$$T = \exp\left[-(h^2\beta_{11} + k^2\beta_{22} + l^2\beta_{33} + 2hk\beta_{12} + 2hl\beta_{13} + 2kl\beta_{23})\right].$$

Table 1.8 lists the final positional and thermal parameters of the heavy atoms along with their standard deviations. In Table 1.9 are the refined positional and isotropic thermal parameters and their standard deviations for the hydrogen atoms. Standard deviations given were obtained from the inverse matrix of the final least-squares refinement cycle. Figures 1.72 and 1.73 show the molecular and crystal structures, respectively. An indication of the directions and root-mean-square amplitudes of vibration for the atoms refined anisotropically is provided by these figures. The bond lengths and the

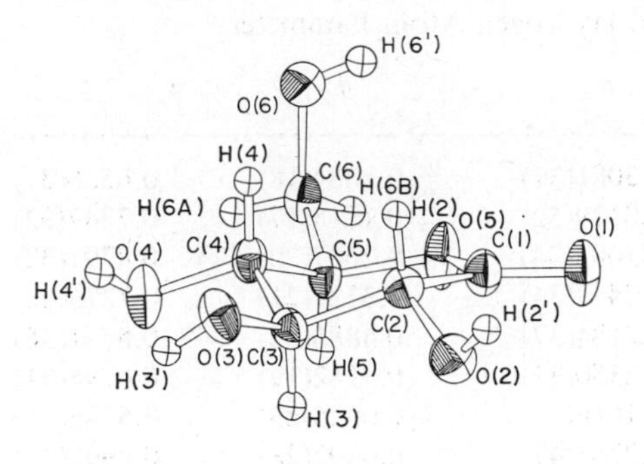

Fig. 1.72 Molecular structure of D-glucono-(1-5)-lactone.

bond angles with standard deviations are given in Table 1.10. The estimated standard deviations were calculated using the variance-covariance matrix.

The best least-squares plane through the δ-lactone carbonyl group was calculated from the final positional parameters. The equation of the least-squares plane and the displacement of the heavy atoms from this plane are given in Table 1.11.

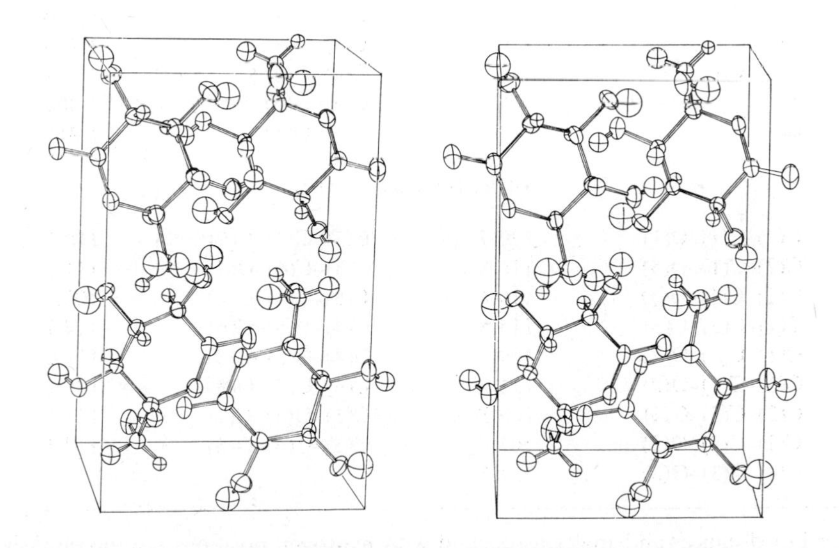

Fig. 1.73 Stereographic drawing showing the molecular packing in the unit cell.

Table 1.9 Refined Hydrogen Atom Parameters[a]

Atom	x	y	z	B
H(2)	0.3081(31)	0.1084(18)	0.6536(31)	1.35(48)
H(3)	−0.0129(32)	0.1718(17)	0.7847(32)	1.16(50)
H(4)	0.3083(34)	0.2607(20)	0.8791(32)	2.13(56)
H(5)	0.0443(34)	0.3727(21)	0.7276(38)	2.08(56)
H(6A)	0.2184(37)	0.4880(22)	0.8771(35)	2.76(63)
H(6B)	0.2350(32)	0.5142(19)	0.6796(31)	1.30(49)
H(2′)	0.1616(37)	−0.0025(23)	0.5086(41)	3.87(85)
H(3′)	0.0767(47)	0.0892(28)	0.9997(48)	4.99(99)
H(4′)	0.1468(41)	0.2769(24)	1.1000(38)	2.85(76)
H(6′)	0.4732(45)	0.4329(27)	0.6956(47)	4.25(93)

[a] Numbers in parentheses are estimated standard deviations in the last significant digits.

Table 1.10 Interatomic Distances and Angles[a]

	Distance (Å), e.s.d. = 0.003 Å		
C(1)–C(2)	1.527	C(1)–O(1)	1.208
C(2)–C(3)	1.508	C(1)–O(5)	1.324
C(3)–C(4)	1.512	C(2)–O(2)	1.414
C(4)–C(5)	1.525	C(3)–O(3)	1.419
C(5)–C(6)	1.491	C(4)–O(4)	1.427
		C(5)–O(5)	1.468
		C(6)–O(6)	1.418
	Angles (°), e.s.d. = 0.2°		
C(2)–C(1)–O(1)	120.7	C(3)–C(4)–C(5)	110.7
C(2)–C(1)–O(5)	119.8	C(3)–C(4)–O(4)	111.9
O(1)–C(1)–O(5)	119.4	C(5)–C(4)–O(4)	105.1
C(1)–C(2)–C(3)	113.5	C(4)–C(5)–C(6)	114.7
C(1)–C(2)–O(2)	109.2	C(4)–C(5)–O(5)	111.0
C(3)–C(2)–O(2)	111.2	C(6)–C(5)–O(5)	106.1
C(2)–C(3)–C(4)	108.8	C(5)–O(5)–C(1)	124.1
C(2)–C(3)–O(3)	107.8	C(5)–C(6)–O(6)	113.1
C(4)–C(3)–O(3)	111.5		

[a] For distances and angles associated with hydrogen positions see the published paper.

Table 1.11 Least-Squares Plane

Description (*):

C(1), C(2), O(1), O(5).

Equation Relative to a, b, c:

$$0.9941X - 0.0372Y + 0.1021Z - 1.910 = 0.$$

Distance from Plane (Å):

Atom	Distance	Atom	Distance
C(1)*	−0.016	O(1)*	0.003
C(2)*	0.005	O(2)	−0.693
C(3)	−0.536	O(3)	−0.216
C(4)	0.088	O(4)	−0.408
C(5)	−0.277	O(5)*	0.003
C(6)	0.456	O(6)	1.865

Computational Considerations

The programs used for these calculations included ALS [80] for superposition computation, ALF [81] for the Patterson, electron density, and difference electron density calculations, ORFLS [82] for the least-squares refinements, ORFFE [83] for computation of bond lengths, bond angles, and their standard deviations, and ORTEP for the production of Figs. 1.72 and 1.73. The ORFLS program requires the largest amount of core, 160K bytes on our IBM 360, Model 65 computer, for the 149 variables involved in this calculation. Approximately 14 minutes of computer time was required for each cycle of refinement, and 2 minutes was required for each electron-density map calculation.

References

1. J. C. Kendrew, *Science* **139,** 1259 (1963).
2. M. F. Perutz, M. G. Rossmann, A. F. Cullis, H. Muirhead, G. Will, and A. T. C. North, *Nature* **185,** 416 (1960).
3. M. F. Perutz, H. Muirhead, J. M. Cox, L. C. G. Goaman, F. S. Mathews, E. L. McGandy, and L. E. Webb, *Nature* **219,** 29 (1968).
4. C. C. F. Blake, D. F. Loenig, G. A. Main, A. T. C. North, D. C. Phillips, and V. R. Sharma, *Nature* **206,** 4986 (1965).
5. W. N. Lipscomb, J. C. Coppola, J. A. Hartsuck, M. L. Ludwig, H. Muirhead, J. Searl, and T. A. Steitz, *J. Mol. Biol.* **19,** 423 (1966).
6. G. Kartha, J. Bello, and D. Harker, *Nature* **213,** 862 (1967).

7. H. P. Avey, M. O. Boles, C. H. Carlisle, S. A. Evans, S. J. Morris, R. A. Palmer, and B. A. Woodhouse, *Nature* **213,** 557 (1967).
8. D. R. Davies, *Ann. Rev. Biochem.* **36,** 321 (1967).
9. A few cases of homometric structures have been found that have exactly the same array of three-dimensional vectors (and therefore X-ray intensities) from different atomic arrangements. An example is the mineral bixbyite. [L. Pauling and M. D. Shappell, *Z. Krist.* **75,** 128 (1930)].
10. C. E. Wilkes and R. A. Jacobson, *Inorg. Chem.* **4,** 99 (1965).
11. J. Waser, *J. Am. Chem. Soc.* **65,** 1451 (1943); *ibid.* **66,** 2035 (1944).
12. D. C. Hodgkin, J. Pickworth, J. H. Robertson, K. N. Trueblood, R. J. Posen, and J. G. White, *Nature* **176,** 325 (1955).
13. A. L. Patterson, *Z. Krist.* **A90,** 517, 543 (1935).
14. D. Wrinch, *Phil. Mag.*, (7) **27,** 98 (1939); M. J. Buerger, *Acta Cryst.* **3,** 87 (1950); *ibid.* **4,** 531 (1951); J. H. Robertson and C. A. Beevers, *ibid.* **4,** 270 (1951); J. Clastre and R. Gray, *Compt. Rend.* **230,** 1876 (1950); J. Garrido, *ibid.* **230,** 1878 (1950); *ibid.* **231,** 297 (1951); D. McLachlan, *Proc. Natl. Acad. Sci. U.S.* **37,** 115 (1951).
15. J. Rodgers and R. A. Jacobson, *J. Chem. Soc.*, **A1970** (1826).
16. R. A. Pasternak, *Acta Cryst.* **9,** 341 (1956).
17. R. A. Jacobson and W. N. Lipscomb, *J. Chem. Phys.* **31,** 605 (1959).
18. R. A. Jacobson, J. A. Wunderlich, and W. N. Lipscomb, *Acta Cryst.* **14,** 598 (1961).
19. S. H. Kim and R. D. Rosenstein, *Acta Cryst.* **22,** 648 (1967).
20. J. H. Derissen, H. J. Endeman, and A. F. Peerdeman, *Acta Cryst.* **B24,** 1349 (1968).
21. H. M. Berman, G. A. Jeffrey, and R. D. Rosenstein, *Acta Cryst.* **B24,** 442 (1968).
22. A. W. Schlueter and R. A. Jacobson, *J. Chem. Soc.* **A1968,** 2317.
23. S. H. Kim, G. A. Jeffrey, R. D. Rosenstein, and P. W. R. Cornfield, *Acta Cryst.* **22,** 733 (1967); T. A. Beineke and L. L. Martin, Jr., *J. Organomet. Chem.*, **20,** 65 (1969).
24. A. D. Mighell and R. A. Jacobson, *Acta Cryst.* **16,** 443 (1963); P. G. Simpson, R. D. Dobrott, and W. N. Lipscomb, *ibid.* **18,** 169 (1965); W. C. Hamilton, *ibid.* **18,** 866 (1965).
25. M. J. Buerger, *Vector Space* (1959) Wiley, New York; R. A. Jacobson, *Trans. Am. Cryst. Assoc.* **2,** 39 (1966); R. A. Jacobson and L. J. Guggenberger, *Acta Cryst.* **20,** 592 (1966).
26. W. Monereif and W. N. Lipscomb, *Acta Cryst.* **21,** 322 (1966).
27. R. Pepinsky and Y. Okaya, *Proc. Natl. Acad. Sci. U.S.* **42,** 286 (1956); W. N. Lipscomb, *J. Chem. Phys.* **26,** 713 (1957).
28. D. Sayre, *Acta Cryst.* **5,** 60 (1952).
29. See for example J. Karle and I. L. Karle, *Acta Cryst.* **21,** 849 (1966).
30. W. Cochran and M. M. Woolfson, *Acta Cryst.* **8,** 1 (1955).
31. F. A. Muller and R. A. Jacobson, *Acta Cryst.*, to be published.
32. H. Hauptman and J. Karle, "Solution of the Phase Problem I. Centrosymmetric Crystal." *Amer. Cryst. Assoc. Monograph No.* 3, Polycrystal Book Service, Pittsburgh, 1953.

33. J. Karle and H. Hauptman, *Acta Cryst.* **9,** 635 (1956).
34. I. L. Karle and J. Karle, *Acta Cryst.* **21,** 860 (1966).
35. Y. Kainuma and W. N. Lipscomb, *Z. Krist.* **113,** 44 (1960).
36. J. Karle, *Acta Cryst.* **B24,** 182 (1968).
37. I. L. Karle and J. Karle, *Acta Cryst.* **17,** 835 (1964).
38. J. Karle and I. L. Karle, *Acta Cryst.* **B24,** 81 (1968).
39. G. N. Reeke, Jr. and W. N. Lipscomb, *J. Am. Chem. Soc.* **90,** 1663 (1968).
40. D. Harker and J. S. Kasper, *Acta Cryst.* **1,** 70 (1948).
41. G. German and M. M. Woolfson, *Acta Cryst.* **B24,** 91 (1968).
42. J. D. Bernal and D. Crowfoot, *Nature* **133,** 794 (1934).
43. C. Bokhoven, J. C. Schoone, and J. M. Bijvoet, *Proc. Acad. Sci. Amsterdam* **52,** 120 (1949); *Acta Cryst.* **4,** 275 (1951).
44. D. W. Green, V. M Ingraham, and M. F. Perutz, *Proc. Roy. Soc.* (London) **A225,** 287 (1954).
45. J. C. Kendrew, G. Bodo, H. M. Dintzis, R. G. Parrish, and H. Wyckoff, *Nature* **181,** 662 (1958).
46. J. C. Kendrew, R. E. Dickerson, B. E. Strandberg, R. G. Hart, D. R. Davies, D. C. Phillips, and V. C. Shore, *Nature* **185,** 4711 (1960).
47. See W. N. Lipscomb, *Structural Chemistry and Molecular Biology*, A. Rich and N. Davidson, eds. Freeman, San Francisco, 1968, p. 38.
48. W. N. Lipscomb, J. A. Hastruck, G. N. Reeke, Jr., F. A. Quiocho, P. H. Bethge, M. L. Ludwig, T. A. Steitz, H. Muirhead, and J. C. Coppola, *Brookhaven Symposia in Biology No.* 21, (1968), p. 24.
49. H. W. Wyckoff, K. D. Hardman, N. M. Allewell, T. Imagami, L. N. Johnson, and F. M. Richards, *J. Biol. Chem.* **242,** 3984 (1967).
50. B. W. Matthews, P. B. Sigler, R. Henderson, and D. M. Blow, *Nature* **214,** 652 (1967).
51. J. Drenth, J. N. Jansonius, R. Koekoek, H. M. Swen, and B. G. Wolthers, *Nature* **218,** 929 (1968).
52. C. S. Wright, R. A. Alden, and J. Kraut, *Nature* **221,** 235, (1969).
53. D. Rabinovich and G. M. J. Schmidt, *Nature* **211,** 1391 (1966).
54. D. E. Williams, *Acta Cryst.* **A25,** 464 (1969).
55. C. E. Nordman and K. Nakatsu, *J. Am. Chem. Soc.* **85,** 353 (1963); C. E. Nordman and S. K. Kumra, *J. Am. Chem. Soc.* **87,** 2059 (1965); C. E. Nordman, *Trans. Am. Cryst. Assoc.* **2,** 29 (1966).
56. W. Hoppe and E. F. Paulus, *Acta Cryst.* **23,** 339 (1967).
57. M. G. Rossmann and D. M. Blow, *Acta Cryst.* **15,** 24 (1962); *ibid.* **16,** 39 (1963).
58. M. G. Rossmann and W. N. Lipscomb, *J. Am. Chem. Soc.* **80,** 2592 (1958).
59. Proof that the total energy P carried through an area A normal to the direction of propagation of a wave $E = E_0 \cos 2\pi(x/\lambda - \nu t + \delta)$ is proportional to E_0^2 is as follows: From electrodynamics we write (in vacuum) $P = (8\pi)^{-1} \int (E^2 + H^2)\, dv = (4\pi)^{-1} \int E^2\, dv = (4\pi)^{-1} \int_0^{t_1} E_0^2\, [\cos 2\pi(x/\lambda - \nu t + \delta)]\, Audt = E_0^2\, Aut_1/(8\pi)$, plus a trigonometric term, $-(1/4\pi\nu)[\sin 4\pi(x/\lambda - \nu t_1 + \delta) - \sin 4\pi (x/\lambda + \delta)]$, which is negligible if $t_1 \gg 1/\nu$.
60. *International Tables for X-ray Crystallography, Volume III*, The Kynoch Press, Birmingham, England, 1962.

61. D. T. Cromer and J. B. Mann, *Acta Cryst.* **A24,** 321 (1968).
62. A. L. Patterson, *Z. Krist.* **90,** 517 (1935); D. Harker, *J. Chem. Phys.* **4,** 381 (1935); G. Albrecht and R. B. Corey, *J. Am. Chem. Soc.* **61,** 1087 (1939); H. Klug and L. Alexander, *ibid.* **66,** 1056 (1944); A. J. C. Wilson, *Nature* **150,** 151 (1942).
63. E. W. Hughes, *J. Am. Chem. Soc.* **63,** 1737 (1941).
64. E. W. Hughes and W. N. Lipscomb, *J. Am. Chem. Soc.* **68,** 1970 (1964).
65. This symmetry, the symmetry of the reciprocal lattice, can be noncentrosymmetric when the structure contains optically active molecules and an X-ray wavelength near an absorption edge of one of the atoms in the crystal is chosen. This method forms the basis for the determination of absolute configuration.
66. *International Tables for X-ray Crystallography, Volume I,* The Knoch Press, Birmingham, England, 1952.
67. See M. J. Buerger, *X-Ray Crystallography,* Wiley, New York, 1942, pp. 69–70.
68. M. J. Buerger, *X-Ray Crystallography,* Wiley, New York, 1942, p. 83; *International Tables for X-Ray Crystallography, Volume I,* The Knoch Press, Birmingham, England, 1952, p. 54.
69. A. J. C. Wilson, *Acta Cryst.* **2,** 318 (1944).
70. E. R. Howells, D. C. Phillips, and D. Rogers, *Acta Cryst.* **3,** 210 (1950).
71. M. Atoji and W. N. Lipscomb, *Acta Cryst.* **7,** 173 (1954).
72. T. B. Reed and W. N. Lipscomb, *Acta Cryst.* **6,** 45 (1953); W. N. Lipscomb, *Norelco Reptr.* **4,** 54 (1957); C. Altona, *Acta Cryst.* **17,** 1282 (1964); G. Abowitz and P. G. Cath. *J. Sci. Instr.* **44,** 156 (1967); G. Abowitz and J. Ladell, *J. Sci. Instr.*, Series 2, **1,** 113, (1968).
73. M. J. Buerger, *X-Ray Crystallography,* Wiley, New York, 1942, pp. 331 ff.; G. H. Stout and L. H. Jensen, *X-Ray Structure Determination,* Macmillan, New York, 1968, pp. 122 ff.; E. W. Nuffield, *X-Ray Diffraction Methods,* Wiley, New York, 1966, Chapter 9.
74. M. J. Buerger, *The Procession Method,* Wiley, New York, 1964; see also Reference 73.
75. R. Wass and J. Donohue, *Acta Cryst.* **10,** 375 (1957).
76. J. Waser, *Rev. Sci. Instr.* **22,** 567 (1951).
77. M. J. Buerger, *X-Ray Crystallography,* Wiley, New York, 1942, Chapter 12; E. W. Nuffield, *X-Ray Diffraction Methods,* Wiley, New York, 1966, Chapter 11; G. H. Stout and L. H. Jensen, *X-Ray Structure Determination,* Macmillan, New York, 1968, pp. 96 ff.
78. P. A. Doyle and P. S. Turner, *Acta Cryst.* **A24,** 390 (1968).
79. R. F. Steward, E. R. Davidson, and W. T. Simpson, *J. Chem. Phys.* **42,** 3175 (1965).
80. C. R. Hubbard and R. A. Jacobson, *ALS, A General Superposition Program in Fortran IV,* Ames Laboratory, Ames, Iowa, 1969.
81. J. Rodgers and R. A. Jacobson, ALF, A General Fourier Program in PLI for Triclinic, Monoclinic and Orthorhombic Space Groups, *Report IS-2155,* Ames Laboratory, Ames, Iowa, 1967.

82. W. R. Busing, K. O. Martin, and H. A. Levy, ORFLS, A Fortran Crystallographic Least-Squares Program, *Report ORNL-TM-305*, Oak Ridge National Laboratory, Oak Ridge, Tenn., 1962.
83. W. R. Busing, K. O. Martin, and H. A. Levy, "ORFFE, A Fortran Crystallographic Function and Error Program," *Report ORNL-TM-306*, Oak Ridge National Laboratory, Oak Ridge, Tenn., 1964.
84. C. K. Johnson, ORTEP, A Fortran Thermal-Ellipsoid Plot Program for the Crystal Structure Illustrations," *Report ORNL-3794*, Oak Ridge National Laboratory, Oak Ridge, Tenn., 1965.

Chapter **II**

ELECTRON DIFFRACTION BY GASES

Lawrence S. Bartell

1 INTRODUCTION

Electron diffraction may be applied to the determination of the structures of molecules in the gaseous state or, less commonly, in the solid state [1]. The application to solids is treated briefly by R. G. Scott in his chapter on electron microscopy. Electron diffraction is frequently used in the study of the arrangement of adsorbed molecules on a variety of substrates [2, 3]. In principle it may even be used to determine the distribution of planetary electrons in atoms and in molecules [4, 5]; to date, however, only a small amount of experimental work has been carried out in this area [4, 6].

Gas-electron diffraction as a structural tool offers certain advantages over the principal alternative methods, molecular spectroscopy and X-ray diffraction. These advantages, together with the main limitations of the technique, are summarized in a later section. Gas-electron diffraction is capable of determining internuclear distances to 0.001 Å or better in favorable cases, permitting, for example, the direct measurement of such subtle quantities as primary and secondary deuterium isotope effects on molecular structure [7]. It also provides useful information on the amplitudes of molecular vibrations.

Since gas-electron diffraction was first introduced by Mark and Wierl in 1930 [8], it has been used to determine the structures of over 1000 volatile compounds. Initial structure analyses were based on visual estimates of intensities of interference rings recorded on photographic plates. Scattered intensities fall off so rapidly with increasing scattering angle, however, that a quantitative measurement of unmodified electron interference patterns is difficult. A considerable advance in precision was made possible by the application of the rotating sector according to the proposals of Debye in 1939 [9] and Hassel in 1940 [10]. In this application the diffracted electrons are selectively masked by the rotating sector in such a way that the electron intensity reaching the photographic plate remains in the optimum range of response of the emulsion. It is possible, with sectored plates, to measure the interference features with considerable sensitivity.

In the last 10 years steady progress has been made in experimental techniques, numerical analyses, and interpretational procedures. The number of laboratories employing the rotating sector (photographic emulsion) microphotometer method of measuring diffracted intensities has increased severalfold, partly because of the increased power of the electron-diffraction method and partly because of the recent availability of commercially manufactured diffraction equipment. Several laboratories have experimented with direct electron counting devices as an alternative to photographic emulsions [11]. Although counting may one day become the preferred method, in practice it is not yet as efficient in recording molecular interference features as the best photographic techniques. Therefore it is not discussed further in this chapter.

2 PREPARATION AND MEASUREMENT OF DIFFRACTION PATTERNS

The interference features arising when electron waves are diffracted by free molecules may be observed by passing a beam of electrons through the vapor under investigation and by recording the scattered electrons on a photographic plate. The production of patterns best suited for interpretation requires that all the electrons have very nearly the same energy, that the beam be brought to a fine focus on the photographic plate, and that the gas molecules intercept the beam along only a very short section of its path [12]. The whole electron path must be enclosed in a high-vacuum chamber because of the relatively high electron-scattering power of air. The diffraction unit consists, then, of an electron gun for producing the beam, an electron optical system for aligning and focusing the beam, a specimen section supporting both the specimen nozzle and the rotating sector, a photographic plate magazine, and auxiliary equipment such as high-vacuum pumps and gages and the high-voltage power supply. The following section describes a simple and sturdy design that has performed well [13].

Diffraction-Camera Construction

In the design in Fig. 2.1 the gun at the top is followed by two pairs of magnetic deflector coils for directing the gun output through the lens aperture, by an electrostatic shutter for exposure control, by one lens that supports a diaphragm with an opening that limits the size of the beam, by two pairs of electrostatic deflector plates for centering the beam, by the nozzle that introduces the gas stream, by the rotating sector, and finally by the photographic plate.

The gun is designed with a self-bias (i.e., the high potential is connected to the grid sheath and thence through a bias resistor to the filament) and produces a fine electron crossover in front of the grid, which is focused on the plate. The cathode provides adjustments for centering the tip of the filament on the axis of the 2.0-mm hole in the grid sheath and for adjusting the depth of the filament behind this hole. The gun is rigidly mounted on the unit for ruggedness and simplicity, unlike most guns in use which can be translated and tilted. Adequate flexibility for aligning the electron optical system is provided by the magnetic and the electrostatic plates referred to above together with a lateral translation of the pole pieces of the magnetic lens. Underneath the gun is the shutter consisting of two circular disks supporting a horizontal diaphragm; batteries supply 300 V for operating the shutter.

The single magnetic lens carries a diaphragm with a 0.45-mm opening serving as a limiting aperture for the electrons. It focuses the electron beam to a spot approximately 0.03 mm in diameter at the photographic plate. Nozzles constructed of platinum or nickel have been used very successfully. A stainless steel nozzle was tried once but proved to be difficult to clean adequately. The nozzle throat is 0.2 to 0.3 mm in internal diameter and about 0.8 mm long. A large diaphragm for intercepting stray electrons is mounted on the nozzle support. Its central section, made of platinum for easy cleaning, is an inverted "roof" with an aperture 0.60 mm in diameter at the roof's apex mounted with its center 0.5 mm above and 0.35 mm to the left of the nozzle tip. The nozzle-to-plate distance can be set at about 7, 11, or 21 cm, and can be measured to within 0.03 mm with a traveling microscope.

The sector is a heart-shaped piece of brass originally made in two halves. Most commonly, the sector's opening is proportional to the cube of the distance from the center of rotation. This has the effect of making the cumulative exposure time greater for the outer part of the photographic plate than for the inner part by the cube of the ratio of the respective distances from the center. For the measurement of the innermost portion of the diffraction pattern a smaller sector with an r^2 angular opening is often used. Knowledge of the precise function represented by the sector opening is not nearly so important as an exact knowledge of the deviation of the function from a smooth curve. Extreme care should be taken in preparing and calibrating the sector. An optical calibration is inadequate at the smaller radii where the slit becomes very narrow. In this range the deviations from smoothness are most easily detected by measuring electron diffraction patterns taken with a monatomic gas sample.

The sector is mounted in a ball bearing with an inner race of internal diameter of 4.5 in. In the inner race is affixed a sleeve with a slanting shoulder against which a rubber drive wheel runs, rotating the sector mount at about 1000 rpm. A slight magnetism of the race can be tolerated if it moves the beam by no more than a few thousandths of a centimeter and if the sector mount is carefully adjusted so that the magnetic displacement in phase with the sector rotation is perpendicular to the sector slit. The sector drive includes a gear train operated from a shaft turning at about 150 rpm through the vacuum wall. The center of the sector is open to allow passage of the electron beam during the alignment operations, but during the recording of diffractions patterns this hole is closed by the centrifugal action of a pivoted mass acting on a small aluminum shutter.

Five photographic plates 4 × 5 in. in a daylight-loading cartridge are brought successively into position by means of a rack and pinion driven by a shaft through the wall of the vacuum chamber. Before recording a diffraction pattern each plate is lifted and pressed firmly against a flange, ensuring that the emulsion surfaces are always in the same position during exposures.

A prerequisite for taking high-quality patterns is an effective trapping of extraneous radiation. Diaphragms for catching stray electrons are mounted at the shutter, between the lens and the electrostatic deflector plates, and on the nozzle. If the aperture on the shutter is the limiting aperture, the cleanest beam is obtained. Some workers place a mask over the sector race to prevent radiation from reflecting from the vertical portions of the race. It is imperative to have an efficient trap on the sector, not only to catch the undeviated beam but also to capture the radiation generated when the beam is stopped. A single, deep, but not very narrow, tube is enormously better than a flat beam stop; nevertheless, such a tube can still allow the secondary radiation directed upward to strike the metal around the nozzle and send back enough radiation to be troublesome at the larger scattering angles. The sector trap should contain baffles to reduce this undesirable radiation. Platinum is an attractive material for nozzles and apertures because of its corrosion resistance and its ease of cleaning. Unfortunately it is a much worse offender from the standpoint of secondary radiation than a low-atomic-number element. An acceptable compromise between quality and convenience can be struck when heavy metals are used by the judicious application of an adherent, conducting carbon coating over the areas of the metals likely to act as a source of secondary radiation.

Two independent diffusion pumps are connected to the gun and the specimen chambers. Because the 0.45-mm hole in the lens diaphragm is the only connection between the two chambers, a high vacuum can be maintained in the gun chamber even during the introduction of a gas sample into the lower chamber. In the design shown in Fig. 2.1 the nozzle directs the gas toward a cold surface very nearly as large as the left-hand wall. This condensing surface consists of a series of thin-walled metal tubes, usually cooled by liquid nitrogen, and is an extremely effective scavenger of condensable

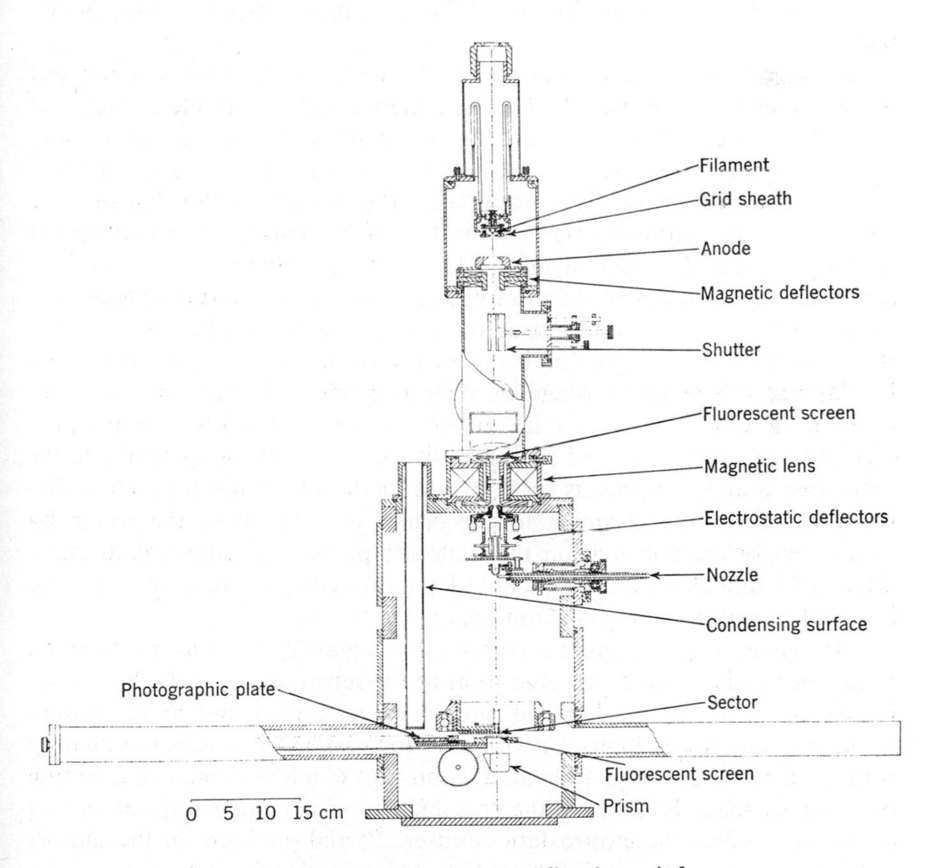

Fig. 2.1 Section through electron-diffraction unit for gases.

samples. A 6-in. vacuum lead at the rear of the specimen chamber conducts out those gases that are not condensed.

Operating Conditions

The electron beam is usually accelerated by 40 kV, corresponding to an equivalent wavelength of about 0.06 Å. The precise relation is

$$\lambda = hc^2(2 \times 10^8 m_0 c^3 eV + 10^{16} e^2 V^2)^{-1/2},$$

in which λ is the wavelength in centimeters, V is the accelerating potential in volts, and h, e, m_0, and c are Planck's constant, the electronic charge, rest mass of the electron, and the velocity of light, respectively. The dc accelerating potential supply is stabilized to 0.02% and is measured to the same precision with the aid of a wire-wound high-resistance voltage divider (mounted in transformer oil for temperature control) and a good potentiometer circuit. The ac component in the high-voltage output is kept below 0.01%.

The beam current can be measured with a probe below the gas nozzle and is usually set at about 0.4 μA. This is achieved with a cathode emission of about 50 μA, depending on the depth of the filament tip behind the opening in the grid (usually near 2 mm) and on the bias resistance (usually 4 MΩ). Adequate concentration of the beam from the cathode is obtained only by operating at the saturation level of the curve of emission versus filament heating current. The alignment of the beam is critical. For mechanical simplicity the positions of the electron gun, the small aperture beneath the lens, and the center of rotation of the sector are rigidly fixed. In aligning the beam the shutter diaphragm is turned out of the way and the nozzle with its diaphragm is removed. Magnetic deflectors under the gun aim the beam at the fixed lens hole to maximize the current received below by a Faraday cage. The lens pole pieces are shifted until the centers of the focused and the unfocused beams coincide, as observed on the lower fluorescent screen. The focused beam is then brought to the center of rotation of the sector by adjustment of the potential on the deflector plates, the shutter diaphragm is returned, and the nozzle is remounted and properly positioned by centering its "roof" aperture on the electron beam.

Unreactive gas specimens may be placed in a glass bulb connected through a stopcock and a standard tapered joint to the tube that supports the nozzle. Corrosive specimens are handled with a Monel Metal and nickel system containing no greased joints, or with a Teflon and Kel-F system terminating with an inert-metal nozzle. Individual exposures of a few seconds to a minute or more are made by opening the specimen valve and triggering the timing circuit controlling the electrostatic shutter. Partial pressures of the sample are chosen to be low enough to avoid serious multiple scattering, yet are high enough to ensure that the specimen scattering is large compared with the extraneous background. Satisfactory results seem to be obtained with sample pressures (torr) given by

$$p = \frac{3 \times 10^4}{\sum_i Z_i^2},$$

where the denominator represents the sum of the squares of the atomic numbers of all the atoms in the molecule. For CCl_4 this represents a pressure of 25 torr. For stable specimens where the recommended pressure cannot be reached at room temperature it is better to use a heated nozzle and a connecting tube rather than to use very long exposure times at much lower pressures.

Eastman Kodak Process Plates and Electron Image Plates have been found suitable when exposed to yield absorbancies between, for example, 0.25 and 0.9. The relationship between exposure, E, and absorbancy, A, has the approximate form [14]:

$$E = A(1 + cA)$$

or, more closely [15]:

$$2cA = (1 - e^{-2cE}),$$

with $c \approx 0.1$. Calibration of individual batches of plates may be made from pairs of plates exposed through a special sector for different times to electrons scattered by a condensable gas. It is essential that plates be agitated adequately during development. A longer and cooler-than-normal development aids uniformity of development [15].

Microphotometer Measurements

The sectored negatives show a series of diffuse concentric diffraction rings. These may be measured with considerable precision by a digital microphotometer. A peripheral mount of the sector is superior to a central mount, for it avoids any shadow of the sector support; this allows the negative to be rotated about the center of the pattern during microphotometering so that the effect of graininess in the emulsion can be averaged out. A recorder trace may be employed to register absorbancies, but a more precise and enormously less tedious measurement may be made with a direct digital device. In this technique the output of the photodiode used to measure the transmitted light is fed into a low-gain, highly stable operational amplifier; the output is measured with a voltage-to-frequency converter, the signal of which is counted and transmitted to a key punch or tape punch. When a digital measurement is made, care should be taken to ensure that the readings are made during an exactly integral number of rotations of the plate to cancel any uneven blackening effects resulting from slight electron-beam misalignments or blackening gradients in development. Readings should be made across the full diameter of a spinning plate so that averages of the right- and left-hand readings cancel any small error in determining the plate center.

When care is taken it is possible to obtain readings smooth to a few parts per ten thousand. It is to be noted, however, that such sensitivity requires that the microphotometer slit be opened widely enough that the area traced out on the spinning plate encompasses nearly 10^8 electrons. Otherwise electron-counting statistics limit the precision. Carefully treated emulsions approach electron-counting statistics in sensitivity. Electron Image Plates with 40-kV electrons are blackened to one absorbancy unit by a current of 5×10^7 to 10^9 electrons/cm^2, depending on development conditions [16]. In Fig. 2.2 is shown the digital microphotometer output for patterns of dimethylcyclopropane, recorded at the three camera geometries. The digital data have been processed by a computer, corrected for the measured sector irregularities and extraneous background, and leveled by dividing by the theoretical atomic intensity as explained later; no further adjustments or smoothing have been made.

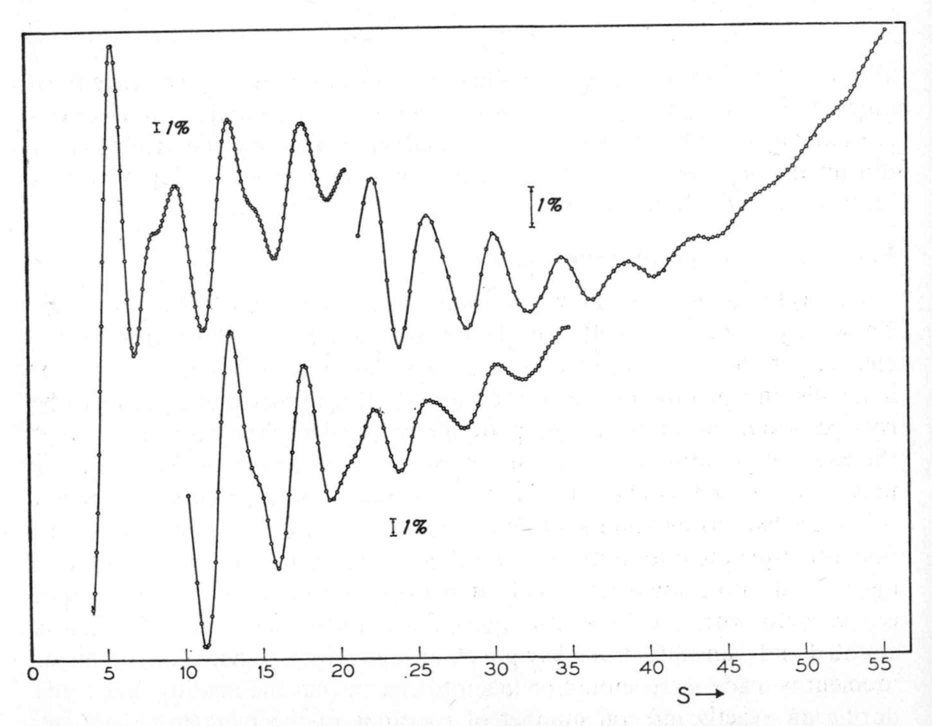

Fig. 2.2 Digital microphometer readings of diffraction patterns of dimethylcylopropane taken at the camera lengths 21 cm, 11 cm, and 7 cm. The vertical arrows represent 1% of the total scattered intensity in each case.

3 INTERPRETATION OF DIFFRACTION PATTERNS

Theoretical Expressions

Total Scattered Intensity

The formulation of quantitatively accurate expressions for the intensity of electrons scattered by gas molecules poses a formidable problem. Effects of electron correlation and chemical-bond formation on the molecular-charge distribution [5], the polarization of the molecular charge by the incident electron [17], the distortion of the incident wave train as it traverses the strong nonuniform field of the molecule [18], relativistic effects [17, 19], and other troublesome factors have been considered in the published literature. They all have appreciable influences in certain regions of scattering angles. Fortunately, however, experience indicates that a reasonably simple expression can be used as the basis of molecular-structure analyses without introducing errors much larger than those expected from the best current experimental techniques. For the remainder of this chapter we confine our attention to this simplified expression, which neglects most of the considerations above and rests on the assumptions that (*a*) each molecule consists of discrete spherical atoms, (*b*) the many molecules contributing to the diffraction pattern exhibit all possible molecular orientations and instantaneous vibrational displacements, (*c*) each diffracted electron is scattered by only one molecule, (*d*) electrons in the incident beam are moving much more rapidly than planetary electrons, and (*e*) wavelets scattered by a molecule may be taken to be the sum of the wavelets scattered independently by the individual atoms. The resultant expression for the total intensity, $I_T(s)$, can be written as

$$I_T(s) = I_A(s) + I_M(s), \tag{2.1}$$

the sum of an atomic intensity, $I_A(s)$, and a molecular intensity, $I_M(s)$, where the natural scattering variable s is $(4\pi/\lambda) \sin (\phi/2)$, in which ϕ is the angle of scattering. The atomic intensity is the sum of an elastic and an inelastic part given by

$$I_A(s) = \frac{N_0 I_0}{R_0^2} \sum_{i=1}^{N} \left[|f_i(s)|^2 + \left(\frac{4}{a^2 s^4}\right) S_i(s) \right], \tag{2.2}$$

and $I_M(s)$ for an N atom molecule is

$$I_M(s) = 2 \frac{N_0 I_0}{R_0^2} \sum_{i<j}^{N} \sum^{N} |f_i| \, |f_j| \cos [\eta_i(s) - \eta_j(s)] \int_0^\infty P_{ij}(r) \frac{\sin sr}{sr} \, dr, \tag{2.3}$$

where N_0 is the number of molecules exposed per unit area of the beam, I_0 is the intensity of the beam, R_0 is the distance from the scattering point to the observation point, $f_i = |f_i| \exp (i\eta_i)$ is the electron elastic scattering factor for the ith atom, where in the limit of the Born approximation $|f_i| = (2/as^2)[Z_i - F_i(s)]$, for which Z is an atomic number and $F(s)$ is an elastic scattering factor for X-rays, $\eta(s)$ represents a phase shift of the scattered wave relative to the shift at the same angle for a wave scattered by an infinitesimal shielded charge, $S(s)$ is an inelastic scattering factor for X-rays, $P_{ij}(r)\,dr$ is the probability that the internuclear distance from the ith to jth atom lies between r and $r + dr$ and $a = h^2/4\pi^2 m\epsilon$ is the Bohr radius. In evaluating a, the relativistic mass of the electron must be used.

For light atoms fairly satisfactory analyses can be performed using the Born approximation $|f|$ with tabulated [20] or analytical [21] $F(s)$ and using the Thomas-Fermi-Dirac $\eta(s)$ tabulated by Ibers and Hoerni [22] and expressed analytically by Bonham and Ukaji [23]. Indeed, the neglect of the $\eta(s)$ would influence derived internuclear distances very little but would lead to apparent amplitudes of vibration that were too large. For heavy atoms, on the other hand, it is imperative if accuracy is desired to use partial-wave (or equivalent) f values derived for Hartree-Fock (or better) atomic fields. The η cannot be neglected because the s-dependence of cos $(\eta_i - \eta_j)$ for an interference term for a bond between a light atom and a heavy atom is of the form to make the interference fringes appear as if the bond consisted of two distinct internuclear distances. At the time of this writing the most satisfactory elastic scattering factors are those tabulated by Bonham and others [24]. The best tabulated inelastic factors (except for a few including electron correlation) are those of Freeman [20], Hanson [25], and Tavard [26].

Two important points can be seen from the above theoretical relations. (*a*) The equations relating molecular structure to gas-diffraction intensities contain the internuclear distances in a molecule (implicit in the $P_{ij}(r)$ distributions) but not the relative orientations of the internuclear vectors; therefore, a full three-dimensional structure can be derived uniquely only when a one-dimensional spectrum of distances, blurred by thermal motion, suffices to establish the structure. (*b*) The term containing the structural information, namely $I_M(s)$ of (2.3), is of such a form, mathematically, that it is susceptible to Fourier analysis.

So many different approaches in structure analyses have been reported that it seems impractical in this chapter to survey them. The more successful variants use analytical functions to represent the $P_{ij}(r)$ and perform least-squares fittings of calculated to experimental intensity functions, or they perform some sort of harmonic analysis on the molecular-diffraction features. Two major schools of analysis exist that might loosely be identified as the Eastern U.S.–Eastern European–Japanese school and the Western U.S.–Western European school. Roughly speaking, the former works with the ratio I_M/I_A, reasoning that certain theoretical errors will cancel, and compensates radial distribution analyses for effects of planetary electrons by additive corrections. The latter [27] works with I_M itself, reasoning that extraneous scattering errors will drop out, and compensates for planetary electrons in radial distribution analyses by multiplicative corrections. Presumably, if appropriate theoretical and experimental precautions are taken, the main differences between the methods reduce to differences of bookkeeping. We shall hereafter describe only the methods of the first-mentioned school.

Reduced Molecular Intensity Function

The theoretical reduced molecular intensity, first employed extensively by the Karles [13, 28], is defined as

$$M(s) = \frac{I_T(s)}{I_A(s)} - 1 = \frac{I_M(s)}{I_A(s)}. \tag{2.4}$$

Practical reasons for focusing attention on this function are as follows. First, only a modest correction is needed to remove the s-dependence resulting from the planetary electrons and leave an expression for the effective nuclear scattering. In addition, certain systematic errors in the theoretical expressions are expected to be similar for both $I_M(s)$ and $I_A(s)$, so that $M(s)$ may be less sensitive to imperfections in theory than $I_M(s)$. Moreover, since nature yields only the total scattering, the resolution of $I_T(s)$ into $I_M(s)$ and $I_A(s)$ requires the somewhat arbitrary step of "drawing" a background $I_B(s)$ through the molecular oscillations to represent $I_A(s)$. Practically, a more sensitive plot can be made to facilitate the drawing of I_B if the $I_T(s)_{\text{exp}}$ experimental function is "leveled" by dividing it by the theoretical $I_A(s)$. Since the shell structure of atoms introduces small undulations into $I_A(s)$, it is apparent that the proper background curve to draw through $I_T(s)_{\text{exp}}$ is a deliberately bumpy curve with just the correct degree of undulation to account for the shell structure. It is tedious to do this precisely while compensating for systematic deviations between $I_T(s)_{\text{exp}}$ and $I_T(s)_{\text{theor}}$. On the other hand, the background to draw through $I_T(s)_{\text{exp}}/I_A(s)_{\text{theor}}$ should be smooth and free of any detail arising from electron shells.

A further simplification of $I_M(s)$ or $M(s)$ requires the specification of the distributions $P_{ij}(r)$. An adequate representation seems to be provided by the distribution function for a Morse oscillator according to which, for a particular atom pair [29],

$$\int_0^\infty P(r)\,\frac{\sin sr}{sr}\,dr \cong \frac{A}{sr_g}\left[\exp\left(-\frac{l_m{}^2s^2}{2}\right)\right]\left[\sin s\left(r_g - \frac{l_m{}^2}{r_g} - \kappa s\right)\right], \tag{2.5}$$

where $A = 1 + (3 - ar_g)(al^2/2r_g) + \cdots$, l_m is an effective root-mean-square amplitude, hereafter called a "mean amplitude," r_g is the mean internuclear distance, a is the Morse asymmetry constant, and $\kappa = (al_m{}^4/6) + \cdots$ is an asymmetry parameter introducing a mild frequency modulation into the molecular interference terms. The corresponding theoretical expression for the reduced intensity is often written [29]:

$$M(s) = \sum_{i<j}^{N}\sum^{N} 2c_{ij}\mu_{ij}(s)\cos(\eta_i - \eta_j)A_{ij}\exp\left(\frac{-l_m^2 s^2}{2}\right)_{ij}$$

$$\times \frac{\{\sin s[r_g - (l_m^2/r_g) - \kappa s^2]\}_{ij}}{(sr_g)_{ij}}, \tag{2.6}$$

where c_{ij} is a constant equal to $Z_iZ_j/\sum_k (Z_k^2 + Z_k)$, A_{ij} is very nearly unity [cf. (2.5)], and μ_{ij} is a slowly varying function tending toward unity at large s given by

$$\mu_{ij}(s) = \frac{|f_i|\,|f_j|\sum_k (Z_k^2 + Z_k)}{Z_iZ_j\sum_k [|f_k|^2 + (4/a^2s^4)S_k]}, \tag{2.7}$$

although some writers absorb η_i into $\mu_{ij}(s)$.

The form of (2.6) is satisfactory as it stands for least-squares comparisons with experimental reduced intensities. If Fourier transformations are to be made, however, certain advantages are to be gained by modifying the reduced intensity function. These modifications are discussed in the following section.

Radial Distribution Function

From the relation

$$sM(s) = \sum_{i<j}^{N}\sum^{N} 2c_{ij}\mu_{ij}\cos(\eta_i - \eta_j)\int_0^\infty \frac{P_{ij}(r)}{r}\sin sr\,dr \tag{2.8}$$

it is readily seen that the Fourier sine transform of $sM(s)$ should display peaks at the internuclear distances. In fact, except for the s-dependency of the coefficients $\mu_{ij}\cos(\eta_i - \eta_j)$, the peaks would be proportional to the true probability distributions. The influence of the variable coefficients is to introduce confusing negative regions and other distortions into the transform. To counteract this, the expedient was introduced many years ago of correcting the experimental $M(s)$ function by an additive correction to a "constant coefficient" function, $M_c(s)$, corresponding to (2.6), with μ replaced by unity [30]. Recent experience with heavy atoms, where μ can deviate markedly from unity has shown that this correction may lead to instabilities. Success has been obtained with the following modification, however, both when heavy atoms are present [31] and when subtle distinctions have to be made in molecules containing only light atoms [32].

A modified reduced intensity function $M_N(s)$ is defined to be identical with the $M(s)$ of (2.6–2.8) except that $\mu_{ij}(s)$ is replaced by $N_{ij}(s)$, where

$$N_{ij}(s) = a_{ij} + b_{ij}\exp(-\beta_{ij}s^2). \tag{2.9}$$

The parameters of the new function $N_{ij}(s)$ are chosen to make $N_{ij}(s)$ similar to $\mu_{ij}(s)$ over the main range of s for which experimental intensities are available. As s goes to 0, however, N_{ij} approaches $a_{ij} + \beta_{ij}$, whereas μ_{ij} approaches 0. Therefore, the Fourier transform of $sM_N(s)$ has an appearance much more nearly like that of a probability distribution curve, and the result is simpler to interpret visually than the transform of $sM(s)$. The conversion of an experimental $M(s)$ to an $M_N(s)$ curve is made with the aid of the theoretical function

$$\begin{aligned}\Delta M_N(s) &= M(s) - M_N(s)\\ &= \sum_{i<j}^{N}\sum^{N} 2c_{ij}(\mu_{ij} - N_{ij}) \cos(\eta_i - \eta_j)\left(\frac{A}{sr_g}\right)_{ij}\\ &\quad \times \exp\left(\frac{-l_m^2 s^2}{2}\right)_{ij}\left\{\sin s\left[r_g - \frac{l_m^2}{r_g} - \kappa s^2\right]\right\}_{ij}. \qquad (2.10)\end{aligned}$$

Since this function depends upon the structure parameters, an analysis based upon it must be recycled to self-consistency. The μ_{ij} and N_{ij} functions pertaining to an analysis [31] of XeF_6 are illustrated in Fig. 2.3 where, among other factors, the difference between the use of Born approximation and more exact scattering factors is portrayed.

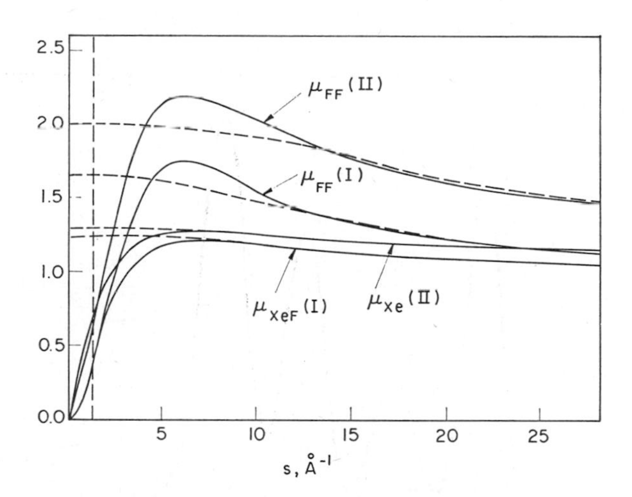

Fig. 2.3 Solid and dashed curves indicate the functions μ_{ij} and N_{ij}, respectively, of (2.6) through (2.10) for the XeF and FF interactions in XeF_6. The vertical dashed line shows the s value at which experimental data start. Curves (I) are calculated using the Born approximation for scattering factors f_i; curves (II) are based on the partial wave scattering factors of Bonham and Cox.

The Fourier transform adopted in analyses is

$$f_N(r) = \int_0^{s_{\max}} sM_N(s)e^{-bs^2} \sin sr\, ds, \tag{2.11}$$

where the modification function e^{-bs^2} serves partly to make the integral insensitive to the upper limit if intensity data do not extend to large s values. A more important function of the modification function in current work, where a very large $s_{\max}$ is attained, is to weight the $sM(s)$ data to reduce effects of noise without seriously changing the aspect of the radial distribution function, $f_N(r)$. In current literature the radial distribution function $f(r)$ [or $f_c(r)$] is more commonly encountered than $f_N(r)$. This function is derived from $M_c(s)$, the reduced intensity for which N_{ij} of (2.9 and 2.10) is taken as unity. When heavy atoms are present, $f_c(r)$ may suffer from an excessive sensitivity to the assumed r_{ij} parameters of 2.10. An experimental $f_c(r)$ function for CF_3Cl is shown in Fig. 2.4.

In earlier, cruder days when it was sufficient to represent the $P_{ij}(r)$ by Gaussian peaks and to neglect the η_i, the integral of (2.11) could have been represented analytically as a simple sum of Gaussians. Even now, if either the anharmonicity or the η_i are neglected, tractable analytical expressions for $f_N(r)$ are available [32]. For molecules with heavy atoms, however, completely satisfactory analytical functions have not yet been published. The only place where highly accurate theoretical expressions of $f_N(r)$ are needed,

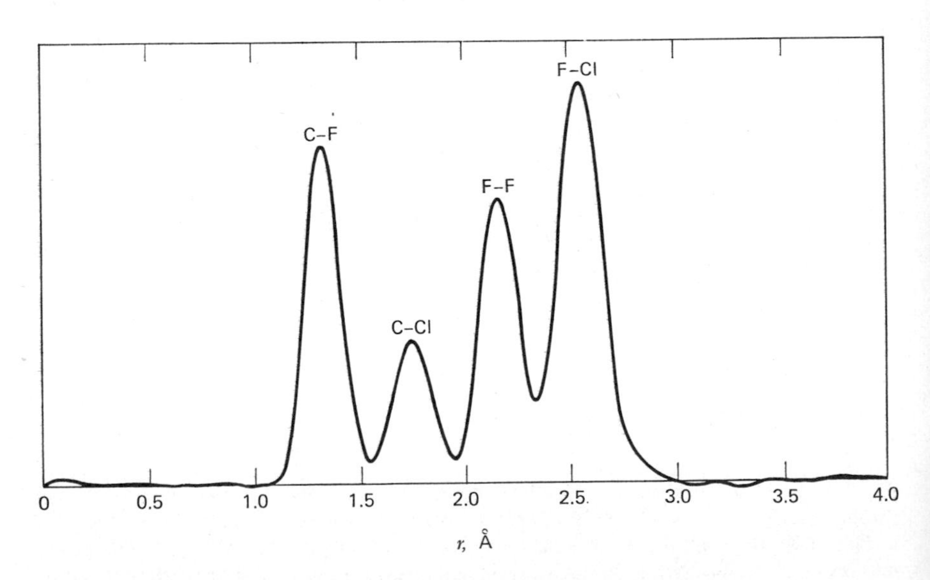

Fig. 2.4 Radial distribution function $f(r)$ for CF_3Cl.

however, is in least-squares analyses of experimental $f_N(r)$ curves. Here it suffices to calculate $f_N(r)$ by numerical integration of $M_N(s)$. Derivatives of $f_N(r)$ with respect to the r_{ij} and l_{ij} parameters adequate for the normal equations of least squares may be obtained from the analytical approximations of the combined Bonham-Ukaji [23] and Kuchitsu-Bartell [29] expressions for Born-corrected anharmonic radial distribution functions.

Geometrical Constraints and Auxiliary Considerations

As mentioned in an earlier section, the information on molecular geometry present in the diffraction intensities consists merely of the distribution of internuclear distances and not of the relative orientations of internuclear vectors. In favorable cases all the internuclear distances can be determined independently. For example, if F_3CCl is considered to have C_{3v} symmetry, it possesses only three independent geometrical parameters but contains four distinct distances, the C–F and C–Cl bond lengths and the F · · · F and F · · · Cl nonbonded distances. Experimentally, as illustrated in Fig. 2.4, four well-resolved peaks are found that turn out to be consistent with C_{3v} symmetry [33]. In many cases, however, some peaks overlap others badly, and some peaks are too weak to be independently determined with accuracy. It is frequently desirable in refining a structure, then, not to attempt to establish all distances independently at once but to impose constraints so that the distances correspond at all times to a geometrically possible structure. Since there are $N(N - 1)/2$ internuclear distances (not necessarily distinct) in an N atom molecule but only $3N - 5$ or $3N - 6$ independent structure parameters (or fewer if elements of symmetry exist), the utility of imposing relationships between the distances is evident. It is often helpful to impose severe simplifications to reduce the independent parameters in initial efforts to unravel the main structural characteristics.

It is also frequently indispensable to augment the electron-diffraction data by adding information from other sources. Symmetries and constraints on certain bond lengths as inferred from partial analyses of molecular spectra can be inserted into a least-squares analysis via the design matrix, **A**, referred to later. Moments of inertia derived from spectra can be included as additional data in the matrix of observations, **M**. Amplitudes of vibration computed from vibrational spectra may be introduced. Before techniques for imposing constraints are discussed, some complicating factors must be aired.

First, perpendicular amplitudes of vibration introduce so-called "Bastiansen-Morino" shrinkage effects [34, 35], tending to make mean nonbonded distances shorter than distances calculated from mean bond lengths and equilibrium bond angles. For example, since any bending motion of the linear CO_2 molecule makes the instantaneous O–C–O angle less than 180°, the mean O · · · O distance is less than twice the C–O bond length even though the equilibrium angle is 180°. For this reason the mean nonbonded internuclear distances in a molecule do not correspond to the set computed for a rigid molecule with the correct mean bond lengths and equilibrium angles. In order to impose geometric constraints properly the effective rigid molecule distances calculated by analytic geometry must be corrected downward if they are to be made rigorously comparable to true *mean* internuclear distances observed in an experiment. Fortunately, a compilation of calculated shrinkage corrections for many molecules has appeared [35]. Tabulated values may serve as guides for estimating corrections for molecules as yet not studied.

A second complication (if it is desired to insert spectroscopic data into the analysis) is that spectroscopic moments of inertia as well as diffraction intensities are influenced appreciably by intramolecular motions. Since the spectroscopic and the diffraction observables represent different types of averages over vibrations, some account must be made of the differences if a precise analysis is desired. This problem and its solution have been discussed in detail, particularly by the Tokyo structural group [36, 37]. Suffice it to say here that both diffraction and spectroscopic analyses can be interpreted in simple cases in terms of the *mean structure*. For this mean structure neither do the associated moments of inertia correspond exactly to uncorrected rotational constants nor do the distances, r_z, between the mean positions of atoms equal the mean r_g distances.

The main idea in the handling of constraints is as follows. Let $F(\rho_i, r_l)$ represent the theoretical dependency of the data on the internuclear distances, r_l. If F represents molecular intensity, or its Fourier transform, ρ_i represents the scattering variable s, or the distance variable r, respectively. Each of the internuclear distances depends upon the set of *independent* parameters, z_m, selected to specify the structure. The derivatives, $\partial F(\rho_i)/\partial z_m$, constitute the elements of the design matrix $\mathbf{A}$ needed in least-squares analyses, as discussed later. Since F can be expressed as the sum of components,

$$F(\rho_i, r_l) = \sum_k F_k(\rho_i, r_k),$$

each internuclear distance contributing one term, the required derivatives are

$$\frac{\partial F}{\partial z_m} = \sum_k \frac{\partial F_k(\rho_i, r_k)}{\partial r_k} \cdot \frac{\partial r_k(z_1 \cdots z_m \cdots)}{\partial z_m}.$$

Constraints may be imposed by Lagrange's method of undetermined multipliers or by building the constraints into the analytical functions expressing how each internuclear distance depends on each *independent* variable, z_m. Obviously, the greater the number of constraints, the fewer the variables, z_m.

One convenient method of treating the problem has been introduced by Hedberg and Iwasaki [27], who employ a "J-matrix" to handle the necessary derivatives. A different subroutine is required for each molecular type. Another technique introduced by Boates [32] utilizes Eyring's transformations as adapted by Bonham and Thompson [38] to express all internuclear distances in terms of a chosen set of independent bond distances, bond angles, and torsion angles. This scheme handles most geometric types with a single program. For technical details of these somewhat involved methods consult the literature [27, 32, 38, 39].

Steps in Interpreting Data

Reduction of Microphotometer Data

The following steps are involved in converting microphotometer readings to reduced intensities suitable for analysis. Photodiode voltages corresponding to light transmitted through diffraction patterns are read (usually at even radial intervals corresponding to steps of about $\frac{1}{8}$ to $\frac{1}{3}$ s units) and input into a computer. It is convenient to program the computer to make instrumental drift corrections, to perform error checks (right side of pattern versus left side, and significant deviations from smoothness), to convert voltages to absorbancies to intensities, to correct for sector irregularities and extraneous scattering, to introduce the flat-plate correction factor $\sec^3 \phi$ for thick emulsions or $\sec^2 \phi$ for thin emulsions, and to divide by an analytical representation of the theoretical atomic scattering. This results in a "leveled intensity" curve, $I_0(s)$. Frequently the scattering variable $q = 10s/\pi$ is used instead of s. Figures 2.5 and 2.6 are plots of leveled experimental intensity data for the molecule $B(CH_3)_3$.

Some workers work with the original intensity points directly, but most prefer to convert by graphical or (better) by computer interpolation schemes to uniform intervals in q or s space. This simplifies plotting and data processing but the attendant smoothing of data introduces correlations among readings that should be, but are not always, taken into account in error analyses.

If theory and experiment were perfect, the atomic background function, $I_B(s) = I_0(s)/[1 + M(s)_{\text{exp}}]$ associated with the leveled total intensity, would pass through the oscillations of $I_0(s)$ as an uncurved horizontal line. Although observed experimental backgrounds betray discrepancies between experiment and theory, if care is taken they may deviate from constancy by only a few percent over a substantial range of scattering angle. A valuable check of the quality of the data and the reliability of scattering factors (a real problem with heavy atoms) is provided in this test. The background function frequently curves markedly in the region below $s \approx 7$. This may be due partly to effects

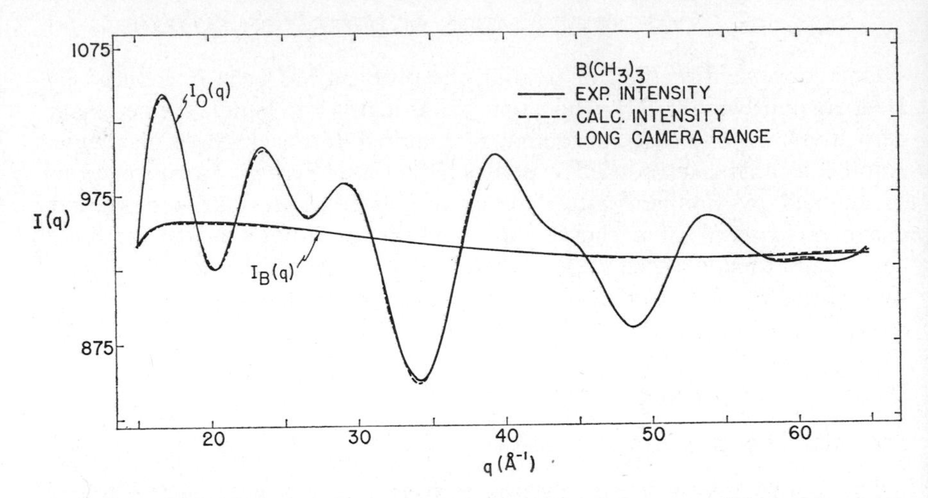

Fig. 2.5 Comparison experimental and calculated (least-squares) leveled intensity curves for $B(CH_3)_3$. The solid background curve $I_B(q)$ was selected to minimize negative regions in $f(r)$, and the dashed background is an analytical function selected by the computer according to the criterion of (4.16), (21-cm camera).

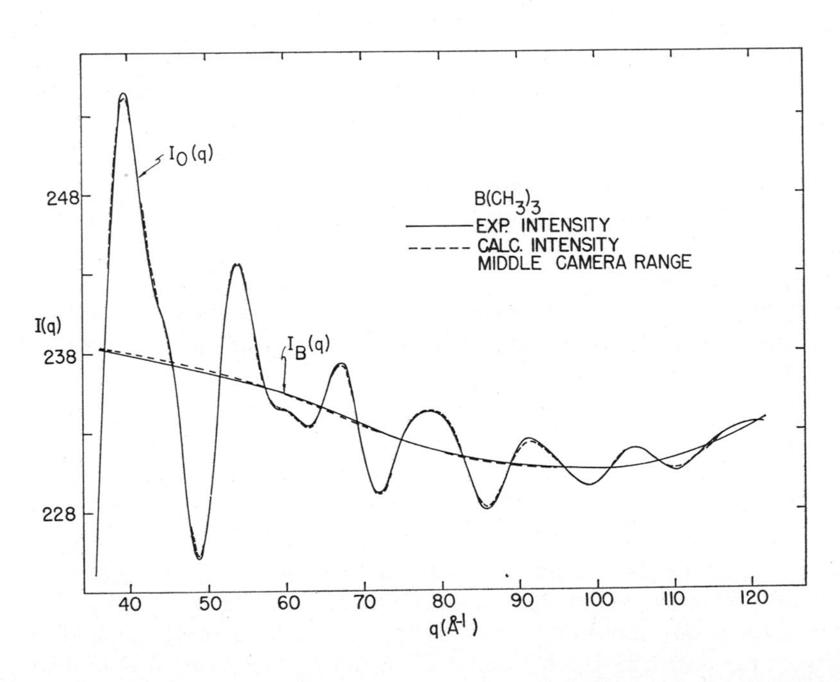

Fig. 2.6 Comparison of experimental and calculated leveled intensities for $B(CH_3)_3$, (11-cm camera).

of electron correlation and chemical bonding that are not included in the theoretical expressions.

The initial trial background function may be drawn by visual estimate to cleave the molecular oscillations more or less evenly, or it may be derived by a polynomial fit with a computer as discussed elsewhere [40]. It is desirable to refine the initial trial background function as the analysis progresses, according to criteria proposed by the Karles [13, 28], namely, that the background must be kept smooth and that the derived $f_c(r)$ radial distribution function be everywhere nonnegative. In this respect it is convenient to reinvert any improper regions detected in the $f(r)$ function in order to diagnose where the improper regions in the background function may lie.

Once a trial background function has been selected, it is possible to compute a trial reduced intensity function $M(s)_{\text{exp}} = \{[I_0(s)_{\text{exp}}/I_B(s)] - 1\}$. Not infrequently a better fit between experiment and calculated curves may be obtained if a factor R, known as the "index of resolution," is introduced, where

$$M(s)_{\text{exp}} = RM(s)_{\text{calc}}. \tag{2.12}$$

Certain types of experimental flaws will lead to an index of resolution that is different from unity, and, moreover, is nonconstant over the range of scattering angle. When this occurs there is no unique method of deriving $R(s)$ from the observed $I_0(s)$ unless all molecular parameters are known from a previous experiment. If such behavior is suspected, and especially if R differs by more than a few percent from unity, the experiment itself should be improved. In cases where R is expected to deviate from unity be only a small, more or less constant amount [41], however, it is very handy to allow R to accommodate the misfit. The R is correlated with the mean amplitudes, l_m, much more strongly than might be expected. If R and the l_m are allowed to vary independently from the outset of analyses, when trial values are poor, the intermediate solutions may meander rather far from the proper values and may converge very slowly indeed. Our best results appear to have been obtained by first fixing R at unity until provisional convergence has been obtained and by then allowing R to vary. Obviously this may bias R, but in our best experiments R does appear to be close to unity. Analyses with arbitrarily fixed backgrounds are not as sensitive to the R, l_m correlation as analyses in which the background is refined simultaneously with the structure parameters. This suggests that derived parameters may be biased significantly by a subjectively chosen, fixed background.

Analysis of Radial-Distribution Function

Most workers find the radial distribution function by all odds the simplest and most direct way to obtain initial insight into the structural feature present in a molecule. Frequently, they even obtain their final structure from least-squares analyses of radial distribution functions. Because it is not practical to obtain intensities down to $s = 0$, it is common to graft calculated intensities onto experimental intensities and to represent the integral of (2.11) by the summation

$$f_N(r) = \sum_{0}^{(s_m - \Delta s)} sM_N(s)_{\text{calc}} e^{-bs^2} \sin sr\, \Delta s + \sum_{s_m}^{s_{max}} sM_N(s)_{\text{exp}} e^{-bs^2} \sin sr\, \Delta s. \quad (2.13)$$

Instead of calculated and experimental data being joined at the minimum s value s_m they may be blended smoothly into each other in some overlap region as a means of weighting the data at small s. If constant intervals Δs are chosen, the right side of (2.13) becomes a Fourier series. Its half-period extending from $r = 0$ to $r = \pi/\Delta s$ is the only region of physical significance. If information is desired beyond, say 8 Å, it is necessary to use an interval in s smaller than $\pi/8$. A coarse grid in s does not necessarily lead to a loss of information in $f_N(r)$ inside $\pi/\Delta s$. If an *exact* $M_N(s)$ is used, the difference between the summation and the integration is utterly negligible until the half-period is closely approached [42]. In assessing errors, however, it is probably better to compute $f(r)$ from all of the original nonequally spaced intensity readings (using an appropriate variable Δs) than to smooth and interpolate the intensities to a coarse but even Δs.

In Fig. 2.7 is illustrated a radial distribution curve for $B(CH_3)_3$ [40, 43]. That the B–C skeleton is a planar, equilateral figure is deduced from the fact that, excepting for a very small "shrinkage" shift, the (C · · · C)/(B–C) distance ratio is $3^{1/2}$. Note that the 3 to 4 Å region contains the C · · · H and H · · · H distributions, which reveal that the methyl groups are rotating freely. A significantly hindered rotation about the B–C bond would have led to a distinctly different distribution. It seems more direct and simpler to establish such behavior from an inspection of $f(r)$ than from $M(s)$.

The b value in the effective weight function se^{-bs^2} should be chosen with care. An excessively small (large) value underweights the intensities at small (large) s values. The optimum value depends upon the distribution of noise in the data, in the minimum and maximum values of s, and on the information desired from $f(r)$. We shall suggest only the crudest sort of guide, leaving the burden of optimization for particular data on the user. If intensities extend to large s and it is desired to determine with accuracy the centers of gravities of nonoverlapping peaks, a value of the order of $(l^2/2)$ may be useful, where l^2 is the mean square amplitude of vibration of the peaks. Large amplitudes require larger b values than small amplitudes if the highest accuracy is desired, although, fortunately, the curve of standard error versus b is quite a flat function. If, on the other hand, it is desired to resolve close distances in overlapping peaks without the aid of information derived from other related peaks, a much smaller value of b is appropriate. Even a negative

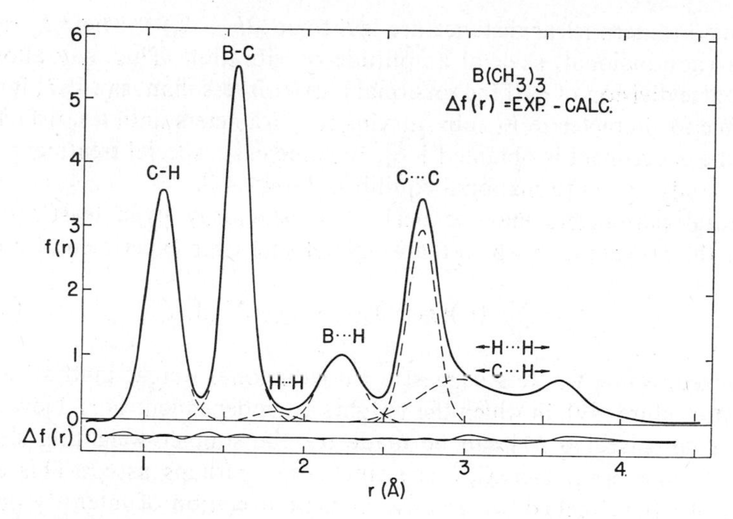

Fig. 2.7 Comparison of experimental and calculated radial distribution functions for $B(CH_3)_3$ with $b = 0.00145$ Å. The calculated H · · · H and C · · · H distributions were computed for freely rotating methyl groups neglecting "shrinkage" corrections. Such corrections would have appreciably lessened $\Delta f(r)$ in the 3 to 4 Å range.

value may be used to sharpen the peaks if the intensities at large s are of high quality.

In cases where internal rotation does not broaden peaks in $f(r)$, the individual components are usually nearly Gaussian peaks. If anharmonicity [29], integral termination corrections [44], and the Born approximation failure [18, 23, 24] are neglected, and if the $N_{ij}(s)$ of (2.9) are taken as unity, the radial distribution curve can be expressed as

$$f(r) = \sum_{i<j}\sum \left[\frac{\pi}{4b + 2l_{ij}^2}\right]^{1/2} \frac{c_{ij}}{r_{ij}} \exp \frac{-(r - r_{ij})^2}{4b + 2l_{ij}^2}. \tag{2.14}$$

If a more accurate curve is needed it can be calculated as discussed in Section 3, pp. 138, 139.

Analytical expressions exist to handle distributions broadened by internal rotation [45]. Usually it is simpler, however, to divide the torsional coordinate α into equal increments, to calculate structures and $f(r)$ functions at each value of the coordinate, and then to compute the net curve as the sum of the individual curves, each curve weighted by the amount expected for the rotational distribution function. If the torsional energy levels are spaced more closely than kT, it may be adequate to adopt the classical distribution $P(\alpha) = C \exp[-V(\alpha)/kT]$, where $V(\alpha)$ is the potential energy as a function of α. Torsional increments should be small enough so that the various increments in internuclear distance are less than $(2b + l_{ij}^2)^{1/2}$, where l_{ij} is the expected (nontorsional) skeletal amplitude of vibration. This may allow a rather coarse division of α. If the torsional barrier is less than, say $4kT$, it may be possible to characterize $V(\alpha)$ by varying its parameters until a satisfactory fit with the experiment is obtained [46]. In some cases special treatments are useful in studying conformational equilibria [7, 45–48].

Structural parameters may be derived satisfactorily by a least-squares fitting of the $f(r)$ curve in which the weighted sum over experimental points

$$Y = \sum_i w(r_i)[f_N(r_i)_{\text{calc}} - f_N(r_i)_{\text{obs}}]^2 \tag{2.15}$$

is minimized. Some workers suggest a "nondiagonal weight matrix" least-squares procedure [49], in which the weights are independent of r_i. Nevertheless, a strong case [50, 51] can be made for the simpler weighting shown above, in which $w(r_i)$ increases as r_i increases, perhaps as r_i^2. This anti-damping of $w(r_i)$ is called for because of the correlation of intensity points with each other; that is, errors are not δ functions in s-space; they are spread over a range of s and their Fourier transform damps in r-space.

In least-squares analyses allowed corrections in parameters should be limited sharply in any one cycle because the linearized normal equations are invalid for large corrections. It is helpful to adopt the relation [40]

$$x_{\text{used}} = [x_{\text{norm eq}}^{-1} + x_{\text{max}}^{-1}]^{-1},$$

where x represents Δr or Δl. Usually Δr_{max} is taken as 0.03 Å and Δl_{max} as 10% of l_m. Also, if the Born approximation correction is neglected, the resultant l_m values may be seriously in error. Even for pairs of light atoms, such as carbon and hydrogen, the influence on l_m may be over 0.002 Å. If differences between atomic numbers are greater, the effect may be considerable.

Analysis of Intensity Function

After establishing the approximate structure with the aid of the radial distribution function, many workers prefer to refine using intensities rather than $f(r)$. Least-squares procedures include (*a*) the "total intensity method" [40], (*b*) the "smooth background method" [52], and (*c*) the conventional "reduced intensity method." Methods (*a*) and (*b*) are, in principle, the most direct methods possible. The computer program automatically refines the background functions simultaneously with the molecular-structure parameters until the best fit is obtained between experimental and calculated intensities. In variant (*a*) a flexible but smooth polynomial function (or, sometimes, a more complex function [40]) is used to represent the background function through the leveled intensity curve. The criterion for refinement is, in matrix notation, the minimization of

$$Y = \mathbf{V}'\mathbf{W}\mathbf{V}, \tag{2.16}$$

with respect to both background and structure parameters, where the column matrix of residuals **V** has elements

$$V_i = I_0(s_i)_{\text{calc}} - I_0(s_i)_{\text{obs}},$$

and where

$$I_0(s)_{\text{calc}} = I_B(s)[1 + RM(s)_{\text{calc}}],$$

in which R may be varied, and $M(s)_{\text{calc}}$ is given by (2.6). The optimum weight matrix **W** is nondiagonal if the data points are correlated [53], but schemes exist that work almost as well if a diagonal **W** is used [51]. Diagonal elements, W_{ii}, should be inversely proportional to the squares of the expected errors in $I_0(s_i)$. The greatest danger in using a diagonal matrix is its careless application, which can lead to excessively small estimates of errors. The results of applying method (*a*) to $B(CH_3)_3$ are depicted in Figs. 2.5 and 2.6 [43].

In variant (*b*), the smooth background method [52], the experimental total intensity $I_0(s)$ is equated to a calculated intensity, and molecular parameters are adjusted automatically to give the smoothest implied background line,

$$I_B(s) = I_0(s)_{\text{obs}}[1 + RM(s)_{\text{calc}}]^{-1}.$$

The smoothness criterion can be formulated in terms of incremental slopes or curvatures. The direct comparisons (*a*) and (*b*) above are often very effective in establishing optimum molecular parameters after the structure is quite well worked out. They are more likely, however, when carelessly used to lead to spurious solutions than the radial distribution method (*c*).

Method (*c*) is a least-squares comparison of experimental and calculated $M(s)$ functions after the background function $I_B(s)$ has been established by trial and error. This approach differs from method (*a*) mainly in requiring a previous determination of the background function. In a sense this method is more subjective than method (*a*), but in the hands of an experienced operator it leads to better numerical stability.

Error Analysis

Error analyses at the time of this writing are still in a somewhat primitive and unsatisfactory state. The greatest difficulties are caused by the correlation of the observed intensity points and by the failure of the least-squares residuals to conform to a simple mathematical model of statistically distributed errors (by virtue of contributions from systematic errors of theory and experiment, and arbitrary, subjective manipulations by the operator).

The correlation problem is orders of magnitude worse in gas-electron diffraction than in X-ray crystallography because the gas-phase molecular pattern is a continuous pattern instead of a pattern of isolated spots, and it is greatly attenuated by the random orientation of molecules in the sample. Accordingly, the gas-phase pattern must be read at close intervals with a sensitivity of a part per thousand or better inasmuch as the molecular pattern itself damps rapidly to parts per thousand of the total intensity. Correlations of seemingly minor import in total intensities can exert a significant influence in the molecular interference terms.

If systematic effects are neglected, errors can be inferred from least-squares analyses [54]. Let **M** be the column matrix whose n elements represent the n observations (which may be $I_0(s_i)$, $M(s_i)$, $s_iM(s_i)$, or $f_N(r_i)$, to name the more commonly used quantities); let $\mathbf{M}_F$ represent the correct matrix of error in observation and $\mathbf{M}_f$, the matrix whose elements are the expectation values of the products V_iV_j of least-squares residuals (with elements of $\mathbf{M}_F$ presumably $n/(n-m)$ times larger than elements of $\mathbf{M}_f$, where m parameters $\theta_1, \ldots, \theta_m$ are deduced from the n observations); let **F** be the column matrix representing the values of the function used to fit the observations; let **A** be the design matrix with elements $\partial F_i/\partial\theta_l$; let **W** be a weight matrix with *arbitrary* elements; let $\mathbf{B} = \mathbf{A'WA}$. If initial guesses $\theta_l{}^0$ are made for the parameters, the least-squares solutions $\theta_l = \theta_l{}^0 + x_l$ are

$$\mathbf{X} = \mathbf{B}^{-1}\mathbf{A'W}(\mathbf{M} - \mathbf{F}^0), \tag{2.17}$$

provided **F** is linear in the x_l, where $\mathbf{F}^0$ is **F** evaluated at the $\theta_l{}^0$, and $(\mathbf{X})_l = x_l$. The smallest errors are obtained if the weight matrix **W** is chosen to be the optimum weight matrix $\mathbf{P} = k\mathbf{M}_f^{-1}$, where k is any constant. If **W** is a reasonably good weight matrix, a perfectly serviceable solution **X** may be obtained and a bona fide error matrix $\mathbf{M}_x$ for parameters may be derived according to this weighting from

$$\mathbf{M}_x = \frac{n}{(n-m)}\, \mathbf{B}^{-1}\mathbf{A}\, \mathbf{W}\mathbf{M}_f\mathbf{W}\mathbf{A}\mathbf{B}^{-1} \tag{2.18}$$

even if **W** is diagonal. The diagonal elements of $\mathbf{M}_x$ are $\sigma^2(\theta_l)$, the squares of the standard errors of parameters. A particularly important function of $\mathbf{M}_x$, apart from furnishing error estimates, is to provide information for a still better set of parameters if at some future time a much better value of one of them becomes available from some other source. Thus if the x_l represent the least-squares solutions and if parameter m is later reliably found to have the value $\theta_m = \theta_m{}^0 + x_m{}^t$, a reoptimized set of parameters corresponds to

$$x_l' = x_l + \left[\frac{(\mathbf{M}_x)_{lm}}{(\mathbf{M}_x)_{mm}}\right](x_m{}^t - x_m). \tag{2.19}$$

Unfortunately, (2.18) is too cumbersome for general use even if it is assumed that $\mathbf{M}_f$ is known (it has been determined in virtually none of the hundreds of published studies to date but obviously should be in future studies). The standard formula used previously was

$$\mathbf{M}_x = \frac{\mathbf{B}^{-1}\mathbf{V}'\mathbf{P}\mathbf{V}}{(n - m)}, \tag{2.20}$$

a result that follows immediately from (2.18) if $\mathbf{W}$ is replaced by the optimum weight matrix $\mathbf{P}$. Equation (2.20) is invalid if $\mathbf{P}$ is taken, as it invariably has been, as a plausible but arbitrary weighting. The careless use of (2.20) can lead to very poor results. Also, it should not be forgotten that background functions, whether analytical or manually drawn, correspond to many arbitrary parameters that should be estimated and totaled in the m of (2.20). A very important step in the resolution of the data correlation problem was made by Murata, Morino, and Kuchitsu [49, 53]. In the following we suggest a preliminary, alternative approach [51] that retains the convenience of working with diagonal weight matrices $\mathbf{W}$.

Let us define a zeroth-order error matrix,

$$\mathbf{M}_x{}^o = \frac{\mathbf{B}^{-1}\mathbf{V}'\mathbf{W}\mathbf{V}}{(n - m)}, \tag{2.21}$$

where n remains the number of *intensity* points used in the analysis whether the analysis is upon the intensity or upon the radial distribution function. Equation (2.18) can be expressed by

$$(\mathbf{M}_x)_{ab} = \mathscr{R}_{ab}(\mathbf{M}_x{}^0)_{ab}, \tag{2.22}$$

where the $\mathscr{R}_{ab}$ can be deduced from the characteristics of $\mathbf{M}_f$. If the analysis is performed on intensities, let us envision the continuous function $M_f(s, s')$ corresponding to the expected mean square error in observations at s if $s = s'$, and to the steady decrease in correlation between observations at s and s' as s' becomes more remote from s. It can be shown in electron-diffraction-intensity analyses that the diagonal factors $\mathscr{R}_{aa}$ are given approximately by

$$\mathscr{R}_{aa} = 2(\langle M_f(s, s)\rangle \Delta s)^{-1} \int_0^{s_M} \langle M_f(s, s')\rangle \cos [r_a(s - s')] \, d(s - s') \tag{2.23}$$

for distance and amplitude parameters in a peak at r_a, where Δs is the interval between observations. If $\langle M_f(s, s')\rangle / \langle M_f(s, s)\rangle$ can be represented by the function $\exp(-\gamma |s - s'|)$, the ratio $\mathscr{R}_{aa}$ between the proper and the zeroth-order error elements is

$$\mathscr{R}_{aa} \approx 2\gamma[\Delta s(\gamma^2 + r_a{}^2)]^{-1} \tag{2.24}$$

if Δs is much less than γ^{-1}. Highly preliminary studies suggest that the correlation parameter γ may be of the order of 1 Å in several laboratories [51, 53, 55]. It should be noted that the commonly used zeroth-order elements of $\mathbf{M}_x{}^0$ tend to be proportional to Δs if $\mathbf{W}$ is diagonal. This dependency is cancelled in $\mathbf{M}_x$, if Δs is small, by the $(\Delta s)^{-1}$ factor in $\mathscr{R}$. It should also be noted that the decrease in $\mathscr{R}_{aa}$ with increasing r_a is a consequence of the breadth of errors in s-space due to correlations between the data points. Fourier transforms of broad error signals damp with increasing r in $f(r)$ curves. The weight function $w(r_i)$ of (2.15) consistent with the intensity correlation function $\exp(-\gamma\,|s - s'|)$ is

$$w(r_i) \approx (\gamma^2 + r_i{}^2). \tag{2.25}$$

If the analysis is upon the radial distribution function in which the intensity correlations are taken into account by a suitable weighting function such as that of (2.25), the zeroth-order error matrix of (2.21) can still be used. The factor $\mathscr{R}$ required to convert the diagonal elements into appropriate squares of standard errors depends on the distribution of noise in the intensities. If, for example, the noise in $sM(s)$ is approximately constant over the range of s, the factor becomes

$$\mathscr{R}_{aa} \approx \left[\frac{(8\pi b)^{1/2} n}{r_{\max} - r_{\min}}\right]\left[\frac{\mathscr{L}_a{}^2}{\mathscr{L}_a{}^2 + 2b}\right]^{p/2}, \tag{2.26}$$

where $p = 3$ for analyses of distances and $p = 5$ for amplitudes, and

$$\mathscr{L}_{ij}^2 = 2b + \delta_{ij}^2 + l_{ij}^2,$$

in which δ_{ij} is the value of $[\eta_i(s) - \eta_j(s)]/s$ at $s = 1.2/l_{ij}$. The relation above was derived from the assumption that $f(r)$ was constructed from intensity data extending to $s \geq (2/b)^{1/2}$ and that atomic-number differences are not huge. It should be noted that the interval Δr in the $f(r)$ curve may be taken to be as small as desired since $[\mathscr{R}(M_x{}^0)]_{aa}$ becomes independent of Δr.

Systematic errors may be appreciable. Errors in electron wavelength and sample to-plate distance influence the internuclear distance scale factor and can be kept at less than one part per thousand. If the microphotometer is provided with a precision screw, radial scanning errors can be kept similarly low. Unsuspected impurities in the sample can cause serious trouble. Errors in the sector shape can contribute appreciably if not taken carefully into account. The uncertainty in drawing the background may sometimes decrease the accuracy of the experiment more than the above random and systematic errors combined [56]. Systematic errors in the emulsion calibration curve and extraneous scattering from stops, apertures, and residual gases can have a marked influence on radial distribution peak shapes and hence on mean amplitudes and on derived separations of closely spaced internuclear distances. Imperfections in scattering theory introduce errors, also. For example, observed peaks in $f(r)$ corresponding to bonds between heavy and light atoms are almost always more asymmetric than curves calculated according to current theoretical expressions. The above sources of uncertainty do not constitute an exhaustive list. Errors arising from systematic factors must be included along with the errors derived from least-squares analyses in the final error assignments. Unfortunately, experience has shown that experimental imperfections may be so well hidden that a derived structure is substantially more in error than error analyses have indicated. Caution is urged.

A criterion to test the often unduly optimistic error estimates resulting when effects of systematic errors are neglected in least-squares analyses is suggested. Imperfections in emulsion development and calibration, scattering factors, and extraneous scattering corrections can be absorbed into a function $T(s)$ such that

$$M(s)_{\text{obs}} = T(s)M(s)_{\text{ideal}}, \tag{2.27}$$

where, for sake of analysis, $T(s)$ may be expressed for a given camera range as

$$T(s) = 1 + \sum_n A_n \cos \frac{n\pi s}{2s_m},$$

in which n ranges from unity to 3 or 4, and s_M is the maximum value of s for that camera range. In current work the A_n might be expected to be several hundredths or even as high as 0.1. As a final refinement it would be prudent to perform a least-squares fitting of the data according to (2.27), in which the A_n were derived simultaneously with the molecular parameters subject to the constraint that the A_n not exceed plausible values. Damaging correlations between systematic errors and molecular parameters would be revealed by the error matrix and would lead to appropriately increased computed standard errors.

As an illustration of magnitudes of errors to be expected in a routine careful analysis of a simple molecule [40, 43], Tables 2.1 and 2.2 show partial results for $B(CH_3)_3$ based on analyses of both intensity (method *a*) and radial distribution functions. Diagonal weight matrices were used, but intensity correlations were accounted for as described above. In general, it is desirable to publish the entire error matrix.

4 APPLICABILITY AND LIMITATIONS OF THE METHOD

The principal application of gas-phase electron diffraction is to determine structures of pure compounds. Valuable additional information on molecular force fields can be derived, however [7]. Mean amplitudes of vibration taken together with spectroscopic frequencies may serve to establish a more complete set of quadratic force constants than can be deduced from the frequencies

Table 2.1 Comparison of $B(CH_3)_3$ Molecular Distances and Amplitudes of Vibration Derived from Various Least-Squares Analyses (in Angstroms)

Parameter	$I_o(s)$ 21-*cm* *Camera*	$I_o(s)$ 11-*cm* *Camera*	$f(r)$ 11- *and* 21-*cm* *Data Merged*
r(B–C)	1.5764	1.5772	1.5758
r(C · · · C)	2.7233	2.7226	2.7233
r(C–H)	1.1072	1.1082	1.1088
r(B · · · H)	2.2403	2.2393	2.2377
1(B–C)	0.0522	0.0537	0.0541
1(C · · · C)	0.0769	0.0756	0.0765
1(C–H)	0.0834	0.0794	0.0812
1(B · · · H)	0.1238	0.1265	0.1273

Table 2.2 Comparison of Standard Errors Computed from Least-Squares Analyses of $B(CH_3)_3$ Data by Various Methods.[a] Systematic Errors Neglected in All Columns except the Last (in Thousandths of an Angstrom).

Parameter	$I_o(s)$ 21-*cm* *Camera*	$I_o(s)$ 11-*cm* *Camera*	$f(r)$ *Merged* *Data*	*Estimated* *Total* *Standard* *Error*
r(B–C)	0.9	0.6	0.9	1.2
r(C · · · C)	1.0	0.7	1.0	2.7[b]
r(C–H)	1.6	1.4	2.3	1.8
r(B · · · H)	3.8	5.3	4.9	5.0
1(B–C)	1.3	0.5	0.9	1.3
1(C · · · H)	1.0	0.5	0.9	2.2[b]
1(C–H)	1.6	1.0	1.9	2.2
1(B · · · H)	3.0	3.0	4.0	4.1

[a] Revised from original analyses of Ref. 40 which neglected intensity correlations. The intensity correlation parameter γ was taken as 1 Å^{-1}.
[b] Radial distribution peak overlapped by C · · · H peak.

alone [57, 58]. Shifts of internuclear distances with temperature [58] or isotopic substitution [56, 59] and analyses of asymmetries of radial distribution peaks provide a measure [59], though not as yet a very delicate measure, of cubic force constants.

Because scattered intensities are proportional to concentrations of molecules, the composition of a mixture of substances can be determined if the components have sufficiently different radial distribution functions and if there are not very many components. The components may be rotational isomers or chemically distinct substances. If components are in equilibrium with each other it may be possible to infer gas-phase equilibrium constants. A temperature dependency would give a fairly detailed account of thermodynamic properties. Although this approach has been applied [60], it is handicapped by the lack of knowledge of accurate temperatures, pressures, and nonequilibrium relaxation effects. Indeed, gas diffraction may turn out to yield useful information on relaxation rates in the microsecond range in gas jets cooling by free expansion [61, 62].

A few words on the relative advantages and disadvantages of gas-phase electron diffraction are appropriate. Certain aspects of the distribution of planetary electrons being studied in current research have not been forthcoming from other techniques (except theory). Insofar as molecular geometry is concerned, there is some advantage of studying free gas-phase molecules instead of molecules in a crystal. Crystals do not permit studies of conformational equilibria. Moreover, stable geometries in crystals are sometimes significantly different from those in the gas phase. The molecules B_2Cl_4 [48] and biphenyl [63] are twisted in the gas phase but are planar in the solid state [64]. Hydrogen fluoride consists of infinite planar hydrogen-bonded chains in the crystal state [65] but of monomers and puckered cyclic hexamers in the vapor state [66]. Xenon hexafluoride is a peculiarly undulating, deformed octahedral monomer in the gas state [67] but a tetramer in the solid state [68]. All true salts have markedly different interionic distances in the two states [69]. Unstable free radicals that do not exist pure, in crystals, have been studied in the gas phase [70].

Electrons enjoy an advantage over X-rays in that they "see" well-localized nuclei and not just the diffuse clouds of planetary electrons. For this reason electron-diffraction bond lengths are often more precise than X-ray bond lengths. Furthermore, certain scattered phase relationships revealing correlated motions of atoms in a given molecule survive in gas-diffraction intensities but not in crystal intensities [67]. These advantages must be weighed against a severe disadvantage. Gas-diffraction data yield only "one-dimensional" distribution functions and cannot lead to unique structures for molecules of any great complexity. Crystallographic data contain three-dimensional information suitable for unique structure determinations of even such molecules as proteins!

In comparison with gas-phase spectroscopic methods, and microwave spectroscopy in particular, electron diffraction offers several advantages. Chief among these are the ability to work with nonpolar molecules, the ability of getting partial structures for some molecules too complex for spectroscopy (e.g., certain higher alkanes [47]), and the somewhat simpler physical significance of the derived bond lengths for complex molecules [37]. On the other hand, the precision of the diffraction data is considerably less than the precision of microwave data. For diatomic molecules and simple polyatomic molecules where suitable corrections for molecular vibrations can be made, the spectroscopic structures may be much more accurate than corresponding diffraction structures. Under the most favorable conditions current diffraction methods yield distances to within 0.001 Å or so. Spectroscopic distances for complex molecules are usually not this accurate (in terms of physically understood internuclear distances).

To date only a few molecules have been studied in exhaustive detail by both electron diffraction and molecular spectroscopy. In those cases where full vibrational corrections have been carried out the agreement has been excellent. Such cases include a handful of diatomic and triatomic molecules, BF_3 [71], and protiated and deuterated species of ammonia, methane, ethane, and ethylene [59]. Considerably more complex molecules than these have proven tractable, and the limitations enunciated in the previous edition of this treatise have been drastically relaxed. New automated techniques to measure and to process diffraction data have enormously relieved the tedium as well as increased the accuracy. The coming decade promises to be yet more productive than the last.

General References

BOOKS

Pirenne, M. H., *The Diffraction of X-Rays and Electrons by Free Molecules*, Cambridge University Press, London, 1946.

Pinsker, Z. G., *Electron Diffraction*, Butterworth, London, 1953.

REVIEW ARTICLES

Bastiansen, O. and P. N. Skancke, *Advan. Chem. Phys.*, **3,** 323 (1960).

Spiridonov, V. P., N. G. Rambidi, and N. V. Alekseev, *Zhur. Strukt. Khimi*, **4,** 779 (1963).

Almenningen, A., O. Bastiansen, A. Haaland, and H. M. Seip, *Angew. Chemie* Internat. ed., **4,** 819 (1965).

Bartell, L., *Trans. Amer. Crystallogr. Assoc.*, **2,** 134 (1966).

Hilderbrandt, R. L. and R. A. Bonham, *Ann. Rev. Phys. Chem.*, **22,** (1971).

REVIEW CHAPTERS

Akishin, P. A., Rambidi, N. G., and V. P. Spiridonov, in The *Characterization of High Temperature Vapors* (J. L. Margrave, ed.), Chapter 12, John Wiley and Sons, New York, 1967.

Bauer, S. H., in *Physical Chemistry, An Advanced Treatise* (D. Henderson, ed.) Volume IV, Chapter 14, Academic Press, New York, 1970.

Karle, J., in *Determination of Organic Structures by Physical Methods* (F. C. Nachod and J. J. Zuckerman, ed.) Volume V, Academic Press, New York, 1971 (in press).

References

1. Z. G. Pinsker, *Electron Diffraction*, Butterworths, London, 1953.
2. L. H. Germer and K. H. Storks, *Proc. Natl. Acad.*, **23**, 390 (1937); *J. Chem. Phys.*, **6**, 280 (1938); *Phys. Rev.*, **55**, 648 (1939); W. C. Bigelow and L. O. Brockway, *J. Colloid. Sci*, **11**, 60 (1956).
3. *Proc. Symp. Low-Energy Electron Diffraction: Trans. Amer. Crystallogr. Assoc.*, **4**, 1–144 (1968).
4. L. S. Bartell and L. O. Brockway, *Phys. Rev.*, **90**, 833 (1953).
5. T. Iijima and R. A. Bonham, *J. Phys. Chem.*, **67**, 2769 (1963); L. S. Bartell and R. M. Gavin, Jr., *J. Amer. Chem. Soc.*, **86**, 3493 (1964); C. Tavard, M. Rouault, and M. Roux, *J. Chim. Phys.*, **62**, 1410 (1965); C. Tavard, *Cahiers. Phys.*, **20**, 397 (1966). D. A. Kohl and L. S. Bartell, *J. Chem. Phys.*, **51**, 2896 (1969).
6. R. A. Bonham and J. Iijima, *J. Chem. Phys.*, **42**, 2612 (1965); D. A. Kohl and R. A. Bonham, *J. Chem. Phys.*, **47**, 1634 (1967); R. A. Bonham and H. L. Cox, Jr., *J. Chem. Phys.*, **47**, 3508 (1967); R. A. Bonham, M. Fink, D. A. Kohl, and E. M. A. Piexoto, *Intern. J. Quantum Chem.*, 1969 Sanibel Island Symp. (to be published). J. Geiger, *Z. Phys.*, **181**, 413 (1964).
7. K. Hedberg, *Trans. Amer. Crystallogr. Assoc.*, **2**, 79 (1966); L. S. Bartell, *ibid.*, **2**, 134 (1966).
8. H. Mark and R. Wierl, *Naturwissenschaften*, **18**, 205 (1930).
9. P. P. Debye, *Physik. Z.*, **40**, 66, 404 (1939).
10. O. Hassel and T. Taarland, *Tidsskr. Kjemi, Bergvesen Met.*, **20**, 167 (1940); C. Finbak, O. Hassel, and B. Ottar, *Arch. Math. Naturvidenskab*, **44**, No. 13, 1 (1941).
11. H. D. Fetzer, doctoral dissertation, University of Texas, 1965; H. D. Fetzer, R. Pohler, B. Turman, and H. P. Hanson, *Trans. Amer. Crystallogr. Assoc.*, **2**, 129 (1966); H. B. Thompson, *Bull. Am. Phys. Soc.*, **13**, 835 (1968); M. Fink and R. A. Bonham, *Rev. Sci. Instr.*, **41**, 389 (1970).
12. Some units designed for low-pressure samples utilize a longer electron path through the sample and compensate by a special lens system. See S. H. Bauer and K. Kimura, *J. Phys. Soc. Japan*, Supplement BII, **17**, 300 (1962).

13. For other designs see ref. 12 and I. L. Karle and J. Karle, *J. Chem. Phys.*, **17,** 1052 (1949); *ibid.*, **18,** 565 (1950); L. O. Brockway and L. S. Bartell, *Rev. Sci. Instr.*, **25,** 569 (1954); O. Bastiansen, O. Hassel, and E. Risberg, *Acta Chem. Scand.*, **9,** 232 (1955); P. A. Akishin *et al.*, *Pribory i Tekhn. Eksperim.*, **2,** 70 (1958); H. C. Corbet *et al.*, *Rec. Trav. Chim.*, **83,** 789 (1964); S. Shibata, *Japan. J. Appl. Phys.*, **3,** 530 (1964); W. Zeil *et al.*, *Z. Instrumentenk.*, **74,** 84 (1966); Y. Murata *et al.*, *Japan. J. Appl. Phys.*, **9,** 591 (1970).
14. L. S. Bartell and L. O. Brockway, *J. Appl. Phys.*, **24,** 656 (1953).
15. R. A. Bonham and H. R. Foster, private communication; R. C. Valentine, *Advances in Optical and Electron Microscopy*, vol. 1, R. Barer and V. E. Cosslett, Eds. Academic, New York, 1966, pp. 180–203.
16. Kodak Pamphlet No. P-140, 1968.
17. N. F. Mott and H. S. W. Massey, *The Theory of Atomic Collisions*, 3rd ed., Oxford University Press, Oxford, 1965; R. A. Bonham, *J. Chem. Phys.*, **43,** 1933 (1965).
18. V. Schomaker and R. Glauber, *Nature*, **170,** 290 (1953); R. Glauber and V. Schomaker, *Phys. Rev.*, **89,** 667 (1952).
19. R. A. Bonham, *J. Chem. Phys.*, **43,** 1434 (1965); A. C. Yates and R. A. Bonham, *ibid.*, **50,** 1056 (1969).
20. *International Tables for X-Ray Crystallography*, Kynoch Press, Birmingham, England, 1952.
21. T. G. Strand and R. A. Bonham, *J. Chem. Phys.*, **40,** 1686 (1964); R. A. Bonham and T. G. Strand, *J. Chem. Phys.*, **39,** 2200 (1963).
22. J. A. Ibers and J. A. Hoerni, *Acta Crystallogr.*, **7,** 405 (1954).
23. R. A. Bonham and T. Ukaji, *J. Chem. Phys.*, **36,** 72 (1962).
24. R. A. Bonham and H. L. Cox, Jr., *J. Chem. Phys.*, **47,** 2599 (1967); *International Tables for X-Ray Crystallography*, revised edition, to be published.
25. R. F. Pohler and H. P. Hanson, *J. Chem. Phys.*, **42,** 2347 (1965).
26. C. Tavard, D. Nicholas, and M. Rouault, *J. Chim. Phys.*, **64,** 540 (1967).
27. See, for example, K. Hedberg *et al.*, *Acta Cryst.*, **17,** 529, 533, 538 (1964); H. Seip *et al.*, *Acta Chem. Scand.*, **19,** 1955 (1965); *ibid.*, **20,** 385, 1535, 2698, 2711 (1966). B. Anderson, H. M. Seip, T. G. Strand, and R. Stölevik, *Acta Chem. Scand.*, **23,** 3224 (1969).
28. J. Karle and I. L. Karle, *J. Chem. Phys.*, **18,** 957 (1950).
29. L. S. Bartell, *J. Chem. Phys.*, **23,** 1219 (1955); K. Kuchitsu and L. S. Bartell, *ibid.*, **35,** 1945 (1961).
30. L. S. Bartell, L. O. Brockway, and R. H. Schwendeman, *J. Chem. Phys.*, **23,** 1854 (1955); R. A. Bonham and L. S. Bartell, *ibid.*, **31,** 702 (1959).
31. R. M. Gavin, Jr. and L. S. Bartell, *J. Chem. Phys.*, **48,** 2460, 2466 (1968).
32. T. L. Boates, doctoral dissertation, Iowa State University, 1966.
33. L. S. Bartell and L. O. Brockway, *J. Chem. Phys.*, **23,** 1860 (1955).
34. O. Bastiansen and M. Traetteberg, *Acta Cryst.*, **13,** 1108 (1960); Y. Morino, *ibid.*, 1107 (1960).
35. S. J. Cyvin, *Molecular Vibrations and Mean Square Amplitudes*, Elsevier, Amsterdam, 1968.

36. T. Oka, *J. Phys. Soc. Japan*, **15,** 2274 (1960); Y. Morino, K. Kuchitsu, and T. Oka, *J. Chem. Phys.*, **36,** 1108 (1962); K. Kuchitsu and L. S. Bartell, *ibid.*, **36,** 2460, 2470 (1962); D. R. Herschbach and V. W. Laurie, *ibid.*, **37,** 1668, 1687 (1962).
37. K. Kuchitsu, *J. Chem. Phys.*, **49,** 4456 (1968).
38. H. B. Thompson, *J. Chem. Phys.*, **47,** 3407 (1967).
39. H. C. Corbet, G. Dallinga, F. Oltmans, and L. H. Toneman, *Rec. Trav. Chim.*, **83,** 789 (1964); G. Dallinga and L. H. Toneman, *ibid.*, **86,** 171 (1967); *J. Mol. Struct.* **1,** 11 (1967).
40. L. S. Bartell, D. A. Kohl, B. L. Carroll, and R. M. Gavin, Jr., *J. Chem. Phys.*, **42,** 3079 (1965).
41. Such as the following cases: (*a*) If the emulsion calibration constant c in $E = A(1 + cA)$ has not been measured accurately and if A is roughly constant over the plate. (*b*) If a little of the gas sample is more or less uniformly delocalized in the diffraction chamber.
42. T. Ino, *J. Phys. Soc. Japan*, **12,** 495 (1957); K. Kuchitsu (unpublished work).
43. L. S. Bartell and B. L. Carroll, *J. Chem. Phys.*, **42,** 3076 (1965).
44. L. S. Bartell and L. O. Brockway, *J. Chem. Phys.*, **32,** 512 (1960).
45. J. Karle, *J. Chem. Phys.*, **22,** 1246 (1954); D. A. Swick and J. Karle, *ibid.*, **23,** 1499 (1955); J. Karle, *ibid.*, **45,** 4149 (1966).
46. K. Kuchitsu, *Bull. Chem. Soc. Japan*, **30,** 391 (1957); Y. Morino and E. Hirota, *J. Chem. Phys.*, **28,** 185 (1958).
47. L. S. Bartell and D. A. Kohl, *J. Chem. Phys.*, **39,** 3097 (1963); L. S. Bartell and J. P. Guillory, *ibid.*, **43,** 647, 654 (1965).
48. J. V. Patton and K. Hedberg, *Bull. Amer. Phys. Soc.*, **13,** 831 (1968).
49. Y. Morino, K. Kuchitsu, and Y. Mirata, *Acta Cryst.*, **18,** 549 (1965).
50. R. A. Bonham and L. S. Bartell, *J. Chem. Phys.*, **31,** 702 (1959).
51. L. S. Bartell, *Acta Cryst.*, **A25,** S76 (1969).
52. S. Shibata, L. S. Bartell, and R. M. Gavin, Jr., *J. Chem. Phys.*, **41,** 717 (1964); *ibid.*, **42,** 1147 (1965). S. Shibata and L. S. Bartell, J. Mol. Struct. (in press).
53. Y. Murata and Y. Morino, *Acta Cryst.*, **20,** 605 (1966).
54. W. C. Hamilton, *Statistics in Physical Science*, Ronald, New York, 1964.
55. R. Stölevik (to be published).
56. S. Shibata and L. S. Bartell, *J. Chem. Phys.*, **42,** 1147 (1965).
57. Y. Morino, Y. Nakamura, and T. Iijima, *J. Chem. Phys.*, **32,** 643 (1960); Y. Morino and Y. Murata, *Bull. Chem. Soc. Japan*, **38,** 104 (1965); Y. Morino and H. Uehara, *J. Chem. Phys.*, **45,** 4543 (1966); Y. Morino, T. Ukaji, and T. Ito, *Bull. Chem. Soc. Japan*, **39,** 64, 71 (1966); S. Konaka, Y. Murata, K. Kuchitsu, and Y. Morino, *ibid.*, **39,** 1134 (1966); T. Ukuji and K. Kuchitsu, *ibid.*, **39,** 2153 (1966).
58. K. Hedberg and M. Iwasaki, *J. Chem. Phys.*, **36,** 589, 594 (1962).
59. L. S. Bartell, K. Kuchitsu, and R. J. deNeui, *J. Chem. Phys.*, **35,** 1211 (1961); L. S. Bartell and H. K. Higginbotham, *ibid.*, **42,** 851 (1965); L. S. Bartell, E. A. Roth, C. D. Hollowell, K. Kuchitsu, and J. E. Young, Jr., *ibid.*, **42,** 2683 (1965); K. Kuchitsu, J. P. Guillory, and L. S. Bartell, *ibid.*, **49,** 2488 (1968).

60. L. S. Su, doctoral dissertation, Indiana University, 1967. R. A. Bonham and L. S. Su, *Bull. Amer. Phys. Soc.*, **13,** 831 (1968).
61. P. Audit and M. Rouault, *Compt. Rend.*, **265B,** 1100 (1967); P. Audit, *Bull. Amer. Phys. Soc.*, **13,** 834 (1968).
62. J. B. Anderson, R. P. Andres, and J. B. Fenn, *Advan. Chem. Phys.*, **10,** 275 (1966); T. A. Milne and F. T. Greene, *J. Chem. Phys.*, **47,** 4095 (1967).
63. O. Bastiansen, *Acta Chem. Scand.*, **3,** 408 (1949).
64. J. Dhar, *Indian J. Phys.*, **7,** 43 (1932); J. Trotter, *Acta Cryst.*, **14,** 1135 (1961); A. Hargreaves and S. H. Rizvi, *ibid.*, **15,** 365 (1962).
65. M. Ajoti and W. N. Lipscomb, *Acta Cryst.*, **7,** 173 (1954).
66. J. Jansen and L. S. Bartell, *J. Chem. Phys.*, **50,** 3611 (1969).
67. L. S. Bartell and R. M. Gavin, Jr., *J. Chem. Phys.*, **48,** 2466 (1968).
68. P. A. Agron, C. K. Johnson, and H. A. Levy, *Inorg. Nucl. Chem. Lett.* **1,** 145 (1965).
69. See, for example, P. A. Akishin and V. P. Spiridonov, *Kristallographia*, **2,** 475 (1957); P. A. Akishin and H. G. Rambidi, *Zhur. Neorg. Khimi*, **5,** 23 (1960); V. M. Tatevskii, V. P. Spirdonov, and P. A. Akishin, *Dokl. Akad. Nauk SSSR*, **138,** 621 (1961); L. V. Vilkov, N. G. Rambidi, and V. P. Spiridonov, *Zhur. Strukt. Khimi*, **8,** 786 (1967).
70. P. Andersen, *Acta Chem. Scand.*, **19,** 629 (1965); B. Andersen and P. Andersen, *Trans. Amer. Cryst. Assoc.*, **2,** 193 (1966); L. Schafer, *J. Amer. Chem. Soc.*, **90,** 3919 (1968).
71. K. Kuchitsu and S. Konaka, *J. Chem. Phys.*, **45,** 4342 (1966).

Chapter **III**

NEUTRON SCATTERING

Julius M. Hastings and Walter C. Hamilton

1 INTRODUCTION

The availability of powerful neutron sources has provided still another tool for the study of matter, in particular the atomic and magnetic structures as well as energy levels of the material under study. That one could use neutrons for this purpose has been recognized ever since their discovery by Sir James Chadwick in 1932. There has always been a practical difficulty of producing sufficiently intense beams. In 1936 Mitchell and Powers [1] succeeded in demonstrating the diffraction of neutrons using a radium–beryllium source, but it was only with the advent of nuclear reactors that it became possible to use neutrons for research in the liquid and the solid states. About a quarter of a century has passed since the pioneering work of Shull and Wollan [2], and a very extensive literature has accumulated. In this chapter the appropriate formalism together with a number of applications is reviewed.

2 SCATTERING OF NEUTRONS

General Properties of Neutrons

The similarity between neutron and X-ray diffraction is far-reaching, but there are sufficient differences so that the information one technique cannot furnish may often be obtained with the other. The geometric aspects of the two methods are identical, but the interaction of the neutron with the scatterer is very different from that of the X-ray photon. In this section we discuss the nature of the neutron interaction with the scatterer, but before doing so we point out some important characteristics of neutrons produced in a reactor.

The neutrons coming from a nuclear reactor are in thermal equilibrium with the moderator and thus have a velocity spectrum that corresponds roughly to a Maxwellian distribution at the temperature of the moderator. If this temperature is 300°K, the mean energy E of the neutrons is approximately 4.2×10^{-14} ergs, or 0.025 eV. From the relation

$$E = \frac{\hbar^2 k^2}{2m}, \tag{3.1}$$

where k is the wave number and m is the mass of the neutron, it follows that the mean wave number is approximately 3.5 Å^{-1}. The wavelength λ corresponding to this wave number is

$$\lambda = \frac{2\pi}{k} = 0.286E^{-1/2}\,(\text{eV}) \simeq 1.8\ \text{Å}. \tag{3.2}$$

This wavelength is just about equal to a typical interatomic spacing, and hence one obtains interference effects similar to those obtained with X-rays of comparable wavelength. There is, however, an important difference when the neutrons or X-rays exchange energy with the scatterer, so-called inelastic scattering. Typical vibrational excitation energies in a solid range from a few milli-electron volts (meV) to several tens of meV and are therefore of the same order of magnitude as the neutron energy. These excitations are therefore readily observable. In the X-ray case the energy of the photon is of the order of 10^7 meV, so that the energy change is essentially unobservable. The possibility of observing inelastic scattering with neutrons has been exploited in a number of ways.

Cross Section

The scattering of neutrons can be conveniently treated by the Born approximation [3]. In this approximation the differential scattering cross section $d\sigma/d\Omega$, which is defined as the ratio of the number of neutrons scattered per unit of time into solid angle $d\Omega$ to the incident flux is given by

$$\frac{d\sigma}{d\Omega} = \frac{k'}{k}\left(\frac{m}{2\pi\hbar^2}\right)^2 \left|\langle s'q'| \int e^{i\mathbf{K}\cdot\mathbf{r}} V(\mathbf{r})\, d\mathbf{r}\, |sq\rangle\right|^2 \tag{3.3}$$

for a process in which the scatterer goes from an initial state q to some final state q', while the neutron changes from wave vector $\mathbf{k}$ and spin state s to wave vector $\mathbf{k}'$ and spin state s'. $V(\mathbf{r})$ is the interaction potential between a neutron with position vector $\mathbf{r}$ and an atom at the origin. $\mathbf{K}$, the momentum transfer in the scattering process, is defined by $\mathbf{K} = \mathbf{k} - \mathbf{k}'$ (see Fig. 3.1) where $\mathbf{k}'$ is subject to the energy conservation condition

$$\frac{\hbar^2 k^2}{2m} + E_q = \frac{\hbar^2 k'^2}{2m} + E_{q'}, \tag{3.4}$$

E_q being the energy of the scatterer in the state q. This latter condition can be

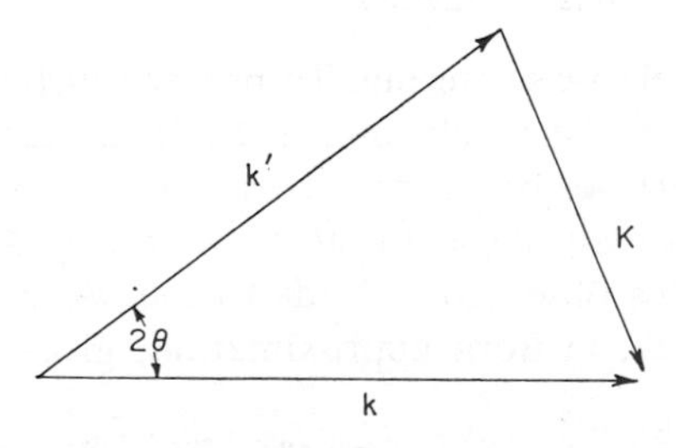

Fig. 3.1 Relationship of the momentum transfer **K** to the wave vectors of the incident and scattered neutrons, **k** and **k**$'$. 2θ is the angle of scattering.

conveniently incorporated into the cross-section formula by multiplication of (3.3) with a δ function of energy. The total cross section is obtained by averaging over all initial states $|qs\rangle$ and summing over all final states $\langle q's'|$. The averaging over initial states involves the probabilities P_q and P_s that the scatterer will be in a state q and that the neutron will have spin s. Usually P_q is given by a Boltzmann distribution and P_s is equal to $\frac{1}{2}$ for an unpolarized beam. The final cross section then appears as

$$\frac{d^2\sigma}{d\Omega\, dE} = \sum_{qs} P_q P_s \sum_{q's'} \frac{k'}{k} \left(\frac{m}{2\pi\hbar^2}\right)^2 \left|\langle q's'| \int e^{i\mathbf{K}\cdot\mathbf{r}} V(\mathbf{r})\, d\mathbf{r}\, |qs\rangle\right|^2 \delta\left(\frac{\hbar^2 k'^2}{2m} + E_{q'} - \frac{\hbar^2 k^2}{2m} - E_q\right). \tag{3.5}$$

The interaction potential $V(\mathbf{r})$ between the neutron and an atom consists of two parts: (*a*) the neutron is scattered by nuclei through nuclear forces and (*b*) the neutron is scattered by electrons in atoms having a magnetic moment through magnetic forces. The interaction of the neutron magnetic moment with the atomic magnetic moment is much weaker than the nuclear interaction, but it has a much longer range and thus the overall cross sections for the two interactions are comparable. The range of the magnetic interactions is given by the size of the orbits of the unpaired electrons, whereas the nuclear interaction is confined to nuclear dimensions. The total $V(\mathbf{r})$ is the sum of these two interactions.

Nuclear Scattering

Effect of Isotopes and Nuclear Spin

Let us consider the nuclear interactions first and treat the case of a single nucleus with zero nuclear spin bound at the origin. Since a fixed nucleus cannot exchange energy with a neutron, (3.5) reduces to

$$\frac{d\sigma}{d\Omega} = \left(\frac{m}{2\pi\hbar^2}\right)^2 |e^{i\mathbf{K}\cdot\mathbf{r}} V(\mathbf{r})\, d\mathbf{r}|^2, \tag{3.6}$$

the cross section for elastic scattering. To proceed further one would have to know the form of $V(\mathbf{r})$ for which, unlike the X-ray case, there is no theory presently available. However, from experiment it is known that $V(\mathbf{r})$ is a very short range potential, so short that the scattering is completely isotropic (no form factor). This observation leads to the use of the so-called Fermi pseudopotential which, in Born approximation, gives the correct isotropic scattering, that is,

$$V(\mathbf{r}) = \frac{2\pi\hbar^2}{m}\, b\, \delta(\mathbf{r}), \tag{3.7}$$

where b has the dimensions of length from which the term "scattering length" arises. If now one inserts this form of $V(\mathbf{r})$ into (3.6), one obtains

$$\frac{d\sigma}{d\Omega} = |b|^2. \tag{3.8}$$

This gives a total cross section $\sigma = 4\pi\ |b|^2$. It must be understood that the Fermi pseudopotential (3.7) is not a true potential, but that the cross section is given correctly by (3.8). One may question the purpose of introducing the formalism of the Born approximation. The utility of this scheme becomes evident when one deals with dynamical effects in the scatterer, such as motion of the nuclei arising from thermal vibrations in a lattice.

Since, as noted, no theory is available to calculate the scattering lengths, b, they must be treated as empirical constants derived from experiment. Normally the neutron energies used are so low compared with nuclear energies that the variation of b with neutron energy is completely negligible. The scattering length as defined by (3.7) is that for a bound (stationary) nucleus. The scattering length of a free atom is given in terms of the bound length by

$$b_{\text{free atom}} = \frac{b}{[(M + m)/M]}, \tag{3.9}$$

where M is the mass of the scatterer. It should be remarked that b can be complex and also show a variation with wavelength when the absorption is large.

Now consider the scattering from a crystal containing one atom per unit cell. The $V(\mathbf{r})$ to be used in (3.5) is the sum of the individual contributions so that

$$V(\mathbf{r}) = \frac{2\pi\hbar^2}{m} \sum_{\mathbf{R}} b_{\mathbf{R}}\ \delta(\mathbf{r} - \mathbf{R}), \tag{3.10}$$

where the nuclear positions $\mathbf{R}$ are the points on a lattice. If, as in the case of the scattering from a single nucleus, we restict ourselves to nonmagnetic materials and ignore lattice vibrations (fixed nuclei), then the quantum numbers q and q' of the scatterer just refer to isotopic composition and nuclear spin orientations. Since, to a very good approximation, the energy of the crystal is independent of either of these, the scattering is, once again, elastic, and the cross section becomes

$$\frac{d\sigma}{d\Omega} = \sum_{\mathbf{R},\mathbf{R}'} e^{i\mathbf{K}\cdot(\mathbf{R}-\mathbf{R}')} \langle b^*_{\mathbf{R}'}\ b_{\mathbf{R}} \rangle, \tag{3.11}$$

where the angular brackets mean average value. The scattering length $b_{\mathbf{R}}$ depends on the isotope as well as on the nuclear spin orientation of the nucleus occupying lattice site $\mathbf{R}$. The value of $b_{\mathbf{R}}$ is uncorrelated with $b_{\mathbf{R}'}$ since the isotope distribution is random so that

$$\begin{aligned}\langle b^*_{\mathbf{R}'} b_{\mathbf{R}} \rangle &= \langle b^*_{\mathbf{R}'} \rangle \langle b_{\mathbf{R}} \rangle = |\langle b \rangle|^2 \qquad \mathbf{R}' \neq \mathbf{R} \\ &= \langle |b|^2 \rangle \qquad \mathbf{R}' = \mathbf{R}.\end{aligned} \tag{3.12}$$

Thus

$$\langle b^*_{\mathbf{R}'} b_{\mathbf{R}} \rangle = |\langle b \rangle|^2 + (\langle |b|^2 \rangle - |\langle b \rangle|^2)\, \delta(\mathbf{R} - \mathbf{R}') \tag{3.13}$$

and substitution into (3.11) gives

$$\frac{d\sigma}{d\Omega} = |\langle b \rangle|^2 \left| \sum_{\mathbf{R}} e^{i\mathbf{K}\cdot\mathbf{R}} \right|^2 + N(\langle |b|^2 \rangle - |\langle b \rangle|^2), \tag{3.14}$$

where N is the number of lattice points.

There is a very striking distinction to be noted between the two terms in the cross section expression given by (3.14). The first term, *defined as* the coherent scattering cross section,

$$\frac{d\sigma^{\text{coh}}}{d\Omega} = |\langle b \rangle|^2 \left| \sum_{\mathbf{R}} e^{i\mathbf{K}\cdot\mathbf{R}} \right|^2, \tag{3.15}$$

contains an interference term ($|\sum_{\mathbf{R}} e^{i\mathbf{K}\cdot\mathbf{R}}|^2$), and is proportional to the average scattering length. The second term in (3.14) is isotropic with no interference effects. This term, the so-called "incoherent scattering," depends on the deviations of the scattering length from the mean, that is,

$$\langle |b|^2 \rangle - |\langle b \rangle|^2 = \langle |b - \langle b \rangle| \rangle^2. \tag{3.16}$$

The values of $\langle |b|^2 \rangle$ and $|\langle b \rangle|^2$ depend on the isotopic composition as well as on nuclear spin. Consider first the case in which the nuclear spin is zero, and label the isotopes with subscripts i and the concentration of the ith isotope with c_i. We then have $\langle b \rangle = \sum_i c_i b_i$ and $\langle |b|^2 \rangle = \sum_i c_i |b_i|^2$. Now reverse the situation and consider the case of a single isotope, but with nuclear spin I. Since the neutron has a spin of $\frac{1}{2}$, the total spin of the interacting system is either $I + \frac{1}{2}$ or $I - \frac{1}{2}$. These states will have associated scattering lengths of b^+ and b^-. According to a familiar quantum mechanical rule, there are $2J + 1$ states of angular momentum J and therefore $2I + 2$ states of spin $I + \frac{1}{2}$ and $2I$ states of spin $I - \frac{1}{2}$. The probability that the state will have spin $I + \frac{1}{2}$ is thus

$$\frac{2I + 2}{(2I + 2) + 2I} = \frac{I + 1}{2I + 1}, \tag{3.17}$$

and for the state $I - \frac{1}{2}$ the probability is $I/(2I + 1)$. The expression for the average scattering length for a single isotope with spin I is therefore

$$\langle b \rangle = \frac{I+1}{2\ \ +1} b^{+} + \frac{I}{2I+1} b^{-} \tag{3.18}$$

and

$$\langle |b|^2 \rangle = \frac{I+1}{2I+1} |b^{+}|^2 + \frac{I}{2I+1} |b^{-}|^2. \tag{3.19}$$

The extension to the situation of several isotopes possessing nuclear spins is straightforward, and the average scattering length is given by

$$\langle b \rangle = \sum_i c_i \left(\frac{I_i + 1}{2I_i + 1} b_i^{+} + \frac{I_i}{2I_i + 1} b_i^{-} \right) \tag{3.20}$$

and

$$\langle |b|^2 \rangle = \sum_i c_i \left(\frac{I_i + 1}{2I_i + 1} |b_i^{+}|^2 + \frac{I_i}{2I_i + 1} |b_i^{-}|^2 \right). \tag{3.21}$$

The presence of either spin or isotope effects gives rise to incoherent scattering. A rather striking example of this occurs in the case of hydrogen, with $I = \frac{1}{2}$ and $b^{+} = 1.07 \times 10^{-12}$ cm and $b^{-} = -4.68 \times 10^{-12}$ cm. These values are taken from the latest compilation [4] of critically evaluated scattering lengths published by the Commission on Neutron Diffraction of the International Union of Crystallography. Substituting into (3.20) and (3.21) gives

$$\begin{aligned}
\langle b \rangle &= \tfrac{3}{4} \times 1.07 \times 10^{-12} - \tfrac{1}{4} \times 4.68 \times 10^{-12} = -0.367 \times 10^{-12}, \\
\sigma_{\mathrm{coh}} &= 4\pi\, |\langle b \rangle|^2 = 1.78 \times 10^{-24}\ \mathrm{cm}^2 \text{ or } 1.78 \text{ barns}, \\
\langle |b|^2 \rangle &= (\tfrac{3}{4} \times |1.07 \times 10^{-12}|^2 + \tfrac{1}{4}\, |4.68 \times 10^{-12}|^2) = 6.334 \times 10^{-12}, \\
\sigma_{\mathrm{incoh}} &= 4\pi \langle |b|^2 \rangle = 79.6 \text{ barns.}
\end{aligned} \tag{3.22}$$

In this case the scattering lengths for the two spin states have opposite signs. In most cases the scattering length or amplitude is positive (corresponding to a 180° phase shift of the scattered beam relative to the incoming one), but in a few cases where there is a nuclear resonance level close by, the amplitude is negative.

Coherent Scattering

The intensity of neutrons coherently scattered from a crystal in an arbitrary direction is given by the square of the amplitude of the resultant wave obtained by summing up the amplitudes of the wavelets coming from the individual scattering centers. As seen from (3.15), the contribution of an individual scatterer is $be^{i\mathbf{K}\cdot\mathbf{R}}$, and the total scattering amplitude from a crystal made up of N unit cells with one atom per unit cell is $\sum_{n=1}^{N} be^{i\mathbf{K}\cdot\mathbf{R}_n}$.

This sum has its exact counterpart in X-ray scattering and leads to interference maxima or Bragg peaks. There are N terms in this sum, and for all except certain special values of $\mathbf{K}$ each term will have a different phase, some being positive and some negative, leading to a cancellation. For N large this cancellation is almost perfect. For the special values of the momentum transfer $\mathbf{K}$, however, all the phases will be the same, namely zero, and the sum will equal Nb. The condition for all the phases to be zero can be treated best in terms of the reciprocal lattice. A reciprocal lattice vector $\boldsymbol{\tau}$ is defined by the condition

$$e^{2\pi i \boldsymbol{\tau}\cdot\mathbf{R}_n} = 1 \qquad \text{for} \quad \text{all } n. \tag{3.23}$$

The vector so defined is perpendicular to a set of planes in the original or direct lattice, and its magnitude is n/d_{hkl}, where n is an integer and d_{hkl} is the interplanar spacing of a set of planes with Miller indices hkl. In terms of the reciprocal lattice the condition for an interference maximum then becomes

$$\mathbf{K} = 2\pi\boldsymbol{\tau}. \tag{3.24}$$

For $\mathbf{K} \neq \boldsymbol{\tau}$, the sum $\sum_{n=1}^{N} be^{i\mathbf{K}\cdot\mathbf{R}_n}$ rapidly approaches zero as the number of unit cells increases.

Equation 3.24 is a less familiar form of the well-known Bragg condition for a diffraction maximum. It can be transformed into its usual form by noting that $K = 2k \sin\theta$, where 2θ is the angle between the incoming and the outgoing wave vectors, that is, the scattering angle (see Fig. 3.1), $\tau = n/d_{hkl}$ and $k = 2\pi/\lambda$. Substitution into (3.24) then gives

$$n\lambda = 2d_{hkl}\sin\theta, \tag{3.25}$$

which is the usual form of Bragg's law.

The discussion of coherent scattering until now has been restricted to a system with one atom per unit cell. For the case of complex structures with more than one atom per unit cell the expression for the scattering amplitude from an individual scatterer, $be^{i\mathbf{K}\cdot\mathbf{R}}$, must be generalized to include all the scattering from a single repeat unit, that is, the unit cell, taking into account the phase relations among the individual scatterers in the unit cell. Consider the situation of a crystal having a unit cell volume V_0 wth N unit cells, each containing j atoms with position vectors $\mathbf{r}_1, \mathbf{r}_2, \ldots, \mathbf{r}_j$ measured relative to some origin in the unit cell. Then the scattering amplitude for the unit cell, $F(\boldsymbol{\tau})$, is given by

$$F(\boldsymbol{\tau}) = \sum_{s=1}^{j} \langle b_s\rangle e^{2\pi i\boldsymbol{\tau}\cdot\mathbf{r}_s}. \tag{3.26}$$

$F(\boldsymbol{\tau})$ is just the familiar structure factor which appears in the X-ray case as well. The cross section for coherent scattering (3.15) then becomes

$$\frac{d\sigma^{\text{coh}}}{d\Omega} = \left|\sum_{\mathbf{R}} e^{i\mathbf{K}\cdot\mathbf{R}}\right|^2 |F(\boldsymbol{\tau})|^2$$
$$= \frac{2\pi^3}{V_0} N \sum_{\boldsymbol{\tau}} \delta(\mathbf{K} - 2\pi\boldsymbol{\tau})\,|F(\boldsymbol{\tau})|^2, \tag{3.27}$$

which is also the standard X-ray result.

Magnetic Scattering

Cross Section

The interaction potential, $V(\mathbf{r})$, which appears in the cross section formula given by (3.5), has a part that is magnetic in origin. The neutron has a magnetic moment and is thus capable of interacting with atoms possessing permanent magnetic moments. The theory of neutron scattering by such atoms requires the use of the Dirac equation and has been given in detail by Halpern and Johnson [5]. When the atomic magnetic moment arises solely from spin [6] the *magnetic* scattering length, p, is given by

$$p = \frac{2e^2\gamma}{m_e c^2} f(\mathbf{K})\{\mathbf{s}\cdot\mathbf{S} - (\mathbf{s}\cdot\hat{\mathbf{K}})(\mathbf{S}\cdot\hat{\mathbf{K}})\}, \tag{3.28}$$

where $\mathbf{s}$ and $\mathbf{S}$ are the neutron and atomic spin vectors, $\hat{\mathbf{K}}$ is a unit vector along $\mathbf{K}$ (^ denotes a unit vector), and $f(\mathbf{K})$ is the magnetic form factor of the spin density. The constant, $2e^2\gamma/m_e c^2$, involves the product of the nuclear and Bohr magnetons. The neutron magnetic moment is γ, and m_e is the mass of the electron. We note two differences between the nuclear and the magnetic scattering lengths. Unlike the nuclear scattering length, the magnetic scattering length can be calculated theoretically. Secondly, the magnetic amplitude is not isotropic, since it depends on the scattering angle through the form factor $f(\mathbf{K})$.

The dependence on the scattering angle arises from the size of the scattering center which, in the magnetic case, is comparable with the wavelength of the neutrons. As a result there is a considerable difference in path length, hence degree of destructive interference, as one varies the angle. Referring to Fig. 3.2, the difference in phase between P_1P_2 and $P_1'P_2'$ is given by $2\pi/\lambda \times$ (path difference) $= 4\pi r \sin\theta/\lambda$, and thus it depends on the scattering angle. The form factor, $f(\mathbf{K})$, just as in the X-ray case, is given by the Fourier transform of the scattering density

$$f(\mathbf{K}) = \int e^{i\mathbf{K}\cdot\mathbf{r}}\,|\psi(\mathbf{r})|^2\,d\mathbf{r}, \tag{3.29}$$

where $\psi(\mathbf{r})$ is the wave function for the unpaired electrons. In principle one

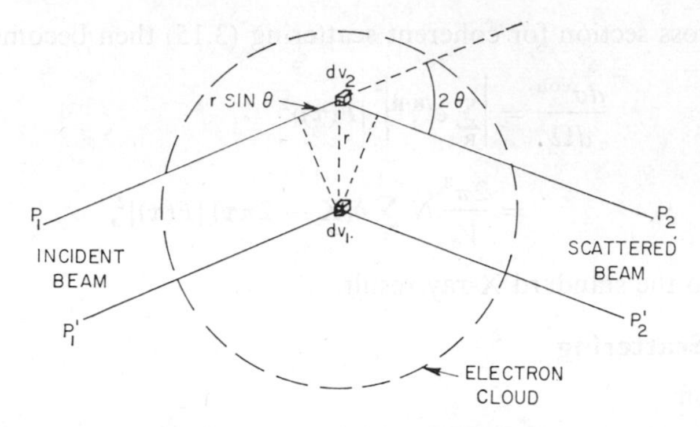

Fig. 3.2 Analysis of scattering by magnetic electrons at Bragg angle θ for two small volume elements, dv_1 and dv_2, separated by a distance r.

could calculate the form factor, but in the important case of the $3d$ transition metals $f(\mathbf{K})$ is treated as an experimental quantity, since the wave functions are not sufficiently well characterized. By Fourier-inverting (3.29), using the empirically determined $f(\mathbf{K})$, one can obtain an experimental magnetic electron distribution function.

Using the scattering length given by (3.28) and noting that there is no dependence of the energy on the neutron polarization, the magnetic scattering cross section for an unpolarized incident beam becomes, according to (3.5),

$$\frac{d^2\sigma}{d\Omega\, dE} = \left(\frac{\gamma e^2}{m_e c^2}\right)^2 \frac{k'}{k} \sum_{\alpha,\beta} (\delta_{\alpha\beta} - \hat{K}_\alpha \hat{K}_\beta) \sum_{\substack{q,q' \\ \mathbf{n},\mathbf{m}}} P_q\, e^{i(\mathbf{K}\cdot\mathbf{n}-\mathbf{m})} |f_{\mathbf{m}}(\mathbf{K})|\,|f_{\mathbf{n}}(\mathbf{K})|$$
$$\times \langle q|\, S_{\mathbf{m}}{}^{\alpha}\, |q'\rangle\langle q'|\, S_{\mathbf{n}}{}^{\beta}\, |q\rangle\, \delta\,(\text{energy}), \quad (3.30)$$

where α and β are to be summed over a set of Cartesian coordinate directions and $\mathbf{m}$ and $\mathbf{n}$ are the position vectors of the scatterers. This is the general expression for the magnetic scattering cross section.

Coherent Scattering

Consider the case of elastic scattering from a system in which all the moments are aligned parallel to a given direction $\hat{\boldsymbol{\varkappa}}$, for example, a ferromagnet. If one defines a vector $\mathbf{q}$

$$\mathbf{q} = \hat{\boldsymbol{\varkappa}} - \hat{\boldsymbol{\epsilon}}(\boldsymbol{\epsilon} \cdot \hat{\boldsymbol{\varkappa}}); \qquad |\mathbf{q}|^2 = \sin^2 (\hat{\boldsymbol{\varkappa}}, \hat{\boldsymbol{\epsilon}}), \quad (3.31)$$

where $\hat{\boldsymbol{\epsilon}}$ is the unit scattering vector ($\hat{\boldsymbol{\epsilon}}$ is equivalent to $\hat{\mathbf{K}}$), the magnetic scattering amplitude (3.28) expressed in terms of this vector becomes

$$p = \frac{\gamma e^2}{m_e c^2} S f(\mathbf{K})\mathbf{q} = p'\mathbf{q}$$
$$p' = 0.539 S f(\mathbf{K}) \times 10^{-12}\ \text{cm}, \tag{3.32}$$

where S is the magnitude of the spin vector. If the direction of polarization of the neutron is $\hat{\boldsymbol{\lambda}}$, the total scattering length, which is the sum of the nuclear and magnetic contributions, can be written as $b + p'\mathbf{q}\cdot\hat{\boldsymbol{\lambda}}$. The coherent cross section for a system with one magnetic atom per unit cell is then

$$\frac{d\sigma^{\text{coh}}}{d\Omega} = \frac{(2\pi)^3}{V_0} N \sum_{\boldsymbol{\tau}} \delta(\mathbf{K} - 2\pi\boldsymbol{\tau})(\langle b\rangle^2 + p'^2|\mathbf{q}|^2 + 2\langle b\rangle p'\hat{\boldsymbol{\lambda}}\cdot\mathbf{q}). \tag{3.33}$$

For an unpolarized beam, which is the usual case, the last term in (3.33) averages to 0. From (3.33) it follows that the magnetic scattering for a ferromagnet consists of an independent set of magnetic Bragg peaks that is superimposed on the set produced by the nuclear scattering. One sees, in addition, that the magnitude of the magnetic Bragg scattering depends on $|\mathbf{q}|^2$ and thus on the relative orientation of the spin and scattering vectors. An unmagnetized ferromagnetic crystal consists of randomly oriented magnetic domains and in any one of these domains the spins are parallel to a given direction ($\hat{\boldsymbol{\varkappa}}$). One must therefore perform an average over the domain orientations in calculating $|\mathbf{q}|^2$. For spherical or cubic symmetry

$$\overline{|\mathbf{q}|^2} = \overline{\sin^2(\hat{\boldsymbol{\varkappa}}, \hat{\boldsymbol{\epsilon}})} = \tfrac{2}{3}. \tag{3.34}$$

When there is more than one atom per unit cell, one must calculate the scattering from the unit cell and then sum over the unit cells just as in the nuclear case. The result is exactly the same as in the nuclear case, in which the amplitude was replaced by a structure factor. The structure factor, $F(\boldsymbol{\tau})$, for a crystal with magnetic atoms is

$$F(\boldsymbol{\tau}) = \sum_{s=1}^{j} (b_s + p_s'\hat{\boldsymbol{\lambda}}\cdot\mathbf{q}_s)\, e^{2\pi i\boldsymbol{\tau}\cdot\mathbf{r}_s}, \tag{3.35}$$

where the sum is taken over the j atoms, both magnetic and nomagnetic, within the unit cell. This form of the structure factor is valid for all ordered magnetic systems, be they ferromagnetic, antiferromagnetic, or ferrimagnetic.

Paramagnetic Scattering

In an ideal paramagnet there is no interaction between the spins, and therefore the energy of the system is independent of the spin orientation. Under these circumstances the scattering is elastic, and the sum over final states, q', in the cross section formula (3.30) can be done by closure so that

$$\frac{d\sigma}{d\Omega} = \left(\frac{\gamma e^2}{m_e c^2}\right)^2 |f_{\mathrm{m}}(\mathbf{K})|\,|f_{\mathrm{n}}(\mathbf{K})| \sum_{\alpha,\beta} (\delta_{\alpha\beta} - \hat{K}_\alpha \hat{K}_\beta) \sum_{\mathrm{m,n}} e^{i\mathbf{K}\cdot(\mathbf{n}-\mathbf{m})} \sum_q P_q \times \langle q|\, S_{\mathrm{m}}{}^\alpha S_{\mathrm{n}}{}^\beta \,|q\rangle. \quad (3.36)$$

The weighted sum of matrix elements $\sum_q P_q \langle q|\, S_{\mathrm{m}}{}^\alpha S_{\mathrm{n}}{}^\delta \,|q\rangle$, is just the thermal average which we designate by

$$\langle S_{\mathrm{m}}{}^\alpha S_{\mathrm{n}}{}^\beta \rangle_T. \quad (3.37)$$

For the ideal paramagnet there is no correlation between spins so that

$$\langle S_{\mathrm{m}}{}^\alpha S_{\mathrm{n}}{}^\beta \rangle_T = \tfrac{1}{3} S(S+1)\, \delta_{\alpha\beta}\, \delta_{mn}, \quad (3.38)$$

since the quantum mechanical expectation value of S^2 is $S(S + 1)$. The cross section for paramagnetic scattering becomes

$$\frac{d\sigma^{\mathrm{para}}}{d\Omega} = \left(\frac{\gamma e^2}{m_e c^2}\right)^2 N\, |f(\mathbf{K})|^2\, \tfrac{2}{3} S(S+1). \quad (3.39)$$

There is therefore no coherent scattering since the spins are randomly oriented, but there is a dependence on angle given by the magnetic form factor.

If there is a departure from ideality in the paramagnet, an inelastic component to the scattering will result. This inelastic component to the scattering is of particular interest in the study of the properties of magnetic materials near the ordering transition, the so-called critical region. Still another type of inelastic scattering can occur if the internal state of the paramagnetic ion is changed during the scattering event. If there are low-lying magnetic states comparable to the energy of the incoming neutron, these can be excited, and the scattered neutrons will show discrete energy changes corresponding to these states.

Temperature Effects

In the previous discussions of both nuclear and magnetic scattering, the scatterer was assumed to be rigid. Because of thermal motion, however, the nuclei at any instant are displaced a small amount, $\mathbf{u}$, from their equilibrium (lattice) positions $\mathbf{R}^0$ so that the instantaneous position vector of the nth nucleus is $\mathbf{R}_n{}^0 + \mathbf{u}_n$. In order to take the thermal motion into account, the matrix element in the general expression for the cross section must be modified. If we, for example, consider the case of nuclear scattering, the appropriate matrix element becomes

$$\langle q's'|\sum_n b_n\, e^{i\mathbf{K}\cdot\mathbf{R}_m{}^0} e^{\,i\mathbf{K}\cdot\mathbf{u}_n}\, |qs\rangle. \quad (3.40)$$

In the harmonic oscillator approximation the displacements will have a Gaussian distribution about $\mathbf{R}_n{}^0$, and the elastic scattering cross section (3.14) becomes

$$\frac{d\sigma}{d\Omega} = |\langle b\rangle|^2 \left|\sum_n e^{i\mathbf{K}\cdot\mathbf{R}_n{}^0} e^{-W_n}\right|^2 + N(\langle|b|^2\rangle - |\langle b\rangle|^2)\, e^{-2W}, \qquad (3.41)$$

where the Debye-Waller factor, W, is

$$W_n = \overline{(\mathbf{K}\cdot\mathbf{u}_n)^2/2}. \qquad (3.42)$$

Weinstock [7] has shown that the Debye approximation can be used to evaluate W so that for a crystal with one atom per unit cell

$$W = \frac{6h^2}{Mk_B\Theta}\left\{\frac{\phi(X)}{X} + \frac{1}{4}\right\}\frac{\sin^2\theta}{\lambda^2}, \qquad (3.43)$$

where Θ is the Debye characteristic temperature, M is the mass of the scatterer, k_B is the Boltzmann constant, and T is the absolute temperature of the sample. The $\phi(X)$ with $X = \Theta/T$, is an integral that has been tabulated by Debye [8]. This angularly dependent reduction in intensity has an exact counterpart in X-ray scattering. It affects all scattering, nuclear as well as magnetic, elastic as well as inelastic. Thus all the expressions for the cross section that have been given must be multiplied by the factor e^{-2W} if one relaxes the condition of a rigid lattice. It will henceforth be assumed that the structure factor F always contains the Debye-Waller correction e^{-W}.

Equation 3.43 is strictly applicable only to cubic crystals containing one type of nucleus. In a complex crystal, in the harmonic approximation, each atom has an anisotropic Debye-Waller factor, and the relationship to a characteristic temperature for the crystal is not simple.

3 METHODS OF MEASUREMENT FOR BRAGG SCATTERING

Powders

The basic equations for the elastic coherent scattering cross section were presented in Section 2, but these have to be adapted to the type of specimen used as well as to the particular scheme of data collection. The significant experimental quantity to be measured is the integrated intensity of a Bragg reflection, which is defined in terms of the total power E reflected by the crystal as it rotates through the reflecting region, that is, Bragg angle θ_B, with an angular velocity ω about an axis parallel to the scattering planes. The integrated intensity, I, is then

$$I = \frac{E\omega}{I_0}, \tag{3.44}$$

where I_0 is the incident neutron flux (neutrons/unit area/unit time). If one considers a very small block of crystal, Δv, for which absorption can be neglected, then the integrated intensity is proportional to Δv:

$$I = Q\Delta v. \tag{3.45}$$

The proportionality constant Q in 3.45 is, except for the polarization factor, exactly the same as one finds in the X-ray case:

$$Q = \frac{N_0{}^2\lambda^3}{\sin 2\theta_B} |F|^2, \tag{3.46}$$

with N_0 equal to the number of unit cells per unit volume.

The appearance of an angular velocity in the integrated intensity, convenient as it may be for deriving (3.44), is nevertheless rather artificial. The integrated intensity can be defined without the use of ω as follows. Let $P(\theta)$ be the reflecting power of the very small block of crystal in the direction θ so that the total power in the scattered beam is

$$E = \int P(\theta) I_0 \frac{d\theta}{\omega} \quad \text{or} \quad \frac{E\omega}{I_0} = \int P(\theta)\, d\theta, \tag{3.47}$$

where the integration is over the reflecting region. Using (3.45) we find

$$\int P(\theta)\, d\theta = Q\Delta v, \tag{3.48}$$

and it is this form of the integrated intensity that is most useful for deriving the formulas for reflection by polycrystalline samples.

In a powder sample that contains a sufficiently large number of crystallites so that one may consider the orientations as having a continuous random distribution, the diffracted rays form a series of cones of semivertical angle $2\theta_B$ about an axis fixed by the incident beam. This is shown schematically in Fig. 3.3. A fraction, $\frac{1}{2} N \cos\theta\, d\theta$, of the total number of crystallites N will have scattering vectors so oriented as to be able to scatter into a given cone. The power in this cone is therefore $\frac{1}{2} N I_0 \bar{P}(\theta) \cos\theta\, d\theta$, where $\bar{P}(\theta)$ is the average value of the reflecting power of a crystallite. The total power I is obtained by integrating over a small range $d\theta$ in the neighborhood of θ_B. Over this interval $\cos\theta$ can be considered to be constant and equal to $\cos\theta_B$ so that we find, using (3.47), that

$$I = \tfrac{1}{2} N I_0 \cos\theta_B \int \bar{P}(\theta)\, d\theta = \tfrac{1}{2} N I_0 \cos\theta_B \overline{Q\Delta v}, \tag{3.49}$$

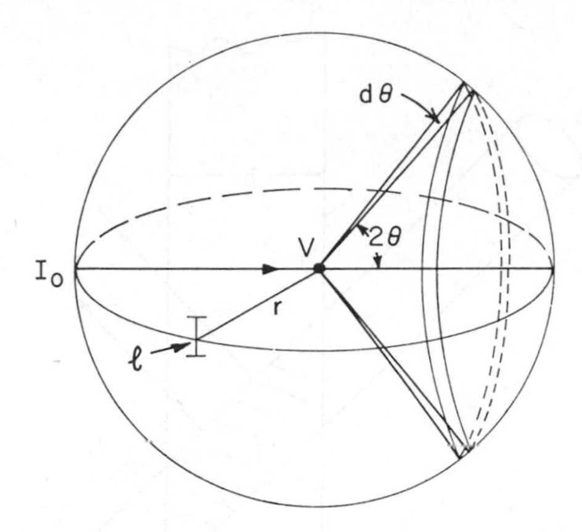

Fig. 3.3 Origin of Debye-Scherrer rings. The detector slit of height l is located at a distance r from the powder sample whose volume is V. The I_0 is the incident beam and 2θ is the scattering angle.

and finally

$$I = \tfrac{1}{2} I_0 \cos \theta_B Q V \tag{3.50}$$

since $N\overline{\Delta v}$ is the total sample volume V.

One must also consider the number of reflecting planes with different Miller indices that have the same interplanar spacing, since they all contribute to the same cone. This is called the multiplicity factor f_{hkl}. In addition, the detecting system records the intensity in a small segment of the halo. If the height of the detector slit is l at a distance r from the sample (see Fig. 3.3), the fraction of the halo that is viewed is $l/2\pi r \sin 2\theta_B$. This gives, using 3.50,

$$I = I_0 \frac{f_{hkl} l}{8\pi r \sin \theta_B} Q V. \tag{3.51}$$

The final factor to be considered is the geometry of the sample arrangement. Consider a slab of polycrystalline material of thickness h with a cross-sectional area that is large compared to that of the incident beam and is arranged for symmetrical transmission as shown in Fig. 3.4. In this configuration the total path length through the crystal (incident + scattered) of any diffracted ray is the same and equal to $h \sec \theta_B$, and the total scattering volume is $(\rho'/\rho)Ah \sec \theta_B$, where ρ' and ρ are the powder and bulk densities, respectively, and A is the cross-sectional area of the incident beam. This then gives, together with (3.51),

$$I = I_0 \frac{f_{hkl} l h}{4\pi r \sin 2\theta_B} \frac{\rho'}{\rho} e^{-\mu h \sec \theta_B} Q, \tag{3.52}$$

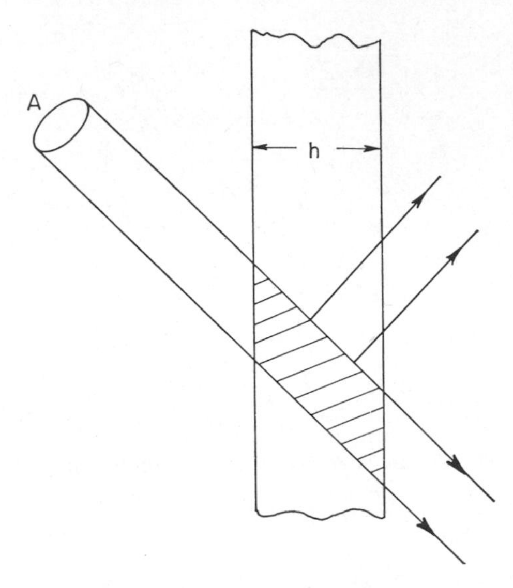

Fig. 3.4 Scattering by a slab of crystal powder of thickness h. The cross-sectional area of the incident beam is A.

where $I_0' = I_0 A$ is the incident power (neutrons/unit time) and μ is the linear absorption coefficient. Using the form of Q given in (3.46) and rearranging terms we get the final expression:

$$I_{hkl} = \left(\frac{I_0'\lambda^3 l}{4\pi r}\right)\left(N_0{}^2 h\,\frac{\rho'}{\rho}\right)\left(\frac{e^{-\mu h \sec\theta_B}}{\sin^2 2\theta_B}\right) j_{hkl}|F|^2. \tag{3.53}$$

We note that the first term in (3.53) is an instrumental constant and that the second term is a constant for a given sample.

Considerations similar to those we have just discussed lead to the following expression for the integrated intensity for a cylindrical sample that is entirely bathed by the incident beam:

$$I_{hkl} = \left(\frac{I_0\lambda^3 l}{4\pi r}\right)\left(N_0{}^2\,\frac{V}{a}\,\frac{\rho'}{\rho}\right)\left(\frac{A^*}{2\sin\theta_B\sin 2\theta_B}\right) f_{hkl}|F|^2. \tag{3.54}$$

The V is the volume of the cylinder and a is its cross-sectional area. The transmission factor A^* is a tabulated function [9] of the linear absorption coefficient.

Single Crystals

The integrated intensity is, as has been noted, the same as that given for the X-ray case with the polarization term omitted. Thus for example, for the case where the single crystal is rotated about an axis making an angle ϕ with the reflecting planes (the rotating crystal method) [10],

$$I = \frac{N_0^2 \lambda^3 \cos\theta}{\sin^2\theta(\cos^2\phi - \sin^2\theta)^{1/2}} A\,|F|^2 V, \tag{3.55}$$

which reduces to

$$I = \frac{N_0^2 \lambda^3}{\sin 2\theta} A\,|F|^2 V \tag{3.56}$$

when the axis of rotation is parallel to the reflecting plane.

Extinction

The preceding discussion of the integrated intensity was based on the assumption that there is no interaction between the incident and the scattered radiation and that there is no multiple scattering. These conditions will be satisfied if the perfect crystal region is sufficiently small, as is the case, for example, with fine powders. It is also true for a thin ideal mosaic single crystal. The ideal mosaic crystal model as an approximation for real crystals was first introduced by Darwin [11]. He assumed that a real crystal was an aggregate of perfect crystal domains with boundaries that represented discontinuities in the periodicity. These domains are assumed to be slightly misoriented with respect to one another, and thus there is no coherence between the scattering from any two domains. If the crystal contains a large number of domains, one may assume that the misalignment is given by a Gaussian distribution about the mean direction. With these assumptions, Darwin showed that the intensity depends on the mean radius, r, and the mosaic spread parameter, η (η is the standard deviation of the Gaussian distribution) of the perfect crystal regions (mosaic blocks). Experience shows that most crystals are characterized by $r \sim 10^{-4}$ cm or less, and η is characterized by the order of 1 to 10 minutes of arc.

In the ideal mosaic crystal we must distinguish between two forms of extinction: primary and secondary. Primary extinction refers to the reduction in the incident beam power within a single mosaic block because of Bragg scattering in that block. Secondary extinction is the decrease in power of the incident beam striking a given mosaic block arising from coherent scattering by similarly oriented blocks previously traversed by the incident beam.

For most real crystals primary extinction can be neglected in the neutron case. This is not true, however, of secondary extinction, which is far more important with neutrons than with X-rays. The depth of penetration of the incident beam is limited by high absorption with X-rays. Secondary extinction is consequently small, if present at all. Linear-absorption coefficients for neutrons with wavelengths of about 1 Å are considerably smaller than those for X-rays, typical values being about 0.3 cm^{-1} for neutrons and 100 cm^{-1} for X-rays. Thus the neutron beam may be expected to reach to a much greater depth, and adequate account must be taken of the shielding of the inner mosaic blocks by those nearer the surface.

Zachariasen [12] has analyzed the problem of secondary extinction in absorbing mosaic crystals and has developed a universal formula for the integrated intensity, which is valid over the entire range—from the perfect crystal to the ideal mosaic crystal. A summary of the Zachariasen treatment together with an extension to the case of anisotropic extinction has been given by Coppens and Hamilton [13].

4 APPARATUS

Two techniques have been used for obtaining diffraction data. The first of these uses a crystal to produce a well-collimated, nearly monochromatic beam by Bragg scattering from the impinging reactor spectrum. The other method uses the entire spectrum, which is chopped to produce a well-defined burst of polychromatic neutrons, and the times for these neutrons to travel from the source of the burst to a detector are measured. This so-called "time-of-flight" technique has rarely been used for structural studies, and we shall therefore concentrate on the crystal-diffraction method. An excellent discussion of time-of-flight techniques appears in an article by Brugger [14].

The crystal spectrometer has three main sections: a monochromator for producing an incident beam, a sample table with room for auxiliary equipment such as dewars, magnets, furnaces, and so on, and a detector-recorder system for scanning the diffracted radiation. In general, because of the intense radiation coming from the reactor and the comparatively large beam sizes used, neutron spectrometers are massive compared with equivalent X-ray units. Figure 3.5 is a diagram of a spectrometer in use at the High Flux Beam Reactor (HFBR) at Brookhaven National Laboratory.

The in-pile section of the spectrometer has a rotating drum that has four fixed positions which can be set remotely. In three of these positions there are collimators of various angular resolution (10′, 20′, and 40′), while the fourth position interposes a 36-in.-long block of steel that acts as a shutter for the main beam. The monochromator section forms the central cylindrical core of a large rotating shield weighing about six tons. The monochromator mount allows three independent motions of the crystal: rotation, translation, and tilt. These motions are controlled manually from the top of the monochromator assembly. The entire assembly is contained within another concentric rotating cylinder that, in turn, is coupled by means of a 2:1 angle divider to the shield. Thus once the monochromator has been oriented, rotation of the shield in order to change wavelength requires no readjustment of the monochromator since the Bragg condition is automatically satisfied at the new wavelength. Provision is made for attaching an optical encoder to the outer rotating cylinder of the monochromator housing, so that the angular setting of the monochromator and hence the wavelength may be recorded.

The sample table as well as the detector arm are supported by a cantilever arm that is attached to the rotating shield. The center of rotation of the sample table is located rather precisely along the center line of the exit collimator in the rotating shield. This arrangement permits a change in wavelength without the necessity of realigning the sample table. The entire sample table-detector combination can be translated parallel to the exit beam to provide a variable monochromator-sample distance. The detector arm rotates about the center of rotation of the sample table. Both the sample and the detector arm have individual optical encoders for reading angular positions.

A number of these spectrometers at the HFBR have provision for energy analysis of the scattered beam. Energy analysis is accomplished by mounting an additional arm on the existing diffractometer arm, as shown in Fig. 3.6. This is then known as a "three-crystal spectrometer," and it finds extensive use in the study of inelastic scattering phenomena.

The choice of monochromator crystal is governed by several factors that influence the overall efficiency. In general, the mosaic spread of the monochromator must be comparable to the angular resolution of the collimation system. If the mosaic spread is much less than the angular resolution, then the monochromator is not capable of scattering all those neutrons that are allowed by the geometry of the arrangement. On the other hand, if the mosaic spread is too large, then the wavelength spread of the nearly monochromatic beam is increased. Other considerations involve the cross section for elastic scattering, which should be high, and a low absorption factor to ensure a high efficiency. In most cases these conditions are best satisfied with metal crystals. Ionic crystals such as NaCl or KBr are usually too perfect, having mosaic spreads of a few minutes of arc. At the HFBR we have been successful in deforming very perfect germanium crystals so that the final mosaic spreads range from 5 to 25 min. A very desirable feature is associated with the use of germanium; this is related to its crystal structure. Germanium has the diamond structure so that if one uses the (111) as the reflecting plane, the second order ($\lambda/2$) contamination is eliminated since the (222) is a forbidden reflection. Copper, lead, magnesium, aluminum, and beryllium have also been used successfully as monochromators. Recently large and reasonably perfect ($\sim 0.5°$) graphite crystals [15] have become available; they seem suited for use as monochromators under special circumstances.

The most widely used detector continues to be the BF_3 proportional counter. These counters have efficiencies of 70% or more and are extremely

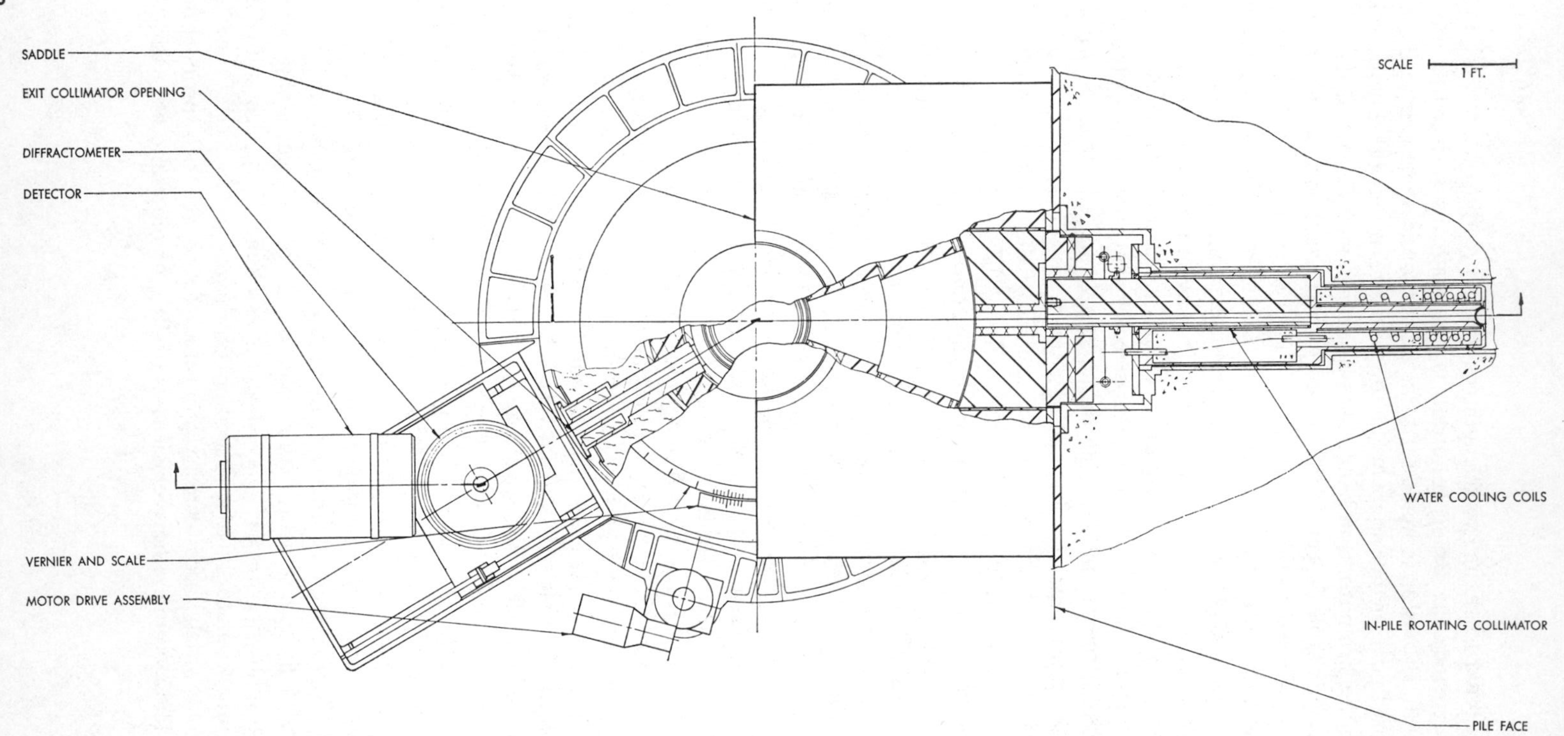

Fig. 3.5 Schematic diagram of 2-axis spectrometer in use at the Brookhaven High Flux Beam Reactor.

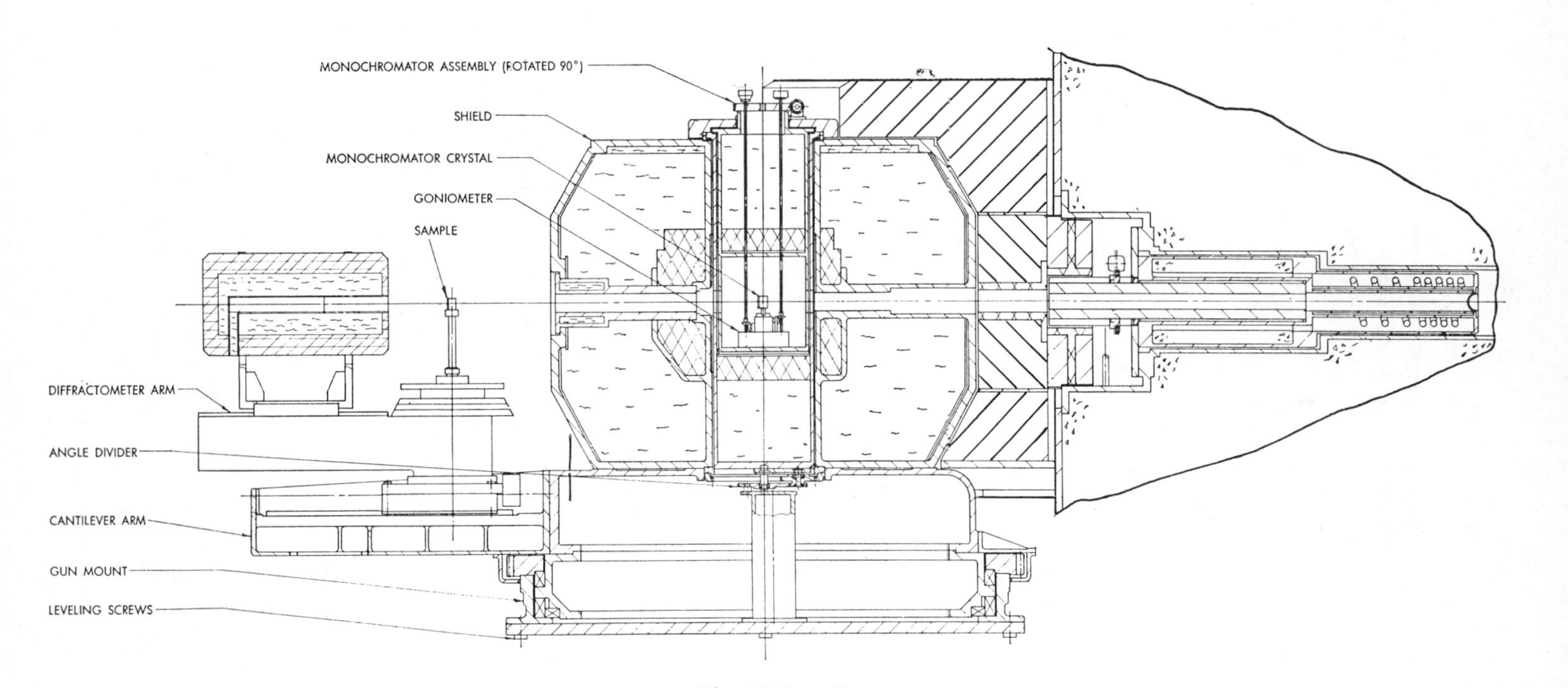
MONOCHROMATOR ASSEMBLY (ROTATED 90°)
SHIELD
MONOCHROMATOR CRYSTAL
GONIOMETER
SAMPLE
DIFFRACTOMETER ARM
ANGLE DIVIDER
CANTILEVER ARM
GUN MOUNT
LEVELING SCREWS

Fig. 3.5 *(contd.)*

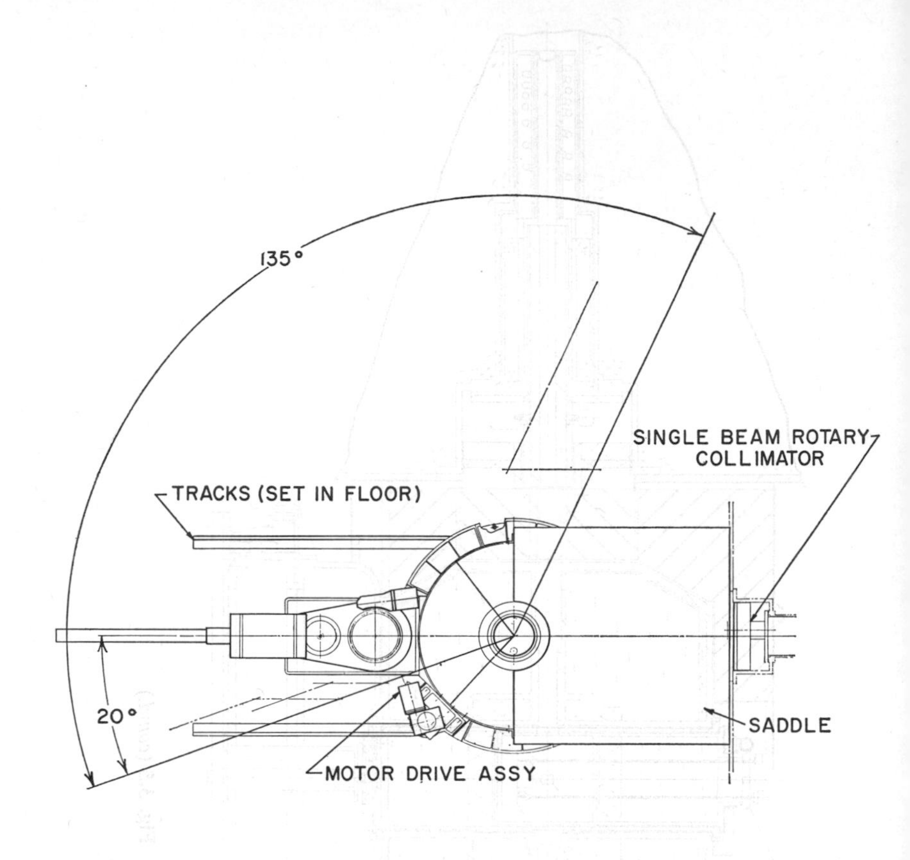

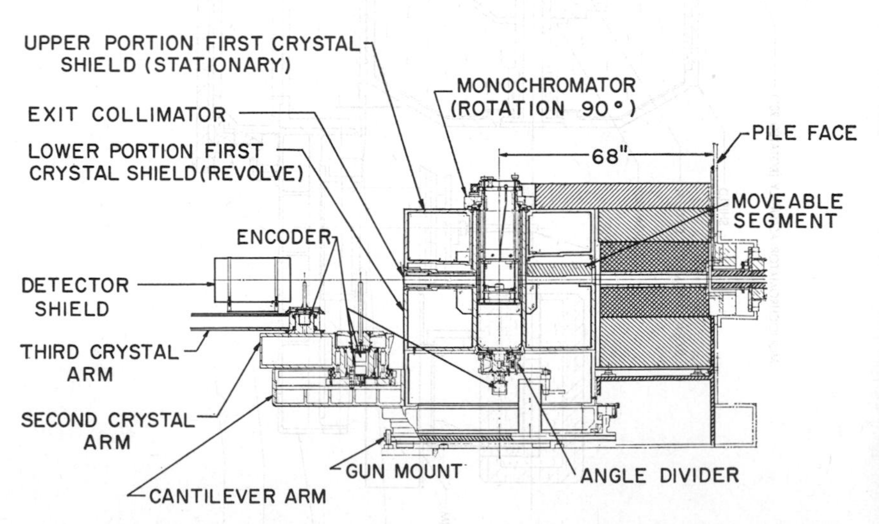

Fig. 3.6 Schematic diagram of 3-axis spectrometer installed at the Brookhaven HFBR.

reliable. The main objection is their bulk, which necessitates a large, heavy shield. Attempts have been made to use lithium-doped ZnS scintillators and other types of small solid-state detectors, but these have all failed either because the efficiency was too low or because the background (from gamma rays) was too high.

The final component, the recording system, has undergone continuous improvement, paralleling very closely the changes that have also taken place in the X-ray case. Almost all neutron spectrometers are now computer controlled. At the HFBR a medium-sized computer (SDS-920) has been installed to operate, control, and monitor nine spectrometers [16]. This system provides a very flexible scheme for data collection and simple data reduction.

5 STRUCTURE DETERMINATION BY NEUTRON DIFFRACTION

The determination of crystal structures by neutron diffraction is in most respects analogous to the determination by X-ray diffraction. The geometries of the experiments, the methods for determining the trial structures, and the refinement procedures are identical in principle. The experiments may differ in detail, and structure solution and refinement may exhibit some interesting variation because of the differences in scattering amplitudes for the two radiations.

Geometry of the Experiments

Most neutron-diffraction data for structural work are collected on a four-circle diffractometer, typical of those used in X-ray diffraction work. X-ray diffractometers are in fact often adapted for neutron-diffraction work. A number of special diffractometers have also been built; a typical example is shown in Fig. 3.7; the large χ circle permits the convenient mounting of cryostats and magnets for special experimental problems. A cryostat for the collection of three-dimensional diffraction data down to liquid helium temperature is shown in Fig. 3.8.

The data are generally collected as the crystal is moved into diffracting position in a monochromatic neutron beam, and the integrated intensity is obtained by a theta–two theta scan using a single BF_3 detector. The data collection procedure is usually automated by a special-purpose programmer or by interfacing to a digital computer [16].

The use of the Laue technique for the collection of intensity data is not common in X-ray diffraction experiments but may be more attractive in neutron-diffraction experiments. Although the technique is not now widely used in neutron diffraction, that it is a method of considerable power and

Fig. 3.7 Single-crystal neutron-diffractometer at the Brookhaven HFBR. The highly shielded diffracted beam monochromator crystal may be used in inelastic scattering experiments or to separate the purely elastic scattering from the diffracted beam. The crystal in the sample position is much larger than the 1 mm^3 normally used.

accuracy has been demonstrated [17]. In the Laue technique a monochromatic incident beam is not used; rather a broad band of wavelengths is allowed to fall on the crystal and to diffract. Many more Bragg reflections can occur for a single crystal setting than if a single wavelength is used (Fig. 3.9). Analysis of the diffracted beams may be carried out by time-of-flight techniques to resolve the diffracted intensity at a single angle into reflections arising from various incident wave lengths and Bragg reflections.

The Debye–Scherrer powder diffraction method is also often used for simple problems with good results. Again, the techniques are identical to those used in X-ray work. A polychromatic beam with time-of-flight analysis can also be used in this case [18].

In the typical monochromatic neutron beam used with a four-circle diffractometer, the incident monochromatic (to 1 to 2%) neutron flux may be on the order of 10^6 to 10^8 neutrons-cm^{-2}-sec^{-1}. This is a few orders of magnitude smaller than the typical flux of photons from an X-ray diffraction tube. A typical value is 10^9 to 10^{10} photons-cm^{-2}-sec^{-1}. Sample sizes required to obtain good counting statistics are thus considerably larger for neutron-diffraction work. At the Brookhaven National Laboratory High-Flux Beam Reactor (40 mW, 7×10^{14} total thermal flux at the end of the beam tubes) single crystals with a volume of 1 mm^3 are sufficient for good intensity data. With a crystal of this size the intensities of 300 reflections per day can be

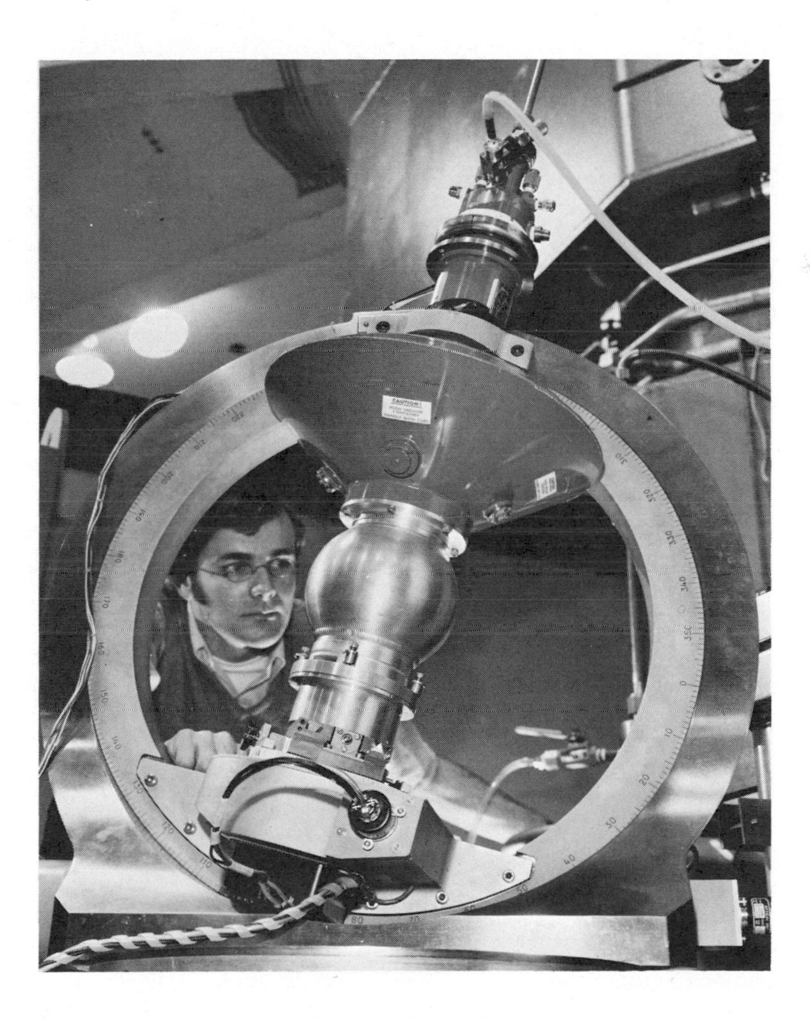

Fig. 3.8 Orientable cryostat used for single-crystal data collection at temperatures down to 4.2° K.

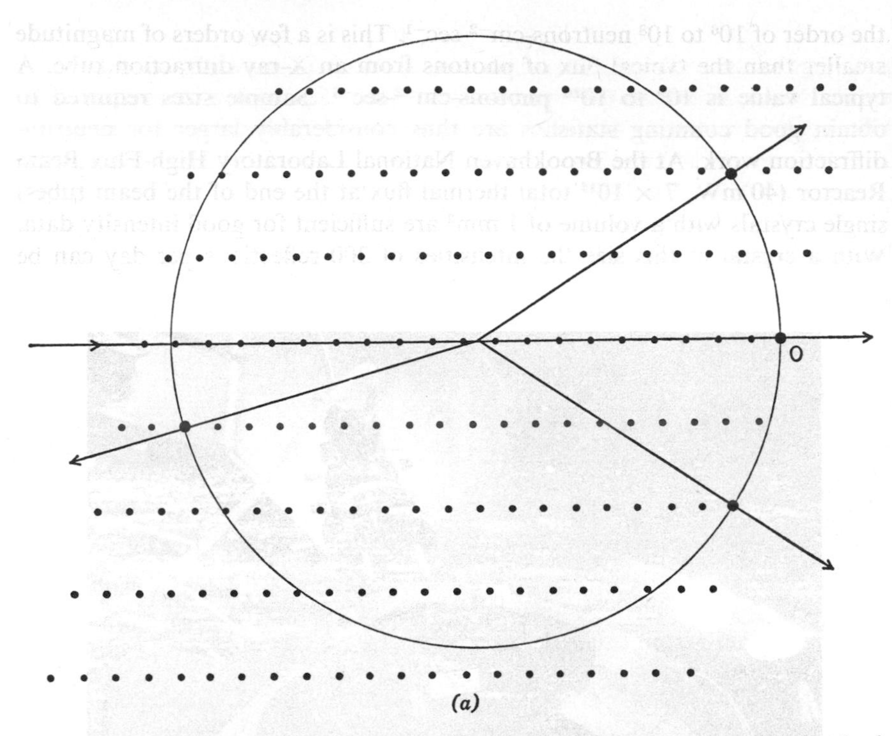

Fig. 3.9 (*a*) Geometry of a diffraction experiment when only one wavelength is involved. The direction of the incident beam is from the left through the center of the circle of radius $1/\lambda$ through the origin of the reciprocal lattice 0. Diffraction occurs only for reciprocal lattice points that lie on the circle (a sphere—the Ewald sphere—in three dimensions). Only a few reflections can occur at once, these in different directions, as indicated by the arrows that give the directions of the diffracted beams. (*b*) Geometry of the diffraction experiment when a band of wavelengths are involved: the Laue method. Wavelengths corresponding to all circles (spheres) lying within the shaded area and passing through the origin 0 are present. All reciprocal lattice points in the shaded area simultaneously diffract, and many in approximately the same direction. (All the diffracted beams are not shown; each originates at the center of a circle passing through the origin and the appropriate reciprocal lattice point.)

measured with sufficient accuracy for a high-quality structural investigation. The time required for an experiment is thus comparable to that required for an X-ray diffraction experiment, provided that good-quality crystals of the requisite size are obtained. Similar data could be obtained in an X-ray diffraction experiment on a crystal 1000 times as small.

As in any diffraction experiment, reliable quantitative results depend upon experiments that have been carried out in such a way as to minimize systematic errors. Although many of the same sources of error exist in X-ray

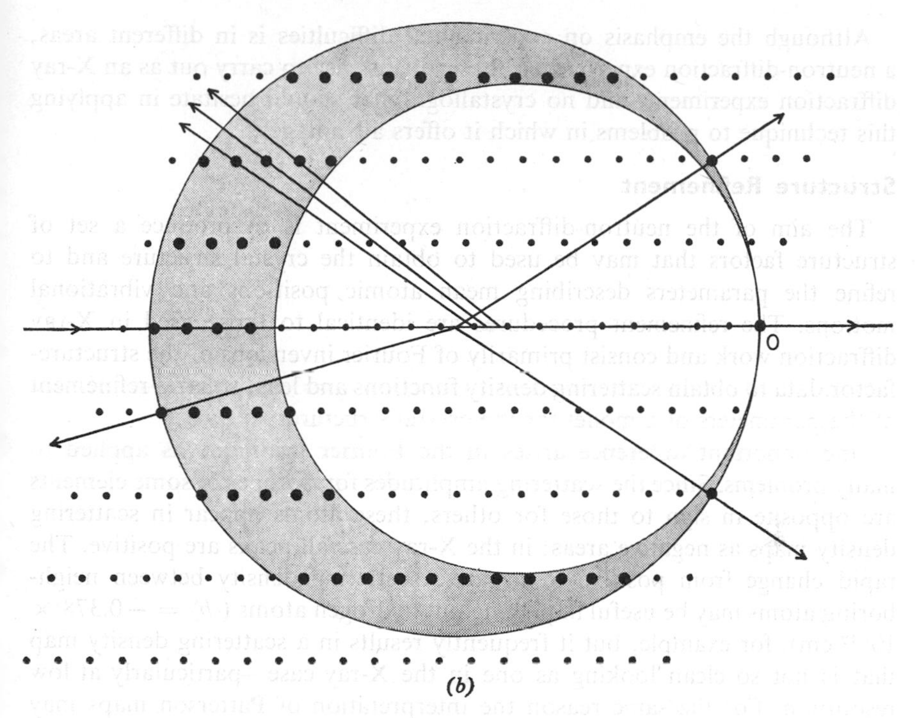

(*b*)

Fig. 3.9 (*contd.*)

and neutron-diffraction experiments, their importances may differ widely between the two techniques. The larger size of crystal required for neutron diffraction usually means that more care must be taken with crystal growth and assessment of crystal perfection. The crystal size, coupled with the relatively large range of wavelengths in the incident beam and in the beam divergence, require that great care be taken so that the same volume of crystal is interacting with a uniform incident flux for every Bragg reflection [19].

Because the neutron energy is comparable to vibrational energy differences in crystals, the cross section for inelastic scattering is large. The inelastic scattering can occur with substantial intensity at the same scattering angle as the Bragg reflections. Correction for this effect is difficult for any but the simplest crystals. Since it may represent an important source of systematic error, it should be considered in every careful neutron-diffraction experiment [20].

As noted in Section 3, primary and secondary extinction can drastically affect the proportionality between the measured intensity and the square of the structure factor.

Although the emphasis on experimental difficulties is in different areas, a neutron-diffraction experiment is basically as easy to carry out as an X-ray diffraction experiment, and no crystallographer should hesitate in applying this technique to problems in which it offers advantages.

Structure Refinement

The aim of the neutron-diffraction experiment is to produce a set of structure factors that may be used to obtain the crystal structure and to refine the parameters describing mean atomic positions and vibrational motions. The refinement procedures are identical to those used in X-ray diffraction work and consist primarily of Fourier inversions of the structure-factor data to obtain scattering density functions and least-squares refinement of the parameters of a model for the crystal structure.

One important difference arises in the Fourier technique as applied to many problems. Since the scattering amplitudes for isotopes of some elements are opposite in sign to those for others, these atoms appear in scattering density maps as negative areas; in the X-ray case all peaks are positive. The rapid change from positive to negative scattering density between neighboring atoms may be useful in picking out hydrogen atoms ($\langle b \rangle = -0.378 \times 10^{-12}$ cm), for example, but it frequently results in a scattering density map that is not so clean looking as one in the X-ray case—particularly at low resolution. For the same reason the interpretation of Patterson maps may be more complex [21].

For many crystals the peaks in neutron-diffraction scattering density maps tend to be sharper than those in X-ray diffraction maps; the lack of a scattering-factor decrease with increasing angle results in a transform that is nearly like a δ function. It is of course modified by thermal motion, but the peaks in a map based on a complete set of data should give an accurate picture of the thermal motion alone—separated from the smearing due to the diffuse electron cloud in the X-ray diffraction case.

Least-squares refinement is usually the last step in a crystal-structure determination. One of the results of such refinement is a set of estimated variances for the parameters of the refined model. These estimates of precision in any kind of experiment do not reveal systematic errors that affect the true accuracy; they are nevertheless universally quoted as estimated probable errors in the refined parameters. The precision of good neutron and X-ray diffraction experiments as assessed in this way is typically 0.003 Å in atomic positions and in root-mean-square amplitudes of thermal vibration. The precision will of course be better for atoms with larger scattering amplitudes. For this reason the precision of hydrogen-atom determination by X-ray diffraction is not good; typically 0.1 Å in the best work. In neutron diffraction, as is discussed below, the precision of hydrogen-atom description approaches that for other atoms in typical medium-sized molecules.

Thus we see that neutron-diffraction data can be collected in times comparable with X-ray diffraction data and that results of comparable precision may be obtained. The experiments—because of the need for nuclear reactors—are considerably more expensive, and one must therefore ask which problems justify such expense. What can be learned that cannot be learned in an X-ray experiment? Several examples are given below.

Special Techniques in Structure Determination by Neutron Diffraction

By structure solution we understand the obtaining from the diffraction data of a structural model close enough to the true structure that an iterative, nonlinear, least-squares procedure leads to estimated true values for all the parameters.

As noted above, the techniques for structure solution by neutron diffraction are identical in principle to those for X-ray diffraction. Of course some differences in emphasis arise because of the differences in scattering amplitudes.

In X-ray diffraction studies the *heavy-atom* technique is widely used. If a few atoms dominate the scattering, they may be easily located by Patterson or *ad hoc* methods. The phases of the structure factors may be well-determined by the contributions of these atoms, and the remaining atoms are found by calculation of a difference Fourier synthesis,

$$\Delta\rho(\mathbf{r}) = \sum_{\boldsymbol{\tau}} [F_{\text{obs}}(\boldsymbol{\tau}) - F_{\text{heavy}}(\boldsymbol{\tau})]e^{-2\pi i\boldsymbol{\tau}\cdot\mathbf{r}}, \qquad (3.57)$$

where the phases of $F_{\text{heavy}}(\boldsymbol{\tau})$ are assigned to $F_{\text{obs}}(\boldsymbol{\tau})$. The technique is less applicable in neutron diffraction, for amplitudes for neutron scattering do not show such a wide variation as amplitudes for X-ray scattering; hence few structures contain a single or a very few atoms that dominate the scattering. In a special sense the heavy-atom approach is used in neutron-diffraction work. In many studies the positions of the heavier atoms in a structure are known from prior X-ray diffraction work. Difference Fourier syntheses have then been used to reveal hydrogen-atom positions; thus

$$\rho_{\text{hydrogen}}(\boldsymbol{\tau}) = \sum_{\boldsymbol{\tau}} [F_{\text{obs}}(\boldsymbol{\tau}) - F_{\text{heavy}}(\boldsymbol{\tau})]e^{-2\pi i\boldsymbol{\tau}\cdot\mathbf{r}}. \qquad (3.58)$$

This technique has been used with good results, even in cases where the hydrogen atoms contribute more than 50% of the scattering.

The use of the so-called direct methods of crystal-structure determination (the derivation of phases from the intensity data by probability methods) has spread widely in the past few years. Most of the non-heavy-atom X-ray structures presently being reported have been solved in this way. The method has also been used in some neutron-diffraction problems [22, 23]. One

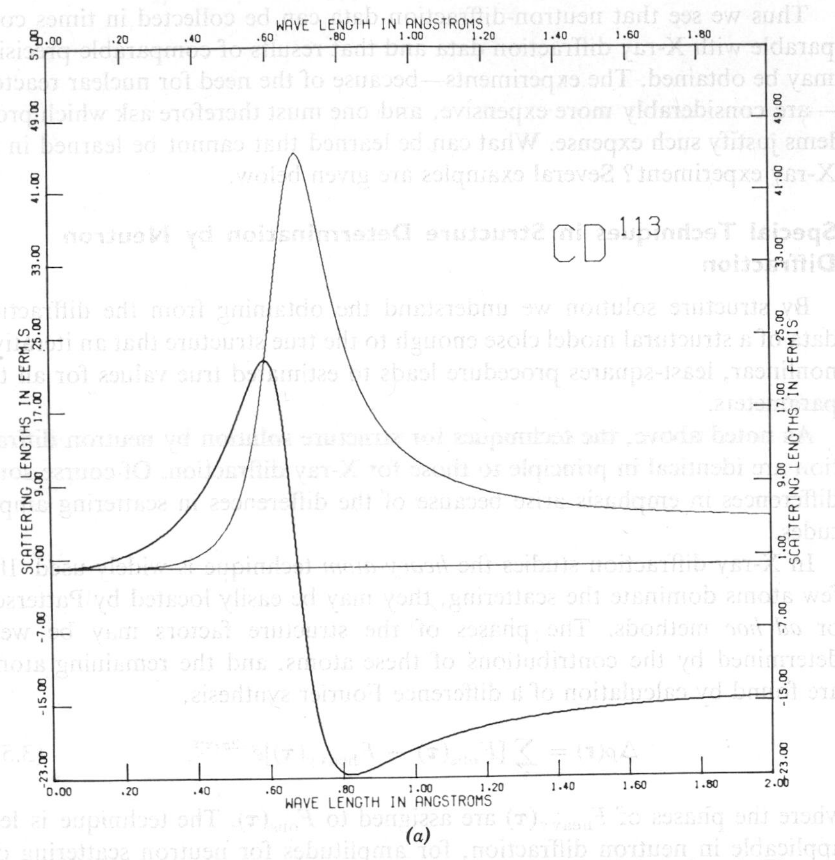

(a)

Fig. 3.10 Anomalous scattering of two isotopes as a function of wavelength. The real part changes sign; the imaginary part f'' reaches a maximum at the resonant wavelength. The large change in these quantities with wavelength can be used in solution of the phase problem. *(a)* ^{113}Cd *(b)* ^{149}Sm.

difficulty is that most of these methods depend on the fact that in an X-ray diffraction experiment the scattering density and hence the Patterson function are everywhere nonnegative. This is not true in neutron diffraction for crystals containing a significant contribution from hydrogen, and it is not at all clear that the methods ought to work so well. Sikka [24] has recently shown that direct methods should work satisfactorily when the total contribution of the hydrogen-atom scattering is less than 25% on an F^2 basis.

In cases where direct methods in the conventional formulation cannot be used, Karle has suggested a *squared-structure* approach [25]. The squared structure is defined as

$$\rho_{\mathrm{sq}}(\mathbf{r}) = \rho^2(\mathbf{r}), \tag{3.59}$$

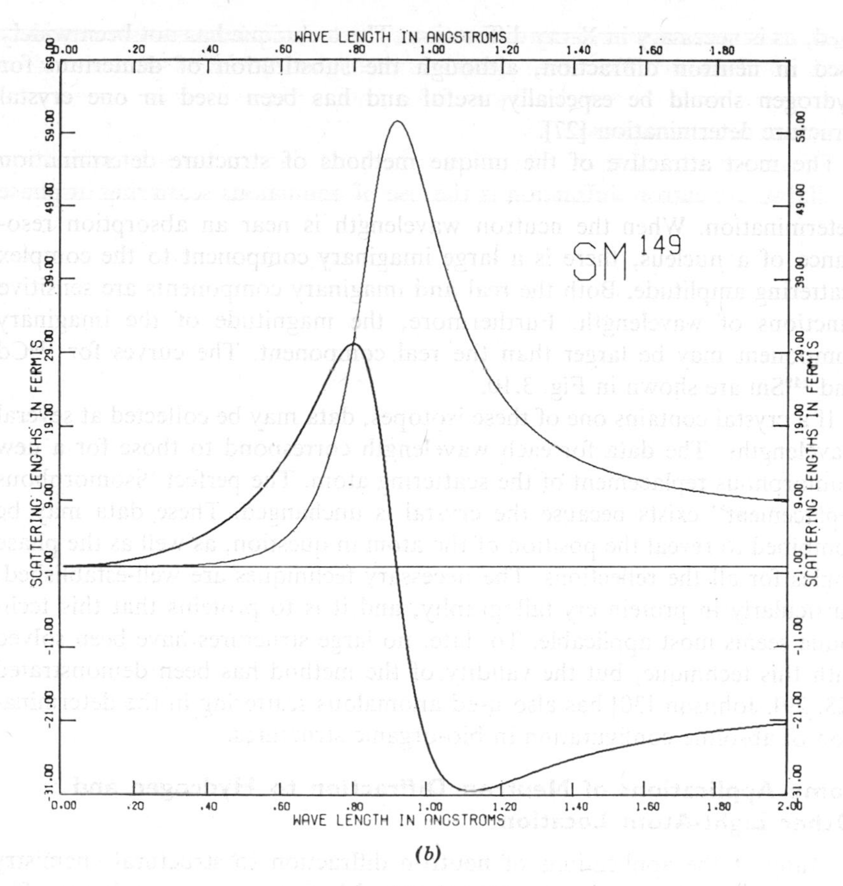

(b)

Fig. 3.10 (*contd.*)

and is thus everywhere positive. Karle develops probability formulas for the structure amplitudes of the squared structure in terms of those observed:

$$\{|F_{\text{obs}}(\boldsymbol{\tau})|\} \Rightarrow \{|F_{\text{sq}}(\boldsymbol{\tau})|\}. \tag{3.60}$$

These probability-generated $\{|F_{\text{sq}}(\boldsymbol{\tau})|\}$ are then used in the direct-methods procedures in the usual way. This method has been successfully used by Ellison and Levy [26] in the crystal-structure determination of glycollic acid.

Different isotopes of an element have different scattering amplitudes. Since the substitution of one isotope for another in a site usually gives rise to isomorphous crystals, the technique of isomorphous replacement for crystal-structure determination is ideally suited for neutron diffraction. The isomorphism may in fact be more perfect than if entirely different elements are used, as is necessary in X-ray diffraction. The technique has not been widely used in neutron diffraction, although the substitution of deuterium for hydrogen should be especially useful and has been used in one crystal structure determination [27].

The most attractive of the unique methods of structure determination available to neutron diffraction is the use of anomalous scattering in phase determination. When the neutron wavelength is near an absorption resonance of a nucleus, there is a large imaginary component to the complex scattering amplitude. Both the real and imaginary components are sensitive functions of wavelength. Furthermore, the magnitude of the imaginary component may be larger than the real component. The curves for ^{113}Cd and ^{149}Sm are shown in Fig. 3.10.

If a crystal contains one of these isotopes, data may be collected at several wavelengths. The data for each wavelength correspond to those for a new isomorphous replacement of the scattering atom. The perfect "isomorphous replacement" exists because the crystal is unchanged. These data may be combined to reveal the position of the atom in question, as well as the phase angles for all the reflections. The necessary techniques are well-established, particularly in protein crystallography, and it is to proteins that this technique seems most applicable. To date, no large structures have been solved with this technique, but the validity of the method has been demonstrated [28, 29]. Johnson [30] has also used anomalous scattering in the determination of absolute configuration in bio-organic structures.

Some Applications of Neutron Diffraction to Hydrogen and Other Light-Atom Location

Many of the applications of neutron diffraction to structural chemistry have been in the precise determination of hydrogen-atom positions. The average contribution of an atom to the intensity of a Bragg reflection in a diffraction experiment is proportional to the square of the scattering amplitude for that atom. Consider a typical organic compound containing equal numbers of hydrogen and carbon atoms (benzene, for example). The X-ray scattering amplitudes at zero scattering angle are one and six electrons. The neutron-scattering amplitudes are -0.378 and $+0.66 \times 10^{-12}$ cm. Thus for X-rays the average contribution of hydrogen to the scattering is $1^2/(1^2 + 6^2) = 1/37$, or 2.7%. For neutrons it is $0.378^2/(0.378^2 + 0.66^2) = 24.7\%$. If the hydrogen atoms are substituted by deuterium ($b = 0.66$), the ratio is even more favorable; the average contribution of hydrogen is 50%. The value of 3% for X-rays is just about at the noise level of the best X-ray experiments; of course some reflections contain contributions of much more than the average for hydrogen, so that hydrogen-atom positions can be determined and refined from X-ray data. Nevertheless, when accurate hydrogen positions are needed in other than the very simplest compounds, neutron diffraction offers the only satisfactory method. Figure 3.11 shows a comparison of hydrogen-scattering density maps for methyl glyoxal *bis*-guanylhydrazone dihydrochloride monohydrate ($C_5H_{14}N_8Cl_2 \cdot H_2O$) [31]. These maps have

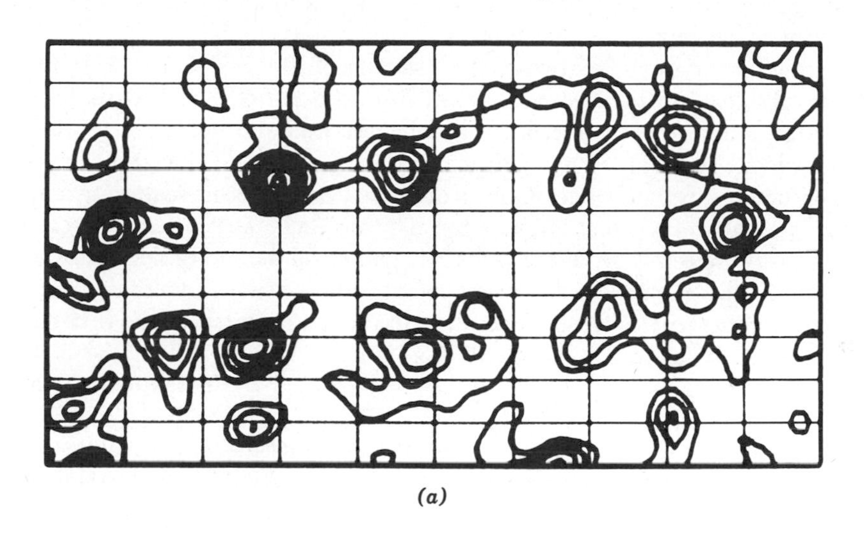

(a)

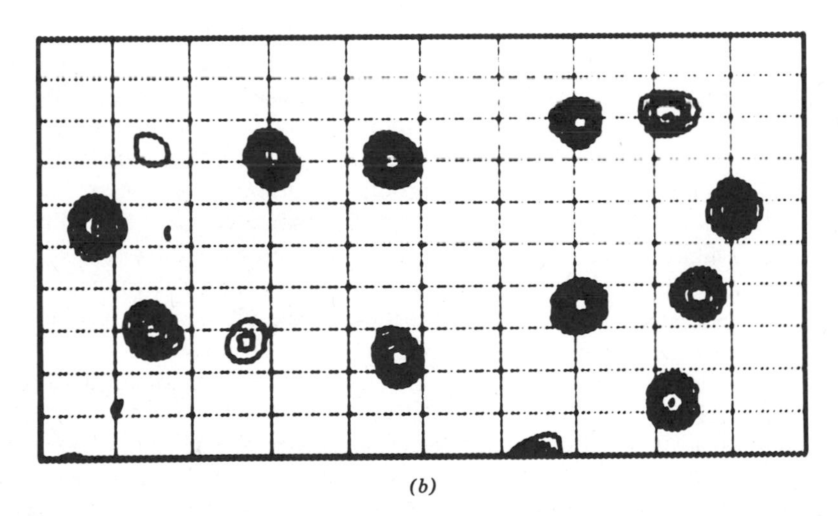

(b)

Fig. 3.11 Scattering density maps for hydrogen atoms in methyl glyoxal *bis*-guanylhydrazone. (*a*) X-ray (*b*) neutron.

been calculated by taking the observed X-ray or neutron-structure factors and subtracting the structure factors calculated for the heavy atoms. The hydrogen atoms are identifiable in the X-ray map, although the noise level in other parts of the map is almost as high as the peaks for the hydrogen atoms. The neutron map is, on the other hand, very clean. All hydrogen-atom peaks are well-shaped and were refinable to a high degree of precision in the least-squares refinement of the structure.

In another more extreme case, the structure of $Mn(CO)_5H$ [32, 33], it was impossible to find the hydrogen atom from the X-ray data because of the very large difference between the manganese and the hydrogen scattering amplitudes. In the neutron-diffraction case manganese has the smallest amplitude in absolute value of any atom in the structure; its scattering amplitude is negative, as is that of hydrogen. The molecular structure and a scattering-density map in one of the principal planes of the molecule is shown in Fig. 3.12. The manganese and hydrogen atoms are readily identified; a least-squares refinement of the data resulted in a determination of the hydrogen-atom position to a precision of 0.02 Å.

The structure of the complex, *bis*(2-amino-2-methyl-3-butanone oximato) nickel(II) chloride monohydrate (Fig. 3.13) has been refined by neutron diffraction [34]. The structure refinement was based on the measurement of 3000 Bragg reflections collected over a two-week period from a crystal with a volume of 28 mm^3. There are 25 hydrogen atoms and 19 heavy atoms. The precision of the determination of molecular geometry and r.m.s. amplitudes of vibration is indicated by some selected values in Table 3.1.

One reason that very precise determinations of hydrogen-atom parameters is desirable is because the examination of the parameters describing the thermal motion can help in an understanding of the potential-energy surfaces governing atomic motions; such studies have been of great importance in hydrogen bonding and in the consideration of motions of methyl groups and ammonium ions in solids. In the nickel amine oxime complex the following interesting analysis could be obtained only from neutron-diffraction data.

The r.m.s. amplitudes of vibration of all atoms were determined to the precision indicated in Table 3.1. Reduction of the vibration ellipsoids to a principal axis description for each methyl group showed that the largest principal axis in each case lies nearly in the direction corresponding to hindered rotation of the methyl group (Fig. 3.14). The agreement between the amplitudes for the three hydrogen atoms on each methyl group gives some assessment of the quality of the data and the validity of a rigid-model description of the methyl group motion. Subtraction of the motion of the neighboring carbon atom (representing the rigid-body motion of the molecule) resulted in mean amplitudes of rotation for the six methyl groups of 35°, 22°, 14°, 14°, 13°, and 13°, corresponding to potential barriers hindering

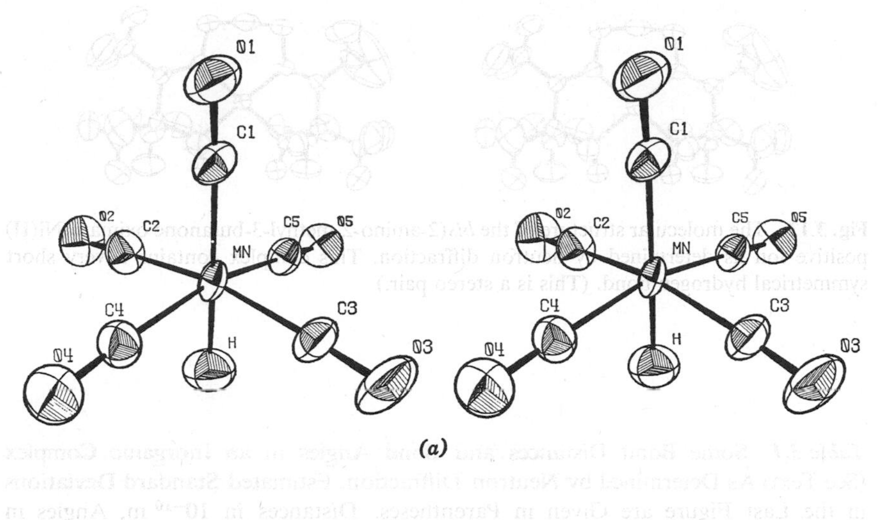

(a)

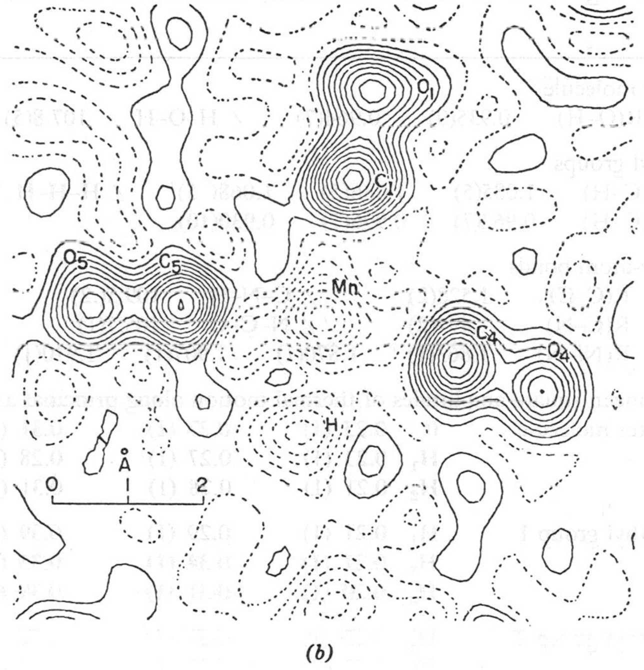

(b)

Fig. 3.12 The structure of $Mn(CO)_5H$ as determined by neutron diffraction. (*a*) The molecular structure; the lengths of the major axes of the ellipsoidal atoms are proportional to the amplitudes of vibration in these directions. (This is a stereo pair.) (*b*) Scattering-density map in one of the vertical mirror planes of the molecule. The manganese and hydrogen density contours are shown as dotted lines; these atoms scatter with a reverse phase.

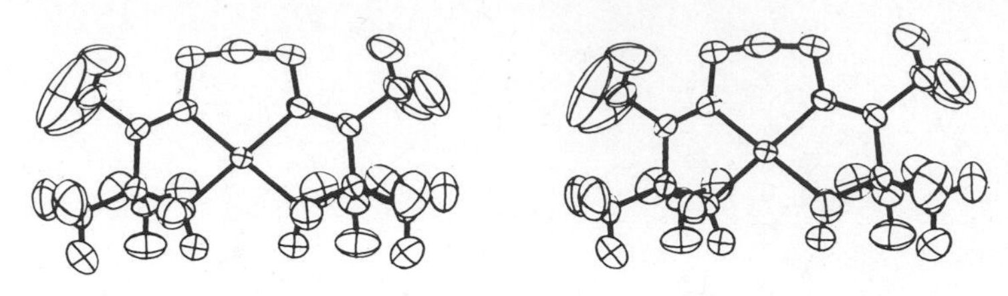

Fig. 3.13 The molecular structure of the *bis*(2-amino-2-methyl-3-butanone oximato)Ni(II) positive ion as determined by neutron diffraction. This complex contains a very short symmetrical hydrogen bond. (This is a stereo pair.)

Table 3.1 Some Bond Distances and Bond Angles in an Inorganic Complex (See Text) As Determined by Neutron Diffraction. Estimated Standard Deviations in the Last Figure are Given in Parentheses, Distances in 10^{-10} m, Angles in Degrees.

Water molecule					
R(O–H)	0.935(9)	0.964(7)	∠ H–O–H	107.8(5)	
Methyl groups					
(1) R(C–H)	1.085(5)	1.086(5)	1.068(5)	∠ H–H–H	57.8(7)
(2) R(C–H)	0.962(7)	0.996(7)	0.930(10)		
Heavy-atom bonds					
R(C–C)	1.528(2)	∠ C–N–H	109.5(.2)		
R(C–N)	1.494(2)	∠ N–C–C	105.7(1)		
R(Ni–N)	1.863(1)	1.870(1)	1.907(1)	1.910(1)	

Root-mean-square amplitudes of thermal motion along principal axes				
Water molecule	0	0.23 (1)	0.27 (1)	0.31 (1)
	H_1	0.23 (1)	0.27 (1)	0.28 (1)
	H_2	0.21 (1)	0.28 (1)	0.31 (1)
Methyl group 1	H_1	0.21 (1)	0.29 (1)	0.39 (1)
	H_2	0.21 (1)	0.34 (1)	0.35 (1)
	H_3	0.20 (1)	0.31 (1)	0.39 (1)
Methyl group 2	H_1	0.20 (1)	0.31 (1)	0.71 (2)
	H_2	0.21 (1)	0.31 (1)	0.62 (2)
	H_3	0.20 (1)	0.37 (1)	0.66 (2)
Ni		0.165(1)	0.180(1)	0.190(1)
N		0.178(2)	0.182(2)	0.210(2)
C		0.172(2)	0.190(2)	0.200(2)

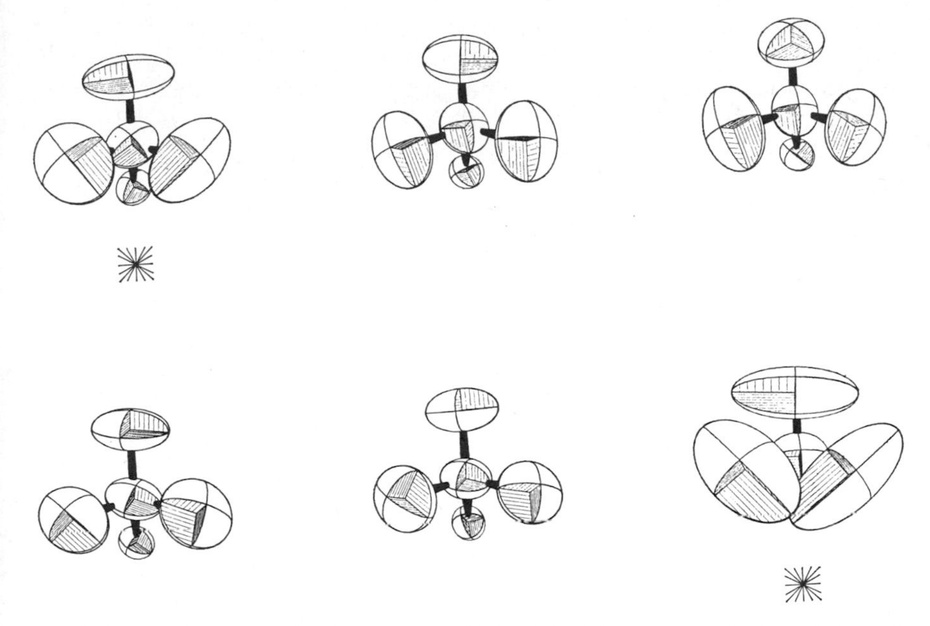

Fig. 3.14 Methyl groups in the complex of Fig. 3.13. The large amplitudes of thermal motion perpendicular to the bonds are indicative of high-amplitude hindered rotation. This is especially marked for the starred groups, which are bonded to sp^2 carbons with no neighboring methyl groups.

rotation of 0.4, 0.9, 2.2, 2.2, 2.2, and 2.2 kcal/mole. Note that the apparent C–H bond lengths for these methyl groups (Table 3.1) are different in a way that beautifully illustrates the apparent shortening of bond lengths that result when an inadequate quadratic model of thermal motion is used to describe a wide-amplitude, hindered rotation. The amplitude of motion is considerably less for methyl group 1 than for methyl group 2, and there is a marked difference in apparent bond lengths.

Even for heavier atoms neutron diffraction is superior to X-ray diffraction for measurements of atomic motions. Other light-atom, heavy-atom problems in addition to hydrogen-atom location have been attacked by neutron-diffraction methods. Crystals of lithium tantalate, $LiTaO_3$, are ferroelectric. The ferroelectric transition is accompanied by large motions of the atoms relative to one another. In particular the lithium and tantalum atoms undergo large motions with respect to the oxygen framework. A detailed description of these motions is necessary for an understanding of the mechanism of the ferroelectric transition. In the X-ray diffraction experiment, the tantalum atom dominates the scattering to the point where the lithium atom is all but invisible. In the neutron-diffraction experiment, on the other hand, the lithium atom plays a larger role in the scattering—although this role is by no means dominant. The oxygen atoms made the most important contribution. In the structures at room temperature X-ray [35] and neutron-diffraction data [36] were collected to equivalent degrees of precision. (The final weighted agreement ratio was 0.05.) The positions of the atoms were determined to the precision indicated in Table 3.2.

Table 3.2 Comparison of Precision of Atomic Position Parameters in X-Ray and Neutron-Diffraction Studies of Lithium Tantalate. Parameters in 10^{-10} m.

	X-Ray	*Neutron*
Li(z)	3.888 ± 0.040	3.845 ± 0.023
O(x)	0.275 ± 0.014	0.258 ± 0.003
O(y)	1.750 ± 0.016	1.771 ± 0.002
O(z)	0.958 ± 0.007	0.947 ± 0.004

Precision Analysis of Thermal Motion and Electron Density

In an X-ray diffraction study the structure-factor expression for a Bragg reflection at a reciprocal lattice point $\boldsymbol{\tau}$ is

$$F(\boldsymbol{\tau}) = \sum_i f_i(\boldsymbol{\tau})T_i(\boldsymbol{\tau}) \exp(2\pi i\boldsymbol{\tau} \cdot \mathbf{r}_i), \tag{3.61}$$

where $f_i(\boldsymbol{\tau})$ is the atomic-scattering amplitude for the atom at rest, $T_i(\boldsymbol{\tau})$ is the Debye-Waller factor, the Fourier transform of which describes the thermal motion of atom i, and $\mathbf{r}_i$ is the mean position of atom i. Because of the changes in electron density when chemical bonding or crystal interaction takes place, $f_i(\boldsymbol{\tau})$ is slightly different from the function that would be characteristic of a free atom. The deviations from the free-atom description may in fact be of considerable theoretical interest. Similarly, the form of $T_i(\boldsymbol{\tau})$ is also of interest in the derivation of information concerning molecular potential functions. If the functional forms of f_i and T_i were known with certainty, then the parameters in both f_i and T_i could be refined in the usual least-squares procedure. However, the functional form is not usually precisely known, although reasonably good models certainly exist. In conventional X-ray diffraction studies the free-atom approximation is usually used for f_i and the harmonic vibration approximation is used for T_i. Parameters for the thermal vibration are refined and reported.

Since f_i and T_i enter into the structure-factor expression in a multiplicative way, any deviations from the assumed model in either affects the refined parameters of the other. In particular, deviations from free-atom f_i's are of the same order of magnitude as the effects of anharmonicities in T_i, and the X-ray diffraction experiment alone is hard-pressed to make a meaningful determination of either. With neutron diffraction, on the other hand, the scattering amplitude $f_i(\boldsymbol{\tau})$ is a constant independent of $\boldsymbol{\tau}$ but characteristic only of the atomic species i. From neutron-diffraction data one may thus determine values for the parameters of T_i which are independent of any assumptions regarding electron distribution in the atom at rest. For definitive determinations of atomic motion parameters, then, neutron diffraction is ideal.

There has been much interest lately in the determination of bonding-electron densities in crystals by X-ray diffraction studies. The interaction of the thermal motion with the electron-density changes makes these studies difficult. One of the most successful ways of avoiding this difficulty has been vigorously pursued by Coppens and co-workers [37, 38]. Neutron-diffraction and X-ray diffraction data are obtained from the same crystal. The neutron-diffraction data are used to refine the positional and thermal parameters, and the X-ray diffraction data are then used to refine parameters describing the bonding electrons. Two approaches have been taken in presenting the results of such measurements. The functions T_i and the atomic positions $\mathbf{r}_i$ may be derived from the neutron-diffraction data. These are used to calculate a set of X-ray structure factors, using the free atom scattering factors f_i^o. A Fourier inversion of these structure factors should give the electron density ρ_{calc}, which would be characteristic of a structure in which no electron redistribution has taken place because of chemical bonding:

$$\rho_{\text{calc}}(\mathbf{r}) = \sum_{\boldsymbol{\tau}} F_{\text{calc}}(\boldsymbol{\tau}) e^{-2\pi i \boldsymbol{\tau} \cdot \mathbf{r}}, \tag{3.62}$$

where F_{calc} is based on the neutron parameters T_i^N and $\mathbf{r}^N$ and the free-atom scattering factors:

$$F_{\text{calc}} = \sum_i f_i^0 T_i^N(\boldsymbol{\tau}) \exp(2\pi i \boldsymbol{\tau} \cdot \mathbf{r}^N). \tag{3.63}$$

The observed electron density is calculated in the usual way:

$$\rho_{\text{obs}}(\mathbf{r}) = \sum F_{\text{obs}}(\boldsymbol{\tau}) e^{-2\pi i \boldsymbol{\tau} \cdot \mathbf{r}} \tag{3.64}$$

The difference between these two,

$$\Delta\rho(\mathbf{r}) = \rho_{\text{obs}}(\mathbf{r}) - \rho_{\text{calc}}(\mathbf{r}) = \sum_{\boldsymbol{\tau}} (F_{\text{obs}}(\boldsymbol{\tau}) - F_{\text{calc}}(\boldsymbol{\tau})) \exp(-2\pi i \boldsymbol{\tau} \cdot \mathbf{r}), \tag{3.65}$$

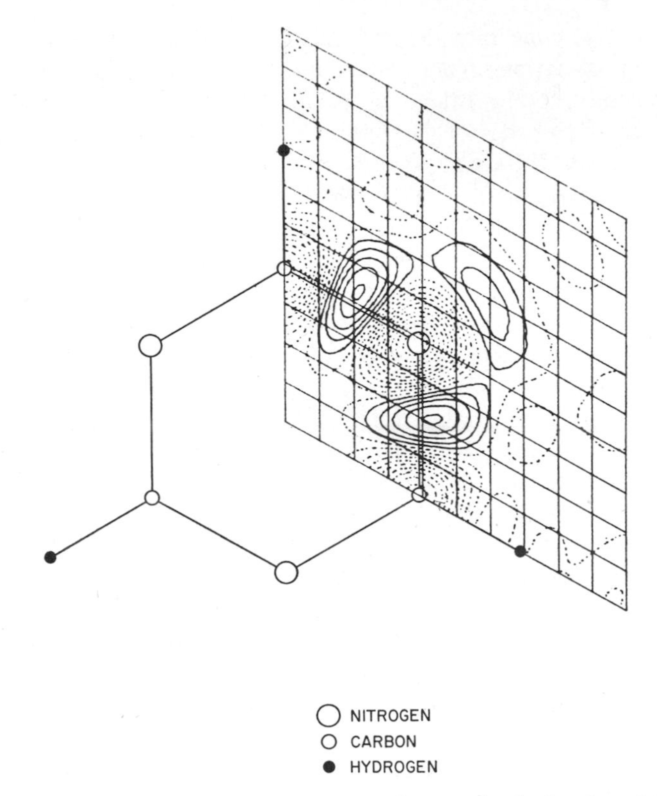

Fig. 3.15 Bonding electron density in *s*-triazine. Electrons in the bonds and lone pair on nitrogen are clearly seen [37].

then gives the deviation of the observed electron density from that which would be characteristic of free atoms. Thus $\Delta\rho$ may be described as the *bonding* electron density. Examples of bonding electron density maps derived in this way are shown in Figs. 3.15 and 3.16.

In another approach the form of $f_i(\boldsymbol{\tau})$ is specified; it contains parameters that are related to molecular obital parameters describing overlap densities in bonds. These parameters may then be refined in the usual least-squares procedures [39].

The possibility of neutron-diffraction determination of protein structures has been demonstrated [40]. Proteins, as do many organic molecules, consist of a core of carbon, nitrogen, and oxygen atoms surrounded by hydrogen atoms. Since hydrogen atoms have a negative scattering density, a rather sharp demarcation exists between the areas in a scattering density map corresponding to the heavy atoms and to the hydrogen atoms. This demarcation

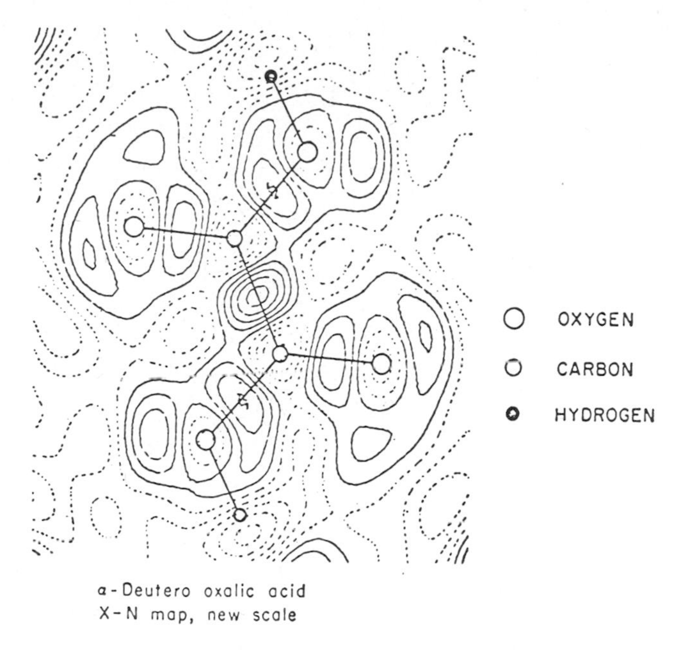

Fig. 3.16 Bonding electron density in oxalic acid dihydrate. Bonding electrons, and one and two lone pair regions on the two oxygens are obvious [38].

may make it easier to pick out the backbone of the molecule than in X-ray diffraction. Furthermore, deuterium substitution may be of use in clearly defining the areas of the crystal that contain replaceable water. A comparison of an X-ray and a neutron-diffraction map for one section of a deuterated myoglobin crystal is presented in Fig. 3.17. At 2.8 Å resolution the maps look very much the same. The phases from an X-ray diffraction study have been used in calculating the neutron-scattering density map. Future use of neutron diffraction for the study of biologically interesting macromolecules may be stimulated by the use of anomalously scattering atoms in the phase determination step and by the development of more efficient means of data collection, including the Laue method and the use of multiple detector systems. Protein crystals are at best difficult to grow. To obtain crystals with a volume of several cubic millimeters—the size necessary for neutron-diffraction studies on materials with large unit cell volumes—has not been possible for more than a few examples. This difficulty may be far outweighed by the fact that protein crystals do not undergo radiation damage in the neutron beam nearly as fast as in the X-ray beam.

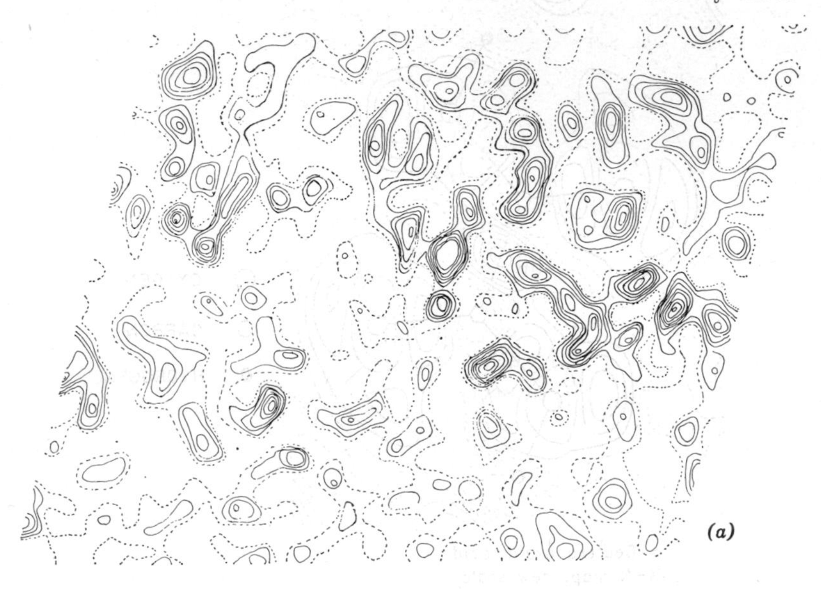

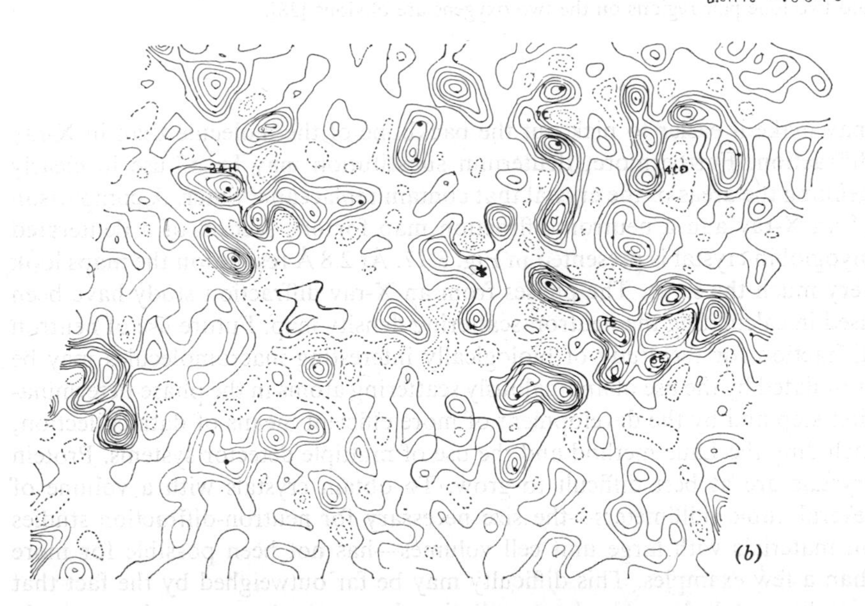

Fig. 3.17 Comparison of sections of three-dimensional scattering density maps for myoglobin structure at a resolution of 2.0 Å [40]. (*a*) X-ray (*b*) neutron.

6 MAGNETIC STRUCTURES

The overlap in information that can be obtained from nuclear scattering of neutrons with that derived from X-rays has been pointed out several times. In the case of crystal structures the choice of method is simply a matter of which technique will furnish more reliable information. The situation in the case of magnetic scattering of neutrons, on the other hand, is qualitatively different; there is *no* analog in the X-ray case. As a matter of fact, neutron magnetic scattering is the only method presently available for the direct determination of the ordered arrangement of spins in a magnetic crystal, the so-called magnetic structure. As was noted in deriving the coherent magnetic scattering cross section (3.33), the amplitude depends on the direction as well as on the magnitude of the atomic magnetic moments so that, in principle, a complete determination of the magnetic structure is possible. This aspect of neutron scattering has been exploited rather widely, and a very large variety of magnetic materials have been investigated. In this section we restrict our discussion to a few interesting examples that should serve as representatives of the kind of information that can be obtained from magnetic scattering.

The general structure factor, including both magnetic and nuclear scattering, is (see 3.35)

$$F(\boldsymbol{\tau}) = \sum_{s=1}^{j} (\langle b_s \rangle + p_s' \hat{\boldsymbol{\lambda}} \cdot \mathbf{q}_s) e^{2\pi i \boldsymbol{\tau} \cdot \mathbf{r}_s} e^{-W}. \tag{3.66}$$

The simplest situation is a monatomic ferromagnet, such as the $3d$ transition metals. In the ordered state (i.e. below the Curie temperature) the magnetic scattering gives rise to a series of Bragg peaks which, according to (3.66), are superimposed on the nuclear Bragg peaks. Several methods are available to disentangle the two sources of scattering. One of these involves placing the sample in a magnetic field sufficient to orient the atomic spins parallel to the scattering vector, $\hat{\boldsymbol{\epsilon}}$, which makes the magnetic contribution vanish since $\mathbf{q} = [\hat{\boldsymbol{\varkappa}} - \hat{\boldsymbol{\epsilon}}(\hat{\boldsymbol{\epsilon}} \cdot \hat{\boldsymbol{\varkappa}})]$ goes to 0 for $\hat{\boldsymbol{\varkappa}} \parallel \hat{\boldsymbol{\epsilon}}$. A second method involves measuring the scattering of a beam of neutrons that is polarized antiparallel and then parallel to the direction of magnetization in the sample. Under ideal experimental conditions the ratio of the two intensities is, from (3.35), $(1 + \nu)^2/(1 - \nu)^2$, where $\nu = p'/b$. In a series of elegant experiments utilizing this technique Shull and co-workers [41] have made very precise measurements of the magnetic form factor of iron, hexagonal cobalt, and nickel. The results were compared with free-atom calculations of the form factor, and in all cases the agreement was good provided they included a uniform *negative* contribution to the magnetization. One cannot determine the origin of this negative contribution from the neutron-diffraction data, but a good possibility is that the $4s$ electron spins are polarized opposite to the $3d$ electron spins.

Another relatively simple case is that of the collinear antiferromagnet, that is, an ordered system of spins, half of which are aligned parallel to a given direction and the other half of which are antiparallel. As an example let us consider the situation in MnS [42], which can be crystallized in three different polymorphic forms. One of these is the NaCl type, and at room temperature the neutron diffraction for this polymorph consists of a set of Bragg peaks arising solely from nuclear scattering. If the sample is cooled below the ordering temperature (the so-called Néel temperature for an antiferromagnet), an additional set of superlattice peaks appear that can be indexed with h, k, and l all odd, using a cubic unit cell with a cube edge equal to twice that of the chemical unit cell. This observation is consistent with a magnetic structure made up of ferromagnetic (111) sheets that alternate in sign, that is, they are antiferromagnetically coupled. The observed intensities require the spins to lie in the ferromagnetic layers, but the direction cannot be specified any further on the basis of powder-diffraction data. The overall structure is shown in Fig. 3.18, in which the spin direction is arbitrarily chosen to be $\pm[1\bar{1}0]$, which fulfills the condition of making the moments lie in the ferromagnetic (111) sheets. A most remarkable feature of the magnetic structure is revealed if one considers the distribution of spin

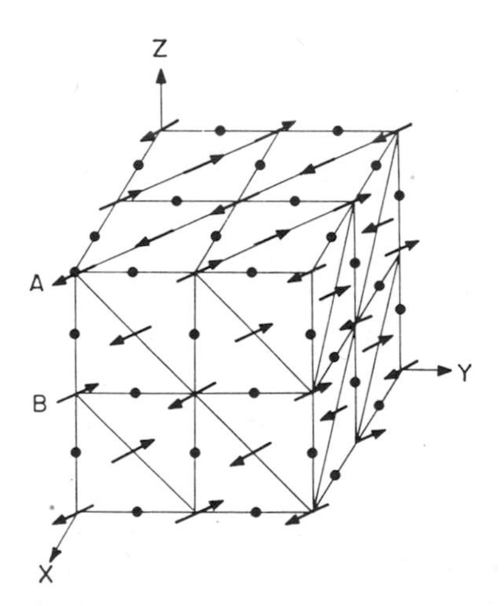

Fig. 3.18 Magnetic structure of MnS (rock salt form) corresponding to ordering of the second kind. The arrows drawn at the positions of the magnetic ions indicate the direction of the spins at those sites. The small circles represent the intervening sulfide ions. The alternating ferromagnetic sheets (parallel to the (111) plane) are indicated by A and B.

directions about a given Mn^{++} spin. The 12 nearest neighbor spins are arranged half-parallel and half-antiparallel to the central spin, but the 6 *next*-nearest neighbors are all antiparallel. Thus the dominant antiferromagnetic coupling is between next-nearest neighbors. These neighbors, however, are separated by intervening S^{2-} anions, so that the coupling mechanism must involve the overlap of anion orbitals with those of the magnetic cations. Kramers [43] some 35 years ago, in trying to understand the early experiments on adiabatic demagnetization, which showed that small exchange couplings existed even between ions separated by diamagnetic groups (such as waters of crystallization), provided the clue. He pointed out that the magnetic ions could cause spin-dependent perturbations in the wave functions of the intervening ions and thus transmit the exchange effect to the next magnetic ion. This mechanism, which is called superexchange, was first demonstrated by neutron-diffraction experiments on MnO [44]. The ideas of Kramers have been clarified and elaborated by Anderson [45] and others. The existence of the three polymorphs in MnS permitted a study of the dependence of superexchange coupling on bonding.

Let us consider another example of an antiferromagnet compound, $MnTe_2$, which crystallizes with the pyrite structure, which is a NaCl-like arrangement of Mn^{++} and Te_2^{2-} "molecules." Just as in the previous example, the magnetic ions are situated on a fcc lattice. The original magnetic structure [46] proposed for this material consisted of ferromagnetic (100) layers alternating in sign in a direction perpendicular to these sheets. The spins, based on the observed intensities, were parallel to the ferromagnetic layers. This ordering scheme, commonly called "ordering of the first kind," does not require an enlargement of the chemical unit cell. The additional magnetic peaks that appear are, however, separated from the nuclear reflections since they have mixed indices (two odd and one even or two even and one odd). Later Mössbauer measurements [47], which give information on the orientation of the internal magnetic field at a tellurium nucleus, were incompatible with this structure. One can overcome this contradiction by suitably generalizing [48] ordering of the first kind in a way that preserves the characteristic neighbor relationships and at the same time accounts for the Mössbauer observations. This is accomplished by relaxing the restriction of collinearity of spins and by analyzing the neutron data in terms of the most general noncollinear model. Let us define a vector $\boldsymbol{\varkappa}(\boldsymbol{\tau})$ such that

$$\boldsymbol{\varkappa}(\boldsymbol{\tau}) = \sum_{s=1}^{j} \hat{\boldsymbol{\varkappa}}_s e^{2\pi i \boldsymbol{\tau} \cdot \mathbf{r}_s}, \tag{3.67}$$

where, as before, $\hat{\boldsymbol{\varkappa}}_s$ is a unit vector in the direction of the spin at the sth atom in the unit cell and p_s' has been set equal to unity for convenience. The sum is over all magnetic atoms in the unit cell. In terms of this vector the *magnetic* structure factor becomes

$$\mathbf{F}(\boldsymbol{\tau}) = \boldsymbol{\varkappa}(\boldsymbol{\tau}) - \hat{\boldsymbol{\epsilon}}(\hat{\boldsymbol{\epsilon}} \cdot \boldsymbol{\varkappa}(\boldsymbol{\tau})). \tag{3.68}$$

This form of structure factor is valid for *any* ordered collection of spins that can be described in terms of a magnetic unit cell. Analysis of the diffraction data for $MnTe_2$ using this formalism yields an infinite number of possible structures. These structures are generated by alternating the x-component of the Mn^{++} spin each half-unit cell translation in either the z or the y directions, and similarly for the z component in the y or the x directions. The original collinear model for ordering of the first kind can be reconstructed in a variety of ways with these rules; for example, let the x and y components have equal magnitudes and alternate in the y and x directions, respectively, while the z component is set equal to 0. There is one structure in this infinite set, belonging to space group Pa3, the space group of the chemical structure as well, which is consistent with the Mössbauer findings. The possibility of generalization to noncollinear structures introduces an additional complexity in deducing magnetic structures. As can be seen from the example just cited the uniqueness of a proposed structure may be in doubt. This ambiguity is, in most cases, not a matter of principle. The scattering from a *single magnetic domain* single crystal should be sufficient to determine a unique structure. It is very difficult, however, especially in the case of antiferromagnets, to prepare a single domain single crystal. If the single crystal sample is multi-domain, then from the point of view of its magnetic scattering, it is equivalent to a powder specimen.

Another interesting class of magnetic structures is one characterized by the appearance in the diffraction pattern of magnetic peaks that cannot be indexed on the basis of a magnetic unit cell that is some simple multiple of the original chemical unit cell. In other words, the reciprocal lattice vectors associated with the magnetic scattering are not commensurate with the reciprocal lattice obtained by any reasonable enlargement of the chemical cell. Structures of this type can be generated by starting with a simple ferro- or antiferromagnetic arrangement of spins and modulating the spins in direction, or amplitude, or both. The form and wavelength of this periodic modulation are arbitrary. It may, for example, be a simple sine wave or spiral, or it may be of a more complex nature. This class of modulated structures was discovered independently by several groups [49] and has since been found in a variety of materials. In particular the rare-earth metals and alloys [50] show a wide variety of modulated structures.

7 NEUTRON SPECTROSCOPY

The discussion of neutron diffraction above has been restricted to the applications of elastic scattering. The new generation of high-flux reactors has opened up a vast new field of inelastic scattering experiments that can be used to study the dynamic properties of solids and liquids. In particular, measurement of the coherent nuclear *inelastic* scattering can yield both the wave vectors and the polarization vectors of the normal modes of vibration. Similar measurements on ordered magnetic systems can be used to determine the dispersion relations (frequency versus wave vector) of the spin waves. In the case of magnetic scattering one can also make measurements in the neighborhood of the ordering temperature, which can then be compared with the predictions of the general theory of second-order phase transitions. Neutron scattering has also been applied to liquids [51] and gases [52]. The variety of these applications gives an indication of the usefulness of neutron scattering.

In this final section we discuss some of the details of neutron spectroscopy as applied to crystalline solids.

Neutron inelastic scattering may be used as a powerful tool in the study of vibrational energy levels in crystalline and molecular systems. Because of the differences in cross section and the possibility of measuring momentum as well as energy changes, the technique is an important complement to infrared spectroscopy. The method is basically simple; a neutron with energy E_0 and momentum $\mathbf{k}$ is scattered to a state with energy E_s and momentum $\mathbf{k}'$. The energy difference

$$\Delta E(\text{neutron}) = E_s - E_0 \tag{3.69}$$

must be equal to the difference between the energy levels of two states of the scattering system

$$\Delta E(\text{scatterer}) = -\Delta E(\text{neutron}). \tag{3.70}$$

The change in the neutron momentum may, as we shall see below, be used to determine the polarizations of the vibrational modes being studied.

Because neutron inelastic scattering is usually applied to condensed phases, the interpretation of the spectra is attendant with all the difficulties inherent in the interpretation of the vibrational modes of any solid or liquid. We restrict our attention here to the spectroscopy of solids. To understand the characteristics of these neutron spectra we remind the reader of a few of the important characteristics of the vibrational properties of solids. We state the formalism for the assumption of harmonic motion for all vibrational modes. Although many low-frequency vibrational modes in solids may be quite anharmonic, a qualitative understanding of neutron spectra may be obtained from the expressions for cross sections in the harmonic approximation.

Let us consider a crystal containing N unit cells, each unit cell containing n atoms. For any harmonic mode with coordinate q_j, the displacement vector $\Delta\mathbf{r}_i$ of the ith atom of mass m_i may be expressed in terms of the normal mode

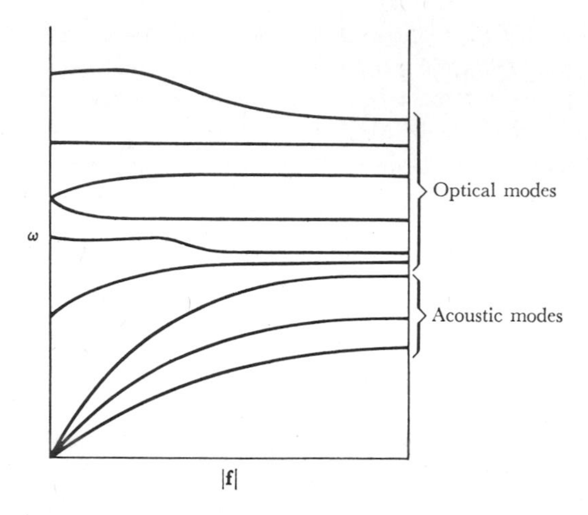

Fig. 3.19 Dispersion curves for a hypothetical crystal with three atoms in the unit cell. The frequency is plotted against the absolute value of the wave vector *f* in one direction of the reciprocal lattice. The curves would be different for different directions. If the dispersion curve for one of the optical modes is nearly flat, this means that there is very little intermolecular coupling.

displacement vectors $\mathbf{C}_{ij}$:

$$\Delta\mathbf{r}_i = (Nm_i)^{-1/2} \sum_j \mathbf{C}_{ij} q_j. \tag{3.71}$$

The $\mathbf{C}_{ij}$ (each with three components) are normalized so that for each mode j the sum over the unit cell of the crystal is

$$\sum_i |C_{ij}|^2 = 1. \tag{3.72}$$

The crystal has $3Nn$ motional degrees of freedom. These are conveniently analyzed in the following way.

Consider the $3n$ normal modes corresponding to the n atoms of a single unit cell. Corresponding to each of these there are in the crystal N modes for which the normal amplitude vector for atom i in the unit cell with origin $\mathbf{R}_m$ is given by

$$\mathbf{C}_{ij}(\mathbf{R}_m) = \mathbf{C}_{ij}(\mathbf{R}_0) \exp (i\mathbf{f} \cdot \mathbf{R}_m), \tag{3.73}$$

where the vector $\mathbf{f}$ takes on the N values

$$\mathbf{f} = 2\pi \sum_{k=1}^{3} \frac{n_{ki}}{N_k} \mathbf{a}_k^* \qquad n_{ki} = 0, 1, \ldots, N_k - 1, \tag{3.74}$$

with $\mathbf{a}_i^*$ being the reciprocal lattice vectors for the crystal and the N_i being the number of unit cells in each direction. For $\mathbf{f} = 0$ the motions of all the unit cells are in phase; three of the modes (the translations) have 0 frequency; the remaining $3n - 3$ are the vibrational modes, which may be studied by conventional infrared and Raman spectroscopy. For $\mathbf{f} \neq 0$, the modes may still be identified with the same intracell vibrations, but there is now a non-zero phase relationship between the motions in different unit cells. The frequency of each of the $3n$ modes is a function of the wave vector

$$\omega_j = \omega_j(\mathbf{f}). \tag{3.75}$$

This relation is known as a dispersion relation. Hypothetical dispersion curves for a crystal with three atoms in the unit cell are shown in Fig. 3.19.

In infrared spectroscopy, transitions are forbidden if $\mathbf{f}$ is not approximately 0; there is no such restriction in neutron spectroscopy, and it is here that one of the important differences lies. If an experiment is carried out in such a way that $\mathbf{f}$ cannot be identified, the neutron spectrum will be a projection of the dispersion curves (properly weighted) on the $\mathbf{f} = 0$ axis (Fig. 3.20). Peaks

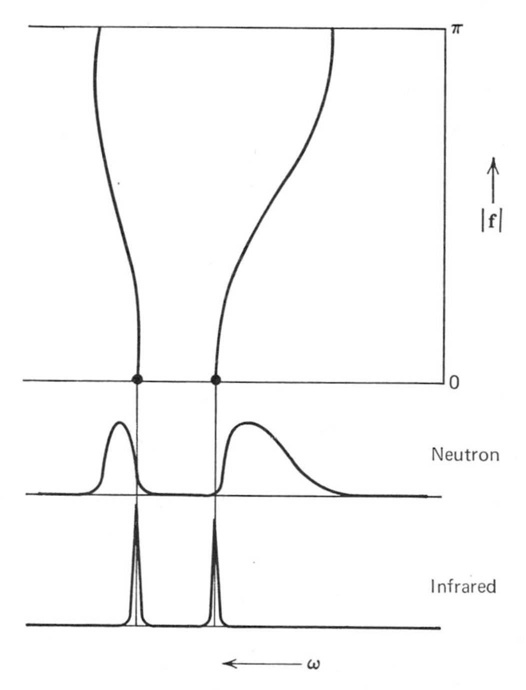

Fig. 3.20 Infrared spectroscopy measures a frequency for $\mathbf{f} = 0$; the line is moderately sharp. In an incoherent scattering experiment from a polycrystalline material there is no resolution of the momentum transfer; hence a broadened spectrum that is the projection of the dispersion curve on the $\mathbf{f} = 0$ axis is observed. Averaging over all populated levels also occurs.

thus tend to be broader than in the infrared experiment. On the other hand, if an experiment is carried out for a particular value of **f** (as it can), much additional information may be obtained about intermolecular forces.

For scattering involving an exchange of one quantum of vibrational energy (one-phonon scattering), cross sections for the coherent and incoherent scattering are as follows:

$$\frac{d\sigma_{\text{coh}}}{d\Omega} = \frac{(2\pi)^3}{V_0}\frac{k'}{k}\left|\sum_i \frac{b_i^{\text{coh}} T_i(\mathbf{K})}{m_i^{1/2}}(\mathbf{K}\cdot\mathbf{C}_{ik})\exp(i\mathbf{K}\cdot\mathbf{r}_i^{\,0})\right|^2$$

$$\times \frac{n\hbar^2}{2\,|\Delta E|}\,\delta(\mathbf{K}-\mathbf{f}-2\pi\boldsymbol{\tau}) \tag{3.76}$$

$$\frac{d\sigma^{\text{inc}}}{d\Omega} = \frac{k'}{k}\sum_i \left| b_i^{\text{inc}} \frac{T_i(\mathbf{K})}{m_i^{1/2}}(\mathbf{K}\cdot\mathbf{C}_{ik})\right|^2 \frac{n\hbar^2}{2\,|\Delta E|}, \tag{3.77}$$

where n is the quantum number of the upper of the two states involved in the transition. The squared sum in (3.76) is the *inelastic structure factor*. For both the coherent and the incoherent cross sections, large contributions arise when the momentum transfer **K** is parallel to the atomic motion as expressed by $\mathbf{C}_{ik}$.

For a given experiment, (3.76) and (3.77) must be averaged over all thermally excited initial states, over all crystalline orientations for a polycrystalline specimen, over all incident and scattered energies if monochromaticity is lacking in either case, and over **f** if momentum transfer is not measured.

The most widespread applications of neutron spectroscopy to complex chemical systems have been based on the application of (3.77) to hydrogenous systems. Because b_i^{inc} for hydrogen is very large, because the mass of hydrogen is small, and because $|\mathbf{C}_{ik}|$ for hydrogen is usually large, vibrational modes involving hydrogen tend to dominate the spectrum.

Many of the experiments of this type have been carried out by time-of-flight methods, where an almost monochromatic beam of very-low-energy neutrons is allowed to fall in bursts on a sample, and the energy of the scattered neutrons is measured by measuring the time required for a given flight path. Typical incident energies may be 25 cm^{-1} with a resolution of 1 cm^{-1} for low-energy transitions. Because of the nearly-zero energy of the incident neutron, this is an experiment in which the neutron gains energy by de-exciting thermally populated levels of the solid, and the technique is thus limited at ordinary temperatures to transitions with energies of less than a few hundred cm^{-1}. It is this region however that is often difficult to reach by infrared methods and it may be difficult to sort out hydrogen modes from other modes.

The method has been vigorously pursued by Rush and co-workers [53] in the investigation of hydrogen-bonded systems. Figure 3.21*a* shows the neutron spectrum from $CuSO_4 \cdot H_2O$. The spectrum is almost entirely due to the vibrations of the water molecule. The peak at 750 cm^{-1} has been assigned to rocking and wagging motions of the molecule, while that at 450 cm^{-1} may be predominantly due to a copper–water stretching vibration and possibly torsional vibrations of the water molecule about its symmetry axis. The lower-frequency peaks may well be due to hydrogen-bond stretching frequencies, superimposed on broad bands of lattice vibrations. The scattered intensity rapidly falls to 0 at high frequencies because of the sparsely populated higher vibrational states. Another example is the work of Reynolds and White [54] on a number of molecular crystals. Figure 3.21*b* shows the neutron spectrum from scattering by crystals of *p*-diiodobenzene. The peak positions and intensities are in good agreement with those calculated from a normal-mode analysis of optical spectroscopic data. Since hydrogen modes do dominate, neutron spectroscopy can be of great help in the assignment of modes in optical spectra in complex cases.

The energy-gain spectroscopy used in the two examples just cited has the disadvantage that high-frequency transitions cannot be seen, as one must depend on thermal population of the excited molecular states. Energy-loss spectroscopy does not have this disadvantage and can also be carried out conveniently and with relatively high resolution. One method is to vary the incident energy of a monochromatic beam by changing the scattering angle at a monochromatizing crystal. A beryllium filter in front of the counter allows scattered neutron with only very low energy (less than 42 cm^{-1}) to enter the counter. The energy change is thus the difference between the incident energy and the mean energy of those neutrons counted (about 25 cm^{-1}). A typical spectrum is shown in Fig. 3.22. This spectrum was recorded at 77°K for formic acid by Collins and Haywood [55].

The large peak at 517 cm^{-1} does not appear in the infrared spectrum of the vapor and is probably due to a number of lattice modes involving significant hydrogen-atom motions. The peak at 717 cm^{-1} is due to the nominal O–C–O bending vibration, which also includes significant contributions from hydrogen motion. The peak at 992 and 1087 cm^{-1} have been assigned to the out-of-plane C–H and O–H bends, respectively, while the small peaks at higher frequency involve C–O stretching and O–H–O bending vibrations.

The shapes of dispersion curves are sensitive to intermolecular force constants; the ability of coherent neutron inelastic scattering to measure energy changes as a function of $\mathbf{f}$ and thus obtain dispersion curves has been of great importance in solid-state studies. For this purpose single crystals must be used, and in the general case the spectrum must be taken as a function of energy at each point in one unit cell of reciprocal space. In addition, the

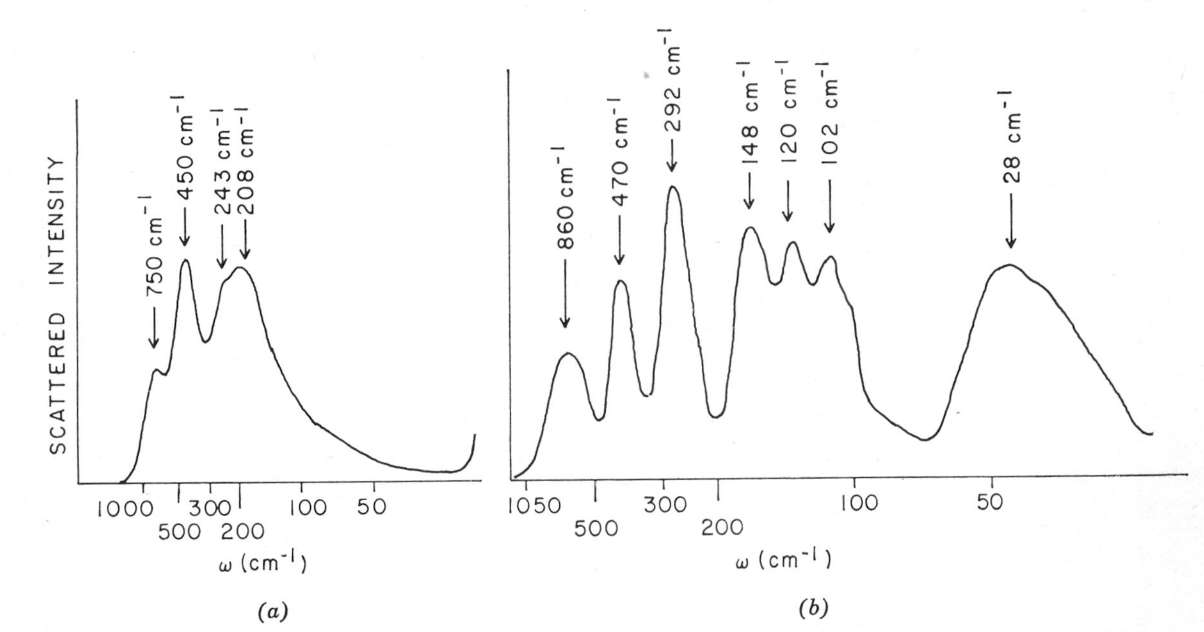

Fig. 3.21 (*a*) Neutron energy-gain spectrum from $CuSO_4 \cdot H_2O$ [53]. (*b*) Neutron energy-loss spectrum of *p* diiodobenzene [54].

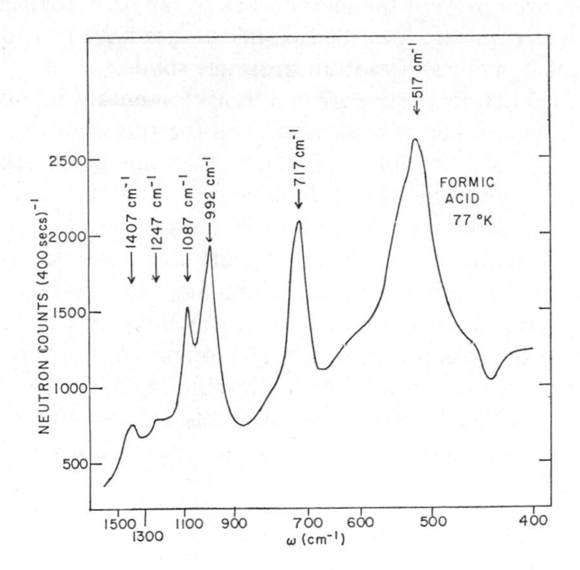

Fig. 3.22 Neutron energy-loss spectrum from crystalline formic acid at 77°K [55].

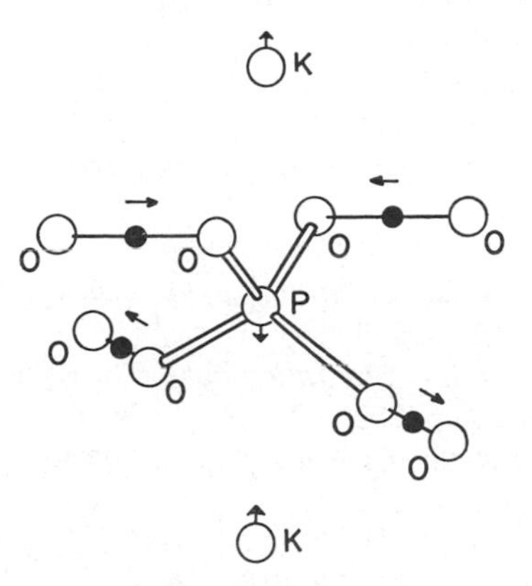

Fig. 3.23 Postulated vibrational mode in ferroelectric potassium dihydrogen phosphate leading to order-disorder transition. Normal mode analysis of phonon dispersion curves measured by neutron scattering has shown that there are also significant motions of the hydrogen atoms in the **c**-direction as well as significant distortions of the oxygen tetrahedron [56].

counting rates are low, and the method has so far been applied to only very simple crystal structures. Nevertheless, the studies have provided significant insight into the dynamics of motion in simple solids.

By measuring not only the energy of a transition but the intensity and hence the inelastic structure factor associated with the transition, the polarization components $\mathbf{C}_{ij}$ can be obtained. Thus the individual atomic motions associated with a given mode can be analyzed. Dramatic use of this possibility has been provided by the analysis of the vibrational modes associated with ferroelectricity in a number of materials. For example, Skalyo, Frazer, and Shirane [56] have shown by this technique that the vibration of ferroelectric potassium dehydrogen phosphate, which collapses into a disordered structure at the ferroelectric transition, is composed not only of hydrogen, potassium, and phosphorus motions corresponding to the so-called Cochran mode, as shown in Fig. 3.23, but also of significant distortions of the oxygen tetrahedra and displacements of the hydrogen atoms in the **c**-direction.

References

1. D. P. Mitchell and P. N. Powers, *Phys. Rev.*, **50,** 486 (1936).
2. E. O. Wollan and C. G. Shull, *Phys. Rev.*, **73,** 830 (1948); C. G. Shull and E. O. Wollan, *Phys. Rev.*, **81,** 527 (1951).
3. See, for example, A. Messiah, *Quantum Mechanics,* North-Holland Publishing Co., Amsterdam, 1961, Chapter 19.
4. Neutron Diffraction Commission, *Acta Cryst.*, **A25,** 391 (1969).
5. O. Halpern and M. H. Johnson, *Phys. Rev.*, **55,** 898 (1939).
6. For an extension to orbital magnetic scattering see G. Trammell, *Phys. Rev.*, **92,** 1387 (1953).
7. R. Weinstock, *Phys. Rev.*, **65,** 1 (1944).
8. *International Tables for X-Ray Crystallography*, The Kynoch Press, Birmingham, England, 1959, Volume 2, p. 264.
9. *International Tables for X-Ray Crystallography*, The Kynoch Press, Birmingham, England, 1959, Volume 2, p. 292.
10. *International Tables for X-Ray Crystallography*, The Kynoch Press, Birmingham, England, 1959, Volume 2, p. 265.
11. C. G. Darwin, *Phil. Mag.*, **27,** 315 675 (1914).
12. W. H. Zachariasen, *Acta Cryst.*, **23,** 558 (1967).
13. P. Coppens and W. C. Hamilton, *Acta Cryst.*, **A26,** 71 (1970).
14. R. M. Brugger, Chapter 2 in *Thermal Neutron Scattering*, P. A. Egealstaff, ed., Academic, London, 1965.
15. Available from Carbon Products Division, Union Carbide Corp.
16. D. R. Beaucage, M. A. Kelley, D. Ophir, S. Rankowitz, R. J. Spinrad, and R. Van Norton, *Nucl. Instr. Methods*, **40,** 26 (1966); W. C. Hamilton, *J. Comp. Phys.*, **2,** 417 (1968).

17. B. Buras and J. Liciejewicz, *Phys. Stat. Sol.*, **4**, 349 (1964). C. R. Hubbard, C. O. Quicksall and R. A. Jacobson, *Acta Cryst.* (to be published).
18. M. J. Moore, J. S. Kasper, and J. H. Menzel, *Nature*, **219**, 848 (1968).
19. M. J. Cooper, *Acta Cryst.*, **A24**, 624 (1968).
20. M. J. Cooper and K. D. Rouse, *Acta Cryst.*, **A24**, 405 (1968).
21. W. C. Hamilton, *Trans. Amer. Cryst. Assoc.*, **2**, 53 (1966).
22. P.-G. Jønssen and W. C. Hamilton, to be published.
23. S. K. Sikka, Thesis, Bombay (1969).
24. S. K. Sikka, *Acta Cryst.*, **A25**, 539 (1969).
25. J. Karle, *Acta Cryst.*, **20**, 881. (1966).
26. R. D. Ellison and H. A. Levy, *Chem. Div. Ann. Progress Reports*, *ORNL*-4164, Oak Ridge National Laboratory, Oak Ridge, Tenn., p. 126.
27. C. K. Johnson, *Chem. Div. Ann. Progress Report*, *ORNL*-4164, Oak Ridge National Laboratory, Oak Ridge, Tenn., p. 115.
28. A. C. McDonald and S. K. Sikka, *Acta Cryst.*, **B25**, 1804 (1969).
29. S. K. Sikka, *Acta Cryst.*, **A25**, 621 (1969).
30. C. K. Johnson, *J. Amer. Chem. Soc.*, **87**, 1802 (1965).
31. W. C. Hamilton and S. J. La Placa, *Acta Cryst.*, **B24**, 1147 (1968).
32. S. J. La Placa, W. C. Hamilton, and J. A. Ibers, *Inorg. Chem.*, **3**, 1491 (1964).
33. S. J. La Placa, W. C. Hamilton, J. A. Ibers, and A. Davison, *Inorg. Chem.*, **8**, 1928 (1969).
34. E. O. Schlemper, W. C. Hamilton, and S. J. La Placa, *J. Chem. Phys.*, **54**, 3990 (1971).
35. S. C. Abrahams and J. L. Bernstein, *J. Phys. Chem. Solids*, **28**, 1685 (1967).
36. S. C. Abrahams, W. C. Hamilton, and A. Sequeira, *J. Phys. Chem. Solids*, **28**, 1693 (1967).
37. P. Coppens, *Science*, **158**, 1577 (1967).
38. P. Coppens, T. M. Sabine, R. G. Delaplanc, and J. A. Ibcrs, *Acta Cryst.* **B25**, 2451 (1969).
39. P. Coppens, *Science*, **167**, 1126 (1970).
40. B. P. Schoenborn, *Nature*, **224**, 143 (1969). Private communication.
41. C. G. Shull and Y. Yamada, *J. Phys. Soc. Japan*, **17**, Suppl. B-III, 1 (1962); R. M. Moon, *Phys. Rev.*, **136**, A195 (1964); C. G. Shull and H. A. Mook, *Phys. Rev. Lett.*, **16**, 184 (1966); H. A. Mook, *Phys. Rev.*, **148**, 495 (1966).
42. L. M. Corliss, N. Elliott, and J. M. Hastings, *Phys. Rev.*, **104**, 924 (1956).
43. H. R. Kramers, *Physica*, **1**, 182 (1934).
44. C. G. Shull and J. S. Smart, *Phys. Rev.*, **76**, 1256 (1949).
45. P. W. Anderson in *Solid State Physics*, Volume 14, F. Seitz and D. Turnbull, Eds., Academic, New York, 1963.
46. J. M. Hastings, N. Elliott, and L. M. Corliss, *Phys. Rev.*, **115**, 13 (1959).
47. M. Pasternak and A. L. Spijkervet, *Phys. Rev.*, **181**, 574 (1969).
48. J. M. Hastings, L. M. Corliss, M. Blume, and M. Pasternak, *Phys. Rev.*, **B1**, 3209 (1970).
49. A. Yoshimori, *J. Phys. Soc. Japan*, **14**, 807 (1959); A. Herpin, P. Meriel, and J. Villain, *Compt. Rend.*, **249**, 1334 (1959); L. M. Corliss, J. M. Hastings, and R. J. Weiss, *Phys. Rev. Lett.*, **3**, 241 (1959).

50. For a comprehensive summary see H. R. Child, *Oak Ridge National Laboratory Report ORNL-TM*-1063, Oak Ridge, Tenn., 1965.
51. K. E. Larsson, Chapter 8 in *Thermal Neutron Scattering*, P. A. Egelstaff, Ed., Adademic, London, 1965.
52. J. A. Janik and A. Kowalska, in *Thermal Neutron Scattering*, P. A. Egelstaff, Ed., Academic, London, Chap. 10, 1965.
53. J. J. Rush, J. R. Ferraro, A. Walker, *Inorg. Chem.*, **6,** 346 (1967).
54. P. A. Reynolds and J. W. White, *Discussions Faraday Soc.*, **48,** 30 (1970).
55. M. F. Collins and B. C. Haywood, *J. Chem. Phys.*, **52,** 5740 (1970).
56. J. Skalyo, B. C. Frazer, and G. Shirane, *Phys. Rev.*, **B1,** 278 (1970).

Chapter **IV**

EXPERIMENTAL ASPECTS OF MÖSSBAUER SPECTROSCOPY

Rolfe H. Herber and Yehonathan Hazony

---

1 INTRODUCTION AND BRIEF SUMMARY OF THEORY

Nuclear Aspects

In the approximately dozen years that have elapsed since its discovery by the German physicist Rudolph L. Mössbauer in 1958 [1] recoilless emission and absorption of γ-radiation—commonly called the Mössbauer effect—has developed into a spectroscopic technique that can make significant contributions to the solution of a wide variety of chemical problems.

In view of the large number of excellent surveys that have appeared in the literature [2–10, 11] describing the fundamental physics behind the Mössbauer effect, it is neither appropriate nor necessary to cover this subject in detail in the present discussion. In order to be able to discuss the experimental aspects of Mössbauer spectroscopy in an intelligent and useful manner, however, we review very briefly some of the physical principles that underlie such measurements. No attempt is made to be either rigorous or exhaustively thorough in this review, and the interested reader is referred to the appropriate literature for a more detailed treatment of this subject.

The law of momentum conservation requires that a free atom of mass m, which emits a photon of energy $h\nu$, must recoil in a direction opposite to that of the emitted photon with an energy given by

$$E_R = \frac{E\gamma^2}{2mc^2}. \tag{4.1}$$

In the case of photons lying in the optical region of the electromagnetic spectrum, this recoil energy is generally small compared to the natural line width, Γ_{nat}, of the photon. This line width is derived from the Heisenberg uncertainty principle to be

$$\Gamma \leq \frac{\hbar}{\tau}. \tag{4.2}$$

where τ is the mean lifetime of the excited state to or from which the transition occurs. In the case where the energy of the photon lies in the region of X-ray or nuclear transitions (i.e., above the kilovolt region), the recoil-energy, E_R, can be very large compared to Γ_{nat}, provided that the excited state lifetime is sufficiently long. Moreover, it is worth noting that in the case of free-atom emission and absorption of radiation, the energy of the emitted photon is given by

$$E_s = E_t - E_R, \tag{4.3}$$

and that of the absorbed photon is given by

$$E_a = E_t + E_R, \tag{4.4}$$

where E_t is the transition energy between the two defined states. Resonant pumping (i.e., the population of the ground state–excited state transition in an absorber by absorption of a photon emitted in the excited state–ground state transition in the source) is proportional to the overlap between the energies of the two quanta. This overlap is shown schematically in Fig. 4.1*a* for the case in which $E_R \ll \Gamma_{\text{nat}}$ (optical case) and in Fig. 4.1*b* for the case in which $E_R \gg \Gamma_{\text{nat}}$ (nuclear case).

Mössbauer's principal discovery was that if the emitting and absorbing atoms are sufficiently tightly bound into a solid lattice, the overlap of the emitted line and the absorbed line could be increased very significantly, and hence resonant pumping could be increased very appreciably. In classical terms this means that the recoil energy is given by

$$E_R = \frac{E\gamma^2}{2Mc^2}, \tag{4.5}$$

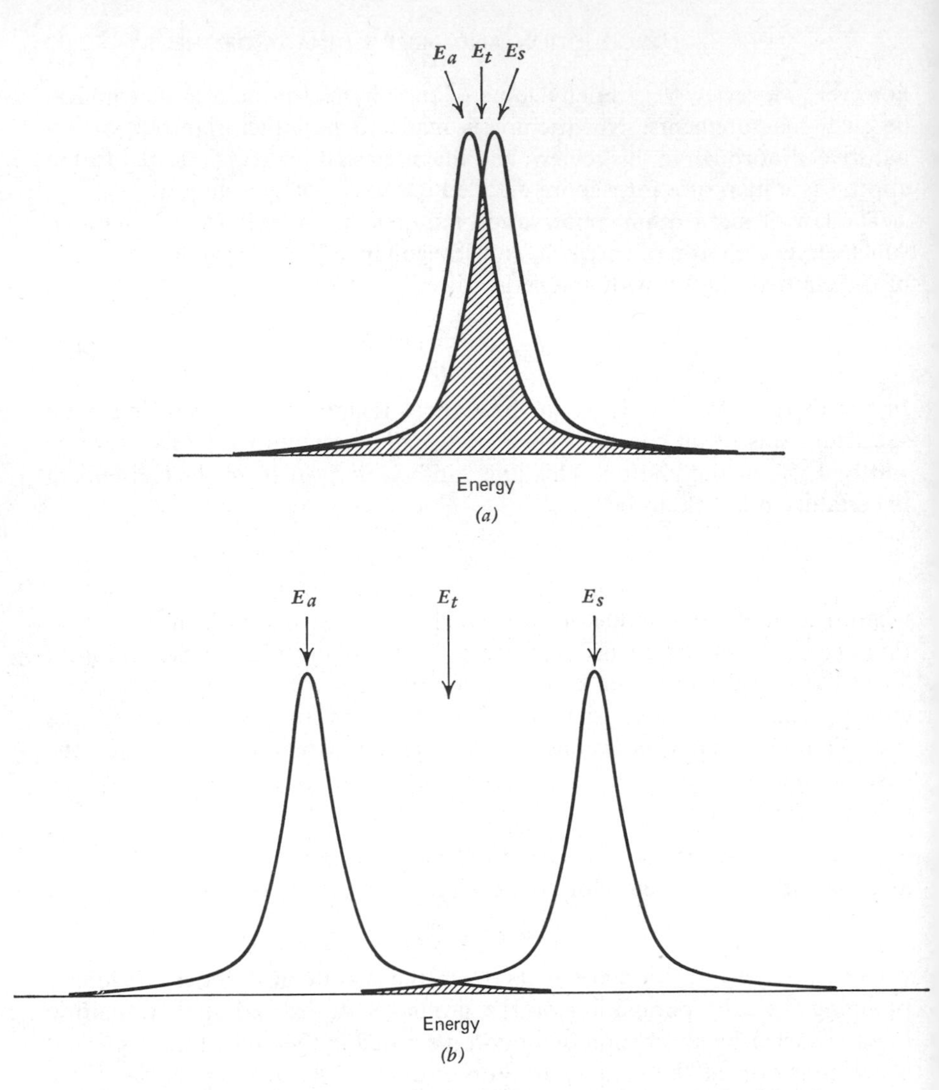

Fig. 4.1 Relationship between transition energy (E_t), the energy emitted by the source that is free to recoil ($E_s = E_t - E_R$), and the energy required by the absorber that is free to recoil ($E_a = E_t + E_R$). (*a*) is for the case where E_R is small compared to the line width; (*b*) is for the case when E_R is large compared to the line width. The resonance effect depends on the overlap of the two lines, shown by the shaded area.

where M is now the mass of the whole lattice (or at least a domain of that lattice containing a large number of atoms). In quantum mechanical terms it means that the probability of observing zero phonon events (i.e., where there is no energy transfer to or from the lattice because the accessible lattice modes lie too high in energy) has been increased appreciably over one phonon, two phonon, and more, events.

On the basis of the brief description above it can readily be seen why the Mössbauer effect is observed for only a limited number of elements. First, the γ-radiation energy must lie within a relatively narrow range. If this is too low, only a small fraction of the radiation escapes from the source, and the absorbers to be studied (see below) will have to be extremely thin for any of the radiation to penetrate and reach the detector. On the other hand, if the γ-ray energy is too large, even cooling of lattices (in which the Mössbauer atom is tightly bonded) to cryogenic temperatures is insufficient to prevent momentum transfer to the lattice. In practical terms the γ-ray energies that satisfy these conditions lie in the range 5 keV $< E_t <$ 150 keV.

Second, in order to observe the Mössbauer effect (resonant pumping) it is necessary to consider only those transitions in the source that decay directly *to the ground state*. The reason for this is that the transition in the absorber (which is presumably originally in the ground state) must involve the identical transition (except for direction) *from the ground state*. This does not mean that the ground state in the absorber necessarily has an infinitely long lifetime, since long-lived nuclear states (e.g., the 1.6×10^7 year ground state of ^{129}I or the 2.14×10^6 year state in ^{237}Np, both of which are suitable as Mössbauer absorbers) have been successfully used in such experiments. It does mean, however, that the source gamma transition energy range referred to above must involve a transition to a state that is preferentially populated over all others in the absorber. Third, the Heisenberg uncertainty principle already referred to imposes some limitations on the lifetime of the nuclear state *from* which the transition occurs in the source and *to* which the analogous transition occurs in the absorber. If this lifetime is too short, the resonance lines will be so broad that the effect will be difficult to observe (or if it is observable the "resolution" of those hyperfine effects that contain most of the "chemical information" will not be seen). On the other hand, if the excited-state lifetime is too long, the line width will be so narrow that very minute accidental vibrations, small temperature differences, or limitations on mechanical portions of Mössbauer spectrometers will make it difficult or impossible to observe the resonance effect. In practical terms the excited-state lifetimes that are useful in Mössbauer transitions fall in the range 10^{-10} sec $< t_e < 10^{-6}$ sec, although there are some exceptions to these limits.

Table 4.1 Mössbauer Nuclides and Some of Their Properties Pertinent to the Observation of the Mössbauer Effect.[a]

Isotope	E_γ *(keV)*	$t_{1/2}$ *(sec)*	w_0 *(mm/sec)*	σ_0 *(cm²)*
^{107}Ag	93	44	6.685×10^{-11}	5.388×10^{-20}
^{197}Au	77.3	1.8×10^{-9}	1.966	4.355×10^{-20}
^{133}Cs	81.0	6.28×10^{-9}	0.5378	1.064×10^{-19}
^{160}Dy	86.8	2.00×10^{-9}	1.576	2.894×10^{-19}
^{161}Dy	25.7	2.87×10^{-8}	0.3716	1.062×10^{-18}
^{161}Dy	74.6	3.10×10^{-9}	1.183	3.216×10^{-20}
^{166}Er	80.6	1.82×10^{-9}	1.865	2.296×10^{-19}
^{151}Eu	21.6	8.8×10^{-9}	1.439	2.330×10^{-19}
^{153}Eu	97.4	1.6×10^{-10}	17.55	1.817×10^{-19}
^{153}Eu	103.2	3.86×10^{-9}	0.6868	5.972×10^{-20}
^{57}Fe	14.4	9.77×10^{-8}	0.1943	2.355×10^{-18}
^{57}Fe	136.4	8.7×10^{-9}	0.2305	4.214×10^{-20}
^{155}Gd	60.0	2.4×10^{-10}	19.00	1.199×10^{-19}
^{155}Gd	86.5	6.15×10^{-9}	0.5142	2.164×10^{-19}
^{156}Gd	89.0	2.19×10^{-9}	1.404	3.042×10^{-19}
^{158}Gd	79.5	2.40×10^{-9}	1.434	2.788×10^{-19}
^{73}Ge	67.0	1.62×10^{-9}	2.520	4.434×10^{-19}
^{177}Hf	113.0	5.0×10^{-10}	4.843	6.906×10^{-20}
^{127}I	57.6	1.86×10^{-9}	2.553	2.070×10^{-19}
^{129}I	27.8	1.61×10^{-8}	0.6123	3.971×10^{-19}
^{191}Ir	129.4	1.28×10^{-10}	16.515	5.452×10^{-20}
^{193}Ir	73	6.0×10^{-9}	0.6245	3.279×10^{-20}
^{40}K	29.4	3.9×10^{-9}	2.386	1.631×10^{-18}
^{83}Kr	9.3	1.47×10^{-7}	0.2001	1.886×10^{-18}
^{61}Ni	67.4	5.3×10^{-9}	0.7658	7.212×10^{-19}
^{237}Np	59.5	6.3×10^{-8}	7.293×10^{-2}	2.917×10^{-19}
^{186}Os	137.2	8.4×10^{-10}	2.374	2.888×10^{-19}
^{188}Os	155.0	7.1×10^{-10}	2.485	2.766×10^{-19}
^{141}Pr	145.4	1.92×10^{-9}	0.9797	1.064×10^{-19}
^{195}Pt	98.8	1.6×10^{-10}	17.30	5.248×10^{-20}
^{195}Pt	129	6.2×10^{-10}	3.420	7.746×10^{-21}
^{187}Re	134.2	1.0×10^{-11}	203.8	5.403×10^{-20}
^{99}Ru	90	2.02×10^{-8}	0.1505	1.007×10^{-19}
^{121}Sb	37.2	3.5×10^{-9}	2.104	2.037×10^{-19}
^{149}Sm	22	7.6×10^{-9}	1.636	3.888×10^{-19}
^{119}Sn	23.9	1.84×10^{-8}	0.6227	1.321×10^{-18}
^{181}Ta	6.25	6.8×10^{-6}	6.436×10^{-3}	1.702×10^{-18}

Table 4.1 (Continued)

Isotope	E_γ *(keV)*	$t_{1/2}$ *(sec)*	w_0 *(mm/sec)*	σ_0 *(cm²)*
^{159}Tb	58.0	1.3×10^{-10}	36.28	9.827×10^{-20}
^{125}Te	35.5	1.48×10^{-9}	5.209	2.718×10^{-19}
^{169}Tm	8.41	3.9×10^{-9}	8.340	2.122×10^{-19}
^{182}W	100.1	1.37×10^{-9}	1.995	2.456×10^{-19}
^{183}W	46.5	2.0×10^{-10}	29.43	2.265×10^{-19}
^{183}W	99.1	6.0×10^{-10}	4.601	6.855×10^{-20}
^{184}W	111.2	1.26×10^{-9}	1.952	2.638×10^{-19}
^{186}W	122.5	1.05×10^{-9}	2.127	2.997×10^{-19}
^{129}Xe	39.6	9.85×10^{-10}	7.016	2.603×10^{-19}
^{131}Xe	80.2	4.5×10^{-10}	7.583	6.141×10^{-20}
^{170}Yb	84.3	1.59×10^{-9}	2.042	2.237×10^{-19}
^{171}Yb	66.7	5×10^{-10}	8.199	9.989×10^{-20}
^{67}Zn	93	9.4×10^{-6}	3.129×10^{-4}	1.179×10^{-19}

[a] Taken from ref. 11, A. H. Muir et al., by permission of the authors.

[b] Additional isotopes in which a resonance has been observed subsequent to the completion of the present compilation: ^{164}Dy, ^{168}Er, ^{160}Gd, ^{176}Hf, ^{180}Hf, and ^{174}Yb.

Finally, as a matter of practical importance, it is desirable to carry out Mössbauer studies on systems in which the transition in the source can be fed from a long-lived precursor state (having a lifetime of at least an hour or so, although coulomb excitation "in-beam" experiments [12–15] in which this condition is not satisfied have been successfully carried out), and in which the suitable absorber atom has an adequate natural abundance (or can be obtained by stable isotope enrichment techniques at costs that are not prohibitive).

The nuclides that meet the above requirements and in which the Mössbauer effect has been successfully observed are summarized in Table 4.1, together with some of the more important parameters characteristic of the Mössbauer transition involved. Although it is likely that this list will have some further additions as sophisticated experimental problems are solved, the absence of representatives of the light elements is striking. This absence is primarily due to the high energies of ground-state to first excited-state transitions in such nuclei and the concomitant short lifetime of the first excited states.

Returning now to the fundamentals of the Mössbauer effect, it is important to consider the experimental consequences of the resonant pumping phenomenon on which the effect is based. In the most common kind of Mössbauer experiment the arrangement of source, the absorber (experimental sample), and the detector is that shown in Fig. 4.2, in which the radiation leaving the source and penetrating through the absorber is observed in an appropriate detector sensitive to the γ-radiation. When resonant pumping occurs in the absorber, the cross section (probability) for absorption of the γ-radiation increases very sharply. This absorption is followed (in a time characteristic

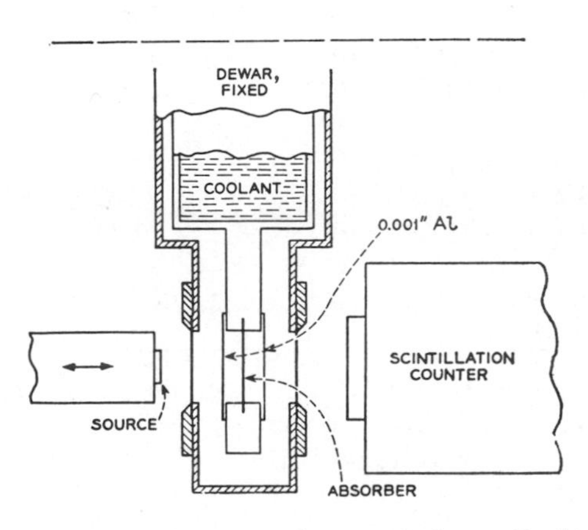

Fig. 4.2 Transmission geometry for a Mössbauer experiment. In this case the stationary sample is cooled in the dewar, and the moving source is held at room temperature.

of the mean life of the excited nuclear state) by an isotropic reemission, so that the absorber acts as an isotropic scatterer. Hence in conditions under which resonant pumping occurs the intensity of the radiation reaching the detector will be significantly reduced compared to the intensity when no resonant pumping in the absorber takes place.

Before this phenomenon can be exploited as a spectroscopic technique, it must be possible to modulate the γ-ray energy sufficiently so that the resonance condition can be swept through in a systematic fashion. This energy modulation can be obtained by virtue of the Doppler effect, by imparting a relative velocity to the source-absorber system. The Doppler energy is given by the equation

$$E_D = \frac{v}{c} E_t, \tag{4.6}$$

where v is the relative source-absorber velocity and c is the speed of light. An order-of-magnitude calculation of v can be effected in the following way:

If the line width calculated from (4.2) is Γ_{nat}, then the minimum experimental line width is $2\Gamma_{nat}$, since the line width of both the radiation emitted by the source and that absorbed by the scatterer are involved. Thus

$$\Gamma_{exp} \geq 2\Gamma_{nat} > \frac{2\hbar}{\tau} = \frac{1.316}{\tau} \times 10^{-15}\ \text{eV sec}$$

$$= \frac{4.562 \times 10^{-16}}{t_{1/2}}\ \text{eV sec.} \tag{4.7}$$

Considering a typical case, the lifetime of the 14.4-keV state in ^{57}Fe is 9.77×10^{-8} sec, so that from (4.7) $\Gamma_{exp} \geq 4.67 \times 10^{-9}$ eV. To modulate the γ-ray energy by the Doppler phenomenon so that an energy shift of Γ_{exp} is observed one sets $E_D = \Gamma_{exp}$ so that for ^{57}Fe

$$v = c\frac{E_D}{E_t} = c\frac{\Gamma_{exp}}{E_t} \geq 0.194\ \text{mm/sec.} \tag{4.8}$$

Thus the relative source-absorber velocities required for a typical Mössbauer experiment in which the γ-ray energy is shifted by one line width must be in the mm/sec range. In order to be able to study phenomena such as magnetic hyperfine effects or Mössbauer transitions involving excited states with lifetimes an order of magnitude smaller than that of ^{57}Fe, most Mössbauer spectrometers have transducers that cover relative velocities of up to 10 cm/sec. The practical aspects of obtaining such velocities in a variety of ways and of incorporating the appropriate velocity transducers in a complete Mössbauer spectrometer are examined in greater detail in Section 4.

At this point it is appropriate to consider briefly the most important parameters that can be extracted from a Mössbauer spectrum, or a series of such spectra run as a function some experimentally variable parameter such as temperature, pressure, chemical composition, magnetic or electric field effects, sample thickness and orientation, and others. Recalling that a Mössbauer spectrum is obtained by determining the intensity of the γ-radiation that reaches the detector as a function of the Doppler modulating velocity, we see that such a spectrum can be represented by a plot of intensity *versus* velocity. Under ideal conditions the velocity-dependent intensity is given by

$$I(v) = \frac{(\Gamma/2)^2 I(0)}{(v - v_0)^2 + (\Gamma/2)^2}, \tag{4.9}$$

in which $I(0)$ is the intensity of the resonance effect when the energies of the emitted and the absorbed gamma quanta match precisely, v is the Doppler velocity, v_0 is the Doppler velocity at the resonance maximum, and Γ is the characteristic line width of the resonance line (equal to $2\hbar/\tau = 2\Gamma_{nat}$ under ideal conditions). The value of $I(0)$ can in turn be related to the resonant absorption cross section, σ_0, which has a value given by

$$\sigma_0 = \frac{2.45 \times 10^9}{E_t^2} \frac{2I_e + 1}{2I_g + 1} \frac{1}{1 + \alpha_t} \text{ barns,} \tag{4.10}$$

in which E_t is the gamma-transition energy in keV, I_e and I_g are the excited- and ground-state nuclear spins, and α_t is the total internal conversion coefficient of the gamma transition.

Returning to (4.9) it is seen that when $v = v_0$, $I(v) = I(0)$, and when $I(v) = \frac{1}{2}I(0), v - v_0 = \Gamma/2$, so that $\Gamma = 2(v - v_0)$. From the latter expression it is seen that the line width Γ is equal to twice the distance from the velocity of resonance maximum to the velocity at which the resonance effect has fallen to just one-half its maximum value. The shape of this resonance line (under ideal conditions) is that of a Lorentzian function, and is represented schematically in Fig. 4.3. In an actual experiment the observed line width, Γ_{exp}, very often is significantly larger than $2\Gamma_{nat}$ (the minimum possible line width arising from the combination of the two Lorentzian lines, one of the source, the other of the absorber) due to the effects of the hyperfine interactions (to be discussed below) and the presence of a finite number of

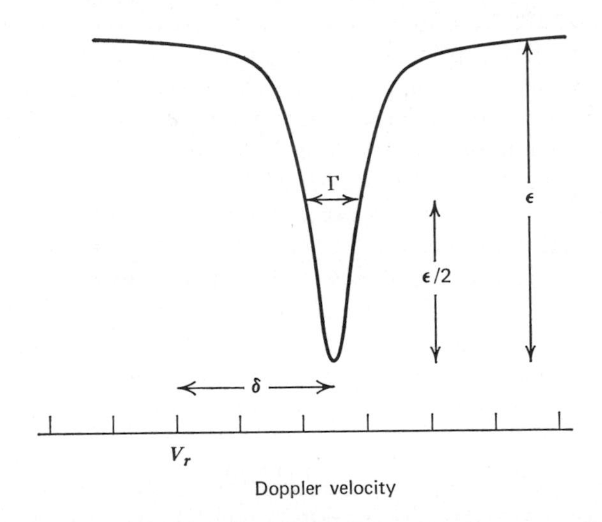

Fig. 4.3 Parameters extractable from a simple Mössbauer spectrum include the effect magnitude (ϵ), the full width at half maximum (Γ) and the isomer shift (δ). The vertical axis represents an arbitrary intensity scale. V_r is the reference point for the isomer shift scale (see p. 227).

absorber (and/or source) nuclei in the optical path. This latter sample thickness broadening is usually expressed in terms of a thickness parameter, t, which is given [16, 17] by the expressions

$$t_s = n_s \sigma_0 f_s \quad \text{and} \quad t_a = n_a \sigma_0 f_a, \tag{4.11}$$

where the subscripts s and a refer to source and absorber, respectively, σ_0 is the resonance cross section defined by (4.10), and f is the recoil-free fraction (also called the Mössbauer-Lamb fraction), which can be related to the dynamics of the Mössbauer atom in the lattice, at least in the case of ideal solids. In its simplest form the recoil-free fraction is given by

$$f = \exp\left\{\frac{-\langle x^2\rangle}{\lambda\!\!\!^{-2}}\right\}, \tag{4.12}$$

where $\langle x^2\rangle$ is the mean-square amplitude of vibration of the Mössbauer atom in the direction of the gamma-quantum emission or absorption and $\lambda\!\!\!^{-} = \lambda/2\pi$, where λ is the characteristic wavelength of the gamma quantum, given by

$$\lambda = \frac{hc}{E_t}. \tag{4.13}$$

For ^{57}Fe,

$$E_t = 1.438 \times 10^4 \text{ eV} \qquad \text{so that}$$

$$\lambda = \frac{(6.6255 \times 10^{-27} \text{ erg sec}) \times (2.997 \times 10^{10} \text{ cm/sec})}{(1.438 \times 10^4 \text{ eV}) \times (1.6021 \times 10^{-12} \text{ erg/eV})}$$

$$= 8.63 \times 10^{-9} \text{ cm} = 0.863 \text{ Å},$$

from which $\lambda\!\!\!^{-} = 0.137$ Å. Thus from (4.12) it is seen that if the mean amplitude of vibration (or linear displacement during the characteristic lifetime of the Mössbauer excited state) is comparable to atomic diameters, f will become vanishingly small; hence the resonance effect will be unobservable. This generalization implies that the Mössbauer technique is applicable only to chemical studies in the solid state, although some results for very viscous liquids have been reported [18, 19].

The parameters that can be deduced from a Mössbauer experiment (or from a related series of such experiments) may be divided roughly into two groups. First, there are those parameters derived from the isomer shift, and quadrupole and magnetic splitting of the spectra. These interactions reflect the details of the electronic charge and the spin-density distributions. Usually in order to evaluate these parameters appropriately it is necessary to carry out experiments over a temperature range. While doing this, it is possible to derive the parameters classified in the second category: the Debye-Waller factor and the thermal shift. The obvious application of these two parameters is in studies in lattice and in atomic dynamics. However, they may also be discussed in terms of effective force constants that are relevant to the details of the electronic-charge density distribution and may therefore be used in some special cases to elucidate problems of chemical bonding.

The Isomer Shift (IS)

Returning now to the implications of (4.9) it might be expected that the resonance maximum should be observed when the Doppler velocity is just equal to zero; that is when the source and the absorber atom are at rest with respect to each other. In fact, this is in general not at all what is observed in a Mössbauer spectrum; rather the resonance maximum is observed at some finite (nonzero) velocity. This shift of the resonance maximum from $v = 0$ (or from some arbitrarily defined reference point, see below) is called the isomer shift (or chemical shift, or even resonance shift; however, in the present discussion the term isomer shift—or the symbol δ—is used exclusively in consonance with recommended conventions. See the Appendix.)

The isomer shift arises from the fact that the nucleus is not an infinitesimally small point in space, but does in fact occupy a significant volume. Quantum theory predicts a finite probability that this same volume will be penetrated by atomic *s*-electrons. The electrostatic interaction between the nuclear-charge density distribution and the penetrating *s*-electron distribution will slightly change the energies of the nuclear levels. If the nuclear-charge density distribution varies from the nuclear ground state to the first excited state, this electrostatic contribution will be different and will therefore shift the energy of the transition between the two levels. The atomic-charge density distribution is affected directly or indirectly (via shielding effects) by chemical bonding. The probability of *s*-electron penetration into (or overlapping with) the nucleus depends on the details of the chemical bonding. Consequently, if the Mössbauer emitter and the absorber atoms are embedded in different chemical environments, the energies of the corresponding nuclear transitions will be slightly shifted relative to each other, reflecting the difference in chemical bonding in the two matrices.

The isomer shift was first correctly identified by Kistner and Sunyar [20] in 1960, in connection with their study of the Mössbauer effect in Fe_2O_3, in which they used a source of ^{57}Co diffused into stainless steel. The isomer shift can be adequately represented by the expression

$$\text{IS} = \frac{2\pi}{5} e^2 Z \, \Delta R^2 [\psi(\text{o})_a{}^2 - \psi(\text{o})_s{}^2], \tag{4.14}$$

in which Z is the nuclear charge (atomic number), $\Delta R^2 = R_e{}^2 - R_g{}^2$ (the difference in the square of the nuclear radii in the excited and ground states, respectively), and $\psi(\text{o})_a{}^2$ and $\psi(\text{o})_s{}^2$ are the electron densities (squares of the wave functions) evaluated within the nuclear volume in the absorber and the source, respectively. Noting that $\Delta R^2 \approx 2R\Delta R$, (4.14) becomes

$$\text{IS} = \left(\frac{4\pi}{5}\right) e^2 Z R^2 [\psi(\text{o})_a{}^2 - \psi(\text{o})_s{}^2] \left(\frac{\Delta R}{R}\right), \tag{4.15}$$

where $\Delta R/R$ is the fractional increase in nuclear radius between the ground state and the excited nuclear state.

There are several important consequences of the relationship expressed in (4.15). For those nuclei for which $\Delta R \neq 0$ for the two nuclear states involved in the Mössbauer transition, the isomer shift will have a nonzero value that can be related to the electronic environment around the Mössbauer atom, and from which, in consequence, chemical information can be extracted. If the chemical form of the source atom is exactly identical to the chemical form of the absorber atom, $\psi(\text{o})_a{}^2 = \psi(\text{o})_s{}^2$, and the isomer shift is identically equal to 0 (assuming that source and absorber are at the same temperature). This situation can be realized in the cases in which the long-lived precursor is an isomeric state of the Mössbauer ground state, as for example in ^{119}Sn (but not in the case of the ^{57}Co precursor of the 14.4-keV transition in ^{57}Fe, or the ^{129}Te precursor of the 27.7-keV transition in ^{129}I). Thus it is possible to determine the position of zero relative velocity (a particular problem in constant acceleration spectrometers, see below) by determining the resonance maximum position for an SnO_2 absorber using an $^{119m}SnO_2$ source.

Finally, it should be noted that the isomer shift as defined by (4.15) always involves the difference in electron densities at the nucleus in two distinct environments: the source and the absorber. Hence absolute values of the IS for a series of absorbers will have little significance since they will also reflect the chemical details of the source, and it has become general practice to refer all isomer shifts (where possible) to the experimentally observed resonance maximum for a reference absorber. For ^{57}Fe this reference point is usually that observed at room temperature for metallic iron (or sodium nitroprusside, $Na_2Fe(CN)_5NO{\cdot}2H_2O$) (see the Appendix); for ^{119}Sn the reference points that have been suggested are metallic tetrahedral tin, or SnO_2 (identical to that observed for $BaSnO_3$ within the presently quoted experimental errors) and I^- for ^{129}I. For other Mössbauer nuclides no generally agreed-upon isomer-shift standards have been adopted at this writing.

A number of correlations and systematics relating the isomer shift to electron configuration, oxidation state, covalency, ligand electronegativity, and other parameters of interest to the chemist have been reported. Because the present discussion is limited to the experimental aspects of Mössbauer spectroscopy rather than to its application to the solution of particular chemical problems, it is not possible to comment on these correlations and systematics in any detail; hence the reader is referred to the current scientific literature for more penetrating and exhaustive discussions of the relationship between the IS and the chemical aspects of the problems that are of particular interest. Some of these relationships have been recently reviewed [21], while additional discussions appear elsewhere in the scientific literature [22–25].

Quadrupole Splitting (Q.S.)

The second type of hyperfine interaction observable in a Mössbauer spectrum arises from the interaction between the nuclear quadrupole moment and the gradient of the electrostatic field at the Mössbauer nucleus.

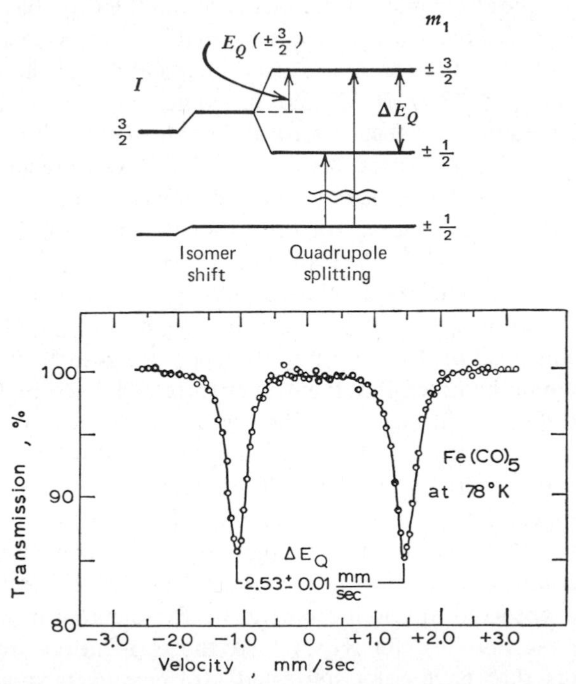

Fig. 4.4 Mössbauer spectrum of $Fe(CO)_5$ at 78°K showing quadrupole splitting of the resonance line. This splitting arises from the interaction between the nuclear quadrupole moment, Q, ($\sim +0.3$ barns in ^{57}Fe for the 14.4-keV transition) and the field gradient at the iron nucleus lattice site, q. From [4].

This interaction is referred to as the quadrupole coupling, or more commonly, the quadrupole splitting, symbolized by QS or Δ.

The nuclear quadrupole moment arises from the deviation from spherical symmetry of the nucleus. The convention relating the quadrupole moment to the shape of the nucleus is that an oblate (flattened) nucleus has a negative quadrupole moment, while a prolate (elongated) nucleus has a positive quadrupole moment. Nuclei whose spins are 0 (even A) or $\frac{1}{2}$ (odd A) are spherically symmetric and have zero quadrupole moment. Thus the ground states of some common Mössbauer nuclides with spins of $\frac{1}{2}$ (e.g., ^{57}Fe, ^{119}Sn, ^{169}Tm, ^{171}Yb, among others) or spins of 0 (e.g., ^{150}Nd, ^{152}Sm, ^{154}Sm ^{154}Gd, ^{174}Yb, among others) cannot interact with the electrostatic field in terms of a quadrupolar interaction and do not contribute to such hyperfine effects. The degeneracy of nuclear levels with spin greater than $\frac{1}{2}$ can, however, be removed in an asymmetric electrostatic field.

In the cases of ^{57}Fe, ^{119}Sn, ^{169}Tm, ^{171}Yb, and others the situation is particularly simple and susceptible to ready correlation with molecular structure and symmetry. For such nuclides the upper ($\frac{3}{2}$) state (in the case of ^{131}Xe this is the lower state) is split under the influence of a noncubic electrostatic field into two levels, which can be designated as $\pm\frac{3}{2}$ and $\pm\frac{1}{2}$, as shown in Fig. 4.4. Whether the $\pm\frac{3}{2}$ level moves up (with respect to the degenerate level energy) or down depends again on the shape of the electrostatic field; that is, whether the latter is oblate or prolate. In either case it is possible to observe two transitions to (in the case of absorbers) or from (in the case of sources) the excited nuclear state, and hence the Mössbauer spectrum will consist (ideally) of two absorption lines, as shown in Fig. 4.5. Whether the two resonance lines can, in fact, be resolved from each other depends on the magnitude of the hyperfine interaction (see below) and on the other experimental conditions, such as the resonance effect magnitude, magnetic or relaxation effects, and so on.

For nuclides in which the Mössbauer transition involves $\frac{1}{2}$ and $\frac{3}{2}$ spin states, the separation between the two resonance lines is given by

$$\mathrm{QS} = \tfrac{1}{2}e^2 qQ[1 + (\tfrac{1}{3})\eta^2]^{1/2}, \tag{4.16}$$

where q and η represent the trace of the electric field gradient (EFG) tensor and are given by $q = V_{zz}$ and $\eta = (V_{xx} - V_{yy})/V_{zz}[V_{zz} = \partial^2 V/\partial z^2$, $V_{xx} = \partial^2 V/\partial x^2$, and $V_{yy} = \partial^2 V/\partial y^2]$, Q is the nuclear quadrupole moment, and e the electronic charge unit.

Generally it is possible to distinguish two separate contributions to the EFG. One is due to the deformation of complete electronic shells induced by the disposition of the effective charges on the ligands, and the second is due to the orientation of electronic charge densities in incomplete electronic shells. The two mechanisms may be exemplified by the groups of high-spin

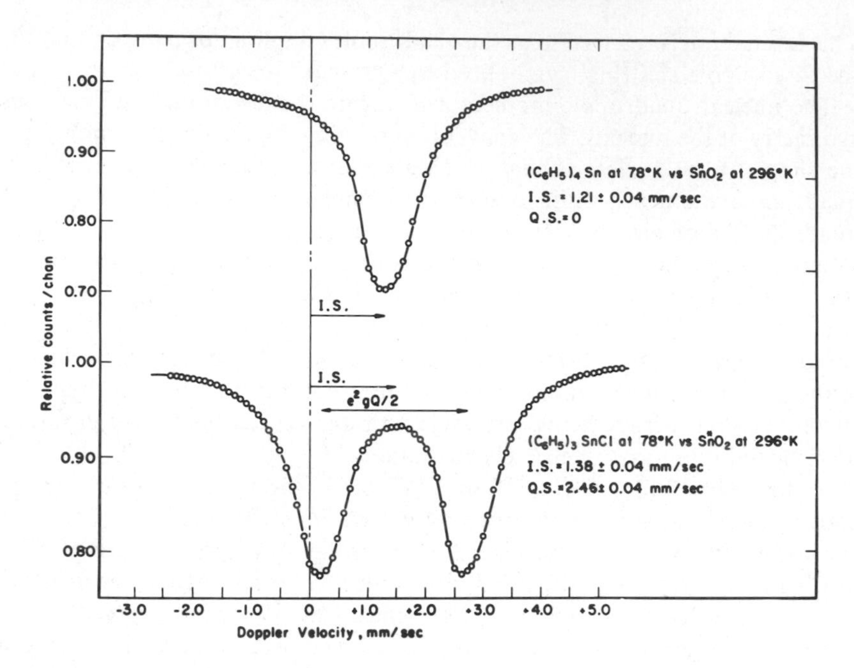

Fig. 4.5 Mössbauer spectra of two organo-tin compounds showing the effect of changing the electric charge symmetry around the ^{119}Sn atom. The unsplit resonance observed for the tetrahedral $(C_6H_5)_4Sn$ is split into a double resonance in $(C_6H_5)_3SnCl$. The magnitude of the interaction between the nuclear quadrupole moment, Q (-0.7 barns in ^{119}Sn for the 23.8-keV transition) and the field gradient, q, is given by the separation between the two components of the doublet.

ferric (Fe^{3+}) and ferrous (Fe^{2+}) compounds of nearly octahedral or tetrahedral symmetry. The QS values observed in the first group are very small (except for cases of very strong distortion from cubic symmetry) and temperature independent, while the QS values observed for the second group are large and in most cases strongly temperature dependent.

In the case where the only contribution to QS is due to the distortion of complete electronic shells, q, the z component of the EFG at the nucleus may be written in terms of the effective charges on the ligands and an antishielding factor $(1 - \gamma_\infty)$, giving $q = (1 - \gamma_\infty)q_{1\mathrm{at}}$. The number of parameters involved in such a computation (effective charges on the ligands as well as the antishielding factor) is large, and the usefulness of such an analysis for a single compound is therefore limited.

The practical consequences of this description can be related to the molecular (ligand or nearest neighbor) symmetry around the Mössbauer atom.

If this atom is in a site of cylindrical symmetry (that is, if a fourfold symmetry axis passes through the nucleus), then a 90°-rotation will leave the field gradient unchanged, so that $V_{xx} = V_{yy}$; hence $\eta = 0$ and the field gradient is completely specified by its z component; that is, for cylindrical symmetry

$$\mathrm{QS} = \tfrac{1}{2}e^2 q_{\mathrm{latt}} Q(1 - \gamma_\infty) \tag{4.17}$$

$$= \tfrac{1}{2}e^2 \frac{\partial^2 V_{\mathrm{latt}}}{\partial z^2} Q(1 - \gamma_\infty). \tag{4.18}$$

(It should be noted that an n-fold symmetry axis through the Mössbauer atom, which leaves the field gradient unchanged on $360°/n$ rotation, also satisfies the above requirement and leads to $\eta = 0$ for $n \geq 3$.)

Moreover, if the Mössbauer atom occupies a site so that two mutually perpendicular axes of threefold or higher symmetry pass through it, the field gradient will vanish, and thus QS will equal 0. This situation may (but does not necessarily) obtain for Mössbauer atoms occupying sites of tetrahedral (e.g., FeO_4^{-2} or $(CH_3)_4Sn$) or octahedral (e.g., $Fe(CN)_6^{-4}$ or SnF_6^{-2}) symmetry, and under such conditions a single resonance line will be observed.

In the discussion of the quadrupole splitting parameter for $\frac{1}{2} \leftrightarrow \frac{3}{2}$ spin-state transitions it is appropriate to note that while the magnitude of QS is readily extracted from the separation between the two resonance lines, the sign of the field gradient (that is whether oblate or prolate, whether the $\pm\frac{3}{2}$ level moves up or down in energy with respect to the degenerate level) cannot be determined. In order to determine this sign it is necessary to work either with single crystals of known orientation with respect to the optical axis or to carry out experiments in a magnetic field. Such latter studies, which were first suggested by Ruby and Flinn [26], have been discussed in detail by Collins [27], and a number of interesting applications to molecular structure studies have been published [28, 29].

The case in which a significant contribution to QS is due to incomplete electronic shells may be illustrated by octahedral (or tetrahedral) high-spin Fe^{2+} compounds or low-spin Fe^{3+} complexes. It appears that the contribution from the single electron outside (or hole in) the closed subshell dominates, and the lattice contribution may be ignored as long as the distortion from cubic symmetry is small [30]. A rather small distortion suffices to remove the threefold degeneracy of the t_{2g} ground level and thus contribute to QS the full value of q associated with a single electron at the lowest t_{2g} level, namely, $\frac{4}{7}\langle r^{-3}\rangle_{3d}$ [31]. Any admixture of these three levels reduces this value. Such an admixture may be caused by the spin-orbit interaction if the energy involved is comparable with the separation between the nondegenerate t_{2g} levels. A different kind of admixture is caused by thermal population of the upper t_{2g} levels when the temperature of the experiment is comparable with the splitting of the t_{2g} levels. Two separable covalency effects cause the reduction of QS via the expansion of the $3d$ orbital. π-bonding mixes the t_{2g} orbitals with combinations of ligand orbitals of appropriate symmetry, that is, forming molecular orbitals (MO). The contribution to QS is determined by the MO mixing parameter α_{MO}^2 according to $\langle r^{-3}\rangle_{\mathrm{MO}} = \alpha_{\mathrm{MO}}^2 \langle r^{-3}\rangle_{3d}$. Additionally, the effective charge on the central ion is significantly reduced below its nominal value because of charge donation from the ligands via both σ and π bonding. The radial expansion of the $3d$ wave function, caused by the strong reduction of the effective charge on the central ion, is called the *nephelauxetic effect*, which may be described by a radial scaling factor [32]. If the radial scaling factor is denoted by β_0, then the density of the $3d$ wave function in the solid will be related to that of the free ion by $\langle r^{-3}\rangle_{3d} = \beta_0^3\langle r^{-3}\rangle_{\mathrm{ion}}$, and to include the effect of MO mixing, $\langle r^{-3}\rangle_{\mathrm{solid}} = \alpha_c^2 \langle r^{-3}\rangle_{\mathrm{ion}}$, where $\alpha_c^2 = \alpha_{\mathrm{MO}}^2 \cdot \beta_0^3$ [23]. While the MO mixing parameter may vary between $\frac{1}{2} < \alpha_{\mathrm{MO}}^2 < 1$, the corresponding range for α_c^2 is $0 < \alpha_c^2 < 1$ because of the additional reduction due to β_0^3. This may explain the very large range of values observed for QS in Fe^{2+} compounds; that is $QS_{FeS} \sim 0.3$ mm/sec or $QS_{FeO} \sim 0.4$ mm/sec compared with $QS_{FeSO_4} \sim 3.6$ mm/sec.

The quadrupole splitting in ferrous compounds can be expressed [33] by the equation

$$QS = \tfrac{2}{7}e^2Q(1 - R)\alpha_c^2 \langle r^{-3}\rangle_{\mathrm{ion}} F(\Delta_1, \Delta_2, \alpha_c^2\lambda_0, T), \qquad (4.19)$$

where α_c^2 has been discussed above, $(1 - R)$ is an antishielding factor and the function F, which depends on the t_{2g} splitting (Δ_1 and Δ_2), the spin-orbit coupling constant λ_0, and the temperature, represents the quenching of QS due to spin-orbit mixing and thermal population of the t_{2g} levels.

As in the case of only lattice contribution to QS, the number of parameters involved in (4.19) is large, and any meaningful discussion requires the analysis of several interrelated compounds and the comparison with IS results and if possible with other physical results [34].

When the Mössbauer transition involves spin states larger than $\frac{3}{2}$, the situation is at the same time more complex and more susceptible to detailed analysis than the situation described above. For ^{129}I spectroscopy, for example, the spin states are $+\frac{7}{2}$ (ground state) and $+\frac{5}{2}$ (excited state). Nuclear gamma spectroscopic selection rules in this case permit transition where the change in the spin state is either ± 1 or 0; hence a total of eight transitions are possible, as shown in Fig. 4.6. The (ideal) relative intensity of these transitions is proportional to the square of the Clebsch-Gordan coefficients, so that a characteristic ^{129}I Mössbauer spectrum is not symmetric, but has a "handedness," as shown in Fig. 4.6. From the appearance of such a spectrum (and experimentally obtained values for the ratio of nuclear quadrupole moments and results for model compounds) it is immediately possible to

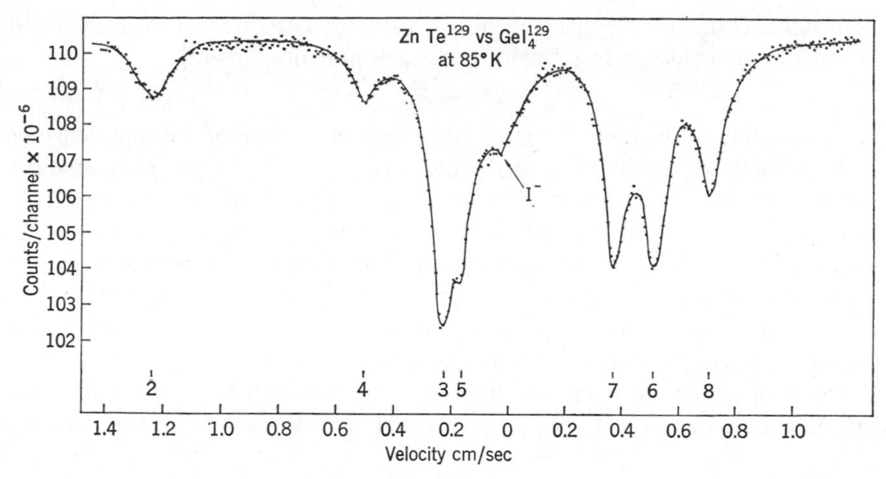

Fig. 4.6 Mössbauer spectrum of the 27.8-keV resonance of ^{129}I in GeI_4. Seven of the eight resonance lines can be resolved from the spectrum. The "handedness," typical of ^{129}I spectra, is readily discernable from such data.

calculate not only the magnitude of the nuclear quadrupole coupling constant, but also the sign of the field gradient. The results of such an analysis can be compared directly to the data obtained in a nuclear quadrupole resonance experiment, and such a comparison has, in general, led to good agreement between the two techniques. Detailed summaries of the results of quadrupole splitting measurements in ^{129}I have been reported by Perlow [35], Hafemeister *et al.* [36], Pasternak [37], Bukshpan [38], and others.

Not all high-spin situations are as favorable as the ^{129}I case noted above. In ^{83}Kr, for example, the two spin states are $+\frac{7}{2}$ and $+\frac{9}{2}$, leading to a total of 11 possible transitions. Only in the most asymmetric compound yet studied, krypton difluoride, KrF_2 [39], could some of these 11 lines be resolved, and thus detailed information concerning the magnitude (or sign) of the field gradient in this compound could be extracted from the available data.

Magnetic Hyperfine Interaction

The last hyperfine effect briefly mentioned in this summary concerns the interaction between the Mössbauer nucleus and a magnetic field, the origin of which may be either internal or external to the sample.

The crystal field information, which can be obtained by applying an external magnetic field to a sample showing quadrupole splitting, has been mentioned in the preceding section. The chemical information one can obtain from the interaction of the Mössbauer nuclear spin with an internal magnetic field stems from the physical origin of this field. The internal magnetic field is generally considered in terms of three components [40–42],

$$H_0 = H_c + H_L + H_D, \tag{4.20}$$

corresponding to the contributions from the polarization of the spin densities of the *s*-electron distributions at the nucleus, from the coupling between the angular momentum of outer electrons with the spin of the nucleus and from the coupling of the nuclear spin with the moment of the spin-density distribution of the outer electrons, respectively. Each of these components depends differently on the electronic charge density distribution in the solid, and in conjunction with the isomer shift and quadrupole interaction, may provide valuable information about bonding [41, 42].

The interaction between the nuclear magnetic dipole moment and the magnetic field gives rise to energy levels that can be specified by the relationship

$$E_m = g\mu_n H m_I, \tag{4.21}$$

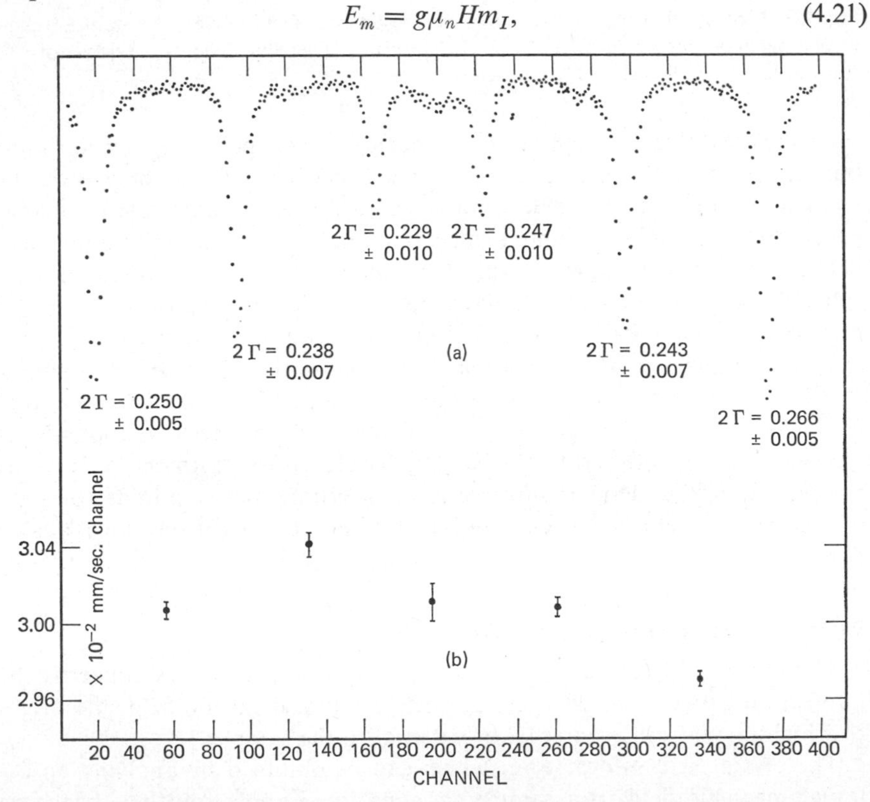

Fig. 4.7 The Mössbauer absorption spectrum of metallic iron at room temperature, taken with an unsplit line source. The line positions and widths are shown in (*a*) and the spectrometer calibration constants (mm/sec. channel) calculated from these data are shown in (*b*).

where g is the nuclear g-factor, μ_n is the nuclear magneton (equal to $eh/4\pi mc$, where m is the nucleon mass), H is the magnetic field strength, and m_I is the spin-state quantum number, which can have values of $I, I-1, \ldots, -I$. Thus there are $2I+1$ magnetically distinct levels present. These levels are degenerate in the absence of a magnetic field, and their presence can be determined only when $H \neq 0$. However, not all possible transitions are actually observed, since nuclear selection rules again come into play. These selection rules depend, in part, on the multipolarity and the parity change involved in the nuclear transition; that is, whether the transition is $E1$ (electric dipole, parity change), $M1$ (magnetic dipole, no parity change), $E2$ (electric quadrupole, no parity change), and so on. In the case of ^{57}Fe, the allowed values of Δm are 0 and ± 1, so that of the eight possible transitions in a magnetic field, six are actually observed, as shown in Fig. 4.7. The relative transition probabilities are given by the Clebsch-Gordan coefficients and are summarized in Table 4.2, together with their angular dependence.

Table 4.2 Transitions between Magnetic Hyperfine States in ^{57}Fe.*

Transition		Δm	*Relative intensity*	*Angular dependence*
I_e	I_g			
$\frac{3}{2}$	$\frac{1}{2}$	-1	3	$(\frac{3}{2})(1 + \cos^2\theta)$
$-\frac{3}{2}$	$-\frac{1}{2}$	1		$(\frac{3}{2})(1 + \cos^2\theta)$
$\frac{1}{2}$	$\frac{1}{2}$	0	2	$2\sin^2\theta$
$-\frac{1}{2}$	$-\frac{1}{2}$	0		$2\sin^2\theta$
$\frac{1}{2}$	$-\frac{1}{2}$	-1	1	$(\frac{1}{2})(1 + \cos^2\theta)$
$-\frac{1}{2}$	$\frac{1}{2}$	1		$(\frac{1}{2})(1 + \cos^2\theta)$

* The angle θ is that between the magnetic and optical axes in the experiment.

For a sample made up of randomly oriented magnetic domains the angular dependence must be integrated over a full sphere, so that six lines of relative intensity 3:2:1:1:2:3 are observed. This six-line, hyperfine pattern (Fig. 4.7) is frequently used in velocity calibrations [43], since the magnetic moment of the ground state in ^{57}Fe can be determined by non-Mössbauer methods [44]. The total splitting observed in an iron spectrum (i.e., the separation between the $+\frac{1}{2} \rightarrow +\frac{3}{2}$ transition, which occurs at the highest Doppler energy, and the $-\frac{1}{2} \rightarrow -\frac{3}{2}$ transition, which occurs at the lowest

Doppler energy) can be calibrated by the imposition of external magnetic fields of variable strength. In this way the internal field in metallic iron (that is, the magnetic field acting on an iron nucleus in the absence of an externally applied magnetic field) is found to be −342 kO*e* at 0°K and −333 kO*e* at 300°K. The negative sign indicates that in metallic iron the direction of the field is opposite to that of the externally applied field [45]. Once this calibration has been effected, the magnitude of the internal fields in other magnetic environments of the ^{57}Fe nucleus can be readily obtained from a Mössbauer

Table 4.3 Magnetic Hyperfine Interactions in Metallic Iron and Inorganic Iron Compounds

Matrix	*Nearest neighbor ligands*	*Magnetic field* kO*e*	*Reference*
Fe metal	Fe	−333	*a*
FeF_3	$6F^-$	622	*b*
$(NH_4)Fe(SO_4)_2 \cdot 12H_2O$	$6H_2O$	584	*c*
Fe_2O_3	$6O^{-2}$	540	*d*
$FeCl_3$	$6Cl^-$	487	*e*
FeF_2	$6F^-$	340	*f*
$FeCl_2$	$6Cl^-$	+2.8	*g*
$FeBr_2$	$6Br^-$	+29.6	*g*
FeI_2	$6I^-$	+74.0	*g*
$FeSiF_6 \cdot 6H_2O$	$6H_2O$	550(z), 250(x)	*h*
$FeCl_2 \cdot 4H_2O$	$4H_2O$, $2Cl^-$	266(z), 440(x)	*h*
$FeCl_2 \cdot 2H_2O$	$2H_2O$, $4Cl^-$	260	*h*
$Fe_3(PO_4)_2 \cdot 8H_2O$	$2H_2O$, $4PO_4^{-3}$	135(z), 270(x)	*i*
$Ni_3[Fe(CN)_6]_2$	$6CN^-$	269	*j*

[a] S. S. Hanna, J. Heberle, G. J. Perlow, R. S. Preston, and D. H. Vincent, *Phys. Rev. Lett.*, **4,** 513 (1960).
[b] D. N. E. Buchanan and G. K. Wertheim, *Bull. Amer. Phys. Soc.*, **117,** 227 (1962).
[c] L. E. Campbell and S. DeBenedetti, *Phys. Lett.*, **20,** 212 (1966).
[d] T. Nakamura and S. Shimizu, *Bull. Inst. Chem. Res. Kyoto Univ.*, **42,** 299 (1964).
[e] C. E. Kocher, *Phys. Lett. A*, **24,** 93 (1967).
[f] G. K. Wertheim, *Phys. Rev.*, **121,** 63 (1961).
[g] D. J. Simkin, *Phys. Rev.*, **177,** 1008 (1969).
[h] C. E. Johnson, *Symp. Faraday Soc.*, **1,** 7 (1967).
[i] U. Gonser and R. W. Grant, *Phys. Stat., Sol.*, **21,** 331 (1967).
[j] B. Sawicki, J. Sawicki, and A. Z. Hrynkiewicz, *Soviet Phys.–Solid State*, **9,** 1100 (1967).

experiment, since this will be directly proportional to the splitting between the two "outside lines" that are observed.

It has been shown recently that the magnitude as well as the sign of the internal magnetic field in Fe^{2+} compounds vary strongly with covalency [41, 42]. The sign of the internal magnetic field in a variety of compounds could be determined by using an external magnetic field. A brief listing of the values of some internal magnetic fields observed in iron compounds is given in Table 4.3. A more extensive discussion of such measurements is found in the references already cited [41, 42, 45].

Similar measurements can be carried out with other Mössbauer nuclides, using either internal or external magnetic fields. In the case of the former the Mössbauer atom may be present as the constituent of a ferri-, ferro-, or antiferromagnetic material, or it can be incorporated into the experimental lattice as an impurity "probe" atom. In the case in which external magnetic fields are applied, the fields required to effect separation (and hence resolution) of the individual lines are so large that superconducting fields are almost always necessary, and hence the experiments are carried out at liquid helium temperature. (See however Oosterhuis and Lang [45a].) Experiments have been reported for Mn_2Sn, Mn_4Sn, $FeSn_2$, Fe–Sn garnets, alloys, and related compounds involving the Mössbauer transition in ^{119}Sn [46], and an extensive series of measurements involving the rare-earth nuclides. A summary of the latter is in the comprehensive review by Ofer, Nowik, and Cohen [47].

Atomic Dynamics

Atomic dynamical data from Mössbauer experiments have been used to elucidate problems in chemical bonding in special cases, such as solid compounds of xenon [48] and krypton [49, 50] and in molecular solids [51–53]. The effective forces acting on an atom are relevant to the chemical bonding in the same sense as thermochemical data and the results of infrared and Raman spectroscopy. Since force constants in a solid are determined by the details of the charge-density distribution, the full accounting for the atomic dynamics should be relevant to the detailed discussion of the hyperfine interactions mentioned above.

Mössbauer experiments provide simultaneous information about two independent dynamical variables: the thermal variation of the mean square displacement $\langle x^2 \rangle$ of the probe atom from the Mössbauer-Debye-Waller factor (or in practice, from the temperature dependence of the area under the resonance line) (4.12) and the mean square velocity, $\langle v^2 \rangle$, from the thermal shift (or second-order Doppler shift) of the resonance line

$$\delta_{\text{Doppler}} = -\frac{1}{2}\frac{\langle v^2 \rangle}{c^2} E_\gamma.$$

The Debye-Waller factor may be obtained from X-rays and neutron-scattering experiments, and the Mössbauer effect provides an alternative—and sometimes more convenient—way of measuring it. The thermal shift, on the other hand, provides a unique way of measuring $\langle v^2 \rangle$ of the probe atoms in the solid. The Mössbauer technique is therefore not only alternative but complementary to other techniques used for lattice dynamical studies.

In spite of this unique property of the thermal shift, most of the available Mössbauer lattice dynamical studies are concerned with the Debye-Waller factor. This is so because even with the most convenient probe atom, ^{57}Fe, meaningful thermal-shift studies require a precision on the order of 1% of the natural line width in the determination of the position of the resonance line. Although this may sound prohibitive, it is clearly within the state of the art.

It is advantageous, of course, to measure the Debye-Waller factor simultaneously with the thermal shift rather than in separate experiments. In a real solid, unlike in an Einstein solid, the two dynamical variables, $\langle v^2 \rangle$ and $\langle x^2 \rangle$, are independent, and a comparison between them provides information about the frequency distribution function as well as anharmonic effects. The relevance of anharmonic effects to chemical bonding becomes apparent when they are discussed in terms of the shape of the interatomic potential and the ionic radii [54, 55].

The Mössbauer fraction may be rewritten as

$$\begin{aligned} \ln f &= -\frac{\langle x^2 \rangle}{\bar{\lambda}^2} \\ &= -\frac{E_\gamma^2}{Mc^2} \int_0^{\omega_{\max}} G(\omega)(\hbar\omega)^{-1}(\tfrac{1}{2} + \eta)\, d\omega, \end{aligned} \tag{4.22}$$

where $G(\omega)$, the frequency distribution function, is normalized by

$$\int_0^{\omega_{\max}} G(\omega)\, d\omega = 1, \tag{4.23}$$

and $\eta = [\exp(\hbar\omega/kT) - 1]^{-1}$.

The thermal shift may be written as

$$\begin{aligned} \delta_{\text{Doppler}} &= -\frac{1}{2}\frac{\langle v^2 \rangle}{c^2} E_\gamma \\ &= -\frac{3}{2}\frac{E_\gamma}{Mc^2} \int_0^{\omega_{\max}} G(\omega)\hbar\omega(\tfrac{1}{2} + \eta)\, d\omega. \end{aligned} \tag{4.24}$$

The difference between (4.22) and (4.24) is better illustrated in their high-temperature limits. For $T > (1/2)(\hbar/k)\omega_{\max}$ one may substitute the integral expressions by the first two terms of the Thirring expansions [56], giving:

$$\delta_{\text{Doppler}} = -\frac{3}{2}\frac{E_\gamma kT}{Mc^2}[1 + \tfrac{1}{12}\theta^2(2)T^{-2} - \cdots] \tag{4.25}$$

$$\ln f = -\frac{E_\gamma{}^2 T}{Mc^2 k}[\theta^{-2}(-2) + \tfrac{1}{12}T^{-2} - \cdots] \tag{4.26}$$

The frequency moments

$$\theta(n) = \left(\frac{\hbar}{k}\right)\left[\int_0^{\omega_{\max}} G(\omega)\omega^n \, d\omega\right]^{1/n} \tag{4.27}$$

are weighted averages of $G(\omega)$, where the weight factors for $\theta^n(n)$ are the nth powers of ω. The difference between the weight factors for $\theta^2(2)$ and $\theta^{-2}(-2)$, and therefore between (4.25) and (4.26), is in the fourth power of ω. Equation (4.25) is strongly weighted by the upper end of $G(\omega)$, while expression (4.26) is weighted by the lower end, which in the case of a molecular solid represents the intra- and intermolecular forces, respectively. Equations (4.22) and (4.24) are often written in terms of the Debye frequency moments $\theta_D(n)$ (or "Mössbauer temperatures"), which are related to $\theta(n)$ by

$$\theta_D(n) = \left[\frac{3+n}{3}\right]^{1/n}\theta(n), \tag{4.28}$$

and are independent of n for the ideal Debye solid only.

It is advantageous to analyze the experimental results in terms of the frequency moments, $\theta(n)$, rather than in terms of the commonly used Debye temperature θ_D or the Mössbauer temperature θ_M. The reason is that when the frequency moments $\theta(\pm 2)$ are derived from (4.25) and (4.26) over the appropriate temperature range they are independent of the details of $G(\omega)$, whereas θ_D is model-dependent and θ_M is not well-defined. One should be especially careful in the use of θ_D in the comparison with the results of other experimental techniques. Such a comparison may be most sensitive to the deviation of $G(\omega)$ from that of the ideal Debye solid.

2 SOURCES

General Considerations

The optimal source for most Mössbauer experiments is one that combines the absence of hyperfine effects with a large recoil-free fraction at the temperature of the experiment, chemical inertness, and ease of preparation. If the Mössbauer transition energy is sufficiently low, it is possible to use appropriate sources at room temperature, thus eliminating the requirement of source cooling, which significantly simplifies the experimental problems. Such room-temperature sources are routinely used for ^{83}Kr (9.3 keV), ^{57}Fe (14.4 keV), and ^{119}Sn (23.8 keV) experiments, among others.

The source strength appropriate to a typical Mössbauer experiment can be estimated from the following order of magnitude calculation: The rate at which a Mössbauer spectrum is accumulated is determined by the nuclear counting system. A nuclear channel of contemporary design (employing, for example, a CO_2–Xe filled counter followed by an FET preamplifier, a bipolar main amplifier, and a single-channel analyzer—all employing silicon transistors) can easily handle up to approximately 10^4 counts per second of 14.4-keV γ-rays of the ^{57}Co Mössbauer source in the channel. Nuclear counting systems capable of handling much higher counting rates have become also commercially available [57]. Basing our estimate on a counting rate of about 10^4 cps in the channel it is necessary to consider also the internal conversion coefficient (the ratio between the total disintegration rate, including the deexcitation of the Mössbauer level via internal conversion electrons, to the emission rate of the γ-rays of interest), which is about 10 for the ^{57}Fe source. For a typical detector area of 10 cm^2 a source-to-detector distance of about 20 cm and an attenuation factor (due to the thickness of the Mössbauer absorber as well as the window of the detector) of about 5, the appropriate source strength is

$$S = 10^4 \times 10 \times \frac{4\pi 20^2}{10} \times 5 \simeq 2.5 \times 10^8$$

disintegration per second, or roughly 10 mCi of ^{57}Co or ^{119}Sn sources. For many experiments, which do not require very fast data-collection times, such sources are useful for periods of two to three half-lives of the precursor nuclides (270 and 245 days for ^{57}Co and ^{119m}Sn, respectively). The effect of source decay can often be compensated for by decreasing the source-detector distance as decay of the precursor proceeds. For short half-life precursors, such as those used for ^{85}Kr and ^{129}I spectroscopies, neutron irradiation to give $\sim$75% of saturation activity (two half-lives) is usually employed.

The preparation of some of the more common Mössbauer sources are briefly considered below. It should be remembered, however, that a number of commercial suppliers [58] can now furnish Mössbauer effect sources—often with "guaranteed" resonance effect magnitudes under stated conditions—and thus each investigator will have to decide for himself whether or not it is worth his time to prepare his own sources. In the case of some Mössbauer nuclides, of course, on-site preparation cannot be avoided, and the experimenter has little choice but to go through trial-and-error work (based on available information in the literature) to find the best method for preparing the sources appropriate to his needs.

^{57}Co

This nuclide, which is the precursor of the 14.4-keV gamma transition used in ^{57}Fe spectroscopy, is produced by the (d, p) reaction on ^{56}Fe (91.52%). The iron cyclotron target is dissolved in acid and the ^{57}Co is freed of other radioactive impurities by ether extraction of iron and anion-exchange purification of cobalt by methods amply described in the literature [59–61]. Most presently-used ^{57}Co sources consist of a metal foil into which the cobalt activity has been diffused to an appropriate depth by high-temperature treatments [62–66]. Typical host lattices include copper, palladium, platinum, chromium, and stainless steel (the latter was used extensively in early Mössbauer work but has been largely abandoned because of the wide line widths observed). Isomer shifts for ^{57}Co in typical host lattices are summarized in Table 4.4. Although it is possible in principle to use suitable cobalt compounds, such as acetyl acetonates, cyanides, or oxides, as well as ^{57}Co-doped alkali metal halides or zinc fluoride, such matrices are in general unsatisfactory because of the hyperfine interactions observed as a result of the nuclear decay processes (electron capture and isomeric transition) that precede the Mössbauer transition [67].

The ^{57}Co activity can be applied to the metal surface prior to annealing, in the form of a solution of a cobaltous salt, which is then reduced in a hydrogen atmosphere, or it can be electroplated in the metallic form with an appropriate electrochemical procedure [68]. The resulting annealing history has to be monitored very carefully so that an optimum compromise is made between two opposing effects: insufficient annealing leads to local high concentrations of domains of cobalt metal with its unwanted magnetic hyperfine effects on the resulting Mössbauer spectral line. Excess annealing leads to too great a diffusion depth and hence to excessive nonresonant scattering (photoelectron and compton), especially in the case of high-Z elements, such as platinum and palladium.

Carefully prepared platinum and palladium matrices for the ^{57}Co activity can be used at room temperature for several half-lives of the activity, and there is no evidence that such sources degrade with time, either by chemical action or by radiation damage. Sources of ^{57}Co diffused into copper are quite readily prepared (because of the ease of electroplating and diffusion), but they must be protected against laboratory atmospheres by a covering of a thin layer of a plastic spray or by similar means. Unprotected sources react with H_2S (and more slowly with O_2) in the air, and the resultant migration of the chemical species so-formed into the lattice tends to lead to progressively wider lines and smaller recoil-free fractions as the use of such sources is continued.

Finally, it should be mentioned in connection with sources for ^{57}Fe spectroscopy that sources of ^{57}Co in cobalt oxide, which at first appeared

Table 4.4 Isomer Shifts for ^{57}Fe and ^{119}Sn in Representative Host Matrices[a]

Absorber matrix	*Isomer shift, mm/sec* ^{57}Fe wrt Fe(O)	^{119}Sn wrt α-Sn
Au	+0.619	+0.15
Ag	+0.513	+0.20
Pt	+0.349	−0.42
Cu	+0.226	−0.13
Ir	+0.214	—
Pd	+0.183	−0.64
W	+0.154	—
Rh	+0.103	—
Mo	+0.049	—
Ta	+0.022	—
Fe	0	−0.42
Nb	−0.026	—
Ti	−0.041	—
$K_4Fe(CN_6)\cdot 3H_2O$	−0.045	—
Vacromium	−0.078	—
302, 310, 316 ss	−0.097	—
V	−0.176	−0.27
Cr	−0.183	—
$Na_2[Fe(CN)_5NO]\cdot 2H_2O$	−0.258	—
In	—	+0.96
Tl	—	+1.01
Pb	—	+1.20
SnO_2	—	−2.10
$BaSnO_3$	—	−2.10
Mg_2Sn	—	−0.30
α (Gray, tetrahedral)	—	0
β (White, tetragonal)	—	+0.55
$(CH_3)_2\,SnF_2$	—	−0.77
$(CH_3)_3\,SnF$	—	−0.80
SnF_4 (80°K)	—	−2.35
Ca_2SnF_6 (80°K)	—	−2.45

[a] These data are taken largely from Table 2.1 in Ref. 5, and are reliable to about ±0.010 mm/sec. The ^{57}Fe values have been calculated with respect to the centroid of a room-temperature metallic iron reference absorber, those for ^{119}Sn with respect to the center of the (single line) absorption in metallic (α) tin. All values, except as noted, refer to both source and absorber at room temperature.

to be superior (with respect to line width and recoil-free fraction) to more commonly used metallic host matrices, are unsuitable. The preparation of such cobalt oxide sources has been reviewed in detail by Mullen and Ok [69, 70], who have ascribed the observation by Wertheim and others [71] of anomalous charge states in ^{57}CoO to nonstoichiometry effects. In fact, the cobaltous oxide sources prepared by Mullen and Ok suffer a severe deterioration with time, as evidenced by the increasing contribution of various hyperfine interactions as these sources age. All factors considered, it appears that the best sources for ^{57}Fe spectroscopy available at the present time are the metallic matrices (principally palladium and platinum referred to above.

^{119}Sn

This nuclide is conveniently produced by the (n, γ) reaction on ^{118}Sn (24.01% natural abundance), using the thermal neutron flux of a nuclear reactor. In the preparation of such sources it is preferable to start with tin enriched in the ^{118}Sn nuclide, not only to increase the specific activity (decays per second per gram) of the ^{119m}Sn, but more importantly to decrease the amount of ^{119}Sn (8.58% natural abundance) in the matrix to reduce self-absorption and line-broadening effects in the source.

Although neutron-irradiated SnO_2 is suitable for use as a Mössbauer source without any additional purification or annealing steps, the inherent broad resonance lines observed with this material due to the noncubic nature of the tin nearest neighbor environment [72], make such sources of limited utility in high-resolution Mössbauer spectroscopy.

A more satisfactory source for ^{119}Sn work suitable for use at room temperature is barium stannate, $BaSnO_3$ [73]. This material can be prepared from neutron-irradiated metallic tin (or from metallic tin obtained by reduction of irradiated SnO_2) by dissolving the metal in concentrated HCl, by passing in an excess of chlorine gas, and then by making the solution strongly basic with NaOH in the presence of sufficient $BaCl_2$ carrier to precipitate all of the tin. The resulting white precipitate is washed copiously with water, dried, and pressed into an appropriate disk for use as a Mössbauer source. Such sources have typically a recoil-free fraction of $\sim$0.8 at 77°K and $\sim$0.55 at 300°K.

The use of magnesium stannide, Mg_2Sn, was first suggested by Flinn and Ruby [74], and was studied in detail by Bryukhanov and others [75], who showed that such sources give rise to a single resonance line with a recoil-free fraction of 0.77 ± 0.08 at 77°K and 0.28 ± 0.03 at 290°K. The relative smallness of the latter value and the sensitivity of Mg_2Sn to moisture have led to the abandonment of this material for most purposes, but such sources are of some utility where source cooling can be conveniently accomplished.

Somewhat more useful are ^{119}Sn Mössbauer sources, in which the activity is incorporated into a metallic lattice. Tetrahedral (gray, alpha) tin is not very widely used as a Mössbauer source because of the small recoil-free fraction observed at room temperature ($f_s \sim 0.039$ at 300°K [76]), although the lines are in general quite narrow and such sources are readily prepared by reduction of irradiated SnO_2 and rolling of the resulting metal. Useful ^{119}Sn sources have been prepared from palladium-tin alloys [77–79], which incorporate up to 10% Sn into the Pd matrix and give narrow resonance lines and moderately large resonance effects. A minor drawback of such sources is their poor reproducibility, since minor differences in the palladium–tin ratio can have a significant effect on the isomer shifts observed. However, the use of standard reference materials for reporting isomer shift data (see the Appendix) overcomes this limitation of palladium–tin sources. Isomer shifts for ^{119}Sn in typical host matrices are summarized in Table 4.4.

^{129}I

Of the two nuclides of iodine suitable for Mössbauer work, ^{127}I and ^{129}I, the latter is considerably more useful because of its narrower resonance lines, despite the fact that the ground state of ^{129}I required for the absorbers to be studied is itself radioactive, with a half-life of 1.6×10^7 years. Although the Mössbauer effect in ^{129}I was first studied [80] using a source of 33-day ^{129m}Te, which had been produced by the (n, γ) reaction on ^{128}Te (32% natural abundance), most of the more recent work in the literature [81] has been done using the ^{129}Te ground-state precursor. This 70-minute activity is obtained by thermal neutron irradiation of ^{66}Zn^{129}Te for about 1 to 2 hours. The zinc telluride can be prepared by mixing stoichiometric amounts of ^{66}Zn (obtained from ORNL and chosen so as to minimize the production of interfering neutron-induced activities) and ^{129}Te, heating to red heat in an evacuated ampul and sedimenting the resulting powder in benzene containing a small amount of polystyrene. The air-dried material can be irradiated repeatedly without significant degradation of the sources. The recoil-free fraction of ZnTe for the 26.8-keV Mössbauer radiation is $\sim$0.15 at 80°K, and provision must be made for simultaneous source and absorber cooling in such experiments.

Telluric acid, $Te(OH)_6$, has also been used as a Mössbauer source in ^{129}I experiments [81]. This material can be formed in a two-step oxidation followed by crystallization under conditions in which the cubic (rather than the monoclinic) form is recovered in the procedure. However, studies with such sources (which turned out to be monoclinic rather than cubic when examined after the decay of the induced activity) show that the lines observed in the Mössbauer spectra are considerably wider than those obtained with ZnTe sources, and so would seem of little advantage over the latter. The major use of such sources has been the study of the chemical consequences of beta decay on a fast time scale.

It is also possible to observe the Mössbauer effect in ^{129}I using a source of elemental tellurium. Except for the convenience of preparing such sources (there is no particular necessity of removing the tellurium activity from the quartz irradiation capsule) there is little to recommend them, and they are not as useful for high-resolution spectroscopy as are the ZnTe sources referred to above.

^{83}Kr

Mössbauer sources of ^{83}Kr are most conveniently prepared by irradiating the corresponding hydroquinone clathrates by a method first suggested by Hazony et al. [82]. The cage compound can be prepared by pressurizing an aqueous hydroquinone solution at 100°C with krypton gas under about 35 atmospheres pressure; this system is then cooled slowly under pressure to room temperature. The resulting crystalline compound is quite stable at 300°K and can be repeatedly irradiated with thermal neutrons to give the desired 1.9-hour ^{83}Kr precursor even a few years after their original crystallization. By appropriate combinations of pressure and/or rate of cooling of the aqueous solution it is possible to obtain a krypton content of 3 to 20% by weight (corresponding to 15 to nearly 100% of the cavities being filled with "guest" noble-gas atoms). These sources have a recoil-free fraction of ~8% at 300°K, thus making them practical for use at room temperature. It should be noted, however, that the line widths obtained are about three times wider than natural, probably because of the distortion from spherical symmetry of the clathrate "cage." Because of the lack of krypton compounds available by synthetic means that can be studied as absorbers, the krypton-clathrate compound is mainly used as an absorber in experiments involving krypton sources produced by the decay of rubidium, bromine, and selenium.

It is also possible to study the 9.3 keV-transition in ^{83}Kr by using the 2.40-hour ^{83}Br precursor. This nuclide, which can be formed by beta decay following the (n, γ) process in ^{82}Se (9.19% natural abundance) undergoes beta decay to ^{83m}Kr, and a number of studies have been reported in which alkali metal bromide sources have been used in conjunction with ^{83}Kr Mössbauer-effect measurements [83]. Finally, it is possible to start even one more element further back in the periodic table and observe the Mössbauer effect in ^{83}Kr using ^{83}Se sources [50]. Neutron irradiation of ^{82}Se (enriched to ~80%) in its elementary form or as selenium dioxide forms ^{83}Se *in situ*. This nuclide decays by two beta decays to ^{83}Kr, and such sources have a useful life of several hours before interfering daughter radiations grow in to reduce the fraction of Mössbauer γ-rays to an intolerable level. It should be noted, however, that the use of ^{83}Se sources requires the employment of high-resolution, solid-state gamma detectors (soft X-ray lithium-drifted silicon detectors) since the abundance of interfering gamma radiations is so large that the signal-to-noise ratio observed with proportional or NaI (Tl) detectors is too small to be useful.

Ruby and Selig [84] in their study of the Mössbauer effect in ^{83}Kr, have made use of the long-lived (83-day) ^{83}Rb precursor. This nuclide is obtained by the (α, $2n$) reaction on ^{81}Br (natural abundance 49.46%), and it decays by 6% of its transitions to the $\frac{7}{2}+$ excited state of krypton. Source preparation consists in dissolving the $CaBr_2$ target, which has been irradiated with 30-MeV alpha particles in water, adding NH_4Cl and precipitating $CaCO_3$ by bubbling CO_2 gas into the aqueous solution. The supernate is evaporated to dryness, and the residue (which contains the ^{83}Rb activity) is taken up in 0.3 ml H_2O ($\sim$6 drops) and is transferred to a filter paper disk that is then dried and sealed with Mylar. The experiments of Ruby and Selig show that at liquid-nitrogen temperature (but not at 300°K!) the retention of the krypton daughter is essentially complete. Although such sources for ^{83}Kr experiments have the advantage of long useful lives compared to the krypton-clathrate sources discussed above, the line widths observed are about seven times natural, and hence the resolution of hyperfine effects is difficult. It seems clear that the ideal host matrix for use with ^{83}Kr Mössbauer spectroscopy has yet to be found.

^{129}Xe

The 39.58-keV Mössbauer transition in ^{129}Xe can be observed [81] using sources incorporating the nuclide ^{129}I since this activity undergoes beta decay to populate the $\frac{3}{2}+$ state in ^{129}Xe. Useful iodine compounds that have been used as Mössbauer sources in such experiments are NaI, $NaIO_3$, $Na_3H_2IO_6$, KIO_4, I_2, $KICl_4 \cdot H_2O$, and $KICl_2 \cdot H_2O$.

It should be remembered, however, that ^{129}I is a radioactive nuclide with a half-life of 1.6×10^7 years hence with a specific activity of 6.4×10^6 decays/sec g = 0.18 mCi/g. Consequently appropriate radiochemical procedures must be followed in the preparation of ^{129}I sources for ^{129}Xe spectroscopy. Because of the long half-life of ^{129}I, the amount needed to give acceptable counting rates in a typical Mössbauer experiment is $\sim$1/2 g (too large a quantity leads to excessive self-absorption of the Mössbauer radiation), and consequently it is not necessary to add carrier ^{127}I to the ^{129}I required in the chemical manipulations involved in source preparation procedures.

Large resonance effects using sources in which the ^{129}I is in a high oxidation state have been reported. In conjunction with a xenon clathrate absorber a 20.4% effect has been observed with $Na_3H^{129}IO_6$, and a 10.4% effect has been observed with $K^{129}IO_4$, both at liquid-helium temperature.

Rare-Earth Sources

The most comprehensive review of Mössbauer spectroscopy involving rare-earth nuclides is that of Ofer, Nowik, and Cohen [47], which summarizes much of the work that has been done with elements in this region of the periodic table. The rare-earths as a group form the most prolific part of the periodic table with respect to Mössbauer studies, a fact emphasized by the observation that of the 15 elements involved only cerium and promethium are not suitable for such experiments. In the cerium isotopes there are no first excited state–ground state transitions of appropriate energy, while in the case of promethium (which has only radioactive nuclides) the ground-state stability problems preclude practical application of compounds of this element as absorbers.

Because of the diversity of nuclear properties among the nuclides of the rare earths, a wide variety of problems is encountered in the experimental requirements of source preparation for these elements. (For a comprehensive table of the pertinent properties, including transition energies, lifetimes, spins, abundances, cross sections, recoil-free fractions, and so on, refer to the extensive tabulation in [47].) The most common chemical form of rare-earth Mössbauer sources is the oxide, of which CeO_2 is the only quadrivalent member, the others (in the case of europium, gadolinium, samarium, dysprosium, terbium, and erbium) having the general formula M_2O_3. A typical Debye temperature for rare-earth oxides is $\sim$220°K [85], so that an approximate value of the recoil-free fraction at various temperatures can be calculated from the known transition energies. The largest f values are, of course, observed for the case of the lowest transition energy, and a brief summary of such sources is given in Table 4.5. In this table are also the

Table 4.5 Mössbauer Parameters for Some Rare-Earth Nuclides

Stable nuclide	*Abundance, %*	E_t *keV*	σ_0 10^{-19} *cm*^{2}	f*(77°K)*
^{169}Tm	100	8.4	700	0.98
^{149}Sm	13.8	22	38	0.77
^{151}Eu	47.8	21.6	68	0.77
^{161}Dy	18.9	25.7	37	0.74
^{163}Dy	25.0	26	36	0.74
^{161}Dy	18.9	44	16.9	0.40
^{157}Gd	15.7	54.5	12.4	0.26
^{159}Tb	100	58	10.9	0.23
^{155}Gd	14.7	60	10.2	0.20
^{157}Gd	15.7	64	9	0.20
^{171}Yb	14.3	66.7	11.0	0.16

abundances of the stable nuclides, the Mössbauer transition energy, and the recoil-free scattering cross-section for each nuclide.

It is clear from Table 4.5 that the "easiest" nuclides in the rare-earth region that can be used for Mössbauer effect studies are europium, dysprosium, and thulium since the low transition energies (21.6, 25.6, and 8.4 keV, respectively) make it possible to use the rare-earth oxide sources at room temperature. This fact is reflected by the observation that by far the largest number of reported Mössbauer investigations have been reported for these elements. However, it should be noted that in only a few cases has it been possible to observe resonance lines having the natural line width. In most instances a variety of hyperfine effects, such as a nonvanishing electric field gradient at the Mössbauer atom lattice site, magnetic relaxation phenomena, nonstoichiometry, and imperfections and impurities can lead to significantly broadened lines. Essentially natural line widths have been observed in the case of Gd_2O_3 [86] and Dy_2O_3 [85], while only slightly broadened lines have been reported for Eu_2O_3 [87], Er_2O_3 [88], and Sm_2O_3 [89].

In addition to rare-earth oxides a number of other chemical forms of the Mössbauer source have been reported. Among these are fluoride sources such as CeF_3 [90] and ErF_3 [91, 92], metallic sources such as terbium in copper [93], erbium metal and thulium metal [94], and a variety of intermetallic compounds [95]. In each case the choice of the chemical form of a particular source must be dictated by the nuclear, chemical, and crystallographical parameters that are involved, and a decision concerning the form of the source must be made on an individual basis.

^{197}Au

The 77.3-keV Mössbauer transition in ^{197}Au is populated by 0.67-MeV β^- decay from 21-hour ^{197}Pt, and has been observed [96] in a variety of alloys and intermetallic compounds. Sources for such experiments can be obtained by neutron irradiation of natural platinum metal to give $\sim$2 mCi of the ^{197}Pt precursor. Using a germanium detector to provide the necessary gamma pulse height resolution, such sources will give $\sim 3 \times 10^4$ counts per second in typical geometry. A detailed review of Mössbauer spectroscopy with ^{197}Au has been given by Shirley [97].

3 DETECTORS AND NUCLEAR SPECTROSCOPY

Electronics—Nuclear Counting System

The electronics of the nuclear counting system is illustrated schematically in Fig. 4.8. All detectors mentioned above require a high-voltage power (HV) supply. The power supply is determined by the choice of the detector. Commercially available [98] proportional counters as well as 14-stage

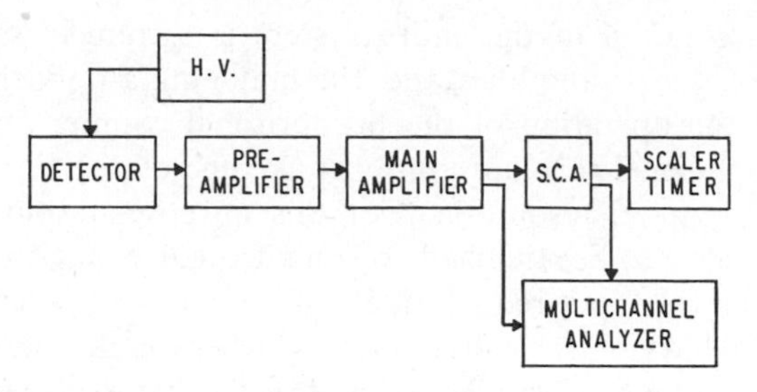

Fig. 4.8 Block diagram for the nuclear electronics required for Mössbauer spectroscopy. The multichannel analyzer operated in the pulse-height mode can be used to set the proper "window" in the single-channel analyzer (SCA) prior to data accumulation as shown. During a Mössbauer spectrum run, data are stored in the MCA, which is operated in the time mode. In this operation the output of the SCA is used to provide input pulses for the MCA.

photomultipliers, used with scintillation counters (NaI(Tl)), generally operate with voltages of up to 2000 V. However, depending on the design and the specifications, they may require even higher voltages (2500 to 3000 V). Proportional counters are generally designed so that the outer cylinder is grounded, and they require positive voltage on the center wire. Photomultipliers, on the other hand, may be mounted to operate with either positive or negative high voltages. Proportional counters require practically no current from the HV supply. The different voltages that are required for the different electrodes in the photomultiplier are tapped off a string of resistors, and, depending on the average current drawn in the final amplification stages, currents as high as 10 ma may be required for high counting rates. This requirement is particularly important for time-dependent, high-counting rates, where too small a current in the resistors' string may cause the gain of the photomultiplier to vary with the counting rate. Si(Li) and Ge(Li) solid-state detectors for high-resolution spectroscopy generally require a high-voltage power supply of few hundreds volts.

Because of the very high gain of the 14-stage photomultiplier its preamplifier may consist of a passive network with a varying capacitor at the anode in order to stretch the output pulses to match the input requirements of the main amplifier. The proportional counter requires a low-noise, charge-sensitive preamplifier with as short a cable or connector between them as possible to reduce the stray capacity at the input. In order to further decrease the noise level of the preamplifier below that of the intrinsic resolution of the Si(Li) detector, the first stage of the charge-sensitive preamplifier is mounted directly in the liquid nitrogen cryostat to minimize both the input capacitance as well as the thermal noise level of the input stage.

The energy resolution of the entire system is extremely sensitive to the pulse shaping of the preamplifier and the matching amplifier. Long pulses are inherent in the operation of the proportional counter and the charge-sensitive preamplifier. The main amplifier is generally designed to reshape long pulses for optimal resolution. For this purpose output pulses of the photomultiplier have to be stretched (by an adequate capacitor at the anode) in order to match the input requirements of the main amplifier. Long pulses in themselves are of course a disadvantage where high counting rates are concerned because they cause pile-ups and drifts that decrease the performance characteristics of the electronic instruments. However, they are required to allow an appropriate reshaping of the signals for optimal performance of the main amplifier and the following single-channel analyzer (SCA). The reshaping of the input signal to the main amplifier may be done in one of two ways. In one method a sequence of differentiations and integrations is applied to produce a bipolar pulse with optimal-time constants; in this method the input pulse is reshaped and followed by a similar signal with opposite polarity, the purpose of which is to discharge the stray capacitors in the amplifier and the following SCA after they were charged by the first pulse. A similar shaping may be done by a "double delay line clipping." The purpose of this technique is of course to minimize drifts and the overloading of the electronic circuits. In a second method the pulse shaping is done by "zero-pole cancellation" and "baseline restoring." The purpose of these two operations is to produce a signal of adequate duration with minimum tail and no overshoot. The two methods of pulse shaping may be provided in the same commercial amplifier as an option at additional cost. It should therefore be emphasized that the two methods operate at cross purposes and that unless the following SCA or MCA (multichannel analyzer) are especially designed to accept either of the two pulse shapes, advantages of either method may be marginal or even nonexistent. Many commercial amplifiers also provide a delayed output, the purpose of which will be immediately apparent.

The purpose of the single-channel analyzer is to sort out of the incoming pulse-height spectrum only those pulses that correspond to the desired radiation. This is accomplished by setting a "lower-level discriminator" or "base line" and an "upper-level discriminator" or "window," which reject pulses that are too small or too large and pass only those pulses of the appropriate size corresponding to the radiation of interest.

A first, but rather good, test for the resolution of the system as well as for the setting of the SCA may be done by monitoring the delayed output of the amplifier by a simple oscilloscope. If the oscilloscope is then triggered by the output of the SCA, only the pulses chosen by the SCA will be displayed. A very complicated pulse-height spectrum may require an MCA operated in coincidence with the output of the SCA for the setting of the window of the SCA. In this mode both the MCA and the SCA are analyzing the output signals of the same detector. Contemporary MCA may have the SCA feature built in, and the proper window is set by adjusting the lower- and upper-level discriminators while operating in live display in the pulse-height-analysis mode. The output of this SCA can be then used to feed signals into the input of the MCA when operated in the time mode as a Mössbauer spectrometer. However, this method, which is less direct than the one using the oscilloscope, may be significantly better only if the input characteristics of the MCA match the output characteristics of the main amplifier and if the spectrum is not distorted. In some of the old MCA models the operation in the coincidence mode would distort or shift the spectrum, and the setting of the SCA this way is unreliable.

The scaler-timer, which is being used when the spectrometer is operated in the constant-velocity mode, incorporates a precision clock and an electronic counter. The electronic counter counts the incoming signals during a preset-time interval determined by the clock. A timer is generally incorporated in the MCA, and the arithmetic scaler is an indispensable component in its operation; consequently, a separate scaler-timer is not required when the spectrometer is operated in the constant acceleration mode. A scaler-timer becomes very useful in such a case if the spectrometer is being used in a long-term experiment that may require a rechecking of the setting of the SCA. This may be done with no interruption of the main experiment by registering the count-rate of the output of the SCA as a function of the setting of its lower level.

Detectors

After a nucleus has been excited by the recoilless absorption of Mössbauer γ-rays, it decays to the ground state by emitting either γ-rays or internal-conversion electrons, and in the latter case, X-rays. The energy of the γ-rays and X-rays encountered in Mössbauer studies varies from about 3.4 keV (L X-rays in ^{119}Sn) to about 200 keV (the highest-energy Mössbauer transition observed so far is 155 keV), and therefore different types of nuclear radiation counters are needed to detect these radiations.

In practically all cases the requirement for energy discrimination to distinguish the Mössbauer radiation from a complex γ-spectrum, coupled with the need for maximum available efficiency, dictates the nature of the detector. Three main types of detectors are most appropriate for these low-energy transitions: NaI(Tl) scintillation counters, proportional counters, Si(Li) and Ge(Li) solid-state diodes.

The effectiveness of these various detectors, which are available commercially, is discussed below; the detailed characteristics of their construction and performance can be found in many texts and review articles on nuclear detectors [99].

Scintillation Counters

For radiation of energy greater than 10 to 15 keV, scintillation crystals mounted on photomultiplier tubes are quite appropriate since they have nearly 100% efficiency. For the lowest energy photons, commercially available (98,100) NaI(Tl) crystals of the order of 0.5 mm thick covered by a 0.002-in.-thick beryllium or 0.001-in.-thick aluminum window integrally mounted on tested photomultiplier tubes [100] are commonly used. As the energy of the radiation increases, proportionately thicker crystals are used. The appropriate thickness is chosen such that the crystal essentially absorbs most of the Mössbauer radiation, but it remains thin and hence inefficient for the higher-energy radiations. The resolution of the detector is determined by the statistics of the conversion of the γ-ray energy to an electronic signal. For a detector with a very efficient reflector at the sides not connected to the photomultiplier and good optical matching to the cathode of the photomultiplier, a very good conversion coefficient is about 0.5 keV per photoelectron. Thus the average number of photoelectrons emitted from the cathode for a 14.4-keV γ-ray from the Mössbauer source ^{57}Co is about 30. The statistical fluctuation of $\pm 30^{1/2}$ results in a spread in the height of the electronic signals of about 35% FWHM (full width at half maximum) in a very good counter assembly. The resolution improves proportionally to the inverse of the square root of the energy of the γ-ray detected.

Proportional Counters

For very-low-energy γ-rays, gas-proportional counters have a marked advantage over scintillation counters in that they exhibit almost the same efficiency but have much better resolution, higher signal-to-noise ratio, and negligible efficiency for the high-energy γ-rays that might be present in the decay and contribute to the background. When followed by a low-noise preamplifier, the resolution of the proportional counter, like the scintillation crystal, is determined by the statistics of the conversion of the energy of the γ-ray to an electrical signal. As a crude estimate one may consider a conversion coefficient of about 40 eV per electron in the primary ionization process in the gas. The statistical fluctuations of about $2 \times N^{-1/2}$, where N is the number of primary electrons, will determine the resolution. For the 14.4-keV γ-radiation the resolution should be approximately 10% FWHM.

Figure 4.9 shows a comparison of the characteristic resolution obtained with an argon-methane proportional counter (at one atmosphere pressure), NaI(Tl), Si(Li), and Ge(Li) detectors.

Gas-proportional counters [101] filled with a mixture of either xenon, krypton, argon, or neon (about 90%) and methane or nitrogen quenching gas (about 10%) have been used most extensively for the isotopes with the lowest energy transitions, such as ^{181}Ta (6.2 keV), ^{169}Tm (8.4 keV), ^{83}Kr (9.3 keV), ^{57}Fe (14.4-keV γ-ray or 6.3 X-ray).

The choice between the various filling gases for the detection of these low-energy radiations is determined partly by the energy of the Mössbauer radiation and partly by the presence and the particular energy of other radiations. A typical proportional counter of 5 cm in diameter has an efficiency of 100% for the 14.4-keV γ-radiation of ^{57}Co if filled with two atmospheres pressure of argon or one atmosphere of krypton. For higher energies xenon filling provides better efficiency. An important consideration is connected with the existence of "escape" peaks. For example, in xenon a secondary peak, owing to the escape from the counter of the xenon K X-rays emitted after the incoming radiation was absorbed by the photoeffect, appears at the full energy minus 34.55 keV, the K-binding energy. Therefore, if there happens to be in the decay another radiation of energy greater than the Mössbauer transition by about 35 keV, the Mössbauer peak will be superimposed on this escape background. For such special cases an argon counter is preferable (K-binding energy = 3.2 keV) since the fluorescence yield of argon (ω = 0.11) is much smaller than that of xenon ($\omega = 0.81$). Proportional counters are relatively easy to construct. However, they are now available commercially [101] in many different shapes, different flow-counter gases, and different windows. Detrimental factors in the operation of a proportional counter are the purity of the gas filling and the cleanliness of the walls and central wire. In the flow counter any contamination of the gas is flushed out continuously. This could be done economically only in argon counters. The commercial availability of long-lived sealed proportional counters with excellent counting characteristics makes this procedure unnecessary. However, the most important advantage of the sealed proportional counter is in its excellent long-term gain stability, since differences in gas-flow rates and diurnal changes in pressure have no effect on performance.

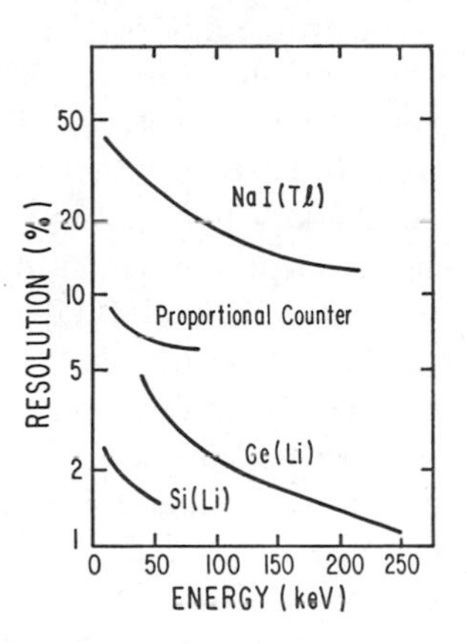

Fig. 4.9 Comparison of nuclear detector response as a function of γ-ray energy.

Solid-State Detectors (Semiconductors)

The commercial availability of Ge(Li) and Si(Li) drifted detectors of relatively large volume and efficiencies have brought about a breakthrough in the problems of energy resolution. Their big advantage stems from the fact that it takes only few electron volts of energy to create an electron-hole pair in a semiconductor. The statistical factor that contributed predominantly to the resolution in the proportional counter is of much less importance here. The lower limits on the resolution of the semiconductor detectors are actually due to preamplifier noise and possible defects in their fabrication.

The problem of resolving low-energy lines in the presence of a background caused by higher-energy X-rays and γ-rays prompted Huntzicker and Rosenblum [102] to apply the lithium-drifted germanium detectors to Mössbauer work. In particular they were interested in observing the Mössbauer effect of the 82.3-keV first excited state in ^{191}Ir. Mössbauer discovered recoilless resonance absorption in his studies [1] of the 129-keV transition that can be clearly resolved with a NaI(Tl) detector, while the weaker 82.3-keV radiation is hidden by the 64.9-keV X-rays and the Compton-scattered higher-energy photons. However, with the very-high-resolution Ge(Li) device the 82.3-keV line is clearly resolved, and the Mössbauer effect was detected for this radiation. The dramatic improvement in resolution obtained with Ge(Li) and Si(Li) detectors over the usual results for NaI(Tl) crystals and proportional counters is displayed in Fig. 4.9. These detectors must be operated at liquid-nitrogen temperature because of the large thermal noise* that is present at room temperature. In order to minimize the contribution of thermal noise in the preamplifier, its first stage, which is generally an FET transistor, is also contained in the cryostat. The fabrication of Ge(Li) detectors has been discussed in great detail in many papers in the literature, and a large variety of detectors as well as cryostats to house them are now commercially available [103]. The availability of Ge(Li) detectors has already opened new possibilities of investigations of isotopes with close-lying or weak γ-branches, such as ^{193}Ir [104]. Furthermore, the number of Mössbauer-effect candidates is being drastically increased by the success of the application of Coulomb excitation and other nuclear reactions to the production of levels otherwise inaccessible by α-, β-, or isomeric decay; in addition, since these new means of excitation in general also produce very large X-ray and γ-ray backgrounds, Ge(Li) detectors might be the only detectors of sufficient resolution for such Mössbauer studies.

The contribution from thermal noise is less important in the Si(Li) detector because of the larger energy gap between the valence and the conductance bands. On the other hand, the stopping power of silicon for γ-rays is much smaller than the stopping power of germanium because of the lower atomic number. The efficiency of the Si(Li) detectors is further reduced by

* Thermal noise in this context is defined as the appearance of spurious pulses due to thermal excitation in the crystal.

the fact that they generally have much smaller volumes than their Ge(Li) counterparts. However, for energies below 50 keV the efficiency of a 3-mm-thick Si(Li) crystal is mainly determined by its area, and this may very often be compensated by an adequate geometry of the Mössbauer experiment. Recently, the present authors have used such a detector, 3 mm thick and with an area of $\sim$30 mm^2, to study the Mössbauer effect in ^{83m}Kr produced by (n, γ) reactions in a reactor using 80%-enriched ^{82}Se compounds as starting materials. The resolution of the detector assembly (Si(Li) detector and the following electronic channel) was about 400 eV. This resolution made it possible to sort the 9.13-keV Mössbauer line of ^{83m}Kr out of a wealth of X-ray activity in the same energy region [50].

Electron Magnetic Spectrometers

In most cases the conversion coefficient of the Mössbauer transition is very large. It would seem therefore that detection in a "scattering" geometry [105] of the conversion electrons emitted from a state that has been excited by recoilless radiation would be a much more efficient tool than the standard absorption measurements. However, because of the extremely low energy of these conversion electrons, very thin "absorbers" must be used to allow escape of the electrons from the sample. Furthermore, the transmission of most electron spectrometers is on the order of a few percent. Therefore, the detection of conversion electrons with a magnetic spectrometer is favorable only in special cases where either specific electronic properties are studied or where no other detection technique is appropriate because of a background that is unresolvable from the radiation of interest by the average scintillation or gas-filled counter.

Solenoid spectrometers adapted for this particular purpose have been used by Kankeleit [106], who detected the conversion electrons from the 100-keV transition in ^{182}W and by Mitrofanov and Shpinel [107], who detected the 19.4-keV L-conversion electrons from ^{119}Sn. Because of the low energy (<100 keV) of the transition in most Mössbauer nuclei, very simple, low-current, air-cooled spectrometers might be designed that would be more suitable to the geometrical requirements of Mössbauer spectroscopy.

Resonant Counters

Mitrofanov and others [108, 109] developed a counter with selective efficiency for recoilless γ-rays. Their counter was used for ^{119}Sn, but it could in principle be used for any other isotope in a compound with a large Mössbauer fraction at room temperature, such as ^{57}Fe [110] and ^{169}Tm [111].

This type of counter is an ordinary Geiger or proportional counter with its inner surface thinly coated with a compound of the isotope under study [107] or with the proper absorber located inside the counter [110]. Recoilless γ-rays are resonantly absorbed by the coating or the internal absorber, and internal-conversion electrons or low-energy X-rays are emitted in the subsequent decay of the excited state. These radiations are now counted by the counter with practically a 100% efficiency and a 2π-geometry. By this method the nonresonant background makes only a very minor contribution to the counting rate. Commercial resonance counters for ^{57}Co have recently become available in the United States [101].

Background Corrections

In Mössbauer experiments that are not carried out in the beam of an accelerator or a nuclear reactor, background radiation arises from the Mössbauer source itself or occasionally from a radioactive absorber, such as ^{129}I [112]. Only experiments involving the absolute measurements of the Mössbauer fraction require a very careful accounting for background radiations and may require sophisticated shielding arrangements. Usually, when the experiment concerns the study of hyperfine interactions, thermal shift, and the relative variation of the Mössbauer fraction with temperature, one may adopt a simpler approach: the shielding is applied in order to improve the signal-to-noise ratio to the extent that the Mössbauer spectrum is easily observed. Several contributions to the radiation background are caused by the Mössbauer source: Compton spectra of higher-energy γ-rays emitted by the source of the Mössbauer radiation, the photopeak of other γ-transitions or X-rays of approximately the same energy, the bremsstrahlung spectrum from high-energy β-transitions, and finally the small but sometimes nonnegligible contribution of nonresonant scattering of recoilless γ-rays.

In dealing with background radiation produced by the source one has to distinguish two separate contributions: one that comes with the beam of relevant radiation, such as β-rays or γ-rays of different energies and X-rays, and a second component that is scattered into the counter from the experimental assembly around the detector; that is, any distribution of masses around the source, absorber, and counter, such as sample holders, cryostat, counter walls, and shielding. For example, the walls of a proportional counter are generally opaque to the soft Mössbauer radiation but may serve as quite an efficient converter for high-energy γ-rays, which eject electrons from the inner walls and may produce a formidable background. For this purpose a single sheet of lead (1 to 3 mm thick) around the detector with a window in the direction of the Mössbauer beam suffices. In some cases, however, backscattering of radiation from the lead or the characteristic X-rays of lead may cause a problem to a similar extent to that of the primary radiation against which the shielding was designed. In such cases the shielding has to be designed in a way to meet the specific background situation. In some cases it may be preferable to have no shielding at all, while in other cases layers of different materials may be sandwiched as an optimum shield. "Graded" shields in which a layer of lead is sandwiched between aluminum absorbers are available commercially [113].

The first step in dealing with the background produced by radiation in the direct beam of the Mössbauer radiation is a proper choice of detector. The magnitude of the Compton contribution from higher-energy γ-rays can be reduced by using a counter that is as efficient as possible for the Mössbauer transition but is inefficient for higher-energy radiations. For example, a NaI(Tl) crystal should be chosen whose thickness is no larger than that needed to stop the radiation under study. NaI(Tl) detectors 0.010 in. thick, are used to detect the 14.4-keV radiation from ^{57}Fe, and the ratio of the peak counting rate to the lowest background point at lower energies is typically between 5 and 10. Similar considerations apply to proportional counters, which have the additional advantage of higher resolution and hence better photopeak-to-Compton background ratio. Very often a double-window proportional counter, that is, one which has beryllium windows both in front of and in the back of the counter, can be used to minimize backscattering of radiation of higher energies from the outer wall of the detector.

With their extremely good resolution and an efficiency higher than that of proportional counters, Ge(Li) detectors are particularly applicable in the medium-energy range from 30 to 100 keV, especially in cases in which the Mössbauer radiation corresponds to a weak transition, such as the 82.3-keV line in ^{191}Ir [114]. Si(Li) detectors, with their superior resolution (Fig. 4.9) but rather small geometrical efficiency, provide a good alternative to the proportional counters at energies below 30 keV when many soft X-rays are present. The cryostat assembly of the solid-state detectors may scatter radiation of higher energies into the detector and may have to be shielded.

The problem of reducing the background contributed by X-rays or γ-rays whose photopeak is completely unresolved from the Mössbauer radiation is much more difficult, and various techniques have been tried. In many cases, where the background radiation is only slightly higher in energy than the Mössbauer radiation, critical absorbers whose K-absorption edge lies between the two radiation energies will be sufficient, since the ratio of absorption coefficients for the two radiations is typically around 3 to 5. However, the thickness, and particularly the position of this critical absorber, must be carefully selected for optimum transmission of the interesting radiation and minimum additional background from the X-rays produced in the critical absorber by the radiation being absorbed. These absorber X-rays will also be unresolved from the other radiations. A dramatic example of the situation above is presented by palladium foil, which absorbs the 25.3-keV K X-ray of ^{119}Sn and thus permits detection of the Mössbauer effect of the 23.8-keV γ-radiation. However, the thickness of the palladium absorber and its position between source and detector [115] must be carefully calculated in order to minimize by self-absorption and unfavorable geometry the detection of the 21.2-keV *K* X-rays of palladium. The absorption coefficients for many elements can be found in the literature [116].

For cases in which the background radiation is lower in energy than in the Mössbauer radiation, absorbers can also be found, which, if their thickness and position are properly selected, will improve the ratio between the Mössbauer line and the background intensity. In general these absorbers will not be the critical absorber for the background radiation but an element of lower Z. For example, cadmium is much more effective in enhancing the 99- and 130-keV transitions of ^{195}Pt in the presence of the 67-keV X-rays than the critical absorber hafnium. Similarly, it is possible to eliminate almost entirely the 6.4-keV X-rays emitted by a ^{57}Co source by using an absorber of about 20 mg/cm^2 of aluminum, while the 14.4-keV Mössbauer radiation is only slightly attenuated.

Coincidence techniques may be used to improve the signal-to-noise ratio, as was demonstrated by Nicholson [117] in the detection of the Mössbauer excitation of the second excited state (at 75 keV) in ^{161}Dy. The 75-keV state decays preferentially by a 25- and 50-keV cascade transition. A conventional experiment would yield a 0.3% effect, while by measuring the coincidence rate between the 25- and 50-keV transitions as a function of the source velocity a factor of eight change has been observed across the resonance. This technique has also been used by Lee et al. [118], who have measured the Mössbauer effect in ^{57}Fe following Coulomb excitation of the 136-keV second excited state. The X-ray background caused by the beam interactions in the target is so large that only by selecting 123:14.4-keV coincidences could the subsequent Mössbauer absorption of the 14.4-keV transition be observed.

Similar effects are achieved by observing either the X-rays produced in the deexcitation of the Mössbauer level with a detector well-shielded from direct radiation from the source, or with the internal conversion electrons. This "scattering" geometry has been extended to the observation of the Mössbauer γ-ray itself, and yields by far the most efficient results at higher energies (>100 keV), where the Mössbauer fractions f and effect ϵ are very small [119] (see the following section).

Another consideration of importance in the usage of proportional counters, NaI(Tl) crystals, and low-temperature dewars with thin beryllium foil windows for work with ^{57}Fe is that the beryllium foils must be iron-free. Spurious results [120] can appear even for very slight contaminations of iron in the beryllium window. Most ordinary aluminum foils contain enough iron impurities to give rise to observable resonance effects. Similarly, ordinary solder flowing on windows can distort ^{119}Sn spectra.

When very strong Mössbauer sources are used, the questions of pile-up and overloading of the detector and the electronic counting system has to be considered. A significant contribution to these problems generally arises from the presence of a very large number of small pulses produced by the background of soft radiation. The intensity of such radiation may be selectively reduced by using low Z absorbers. Even when the overall counting level has been reduced to the level where the system is not overloaded and the radiation of interest may be resolved, the resolution may still be deteriorated because of pile-up (when several low-energy signals overlap and add up in the detector to give signals of similar amplitude to that of the relevant radiation), and the overall counting rate should be further reduced.

Special considerations apply to background radiation of rapidly decaying sources such as ^{129}Te [121] or ^{83}Kr [122], for which the signal-to-background ratio changes with time. For these cases there exists an optimum measuring time during which the statistical accuracy of the experiment reaches a maximum.

In addition to the natural background or the background arising from other radiations in the source, other effects contribute to the background and can lead to errors in the results, particularly in the measurements of recoil-free fractions. For low X-ray energies relatively large-angle Compton scattering can occur with hardly any loss in energy, or at least the energy change is too small to be resolved by the usual detectors. However, such inelastic scattering in the source and in the absorber can be minimized by a propitious choice of geometry and by very high resolution detectors.

Houseley has discussed in detail methods to determine experimentally such a background contribution as well as other subtle effects that are relevant in precision measurements of the Mössbauer absorption intensity [123].

Scattering Geometry

As has already been noted (see Section 1, p. 222), after absorption of a recoilless gamma quantum, the Mössbauer absorber acts as an isotropic scatterer in the deexcitation process in which X-rays and/or conversion electrons are emitted. While the majority of Mössbauer experiments that have been reported in the literature have made use of transmission geometry, because of its great simplicity, it is also possible to observe the Mössbauer effect in a scattering geometry. Detailed discussions of scattering experiments have been presented by Frauenfelder, deBrunner, and Major [105].

Apart from the intrinsic interest in the scattering process, such experiments are used principally in the study of high-energy Mössbauer transitions where scattering configurations can lead to a dramatic increase in the signal-to-noise ratio. Scattering geometry is also useful in the study of the properties of surfaces of solids such as steels [124], in the investigation of corrosion processes [125], and in the field use of geological prospecting devices such as those described by Gol'danskii [126].

4 SPECTROMETERS AND DRIVES

Electronics

A Mössbauer experimental setup consists of an electromechanical transducer and its driving system, a nuclear detecting, sorting, counting and data-handling system, and very often a low-temperature cryostat and/or a high-temperature oven with appropriate associated temperature control and recording units. Such units have become commercially available in increasing sophistication and can often be tailored to individual specific needs. The design of some of the components improve and change significantly in time as a result of the fast evolution in solid-state electronics. Consequently, we shall go into the details of the design of the electronic components only insofar as it seems essential for the understanding of the operation of a complete Mössbauer spectrometer system.

Most of the Mössbauer spectrometers employ electromechanical velocity drives, which are operated either in the constant acceleration or the constant velocity mode. For a velocity range of ± 10 mm/sec, which is adequate for some of the most commonly used Mössbauer isotopes (e.g., ^{57}Fe, ^{119}Sn, ^{129}I), both types of instruments display similar performance, and the choice between them depends on the types of experiments and data handling to be used. For much higher velocity ranges the constant velocity spectrometer becomes impractical, due in large part to the purely mechanical problems associated with stopping and reversing the transducer.

The basic design of both spectrometers is the same: the motion of an electromechanical transducer is enslaved to a velocity reference signal. While the appropriate range of relative velocity between the Mössbauer source and the absorber (or scatterer) is automatically scanned, the transmission of the Mössbauer radiation through (or scattering from) the sample is measured, stored, and recorded as a function of the velocity.

Electromechanical Transducer and Driving Unit

The electromechanical transducer and its driving electronics are shown schematically in Fig. 4.10. A typical velocity transducer consists of a dual-coil loudspeaker, where one coil is being used to drive the moving shaft that carries the source and the second is used as a velocity-sensing coil. The driving coil has to be of low enough impedance (typically a few ohms) to allow the application of adequate power to the drive within the voltage span of the output of the power amplifier. For a straightforward design of the electromechanical drive, the emf generated across the pickup coil should be proportional to the relative velocity between the coil and the magnet. This can be achieved if the magnetic flux intercepted by the coil is linearly related to its relative displacement from the magnet over the entire stroke of the motion, if only a negligible amount of current is drawn from the pickup coil in the process of measuring the velocity and if there is no inductive coupling between the driving and the sensing coils.

It appears that commercially available velocity transducers as well as high-fidelity loudspeakers generally meet the first requirement. The second requirement is met by the design of the feedback network that controls the motion, namely, the pickup coil is a part of a high-impedance circuit. The third requirement (absence of cross talk) may become a problem when the two coils are mounted in the same transducer (double-coil loudspeaker).

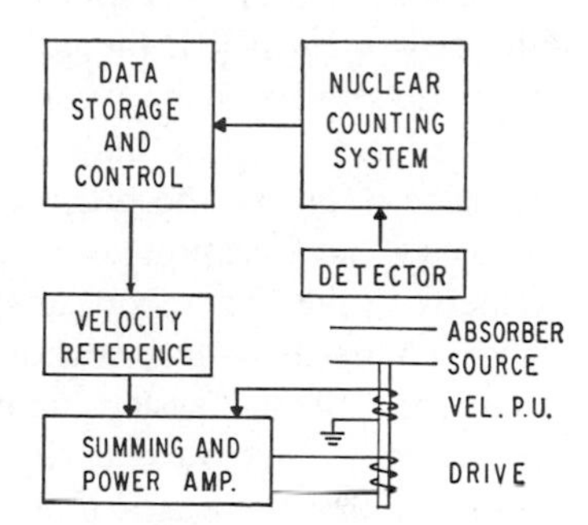

Fig. 4.10 Block diagram of complete electronics for a Mössbauer spectrometer. The velocity pickup signal (VEL. PU) is used to sense the instantaneous velocity of the drive.

Any electromechanical transducer tends to oscillate at a resonance frequency when displaced from its equilibrium position. In principle it is possible to enslave the motion of the transducer to follow any velocity wave form, provided its repetition rate is far off the resonance frequency [127, 128]. To do so the transducer is incorporated into an electromechanical feedback loop, as illustrated in Fig. 4.10.

Since the accumulation and recording of the Mössbauer spectrum is usually done digitally, the most natural way to generate the velocity reference signal (VRS) is by coupling an address scaler to a digital to analog converter (DAC) [129–131]. The wave form of the VRS is determined by programming of the advance of the address scaler, which will be done differently for the constant acceleration and the constant velocity spectrometers. The appropriate wave forms may be also obtained from analog signal generators, which are capable of generating pure wave forms (square, sawtooth, sinusoidal, or triangular) in a reproducible manner [132, 133].

The VRS is compared with the output of the velocity pickup coil by means of a difference (operational) amplifier. The difference between the two signals is amplified and fed through the power amplifier to the driving coil of the transducer. The signal observed across the driving coil is related to the power required to force the drive to follow the desired velocity and has a different appearance from the velocity or displacement wave forms. The quality of the velocity wave form is measured in terms of stability and linearity. The stability is determined by the zero-level drift, the peak-to-peak amplitude changes in the VRS generator, and the offset drift in the difference amplifier. VRS and difference amplifiers are available that allow for a stability better than 10^{-4} of the full velocity sweep, or 10^{-3} mm/sec for a full scale of 10 mm/sec [131]. The linearity of the velocity scale is determined by the linearity of the VRS, the gain of the feedback loop, and the linearity of the response of the pickup coil. While the VRS can be made as linear and the feedback gain as high as necessary, the limiting factor in the linearity is the response of the pickup coil. In both versions of the Mössbauer drive (constant acceleration and constant velocity) the Fourier analysis of the velocity wave form contains high-frequency components associated with large changes in acceleration [127, 128]. A component at the resonance frequency of the system may distort the linearity of the velocity scale. However, an overall linearity of 1% is easily obtained.

In practice there is a limit to the gain of the feedback loop. Too high a gain may cause the entire system to oscillate at a high frequency. The origin of this oscillation is in the finite time required for a perturbation to travel over the entire feedback loop and to return to the summing point. This delay causes a phase shift between the perturbation and the feedback signals which, for high enough frequency, is enough to reverse the sign of the feedback gain. If the gain is larger than unity for this frequency, the system will oscillate. To quench this oscillation the feedback gain can be reduced selectively at the high-frequency end by a capacitative filter.

Constant Velocity

Using a constant voltage as a velocity reference signal, the velocity of the transducer will be constant within its limited stroke. At the limits, which can be marked by either mechanical [134] or optical limit switches [129], a flyback signal is required to bring the drive back to its starting position. During the flyback time the nuclear counting system and the timer are blocked. The flyback-signal generator recognizes the reversal of the direction of motion and provides the flyback signal with the appropriate polarity. After a given counting time at a given velocity (selected by the experimenter to optimize statistics and to minimize drift problems), the total number of counts stored in the electronic scaler is recorded and the counter is reset to zero. The system then advances to the next velocity point, and the cycle is repeated.

The natural way to generate a constant reference voltage is by using an address scaler and a DAC. The velocity range of interest will be covered by programming the advance of the address scaler. Since each data point is measured separately, the spectrum is obtained progressively as the velocity range of interest is scanned. Automatic X-Y plotters can be used to provide a graphic record of the accumulated data.

Technical difficulties may arise at zero velocity because of the reversal of the direction of motion. Depending on the quality of the limit switches, one may observe a small discontinuity in the base-line counting rate at zero velocity that will make it difficult to measure very small effects close to this velocity.

Additional difficulty arises when the stroke of the drive is not negligible in comparison with the distance between the source and the detector. In such a case at the vicinity of zero velocity one would experience a somewhat larger scatter of data points than due to the statistics of the experiment. However, these are minor difficulties, and if necessary it is possible to overcome them by choosing an appropriate chemical form for the source so that the isomer shift for the absorber of interest lies well away from zero relative velocity (see discussion of sources above).

For high-velocity ranges the constant velocity mode becomes impractical when the period of measurement becomes comparable to the flyback interval.

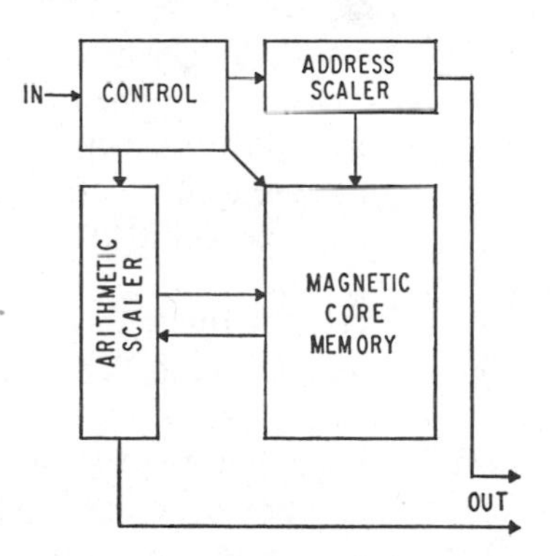

Fig. 4.11 Block diagram of logic circuit for multiscaler operation of a multichannel analyzer.

Constant Acceleration Mode

In a Mössbauer spectrometer operating in the constant acceleration mode the motion of the transducer is enslaved to a VRS having a constant slope. This is achieved by using a symmetric triangular wave form, an asymmetric triangular (sawtoothed) wave form, or a ramp signal. The repetition rate of a few cycles per second is determined by the limited stroke of the transducer. The linearly changing velocity (constant acceleration) is synchronized to the advance of the address scaler of a multichannel analyzer (MCA) operating in the multiscaler mode.

A MCA operating in the multiscaler mode is essentially a small computer in which a magnetic core memory is operating in conjunction with two scalers and some logic circuits (Fig. 4.11). The address scaler chooses a particular memory channel, whereas the arithmetic scaler transfers digital information in (and out) of the particular channel. When a channel advance signal is given, a memory cycle is initiated. The memory cycle consists of the following sequence of events: (1) the input to the arithmetic scaler, which was counting incoming pulses from the counting system, is blocked; (2) the content of the arithmetic scaler is copied by (transferred in parallel into) the passive memory elements of the particular memory channel, n, selected by the address scaler; (3) the address scaler is advanced to the next channel, $n + 1$; (4) the content of the $n + 1$ memory channel is copied by the arithmetic scaler; (5) the input gate to the arithmetic scaler is reopened to the incoming pulses from the counting system.

During the time in which the system dwells at a particular channel, the incoming pulses are counted by the arithmetic scaler and added to the number that was copied from the memory at the end of the last memory cycle. At the end of this counting period the next memory cycle starts by copying the content of the arithmetic scaler into the memory channel $n + 1$. When operated in conjunction with a constant acceleration Mössbauer spectrometer, the address scaler is being advanced at equal time intervals controlled by a precision oscillator. During each time interval the velocity of the Mössbauer drive changes by an equal velocity increment. The pulses arriving from the nuclear counting channel are registered in the memory for each such velocity increment. If the sweep of the address scaler is synchronized with the velocity sweep, the spectrum stored in the MCA will represent the counting rate as a function of velocity.

Constant acceleration spectrometers may be subdivided into three groups.

1. The velocity function is determined by a symmetric triangular velocity reference signal (Fig. 4.12*a*) [132, 133]. In such a case the velocity range is scanned twice during a velocity cycle. The two Mössbauer spectra (which are, ideally, mirror images of each other) obtained during the two halves of the velocity cycle are stored separately in the two halves of the memory of the MCA. The two spectra are stored with reversed velocity scales, and thus cannot be superimposed or simply added to each other. They can, however, be added by appropriate "folding" procedures, provided that the calibration constant (velocity increment per channel increment) is identical in the two halves. The folding of the velocity scale and the addition of the two spectra can be done by a computer in the process of curve fitting to the data. Such an addition increases the statistics of the experiment (by a factor of $\sqrt{2}$), but is desirable only when the inversion of the velocity scale is the only difference between the two spectra. It can be done only when the differential and the integral linearities of the velocity scale are very good in both halves of the velocity cycle, if both halves are free from vibrational disturbances, and in the absence of a significant phase shift in the synchronization between the advance of the address scaler and the velocity function; otherwise folding and adding the data is a disadvantage, since either of the unfolded spectra may be of a better quality than the folded one.

2. With the above reservations, the folding and addition can be done automatically during the accumulation of the data if the access to the memory channels is done with an "up-down" address scaler [135, 136]. In this mode the scanning of the memory address is done from the first channel up to the last in the first half and from the last channel down to the first in the second half of the velocity cycle.

3. It is also possible to use the full memory of the MCA to store only during one-half of the velocity cycle and to compensate for the 50% experimental dead time by doubling the strength of the Mössbauer source. This mode of operation makes it possible to use the full channel capacity of the MCA to record a single spectrum, thus aiding in line resolution or the recording of data when very large hyperfine effects are involved. Under these conditions the folding procedure is, of course, unnecessary.

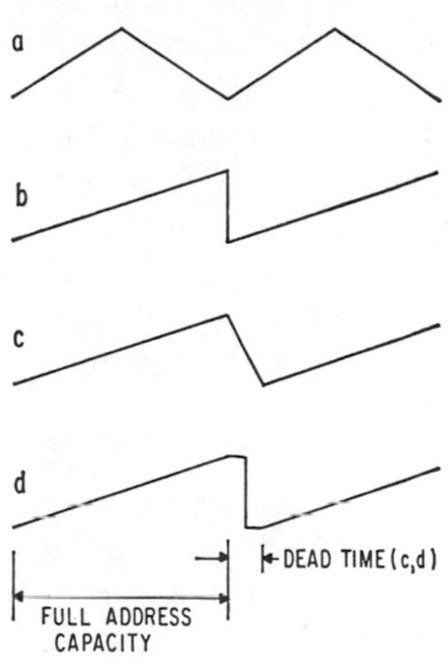

Fig. 4.12 Wave forms for constant acceleration mode operation of the velocity transducer. (*a*) is the symmetric saw-tooth reference signal applied to the input of the summing amplifier for comparison with the velocity pickup; (*b*) is a ramp reference signal; (*c*) is an asymmetric sawtooth reference signal; during the negative slope period of the signal the velocity is reversed from maximum positive to maximum negative while the spectrometer is disabled (dead time); (*d*) is the shape of the reference signal during the dead time that has been used for stabilization of zero velocity of the spectrometer (see [131]).

It is possible to reduce the dead time significantly by shortening the time required to sweep through the second half of the velocity cycle [129–131]. Ideally, this could be achieved by using a ramp as a velocity reference signal [Fig. 4.12*b*]. However, too fast a turnover from the maximum positive to the maximum negative velocities causes a transient signal that distorts the linearity of the velocity scale. This difficulty is overcome by allowing the turnover to occur during a finite time, that is, using a velocity reference signal of asymmetric-triangular or other shapes (Fig. 4.12).

The symmetric triangular wave form used in Fig. 4.12*a* is generally produced by an analog signal generator synchronized with the address scaler of the MCA. In Figs. 4.12*b* and *c* the wave forms may be generated by coupling the address scaler with a digital to analog converter (DAC), which in the case of Fig. 4.12*c* requires some additional circuitry associated with the second half of the velocity cycle. Some MCA's use an "up-down" address scaler; however, unless otherwise spelled out in their specifications, the DAC used for the visual display and the X-Y recording is not good enough to be used as a VRS generator.

Finally it should be mentioned that as in the case of the constant velocity drive, zero velocity may present some problems. The velocity of the drive is zero at the end of the stroke, where the direction of motion is reversed. Any minor strains in the mounting of the drive or friction may significantly perturb the performance of the drive and distort the component of the spectrum near zero velocity. Furthermore, there is no simple way to verify during an experiment which channel in the MCA corresponds to zero velocity, while it is important when the data are folded automaticaly that this be the middle channel.

Constant Acceleration versus Constant Velocity Modes

In the constant acceleration mode the data are accumulated in all velocity channels in each velocity cycle and therefore require the memory of a MCA for their storage. The use of the MCA makes such a spectrometer, in principle, more expensive. However, a MCA is generally more complex and versatile than just a storing device and is useful for many other applications. In the case where a MCA is available for a part-time use as a Mössbauer spectrometer, the construction of an auxiliary constant acceleration drive may be relatively easy and inexpensive. Some of the MCA's incorporate, apart from the detector and preamplifier, most of the nuclear counting electronics, such as high-voltage supplies, single-channel analysis capability, and coincidence circuitry. It may also have an "up-down" address scaler as well as an adequate digital-to-analog converter for the velocity reference signal generator. The use of a constant acceleration system is essential for velocities above ± 2 cm/sec, for the measurements of very small effects, that is, for resonance depth smaller than about 1% of the base-line counting rate and for experiments with decaying Mössbauer sources (such as ^{83}Kr, ^{129}I, and others) that have a time-dependent counting rate. The accumulation of the data at all velocities simultaneously makes this mode of operation insensitive to small drifts in the nuclear counting system, which become prohibitive in the constant-velocity mode at about 1% resonance depth or less, in addition to the difficulties near zero velocity mentioned above.

The constant velocity spectrometer has the obvious advantage of not using a MCA. Furthermore, its advantage becomes apparent in experiments in which the evolution of the Mössbauer spectra is being studied as a function of temperature and/or time. When other physical properties, such as dielectric constant [137], electrical resistivity [138], or others are being continuously monitored while many spectra are being taken and stored for later analysis by computer, it becomes essential that the Mössbauer spectra are also monitored continuously. This is necessary to allow the experimenter to make a visual evaluation of the experiment before proceeding with full computer analysis of his data. The difficulties involved in constant surveillance of the experiment by the experimenter make it essential that this be done automatically. MCA's generally provide one type of output in their automatic mode of operation and require a modification of their automatic cycle in order to provide punched cards or tape output as well as a plotted display. The automatic plotting of many consecutive spectra taken in the constant acceleration mode has to be done on an X-Y recorder with a roll-chart adapter, which shifts the paper by a full frame for each recording. However, such a recorder is incapable of continuous recording of what happens during the run. Consequently, temperature and other physical parameters of interest have to be separately recorded continuously on a multipen strip-chart recorder. In the constant velocity mode, however, the Mössbauer spectrum is plotted and numerically punched continuously during accumulation and therefore may be plotted simultaneously with any other parameter on the same multipen strip-chart recorder. The recording is done for monitoring and not for detailed data analysis and plotting can be done therefore in a very condensed form with a slow movement of the chart. However, if the plotted output is required for graphical analysis of the data, the X-Y recorder with the roll-chart adapter rather than the strip-chart recorder should be used. The reason for this is inherent in the design of the constant velocity spectrometer. The relative dead time associated with the flyback of the drive increases with the velocity due to the increase in the repetition rate of the drive. Consequently, the real time of accumulation at each velocity changes with the velocity and distorts the linear relation between velocity and time.

A constant-velocity spectrometer may be easily designed to provide the option of being used as a constant acceleration spectrometer in conjunction with a MCA. However, for the range of comparable performance it will be more convenient to use it in the constant velocity mode for the following reasons: (1) the time efficiency is increased because only the velocity range of interest is being scanned and (2) it is relatively easy to incorporate the constant-velocity spectrometer in a prolonged automated experiment, during which the MCA becomes available for other applications. These considerations are not valid of course for experiments requiring velocities larger than ± 2 cm/sec, or involving very small effects (the practical lower limit is about 1%), or using decaying radioactive sources with mean life comparable with the time of the experiment.

5 DEWARS AND OVENS

Introduction

Mössbauer experiments are very often associated with changing temperatures, either because many Mössbauer isotopes require low temperatures in order to detect the effect, or, in the case of isotopes that show significant effects at room temperature, many of the parameters studied are derived from the temperature dependence of the Mössbauer spectra. Mössbauer cryostats and ovens differ from conventional ones in that they have to allow the transmission of rather soft γ-radiation. This transmission is obtained by employing vacuum-sealed radiation windows, which may consist of thin beryllium, mylar, or aluminized mylar foils. Many cryostats have been described and reviewed in the literature [140, 141]. They may differ in the structural details because of different experimental requirements as well as differences in prejudices of the designers. The choice of a particular design depends on whether the temperatures desired are above or below that of liquid nitrogen, on whether both source and absorber or only one of them have to be cooled, or on whether the experiment is going to be performed at a single fixed temperature or over a large temperature range. Many commercial cryostats are available [142], and the choice between them has to be made on the basis of compatibility with the experiment in question as well as on the reputation of the manufacturer and the accessibility of service when needed.

Coolants

Liquid nitrogen and helium-4 are the most widely used coolants. Their cryogenic properties are listed in Table 4.6 together with the relevant

Table 4.6 Properties of Some Common Cryogenic Liquids Suitable for Experiments at Low Temperatures

Cryogen	*Normal boiling point, °K*	*Heat of vaporization cal/gm*	*Temperature reached by pumping, °K*	*Density gm/cm³*
Argon	87.2	39.0	—	1.400
Nitrogen	77.3	47.6	63.3	0.808
Neon	27.1	20.5	15.5	1.207
Hydrogen	20.3	108.0	14.0	0.071
Helium (^{4}He)	4.2	4.9	1.2	0.125
Helium (^{3}He)	3.2	—	—	—

properties for neon and hydrogen. The choice between hydrogen and helium is determined in the first place by whether the temperature of interest is above or below 20.4°K. For temperatures above the boiling point of hydrogen but below 78°K a compromise has to be made between the hazards involved in employing hydrogen and its availability and price. The cost of neon is prohibitive in any case, and its use as a cryogenic liquid is seldom justified.

Temperatures below the boiling points of cryogenic liquids may be reached by pumping on the liquid. Temperatures above the boiling point of the coolant are obtained by a combination of thermal insulation from the cold bath and electrical heating.

Glass versus Metal Dewars

Apart from the obvious difference in fragility, almost any metallic configuration of a cryostat may be replaced with a glass counterpart. Glass dewars are cheaper and in principle may be modified to have radiation windows adequate for Mössbauer spectroscopy. The problems of the permeability of borosilicate glass to helium have recently been solved by appropriate coating techniques. Both glass and metallic dewars have been developed to the point that they have become commercially available [142] for Mössbauer experiments.

Windows

Dewars for Mössbauer experiments must have thin radiation windows to allow the γ-radiation of relatively low energy to reach the detector. At the room-temperature end of the vacuum jacket of a metallic dewar a 0.005-in. beryllium window or even thinner mylar or aluminized mylar vacuum sealed windows with an O-ring seal are sufficient. These windows may be glued to the metal with epoxy cement. In cryostat configurations involving a window in the low-temperature end, the soldered window is the most reliable one. The mounting and performance of such windows has been described in many places [143–146]. The use of mylar film coated on one side with a thermoplastic adhesive [147] is considered in Section 6. Such film is very useful for room-temperature vacuum seals and can support atmospheric pressures over an area of $\sim$20 cm^2 or more.

Nitrogen Dewars

The low price and favorable cryogenic properties of liquid nitrogen make it easy to use in homemade improvised dewars using styrofoam for thermal insulation. Such a dewar may be very simple, inexpensive, and useful for occasional experiments at fixed temperature close to the boiling temperature of nitrogen. It is not recommended, however, for extensive and prolonged experiments with accurate and variable temperatures. For such experiments a metallic cryostat is more appropriate. The easy manipulation, reproducibility in experimental arrangement, and time saved make such an investment very worthwhile in the long run.

A typical metallic dewar for liquid nitrogen is schematically illustrated in Fig. 4.13. The liquid nitrogen container, which is usually made of stainless steel, is hung from the top of a second stainless steel container by thin-walled stainless steel tubing. The thermal insulation is provided by the vacuum between the two containers and the low thermal conductivity of the thin-walled stainless steel tubing. The Mössbauer absorber is attached to a cold finger, usually made of copper and thermally attached to the nitrogen reservoir. Such a configuration is most appropriate for a constant-temperature

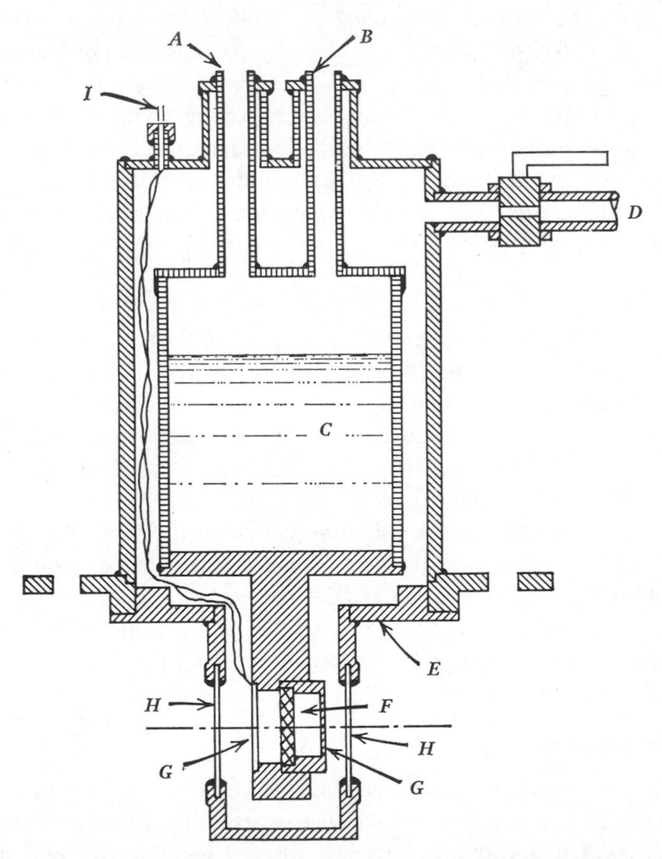

Fig. 4.13 Metal Dewar for liquid nitrogen. The transmission geometry is that used most commonly in conjunction with moving room-temperature sources: *A*, nitrogen fill, *B*, nitrogen vent; *C*, liquid nitrogen; *D*, main pump-out; *E*, detachable tail section; *F*, sample; *G*, aluminum foils; *H*, Mylar window; and *I*, thermocouple.

experiment. For an experimental arrangement with variable temperature it is necessary to control the thermal conductivity between the cold bath and the sample. This may be done as shown in Fig. 4.14. The cold finger is attached to the cold bath via a thin-walled stainless steel tube, while a second tube made of copper is concentrically extended from the sample holder copper block up into the liquid nitrogen container. The thermal conduction between the cold bath and the cold finger is therefore determined by the gas pressure (generally referred to as the exchange gas) between the stainless steel and the copper tubes extending up to the liquid nitrogen level. The pressure of the exchange gas is controlled by a separate vacuum system. The Mössbauer radiation is transmitted through two radiation windows mounted on the external container. The most convenient way to change the absorber is by having a demountable tail section, which requires, of course, the warming up of the dewar for each such operation. In order to change temperatures one may vary the pressure in the exchange tube and/or use an electrical heater mounted on the cold finger while monitoring the temperature by an adequate thermometer. In such a cryostat one could either cool the source or the absorber, but not both because of the necessity of introducing the Doppler motion. Experiments that require both source and absorber to be cooled require also lower temperatures. It is not worthwhile to increase the sophistication of a nitrogen cryostat to allow for cooling of both source and absorber unless the cryostat is designed for liquid helium experiments.

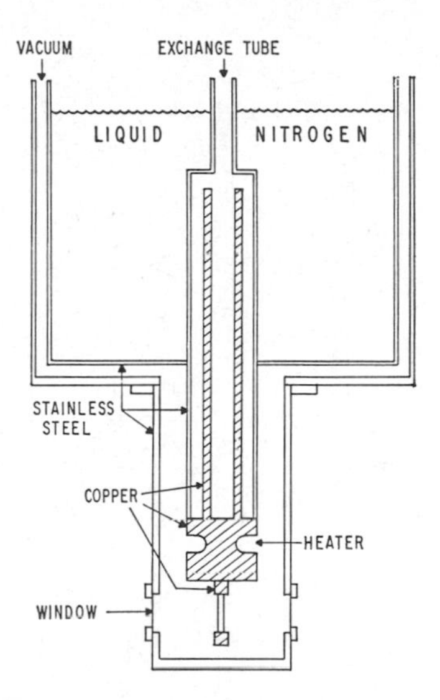

Fig. 4.14 Metal dewar for liquid-nitrogen cooling of a stationary absorber. This dewar is suitable for measurements at any temperature between that of the cryogenic liquid and room temperature. Sample temperatures can be ascertained from appropriate thermocouple or platinum thermometer measurements. Heat transfer between the copper sample block and the nitrogen reservoir is controlled by variation of the pressure in the exchange gas tube.

Helium Dewars

Because of the very low heat of vaporization of liquid helium, a helium cryostat has to be carefully designed with minimum heat leaks. Because of the much larger heat of vaporization, a liquid hydrogen cryostat does not require such care concerning heat leaks, but greater care needs to be exercised to prevent the possibilities of explosion. Apart from these differences the requirements are the same and the design is very similar. In both cases the cold bath must be surrounded by two evacuated regions with a liquid nitrogen shield in between. For experiments in which only the source or the absorber has to be cooled a design similar to that in Fig. 4.14 may be used. In this case the vacuum container serves also as the inner wall of a liquid nitrogen container, used as a radiation shield while being thermally insulated from room temperature by an additional vacuum region (Fig. 4.15). The vacuum-sealed radiation windows are mounted only on the outer container, while the windows on the radiation shield may be made of thin aluminum or aluminized mylar foils taped on the metallic shield. In order to obtain liquid helium temperatures liquid helium has to be in the exchange gas tube. This may be done either by using a controlled leak from the main reservoir or by introducing helium gas into the exchange manifold at a pressure slightly above atmospheric and thus liquifying it. Both methods are frequently used and involve some differences in the details of the design of the cryostat. Temperatures down to about 1.8°K may be easily obtained by pumping on the liquid helium in the exchange gas with a simple rotary vacuum pump. With a large mechanical pump and a well-designed cryostat temperatures as low as 1.2°K may be reached.

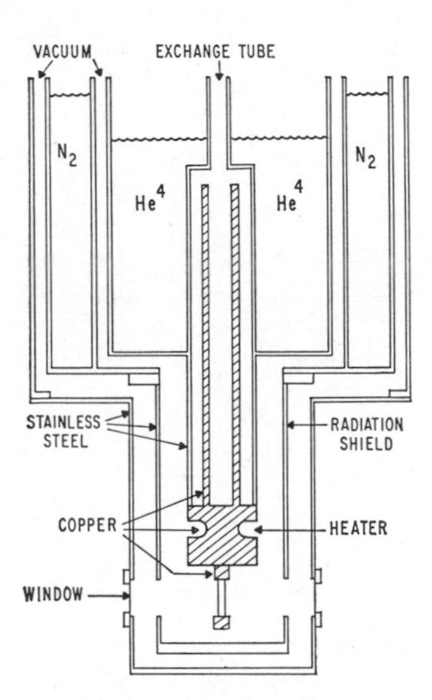

Fig. 4.15 Metal Dewar for liquid helium cooling of a stationary absorber. The operation of this Dewar is similar, in principle, to that shown in Fig. 4.14, with the addition of a helium reservoir and a liquid nitrogen cooled thermal radiation shield.

Typical liquid helium consumption in a routine operation with a Mössbauer cryostat is about 200 cm^3/hr. First cooling to liquid helium temperatures, after precooling by liquid nitrogen, may require an additional amount of about 3 liters while as little as 1 liter may be wasted in subsequent fillings and cooling of the transfer tube. These numbers are based on experience with a commercial cryostat [148]. The numbers vary, of course, depending on the design of the cryostat as well as on the details of the experiment; however, they should give some idea of the quantities involved.

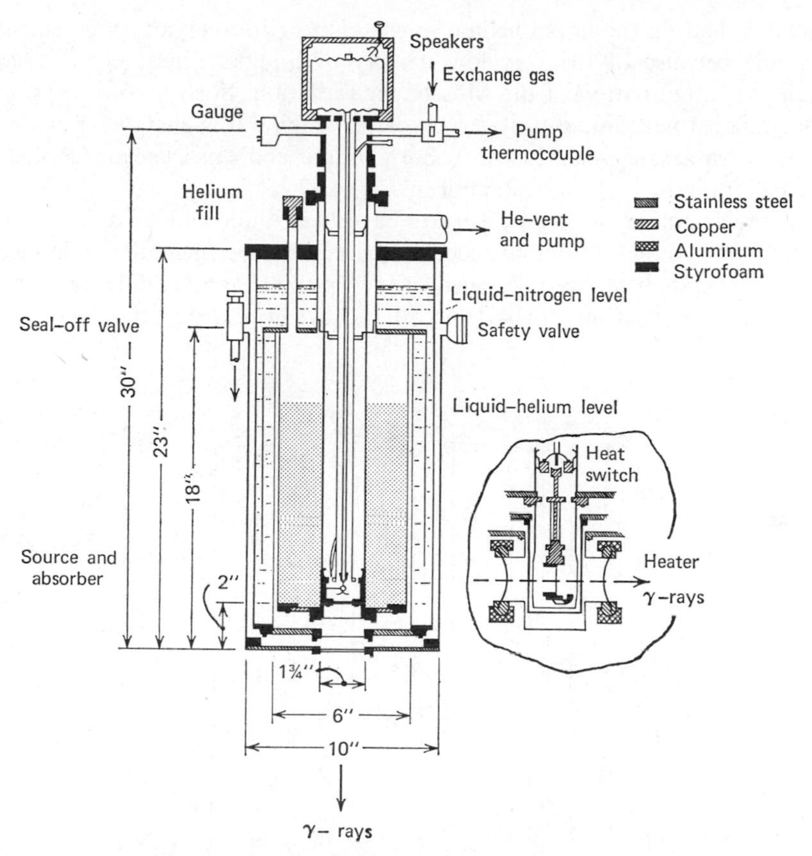

Fig. 4.16 Metal Dewar for liquid helium cooling of a stationary absorber and moving source (or vice versa). This Dewar uses a vertical geometry with the moving element supported longitudinally and a source immersed in the cryogenic (helium) bath. This mode of operation is possible even with soft γ-emitters because of the low absorption coefficient of helium, even in the condensed phase.

A liquid helium cryostat in which both source and absorber are cooled is shown in Fig. 4.16. This cryostat has been designed by Kalvius [140]. The velocity transducer is mounted on the top of the exchange tube driving the source. The versatility of this design allows the mounting of a small heater on the source holder and varying the temperature of the source, while the demountable bottom allows mounting the absorber with thermal insulation from the cold bath and varying its temperature. When the source temperature is to be maintained only at liquid helium temperature, a simpler version of this cryostat may be used. In this case the exchange tube is eliminated and the source is held in the liquid helium in the cold bath. Such an arrangement is possible because of the very low density of liquid helium, giving rise to negligible attentuation of the Mössbauer radiation. Such a configuration is not practical with liquid nitrogen cryostats when a low-energy γ-ray source is used. An arrangement in which both source and absorber are cooled by the helium bath is shown in detail in Fig. 4.17.

A large number of metal cryostats have been built and used successfully in different configurations to meet a variety of experimental requirements. Some cryostats have been designed to incorporate superconducting magnets in Mössbauer experiments [149] and others have been designed to incorporate

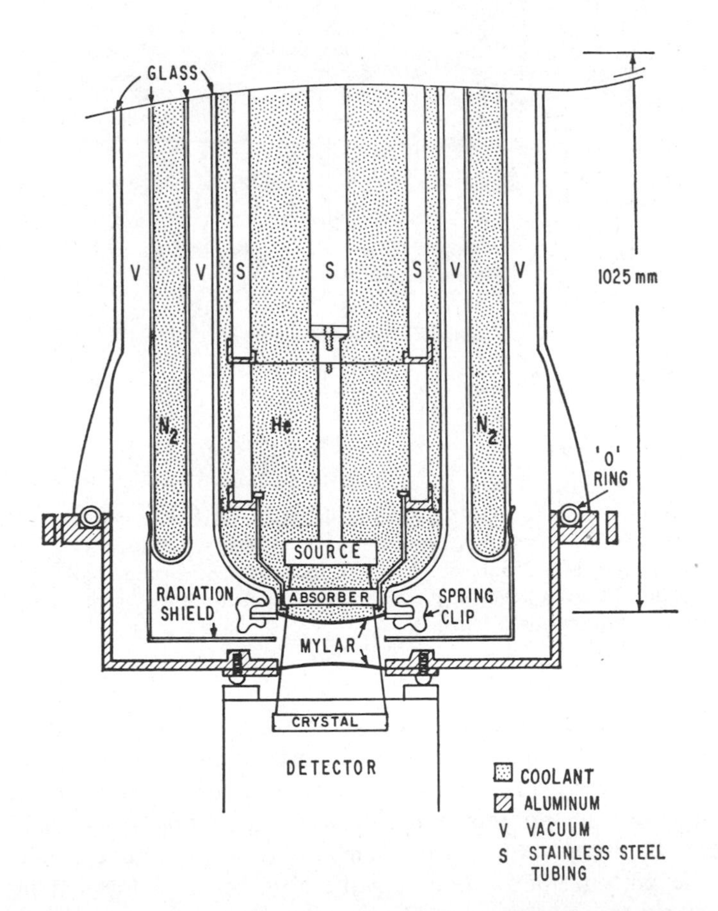

Fig. 4.17 Glass version of the helium Dewar shown in Fig. 16*a* (S. L. Ruby and B. Zabransky, private communication). The acceptance cone seen by the detector crystal is indicated to show the large solid-angle geometry possible in such a design.

a liquid helium Mössbauer experiment in the beams of accelerators [150–154]. Since cryogenics is an art, there are always several good ways to achieve similar goals. The purpose in this summary is not to give an extensive review of the existing literature in this field, but rather to provide an initial entrée into the state of the art to the newcomer to low-temperature experimentation.

Ovens

The investigation of quadrupole and magnetic hyperfine interactions as well as of precision measurements of the isomeric shift and atomic dynamics often requires high temperatures [155–157]. The design of a high-temperature chamber is similar in principle to that of a low-temperature chamber. The sample holder is held at the required temperature, and the heat leak is minimized by vacuum and radiation shields. The choice of materials in the inner hot chamber is of course different. Because of the possibility of high-temperature chemical reaction between the sample holder and the sample, inert low-Z materials such as graphite or boron nitride are preferable to metals such as beryllium and aluminum. Special attention should be paid to possible high-temperature reactions between the sample and the rarified atmosphere around it because of the finite vacuum in the oven. The rate of such reactions may be appreciably reduced by enclosing the sample in a tightly fit ceramic sample holder and cover. In this way the flow rate of gas to and from the sample is significantly reduced, reducing reaction rates as well as sublimation. Another way is to fill the oven or the inner chamber with an inert gas [158, 159]. As in low-temperature cryostats, the problem of the transmission of the Mössbauer radiation exists and is solved in a similar manner. Vacuum-sealed radiation windows are needed only on the outer container, which is at room temperature, and therefore the same kinds of windows can be used. Thin aluminum foils are used in radiation windows on the heat shields. In order to replace samples, ovens have to be demountable, and the vacuum sealing is done by O-rings that may have to be cooled in order to serve properly. The temperature of the O-rings may be maintained close to room temperature by water-cooling the metallic cover on them. Several ovens have been described in the literature [155, 158–160], and some are commercially available [161].

Thermometers for Temperature Measurements and Control

Low as well as high temperatures may be measured by thermocouples. Carbon and germanium resistance thermometers are very sensitive below 20°K [162–166]. A most convenient thermometer for temperatures above 20°K is the platinum resistance thermometer, which has become commercially available. In principle it may be used up to the melting temperature of platinum (2042°K). Calibration charts for standard thermocouples are available in handbooks. A carbon resistance thermometer can be calibrated by measuring its resistance at the boiling temperatures of helium and nitrogen; the interpolation for intermediate temperatures or the extrapolation to lower temperatures is done with the equation:

$$\log R + \frac{K}{\log R} = A + \frac{B}{T}, \tag{4.29}$$

where K, A, and B are empirical constants that must be determined (by measurements at three temperatures) for each thermometer. Calibration curves for germanium and platinum resistors can be supplied by the manufacturers at nominal charges. Because of self-heating problems, no current is allowed to flow through the germanium and carbon resistors, and their resistance has to be determined with a Wheatstone bridge. This problem is unimportant in the temperature ranges in which a platinum resistor may be employed. The reading of a thermocouple may be modified if there is a break in its line at the vacuum feedthroughs because of possible thermal gradients across them. It is therefore necessary to cross check the calibration of the thermometers under the experimental conditions. For cryostats, the boiling points of helium, hydrogen, and nitrogen, as well as room temperature, provide the necessary check points. The calibration of thermometers in high-temperature ovens may be effected by carrying out measurements at the melting points of well-defined standards. Such calibrations cannot in general be carried out under the experimental conditions of the Mössbauer experiment, thus introducing uncertainties in the temperature data. In the future it may be possible to calibrate ovens by measurement of magnetic hyperfine interactions in a Mössbauer experiment, but such reference data are not as yet available. A distinction should be made between the use of a thermometer for temperature measurement and for its control, and it is advisable when possible to use two separate thermometers for the different functions. The reason for this distinction is that while for precision temperature mesurements the thermometer should not be overloaded (no current is expected to flow through a thermocouple or carbon and germanium resistors), for optimum design of temperature controllers the thermometer may often be operated in this condition. An overloaded thermometer may be still used for temperature reading if it is properly recalibrated.

Several temperature controllers have been described in the literature [149, 167, 168]. A simple inexpensive temperature controller is illustrated in Fig. 4.17 [137]. It consists of a closed feedback loop similar to the one used in the electromechanical driving unit described in Fig. 4.10. The output of the thermometer is compared with the setting of a reference current at the summing point of an operational amplifier. The difference between the two dc levels is amplified and fed through the power amplifier into the electrical heater. The heater is generally made of nichrome or tungsten wire, and it is advisable to choose a low-enough resistivity so that the voltage across the heater at full power consumption will not exceed the voltage range of the output of the operational and power amplifiers used. Contemporary, inexpensive, solid-state amplifiers operate with power levels of ± 12 or 15 V, and a compatible heater should require about 20 V at full power. The design shown in Fig. 4.17 has been successfully used in conjunction with a commercial Janis cryostat and with an Elron oven [137]. The power requirements are entirely different in the two cases: while maximum power consumption in the low-temperature cryostat was about 1 W, about 50 W were required to maintain a temperature of about 1000°K in the oven. The temperature stability maintained by such a feedback loop depends on the ratio between the current drift of the operational amplifier and the current drawn from the thermometer and its gain. An iron-constantan thermocouple provided with the commercial oven feeds current into the summing point through a 10 KΩ resistor, maintaining temperature stability of about 1°K, while being overloaded. The calibration of the thermocouple under these conditions showed about 0.5% reduction in the output reading. Improving the temperature stability will require either a better operational amplifier or higher currents drawn from the thermometer. Some strip-chart-recorders used for temperature recording provide the possibility of increasing the current fed into the feedback loop (without overloading the thermocouple) by supplying a potentiometer output synchronized with the motion of the pen. A less expensive way, however, is to use a platinum resistance thermometer. A typical value for the offset current drift of an inexpensive operational amplifier is 10^{-8} A/day, while the amount of current it is possible to draw from a platinum resistor depends on the bias current fed into it. Using a constant bias current of 1 mA through a platinum resistor of sensitivity of 1 Ω/deg and feeding this to the summing point through a 1200-Ω resistor, it was possible to advance the temperature in the cryostat in 0.5°K steps and with a stability of 0.1°K [137].

Such a temperature controller may be easily incorporated in a fully automated Mössbauer spectrometer, where the temperature is changed between consecutive experiments. To do so, the reference voltage is tapped off a chain of resistors via the center tap of a stepping relay [169], the position of which is advanced at the end of each experiment (Fig. 4.18). Such a stepping relay is also provided with additional sets of contacts that may be used for indexing the punched or recorded output.

As in any feedback system, a temperature controller may overshoot or even oscillate if the gain is too high. This may be prevented by reducing the level of maximum power to be no more than a few times the power required to maintain constant temperature. If temperatures are automatically changed over a large temperature range, the level of maximum power may be kept

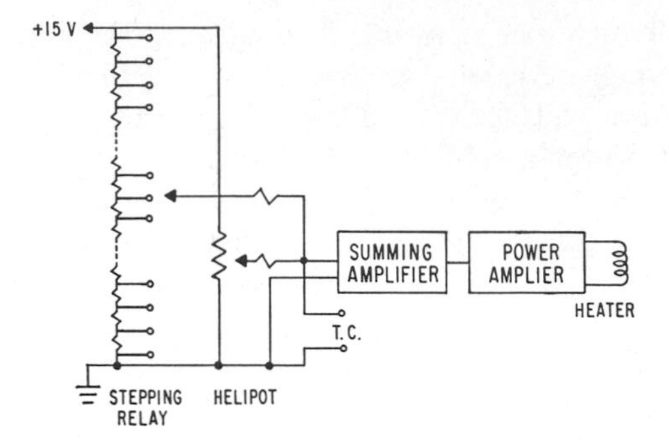

Fig. 4.18 Automatic proportional temperature control circuit. The reference voltage used for control can be varied either manually, using a potentiometer, or automatically by using a clock driven stepping relay. The temperature sensor (TC) can be either a thermocouple or a resistance thermometer as appropriate for the temperature range in question.

the same for several temperatures but it has to be readjusted for each temperature range. This is done by changing a resistor in series with the heater either manually—with a rotary switch—or automatically, using an additional set of contacts on the stepping relay.

6 SAMPLE MANIPULATION

In this section we consider briefly the most important aspects of sample preparation and treatment as they pertain to Mössbauer-effect measurements. In many ways such treatment parallels the sample preparation and handling techniques of optical spectroscopy, with appropriate modifications as required by the nuclear aspects of the method.

Sample Thickness

Reference has been made earlier to the relationship between the experimentally observed line width (Γ_{exp}) and the effective thickness (t_a) of the Mössbauer absorber [see (4.11)]. In most instances the most judicious compromise between this thickness broadening, which lowers the spectroscopic resolution as sample size increases, and the resonance effect magnitude, which increases as sample size increases up to the point where background considerations become dominant, governs the choice of sample thickness.

The relationship between the experimentally observed full width of half maximum, Γ_{exp} and the thickness parameter t has been critically examined by Margulies and Ehrman [170], O'Connor [172], and others [173] and can generally be represented by a relationship of the form

$$\Gamma_{\text{exp}} = \Gamma_s + \Gamma_a + 0.27\Gamma_{\text{nat}}\sigma_0 Nf. \tag{4.30}$$

Substituting $t = \sigma_0 N f$, (4.30) can be rewritten in the form

$$\Gamma_{\text{exp}} = \Gamma_s + \Gamma_a + 0.27\Gamma_{\text{nat}} t. \quad (4.31)$$

A relationship between the experimentally observed line width, Γ_{exp}, and the sample thickness, which takes into account higher order terms in t, has been given by Heberle [174], and takes the form

$$\Gamma_{\text{exp}} = 2\Gamma_{\text{nat}}(1 + 0.1288t + 4.733 \times 10^{-3}t^2 - 9.21 \times 10^{-4}t^3 + 3.63 \times 10^{-5}t^4), \quad (4.32)$$

in which it is assumed that in the absence of thickness-broadening, both source and absorber contribute only the natural line width. In many instances it is possible to choose a chemical and a physical form of the source so that $\Gamma_s \approx \Gamma_{\text{nat}}$. For an absorber in which hyperfine interactions are absent or for one in which the hyperfine effects (e.g., quadrupole coupling constants and magnetic splittings) are so large that the various resonance lines are well-resolved $\Gamma_a \approx \Gamma_{\text{nat}}$ except for thickness effects so that (4.31) can be expressed as

$$\Gamma_{\text{exp}} = \frac{h}{2\pi\tau}[2 + 0.27t]. \quad (4.33)$$

Thus, from the $t = 0$ intercept of a plot of Γ_{exp} versus t it is possible to calculate a lower limit for the excited-state mean lifetime τ. Conversely, if t is known, the recoil-free fraction can be calculated from (4.33) by appropriate numerical methods. In the absence of other complicating factors an appropriate sample thickness is one in which $t_a \approx 1$, so that n_a, the number of absorber nuclei per square centimeter of sample, is given by

$$n_a = \frac{t_a}{\sigma_0 f_a} \approx \frac{1}{\sigma_0 f_a}, \quad (4.34)$$

where σ_0 is the cross section for resonant scattering [given by (4.10)] and f_a is the recoil-free fraction in the absorber. The n_a so calculated is that of the Mössbauer effect nuclide. If the element in question is polyisotopic, the abundance of the Mössbauer active nuclide, a_M, must be taken into account, so that (4.34) becomes

$$n_a^1 \approx \frac{1}{\sigma_0 f_a a_M}, \quad (4.35)$$

where n_a^1 is the number of all atoms of a particular element per unit area of the absorber.

For ^{57}Fe, $\sigma_0 = 2.38 \times 10^{-18}$ cm^2 and $a_M = 2.19\%$. Assuming $f_a \approx 0.5$ gives $n_a^1 = (2.38 \times 10^{-18} \times 2.19 \times 10^{-2} \times 5 \times 10^{-1})^{-1} \cong 3.8 \times 10^{19}$ atoms of iron/cm^2. For a compound such as sodium nitroprusside, $Na_2Fe[(CN)_5NO]\cdot 2H_2O$, (mol. wt $= 297.97$ g/mole, $d = 1.72$ g/cm^2), this is equivalent to a sample thickness of $\sim$1.9 $\times$ 10^{-2} g/cm^2 or $\sim$0.1 mm thick. For metallic iron (d = 7.86 g/cm^3), unit thickness (at $f_a = 0.5$) corresponds to a foil $\sim$4.5 $\times$ 10^{-4} cm thick; that is, $\sim$0.2 mils. Note that velocity calibrations with metallic iron (*vide infra*) are usually carried out with foils $\sim$0.5 mils thick in consonance with this calculation.

For ^{119}Sn, $\sigma_0 = 1.321 \times 10^{-18}$ cm^2 and a_M = 8.58%. Again for $f_a \approx 0.5$ this means that $t_a = 1$ corresponds to $\sim(1.32 \times 10^{-18} \times 8.58 \times 10^{-2} \times 5 \times 10^{-1})^{-1} \sim 1.8 \times 10^{19}$ tin atoms per square centimeter. This figure represents $\sim$3 $\times$ 10^{-5} moles/cm^2, corresponding to a sample thickness of $\sim$9 mg/cm for a compound of molecular weight of 300 g/mole. Since the ^{119}Sn atom in most chemical compounds has an f_a appreciably smaller than 0.5, typical sample thicknesses of several tens of milligrams per square centimeter of tin compounds are normally used as absorbers.

In the above calculation no consideration has been given to the fact that nonresonant scattering increases as sample thickness increases. For relatively-high-energy Mössbauer transitions; that is, those with $E_\gamma \geqslant 50$ keV, the half-thickness is sufficiently large so that this factor can be ignored in most instances. For softer γ-rays, however, this problem may become severe, and must be taken into account. In ^{57}Fe spectroscopy on biologically active materials, for example, molecular weights of 10^5 can be encountered with samples having a single iron atom per molecule [175]. Under such conditions to achieve $t_a \sim 1$ would require a sample thickness of $\sim$6.4 g/cm^2, which is very much larger than the nonresonant scattering half thickness for the 14.4-keV γ-radiation, and hence only a small fraction of the incident γ-rays can penetrate through the sample. To overcome this problem it may be helpful to use enriched ^{57}Fe in absorber preparation. In the case of samples of biological interest this may require the incorporation of enriched ^{57}Fe into the diet or substrate from which an organism (or animal) derives its metabolic requirements, and a number of such experiments have been reported in the literature [176–179].

In the case of other Mössbauer nuclides an appropriate choice of sample thickness, isotopic enrichment, and matrix must be made in the light of the particular nuclear parameters involved. For ^{197}Au and ^{127}I, and others, for example, the isotopic abundance is 100% so that isotopic enrichment is not possible. Similarly, for ^{195}Pt (99 and 129 keV), ^{186}Os (137 keV), and ^{191}Ir (129 keV), and others, the energy of the Mössbauer transition is so high that sample thickness for nonresonant scattering is not an important consideration.

Foils

In many studies, especially where accurate thickness control is important, it is desirable to use foil absorbers. For this purpose iron, tin, gold, platinum, zinc, silver, nickel, iridium, and other foils have been used both for investigations on the pure metal and for the study of a large variety of alloys that can be rolled into appropriate thickness. For most chemical purposes, however, the use of foils as absorbers is usually limited to standard or calibration samples, and the use of such foils (e.g., metallic iron and stainless steel, metallic tin, and so on) is discussed in Section 8.

Powdered Samples

In the majority of chemical studies the sample to be investigated is obtained as a powdered material. Such samples can be prepared for examination by Mössbauer spectroscopy by pressing a well-ground sample of appropriate thickness (see above) between two foils and then mounting this "sandwich" in the optical path of the spectrometer. For ^{57}Fe and ^{119}Sn spectroscopy, especially at low temperatures where good thermal conductivity between the sample and the cryogenic space must be assured, ordinary kitchen aluminum foil has been very widely used. A word of caution is appropriate here, however, since all such foil (usually obtained by rolling of the aluminum billet between steel rollers) contains small but significant amounts of iron (probably as FeO). For the study of small ^{57}Fe resonance effects, such as in organometallic compounds [180, 181] or in biological materials containing iron, the parasitic resonance effect observed for this iron in the aluminum foil can be ruinous and must be avoided. High-purity (99.99%) aluminum foil, which is available commercially [182], can be used in such cases to avoid the parasitic resonance effect. Mylar film, either aluminized or not, of 0.5 to 2 mil thickness, which is available commercially [183], can also be used for the "sandwiching" material.

Mylar film that is coated on one side with a thermoplastic material [184] can be used to good advantage in encapsulating air- and/or moisture-sensitive materials, provided that these can withstand the modest (~130 to 140°C) sealing temperatures of the film. Such films are also useful for heat-sealing to metal absorber holders, as in the case of thin liquid cells (see p. 283).

A problem often encountered in the use of crystalline samples as Mössbauer absorbers has to do with accidental preferential crystal orientation. This orientation can be observed as an asymmetry in the intensity ratio of the two doublets in a quadrupole split spectrum in ^{57}Fe [185] or ^{119}Sn [186], or in a departure from the intensity ratios calculated from Clebsch-Gordan coefficients in ^{129}I [187], for example. In some cases it is possible to overcome this difficulty by finely grinding the Mössbauer sample with quartz or Pyrex glass powder [188] prior to mounting in the absorber holder. Another technique is to mix the absorber material intimately with a nonreactive stopcock grease (such as the Apiezon L, N, or T, Kel-F or silicone greases used in high-vacuum work), and then spreading this "paste" on an appropriate backing foil or film.

Finally, it is worth commenting briefly on the use of self-supporting absorbers in Mössbauer measurements. Frequently the use of such films is demanded by the quantitative aspects of the problem, especially where an accurate measure of the number of Mössbauer atoms per unit area of sample is required. In the case of metallic absorbers such self-supporting samples usually take the form of rolled foils of known thickness. For chemical compounds of the kind usually obtained as micro-crystalline powders it is often possible to obtain self-supporting disks by pressing the material in a hydraulic press (for example, the kind used to prepare KBr pellets for infrared spectroscopy, found in most chemical laboratories) in a suitable stainless steel die. For ultra-thin absorbers or for those in which self-cohesion properties are poor, prior admixture of an inert, low-Z, compressible powder such as Al_2O_3 or benzoic acid often leads to satisfactory self-supporting disks that can be weighed and used as Mössbauer absorbers. Incorporation of the sample of interest in paraffin, polymethyl methacrylate [188, 189], polyethylene, and others, has also been used to obtain self-supporting films suitable for spectroscopic study.

Single Crystals

Reference was made above to the problems associated with preferential crystal orientation with respect to the optical axis of the Mössbauer experiment. In certain experiments, such as those associated with the determination of the sign of the electrostatic field gradient (in $\frac{1}{2} \leftrightarrow \frac{3}{2}$ transitions), and in studies of the anisotropy of the recoil-free fraction (Gol'danskii–Karyagin effect [190]), it is sometimes appropriate to study single crystals oriented in a known manner in the spectrometer. For samples that can be grown into large crystals, which can subsequently be sliced or cleaved (e.g., sodium nitroprusside, iron garnets, rare-earth salts), crystal orientation is effected in a straightforward manner. By the use of rotatable dewar tail sections or sample mounts, such oriented single crystals can be studied as a function of angle, and such studies can shed considerable light on lattice dynamical properties and on magnetization phenomena in appropriate solids. For samples that are not readily obtained as large single crystals but which grow in the form of plates, sheets, or needles, it may still be possible to carry out a known orientation Mössbauer experiment by low-power microscopic examination of the sample and by manual orientation of many small crystals in a matrix such as stopcock grease or on a mildly adhesive surface (pressure-sensitive tape).

Frozen Liquids and Glasses

As indicated in the section above, Mössbauer measurements are almost always carried out on samples in the solid state. This does not, however, restrict the experimenter to the study of neat crystalline materials, since much information can be obtained from the study of frozen solutions [188, 189, 191–193]. For the higher-energy Mössbauer nuclides the construction of liquid cells that can be cooled to cryogenic temperatures is no particular problem, since sample and window penetration problems are minimal. For the softer Mössbauer transitions, however, such as ^{83}Kr (9.3 keV), ^{57}Fe (14.4 keV), ^{119}Sn (23.8 keV), and ^{129}I (27.8 keV) *inter alia*, both sample thickness and window absorption can become a significant problem. The construction of thin (0.3 to 2 mm in depth) sample cells out of copper, brass, stainless steel, or plastic is required for such purposes. Window materials for such cells should be rigid where possible, and should be made of low-Z materials to avoid nonresonant scattering effects. The use of Mylar film coated on one side with a thermoplastic adhesive [184] is particularly suitable for such purposes.

The choice of solvents that can be used for frozen solution experiments depends in large measure on the particular chemical requirements of the problem. A good deal of work has been reported on frozen aqueous solutions [194–196], and because of the unique nature of water as a chemical solvent such studies have shed much light on the nature of the ionic and molecular species in water and on the structure of ice itself (where the Mössbauer atom is used as a probe). Crystalline organic solvents such as benzene have been used to study the nature of the benzene-I_2 complex [197] and the nature of the spectral changes that occur on the solution of iodine in polar and nonpolar solvents.

The use of organic solvents that set to glasses at low temperature is especially useful in the study of conformational changes that accompany the solution process [189], and such measurements are particularly useful in conjunction with studies on organometallic molecules. Solvents that set to transparent glasses at low temperatures which can be used for such studies are methylhydrofuran ($C_4H_3OCH_3$, $d = 0.923$ g/cm^3, bp 65.5°C) tetrahydrofuran (C_4H_8O, bp 64–66°C), 2 methyl pentane (C_6H_{14}, $d = 0.654$ g/cm^3, bp 60°C), and others [198]. It must be remembered, however, that recoil-free fractions in such glassy matrices are usually considerably smaller than they are in materials that are solids at room temperature. Moreover, the solubility limit of the sample in the solvent tends to make the Mössbauer nuclide density (atoms per square centimeter of sample) considerably smaller than for the usual neat solid absorber. These two factors tend to give rise to very small resonance effects (frequently less than 1%) so that considerable care must be taken with respect to optimizing the experimental conditions under which the Mössbauer effect can be observed.

Experiments at High Pressure

A number of studies have been reported in which either the source or the absorber in a Mössbauer experiment has been subjected to high hydrostatic pressures. Many of these results are discussed in terms of chemical bonding. The earliest such results on the effect of pressure on the 14.4-keV transition in ^{57}Fe are those of Pound et al. [199], who made use of the equipment described earlier [200] in the measurements of the effect of gravity on the Mössbauer transition. A source of ~2 mCi of ^{57}Co diffused into metallic iron was enclosed in a beryllium copper pressure bomb equipped with a $\frac{1}{2}$-in.-thick beryllium window and connected by means of a stainless steel tube to a Bridgman press. Measurements at 1 atmosphere, 2000 kg/cm^2, and 3000 kg/cm^2 were reported in this study.

A very extensive series of measurements of the effect of pressure on the properties of materials by means of Mössbauer spectroscopy has been reported by Drickamer's high-pressure group at the University of Illinois. This group has largely adapted the experimental techniques developed earlier in the study of optical spectra at high pressures to the specific requirements imposed by the nuclear parameters of the technique. In an extensive study on metallic iron foils reported by Pipkorn et al. [201] a detailed description of the Illinois experimental equipment is presented. The high-pressure cell used in this work consists essentially of massive carboloy pistons in the shape of truncated cones, jacketed by hardened steel supports. The sample was pressed in a boron gasket (mixed with 15% LiH) to prevent extrusion (for details, see Fig. 4.3 of ref. 201), and the whole system can be described (authors' quote) ". . . as a series of concentric pressure cells each one containing the one inside it and each one strong enough to support the pressure gradient across it. . . ." The γ-rays emitted by the source (^{57}Co electroplated on Armco iron, annealed in hydrogen) are emitted through a window slot (0.030 × 0.530 in.) extending radially from the sample containment anvil. Using this arrangement Pipkorn et al. [201] were able to extend their Mössbauer measurements to about 240 kilobars. In a subsequent study Edge et al. extended these measurements to ^{57}Fe as a dilute solution in titanium, vanadium, and copper, covering the pressure range to 250 kilobars [202].

High-pressure experiments on the ^{57}Fe resonance in powdered samples were effected by Coston et al. [203, 204], using a cell made of a Teflon wafer, into which was inserted a small disk of pyrophyllite that had an 0.010-in. segment ground off one side. The sample to be studied was inserted into the cavity formed by this missing segment, and the whole assembly was fitted into the cell described by Pipkorn et al. [201]. Pressures up to ~250 kilobars could be exerted on the sample maintained in the high-pressure anvil at room temperature. A number of other high-pressure experiments using the ^{57}Fe resonance have been reported [205].

High-pressure experiments on the ^{119}Sn resonance have been reported in which either source or absorber were held at elevated pressures. Herber and Spijkerman [206] studied the effect of pressure on the resonance observed in a stannic oxide (SnO_2) absorber compressed to 150 kilobars. The pressure cell used in this experiment was that originally described by Weir et al. [207] for use in infrared studies to 300 kilobars. The cell itself is made of small (~36 mgm) type-II, gem-cut diamonds that have been lapped optically flat with porous cast-iron lap charged with 5 micron diamond dust to provide a window area of $\sim 2 \times 10^{-4}$ in^2. Although the original infrared measurements were extended to ~300 kilobars, the Mössbauer data were obtained at about $\frac{1}{2}$ this maximal pressure in the SnO_2 study.

Experiments in which a magnesium stannide (Mg_2Sn) source was subjected to pressures up to 100 kilobars and used in conjunction with metallic tin absorbers at 1 atmosphere, have been reported by Möller and Mössbauer [208]. The experimental arrangement used in this study has been subsequently discussed in detail by Möller [208], and consists of a 100-ton hydraulic press that exerted the applied force to a set of tungsten carbide pistons, which in turn transmitted the pressure to the sample through a medium of "Gupalon" (Gus-Solit Hajek & Co., Munich, Germany [as cited by Möller]) in which the powdered source was embedded. In addition to the measurements on Mg_2Sn Möller also discussed results obtained for β–Sn, Sn–Au, $SnSO_4 \cdot 2H_2O$, and Pd–Sn alloy (12% tin by weight) in the pressure range up to 100 kilobars. The results of comparable studies on β–Sn up to 100 kilobars have been published by Panyushkin and Voronov [209], who used a modified high-pressure X-ray cell [210] in their investigations.

7 DATA HANDLING

The data output of a Mössbauer spectrometer may be recorded in analog or in digital form. In the case of a constant velocity spectrometer the analog output is obtained from the same digital to analog converter used to generate the velocity reference signal, and therefore the accuracy of the spectrum recorded on an X-Y recorder may match that of the spectrometer itself and can therefore be used directly for graphical analysis. This is not generally the case for constant acceleration spectrometers. The analog output of a contemporary multichannel analyzer is generally of a lower quality than that of the Mössbauer spectrometer that may be attached to it. In order to use methods of graphical analysis one has either to use a high-quality digital-to-analog converter or reproduce the analog plot point by point from digital output, such as typewriter copy, paper tape, and so forth. In experiments involving a small number of Mössbauer spectra the methods of graphical analysis, that is, the comparison of the spectrum with a number of Lorentzian shapes either manually or by the help of a simple analog computer (such as the Dupont Curve Resolver), may provide a detailed and quite accurate interpretation of the spectrum.

In cases in which the spectrum is very complex, involving many lines, some of which may strongly overlap and may therefore be unresolved, or in cases in which the experiment involves a large number of Mössbauer spectra it is necessary to apply methods of least-mean-square fitting using large digital computers. Programs for least-mean-square fitting are readily available from computer libraries [211] and may be easily adapted to handle Mössbauer data.

In order to use the input facilities of available computers it is necessary to have compatible output-data-recording devices. Commercial spectrometers are generally equipped with either a serial printer, a typewriter, a teletypewriter, a paper tape, or a magnetic tape recorder. The paper tape and magnetic tape output frequently, and the printed output always require conversion into punched cards in order to feed the data into the available large computer for further analysis. Punched cards are the best way to handle data since the card reader is the most commonly used input device. Additionally, punched cards provide a convenient way of storing, cataloguing, and retrieving the data when necessary. Commercial spectrometers are generally not equipped with a punch-card-output because the lack of standardization in the design of the key punches makes it impossible to supply a universal interface unit. It should be noted, however, that the conversion of the available interface between the MCA and a typewriter or teletypewriter to one that will drive an IBM key punch is generally quite straightforward and is a worthwhile investment. It is not possible to give a detailed description of this interface because of the lack of standardization of MCA outputs. However, an IBM key punch is operated by a number of relays that are activated by closing a contact either manually or by remote control. The output of the interface unit has to provide such contacts, and this may be done by an additional set of 12 relays—10 for the numbers 0 through 9, one for the card-release operation, and one for the punch timing signal (Fig. 4.19). Such a set may already be included in the interface that is generally provided for a typewriter. Since key punch consoles generally operate on a much higher power level than ordinary solid-state circuitry, the interface unit should provide maximum noise decoupling between the two units. Such a decoupling may be provided by the additional set of relays as shown in Fig. 4.19 [212]. If the output of the

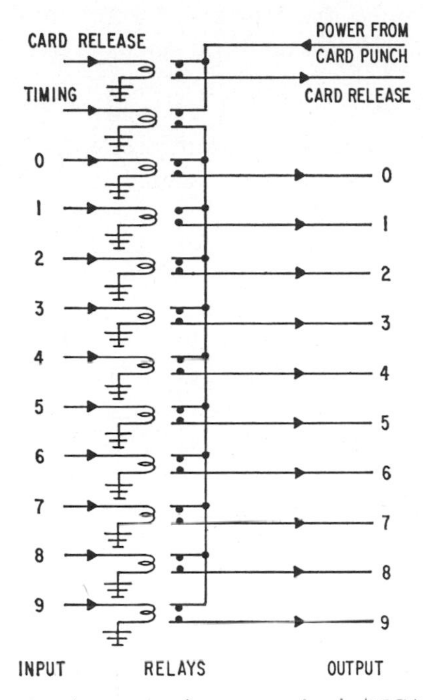

Fig. 4.19 Card-punch interface between pulse storage circuit (MCA) and IBM card punch for automatic data transfer to computer-compatible input format. This circuit allows for isolation of the low-power-level data signals from the high-power signals needed to drive the card punch.

multichannel analyzer is presented in binary rather than decimal form (interface to teletypewriter), it may be easily converted to decimal form by a binary-to-decimal decoder that involves an additional set of relays or a single integrated circuit.

The manual fitting of simple spectra has been discussed [130] and provides an adequate (albeit tedious) method of data reduction where relatively few spectra need to be analyzed.

8 STANDARDS, CALIBRATIONS, ABSORBERS AND REFERENCE

Standard Absorbers

There are a variety of measurements encountered in Mössbauer spectroscopy in which it is desirable to relate the experimental parameter of interest to an appropriate standard. In general such measurements are related to (1) velocity calibrations, (2) isomer shifts, and (3) internal magnetic fields. The orderly development of the Mössbauer effect into a broadly applicable spectroscopy depends in part on the judicious choice of such standards.

Velocity Calibration

For velocity calibration of a constant acceleration spectrometer the most appropriate absorber is a thin foil of metallic iron. The advantages of such an absorber are the fact that the ground-state splitting can be obtained from non-Mössbauer measurements, the six-line hyperfine spectrum is readily resolved, the recoil-free fraction is large at room temperature, and high-purity foils are readily obtained. Although at the time of this writing no certified samples have become available, it has been stated [213] that the National Bureau of Standards will certify metallic iron Mössbauer absorbers for velocity calibrations and that such standards will be issued shortly. The use of such foils in velocity calibrations [214] has become widespread, and it is the most common practical means of effecting such calibrations. It should be noted, however, that for reasonably high-precision work, high-purity (at least 99.9%) iron must be used, since the internal field that is the basis of the calibration is very sensitive to the presence of small amounts of impurities, principally carbon and silicon [215].

One drawback to the use of metallic iron as a means of obtaining a velocity calibration lies in the fact that an independent ^{57}Co source–metallic iron measurement must be made. When the major emphasis of the scientific study involves iron compounds, this is of little consequence since the same source can be used in the subsequent measurements. When these measurements involve other Mössbauer nuclides, however, a change must be made not only in the case of the absorber but in the case of the source as well. If the velocity transducer performance is sensitive to the mass of the source, it is possible to introduce significant systematic experimental error in changing sources between the velocity calibration run and the experimental measuring run, and the experimenter will have to convince himself that this error is either small enough to be safely ignored or that he can make appropriate corrections for it.

If the ground-state ($\frac{1}{2}$) magnetic splitting in metallic iron is designated by g_0, and the excited-state ($\frac{3}{2}$) splitting by g_1, and if use is made of the notation $\alpha = \frac{1}{2}(g_0 - g_1)$, then the positions of the six lines observed in an iron foil Mössbauer spectrum are as shown in Table 4.7, taken from [215]. The "observed" line positions are those obtained using a 1.9 mg/cm^2-thick Armco iron foil of 99.98% purity (0.02% C by weight was the only known noniron constituent present). By using such foils it is found that the line width of the two inner resonance lines ($-\frac{1}{2}$ to $-\frac{1}{2}$ and $-\frac{1}{2}$ to $+\frac{1}{2}$ transition is $\sim$0.236 mm/sec in good agreement with the value calculated from the lifetime of the excited state, taking sample thickness into account. The presence

Table 4.7 Line Positions Observed Using a Metallic Iron Foil Absorber[a]

I_e	I_g	Line	E[b]	Position[c]	Splitting
$-\frac{3}{2}$	$-\frac{1}{2}$	1	E_0	-5.313	
$-\frac{1}{2}$	$-\frac{1}{2}$	2	$E_0 + g_1$	-3.081	g_1 (lines 2–3); g_0 (lines 2–4); 2α (lines 3–4)
$\frac{1}{2}$	$-\frac{1}{2}$	3	$E_0 + 2g_1$	-0.842	g_0 (lines 3–5); H_i (lines 1–6)
$-\frac{1}{2}$	$\frac{1}{2}$	4	$E_0 + g_0 + g_1$	0.842	g_1 (lines 4–5)
$\frac{1}{2}$	$\frac{1}{2}$	5	$E_0 + g_0 + 2g_1$	3.081	
$\frac{3}{2}$	$\frac{1}{2}$	6	$E_0 + g_0 + 3g_1$	5.313	

[a] Because of the cubic symmetry of the iron atom lattice site $(\frac{1}{2})e^2qQ = 0$.

[b] E_0 is the energy difference between the highest lying $(-\frac{1}{2})$ ground-state level and the lowest lying $(-\frac{3}{2})$ excited-state level. The ground-state splitting (g_0) and the excited-state splitting (g_1) are the values of $\mu gH/I_g$ and $\mu eH/I_e$, respectively.

[c] For the purposes of this calculation the numbers $g_0 = 3.9177$ mm/sec and $g_1 = 2.2363$ mm/sec have been taken. These are the currently best available values and are quoted with an experimental error of ± 0.0010 mm/sec. Note that $g_0/g_1 = 3\mu g/\mu_e = 1.754 \pm 0.002$ and that $g_0 + 3g_1 = 10.627$ mm/sec $= 333$ kOe at room temperature.

of small amounts of carbon in the iron absorber significantly decreases the magnetic hyperfine splitting that is observed, and hence will be reflected in the velocity constant calculated from such data. Carbon impurities are often found in ^{57}Fe-enriched samples of metallic iron that have been obtained by reduction of $^{57}Fe_2O_3$ in graphite crucibles. The use of such foils should be avoided where possible.

For most chemically oriented work with ^{57}Fe (and ^{119}Sn, for that matter) the velocity region covered by the four "inner" lines of the metallic iron spectrum is adequate to cover the Doppler shifts of interest. With respect to the center of a metallic iron resonance spectrum, these span a velocity range from $-(g_1 + \alpha)$ to $(g_1 + \alpha)$ (approximately -3.1 to $+3.1$ mm/sec). The spectrometer "calibration constant" for constant acceleration devices, defined in terms of millimeters-per-second channel, can then be calculated from the Mössbauer data by means of the relationship

$$k = \frac{g_0}{n_5 - n_3} = \frac{g_0}{n_4 - n_2}, \tag{4.36}$$

where n_i is the resonance line midpoint (expressed in terms of multichannel analyzer address) of the ith resonance line.

In principle, the same velocity calibration procedure can be followed for constant velocity driven transducers, the only modification being that the

Mössbauer spectrum has now accumulated one velocity point at a time rather than all velocity points being accumulated simultaneously.

Although the ground-state splitting, g_0, from which the velocity constant, k, can be calculated is derived from a non-Mössbauer (nmr) experiment, there is in principle an objection to using a Mössbauer spectrum to calibrate a spectrometer. This objection can be largely overcome by making use of "absolute" optical calibrations, using a laser interferometer technique. The principle of this method consists of reflecting laser light of known wavelength (λ) from a corner-cube reflector mounted on the moving transducer, and counting the interference fringes by means of a stationary photomultiplier. Constructive interference (hence a velocity pulse count in the photomultiplier) occurs for every $2(\frac{1}{2}\lambda)$ travelled by the transducer. If the frequency, v, of the laser line is accurately known, then the wavelength ($\lambda = c/v$) is also accurately known. By determining the number of constructive interferences per transducer excursion cycle that are registered in the photomultiplier, the distance travelled by the transducer, and hence its velocity, can be accurately calculated without reference to Mössbauer data. A laser velocity determining attachment for spectrometer calibration purposes has been used for several years at the National Bureau of Standards [216] and is now available commercially [217].

Finally, in the context of the above discussion it should be emphasized that optical spectrometer calibration is especially useful where work with Mössbauer nuclides other than ^{57}Fe is involved, since no investment need be made in ^{57}Co sources and standard ^{57}Fe absorbers. For those investigators who are largely concerned with the 14.4-keV resonance in iron, the use of N.B.S. calibrated metallic iron foils in velocity constant determinations is probably satisfactory for all but the most demanding ultra-high (state-of-the-art limit) precision measurements.

Isomer-Shift Reference Standards

In order to facilitate the intercomparison of Mössbauer spectroscopic results between different laboratories, and even between sets of measurements in the same laboratory, it has become useful to report isomer-shift measurements with respect to a standard reference absorber [218]. The early Mössbauer literature consists largely of data in which isomer shifts are reported with respect to a given source, and hence meaningful comparison of such data between two or more investigators using different sources introduces large experimental uncertainties.

To overcome this difficulty for ^{57}Fe data it was suggested [218] that sodium nitroprusside [$Na_2Fe(CN)_5NO \cdot 2H_2O$] would be a good reference standard for iron isomer shifts, and a large body of data has been reported with respect to the centroid of the Mössbauer spectrum of this material. More recently, the Mössbauer Spectroscopy Task Group of the A.S.T.M. has recommended (see Appendix 1) that the centroid of the metallic iron spectrum be used instead for such purposes. Some representative isomer shifts for ^{57}Co host matrices and ^{57}Fe standard absorbers, calculated with respect to the center of the metallic iron spectrum, have been summarized in Table 4.4.

For ^{119}Sn spectroscopy two standard reference absorbers have been used in conjunction with isomer-shift data. Metallic (α, gray, tetrahedral) tin has been used, largely by physicists, because this material is well-characterized and readily available in high purity. The only drawback to the use of the single-resonance line of metallic tin as an isomer-shift reference point is the small resonance effect observed at room temperature [176]. However, with good single-line ^{119}Sn sources and fast counting electronics this objection is not a serious one, since even a 1% resonance effect can be accurately determined in a short time under such conditions.

On the other hand, because the Mössbauer transition in ^{119}Sn is populated by an isomeric transition precursor event, it is possible to make use of this nuclide to determine zero velocity, by running a Mössbauer experiment in which source and absorber are chemically identical. The most advantageous material for use in such an experiment is stannic oxide, SnO_2 [219], and a large amount of data have been reported with reference to this isomer-shift calibration point. It should be noted, however, that the tin atom in SnO_2 does not occupy a lattice point having cubic symmetry, and hence $e^2qQ \neq 0$. The small hyperfine effect that is observed (QS $\sim$ 0.43 mm/sec at room temperature [220]) leads to significantly broadened resonance lines ($\Gamma_{exp} \sim 1.2$ to 1.5 mm/sec even for thin absorbers) which cannot, in general, be resolved in an SnO_2 (source)–SnO_2 (absorber) experiment. It has been observed [221], however, that the centroid of a $BaSnO_3$ (source)–SnO_2 (absorber) [and of course a $BaSnO_3$ (source)–$BaSnO_3$ (absorber) experiment] spectrum is identical to that of the stannic oxide spectrum, and hence the SnO_2 and $BaSnO_3$ isomer-shift reference points are identical within experimental error.

There has recently been a good deal of interest in using the doublet observed in the ^{119}Sn resonance of dimethyl tin difluoride [$(CH_3)_2SnF_2$] as a reference spectrum, and calibration experiments are currently under way at the National Bureau of Standards on this material [222]. The advantages here lie in the high recoil-free fraction, the inertness, and the large quadrupole splitting (4.54 mm/sec at 294°K, 4.75 mm/sec at 9°K [223]) observed, which makes the line resolution easier. However, the presence of a large Gol'danskii–Karyagin asymmetry [223] and the general unavailability of this material may militate against its use as an isomer-shift reference standard.

The only other element for which a reference material has been widely used in reporting isomer-shift data is iodine, for which both ^{127}I and ^{129}I isomer-shift data have been tabulated [224] with respect to the iodide ion as a reference point. This choice has been made for the following reasons: (1) the electron configuration of I^- corresponds to (Kr core) $5s^2\ 4d^{10}\ 5p^6$, and is isoelectronic with the electron configuration of elemental xenon; (2) iodide compounds are readily accessible by standard synthetic techniques in high purity on a milligram scale (this is important for work with ^{129}I absorbers); (3) the hypothetical value for the isomer shift for I^- (+0.18 mm/sec with respect to the ZnTe source based on extrapolation of the experimental results for HI = $+0.16 \pm 0.02$, KI and NaI = $+0.14 \pm 0.02$, CsI = $+0.12 \pm 0.02$ mm/sec with respect to a ZnTe source) falls almost half-way between that observed for $Na_3H_2IO_6$ ($+1.02 \pm 0.01$ mm/sec) and that for $KICl_4 \cdot H_2O$ (-1.39 ± 0.05 mm/sec) [224]. Moreover, much of the literature on Mössbauer spectroscopy using the transition in ^{129}I has been reported with respect to I^- as a reference point, and the evaluation of isomer-shift data with respect to the electron configuration in iodine compounds has usually been expressed in terms of this standard.

In the case of xenon Mössbauer spectroscopy the work reported in the literature has come from too few laboratories for the choice of an isomer-shift reference point to have become a subject of significant discussion. It seems probable, however, that future workers will follow the lead set by Perlow [224], who has reported his extensive isomer-shift measurements with respect to xenon clathrate. Perlow points out that this form should be representative of Xe^0, and while there may be some discussion of whether xenon is chemically bound in the clathrate or not [50], the observed value lies within the commonly quoted present experimental error of that observed for elemental xenon.

For the rare earths most isomer-shift data are reported with respect to the rare-earth oxide, M_2O_3. In their extensive review Ofer, Nowik, and Cohen [47] have tabulated isomer shifts for the three Mössbauer transitions in europium (21.6 keV in ^{151}Eu, 97 and 103 keV in ^{153}Eu) with respect to Eu_2O_3, and thus have facilitated a direct comparison of data on identical compounds observed for the three modes of decay. Moreover, using Eu_2O_3 as a standard, these authors have been able to show the linear relationship between the isomer shifts observed in the two ^{153}Eu transitions, and thus have settled an interesting theoretical point on the basis of the elegant data of Atzmony et al. [85].

The extensive data on dysprosium intermetallic compounds reviewed by Ofer et al. [47] have also been referred to Dy_2O_3 as a reference point. Finally, it should be noted that Cohen and Kalvius [226] have proposed a number of isomer shift standards and have discussed criteria which can be used in arriving at suitable choices for such materials.

Reporting of Mössbauer Data

In order to facilitate the evaluation and intercomparison of Mössbauer data between various laboratories a set of ground rules covering nomenclature and units has been suggested by the Mössbauer Spectroscopy Task Group of Committee E-4 (Metallography) of the American Society for Testing and Materials (ASTM). A copy of this report is included as Appendix 1. This report has also been published in the scientific literature [225] and has been revised by the Ad Hoc Panel on Mössbauer Data of the Numerical Data Advisory Board, National Research Council. This panel (under the chairmanship of Prof. J. J. Zuckerman) has submitted a draft copy of its report to the IUPAC Committee on Nomenclature and Standards, for ultimate adoption.

Acknowledgement

The preparation of this chapter has been supported in part by the United States Atomic Energy Commission, the Petroleum Research Fund administered by the American Chemical Society, and the Research Council of Rutgers University. This support is gratefully acknowledged. This report constitutes AEC Document NYO-2472-66 and NYO-2807-69.

APPENDIX I. SUGGESTED NOMENCLATURE AND CONVENTIONS FOR REPORTING MÖSSBAUER DATA

The following persons have made major contributions and suggestions that are incorporated in this Appendix and their assistance is gratefully acknowledged: F. W. Christ, R. L. Collins, J. R. DeVoe, A. M. Ehrlich, P. Flinn, P. R. Girardot, U. Gonser, C. W. Kocher, L. May, A. H. Muir, Jr., C. Seidel, and J. J. Spijkerman.

Text

The text should include the following:

1. Method of sample mounting, sample thickness control (if any), sample confinement.
2. Method of data reduction (visual, computer, X-Y recorder).
3. Absorber form (single crystal, polycrystalline powder, inert matrix if used, evaporated film, folled foil, etc.).
4. Geometry of the experiment (transmission, scattering, etc.).
5. Critical absorbers if used.
6. Detector used and comments about associated electronics (i.e., single-channel window, escape peak, solid-state detector) if appropriate or unusual.

Symbols and Nomenclature

Name	*Symbol*	*Units*	*Definition and Comments*
Isomer shift	δ	mm/sec[a]	Displacement of center[b] of resonance spectrum from reference point
Quadrupole splitting	Δ or ΔE_Q	mm/sec[a] MC[c]	Hyperfine interaction (line splitting) between the nuclear quadrupole moment and the electric field gradient
Line width	Γ or Γ_{exp}	mm/sec	Full width at half-maximum of the experimentally observed resonance line
Natural line width	Γ_{nat}	mm/sec	Usually calculated from $\Gamma_{\mathrm{nat}} = h/2\pi\tau$
Extrapolated line width	Γ_0	mm/sec	The line width (Γ) extrapolated to zero thickness
Resonance effect magnitude	ϵ	%	$100\,[I(\infty) - I(0)]/I(\infty)$[d]
Recoil-free fraction	f	%	The fraction of all γ-rays of the Mössbauer transition that are emitted (f_s) or absorbed (f_a) without recoil
Mössbauer thickness	T	—	The effective thickness of a source (T_s) or absorber (T_a) $= n\sigma_0 af$[e]
Resonance cross section	σ_0	cm^2	The cross section for resonant absorption, related to the nuclear spins of the resonant nuclide and the internal conversion coefficient
Internal magnetic field	H_n	Oe	Magnitude of the magnetic field at the nucleus (from experiment)
External magnetic field	H_e	Oe	Magnitude of the applied magnetic field
Electric field gradient tensor	EFG		A tensor describing the electric field gradient, which is specified by η and V_{zz}[f] in addition to the Euler angles specifying the tensor orientation, if required
Asymmetry parameter	η		$= (V_{xx} - V_{yy})/V_{zz}$

Symbols and Nomenclature (*cont.*)

Name	*Symbol*	*Units*	*Definition and Comments*
Principal component of EFG	V_{zz}	V/cm^2	$= \dfrac{\partial^2 V}{\partial z^2} = eq$ (e is the proton charge)
Nuclear quadrupole moment	Q	barn or cm^2	A parameter that describes the shape of the nuclear charge distribution

[a] Although millimeters-per-second Doppler velocity units have become accepted usage in ^{57}Fe and ^{119}Sn spectroscopy, centimeters-per-second units may be more appropriate for other nuclides or where hyperfine interactions cause large Doppler velocities to be involved in the data.

[b] The center of a Mössbauer spectrum is defined as the Doppler velocity at which the resonance maximum is observed when all hyperfine interactions are (or would be) absent. That is $\delta \equiv \lim (E - E_0)$, $H \to O$, $V_{zz} \to O$. The contribution of the second-order Doppler shift should be indicated, if possible.

[c] Quadrupole splittings are frequently reported in megacycles per second, especially when direct comparison with nmr or nqr data can be effected (*e.g.*, in the case of ^{129}I). If such units are used in conjunction with data derived from Doppler-shift measurements, the conversion units should be stated.

[d] $I(0)$ is the counting rate (or transmission or scattering intensity) at the resonance maximum. $I(\infty)$ is the corresponding rate at a velocity at which the resonance effect is negligible. If corrections for nonresonant γ- or X-rays or other base-line corrections have been made in evaluating I, these should be stated.

[e] For an experiment involving element Z, n is the number of atoms of Z per centimeters squared, σ is the cross section for resonance absorption, a is the fractional abundance of the Z nuclide that show the Mössbauer effect being observed, and f is the recoil-free fraction.

[f] $V_{xx} + V_{yy} + V_{zz} = 0$ independent of the choice of axes. Principal axes are chosen so that the off-diagonal matrix elements vanish, $V_{ij} = 0$ $(i, j = x, y, z;$ $i \neq j)$, and are relabeled such that $|V_{zz}| \geq |V_{yy}| \geq |V_{xx}|$, so that $0 \leq \eta \leq 1$.

$$\text{EFG} = \left(-\frac{\partial^2 V}{\partial x_i \, \partial x_j}\right) \quad \begin{matrix} i = 1, 2, 3 \\ j = 1, 2, 3 \end{matrix}$$

7. Isomer-shift sign convention (commonly in a transmission experiment when source approaches absorber, that is, when the γ-ray incident on the absorber increases in energy, the shift is said to be positive; see Notes 1 and 3).
8. Method of spectrometer calibration (see Note 2).

Notes:

1. Metallic iron has been recommended both as a velocity calibrator for constant acceleration spectrometers (see Note 2) and as an isomer-shift standard for ^{57}Fe data. Thin foils (0.25 to 1.0 mil) of natural iron of at least 99.9% purity (preferably 99.99% and/or NBS Standard Reference Material, if available) are adequate for isomer-shift data. The centroid of the metallic iron absorption (or scattering) spectrum should be calculated from (at least) the resonance maximum positions for the four "inner" lines (i.e., the $|\frac{1}{2}| \leftrightarrow |\frac{1}{2}|$ transitions). Sodium nitroprusside dihydrate

$$[SNP, Na_2[Fe(CN)_5NO]\cdot 2H_2O]$$

has been standardized by the National Bureau of Standards using an absolutely calibrated spectrometer. Samples of SNP are available from the National Bureau of Standards as a Standard Reference Material. For interlaboratory comparison of data, isomer shifts for other standards (e.g., $K_4Fe(CN)_6$, stainless steels, iron oxides) should be determined experimentally using the spectrometer in question, rather than depending on literature values.

2. Constant-acceleration spectrometers to be used for work in the millimeters-per-second range should be calibrated with respect to velocity, using metallic iron of at least 99.99% purity. The ground-state and excited-state splittings reported in the literature [R. S. Preston, S. S. Hanna, and J. Heberle, *Phys. Rev.*, **128,** 2207 (1962); J. G. Dash, R. D. Taylor, D. E. Nagle, P. P. Craig, and M. Visscher, *Phys. Rev.*, **122,** 1116 (1961); S. S. Hanna *et al.*, *Phys. Rev. Lett.* **4,** 177 (1960); J. I. Budnick, L. J. Bruner, R. J. Blume, and B. L. Boyd, *J. Appl. Phys.*, 3**2,** 1205 (1962); Fritz and Schulze, *Nucl. Inst. Methods*, **62,** 317 (1968); H. Shechter, M. Ron, S. Niedzwiedz, and R. H. Herber, *Nucl. Inst. Methods*, **44,** 268 (1966)] are $g_0 = 3.9177 \pm 0.0010$ mm/sec, $g_1 = 2.2363 \pm 0.0010$ mm/sec at room temperature. The quadrupole splitting in sodium nitroprusside $[Na_2Fe(CN)_5NO\cdot 2H_2O]$ should not be used for velocity calibration.

3. There are at the present no generally accepted isomer-shift standards for other nuclides. Stannic oxide (SnO_2) and barium stannate ($BaSnO_3$), which give essentially identical isomer shifts, have been used in ^{119}Sn spectroscopy, as have various forms of metallic tin. The $K^{129m}I$ has been used for ^{129}I spectroscopy. NpO_2 has been used in ^{237}Np spectroscopy. For metals some possible standards are ^{40}K in K, ^{99}Ru in Ru, and ^{197}Au in Au. Whatever reference standard is employed, sufficient data should be reported to permit interconversion to other data.

Numerical (Tabular) Data

Numerical (Tabular) data should include the following:

1. Chemical state of source and absorber.
2. Temperature of source and absorber.

3. Mössbauer parameters in millimeters per second (or centimeters per second, or other appropriate units) with estimated errors.
4. Isomer-shift standard or data for reference absorber.
5. Line widths (full width at half maximum) observed, if pertinent.
6. Line intensities observed, if pertinent.

Figures (Spectra)

1. At least one spectrum (percent transmission or counting rate versus Doppler velocity or isomer shift, or comparable data—not "channel number") should be shown to indicate the "quality" of the data (see Note 1).
2. "Effect" or intensity (i.e., percent transmission or counting rate) axis should be normalized to $V = \infty$ if possible.
3. Statistical counting error or limits should be indicated for at least one data point (see Note 2).
4. Individual data points (rather than only a smoothed curve) should be shown.

Notes:

1. It has become customary to display data obtained in transmission geometry with the resonance maximum "down" and scattering data with the resonance maximum "up." In either case sufficient data should be shown far from the resonance peaks to establish the nonresonant base line.
2. In most instances (where the data are uncorrected counting results) the statistical counting error is given by $\pm N^{1/2}$, where N is the total number of counts stored per channel or velocity point. For corrected data the statistical counting error should be obtained by the usual statistical methods.

References

1. R. L. Mössbauer, *Z. Physik*, **151**, 124 (1958); *Naturwissenschaften*, **45**, 538 (1958); *Z. Naturforsch.*, **14a**, 211 (1959).
2. H. Frauenfelder, *The Mössbauer Effect*, W. A. Benjamin, New York, 1962.
3. V. I. Gol'danskii, "The Mössbauer Effect and Its Application to Chemistry," *Izv. Akad. Nauk SSSR*, Moscow, 1963; *At. Energy Rev.*, **1**, 3 (1963).
4. G. K. Wertheim, *Mössbauer Effect. Principles and Applications*, Academic, New York, 1964.
5. V. I. Gol'danskii and R. H. Herber, eds., *Chemical Applications of Mössbauer Spectroscopy*, Academic, New York, 1968.
6. R. H. Herber, *J. Chem. Educ.*, **42**, 180 (1965).

7a. R. H. Herber, *Ann. Rev. Phys. Chem.*, **17**, 261 (1966).

7b. D. A. Shirley, *Ann. Rev. Phys. Chem.*, **20**, 25 (1969).

8. N. N. Greenwood, *Chem. Brit.*, **3**, 56 (1967).
9. H. Wegner, *Der Mössbauer Effekt und seine Anwendungen in der Physik und Chemie*, Bibliog. Institut AG, Mannheim, Germany, 1965.

10. J. Danon, *Lectures on the Mössbauer Effect*, Gordon and Breach, New York, 1968.
11. The most complete collection of references (as well as useful data on Mössbauer nuclides and parameters for related isotopes) for the period 1958 to 1965 are found in A. H. Muir, Jr., K. J. Ando, and H. M. Coogan, "Mossbauer Effect Data index, 1958–1965 , Wiley-Interscience, New York, 1966 and "Mössbauer Effect Data Index Covering the 1969 Literature," J. G. Stevens and V. E. Stevens, eds., IFI/Plenum, New York, 1970. See also J. R. DeVoe and J. J. Spijkerman, *Anal. Chem.*, **38,** 382R (1966); *ibid.*, **40,** 472R (1968).
12. Y. Lee, P. W. Keaton, Jr., E. T. Ritter, and J. C. Walker, *Phys. Rev. Lett.*, **14,** 957 (1965).
13. D. Seyboth, F. E. Obenshain, and G. Czjzek, *Phys. Rev. Lett.*, **14,** 954 (1965).
14. G. Czjzek, J. L. C. Ford, Jr., F. E. Obenshain, and D. Seyboth, *Physics Lett.*, **19,** 673 (1966).
15. J. Eck, Y. K. Lee, E. T. Ritter, R. R. Stevens, Jr., and J. C. Walker, *Phys. Rev. Lett.*, **17,** 120 (1966).
16. R. L. Mössbauer and W. Wiedemann, *Z. Physik*, **159,** 33 (1960).
17. S. Margulies and J. R. Ehrman, *Nucl. Inst. Methods*, **12,** 131 (1961).
18. D. St. P. Bunbury, J. A. Elliott, H. E. Hall, and J. M. Williams, *Phys. Lett.*, **6,** 34 (1963).
19. P. P. Craig and N. Sutin, *Phys. Rev. Lett.*, **11,** 460 (1963).
20. O. C. Kistner and A. W. Sunyar, *Phys. Rev. Lett.*, **4,** 412 (1960).
21. Such correlations are reviewed in detail for ^{75}Fe, ^{119}Sn, ^{129}I, ^{131}Xe, the rare earths, *inter alia*, in ref. 5.
22. N. E. Erickson, in "The Mössbauer Effect and Its Application in Chemistry," *Advan. Chem. Ser.*, No. 68, American Chemical Society, Washington, 1968.
23. Y. Hazony, R. C. Axtmann, and J. W. Hurley, Jr., *Chem. Phys. Lett.*, **2,** 440 (1968); *ibid.*, **2,** 673 (1968); *J. Chem. Phys.*, **52,** 3309 (1970).
24. L. H. Bowen, J. G. Stevens, and G. C. Long, *J. Chem. Phys.*, **51,** 2010 (1969).
25. G. K. Shenoy and S. L. Ruby, "Mössbauer Effect Methodology," Vol. 5, I. J. Gruverman, Ed., Plenum, New York, 1970.
26. S. L. Ruby and P. A. Flinn, *Rev. Mod. Phys.*, **36,** 351 (1964).
27. R. L. Collins, *J. Chem. Phys.*, **42,** 1072 (1965).
28. V. I. Gol'danskii, V. A. Trukhtanov, M. N. Devisheva, and V. F. Belov, *Phys. Lett.*, **15,** 317 (1965); *Zh. Eksperim. i Teor. Fiz. Pis'ma Redaktsiyu* [1], **1,** 31 (1965), English translation *JETP Lett.*, **1,** 31 (1965); K. P. Belov and I. S. Lyubutin, *Zh. Eksperim i Teor. Fiz. Pis'ma Redaktsiyu* [1], **1,** 26 (1965), English translation *JETP Lett.*, **1,** 26 (1965).
29. R. L. Collins, in *Mössbauer Effect Methodology*, Vol. 3, I. J. Gruverman, Ed., Plenum, New York, 1967.
30. A. J. Nozik and M. Kaplan, *Phys. Rev.*, **159,** 273 (1967).
31. J. Danon, Chapter 3 in Ref. 5.
32. C. K. Jørgensen, *Oxidation Number and Oxidation State*, Springer Verlag, New York, 1969.
33. R. Ingalls, *Phys. Rev.*, **133,** A787 (1964).

34. See, for example, the discussion by V. I. Gol'danskii and E. F. Makarov, Ref. 5, Chapter 1.
35. G. J. Perlow, in *Chemical Applications of Mössbauer Spectroscopy*, V. I. Gol'danskii and R. H. Herber, Eds., Academic, New York, 1968, Chapter 7 and references therein.
36. D. W. Hafemeister, G. de Pasquali, and H. de Waard, *Phys. Rev.*, **135,** B1089 (1964).
37. M. Pasternak, A. Simopoulis, and Y. Hazony, *Phys. Rev.*, **140,** A1892 (1965); M. Pasternak, *Symp. Faraday Soc.*, No. 1, p. 119 ff; M. Pasternak and S. Bukshpan, *Phys. Rev.*, **163,** 297 (1967).
38. S. Bukshpan, Ph.D. dissertation, Weizmann Institute, Rehovoth, Israel, 1968.
39. S. L. Ruby and H. Selig, *Phys. Rev.*, **147,** 348 (1966).
40. R. E. Watson and A. J. Freeman, *Phys. Rev.*, **123,** 2048 (1961).
41. C. E. Johnson, *Symp. Faraday Soc.*, **1,** 7 (1967).
42. D. J. Simkin, *Phys. Rev.*, **177,** 1008 (1969).
43. R. S. Preston, S. S. Hanna, and J. Heberle, *Phys. Rev.*, **128,** 2207 (1962); J. G. Dash, R. D. Taylor, D. E. Nagle, P. P. Craig, and M. Visscher, *Phys. Rev.*, **122,** 1116 (1961); S. S. Hanna et al., *Phys. Rev. Lett.*, **4,** 177 (1960); H. Shechter, M. Ron, S. Niedzwiedz, and R. H. Herber, *Nucl. Instr. Methods*, **44,** 268 (1966).
44. J. I. Budnick, L. J. Bruner, R. J. Blume, and B. L. Boyd, *J. Appl. Phys.*, **32,** 1205 (1961).
45. See, for example, the discussion by J. Danon, Ref. 5, Chapter 3.
45a. W. T. Oosterhuis and G. Lang, *Phys. Rev.*, **178,** 439 (1969).
46. M. Cordey Hayes, see references to Table 5.5, Ref. 5.
47. S. Ofer, I. Nowik, and S. G. Cohen, Ref. 5, Chapter 8.
48. G. J. Perlow, Ref. 5, Chapter 7.
49. Y. Hazony and S. L. Ruby, *J. Chem. Phys.*, **49,** 1478 (1968).
50. Y. Hazony and R. H. Herber, *J. Inorg. Nuclear Chem.*, **33,** 961 (1971).
51. Y. Hazony, *J. Chem. Phys.*, **45,** 2664 (1966); *ibid.*, **49,** 159 (1968); *Disc. Faraday Soc.*, **48,** 148 (1969).
52. H. A. Stöckler, H. Sano, and R. H. Herber, *J. Chem. Phys.*, **45,** 1182 (1966); *ibid.*, **47,** 1567 (1967).
53. H. A. Stöckler and H. Sano, *Phys. Lett.*, **25A,** 550 (1967).
54. D. P. Johnson and J. G. Dash, *Phys. Rev.*, **172,** 983 (1968).
55. Y. Hazony, R. C. Axtmann, and J. W. Hurley, Jr., *Bull. Amer. Phys. Soc.*, **15,** 106 (1970) [HH4].
56. R. M. Housley and F. Hess, *Phys. Rev.*, **146,** 517 (1966).
57. For example, the system designed by Austin Research Associates, Austin, Tex.
58. For example, New England Nuclear Corp., Boston, Mass., or International Chemical and Nuclear Corp., Irvine, Calif.
59. S. B. Tomilov, *Radiokhimiya*, **6,** 377 (1964).
60. J. Bara, A. Z. Hrynkiewicz, and I. Stronski, *Kernenenergie*, **7,** 317 (1964).
61. For detailed references see N. Benczer-Koller and R. H. Herber, in Ref. 5, Chapter 2.
62. J. Stephen, *Nucl. Inst. Methods*, **26,** 269 (1964).

63. J. P. Schiffer, P. N. Parks, and J. Heberle, *Phys. Rev.*, **133,** A1553 (1964).
64. T. A. Kitchens, W. A. Steyert, and R. D. Taylor, *Phys. Rev.*, **138,** A467 (1965).
65. P. P. Craig, D. E. Nagle, W. A. Steyert, and R. D. Taylor, *Phys. Rev. Lett.*, **9,** 12 (1962).
66. R. M. Housley, N. E. Erickson, and J. G. Dash, *Nucl. Inst. Methods,* **27,** 29 (1964).
67. H. H. Wickman and G. K. Wertheim, Ref. 5, Chapter 11; G. K. Wertheim and H. J. Guggenheim, *J. Chem. Phys.*, **42,** 3873 (1965); *Phys. Rev. Lett.*, **20,** 1158 (1968).
68. A. Mustachi, *Nucl. Inst. Methods,* **26,** 219 (1964).
69. H. N. Ok and J. G. Mullen, *Phys. Rev.*, **168,** 563 (1968); J. G. Mullen and H. N. Ok, *Phys. Rev. Lett.*, **17,** 287 (1966).
70. H. N. Ok and J. G. Mullen, *Phys. Rev.*, **168,** 550 (1960).
71. G. K. Wertheim, *Phys. Rev.*, **124,** 764 (1961); W. Trifthauser, and P. P. Craig, *Phys. Rev.*, **162,** 274 (1967).
72. R. H. Herber and J. J. Spijkerman, *J. Chem. Phys.*, **42,** 4312 (1965).
73. H. Sano and R. H. Herber, *J. Inorg. Nucl. Chem.*, **30,** 409 (1968).
74. P. Flinn and S. L. Ruby, *Rev. Mod. Phys.*, **36,** 352 (1964).
75. V. A. Bryukhanov, N. N. Delyagin, and R. N. Kuzmin, *Zh. Eksperim. i Teor. Fiz.*, **46,** 137 (1964), [English translation, *Soviet Phys. JETP,* **19,** 98 (1964)].
76. C. Hohenemser, *Phys. Rev.*, **139,** 185 (1965).
77. V. A. Bryukhanov, N. N. Delyagin, and V. S. Shpinel, *Zh. Eksperim. i Teor. Fiz.*, **47,** 2085 (1964), English translation, *Soviet Phys. JETP*, **20,** 1400 (1965).
78. R. H. Herber and J. J. Spijkerman, *J. Chem. Phys.*, **43,** 4057 (1965).
79. H. A. Stöckler and H. Sano, *Nucl. Instrum. Methods*, **44,** 103 (1966).
80. S. Jha, R. Segnan, and G. Lang, *Phys. Rev.*, **128,** 1160 (1962).
81. See G. J. Perlow, ref. 5, Chapter 7, for a detailed review of the ^{129}I work through 1966; see also D. W. Hafemeister, G. De Pasquali, and H. De Waard, *Phys. Rev.*, **135,** B1089 (1964).
82. Y. Hazony, P. Hillman, M. Pasternak, and S. L. Ruby, *Phys. Lett.*, **2,** 337 (1962).
83. M. Pasternak and T. Sonnino, *Phys. Rev.*, **164,** 384 (1967).
84. S. L. Ruby and H. Selig, *Phys. Rev.*, **147,** 348 (1966).
85. S. Ofer, P. Avivi, R. Bauminger, A. Marinov, and S. G. Cohen, *Phys. Rev.*, **120,** 460 (1960); U. Atzmony and S. Ofer, *Phys. Rev.*, **145,** 915 (1966); U. Atzmony, E. R. Bauminger, I. Nowik, S. Ofer, and J. H. Wernick, *Phys. Rev.*, **156,** 262 (1967); U. Atzmony, E. R. Bauminger, and S. Ofer, *Nucl. Phys.*, **89,** 433 (1966).
86. D. A. Shirley, M. Kaplan, R. W. Grant, and D. A. Keller, *Phys. Rev.*, **127,** 2097 (1962).
87. S. Ofer, E. Segal, I. Nowik, E. R. Bauminger, L. Grodzins, A. J. Freeman, and M. Schieber, *Phys. Rev.*, **137,** A627 (1965); S. Jha, R. Segnan, and G. Lang, *Phys. Lett.*, **2,** 1171 (1962).
88. I. Nowik and S. Ofer, *Phys. Lett.*, **3,** 192 (1963); R. L. Cohen, *Phys. Lett.*, **5,**

177 (1963); M. Kalvius, P. Kienle, H. Eicher, W. Wiedemann, and C. Schuler, *Z. Physik*, **172,** 231 (1963).
89. U. Atzmony, A. Mualem, and S. Ofer, *Phys. Rev.*, **136,** B1237 (1964).
90. R. J. Morrison, *Nucl. Sci. Abstr.*, **19,** No. 6638 (1965).
91. I. Nowik and H. H. Wickman, *Phys. Rev. Lett.*, **17,** 949 (1966).
92. S. Ofer and I. Nowik, *Phys. Lett.*, **24A,** 88 (1967).
93. R. Bauminger, S. G. Cohen, A. Marinov, and S. Ofer, *Phys. Rev. Lett.*, **6,** 467 (1961).
94. F. E. Wagner, F. W. Staneck, P. Kienle, and H. Eicher, *Z. Physik*, **166,** 1 (1962).
95. For detailed references, see ref. 47.
96. D. Nagle, P. P. Craig, J. G. Dash, and R. D. Reiswig, *Phys. Rev. Lett.*, **4,** 237 (1960).
97. D. A. Shirley, ref. 5, Chapter 9; D. A. Shirley, *Phys. Rev.*, **124,** 354 (1961); see also ref. 7b.
98. A comprehensive guide to commercial supplies of nuclear electronics is in the Annual Buyer's Guide issue of *Science*, November 26, 1968, and September 23, 1969. For details of commercially available nuclear scalars and detectors see *Ind. Res.*, November 20, 1969, pp. 86–92.
99. *Alpha, Beta and Gamma Ray Spectroscopy*, K. Siegbahn, ed., North Holland, Amsterdam, 1965.
100. The Harshaw Chemical Co., Cleveland, Ohio 44139; International Chemical and Nuclear Corp., Oakland, Calif. 94612; Nuclear Equipment Chemical Corp., Farmingdale, N.Y. 11735; Quartz Production Corp., Plainfield, N.J. 07061; Semi-Elements Corp., Saxonburg, Pa. 16056.
101. Reuter-Stokes Electronics Components, Inc., Cleveland, Ohio 44139.
102. J. Huntzicker and S. S. Rosenblum, *Nucleonics*, **23,** 62 (1965).
103. See ref. 98 and *Nucleonics Buyers' Guide*, McGraw-Hill, New York, 1966.
104. J. Huntzicker, E. Matthias, S. S. Rosenblum, and D. A. Shirley, *Bull. Amer. Phys. Soc.*, **9,** 435 (1964) [ED8].
105. For a review of Mössbauer experiments using scattering geometry see P. de Brunner and H. Frauenfelder, *Applications of the Mössbauer Effect in Chemistry and Solid State Physics*, I.A.E.A. Tech. Repts. Series No. 50, Vienna, 1966; J. K. Major, in *Mössbauer Effect Methodology*, Vol. 1, I. Gruverman, Ed., pp. 89–95, Plenum, New York, 1965; P. De Brunner, *ibid.*, pp. 96–105.
106. E. Kankeleit, *Z. Physik*, **164,** 442 (1961).
107. K. P. Mitrofanov and V. S. Shpinel, *Zh. Eksperim. i Teor. Fiz.*, **40,** 983 (1961).
108. K. P. Mitrofanov, N. V. Illarionova, and V. S. Shpinel, *Pribory i Tekhn. Eksperim.*, **8,** 49 (1963), English translation *Instr. Exper. Tech. (USSR)*, **3,** 415 (1963).
109. K. P. Mitrofanov, N. V. Illarionova, and V. S. Shpinel, *Proc. Conf. Mössbauer Effect*, Dubna, July 1962; Paper 3, Consultants Bureau, New York, 1963.
110. D. A. O'Connor, N. M. Butt, and A. S. Chohan, *Rev. Mod. Phys.*, **36,** 361A (1964); W. A. Owens, *Bull. Amer. Phys. Soc.*, **9,** 689 (1964) [D6]; N. Hershkowitz, J. S. Eck, and J. C. Walker, *ibid.*, **10,** 577 (1965) [AC12].

111. J. S. Eck, N. Hershkowitz, and J. C. Walker, *Bull. Amer. Phys. Soc.*, **10,** 577 (1965) [AC13].
112. S. Jha, R. Segnan, and G. Lang, *Phys. Rev.*, **128,** 1160 (1962).
113. Austin Research Associates, Austin, Tex.; International Chemical and Nuclear Corp., Oakland, Calif. 94612.
114. See D. A. Shirley, ref. 5, Chapter 9.
115. As a first approximation, geometrical considerations indicate that the critical absorber should be placed at the midpoint.
116. G. W. Grodstein, *Nat. Bur. Std. Cir. 583*, 1957; R. T. McGinnies, *ibid.*, 1959.
117. W. J. Nicholson, *Phys. Lett.*, **10,** 184 (1964).
118. Y. K. Lee, P. W. Keaton, E. T. Ritter, and J. C. Walker, *Phys. Rev. Lett.*, **14,** 957 (1965).
119. See ref. 105.
120. O. C. Kistner and J. B. Swann, *The Mössbauer Effect*, D. M. J. Compton and A. H. Schoen, eds., Vol. 1, Wiley, New York, 1962, p. 270.
121. D. W. Hafemeister, G. De Pasquali, and H. De Waard, *Phys. Rev.*, **135,** B1089 (1964).
122. Y. Hazony, P. Hillman, M. Pasternak, and S. L. Ruby, *Phys. Lett.*, **2,** 337 (1962).
123. R. M. Housley, *Nucl. Instr. Methods*, **35,** 77 (1965).
124. J. J. Spijkerman, *J. Appl. Phys. Lett.* (in press).
125. K. R. Swanson and J. J. Spijkerman, *NBS Technical Note 451*, pp. 29–44, 1969.
126. V. I. Gol'danskii, *Usp. Fiz. Nauk* **89,** 333 (1966); *Angew. Chemie* **79,** 844 (1967), *Internatl. Edit.*, **6,** 830 (1967).
127. R. Zane, *Nucl. Instr. Methods*, **43,** 333 (1966).
128. J. Pahor, D. Kelsin, A. Kodre, D. Hanzel, and A. Moljk, *Nucl. Instr. Methods*, **46,** 289 (1967).
129. F. C. Ruegg, J. J. Spijkerman, and J. De Voe, *Rev. Sci. Inst.*, **36,** 356 (1965); "Application of the Mössbauer Effect in Chemistry and Solid State Physics," *Tech. Rep. Ser. No. 50*, I.A.E.A., Vienna, 1966, pp. 53–58.
130. H. Brafman, M. Greenshpan, and R. H. Herber, *Nucl. Instr. Methods*, **42,** 245 (1966).
131. Y. Hazony, *Rev. Sci. Instr.*, **38,** 1760 (1967).
132. E. Kankeleit, *Rev. Sci. Instr.*, **35,** 194 (1964).
133. R. L. Cohen, P. G. McMullin, and G. K. Wertheim, *Rev. Sci. Instr.*, **34,** 671 (1963); *ibid.*, **37,** 957 (1966); G. K. Wertheim and R. L. Cohen, "Application of the Mössbauer Effect in Chemistry and Solid State Physics," *Tech. Rep. Ser. No. 50*, I.A.E.A., Vienna, 1966. This drive is commonly referred to as the "Bell Laboratories drive" and is the most commonly used transducer in Mössbauer spectroscopy.
134. J. Lipkin, B. Schechter, S. Shtrikman, and D. Treves, *Rev. Sci. Instr.*, **35,** 1336 (1964).
135. Y. Reggev, S. Bukshpan, M. Pasternak, and D. Segal, *Nucl. Instr. Methods*, **52,** 193 (1967).
136. E. Nadav and M. Palmai, *Nucl. Instr. Methods*, **56,** 165 (1967).

137. Y. Hazony, D. E. Earls, and I. Lefkowitz, *Phys. Rev.*, **166,** 507 (1968).
138. I. Pelah and S. L. Ruby, *J. Chem. Phys.*, **51,** 383 (1969).
140. M. Kalvius, "Mössbauer Effect Methodology," I. J. Gruverman, ed., Vol. I, Plenum, New York, 1965.
141. N. Benczer-Koller and R. H. Herber, ref. 5, Chapter 2.
142. See, for example, the listing in *Ind. Res.*, November 20, 1969, pp. 116–117 for metallic dewars. Glass dewars are available from Pope Scientific Inc. Menomonee Falls, Wisconsin and H. S. Martin & Son, Evanston, Ill.
143. J. H. Heberle, *Rev. Sci. Instr.*, **33,** 1476 (1962).
144. J. R. Harris, G. M. Rothberg, and N. Benczer-Koller, *Rev. Sci. Instr.*, **137,** A1101 (1965); *ibid.*, **138,** B554 (1965).
145. R. D. Taylor, W. A. Steyert, and G. A. Fox, *Rev. Sci. Instr.*, **36,** 563 (1965).
146. P. P. Craig, W. A. Steyert, and R. D. Taylor, *Rev. Sci. Instr.*, **33,** 869 (1962).
147. G. T. Schjeldahl Co., Northfield, Minn., GT-300 Mylar film (1 mil adhesive ± 5 mil Mylar).
148. Model 65DT2L helium dewar manufactured by Janis Research Co., Stoneham, Mass.
149. P. P. Craig, *Mössbauer Effect Methodology*, Vol. 1, I. J. Gruverman, ed., Plenum, New York, 1965.
150. Y. K. Lee, P. W. Keaton, Jr., E. T. Ritter, and J. C. Walker, *Phys. Rev. Lett.*, **14,** 957 (1965).
151. S. L. Ruby and R. E. Holland, *Phys. Rev. Lett.*, **14,** 591 (1965).
152. D. Seyboth, F. E. Obenshain, and G. Czjzek, *Phys. Rev. Lett.*, **14,** 954 (1965).
153. G. Czjzek, J. L. C. Ford, Jr., F. E. Obenshain, and D. Seyboth, *Phys. Lett.*, **19,** 673 (1966).
154. R. R. Stevens, J. S. Eck, E. T. Ritter, Y. K. Lee, and J. C. Walker, *Phys. Rev.*, **158,** 1118 (1967).
155. R. S. Preston, S. S. Hanna, and J. Heberle, *Phys. Rev.*, **128,** 2207 (1962).
156. K. Ono, A. Ito, and T. Fujita, *J. Phys. Soc. Japan*, **19,** 2119 (1964).
157. U. Ganiel and S. Shtrikman, *Phys. Rev.*, **177,** 503 (1969).
158. D. E. Nagle, H. Frauenfelder, R. D. Taylor, D. Cochran, and B. T. Matthias, *Phys. Rev. Lett.*, **5,** 364 (1960).
159. R. G. Barnes, R. L. Mössbauer, E. Kankeleit, and J. M. Poindexter, *Phys. Rev.*, **136,** A175 (1964).
160. B. Sharon and D. Treves, *Rev. Sci. Instr.*, **37,** 1252 (1966).
161. Elscint Inc., Edison, N.J. 08817; High Voltage Engineering Corp., Burlington, Mass. 01803.
162. G. K. White, *Experimental Techniques in Low Temperature Physics*, Oxford University Press, London and New York, 1959.
163. R. Berman and D. J. Huntley, *J. Cryog.*, **3,** 70 (1963); R. Berman, J. C. F. Brock, and D. J. Huntley, *ibid.*, **4,** 233 (1963); R. L. Rosenbaum, R. R. Oder, and R. B. Goldener, *ibid.*, **4,** 333 (1964).
164. R. B. Scott, *Cryogenic Engineering*, Van Nostrand, New York, 1959, p. 349.
165. C. R. Barber, "Resistance Thermometers for Low Temperatures," *Progr. Cryog.*, **2,** 147 (1960).

166. N. A. Blum, in *Mössbauer Effect Methodology*, I. Gruverman, ed., Vol. I, p. 159, Plenum, New York, 1965.
167. W. Wiedemann, W. A. Mundt, and D. Kuhlman, *Cryog.*, **5**, 94 (1965).
168. A. J. Nozik and M. Kaplan, *Anal. Chem.*, **39**, 854 (1967).
169. American Relays, Electronics Division, New York.
170. S. Margulies and J. R. Ehrman, *Nucl. Instr. Methods*, **12**, 131 (1961).
171. J. Heberle, *Nucl. Instr. Methods*, **58**, 90 (1968).
172. D. A. O'Connor, *Nucl. Instr. Methods*, **21**, 318 (1963).
173. A. M. Afanas'ev and Yu. M. Kagan, *Zh. Eksperim. i Teor. Fiz.*, **48**, 327 (1965); Yu. M. Kagan and A. M. Afanas'ev, *Zh. Eksperim. i Teor. Fiz.*, **50**, 271 (1966).
174. J. Heberle, *Nucl. Instr. Methods*, **58**, 90 (1968).
175. J. E. Maling and M. Weissbluth, "Application of Mössbauer Spectroscopy to the Study of Iron in Heme Proteins," *Solid State Biophysics*, S. J. Wyard, ed., McGraw-Hill, New York, 1969, Chapter 10.
176. A. J. F. Boyle and H. E. Hall, *Rept. Progr. Phys.*, **25**, 441 (1962).
177. U. Gonser and R. W. Grant, *Biophys. J.*, **5**, 823 (1965).
178. W. S. Caughey, W. Y. Fujimoto, A. J. Bearden, and T. H. Moss, *Biochem. J.*, **5**, 1255 (1966).
179. G. Lang and W. Marshall, *Proc. Phys. Soc.*, **87**, 3 (1966).
180. E. Fluck, in ref. 5, Chapter 4.
181. R. H. Herber, in "Characterization of Organometallic Compounds," M. Tsutsui, ed., Interscience, New York, 1969, Chapter 7.
182. A. D. Mackay Co., New York; United Mineral and Chemical Corp., New York 10013.
183. National Metallizing Division, Standard Packaging Corp., Cranbury, N.J. 08512.
184. E. I. du Pont Film Dept., Wilmington, Del. 19898; M. R. Guedon Co., Audubon, N.J. 08106; see also ref. 147.
185. M. Kalvius, U. Zahn, P. Kienle, and H. Eicher, *Z. Naturforsch.*, **17a**, 494 (1962).
186. A. J. F. Boyle, D. St. P. Bunbury, and C. Edwards, *Proc. Phys. Soc. (London)*, **79**, 416 (1962).
187. See the discussion by G. J. Perlow, in ref. 5, Chapter 7, p. 390.
188. R. H. Herber and S. Chandra, in *Mössbauer Effect Methodology*, Vol. 5, I. J. Gruverman, ed., Plenum, New York, 1970; R. H. Herber and S. Chandra, *J. Chem. Phys.*, **52**, 6045 (1970).
189. R. H. Herber and Y. Goscinny, *Inorg. Chem.*, **7**, 1293 (1968).
190. See the discussion by V. I. Gol'danskii and E. F. Makarov, in ref. 5, Chapter 2, pp. 102–107, and references therein.
191. R. Grubbs, R. Breslow, R. Herber, and S. J. Lippard, *J. Amer. Chem. Soc.*, **89**, 6864 (1967).
192. B. Gassenheimer and R. H. Herber, *Inorg. Chem.*, **8**, 1120 (1969).
193. C. R. Kurkjian and D. N. E. Buchanan, *Phys. Chem. Glasses*, **5**, 63 (1964); H. Pollak, M. de Coster, and S. Amelinckx, in *The Mössbauer Effect*, D. M. J. Compton and A. H. Schoen, eds., Wiley, New York, 1962.

194. S. L. Ruby, P. K. Tseng, and H.-S. Cheng, *Chem. Phys. Lett.*, **2,** 39 (1968).
195. A. J. Nozik and M. Kaplan, *J. Chem. Phys.*, **47,** 2960 (1967).
196. Y. Hazony and S. Bukshpan, *Rev. Mod. Phys.*, **36,** 360 (1964); I. Dezsi, L. Keszthelyi, L. Pocs, and L. Korecz, *Phys. Lett.*, **14,** 14 (1965).
197. S. Bukshpan, C. Goldstein, and T. Sonnino, *J. Chem. Phys.*, **49,** 5477 (1968).
198. For a list of solvents useful for such experiments see G. Adam and J. H. Gibbs, *J. Chem. Phys.*, **43,** 144 (1965).
199. R. V. Pound, G. B. Benedek, and R. Drever, *Phys. Rev. Lett.*, **7,** 405 (1961).
200. R. V. Pound and G. A. Rebka, Jr., *Phys. Rev. Lett.*, **4,** 337 (1960).
201. D. N. Pipkorn, C. K. Edge, P. deBrunner, G. DePasquali, H. G. Drickamer, and H. Frauenfelder, *Phys. Rev.*, **135,** A1604 (1964).
202. C. K. Edge, R. Ingalls, P. deBrunner, H. G. Drickamer, and H. Frauenfelder, *Phys. Rev.*, **138,** A727 (1965).
203. C. J. Coston, R. Ingalls, and H. G. Drickamer, *Phys. Rev.*, **145,** 409 (1966); *J. Appl. Phys.*, **37,** 1400 (1966) (A).
204. R. Ingalls, C. J. Coston, G. DePasquali, H. G. Drickamer, and J. J. Pinajian, *J. Chem. Phys.*, **45,** 1057 (1966).
205. A. R. Champion, R. W. Vaughan, and H. G. Drickamer, *J. Chem. Phys.*, **47,** 2583 (1967): A. R. Champion and H. G. Drickamer, *J. Chem. Phys.*, **47,** 2591 (1967); *ibid.* **47,** 1503 (1967); *ibid.* **47,** 468 (1967); S. C. Fung and H. G. Drickamer, *J. Chem. Phys.*, **51,** 4353 (1969); *ibid.* **51,** 4360 (1969). For a general brief review see J. Danon, *Lectures on the Mössbauer Effect*, Gordon and Breach, New York, 1968, pp. 104–111.
206. R. H. Herber and J. J. Spijkerman, *J. Chem. Phys.*, **42,** 4312 (1965). It is possible that the compression in these experiments was not isotropic, since no independent measurements were made to verify this.
207. C. E. Weir, E. R. Lippincott, A. Van Valkenburg, and E. N. Bunting, *J. Res. N.B.S.*, **63A,** 55 (1959).
208. H. S. Moeller and R. L. Mössbauer, *Phys. Lett.*, **A24,** 416 (1967); H. S. Moeller, *Z. Physik*, **212,** 107 (1968)
209. V. N. Panyushkin and I. F. Voronov, *JETP Pis'ma V Redaktsiyu*, **2,** 153 (1965), English translation *Sov. Phys. JETP Lett.*, **2,** 97 (1965).
210. S. S. Kabalkin and Z. V. Troitskaya, *Dokl. Akad. Nauk USSR* **151,** 1068 (1963), English translation, *Sov. Phys. Doklady* **8,** 800 (1964).
211. I.B.M. Corp., Yorktown Heights, N.Y. 10598; Computer Center, Argonne National Laboratory, Lemont, Ill. See also E. Rhodes, W. O Neal, and J. J. Spijkerman, *NBS Tech. Note*, **404,** 108 (1966).
212. One of us (Y. H.) and others at Princeton University have been using successfully several versions of this interface between two different MCA's (RIDL and TMC), a constant-velocity spectrometer, a molecular-beam experiment, and I.B.M. card-punch consoles, models 026, 526, and 029.
213. J. R. DeVoe, private communication.
214. S. S. Hanna, J. Heberle, C. Littlejohn, G. J. Perlow, R. S. Preston and D. H. Vincent, *Phys. Rev. Lett.*, **4,** 177 (1960). This paper appears to contain the first suggestion of this method.

215. H. Shechter, M. Ron, S. Niedzwiedz, and R. H. Herber, *Nucl. Instr. Methods*, **44,** 268 (1966).
216. J. J. Spijkerman, F. C. Ruegg, and J. R. DeVoe in "Applications of the Mössbauer Effect in Chemistry and Solid State Physics," *Tech. Rep. Series No. 50*, I.A.E.A., Vienna, 1966; see also NBS Technical Notes of the Radiochemistry Division.
217. Austin Science Associates, Austin, Tex. 78712; $1500.
218. R. H. Herber, in *Mössbauer Effect Methodology*, Vol. I, I. Gruverman, ed., Plenum, New York, 1965.
219. This is one form in which ^{119}Sn activity is available from commercial suppliers. Stannic oxide sources are not recommended, however, because of the wide lines observed (vide infra).
220. See Ref. 78. A detailed study of SnO_2 is contained in the Ph.D. dissertation of H. A. Stockler, Rutgers University, 1967; H. A. Stockler and H. Sano, *Nucl. Instr. Methods*, **44,** 103 (1966).
221. R. H. Herber, unpublished results; see also Table 1 of R. H. Herber and H.-S. Cheng, *Inorg. Chem.*, **8,** 2145 (1969).
222. J. J. Spijkerman, private communication.
223. R. H. Herber, Paper 173, Division of Physical Chemistry. New York Meeting, American Chemical Society, September 1969; R. H. Herber and S. Chandra, *J. Chem. Phys.*, **52,** 6045 (1970); **54,** 1847 (1971).
224. G. J. Perlow, ref. 5, Chapter 7.
225. R. H. Herber, in "Mössbauer Effect Methodology," Vol. 6, I. Gruverman, Ed., Plenum Press, New York, 1971, pp. 3–15. See also *Inorg. Chim. Acta*, 1970 ("Notice to the authors of Papers").
226. R. L. Cohen and G. M. Kalvius, *Nuclear Instr. Methods*, **86,** 209 (1970).

Chapter **V**

PHOTOELECTRON SPECTROSCOPY

Royal G. Albridge

1 INTRODUCTION

Photoelectron spectroscopy is the analysis of electrons that are ejected when molecules* are irradiated with X-rays or with ultraviolet light. In this method an electron spectrometer measures the number of photoelectrons as a function of the energy of the photoelectrons. The spectra so obtained give information about the molecular energy levels and thus about the structure and the chemical nature of the sample.

After the photoionization, the molecule can deexcite by one of two secondary processes: fluorescence emission or Auger electron emission. These processes, as well as the absorption of the incident photons themselves, can also give information about the electronic structure of molecules. Table 5.1

Table 5.1 Spectra That Can Be Recorded from a Sample Irradiated with Photons

	Primary processes	*Secondary processes*
Electron spectra	Photoelectron and shake-off	Auger
Photon spectra	Absorption	Fluorescence emission

summarizes this information by classifying the spectra that can be recorded from a sample irradiated with photons.

During the photoelectric process it is possible for the photon energy to be shared between two electrons so that (1) both electrons are ejected or (2) one electron is ejected and the other is excited to an outer level [1]. If two electrons are ejected, one is emitted with an energy nearly equal to that of the photoelectron of the single photoionization process and the other is emitted with little energy. The emission of the low-energy electron is called "electron shake-off." For simplicity we shall refer to both processes (1) and (2) as "double excitation."

Auger electron emission [2] occurs when a vacancy in an inner orbit is filled by an electron dropping from a higher orbit. The energy released in the process can eject an electron (Auger emission) or can release a photon (fluorescence emission). Auger electron emission and shake-off electron emission are not, strictly speaking, a part of photoelectron spectroscopy; however, since they give information about the electronic structure of the

* In many places in this article the word "molecule" can be taken to mean "molecule or atom."

sample and since they are studied by means of the same spectrometers used for photoelectron spectroscopy, they are discussed briefly in this article.

The spectroscopies that involve the study of photons were developed and used to great advantage much earlier than was electron spectroscopy. The early attempts [3] at electron spectroscopy were made by means of low-resolution instruments, which recorded lines with long, low-energy tails that made accurate measurements impossible. These tails are the result of inelastic scattering of the electrons in the sample. The problem of scattering is not as serious in photon spectroscopy because for $h\nu$ less than about 500 keV, photon scattering is less probable than is photoelectric absorption [4]; thus the photons are either recorded with their full energy or are not recorded at all.

The low-energy losses suffered by the electrons in passing through a sample are quantized [5, 6], and the most probable losses have magnitudes of several electron volts; thus a high-resolution instrument can separate the energy-loss tail from the photoelectron line produced by electrons that have escaped the sample without energy loss. This fact was not appreciated, however, until recently (the 1950s), when high-resolution spectrometers were developed and applied to photoelectron spectroscopy. An example of the separation of a line from the energy-loss spectrum is shown in Fig. 5.1, which is the data of Siegbahn et al. [7].

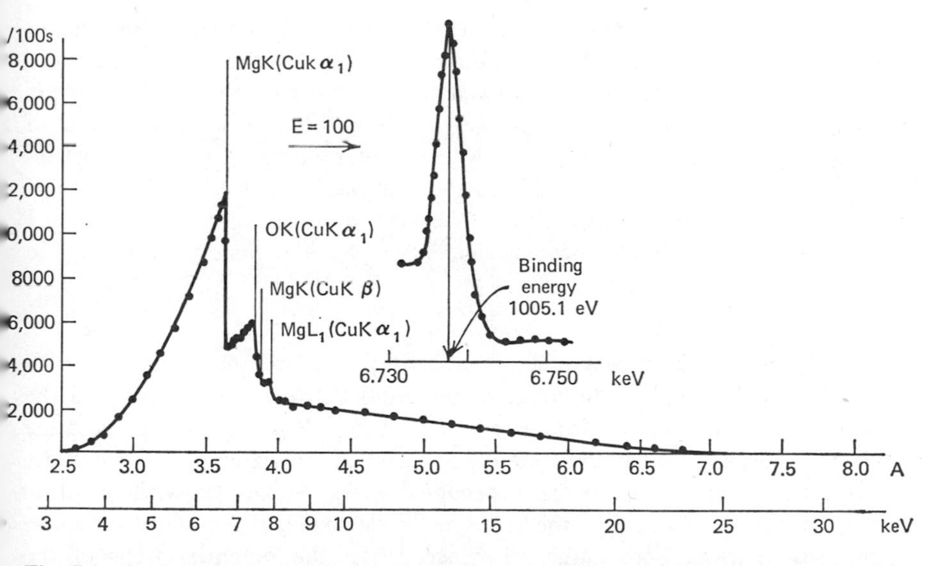

Fig. 5.1 A photoelectron spectrum of magnesium oxide irradiated with copper X-rays showing the separation of lines from the energy-loss spectrum. The narrowness of the lines is shown by the inset, which has an abscissa that is expanded by a factor of 100. This figure is reproduced, with permission, from *ESCA* by Siegbahn et al. [7].

Photoelectron spectroscopy provides an accurate and simple means of measuring the binding energies of electrons in atoms and molecules. All orbitals from the innermost to the outermost can be studied, and the method is applicable to all elements. The measured binding energies and the shifts of binding energy due to chemical bonding can be interpreted and used in a variety of ways, as described in Section 5.

Two groups have been primarily responsible for the development of photoelectron spectroscopy: a Swedish group under the direction of Professor Kai Siegbahn and an English group under the direction of Professor D. W. Turner. The Siegbahn group developed their methods using mostly X-ray excitation, but they now use also ultraviolet excitation and electron impact. The Turner group uses ultraviolet excitation. Both Siegbahn et al. [7, 8] and Turner [9] have recently published reviews of their work. In addition, R. S. Berry [10] has recently written a review article on electron spectroscopy. Hollander and Shirley [10a] have compared the various inner-shell spectroscopics.

2 THE PRINCIPLES OF PHOTOELECTRON SPECTROSCOPY

The energy balance in the photoelectric process is given by

$$h\nu = T + E_B + E_R, \tag{5.1}$$

where T is the kinetic energy of the ejected electron, E_B is the binding energy of the electron in the sample, E_R is the recoil energy of the ion, and $h\nu$ is the energy of the incident photon. The magnitude of the recoil energy, which is determined by the conservation of energy and momentum, is small enough to be neglected except in those cases where higher-energy X-rays impinge on very light elements. (For example, if the commonly used AlK_α X-ray ($h\nu =$ 1486.6 eV) is incident upon lithium, the maximum recoil energy is only 0.1 eV, and it is much less for heavier elements.) Thus the basic equation can be written as

$$h\nu = T + E_B. \tag{5.2}$$

For experimental measurements involving solid samples, (5.2) must be modified to account for the contact potential between the sample and the metal from which the spectrometer is constructed. When the sample and the spectrometer are in electrical contact, electrons in the material with the larger Fermi energy can move to the unoccupied energy states below them in the other material. Because of the transfer of electrons, the materials acquire opposite charges. The charge difference causes the potential wells of the materials to shift until the Fermi surfaces are at the same level, so that there is no reason for further electron transfer. The shifts of the Fermi surfaces are brought about primarily by the charge difference due to a few electrons and not by appreciable changes in the depths of Fermi the seas, which would occur if many electrons transferred. Since the depths of the Fermi seas do not change appreciably, the work functions of the materials are not altered; however, a potential difference (a contact potential) will exist between the two materials.

These facts are represented schematically in Fig. 5.2, where the energy levels of the materials are shown before and after equilibrium for the case of

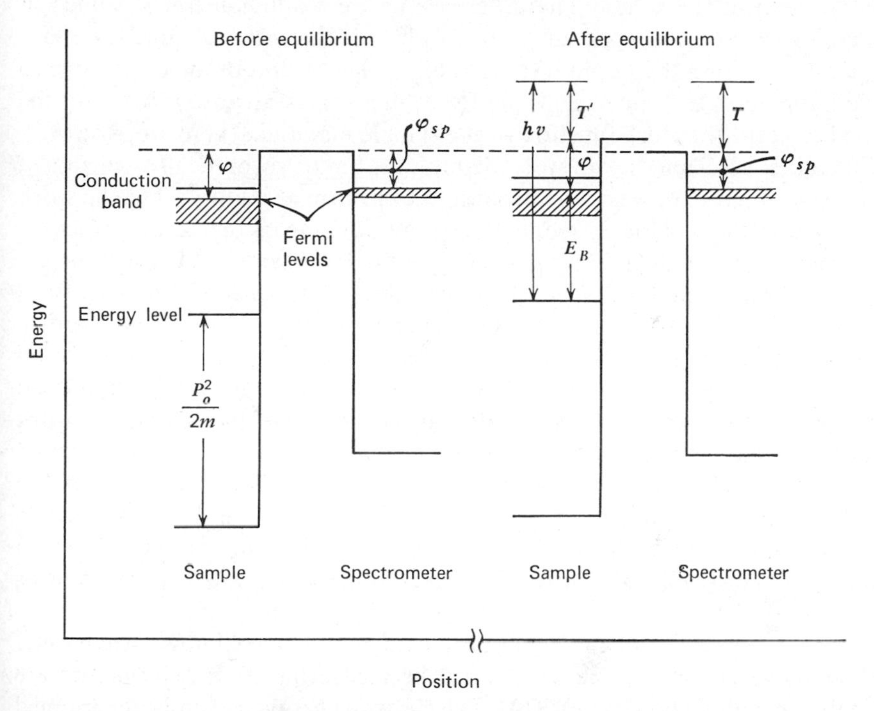

Fig. 5.2 The energy levels of a conducting sample and the spectrometer metal before and after the materials are brought into electrical equilibrium. The Fermi levels adjust themselves to the same potential creating a contact potential between the surfaces. The energy relation is $h\nu = T + E_B + \phi_{sp}$, when E_B is defined relative to the Fermi energy.

a conducting sample. For samples that are insulators it is thought [11, 11a] that the same condition of equilibrium is achieved because of the large number of free electrons produced when the sample is irradiated. The contact potential causes the kinetic energy of the photoelectron, as measured in the spectrometer, to depend upon the spectrometer work function, ϕ_{sp}, rather than on the work function, ϕ, of the sample [11–13]. If the contact potential were not present, the kinetic energy would be T', such that

$$h\nu = T' + E_B + \phi. \tag{5.3}$$

Because of the contact potential we must write

$$h\nu = T + E_B + \phi_{sp}, \tag{5.4}$$

where $T' + \phi = T + \phi_{sp}$ and where E_B is defined relative to the Fermi level.

A straightforward definition of binding energy is "the energy required to move an electron from its orbital to infinity." Note that the definition of E_B in 5.4 is "the energy required to promote the electron from its orbital to the Fermi level of the solid." The difference in the binding energies defined in these two ways is the work function, ϕ, which is the energy required to move the electron from the Fermi level to infinity. The definition of E_B relative to the Fermi level is a convenient one for solid samples because it bypasses the need to know the work function of the sample material. There are, however, difficulties associated with this definition. First, theoretically calculated binding energies are usually referred to the electron at infinity; thus the work function of the material is required anyway, if comparisons to theory are to be made. For metals the Fermi level is well-defined, and work functions are known; however, the Fermi levels of insulators and semiconductors are hard to locate, and work functions are not generally known. Also, it is not known for sure that at the surface of the sample the equilibrium between Fermi levels of the sample and the spectrometer, as shown in Fig. 5.2, is established for nonmetallic samples. And, finally, at the surface of the sample the work function and the position of the Fermi level relative to the electronic energy levels can be altered by surface effects, such as adsorption. Thus, even if one is interested only in comparing binding energies of the same element in different compounds, the uncertainties in the positions of the Fermi levels in the compounds will make the results uncertain to some extent (perhaps by as much as 1 eV).

Since the energies of the commonly used photons are known accurately, the binding energy can be determined by measuring T. X-ray energies are known to within about 0.1 eV [14]. The electron energy, T, can be determined to within about $\frac{1}{10}$ the recorded line width. This width, which has contributions* from the photon width, the level width, the source and detector slit widths, and instrumental aberrations, lies in the range of about 1 to 4 eV for X-ray induced spectra; thus the error limits of T range from about 0.1 to about 0.4 eV, when X-ray excitation is used. The work function of the spectrometer material can be determined experimentally to within about 0.3 eV [15]. The combined uncertainties result in error limits for E_B of the order of tenths of an electron volt.

The characteristics of electron spectrometers are such that the instrumental contribution to line width decreases with decreasing energy. Thus, when the

* Gaseous sources give narrower lines than do solid sources; thus there is also a "solid-state" contribution to the line width.

helium resonance line with an energy of only 21 eV and with a natural width of less than 0.005 eV [16] is used as the exciting radiation in gas-phase studies, very narrow lines ($\Delta E \sim 0.01$ eV) can be obtained. In these cases precision of the order of $\frac{1}{100}$ of an electron volt is possible. Samson [16] has considered several factors (Doppler broadening, Stark broadening, pressure, intrinsic widths, and molecular motion) that contribute to line widths in ultraviolet-excited, gas-phase photoelectron spectroscopy. He concludes that the main contribution arises from the thermal motion of the sample molecules. Ultraviolet radiation has been used almost exclusively for studies of gases since the low-penetrating power of the photons makes measurements of the solid state especially difficult, and the low energy of the photons permits only the band structure of the solid to be studied.

An ionized free molecule can have vibrational and rotational modes of excitation in addition to electronic modes. To account for this fact the photoelectric equation for gases must be written

$$h\nu = T + E_B + E_{\text{vib}} + E_{\text{rot}} + \delta, \qquad (5.5)$$

where E_{vib} and E_{rot} are the vibrational and rotational energies of the ionized molecule. In high-resolution, gas-phase studies of spectra excited by ultraviolet radiation, series of lines corresponding to different vibrational states can be observed. In favorable cases line broadening caused by rotational substructure can be discerned. (See Fig. 5.15.)

In (5.5) δ is a term used to correct for variation of the energy calibration with total gas pressure, X-ray intensity, and type of gas studied. Presumably these variations are due to the buildup of space and surface charges in the sample chamber [8]. As the total pressure has been varied from ~ 0 up to 1 torr, T has been observed to change by as much as 1 eV. [8] Krause and Carlson in a private communication have reported a strong dependence of T on the intensity of the X-ray beam.

3 APPARATUS

Electron spectrometers measure the energies and the intensities of the emitted electron groups. The spectrometers use either electric or magnetic fields (or a combination of both) and can operate on the principle of deflection, retardation, or time-of-flight.

Deflecting-Field Analyzers

Deflecting-field spectrometers focus the photoelectrons onto an exit slit, behind which is a detector. The focusing field is varied to bring electrons of different energies successively into focus at the exit slit, and the spectrum is a plot of detector response versus focusing field. In several respects deflection-type spectrometers are analogous to optical spectrometers, and the term "electron optics" is used in reference to the focusing properties of the instruments. The spectrometer must be capable of separating electrons of different energies (i.e., it must possess energy dispersion), and it must be capable of focusing electrons onto a small detector aperture even though the electrons are emitted at different angles and from different positions on a source of finite extent (i.e., it must possess small aberrations).

The "width," ΔE, of a recorded electron line is defined as the full width of the line, expressed in energy units, at one-half the maximum height of the line. As stated in Section 2, this width has contributions from the photon width, the level width, the source and detector slit widths, and the instrumental aberrations. "Resolution" is defined as the line width divided by E, the energy of the line; and "instrumental resolution" is defined as the instrumental contributions to the line width, ΔE_{instr}, divided by the energy of the line. "Transmission" is the fraction of the total solid angle that is subtended by the spectrometer aperture. A good spectrometer design tends to maximize the transmission and to minimize the instrumental resolution, $\Delta E_{\text{instr}}/E$.

The dispersive and focusing properties of a spectrometer are determined by the shape (i.e., the spatial distribution) of the electric or the magnetic field. The electric field shape can be determined by the arrangement of metal plates or grids held at suitable potentials, and the magnetic field shape can be determined by the arrangement of current-carrying coils. Extensive and complex calculations are required to determine the focusing properties of a given field and to determine the proper plate or coil arrangement required to produce the field; and skill is required to design and construct the instrument within the mechanical tolerances allowed. The detailed theory [17] of the focusing properties of these spectrometers is not appropriate material for this article; however, a general discussion presents useful information and terminology.

Electrostatic Spectrometers*

There are four different types of spectrometers in practice that focus by means of an electric field: parallel-plate, cylindrical, toroidal, and spherical. Because of its simplicity the parallel-plate spectrometer has been used in the past in spite of its crude focusing properties [18]. It can be used to introduce the concepts and the terminology of electron focusing.

The field of a parallel-plate spectrometer is uniform, and any particular electron trajectory in the spectrometer is analogous to that of a stream of water projected in a homogeneous gravitational field, if air resistance is neglected (see Fig. 5.3). That is, in both cases the trajectories are parabolic.

* The author is indebted to Dr. Reimar Spohr for much of the material in this section.

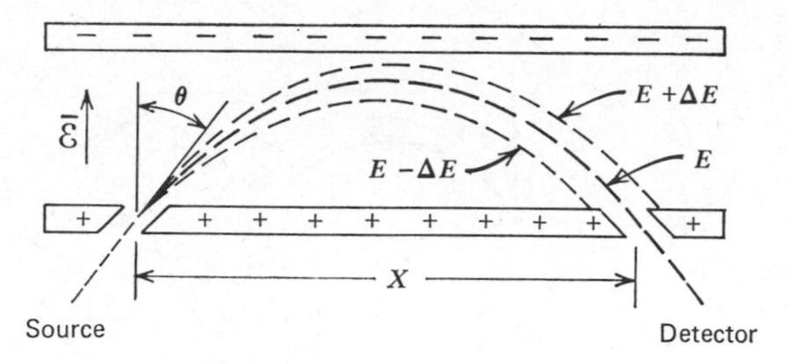

Fig. 5.3 Electron trajectories in a parallel-plate spectrometer.

This spectrometer possesses energy dispersion, since for a given field, E, only electrons of the proper energy, E, will be focused at the detector. It is easily shown that

$$X = kE \sin 2\theta, \tag{5.6}$$

where E is the electron energy and k is a constant. Thus $\partial X/\partial E$, the energy dispersion, is $k \sin 2\theta$, which is zero only for $\theta = 0$ or $90°$ and which is a maximum for $\theta = 45°$.

Children who play with water hoses in their backyards know that the maximum range, X_{max}, is obtained for $\theta = 45°$ and that at this angle small changes in θ have little effect on the range. These facts can be derived from (5.6) by showing that $\partial X/\partial \theta$ vanishes at $\theta = 45°$.

The insensitivity of X to small changes in θ is of fundamental importance in any focusing device since the aperture at the source must accept a range of angles in order to produce a finite intensity at the detector. When the spectrometer fails to focus electrons of the same energy but different emission angles, it is said to possess aberrations. Of course, the dependence of X on θ is not completely determined by $\partial X/\partial \theta$ but by higher derivatives as well. In the present example $\partial^2 X/\partial \theta^2$ does not vanish at $\theta = 45°$; hence this spectrometer possesses first-order focusing but not second-order focusing. The value of $\partial^2 X/\partial \theta^2$ is called "the second order aberration coefficient."

To completely describe in three dimensions the direction of emission of an electron two angles are required. Thus a complete description of the focusing properties of a spectrometer requires that the order of focusing for both angles be given. In the case of the parallel-plate spectrometer a change $d\phi$ in ϕ (Fig. 5.4) causes a change $ds = r\,d\phi$ in the position of focus. Since $ds/d\phi = r$, the focus in ϕ is zeroth order, and the spectrometer has one-dimensional focusing only.

Figure 5.5 illustrates the configurations and the electron trajectories of the several types of electrostatic spectrometers. The radially and axially cylindrical configurations can be thought of as parallel plates bent around an

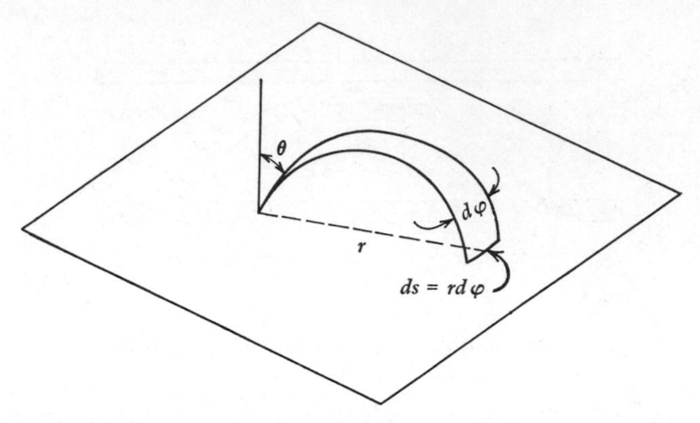

Fig. 5.4 The parallel-plate spectrometer is shown in three dimensions to illustrate that it has only one-dimensional focusing properties.

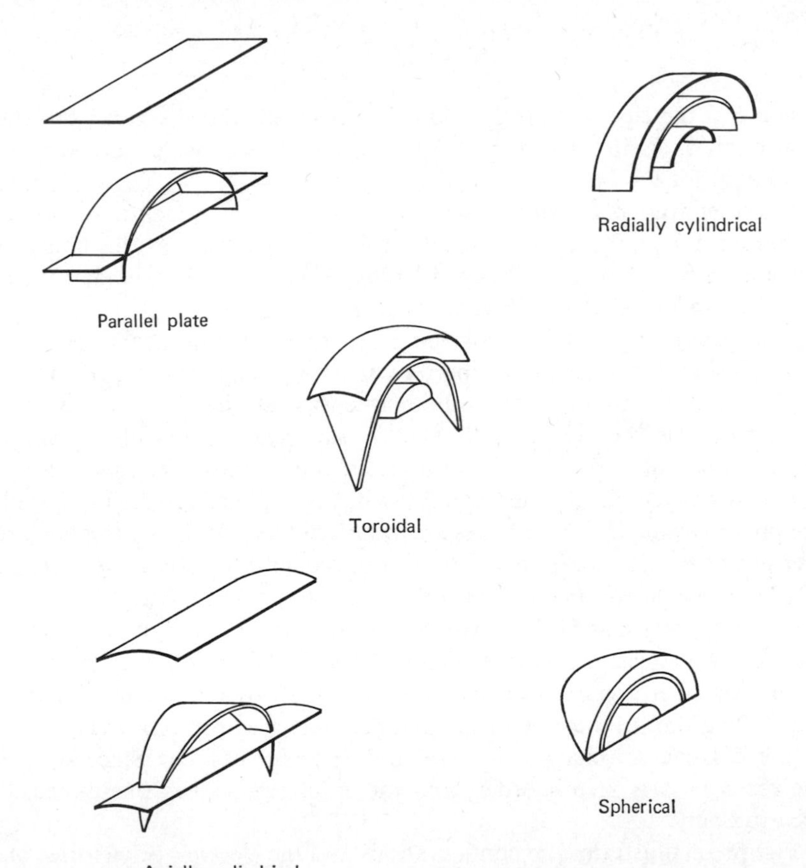

Fig. 5.5 The configurations and the electron trajectories of the several types of electrostatic spectrometers.

axis that is parallel respectively to the short and long dimensions of the plates. The axially cylindrical spectrometer is known commonly as the cylindrical mirror spectrometer. The toroidal configuration consists of plates bent around both axes, and the spherical configuration is a special case of the toroid in which the two radii of curvature are equal to each other.

The radially cylindrical spectrometer has one-dimensional focusing to first order [19, 20]. Both the axially cylindrical [21–24] and the spherical configurations [25–30] have two-dimensional focusing (are "double focusing"). The axially cylindrical spectrometer has been compared to the spherical deflector by Hafner and others [31] who found the focusing properties of the former to be superior. Table 5.2 summarizes the focusing properties of the types discussed.

Table 5.2 A Rough Summary of the Focusing Properties of Electrostatic Spectrometers*

Type	*Potential varies with*	*Order of focus in* θ, ϕ
Parallel plate	z	1, 0
Radially cylindrical	$\ln r$	1, 0
Axially cylindrical	$\ln r$	2, ∞
Spherical	$1/r$	1, ∞

* The order of focus is shown for both angles that describe the electrons' initial direction of motion.

The spherical electrostatic spectrometer has attracted the most widespread interest as a photoelectron spectrometer. Siegbahn et al. [32] have designed and constructed one with sector angles of 158° (horizontal) and 60° (vertical). An instrument of this design is being made commercially [33]. Spectrometers utilizing full spheres [34] or hemispheres [35] are also being made commercially. Radially cylindrical instruments for gas-phase studies are also available commercially [36].

Magnetic Spectrometers

Magnetic instruments have been used extensively for nuclear research to study electrons and beta particles with energies up to several million electron volts. Thus the focusing properties of these instruments have been calculated extensively and have been reported fully in the literature. Also, at the present time the electrostatic instruments have attracted the most interest for photoelectron studies. Therefore we present here only a small amount of basic information on magnetic instruments. For more detailed information see the references cited.

Magnetic spectrometers are of either the lens type or the transverse type. The field lines of the most simple lens spectrometer are parallel to a line connecting source and detector; and the electrons spiral down this line focusing on an annular slit. These are not high-resolution instruments and are not used often for photoelectron spectroscopy.

The field lines of transverse spectrometers are perpendicular to the plane formed by the source, the detector, and the midpoint of the baffle; and the electrons traverse circular paths. Such spectrometers have either a homogeneous field with one-dimensional focusing or a nonuniform field that gives two-dimensional focusing. A field that varies with the radius, r, as $1/\sqrt{r}$ gives double focusing.

A homogeneous-field instrument focuses the beam at an angle of 180° along the circular path. The transmission of these instruments is relatively small; however, the focal plane is quite large so that long photographic plates can be used to record a large portion of the spectrum at one field setting. For this reason uniform-field spectrographs often utilize permanent magnets. Photographic recording is not desirable because intensity measurements are difficult to obtain from photographic film; however, Siegbahn et al. [37] have developed an automatic scanner to count individual electron tracks in nuclear emulsions, and they can thus obtain accurate intensity measurements. When large position-sensitive detectors become available (see Section 3, p. 322), uniform-field instruments can become even more useful tools for photoelectron spectroscopy. In addition to their ability to record large portions of the spectrum at one time, the advantages of these instruments are their simplicity, stability, low cost, and the fact that no external field compensation is required.

The transverse, double-focusing spectrometer focuses the electron beam at an angle of $\pi\sqrt{2}$. Several different types of this spectrometer have been constructed. Some utilize several pairs of coils symmetrically arranged above and below the optic circle [17, 38, 39], while others use two concentric solenoids with the optic circle between them [17, 40, 41].

Fadley and others [42] have designed but not yet built a unique magnetic instrument that creates a double-focusing field by means of coils properly arranged along the length of a horizontal cylinder. Unlike other solenoidal instruments, only the field outside the solenoid is utilized; thus the vacuum chamber is designed to surround the cylinder so that the chamber is easily accessible and can be slid away for high-vacuum bake-out. Presumably, different vacuum chambers could be used interchangeably. The field has been designed to give very high transmission (0.2% of 4π) at high resolution ($\Delta E/E = 0.02\%$) and to give an exceptionally wide focal plane. The wide focal plane could accommodate a large position-sensitive detector (see Section 3, p. 322) to give a factor of 100 increase in the rate of data acquisition. Other double-focusing spectrometers (magnetic and electrostatic) have focal planes sufficiently large to gain some advantage by the use of position-sensitive detectors.

Magnetic instruments have the disadvantage that since they create a magnetic field it is difficult to provide for automatic compensation of external fields. It is necessary to utilize a coil arrangement that produces zero field at some point near the spectrometer for any value of the spectrometer current [40]. The magnetic probe can be securely fixed at that position, and the compensating coils used to null the field at that point.

Crossed-Field Spectrometers (Wien Filters)

In these instruments an electric and a magnetic field are produced at right angles to each other and are adjusted so that the deflection caused by each exactly cancels for electrons of a given energy [43]. The electrons of interest traverse a straight-line path from entrance slit to detector. The spectrum is recorded by sweeping either the electric field or the magnetic field.

Retardation Enhancement

For electrostatic, double-focusing spectrometers the instrumental contribution to the line width, ΔE, is proportional to the electron energy, E. Therefore at constant slit and constant baffle openings the application of a retarding potential to the sample decreases E and reduces ΔE, or if the slit and the baffle apertures are increased in size, ΔE can be held constant and the transmission can be enhanced by a factor of about 10. Helmer and Weichert [44] have discussed this technique in detail. The application of an electrostatic retarding potential to the sample will have a similar effect on the performance of a magnetic double-focusing spectrometer.

Dispersion Compensation

Since the X-ray source emits a continuous Bremsstrahlung spectrum as well as discrete characteristic X-rays, the photoelectron spectrum consists of a continuous background with discrete lines superimposed. Thus the peak-to-background ratio can be increased if the X-ray beam is filtered or crystal-monochromatized. Furthermore if crystal monochromatization is used, the dispersion of the monochromator can be matched to the dispersion of the electron spectrometer in such a way that the contribution of the characteristic X-ray line width to the photoelectron line width can be eliminated [45]. The method is explained in Fig. 5.6.

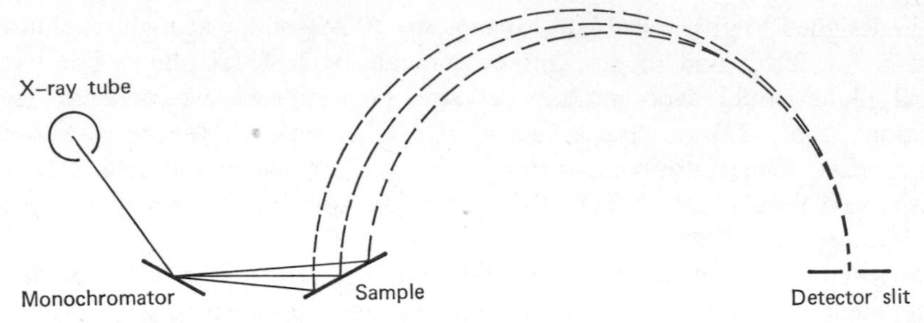

Fig. 5.6 Dispersion compensation. The X-ray beam is a mixture of many wavelengths. The X-rays are dispersed in space according to wavelength by the crystal monochromator, and a characteristic X-ray with a finite energy width strikes the sample. The X-ray line is dispersed such that the shorter wavelengths fall to the left in the figure. The photoelectrons from the sample are focused by the electron spectrometer onto the detector slit. The sample is so positioned that all electrons have a common position of focus, and the finite width of the X-ray line makes no contribution to the width of the photoelectron line.

Retarding-Field Analyzers

Retarding-field analyzers utilize an electrostatic field to retard the passage of electrons to a detector. The electrostatic field lies between two electrodes that can be arranged in the geometry of a parallel-plate [46], cylindrical [9, 47], or spherical condenser [9, 46, 48, 49]. Only those electrons with sufficient energy to overcome the retarding field reach the detector and are counted. The spectra obtained are plots of the detector output as a function of the retarding potential. These are, unfortunately, integral, step-function curves. Differential curves can be computed or can be obtained by electronic differentiation of the detector output [9], but the accuracy is still limited by the integral nature of the primary data.

The greatest disadvantage of retarding-field analyzers is that they measure only the component of energy parallel to the retarding field. Thus, if the electron beam has a finite angular spread, the spectrometer will record a range of energies even if the electron beam is monoenergetic [9, 46]. This difficulty is not encountered if a spherical condenser is used with a point source; however, the use of point sources is not practical. Al-Joboury and Turner [9, 47], using a cylindrical condenser, have obtained resolutions of the order of $\Delta E/E \sim 2\%$ as compared to values of $\sim 0.04\%$ obtained with deflecting-field analyzers [7]. Of course, the geometry of the cylindrical and the spherical condensers gives a much higher collection efficiency than is obtained with the deflecting-field instruments.

Time-of-Flight Analyzers

The time dispersion of electrons with different velocities moving a fixed distance can be used to obtain energy spectra [50]. The technique is a convenient one since it is possible to record all points on the energy spectrum at one time. Since electron beams with energies in the range of 0 to 10 eV are difficult to measure by means of deflecting-field instruments and since at these energies the electron velocities are small enough to be timed by modern electronics, the applications of time-of-flight instruments to date have been in the very-low-energy range. No application to photoelectron spectroscopy has yet been made.

Accessory Equipment

Shielding and Compensation of External Fields

In order to avoid hysteresis effects photoelectron spectrometers (with the exception of permanent-magnet instruments) are constructed iron-free. Thus no magnetic materials should be near the instrument when it is operating, and all external magnetic fields (in particular the Earth's field) must be eliminated. External magnetic fields can be shielded against by high-μ metals or by compensating coils that produce a field in opposition to the external field.

The strictest requirements for shielding occur for low-energy electrons in instruments of large radius. The magnetic field required to focus a 20-eV electron at a radius of 30 cm is ~0.5 gauss. Precision of 10^{-4} requires that the external field be zero to within approximately $\pm 5 \times 10^{-5}$ gauss ($\pm 5\gamma G$). These requirements are encountered in work done by means of ultraviolet radiation. If X-ray excitation is used, the electron energies are of the order of 1000 eV, and fields of ~3 gauss are required to focus the beam with a radius of ~30 cm. In this case precision of 10^{-4} requires that the external field be zero to within about $\pm 30\gamma G$.

An external magnetic field that is constant in time and uniform in space is, of course, easiest to eliminate. Fields that vary in time or space or both present special problems. The Earth's field is approximately constant and uniform; however, time-varying fields and/or fields with large gradients are often present in laboratories because of electric transformers, power lines, iron equipment, building superstructure, and other magnetic equipment.

Magnetic shielding will discriminate against all fields providing the shield is thick enough and covers the instrument adequately. The disadvantage of the shield is the inconvenience of removing it when access to the spectrometer is desired. Magnetic shielding cannot be used conveniently with magnetic-type instruments because of the effect of the shield on the field produced by the spectrometer.

Three pairs of compensating coils at right angles to each other can be used to compensate the three components of the external field [51, 52]. For efficient compensation the coil dimensions should be of the order of at least two or three times the diameter of the spectrometer. One can reduce field gradients by twisting the compensating coils so that they do not lie in a flat plane, while monitoring the external field at the position of the electron trajectory. A trial-and-error method must be used, and many hours can be required to reduce the gradient to within $\pm 10\gamma G$ over the path of the electron. Field gradients can be reduced also by the use of more than three pairs of coil.

Time-varying fields can be reduced by means of a feedback system that controls the current to the compensating coils. A magnetic probe senses the field, and the feedback loop adjusts the compensating current until the magnetometer reads zero [40]. Rapidly varying fields such as those found near large transformers are not eliminated by the feedback device. If these fields are sufficiently large, they cause line broadening. Fortunately, the metal chamber of the spectrometer itself reduces to some extent the magnitude of rapidly alternating fields inside the tank.

Electron Detectors

It is possible to detect either individual electrons striking a detector or the current flowing from a collector electrode. The former method is commonly used with deflecting-field analyzers, the latter method with retarding-field analyzers.

Digital-type detectors include electron-multiplier tubes, continuous-channel multipliers, and position-sensitive detectors. Continuous-channel multipliers work much like electron-multiplier tubes; the incoming electron is accelerated down a tube so that it repeatedly strikes a semi-conducting coating from which it knocks loose secondary electrons [53–57]. The secondary electrons are accelerated and produce a cascade. The gains of these detectors are typically in the range of from 10^5 to 10^8. A typical tube is only about 1 mm in diameter and 6 cm long, and it is bent to prevent regenerative ion feedback [54]. It operates with a potential difference of about 4 keV across its length. Kanayama and others [58] have developed a parallel-plate multiplier that uses the same principle.

Position-sensitive detectors are detectors that indicate not only the arrival of particle at the detector, but also the position of the event along the length of the detector. If such detectors can be developed to give sufficient spatial resolution, their use in spectrometers with wide focal planes can increase the speed of data acquisition by one or two orders of magnitude. There have been developed, for the detection of ions, solid-state detectors that give one voltage pulse to indicate an event and another to give the position of the event [59]. A multichannel detector consisting of an array of closely spaced solid-state detectors is also feasible [60].

Mehlhorn [23] and Al-Joboury and Turner [47] have described detector systems used to detect the current rather than the individual electron pulses.

Spectrometer Systems

A photoelectron spectrometer system consists of the electron spectrometer, a central process controller, the detectors and associated electronics, the

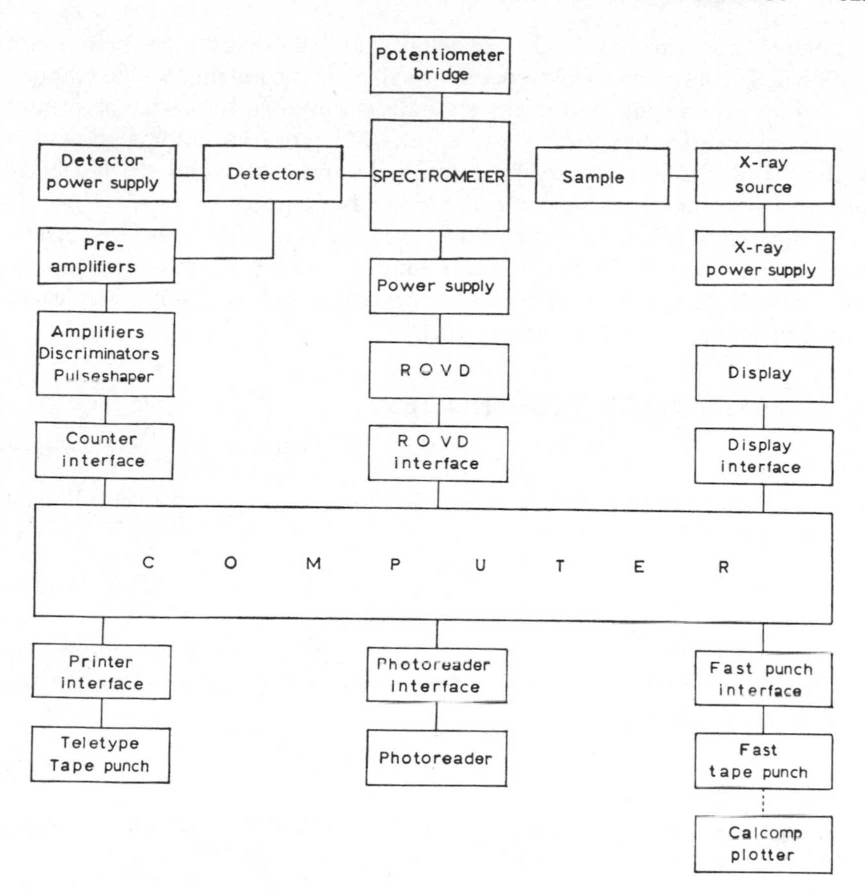

Fig. 5.7 A block diagram of the Vanderbilt photoelectron spectrometer system.

spectrometer power supply with control and calibrator, the data output devices, the external-field compensation unit, the high-vacuum system, and the X-ray, ultraviolet, and/or electron impact units used to excite the sample. A block diagram of the Vanderbilt spectrometer system is shown in Fig. 5.7, as an example [61]. The spectrometer is as solenoidal, split-coil, magnetic instrument designed and built by the Uppsala group. The high-vacuum pump is a turbo-molecular pump that does not require trapping and does not introduce oil into the system. The external field is compensated by three pairs of coils.

The process controller is an 8000-word, 16-bit computer that controls the spectrometer and stores and outputs the data. Spectra are scanned by the stepping of a computer-controlled, relay-operated voltage divider. With the computer program now used, step lengths and dwell time per setting are optional, and as many as five separate portions of a spectrum can be scanned during an experiment. At present the data is outputed to a teletype printer simultaneously in digital form and as an X-Y plot. An automated system such as this one is necessary for continuous operation, and the flexibility offered by a programmable controller is highly desirable.

Another type of automation, which is used principally with electrostatic instruments, sweeps the spectrometer field repeatedly by means of a saw-toothed voltage and records the data by means of a multiscalar that is triggered at the beginning of each sweep [1, 61a].

4 EXPERIMENTAL TECHNIQUES

Calibration

In a spherical electrostatic spectrometer the radius, r_0, of the central orbit is given by

$$\frac{mv^2}{r_0} = k'V, \tag{5.7}$$

where V is the voltage between the plates, m and v are the mass and velocity of the electron, and k' is a constant. Substituting into (5.7) the nonrelativistic kinetic energy, $T = \frac{1}{2}mv^2$, gives

$$T = k''V \quad \text{(nonrelativistic)}, \tag{5.8}$$

where $k'' = \frac{1}{2}k'r_0$.

Substituting into (5.7) the relativistic kinetic energy, $T_r = mc^2 - m_0c^2$, where m_0 is the rest-mass of the electron, gives

$$T_r = kV\left(1 + \frac{T_r}{(2E_0 + T_r)}\right) \quad \text{(relativistic)}, \tag{5.9}$$

where E_0 is the rest-mass energy of the electron. The relativistic correction, $T_r/(2E_0 + T_r)$, for a 1000-eV photoelectron is 0.1%.

In a magnetic spectrometer the radius of the central orbit is given by

$$\frac{mv^2}{r_0} = Bev, \tag{5.10}$$

where B is the magnitude of the magnetic field, e is the charge on the electron, and other symbols are as defined above. Combining $T_r = mc^2 - m_0c^2$ with (5.10) gives

$$T_r = [E_0^2 + e^2c^2(Br_0)^2]^{1/2} - E_0 \quad \text{(relativistic)}. \tag{5.11}$$

In the double-focusing, iron-free, magnetic spectrometer, the field, B, is proportional to I, the current in the coils: $B = K'I$ or, since r_0 is constant,

$Br_0 = KI$. Thus (5.11) can be rewritten as

$$T = [E_0^2 + e^2c^2K^2I^2]^{1/2} - E_0 \quad \text{(relativistic)}. \tag{5.12}$$

Calibration of a spectrometer consists of determining the constant k or K by measuring the position of a line of known energy [11a]. The spectrometer constant depends upon the geometry of the spectrometer, which can change when conditions such as the temperature vary. In addition effects such as source charging or the presence of an external magnetic field can cause an apparent change in the constant. For these reasons it is necessary to use an internal standard if accurate work is to be done.

The Uppsala group has used as a calibration line for solid samples the carbon 1*s* line from hydrocarbon contamination that deposits itself on the sample. It is not known whether in some cases interaction between the oil and the sample will shift the carbon line or whether the sample and the pump oil will maintain the same electrostatic potential. It has been suggested that there are sufficient free-charge carriers in the vacuum around the sample to prohibit charge buildup even if the sample is an insulator. However, with thick insulating samples it is possible to observe shifts as large as 1 eV, which can be interpreted as a source-charging effect. The Vanderbilt group has observed shifts in the carbon 1*s* line of vacuum-system hydrocarbons that they attributed to an interaction between the source and the thin hydrocarbon overlayer [62]. To avoid this difficulty they have used for calibration powdered graphite mixed with the samples [62]. A metal such as gold might serve well as a calibration source. It could, for example, be deposited as a thin layer on top of the sample.

It is usually not obvious whether small changes in k (or K) are real changes due to variation of spectrometer geometry or apparent changes due to source charging or incomplete shielding of the external field. Strictly speaking, one should treat the latter two effects by correcting T and B, respectively; but if the effects are small, treating them by recalculating the spectrometer constant introduces negligible error.

In the gas phase calibration is easier to achieve since the calibration standard can be mixed uniformly with the sample. The intensity of the calibration line is in part determined by the partial pressure of the calibration gas.

Sample Preparation

Solid Samples

Solid samples should be as thin as possible and should be mounted on a conducting backing in order to minimize source charging. The sample should be dense to give high intensity and should be as free as possible from surface contamination.

Common methods of mounting powdered samples include pressing the sample into a copper mesh. A piece of tape on the back side of the mesh provides a sticky surface to which the sample can adhere. The sample can be pressed by means of a metal spatula held in the hand, by means of a mechanical or hydraulic press, or by means of a piece of hard steel that is placed on top of the sample and rapped hard with a hammer. The powder can be sprinkled directly onto a piece of tape, which is mounted on a metal strip, but care must be taken to see that the particles are not "rolled" in such a way that surfaces contaminated by the adhesive of the tape are exposed to the X-ray beam. Also, the powder can be slurried with an inert liquid, painted onto a metal strip, and dried. Highly stable materials such as metals can be vacuum-vaporized onto the backing material.

Large pieces or strips of solids can be mounted by means of tape or clamps or can be pulverized and handled as described above. The latter method is perhaps better, since it creates fresh surface areas. The sample itself can be large since it is never the object of the electron-optical system. Rather, a thin slit immediately in front of the sample defines the object.

Gas Samples

Gas samples are bled into the source chamber at a controlled rate. For higher pressure work (up to 1 torr) differential pumping can be used to maintain the spectrometer vacuum. At higher pressures it is also desirable to utilize an adjustable slit in addition to the source-defining slit in order to limit the gas leakage into the main tank of the spectrometer. Even without differential pumping and extra slits, gases can be introduced into source chambers at pressures up to several microns.

Gas samples can be frozen onto a metal plate cooled by either liquid nitrogen or liquid helium. It is useful to utilize the same chamber for both gas and condensed-gas experiments so that differences in binding energies caused by solid-state effects can be detected.

Liquid Samples

Liquid samples cannot themselves be studied in the vacuum of the spectrometer unless they have extremely low vapor pressures. Liquids can be frozen onto the same metal finger used for gas analysis (subsection above). Vapor from the liquid is bled into the chamber through a capillary tube that terminates near the cold finger. Liquids can also be evaporated for gas-phase studies.

Figure 5.8 is a spectrum obtained at Vanderbilt from a liquid sample of siloxane [62]. The liquid was spread onto the surface of a copper mesh and analyzed in the liquid state. Typical widths of 2.0 eV were obtained for lines from solid samples of silicon compounds compared to widths of 1.6 eV for lines of the liquid sample.

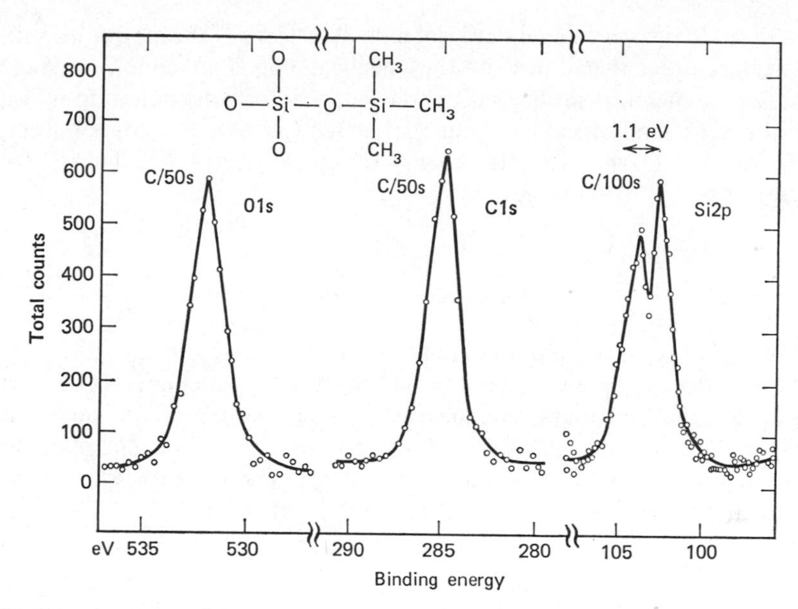

Fig. 5.8 A portion of the photoelectron spectrum of siloxane excited by means of magnesium K_α X-rays. The sample was a liquid spread onto the surface of a copper mesh.

Solutions

Kramer and Klein [63] have frozen water solutions onto gold sample plates in a dry-nitrogen atmosphere for photoelectron measurements. Their tentative results show that the frozen samples were not homogeneous in concentration.

5 APPLICATIONS

Determinations of Molecular Structure

One can use experimentally measured binding energies to deduce certain features of molecular structure. Use is made of empirical correlations between the measured binding energies and some other quantity that is used to describe structure. In photoelectron spectroscopy, binding energies have been correlated with oxidation numbers, calculated charges, and thermochemical data. This approach is similar to that used in nuclear-magnetic resonance spectroscopy, where correlations between chemical shifts and functional groups are utilized [64], and in Mössbauer spectroscopy, where correlations are drawn between isomer shifts and the partial ionic character of chemical bonds [65].

The chemical shifts observed in photoelectron spectroscopy can be interpreted in terms of the charge distributions within the molecules. The charge distributions, in turn, are dependent upon the structures of the molecules and, in particular, upon the nature of the chemical bonds. A simple classical explanation of chemical shifts pictures the valence electrons of an atom as a spherical charge shell of radius r and charge $-q$ [66–68]. The contribution, U_v, of this shell to the potential energy of an electron, with charge $-e$, brought to the shell from infinity is

$$U_v = \frac{e}{4\pi\epsilon_0}\int_{\infty}^{r}\frac{q\,dr'}{r'^2} = \frac{e}{4\pi\epsilon_0}\frac{q}{r}, \tag{5.13}$$

where the arbitrary zero potential energy is taken as $U_{v(\infty)} = 0$.

If the electron is taken inside the valence shell to an inner orbital at radius $r_0 < r$, the contribution of the valence shell to the potential energy is constant at $U_v = (e/4\pi\epsilon_0)(q/r)$. Therefore if for any reason the charge of the valence shell is changed by an amount Δq, the potential energies, U, of all electrons within the valence shell will alter by the amount

$$\Delta U = \frac{e}{4\pi\epsilon_0}\frac{\Delta q}{r}. \tag{5.14}$$

This change in potential energy is observed as a change in binding energy and is related to Δq in such a way that a decrease in q (e.g., the removal of a valence electron) will result in an increase in the binding energy, (i.e., the binding energy becomes more negative).

We can interpret Δq in (5.14) as the net charge on an atom when the atom forms a chemical bond that is partially or wholly ionic in character. This treatment is crude; however, it does predict the proper direction of the observed shifts and the fact that all inner shells are shifted by approximately the same amount. In addition, (5.14) predicts an upper limit of $\sim$14 eV for binding energy shifts caused by the complete removal of an electron from a valence shell of initial radius $r = 1$ Å.

This approach does not account for the fact that in a partially covalent bond the bonding electrons are not removed from the atom but reside in bonding orbitals in the vicinity of the atom; nor does it take into account the shifts in binding energies caused by charges on neighboring atoms. To account for these effects it would be necessary to recast (5.14) in the form

$$\Delta U = \Delta U\ (\text{free ion}) + \Delta U\ (\text{molecular potential}), \tag{5.15}$$

where ΔU (free ion) is the shift given by (5.14) for a free ion and ΔU (molecular potential) is the shift caused by the electronic and nuclear charges distributed throughout the molecule. The free-ion shifts can be calculated also by

Hartree-Fock [68] and Hartree-Fock-Slater [7] methods. Fadley et al. [68] have shown that for the halogens and for europium, the binding-energy shifts given by (5.14) are several electron volts larger than those calculated by a Hartree-Fock program.

If the atom of interest occupies a position in a crystal lattice, it is necessary to write

$$\Delta U = \Delta U \text{ (free ion)} + \Delta U \text{ (crystal potential)}, \tag{5.16}$$

where ΔU (crystal potential) is the shift caused by the charges distributed throughout the crystal lattice. Calculations of ΔU (crystal potential) are similar to calculations of crystal energies and have been discussed in detail by both Siegbahn et al. [7] and Fadley et al. [68]. Unless the crystal structure is known accurately and unless the compound has a high degree of ionic bonding, the calculation of ΔU (crystal potential) gives only an approximate result. Fadley et al. [68] have demonstrated the use of (5.15) for a case in which it was not possible to calculate ΔU (crystal potential) for (5.16).

Equations (5.15) and (5.16) introduce considerable complexity into the notion of binding-energy-versus-charge correlation curves; nevertheless it has been possible for a number of workers to establish correlation curves by using only the concepts embodied in (5.14), that is, by ignoring molecular and crystal potentials. The correlations so obtained are better than one might expect from such an approximate use of a classical approach. Some of these correlations are now discussed as examples of the technique.

Since the "charge on an atom" within a compound is related to the oxidation number of the element in the compound, attempts have been made to correlate observed binding-energy shifts to oxidation numbers. Such correlations have been demonstrated for some sulfur compounds [66, 69, 70] and for some chlorine compounds [70, 71]; however, the oxidation number does not give a good approximation to the charge, since it ignores any covalent character of the chemical bond. If the oxidation number were a true measure of charge, (5.14) would predict a shift of ~14 eV for a unit change in the oxidation number. Such a large shift of inner-shell binding energies per unit oxidation has been demonstrated only for the special case of the oxidation of Eu^{+2} to Eu^{+3}, where shifts of ~10 eV were observed [68]. More typical is the case of iodine, where the inner levels shift by only about 6 eV between the -1 state (I^-) and the $+7$ state (IO_4^-), or by only about 0.8 eV per unit change in oxidation number [68].

Figure 5.9 is a correlation curve established by Nordberg et al. [72] for a series of nitrogen compounds. The charges were calculated by the method of Pauling [73], which uses concepts of electronegativity, partial ionic character, and bond resonance. The electronegativities, χ, used were those established by Pauling on the basis of single-bond energies. The partial ionic character,

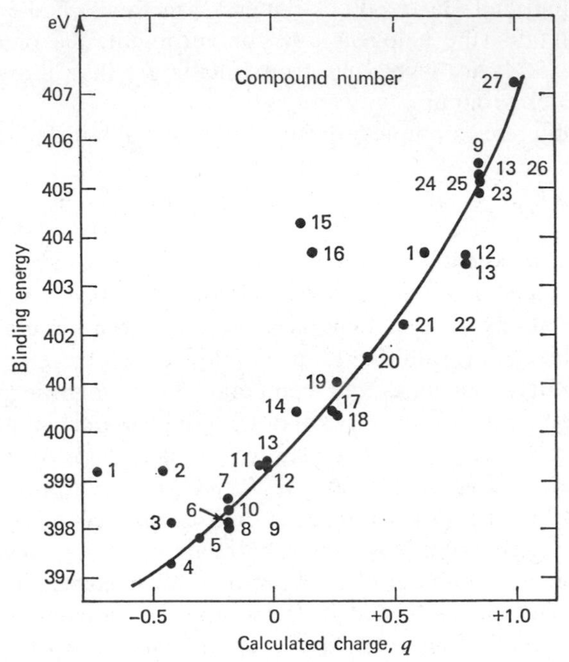

Fig. 5.9 Binding-energy versus charge correlation curve for the 1*s* level in nitrogen. The charges were calculated by the method of Pauling, as explained in the text. The curve is reproduced from [72].

I, of a bond between atom *A* and atom *B* was deduced from Pauling's empirical equation

$$I = 1 - e^{-0.25(\chi_A - \chi_B)^2} \tag{5.17}$$

which is based on the dipole moments of simple halides.

The method of calculation is best illustrated by an example. Compound 20 in Fig. 5.9 is tetramethylammonium chloride, in which the nitrogen has a formal positive charge:

$$\left[\begin{array}{c} H_3C \quad\; CH_3 \\ \diagdown\overset{\oplus}{N}\diagup \\ \diagup\quad\diagdown \\ H_3C \quad\; CH_3 \end{array}\right] Cl^-$$

The electronegativities of carbon, nitrogen, and oxygen are 2.5, 3.0, and 3.5, respectively. The normal positive charge on the nitrogen will give it an even larger electronegativity. Pauling [73] has estimated that a formal charge of one unit will shift the electronegativity two-thirds of the distance to the next element in the periodic table. Thus the electronegativity of nitrogen is

$[3.0 + \frac{2}{3}(3.5 - 3.0)] = 3.3$. The partial ionic character of the C–N bonds is $I = 1 - e^{-0.25(3.3-2.5)^2} = 0.15$. Since there are four such bonds, nitrogen will acquire a charge of $-4(0.15) = -0.60$, which will add to the formal charge of $+1.0$ to give a net calculated charge of $+0.40$.

If different resonance structures are possible, the calculated charge for each structure is determined and the weighted average is taken. The weighting factors are proportional to the contribution each structure makes to the final structure which is determined from concepts such as the electroneutrality rule and the adjacent charge rule [73]. Conversely a measurement of chemical shifts may allow one to determine the contributions of different resonance structures. For example, in the work cited above Norberg et al. [72] were able to estimate the contributions of conjugated or resonance structures to the structure of aniline, 4-amino-4′-nitroazobenzene, 4-(acetamidocarboxy-methyl)-1,2-dithiolane, and to compounds of the general type $RHNSO_2R'$.

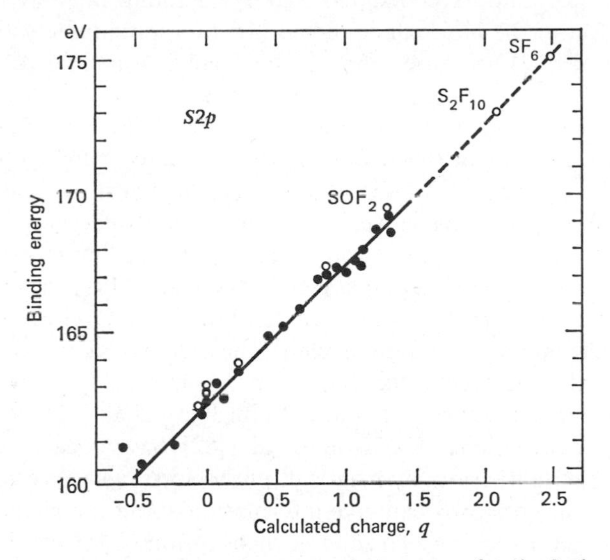

Fig. 5.10 Binding-energy versus charge correlation curve for the $2p$ level of sulfur. The charges were calculated by the method of Pauling. The curve is reproduced from [74].

The correlation curves shown in Fig. 5.9 for nitrogen and in Fig. 5.10 for sulfur [74] illustrate that correlations can be established by means of the concept of electronegativity. However, the fact that the charge calculations are very approximate and the fact that not all of the points fall on the smooth curves have caused investigators to try the use of more sophisticated approaches to the concept of electronegativity and the use of charges obtained by quantum-mechanical calculations.

Nordberg and others [75] have used the concepts of orbital [76, 77] and group [78] electronegativities in their analysis of shifts in the carbon $1s$ binding energy. When the Pauling method of calculating charges was used, the orbital electronegativities of Hinze and Jaffé [76] were used to correct for the considerable s-character of chlorine bound to carbon. Also, the authors were able to fit their data to a correlation curve by assigning group electronegativities and group charges to about 40 chemical groups. The calculated charge on a carbon atom was taken to be the sum of the group charges of the groups attached to the carbon atom. The group charges were assigned to give the best least-squares fit of shift versus calculated charge. Nordberg et al. [75] also attempted to fit the data to a correlation curve by calculating the charges by means of the extended Hückel method; however, the correlation in this case was less good than that obtained when Pauling's method was used, unless the charges were corrected by an additive term representing the molecular potential. The authors concluded that the Pauling method somehow includes the effect of the molecular potential in an approximate way.

Nordberg et al. [72] have used the extended Hückel method to calculate charges for many of the compounds of Fig. 5.9 and have obtained a linear correlation curve with very little scatter. Hollander et al. [79] have shown that correlation data for nitrogen fall on two separate straight lines if the charges are calculated from CNDO molecular-orbital eigenfunctions. Pelavin et al. [80] have measured phosphorus $2p$ electron binding energies for a large number of compounds and have attempted to correlate the data to atomic charges calculated by several methods. Hendrickson et al. [81] have measured for a series of compounds the binding energies of the $1s$ level of boron and the $3p$ level of chromium and have correlated these data to charges calculated by CNDO and extended-Hückel techniques. In the same work [81] the authors were able to correlate the carbon $1s$ data of Nordberg et al. [75] by means of CNDO charge calculations. Karlsson et al. [82] have measured charge distribution in xenon fluorides by means of photoelectron spectroscopy.

Photoelectron spectroscopy can give information about the phenomena of back donation and π-bonding. The charge distribution within certain species are in part determined by back donation of d-electrons, as in $p_\pi - d_\pi$ bonding, and the binding-energy shifts caused by the back donation may be large enough to be detected by photoelectron spectroscopy. At Vanderbilt workers recently established a correlation curve for the silicon $2p$ binding energy in 16 compounds [62]. A straight line was drawn through four points representing compounds for which little $p_\pi - d_\pi$ bonding is expected, and deviations from this line (all but two of the remaining points fall below the line) were explained in terms of back donation to the silicon p orbitals. Some *ab initio* LCOA-MO-SCF calculations were done on silicon, phosphorus, and sulfur atoms from which it was deduced that the back donation effects the silicon $2p$ orbital energy to the extent of $\sim$ 5 eV for each electron back bonded. On the basis of these assumptions the data could be explained in a reasonable and consistent manner. In addition, silicon, phosphorus, and sulfur were compared to one another by plotting the $2p$ binding energies of one of these elements against those of another for series of compounds in which the surrounding atoms are of the same kind and number. These plots were linear with slopes about equal to those obtained when corresponding plots were made of the binding energies derived from the *ab initio* calculations (see Fig. 5.11).

In another study performed at Vanderbilt [82a], evidence for π-bonding in complexes of triphenylphosphine was obtained. The phosphorous $2p$ binding energy in the coordination compounds $M[P(C_6H_5)_3]_2Cl_2$ for M = Ni, Pd, and Cd was measured to be 131.6 ± 0.2 eV, compared to 131.9 ± 0.4 eV for the triphenylphosphine. The fact that the value for the complexes was not greater than the value for the triphenylphosphine was interpreted as evidence of $p_\pi - d_\pi$ bonding.

The simple charge-correlation approach applied to solid samples suffers from the fact that the effect of the crystal potential on the binding energies is not taken into account. Fadley et al. [68] have treated the problem of crystal potential by considering the ionization of an atom within a crystal lattice as the sum of three distinct steps.

1. The atom of interest, A, with its charge, Z, is removed from the crystal lattice. This process requires energy E_1. The charge, Z, is integral for ionic compounds and is determined empirically for other compounds.
2. A core electron in the ith level is removed from the ion A^Z, producing ion A^{Z+1}. The energy for this process is the free-ion binding energy for the ith level, $E_B(A, i, Z)$.
3. The ion A^{Z+1} is placed back into the lattice. This process requires energy E_2.

The sum of these three processes is equivalent to the ionization, in level i, of atom A within the crystal lattice

$$E_B(A, i, \text{in lattice}) = E_B(A, i, Z) + E_1 + E_2.$$

The free-ion binding energies were assumed to be the Hartree-Fock eigenvalues of atomic orbitals expanded in terms of Slater-type functions. The energies E_1 and E_2 were taken to be the point-charge Coloumb energy (the Madelung energy) for the ions A^Z and A^{Z+1}, respectively. The authors [68] concluded that for ionic solids, the Madelung energy must always be considered. Seigbahn et al. [83] have calculated Madelung contributions to binding-energy shifts for some sulfur compounds but saw no experimental evidence for them.

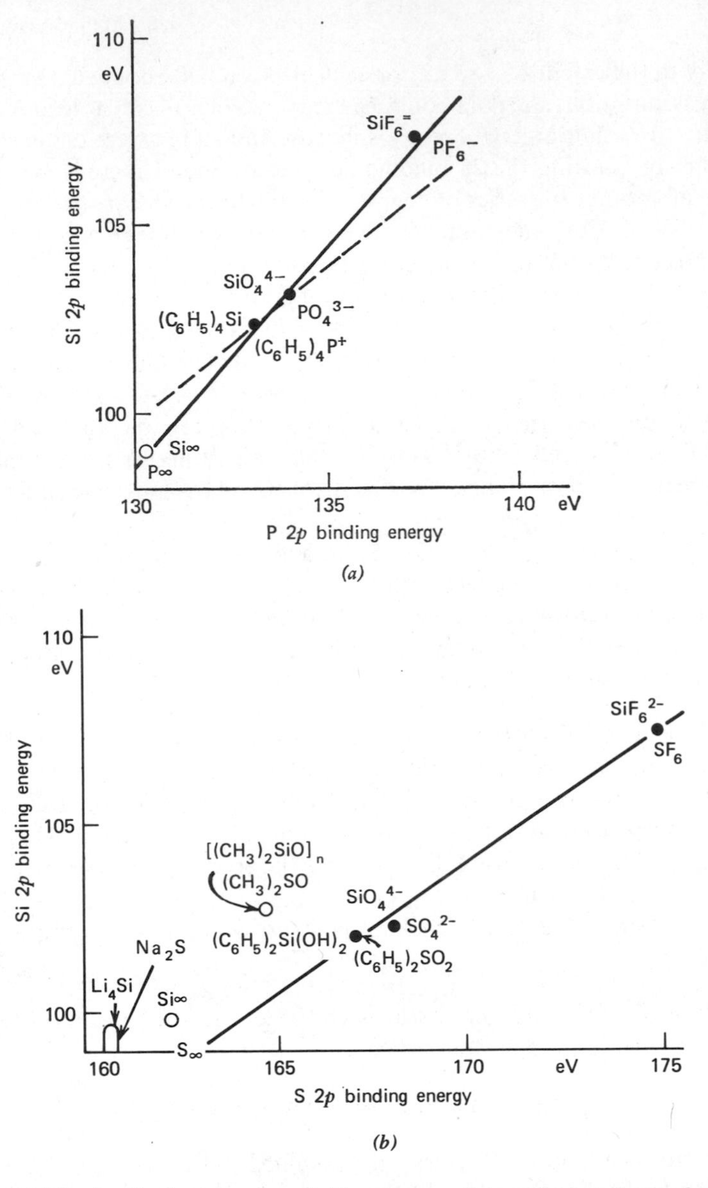

Fig. 5.11 (*a*) The 2*p* binding energies of silicon versus those of phosphorus for compounds having the same number, kind, and arrangement of nearest-neighbor atoms (solid points). The open point in this figure corresponds to a structure that is closely related but does not fulfill all of the specifications for the solid points. The solid line has been drawn through the experimental points, and the dotted line gives the slope corresponding to calculated 2*p* binding energies of various atomic configurations. (*b*) The 2*p* binding energies of silicon versus those of sulfur for compounds having the same number, kind, and arrangement of nearest-neighbor atoms (solid points). The open points correspond to structures that are closely related but do not fulfill all of the specifications for the solid points. The line has the same slope as the line drawn through the points in a similar plot of silicon and sulfur 2*p* binding energies calculated for various atomic configurations.

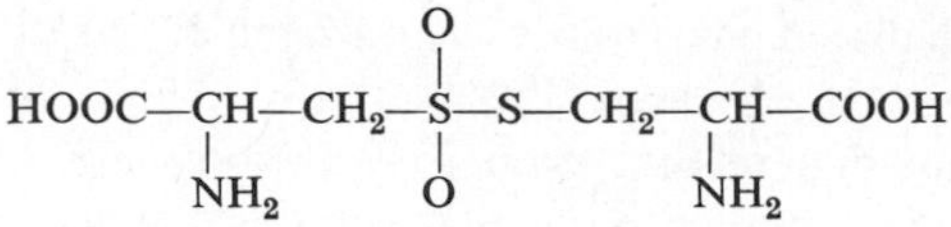

$$\mathrm{HOOC{-}\underset{\displaystyle NH_2}{\underset{|}{C}H}{-}CH_2{-}\overset{\displaystyle O}{\overset{|}{\underset{\displaystyle O}{\underset{|}{S}}}}{-}S{-}CH_2{-}\underset{\displaystyle NH_2}{\underset{|}{C}H}{-}COOH}$$

(*a*)

$$\mathrm{HOOC{-}\underset{\displaystyle NH_2}{\underset{|}{C}H}{-}CH_2{-}\overset{\displaystyle O}{\overset{|}{S}}{-}\overset{\displaystyle O}{\overset{|}{S}}{-}CH_2{-}\underset{\displaystyle NH_2}{\underset{|}{C}H}{-}COOH}$$

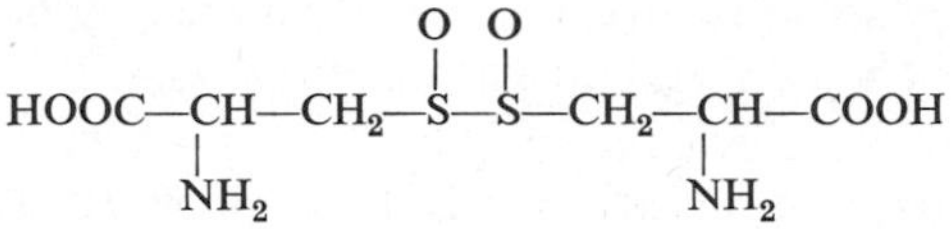

(*b*)

Fig. 5.12 The two possible structures of cystine S-dioxide. (*a*) Thiosulfonate structure. (*b*) Disulfoxide structure.

It is not always necessary for one to establish detailed correlation curves in order to determine certain features of a structure. The number of photoelectron lines observed or the width of a single line can indicate if two atoms of the same element occupy equivalent or nonequivalent position in the molecule. For example, the fact that cystine S-dioxide gives a double peak for the S2*p* photoelectron line allowed Axelson et al. [84] to conclude that the molecule has the thiosulfonate structure with nonequivalent sulfur atoms (Fig. 5.12*a*) instead of the α-disulfoxide structure with equivalent sulfur atoms (Fig. 5.12*b*).

Similarly, Jolly et al. [85] were able to determine the structure of the oxyhyponitrite ion, $N_2O_3^{=}$ by means of photoelectron spectroscopy. The three structures postulated for this ion are shown in Fig. 5.13. The photoelectron spectrum shows two nitrogen 1*s* lines, which rules out structure **1** in which the nitrogen atoms are equivalent. It was shown by both extended-Hückel and CNDO molecular orbital calculations that the charges on the nitrogen atoms in structure **2** are nearly equal to each other, but that for

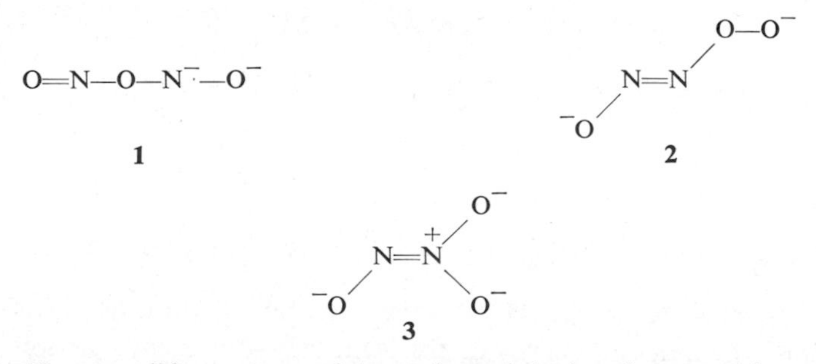

Fig. 5.13 Three possible structures of the oxyhyponitrite ion. Structure **1** has equivalent nitrogens. The calculated charges on the nitrogens in structure **2** are about equal, but in structure **3** the calculated charges differ by an amount sufficient to explain the difference in the energies of the two nitrogen 1*s* lines in the photoelectron spectrum.

structure **3** the calculated charges for the nitrogen atoms differ by an amount sufficient to explain the observed shift.

It should be noted here that even if two atoms are exactly equivalent, delocalization of the pertinent molecular orbitals may lead to nondegenerate orbital energies. For example, Schaad and Kinser [86] have calculated differences of a few tenths of an electron volt for the energies of the innermost orbitals of N_2 and C_2. In general such differences will be too small to be detectable by photoelectron spectroscopy.

Because the charge on an atom, which is not a measurable quantity, is difficult to estimate, Jolly and Hendrickson [87] have attempted to correlate binding energies to measurable thermochemical data. An example of their approach is their consideration of the photoejection of a 1*s* electron from nitrogen in solid sodium nitrite. They describe this process as the sum of three processes:

$$NaNO_2(s) \rightarrow Na^+(g) + NO_2^-(g)$$

$$NO_2^-(g) \rightarrow NO_2^*(g) + e^-(g)$$

$$NO_2^*(g) \rightarrow NO_2^*(s),$$

where (*s*) and (*g*) refer to solid phase and gas phase, respectively, and the asterisk indicates an inner vacancy.

To put these equations into a form that can utilize thermochemical data, Jolly and Hendrickson [87] assumed that atomic cores that have the same charge are chemically equivalent. Thus N^{+*}, which has a core charge of +6 (seven protons and one electron), is assumed to be equivalent to O^+ that also has a core charge of +6 (eight protons and two electrons). In both cases there are five valence electrons and in both cases these valence electrons see the same core charge. Under this assumption NO_2^* is equivalent to O_3, and the equations are written as

$$NaNO_2(s) \rightarrow Na^+(g) + NO_2^-(g)$$

$$O^+(g) + NO_2^- \rightarrow N^{+*}(g) + O_3(g) + e^-(g)$$

$$2O_3(g) \rightarrow 2O_3(s).$$

The sum of these equations is

$$NaNO_2(s) + O^+(g) + O_3(g) \rightarrow 2O_3(s) + N^{+*}(g) + Na^+(g) + e^-(g),$$

which is assumed to be equivalent to the photoionization of $NaNO_2$ in the solid state. The reaction energy of this "thermo-equivalent reaction" can be evaluated from thermochemical data. Since the species $N^{+*}(g)$ and $O^+(g)$ appear in the thermoequivalent equation for any process involving the photoionization of nitrogen, their energies of formation are not required if one considers only differences in reaction energies. Thus the reaction energies, E_T, were calculated by ignoring $N^{+*}(g)$ and $O^{+}(g)$.

Since the measured binding energies, E_B, are referred to the Fermi level and not the vacuum level, the calculated E_T values should also be corrected for the work function, ϕ, of the compound. Since work functions are not known for many compounds, it was assumed that this correction would be small and would be approximately constant for similar materials, and it too was ignored.

Jolly and Hendrickson [87] calculated E_T for several photoionization processes involving nitrogen and measured the corresponding binding energies, E_B. A plot of $E_B - \gamma$ versus $E_T - \beta$ where γ and β are adjustable parameters, gives a straight line with slope unity and with very little scatter to the points. Data were plotted also for a very small number of boron, carbon, and iodine compounds, with similar results.

Investigations of Molecular Orbitals and of Excited States of Singly-Ionized Molecules

Binding energy-charge correlation curves are useful for determinations of certain features of molecular structure; however, a proper description of molecular structure can be written only in the language of quantum mechanics, and it is not until the measured binding energies are related directly to the results of quantum calculations that the full value of the experimental data is utilized. The desired description of molecular structure is the listing of accurate molecular-orbital wave functions, and the determination of these wave functions is a difficult theoretical problem. Fortunately, the application of quantum mechanics to chemical problems has experienced great growth during the last decade because of the availability of large computers to handle complex numerical analysis. (For example, *ab initio* calculations of the ground state of pyridine have been made [88].) However, many calculations and checks against experiment are required before it can be presumed that an adequate approximation to a molecular wave function has been deduced. For simple molecules such quantities as orbital energies, total energies, dipole moments, quadrupole moments, field gradients, and potentials can be calculated by means of molecular orbital theory and can be checked against experimental data [89]. Krauss [90] has published a compendium of *ab initio* calculations of molecular energies and other properties. Thus, the binding energies measured by photoelectron spectroscopy can be used to assess the validity of the theoretical approach.

The binding energy of an electron within a molecule is defined as the negative of the work required to move the electron from its orbital to a position of rest an infinite distance from the molecule. Since the electron in its final state has no kinetic or potential energy, the binding energy is equal to the difference in the energies of the initial and final states of the molecule:

$$E_B = E_{\text{molecule}} - E_{\text{ion}}, \tag{5.18}$$

where E_B is the binding energy, E_{molecule} is the total electronic energy of the molecule, and E_{ion} is the total electronic energy of the molecular ion.

It is possible by means of self-consistent-field methods to calculate the energies E_{molecule} and E_{ion} for simple molecules and for atoms. E_{ion} can be calculated from the wave functions used to determine E_{molecule}, or E_{ion} can be calculated by means of separate SCF calculation. Koopmans' theorem [91–94] states that if the general Hartree-Fock method is used and if the same orbitals used to determine E_{molecule} are used to determine E_{ion}, then the binding energies are equal to the negative of the one-electron eigenvalues for the molecule. Thus within the applicability of Koopmans' theorem, a direct comparison can be made between the binding energies measured by means of photoelectron spectroscopy and the one-electron eigenvalues calculated by theory.

Koopmans' theorem assumes that the same orbitals can be used to describe both the molecule and the molecular ion. It seems intuitively obvious that this assumption is not valid, at least for inner-shell ionization, in which case a screening electron is removed. Lindgren [95], Gianturco and Coulson [96], and Fadley et al. [68] have shown that for atoms and free ions orbital-energy eigenvalues are significantly (a few percent) higher than the differences between initial- and final-state energies determined by separate SCF calculations. The nonapplicability of Koopmans' theorem to outer molecular orbitals of some organic compounds [97] and metal complexes [98] has been reported also.

Even if Koopmans' theorem were applicable or even if separate SCF calculations were made for the ion, the calculated binding energies would not be exactly correct because the general Hartree-Fock method does not include relativistic and correlation effects. Ignoring these introduces no error if the effects are the same size in the molecule and ion; however, in general, such is not the case. Verhaegen et al. [99] and Richards and Wilson [100] have made estimates of differences in correlation energies in specific cases. Hollister and Sinanoğlu [101], and Phillips [102] have attempted to develop general methods for including correlation.

In general the correlation energy of the ion is less than that of the molecule since the ion has one less electron. Therefore the neglect of correlation tends to make calculated binding energies smaller than the true values. On the other hand any error due to the nonapplicability of Koopmans' theorem is in the direction that makes the calculated energy of the ion too large, and as a result, the calculated binding energy is too large. In some cases these two errors may cancel each other, with the result that binding energies calculated on the assumption of Koopmans' theorem agree better with experimental values than do binding energies calculated by separate SCF calculations for molecule and ion [103]. (See Fig. 5.14.) Bagus [104] has reported that for cases involving the removal of an outermost *s*-electron, the correlation energy of the ion can be *larger* than that of the molecule.

For cases in which binding-energy shifts for the inner levels of an element in several different compounds are measured, it might be expected that the

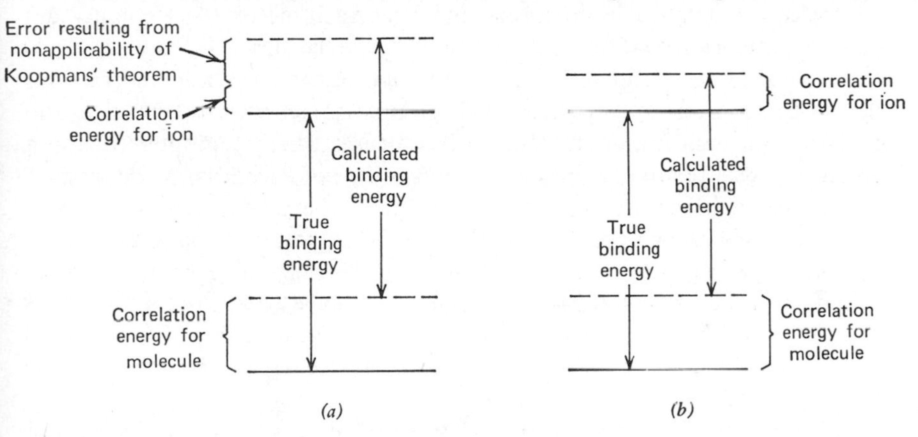

Fig. 5.14 A schematic representation of the true (solid lines) and calculated (dotted lines) energies for a molecule and its ion. The energy of the ion is calculated under the assumption of Koopmans' theorem *(a)* and by means of a separate SCF calculation *(b)*. In *(a)* the error introduced by the nonapplicability of Koopmans' theorem cancels the error from correlation energies with the result that *(a)* gives better agreement between calculated and true binding energies than does *(b)*. Such cancellation can occur in specific cases, but it is not general.

errors due to the nonapplicability of Koopmans' theorem (and perhaps errors due to relativistic effects and correlation) would be the same in all of the compounds, hence would cancel when differences in binding energies are considered. If these errors are associated primarily with "atomic orbitals" localized around the atom of interest, such cancellation might occur; however, if they are associated with the electronic structure of the entire molecule, or entire crystal lattice, the size of the effect might be different for different compounds. Fadley et al. [68] have estimated for some specific iodine compounds that differences related to the nonapplicability of Koopmans' theorem might be of the order of 1 eV, which is not small in comparison to observed chemical shifts. Basch and Snyder [105] have calculated binding-energy shifts for the 1*s* level of carbon, nitrogen, oxygen, and fluorine in some small molecules on the assumption that errors related to correlation, to relativity, and to Koopmans' theorem are independent of an atom's molecular environment. The authors were unable to assess the validity of their approach because of the lack of experimental data for comparison.

The limitations of Koopmans' theorem discussed above apply to all molecules. For the specific case of open-shell molecules additional complications arise because of cross terms in the energy expression and because it is impossible to interpret the theoretical results in clear physical terms [106]. Richards [106] has discussed the applicability of Koopmans' theorem to photoelectron data for both closed- and open-shell molecules. He concludes that separate Hartree-Fock calculations should be made for the molecule and for each state of the ion, and that relativistic and correlation corrections should be estimated and applied before comparisons between theoretical and experimental data are attempted. In those cases where computer time and operating cost are not prohibitive the suggestions of Richards can be attempted; however, in more complicated cases calculations of eigenvalues can be made, and Koopmans' theorem can be used to provide a rough test of the theoretical values. Recently the differences between Koopmans'-theorem values and values calculated from total energies have been expressed in terms of a polarization potential by Hedin and Johansson [106a]. The potential can be calculated and used to correct the binding energies derived under the assumption of Koopmans' theorem.

Another aspect of the photoelectric process that is of interest in regard to the interpretation of photoelectron spectra is the quantum mechanical theory of the sudden approximation. In time-dependent perturbation theory there are two extreme cases: the adiabatic perturbation and the sudden perturbation [107]. In the former case the perturbation is built up so slowly that regardless of the final magnitude of the perturbation, the system is able to adjust itself continuously so that it always remains in an eigenstate of the perturbed system. In the latter case the perturbation acts so quickly that the system finds itself in a nonstationary state that is a superposition of the eigenstates of the final Hamiltonian of the system.

The expulsion of a photoelectron from a molecule alters the potential field of the remaining electrons. If the eigenstates of the original molecule are u_n and the eigenstates of the molecular ion are v_m, then the molecular ion will be formed in a stationary state, v_f; if the process is adiabatic, and in a nonstationary state,

$$\psi = \sum_m C_m v_m e^{-iE_m t/\hbar} \tag{5.19}$$

if the process is sudden. In the latter case the coefficients are given by

$$C_m = \int v_m^* u_0 \, d\tau, \tag{5.20}$$

where u_0 is the initial state. The probability of measuring E_m as the energy of the final state is $|C_m|^2$.

The condition required for the sudden approximation to be valid [107] is

$$\frac{\tau(\epsilon_m - \epsilon_0)}{\hbar} \ll 1, \tag{5.21}$$

where τ is the time interval over which the perturbation occurs and ϵ_0 and ϵ_m are the initial- and final-state energies, respectively. Carlson and Krause [108] have concluded that for the K-shell photoionization of neon, the sudden approximation is valid if $\tau(\epsilon_m - \epsilon_0)/\hbar < 0.2$. The photoelectric process occurs in a time interval, τ, of the order r/v, where r is the atomic radius and v is the velocity of the photoelectron. For a 1000-eV photoelectron $r/v \sim 5 \times 10^{-18}$ sec; therefore, $\tau(\epsilon_m - \epsilon_0)/\hbar$ is less than 0.2, if $(\epsilon_m - \epsilon_0)$ is less than 20 eV. One can conclude that under these conditions the molecular ions can be created with their outer electrons excited by some tens of electron volts, which includes states in the continuum. This is the double excitation (electron shake off) mentioned in the introduction. Shake-off excitation of inner shells also occurs, but the probability is low because the condition for the sudden approximation is not met and because the overlap between the initial bound state and the final excited state is not large. Carlson and coworkers have reported on both the experimental and theoretical aspects of electron shake-off [107–109].

If the molecular ions are formed in different states of electronic excitation, the photoelectrons will be ejected in groups with different kinetic energies. With sufficiently high resolution these groups can be resolved into separate lines in the photoelectron spectrum. Siegbahn et al. [8] have interpreted in this manner spectral lines seen in gas-phase studies.

It can be shown also that the sudden approximation is valid in regard to the vibrational excitations of the molecular ions. Thus in gas-phase photoelectron spectroscopy the ions will be formed in different vibrational states and the photoelectrons will be emitted in groups of different energies, which reflect the vibrational band structure of the molecular ion. This phenomenon is discussed more fully below.

It should be noted that, contrary to what has become common practice, we have discussed the applicability of Koopmans' theorem and the concept of the sudden approximation as quite separate topics. This is because Koopmans' theorem refers only to the nature of the orbitals used to describe the final state, and these orbitals do not depend upon the transition rate. The observed final state of the ion must be an eigenstate of the Hamiltonian of the ion regardless of the transition rate. Koopmans' theorem is concerned with the orbitals used to describe this state, and the sudden approximation deals with the problem of whether the final state will be the ground state of the ion or one of several excited states of the ion.

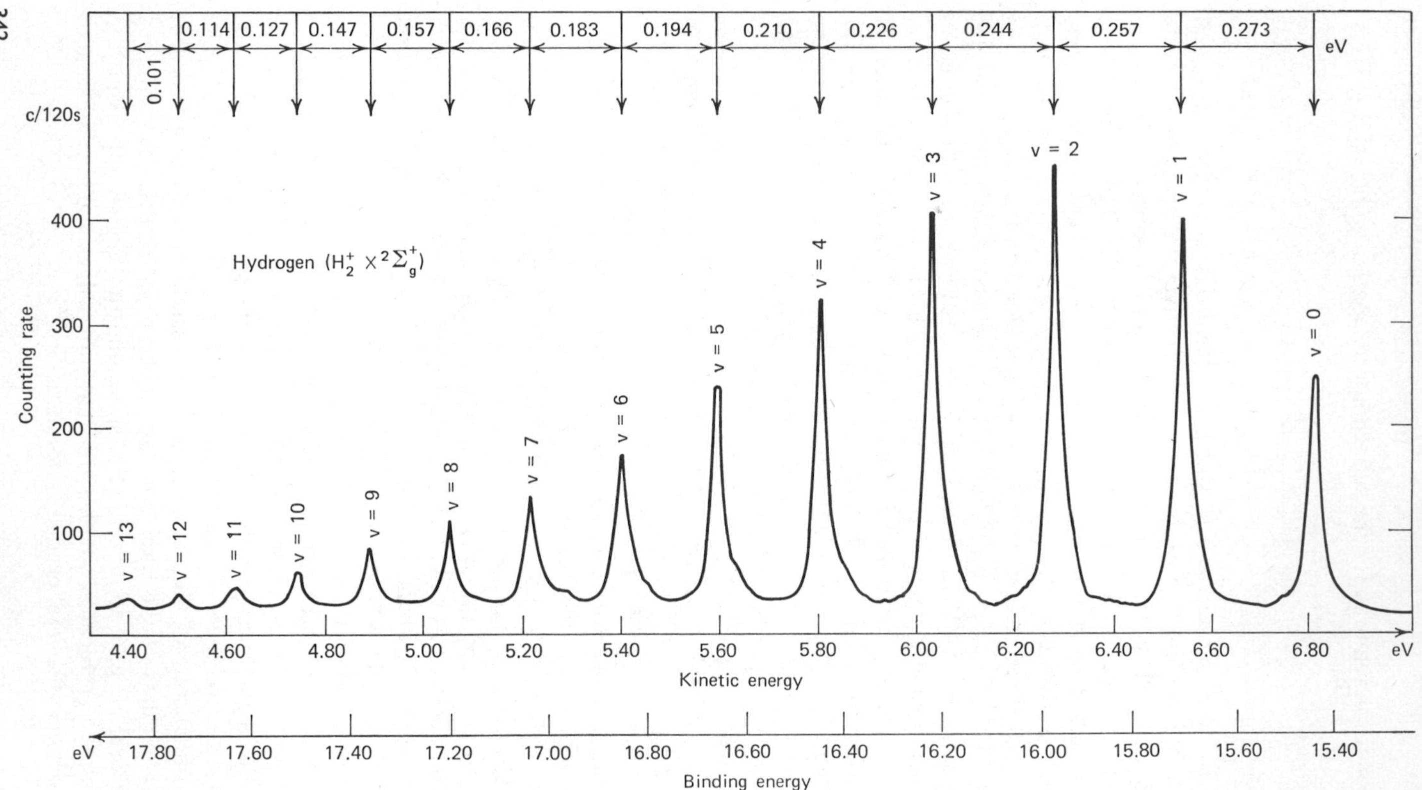

Fig. 5.15 The electron spectrum from molecular hydrogen excited by helium-resonance radiation. The different peaks correspond to different vibrational levels of the molecular ion. The broadening of the lines at the base is rotational structure. This curve is the data of Siegbahn et al. [113].

For the purpose of testing theory the binding energies of all the orbitals of a molecular species should be measured; thus the use of several radiations of different energies is required. A necessary condition for ionization of a level is that $h\nu > E_B$; however, since the instrumental contribution to the line width is proportional to the electron energy, it is an advantage if $h\nu$ is not much greater than E_B. Furthermore, for levels with $E_B < 21$ eV, excitation by means of the helium resonance line ($h\nu = 21.218$ eV) has the added advantage that the intrinsic width of the radiation is less than 0.005 eV [16], and the resulting electron line width is small enough so that in gas-phase studies vibrational structure can be resolved.

It is unfortunate that a convenient source of monochromatic radiation with $h\nu \sim 100$ eV is not available for study of molecular orbitals in the energy region of 20 to 100 eV. The He^+ resonance line with $h\nu = 40.812$ eV has been used [110] but it is of low intensity, and its energy is considerably less than 100 eV. Monochromatized synchrotron radiation [111, 112] is a possible source of excitation. Krause [112a] reports that the Mζ X-rays of yttrium and zirconium (133 and 151 eV, respectively) show some promise.

The helium resonance line has not been used often in solid-phase studies, but has been of great value for gas-phase studies of outer molecular orbitals. The information obtainable differs from that obtained in inner-shell studies in two regards; first, the narrow line widths make possible more accurate energy measurements and second, the vibrational structure that can be observed gives additional information about the nature of the orbits. If some of the excitation energy goes into vibrational degrees of freedom, the photoelectron will escape with less energy and the electron spectrum will show a series of lines, each line corresponding to a different mode of vibration. An example of vibrational spectra recorded by photoelectron spectroscopy is that for H_2^+ published by Siegbahn et al. [113]. It is reproduced in Fig. 5.15.

Figure 5.16 shows the potential energy curves for two diatomic molecules and their molecular ions as a function of the internuclear distance, R. In Fig. 5.16*a* the equilibrium internuclear distance is nearly the same for the molecule and the ion, whereas in Fig. 5.16*b* it is different. Shown also in the figure are some of the vibrational energy levels of the molecules and the ions, and depicted in a schematic way are the probability distributions, $\psi\psi^*$, for the vibrational states. The Franck-Condon principle states that for an electronic transition the electron configuration changes so fast that the positions and velocities of the nuclei do not change appreciably until after the transition. In view of this principle, transitions from the molecule to the molecular ion are presented in Fig. 5.16 by vertical lines (lines of constant R).

The electronic part of the transition probability is assumed to be constant and independent of the internuclear distance; therefore the transitions with

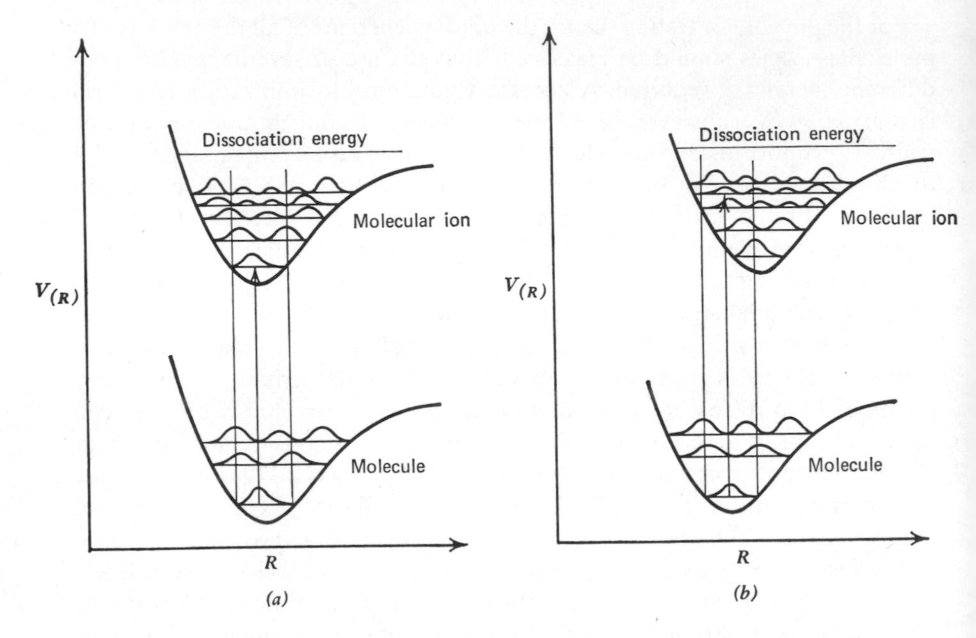

Fig. 5.16 Potential energy curves for diatomic molecules and their molecular ions as a function of the internuclear distance, R. Some vibrational energy levels and their probability distributions are shown. The vertical lines show the regions in which Franck-Condon transitions can occur. In (*a*) the equilibrium internuclear distance for the molecular ion is nearly the same as that for the molecule. In (*b*) the equilibrium distances are different.

the greatest probabilities are those between states whose vibrational eigenfunctions overlap the most. These preferred transitions are represented in diagrams such as Fig. 5.16 by vertical lines drawn from a maximum in the $\psi\psi^*$ distribution of a level on the lower potential curve to a maximum in the $\psi\psi^*$ distribution of a level on the upper curve. In photoelectron spectroscopy the molecule will nearly always be in its lowest vibrational state, for which the maximum occurs in the center.

The vertical lines in Fig. 5.16 show the regions in which Franck-Condon transitions can occur. For the higher vibrational quantum numbers $\psi\psi^*$ has large values only at the extreme values of R; thus the higher vibrational states can be reached only if the potential curve of the molecular ion is displaced appreciably from that of the molecule, as in Fig. 5.16*b*. Furthermore, in this case many vibrational lines are observed, since for the higher levels the density of states is large.

The relative positions of the potential curves along the R-axis are determined by the bonding character of the orbital from which the electron is ejected. If this orbital is nonbonding, one curve lies above the other, as in

Fig. 5.16*a*, and little vibrational structure is observed. If the orbital is bonding, the bond is weakened by the loss of the electron, and the new equilibrium position is toward larger values of R, as in Fig. 5.16*b*. In this case vibrational structure is observed, and the relative intensities of the lines are given by the Franck-Condon factors, $|\int \psi_{v'} \psi_{v''}\, dr|^2$. Here v' and v'' are the vibrational quantum numbers of the levels for the molecular ion and the molecule, respectively. If the orbital is antibonding, the upper curve is shifted to the left and again vibrational structure is expected in the photoelectron spectrum. Thus it is seen that the intensity distribution in the vibrational band structure of the photoelectron spectrum permits one to deduce the relative position of the molecular-ion potential curve, to determine Franck-Condon factors for ionization, and to characterize the orbital of the emitted electron.

If the exciting photon is of sufficient energy, the final state may be in the continuum above the dissociation limit, in which case the molecular ion dissociates and a continuous spectrum is observed. There will be, of course, more than one electronic state for the ionized molecule; thus several vibrational bands may be observed. If two of the electronic states cross, as shown in Fig. 5.17, predissociation may occur. In this case a vibrational level, a', of electronic state A' lies below the dissociation limit of A' but above the dissociation limit of electronic state B'. If certain selection rules are satisfied, a nonradiative transition can occur between a' and the continuum of B',

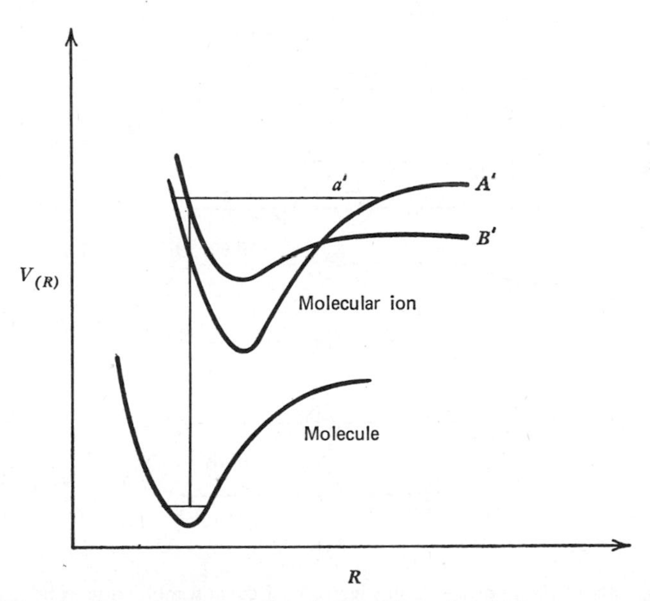

Fig. 5.17 The two electronic states, A' and B', of the molecular ion cross, making possible the phenomenon of predissociation.

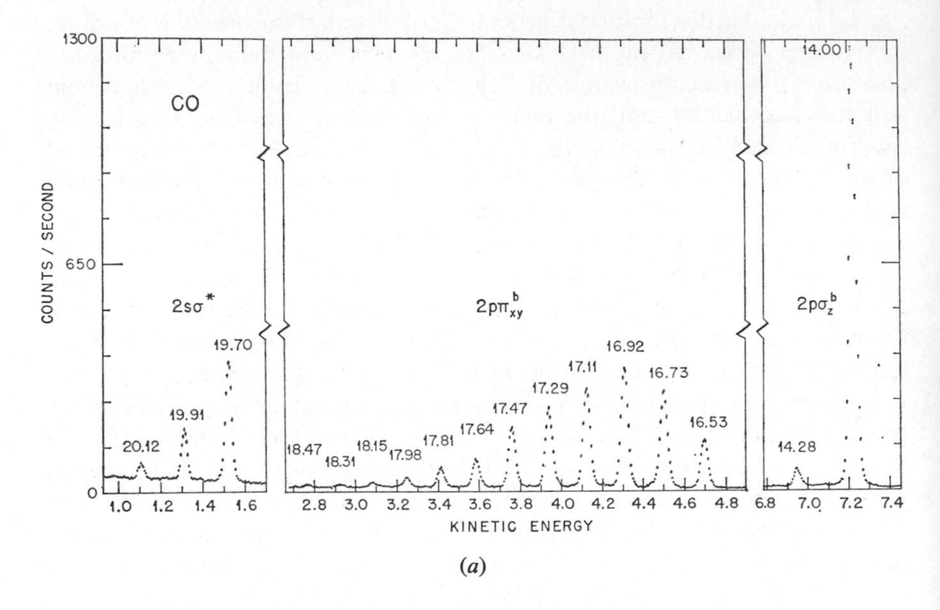

(*a*)

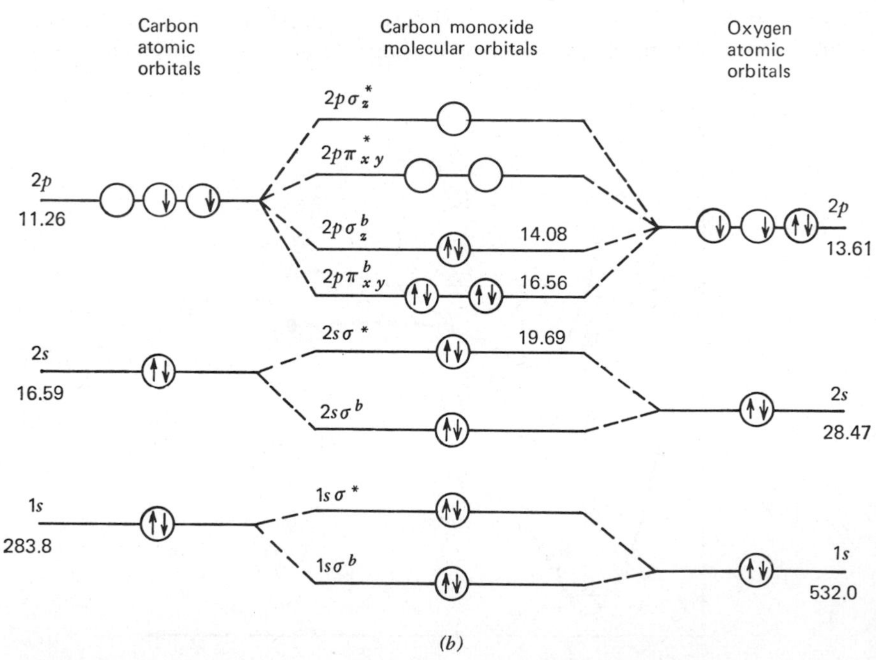

(*b*)

Fig. 5.18 (*a*) Part of the photoelectron spectrum of carbon monoxide excited by helium-resonance radiation, as recorded by Carlson et al. [29]. (*b*) The molecular orbital correlation diagram for carbon monoxide.

followed by dissociation. This additional mode of decay of state a' will shorten the lifetime of a' and, by the uncertainty principle, increase the width of the state and of the observed spectral line.

The basic principles discussed above for diatomic molecules are also true for polyatomic molecules, but the physical situation and the observed spectra are more complicated. Even so, valuable information can be obtained, including binding energies, vibrational structure, and the number, the ordering, the spacing, and the symmetries of molecular orbitals. In many cases the photoionization excites principally only one mode of vibration, and the spectra are less complicated than might be expected.

Examples of the vibrational spectra recorded by means of photoelectron spectroscopy are shown in Figs. 5.18 and 5.19. Fig. 5.18*a* is a portion of the helium-excited carbon monoxide spectrum recorded by Carlson et al. [29]. The molecular orbital correlation diagram for carbon monoxide is shown in Fig. 5.18*b*. The difference in the complexity of the spectra for the $2p\sigma_z^b$, the $2p\pi_{xy}^b$, and the $2s\sigma^*$ orbitals is clearly evident in the electron spectrum. Figure 5.19 is a spectrum of methyl fluoride recorded by Pullen et al. [114]. The broad line with no resolved vibrational structure is probably evidence of dissociation or predissociation [114].

In his review article Turner [9] has discussed in detail the spectra of several diatomic and polyatomic gases.

Observations of Level Splittings

Another interesting application of photoelectron spectroscopy is the measurement of the splitting of degenerate levels caused by electric-field gradients or exchange interactions. Such splittings caused by internal electric-field gradients have been observed in Mössbauer spectroscopy [115] and in internal conversion spectroscopy [116]. More recently, splittings of the $p_{3/2}$ levels in compounds of thorium, uranium, plutonium, and gold have been observed by means of photoelectron spectroscopy [117]. In the case of the gold compounds the splittings observed in the photoelectron spectrum were shown to be directly proportional to the Mössbauer quadrupole splittings [115].

Energy splittings of core-electron levels caused by exchange interaction with unfilled shells have been observed by means of photoelectron spectroscopy. Hedman et al. [118] have observed splittings of ~1 eV in the paramagnetic molecules O_2, NO, and NO_2, but they saw no splittings in diamagnetic N_2. The relative instensities of the two components of the split lines are in agreement with the statistical weighting factor $2j + 1$. Shirley et al. [119] have recently observed similar splittings in the core levels of manganese and iron compounds.

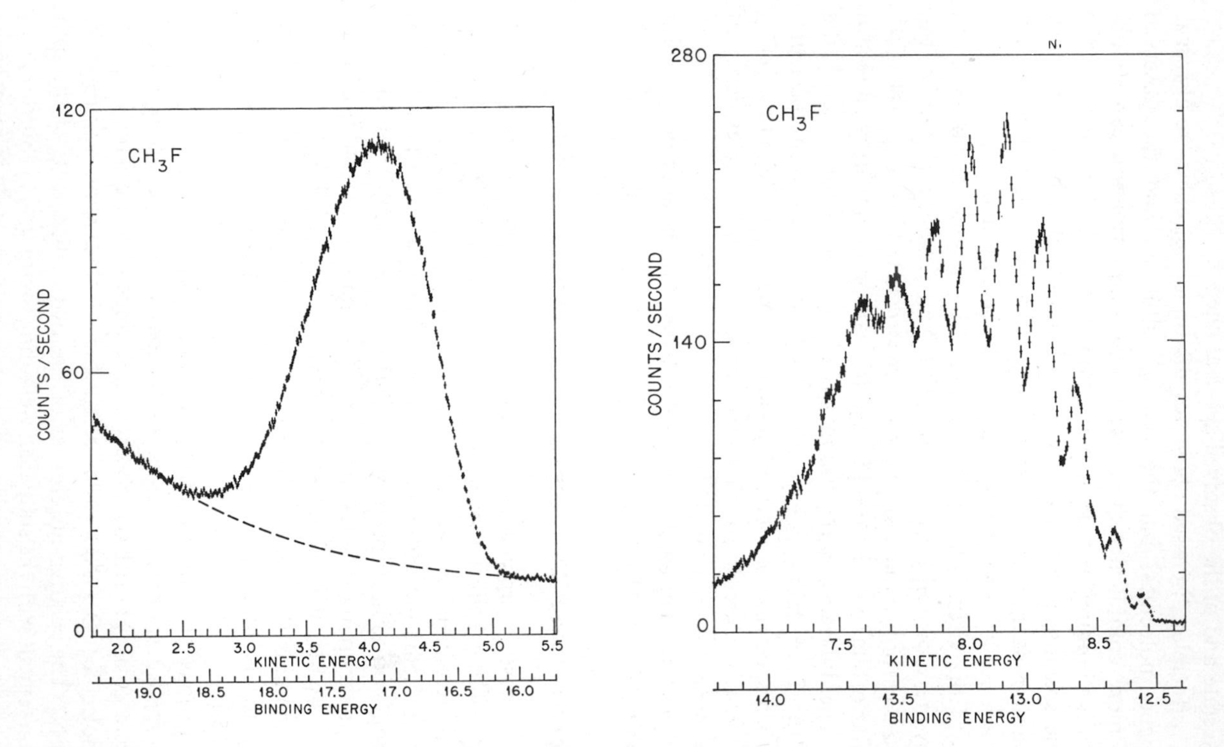

Fig. 5.19 Part of the photoelectron spectrum of methyl fluoride excited by helium-resonance radiation, as recorded by Pullen et al. [114]. The broad line with no resolved structure is probably the result of dissociation or predissociation.

Measurements of Densities of States

Fadley and Shirley [120] have used photoelectron spectroscopy to measure the density of electron states near the Fermi level in iron, cobalt, nickel, copper, and platinum. Their results are in general in good agreement with results of ion-neutralization and X-ray emission measurements, but in some cases they show disagreement with measurements made by means of ultraviolet photoemission spectroscopy.

Studies of Surface and Near-Surface Phenomena

Delgass [121] has shown that for the catalyst FeV_2O_4 the binding energy of the oxygen $1s$ level decreases by about 1 eV, and that of the vanadium level increases by about 1 eV when the catalyst is used to dehydrogenate cyclohexane. The binding energy of the iron $3p$ level does not change. Results such as these indicate that photoelectron spectroscopy may be of use for the examination of the surface states of catalysts and for the acquisition of information about the mechanism of catalysis.

The depth to which a solid is studied depends upon the range of the photoelectron, which in turn depends upon the energy of the ionizing photon and the absorption coefficient of the solid. Electron ranges can vary from a few angstroms for $\sim$300 eV electrons in platinum to 100 Å for $\sim$800-eV electrons in organic materials and to 1000 Å for $\sim$ 1000-eV electrons in light elements. The limited range of the electrons make possible surface studies; however, the situation is not ideal since the range, hence the depth of surface studies, is difficult to determine accurately. Furthermore, it is known that the absorption of gases and oils on the sample affects the results. To date, solids have been studied in vacuum systems capable of 10^{-6}, or at best 10^{-8}, torr, and these vacuums have not been especially clean. It will be interesting to discover what differences are observed in the spectra of solids when vacuums of 10^{-10} or 10^{-11} torr are realized. Other possible solutions to the problem of surface contamination are cleaning by ion bombardment, which has been used by Harris [122] in Auger studies of surfaces, and cleaning by heating in a hydrogen atmosphere, which has been used by Fadley and Shirley [120].

Although there are uncertainties about the surface condition and about the exact depth of surface that is studied, photoelectron spectroscopy may prove to be one of the most powerful tools available for the study of surface and near-surface phenomena. Control and knowledge of the condition of surfaces will continue to be problems encountered in all methods of surface study, and these are only technical difficulties, which presumably can be overcome. The uncertainties about the depth of the surface studied can be eliminated by further investigation. At any rate, the fact that photoelectron spectroscopy measures surface phenomena to a depth of from one to a few hundred angstroms is an advantage over techniques such as microprobe analysis, which probe to depths of about one micron.

Bordass and Linnett [123] have studied adsorbed gases on surfaces by means of photoelectron spectroscopy, using the helium resonance line for excitation. They used grazing incidence of the photon beam in order to minimize the number of photoelectrons originating below the surface. They were able to show for methanol adsorbed on tungsten that the π-type orbital of the methyl group is displaced by ~ 1 eV from its position in gas-phase measurements. The authors concluded, therefore, that this orbital is the bonding orbital in the adsorption process. The photoelectron peak corresponding to the oxygen lone pairs was broadened but not shifted. This preliminary work indicates that it may be possible to identify the orbitals involved in absorption and to investigate catalytic intermediates by means of ultraviolet photoelectron spectroscopy. The ability of the photoelectron method to detect multivalent states certainly makes it of great potential value for the study of catalysis. Mössbauer spectroscopy [124] has been applied toward this purpose but with only limited success.

Applications to Analytical Chemistry

Binding energies of inner orbitals are characteristic of individual atoms; therefore peak positions in photoelectron spectra can be used to identify elements. As discussed above, the binding energies also depend upon the chemical environment, but since the chemical shifts are small, few ambiguities should arise. To the contrary, the observed shifts give additional information about the chemical state of the atom. If uncertainties in identification of peaks are encountered, they can be removed by the measurement of more than one level in the atom or by the use of photons of different energies. The sensitivity of the method is determined by the photoelectric cross section, the absorption of the electrons in the sample, the detector efficiency, the signal-to-noise ratio, the intensity of the ionizing radiation, and the transmission of the spectrometer. Since these factors interrelate in a complicated way, it is not possible to predict accurately the sensitivity for a given element in a given compound. Siegbahn et al. [125] have demonstrated the sensitivity of the method for detecting small amounts of heavier elements in organic compounds by recording cobalt lines from vitamin B_{12}, in which the atom ratio of cobalt to other elements is 1:180.

The fact that photoelectron spectroscopy measures surface phenomena is a handicap if analytical information about the bulk sample is desired, since surfaces are sometimes nonstochiometric. Attempts to analyze solids with more than one phase would pose especially difficult problems. Also, adsorbed gases and oils produce unwanted lines and decrease the intensities of the lines originating in the sample itself. Such problems are particularly undesirable if quantitative information is desired. Nordling et al. [126] have measured the atom ratios of elements in some organic molecules and in some metal alloys with a precision of better than 5%; however, the general applicability of the method for quantitative analysis has not been demonstrated.

The desirable features of the method as an analytical tool are that it is applicable to all elements, with the exception of hydrogen; it has, in general, high sensitivity; and it is basically nondestructive. The problems discussed above in regard to surfaces are not encountered, of course, in gas-phase studies.

Measurements of Angular Distributions of Photoelectrons

Most photoelectron spectrometers are designed to detect photoelectrons ejected from the source at a single, fixed angle; thus it is desirable to know the angular distributions of the photoelectrons so that the optimum angle of detection can be chosen and so that the effect of anisotropies on measurements of relative intensities can be determined. In addition, comparisons between measured and calculated distributions can give information about the angular momenta of the orbitals from which the electrons are ejected.

If it is assumed that the photoelectric effect is a single-electron process, the angular distribution of the photoelectrons from atoms is given by [127]

$$\frac{d\sigma}{d\Omega} = \frac{\sigma}{4\pi}[1 - (\tfrac{1}{2})\beta P_2(\cos\theta)], \tag{5.22}$$

where σ and $d\sigma/d\Omega$ are the total and differential cross sections for the shell of interest, θ is the angle between the incident photon beam and the direction of the outgoing electron, $P_2(\cos\theta) = \frac{1}{2}(3\cos^2\theta - 1)$, and β is an asymmetry parameter. Equation (5.22) is valid in general for any particle ejected in an electric-dipole absorption process, which includes Auger as well as photoelectron processes. Cooper [128], and Manson and Cooper [129], assuming a central potential, have derived an expression for the total photoelectric cross section for a given subshell. The asymmetry parameter, β, depends in a complicated way upon the angular momentum, l, of the orbit from which the electron is ejected [130].

Krause [131] has measured the angular distributions of the 3*s*, 3*p*, and 3*d* photoelectrons for krypton and has determined the relative subshell contributions to the photoelectric cross section of krypton. Cooper and Manson [127] have shown that these experimental results of Krause [131] are in satisfactory agreement with the central-field theory. Lin [132] has calculated differential and total cross sections for the photoionization of various states of the hydrogen atom. Measurements have been made of the angular distributions of photoelectrons ejected from simple molecules [133, 133*a*]. Schneider and Berry [134] have used a pseudo-potential method to calculate total and differential cross sections for the photoionization of N_2. The results are compared to the experimental data of Berkowitz et al. [133]. Tully and others [134*a*] have derived expressions for nonrelativistic differential cross sections for the photoionization of randomly oriented diatomic molecules.

Most of the early data on angular distributions were obtained at a few fixed angles by means of retarding-field spectrometers. Carlson is at present using a high-dispersion, double-focusing, electrostatic spectrometer equipped so that the angle of the incident photon beam can be varied in a continuous manner. He has found for N_2 that the relative intensities of the vibrational structure in the photoelectron spectrum depends upon the angle of the ejected electron and has discussed the results in terms of autoionization states and the Born-Oppenheimer approximation [134b].

6 AUGER ELECTRON SPECTROSCOPY

When an atom or a molecule is ionized in an inner shell, the vacancy is filled by an electron from a higher-energy shell, and the excitation energy is either emitted as photon or is transferred to an orbital electron, which is emitted. The latter process is called the Auger process after its discoverer, Pierre Auger [2, 135, 136].

For an Auger process in which an initial vacancy in the K-shell is filled by an electron from the L_2 shell with the emission of an electron from the L_3 shell (a KL_2L_3 Auger process), the energy of the emitted Auger electron is given for the case of pure *jj* coupling by the difference in binding energies:

$$T_{KL_2L_3} = E_K - E_{L_2} - E'_{L_3}. \tag{5.23}$$

The prime indicates that the binding energy used for the L_3 shell should be that for an atom already ionized in the L_2 shell.

The Auger processes are named according to the shell in which the initial vacancy is produced (e.g., K Auger electrons, L Auger electrons), according to the outer shells involved in the process (e.g., KLL, KLM, LMN, and LMM Auger electrons), and according to the subshells involved (e.g., for *jj* coupling, KL_2L_3, KL_1M_1, $L_1M_2M_3$ Auger electrons). Auger transitions that transfer a vacancy from one subshell to another of the same shell (e.g., LLM and MMN) are given the special name Coster-Kronig transitions.

The assumption that either pure *jj* or pure LS coupling can be used to describe the Auger process is not valid in most cases. For heavier elements the use of *jj* coupling is a good approximation; however, for the intermediate Z elements an intermediate coupling scheme is required that approaches LS coupling for the lightest elements. Furthermore, the energies of the emitted Auger electrons are given accurately by (5.23) only for pure *jj* coupling. In all cases the energies are given correctly by the differences in the energies of the initial atomic or molecular state (with one vacancy) and the final state (with two vacancies). Nevertheless, it is convenient in a general discussion to describe the Auger transitions in terms of the *jj* coupling scheme and to approximate their energies in terms of binding-energy differences.

Auger transitions have been studied for years by nuclear spectroscopists because Auger electrons are emitted as a secondary process in most radioactive decays. Atomic spectroscopists have also studied Auger transitions because Auger-transition data can give information about atomic structure. The data available from studies of Auger transitions that are of interest to the chemists are of the same type as those available from photoelectron spectroscopy studies, namely chemical-shift, molecular-orbital, and analytical information. Also, the Auger process provides information about excited states of doubly-ionized molecules. For these purposes the *KLL* Auger transitions are the most useful and are the only ones discussed here.

Chemical shifts of binding energies will result in a shift in, for example, $T_{KL_2L_3}$ of the order of

$$\Delta T_{KL_2L_3} = \Delta E_K - \Delta E_{L_2} - \Delta E'_{L_3}.$$

Since all shells shift approximately the same,

$$\Delta E_{L_2} \cong \Delta E_{L_3} \cong \Delta E_K$$

and

$$\Delta T_{KL_2L_3} = -\Delta E_K. \tag{5.24}$$

Therefore, the chemical shifts of Auger-electron lines are as large as and as easy to measure as shifts of photoelectron lines. The shifts in the Auger spectra are more difficult to interpret, however, because of the larger number of energy levels involved in the process.

For the first-row elements beryllium through neon the *L* electrons are valence electrons, which participate directly in chemical bonding. Therefore studies of *KLL* Auger transitions in molecules containing these elements can give information about molecular orbitals. Since all of the *KLL* Auger transitions of a given element in a given compound begin with the same *K* vacancy, the energy spacings of the Auger lines are the same as the energy spacings of the molecular orbitals. The relative intensities of the Auger lines provide information about the overlap between the molecular orbitals and the *K* orbitals.

An example of a *KLL* Auger spectrum of a first-row element is shown in Fig. 5.20, which is the data of Carlson et al. [29] for oxygen in CO, CO_2, NO, and O_2. The spectra are markedly different because of the differences in the molecular orbitals of the compounds. The observed differences may also prove to be of use for qualitative analysis.

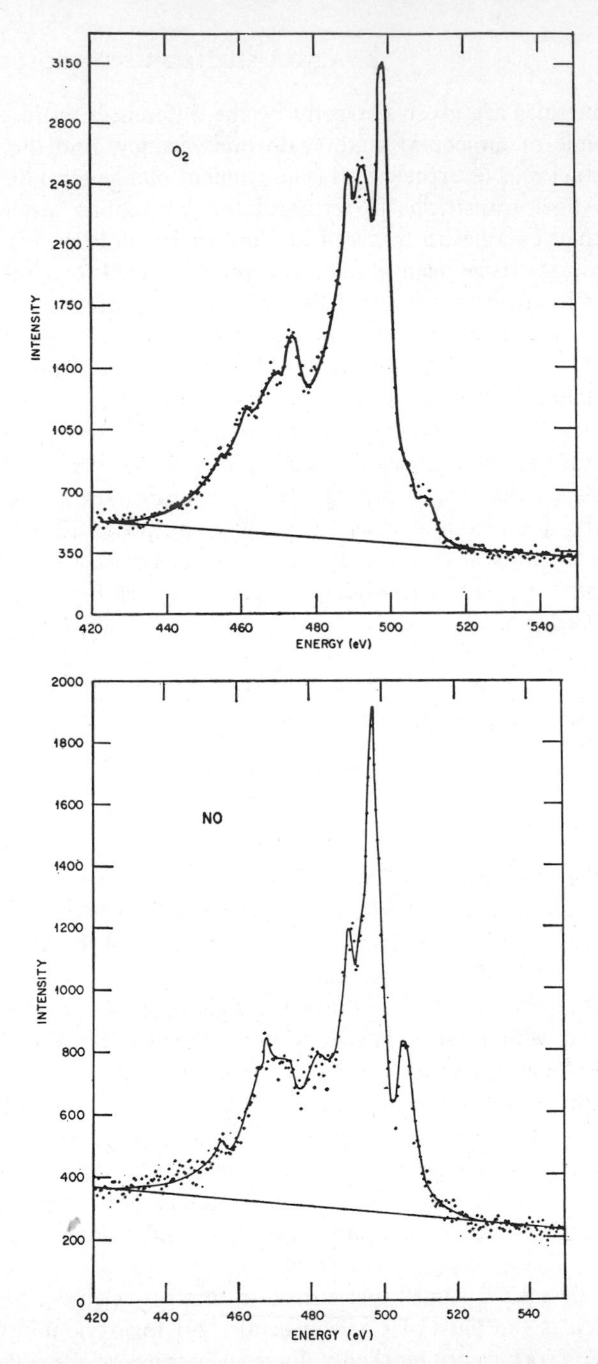

Fig. 5.20 The data of Carlson et al. [29] on the KLL Auger spectra of oxygen in CO, CO_2, NO, and O_2. The Auger spectra were excited by bombardment of the gaseous samples with 2-keV electrons from an electron gun.

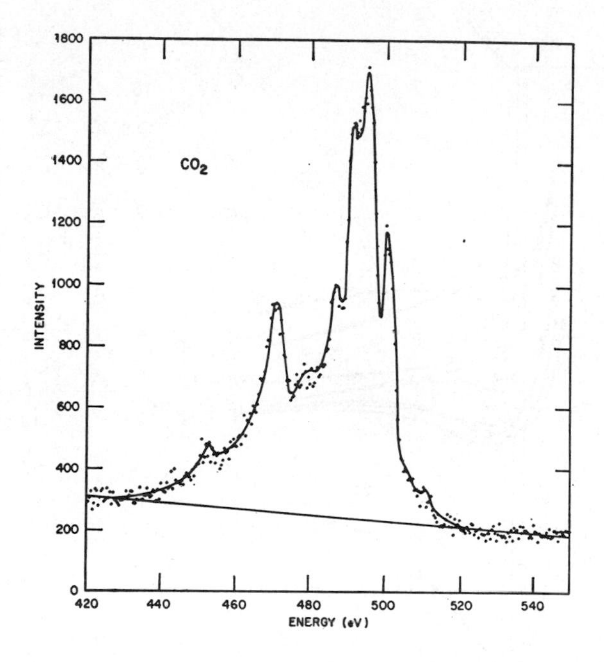

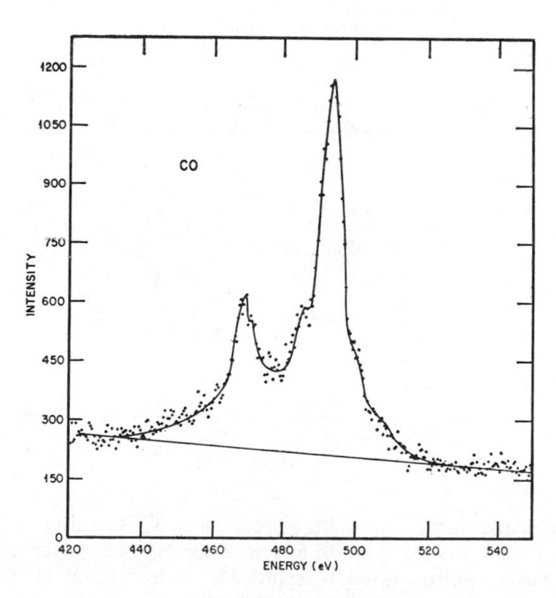

Fig. 5.20 (*continued*)

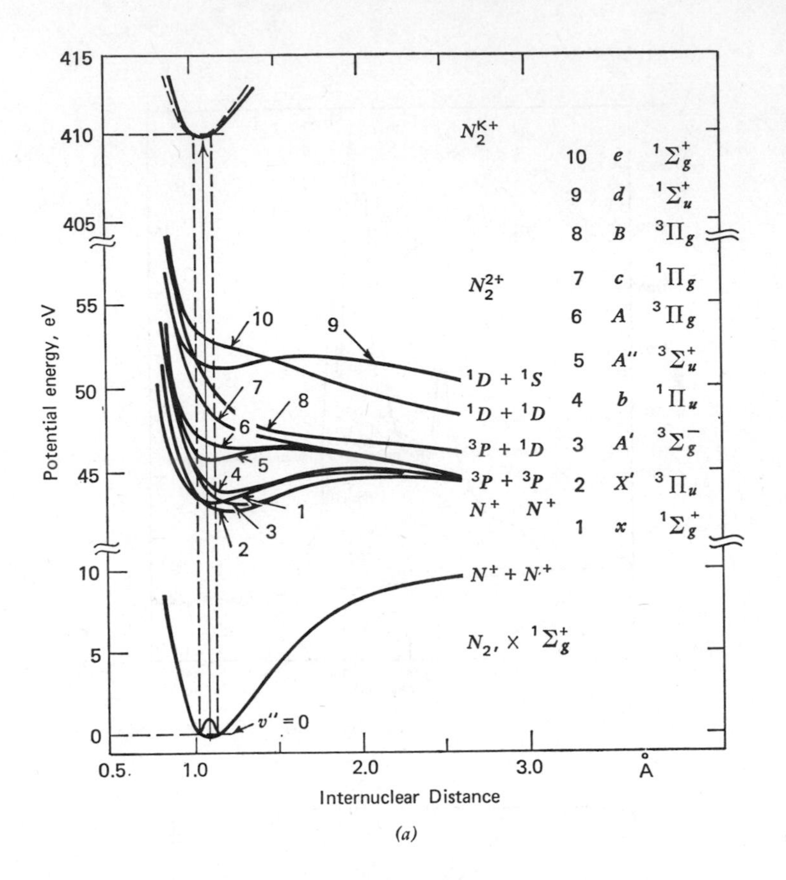

(*a*)

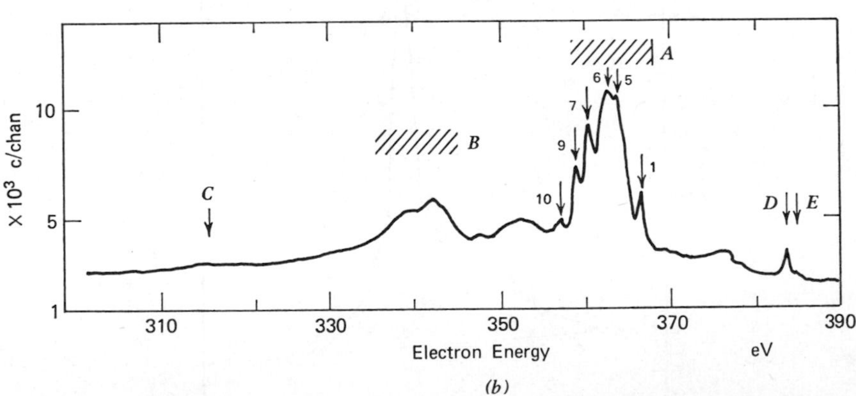

(*b*)

Fig. 5.21 (*a*) The potential energy curves for N_2, N_2^{2+}, and N_2^{K+} as given by Stalherm et al. [137]. A vertical line from the N_2^{K+} curve to one of the N_2^{2+} curves represents an Auger process. (*b*) The K Auger spectrum of N_2 as recorded by Stalherm et al. [137]. The numbers identifying certain peaks refer to the final states of the doubly ionized molecule as identified in (*a*).

Auger spectra give information about excited states of doubly ionized molecules in much the same way that photoelectron spectroscopy gives information about states of singly ionized molecules. An example is shown in Fig. 5.21, which is the data of Stalherm et al. [137] on the *KLL* Auger spectra of nitrogen in N_2. In Fig. 5.21*a* are shown the potential curves for N_2, N_2^{2+} and N_2^{K+}. (The $K+$ indicates a vacancy in a K orbital.) On this diagram the primary ionization is represented by a vertical line from the N_2 curve up to the N_2^{K+} curve, and the Auger process is represented by a vertical line from the N_2^{K+} curve down to one of the N_2^{2+} curves. The lines are drawn vertically under the assumption of the Franck-Condon principle.

The kinetic energy T_{KLL} of the emitted electron is given by the difference in energy of the initial and the final states:

$$T_{KLL} = E_K - E_K(N_2^{2+}). \tag{5.25}$$

The measured values of T_{KLL} combined with the previously measured value of E_K allowed Stalherm and others [137] to determine $E_K(N_2^{2+})$ for several excited states of the doubly ionized nitrogen molecule. Fig. 5.21*b* shows the Auger spectra and the interpretation of some of the stronger lines. The instrumental resolution used was not sufficient to resolve vibrational structure in the spectrum.

The energies of the Auger electrons are characteristic of the element; thus Auger spectra can be used for qualitative chemical analysis. In principle quantitative analysis could also be accomplished but the difficulties in calibration are the same as those for photoelectron analysis, and these difficulties have not yet been overcome in general, although it is possible with extreme care to calibrate quantitatively thin overlayers of elements deposited on clean surfaces [138].

Since high-flux electron guns can be built, electron impact is often used to excite Auger spectra, particularly in gas-phase studies where the low density of the target material makes the high flux particularly advantageous. Electron guns can be specially designed for this purpose, or modified cathode-ray tubes can be used.

Auger studies of solid materials are often made by means of low-energy-electron-diffraction (LEED) systems [139] so that the analysis by Auger spectroscopy can be supplemented by electron-diffraction studies of surface structure [138–140]. Palmberg has reported on the use [141] and the optimization [142] of LEED systems for Auger studies. If low-energy (< 500 eV) *K*- or *L*-Auger electrons are studied, only a few monolayers of the surface of the sample contribute to the recorded spectrum. Cesium and potassium in deposits as thin as 0.1 monolayers [143] and 0.02 monolayers [138] have been detected and identified. The Auger yield decreases with increasing Z; hence the method is most sensitive to elements of low atomic number.

Chemical analysis by means of Auger spectroscopy has been applied to studies of surface contamination [143, 144], migration [144], segregation [144, 145], diffusion [144], and evaporation [146]. Harris [147, 148] and Yin and others [149] have discussed in a broad sense the analytical applications of Auger spectroscopy.

7 ELECTRON IMPACT SPECTROSCOPY

When a photon strikes a molecule, either all or a part of the photon energy can be given to the molecule. These alternatives correspond to the photoelectric and the Compton effect, respectively. For small photon energies the photoelectric effect dominates, and photoionization is characterized by monoenergetic electrons of energies $h\nu - E_B$. Those Compton processes that do occur give rise to a background of electrons with a continuous range of kinetic energy.

Electron-impact spectroscopy differs from photoelectron spectroscopy in that an electron beam, rather than a photon beam, is used to excite the molecules. During ionization by electron impact, an incident electron scatters from a bound electron, and the available kinetic energy can be shared between the two electrons in any manner consistent with the conservation laws. Thus this process is similar to Compton scattering in that it is characterized by a continuous distribution of electron energies. There are, however, three ways in which electron impact can produce discrete lines in the electron spectrum: (1) the Auger processes that follow the ionization (see Section 6), (2) autoionization, and (3) the excitation of electrons to available bound states.

Autoionization is the process by which a transition takes place between an excited state of the neutral molecule and an ionized state of the molecule. When an electron scatters from molecule M and excites M to M^*, the autoionization process is described as $M^* \rightarrow M^+ + e^-$. Measurements of the electrons ejected during autoionization give information about the autoionization states, M^*.

When the scattering electron excites the molecule from M to M^*, it will lose an amount of energy equal to the molecular excitation energy; thus if the incident electron beam is monoenergetic, an energy analysis of the scattered electrons will show energy-loss peaks corresponding to the excitation of bound electrons to available molecular orbitals. Therefore whereas photoelectron spectroscopy gives information about molecular levels that are normally occupied, electron-impact spectroscopy gives information about levels that are normally empty. Electron-impact spectroscopy can be used to measure the energies of excited states, to measure generalized oscillator strengths, to determine Franck-Condon factors, and to determine the multipolarities of transitions. Electron-impact spectroscopy is particularly useful for studying transitions that are optically forbidden. Berry [10] has published recently a detailed review of electron-impact spectroscopy, and Moiseiwitsch and Smith [150] have published a report on the electron-impact excitation of atoms.

Acknowledgment

The author is grateful to the following persons for enlightening conversations regarding this manuscript: Thomas Carlson, Charles Fadley, Jack Hollander, Manfred Krause, Werner Mehlhorn, Ragnar Nordberg, Thomas Pinkston, Larry Schaad, David Shirley, Reimar Spohr, and John Van Wazer, and to Thomas Carlson, Werner Mehlhorn, Kai Siegbahn, Reimar Spohr, and their coworkers for permission to use data and diagrams from their work.

References

1. M. O. Krause, T. A. Carlson, and R. D. Dismukes, *Phys. Rev.*, **170,** 37 (1968).
2. I. Bergström and C. Nordling, in *Alpha-, Beta-, and Gamma-Ray Spectroscopy*, Kai Siegbahn, Ed., North-Holland, Amsterdam, 1965, p. 1523.
3. H. Robinson and W. F. Rawlinson, *Phil. Mag.*, **28,** 277 (1914); E. Rutherford, H. Robinson, W. F. Rawlinson, *Phil. Mag.*, **28,** 281 (1914); K.-F. Hu, *Phys. Rev.*, **11,** 505 (1918); M. de Broglie, *Compt. Rend.*, **172,** 274, 527, 806 (1921); *ibid.*, **173,** 527, 1157 (1921); *ibid.*, **174,** 939 (1922); *J. Phys. Radium*, **2,** 265 (1921); M. de Broglie and L. de Broglie, *ibid.*, **172,** 746 (1921); *ibid.*, **175,** 1139 (1922); R. Whiddington, *Phil. Mag.*, **43,** 1116 (1922); H. Robinson, *Proc. Roy. Soc. (London)*, **104A,** 455 (1923); *Phil. Mag.*, **50,** 241 (1925).
4. C. M. Davisson, in *Alpha-, Beta-, and Gamma-Ray Spectroscopy*, Kai Siegbahn, Ed., North-Holland, Amsterdam, 1965, p. 37.
5. R. D. Birkhoff, *Encyclopedia Phys.*, **34,** 51 (1955).
6. D. Pines, *Elementary Excitations in Solids*, Benjamin, New York, 1963.
7. K. Siegbahn, C. Nordling, A. Fahlman, R. Nordberg, K. Hamrin, J. Hedman, G. Johansson, T. Bergmark, S.-E. Karlsson, I. Lindgren, and B. Lindberg, *ESCA, Atomic, Molecular, and Solid-State Structure Studied by Means of Electron Spectroscopy*, Almqvist and Wiksells, Uppsala, 1967.
8. K. Siegbahn, C. Nordling, G. Johansson, J. Hedman, P. Hedén, K. Hamrin, U. Gelius, T. Bergmark, L. Werme, R. Manne, and Y. Baer, *ESCA Applied to Free Molecules*, North-Holland, Amsterdam, 1969.
9. D. W. Turner, in *Physical Methods in Advanced Inorganic Chemistry*, H. A. O. Hill and P. Day, Ed., Interscience Publishers, London, 1968, p. 74.

10. R. S. Berry, *Ann. Rev. Phys. Chem.*, **20,** 357 (1969).
10a. J. Hollander and D. Shirley, *Ann. Revs. Nuclr. Sci.*, **20** (1970).
11. Reference 7, p. 36; S. Hagström and S.-E. Karlsson, *Ark. Fysik*, **26,** 451 (1964).
11a. C. Fadley, G. Geoffroy, S. Hagström, and J. Hollander, *Nucl. Instr. Methods*, **68,** 177 (1969).
12. E. Sokolowski, *Ark. Fysik*, **15,** 1 (1959).
13. C. Nordling, *Ark. Fysik*, **15,** 397 (1959).
14. J. A. Bearden, USAEC, ORNL, NYO 10586 (available from U.S. Department of Commerce); *Rev. Mod. Physics*, **39,** 78 (1967).
15. A. Fahlman, S. Hagström, K. Hamrin, R. Nordberg, C. Nordling, and K. Siegbahn, *Ark. Fysik*, **31,** 479 (1966).
16. J. A. R. Samson, *Rev. Sci. Instr.*, **40,** 1174 (1969).
17. K. Siegbahn, in *Alpha-, Beta-, and Gamma-Ray Spectroscopy*, Kai Siegbahn, Ed., North-Holland, Amsterdam, 1965, p. 79.
18. J. H. D. Eland and C. J. Danby, *J. Sci. Inst.*, **45,** 406 (1968); *Z. Naturforsch.*, **23A,** 355 (1968); *J. Mass Spectr. Ion Phys.*, **1,** 111 (1968); J. H. D. Eland, *J. Mass Spectr. Ion Phys.*, **2,** 471 (1969).
19. G. C. Theodoridis and F. R. Paolini, *Rev. Sci. Instr.*, **39,** 326 (1968).
20. D. W. Turner, *Proc. Roy. Soc.* (*London*), **A307,** 15 (1968).
21. H. Z. Sar-El, *Rev. Sci. Instr.*, **38,** 1210 (1967).
21a. E. Blauth, *Z. Physik*, **147,** 228 (1957).
22. V. V. Zashkvara, M. I. Korsunskii, and O. S. Kosmachev, *Sov. Phys.–Tech. Phys.*, **11,** 96 (1966).
23. W. Mehlhorn, *Z. Physik*, **187,** 21 (1965).
24. H. Körber and W. Mehlhorn, *Z. Physik*, **191,** 217 (1966).
25. F. R. Paolini and G. C. Theodoridis, *Rev. Sci. Instr.*, **38,** 579 (1967); I. R. Toscano and P. J. Prosciutto, private communication, August 1969.
26. C. E. Kuyatt and J. A. Simpson, *Rev. Sci. Instr.*, **38,** 103 (1967).
27. E. M. Purcell, *Phys. Rev.*, **54,** 818 (1938).
28. M. O. Krause, *Phys. Rev.*, **140,** 1845 (1965).
29. T. A. Carlson, M. O. Krause, W. E. Moddeman, B. P. Pullen, and F. W. Ward, Oak Ridge National Labs Physics Division Annual Reports, ORNL 4395, December 1968; B. Pullen, T. Carlson, W. Moddeman, G. Schweitzer, W. Bull, and F. Grimm, *J. Chem. Phys.*, **53,** 768 (1970).
30. E. N. Lassettre, A. Skerbelle, M. A. Dillon, and K. J. Ross, *J. Chem. Phys.*, **48,** 5066 (1968).
31. H. Hafner, J. A. Simpson, and C. E. Kuyatt, *Rev. Sci. Instr.*, **39,** 33 (1968).
32. Reference 7, p. 198.
33. McPherson Instrument Co., Acton, Mass.
34. Varian, Inc., Palo Alto, Calif.
35. Hewlett-Packard, Palo Alto, Calif.; AEI Ltd., Manchester, England.
36. Perkin-Elmer Corp., Beaconsfield, Bucks, England.
37. Reference 7, p. 217.
38. R. L. Graham, G. T. Ewan, and J. S. Geiger, *Nucl. Instr. Methods*, **9,** 245 (1960).

39. Q. L. Baird, J. C. Nall, S. K. Haynes, and J. H. Hamilton, *Nucl. Instr. Methods*, **16,** 275 (1962).
40. R. Nordberg, J. Hedman, P. F. Hedén, C. Nordling, and K. Siegbahn, *Ark. Fysik*, **37,** 489 (1968); Ref. 7, p. 189.
41. C. D. Vries and A. H. Wapstra, *Nucl. Instr. Methods*, **8,** 121 (1960).
42. C. S. Fadley, C. E. Miner, and J. M. Hollander, *Lawrence Radiation Laboratory Report UCRL–18954*, July 1969; *Appl. Phys. Lett.*, **15,** 223 (1969).
43. H. Boersch, J. Geiger, and W. Stickel, *Z. Physik*, **180,** 415 (1964).
44. J. C. Helmer and N. H. Weichert, *Appl. Phys. Lett.*, **13,** 266 (1968).
45. Reference 7, p. 172.
46. J. A. Simpson, *Rev. Sci. Instr.*, **32,** 1283 (1961).
47. M. I. Al-Joboury and D. W. Turner, *J. Amer. Chem. Soc.*, **1963,** 5141.
48. J. A. R. Samson and R. B. Cairns, *Phys. Rev.*, **173,** 80 (1968).
49. D. C. Frost, C. A. McDowell, and D. A. Vroom, *Proc. Roy. Soc. (London)*, **A296,** 566 (1967).
50. G. C. Baldwin and S. I. Friedman, *Rev. Sci. Instr.*, **38,** 519 (1967).
51. M. Ference, A. E. Shaw, and R. J. Stephenson, *Rev. Sci. Instr.*, **11,** 57 (1940).
52. Reference 7, p. 215.
53. G. W. Goodrich and W. C. Wiley, *Rev. Sci. Instr.*, **33,** 761 (1962).
54. D. S. Evans, *Rev. Sci. Instr.*, **36,** 375 (1965).
55. T. Hayashi and M. Hashimoto, *Rev. Sci. Instr.*, **40,** 1239 (1969).
56. A. Egidi, R. Marconero, G. Pizzella and F. Sperli, *Rev. Sci. Instr.*, **40,** 88 (1969).
57. U. Amaldi, A. Egidi, R. Marconero, and G. Pizzella, *Rev. Sci. Instr.*, **40,** 1001 (1969).
58. M. Kanayama, T. Konno, and S. Kiyono, *Rev. Sci. Instr.*, **40,** 129 (1969).
59. E. Laegsgaard, F. W. Martin, and W. M. Gibson, *Nucl. Instr. Methods*, **60,** 24 (1968).
60. Bendix Mosaic, Sturmbridge, Mass.
61. A. Fahlman, R. G. Albridge, R. Nordberg, and W. M. LaCasse, *Rev. Sci. Instr.*, **41,** 596 (1970).
61a. M. O. Krause, *Phys. Lett.*, **19,** 14 (1965).
62. R. Nordberg, H. Brecht, R. G. Albridge, A. Fahlman, and J. R. Van Wazer, *Inorg. Chem.*, **9,** 2469 (1970).
63. L. Kramer and M. Klein, preprint (1969).
64. D. R. Eaton, in *Physical Methods in Advanced Inorganic Chemistry*, H. A. O. Hill and P. Day, Ed., Interscience, London, 1968, p. 462.
65. J. Danon, in *Physical Methods in Advanced Inorganic Chemistry*, H. A. O. Hill and P. Day, Eds., Interscience, London, 1968, p. 380.
66. A. Fahlman, K. Hamrin, J. Hedman, R. Nordberg, C. Nordling, and K. Siegbahn, *Nature*, **210,** 4 (1966).
67. Reference 7, p. 79.
68. C. S. Fadley, S. Hagstrom, M. P. Klein, and D. A. S. Shirley, *J. Chem. Phys.*, **48,** 3779 (1968).
69. K. Hamrin, G. Johansson, A. Fahlman, C. Nordling, K. Siegbahn, and B. Lindberg, *Chem. Phys. Lett.*, **1,** 557 (1968).

70. Reference 7, p. 98.
71. A. Fahlman, R. Carlsson, and K. Siegbahn, *Arkiv Kemi*, **25,** 301 (1966).
72. R. Nordberg, R. G. Albridge, T. Bergmark, U. Ericson, J. Hedman, C. Nordling, K. Siegbahn, and B. Lindberg, *Arkiv Kemi*, **28,** 257 (1968); Ref. 7, p. 108.
73. L. Pauling, *The Nature of the Chemical Bond*, 3rd ed., Cornell University Press, Ithaca, N.Y., 1960.
74. Reference 7, p. 116.
75. R. Nordberg, U. Gelius, P.-F. Hedén, J. Hedman, C. Nordling, K. Siegbahn and B. Lindberg, UUIP, March 1968.
76. J. Hinze and H. H. Jaffé, *J. Amer. Chem. Soc.*, **84,** 540 (1962).
77. J. Hinze, M. A. Whitehead, and H. H. Jaffé, *J. Amer. Chem. Soc.* **85,** 148 (1963).
78. J. E. Huheey, *J. Phys. Chem.*, **69,** 3284 (1965); *ibid.*, **70,** 2086 (1966); *J. Org. Chem.*, **31,** 2365 (1966).
79. J. M. Hollander, D. N. Hendrickson. and W. L. Jolly. *J. Chem. Phys.*, **49,** 3315 (1968).
80. M. Pelavin, D. N. Hendrickson, J. M. Hollander, and W. L. Jolly, *J. Phys. Chem.*, **74,** 1116 (1970).
81. D. N. Hendrickson, J. M. Hollander, and W. L. Jolly, *Inorg. Chem.*, **9,** 612 (1970).
82. S. E. Karlsson, K. Siegbahn, and N. Bartlett, preprint.
82a. J. R. Blackburn, R. Nordberg, F. Stevie, R. G. Albridge, and M. M. Jones, *Inorg. Chem.*, **9,** 2374 (1970).
83. Reference 7, p. 93.
84. G. Axelson, K. Hamrin, A. Fahlman, C. Nordling, and B. Lindberg, *Spectrochim. Acta*, **23A,** 2015 (1967); Ref. 7, p. 135.
85. W. L. Jolly, private communication, 1969.
86. L. J. Schaad and H. B. Kinser, *J. Phys. Chem.*, **73,** 1901 (1969).
87. W. L. Jolly and D. N. Hendrickson, *Lawrence Radiation Laboratory Report, UCRL–19050* (July 1969); *J. Amer. Chem. Soc.*, **92,** 1863 (1970); see also W. L. Jolly, *ibid.*, **92,** 3260 (1970).
88. E. Clementi, *J. Chem. Phys.*, **46,** 4731 (1967).
89. M. L. Unland, T. H. Dunning, Jr., and J. R. Van Wazer, *J. Chem. Phys.*, **50,** 3208 (1969); M. L. Unland, J. H. Letcher, and J. R. Van Wazer, *J. Chem. Phys.*, **50,** 3214 (1969).
90. M. Krauss, *Compendium of ab initio Calculations of Molecular Energies and Properties*, NBS Technical Note 438 (1967), U.S. Department of Commerce, National Bureau of Standards.
91. T. Koopmans, *Physica*, **1,** 104 (1933).
92. R. S. Mulliken, *Phys. Rev.*, **74,** 736 (1948).
93. W. G. Laidlaw and F. W. Birss, *Theoret. Chim. Acta* (Berl.) **2,** 181 (1964); F. W. Birss and W. G. Laidlaw, *Theoret. Chim. Acta* (Berl.), **2,** 186 (1964).
94. M. D. Newton, *J. Chem. Phys.*, **48,** 2825 (1968).
95. I. Lindgren, *Phys. Lett.*, **19,** 382 (1965); Ref. 7, p. 70.
96. F. A. Gianturco and C. A. Coulson, *Mol. Phys.* **14,** 223 (1968).

97. J. R. Hoyland and L. Goodman, *J. Chem. Phys.*, **33,** 946 (1960); *ibid.*, **36,** 12 (1962).
98. S. Schildcrout, R. Pearson, and F. Stafford, *J. Amer. Chem. Soc.*, **90,** 4006 (1968).
99. G. Verhaegen, W. G. Richards, and C. M. Moser, *J. Chem. Phys.*, **47,** 2595 (1967).
100. W. G. Richards and R. C. Wilson, *Trans. Faraday Soc.*, **64,** 1729 (1968).
101. C. Hollister and O. Sinanoğlu, *J. Amer. Chem. Soc.*, **88,** 13 (1966).
102. J. C. Phillips, *Phys. Rev.*, **123,** 420 (1961).
103. R. S. Mulliken, *J. Chem. Phys.*, **46,** 497 (1949).
104. P. S. Bagus, *Phys. Rev.*, **139,** A619 (1965).
105. H. Basch and L. C. Snyder, *Chem. Phys. Lett.*, **3,** 333 (1969).
106. W. G. Richards, *Int. J. Mass Spectr. Ion Phys.*, **2,** 419 (1969).
106a. L. Hedin and A. Johansson, *J. Phys. B*, **2,** 1336 (1969).
107. D. Bohm, *Quantum Theory*, Prentice-Hall, New York, 1951, Chapter 20.
108. T. A. Carlson and M. O. Krause, *Phys. Rev.*, **140A,** 1057 (1965).
109. T. A. Carlson, C. W. Nestor, Jr., and T. C. Tucker, *Phys. Rev.*, **169,** 27 (1968).
110. C. R. Brundle, M. B. Robin, and G. R. Jones, *J. Chem. Phys.*, **52,** 3383 (1970).
111. Nakamura, Sasanuma, Sato, Watanabe, Yamashita, Iguchi, Eiri, Nakai, Yamaguchi, Sagawa, Nakai, and Oshio, *Phys. Rev.*, **178,** 80 (1969).
112. K. Codling, R. P. Madden, and D. L. Ederer, *Phys. Rev.*, **155,** 26 (1967).
112a. M. O. Krause, private communication.
113. Reference 7, p. 209.
114. B. P. Pullen, T. A. Carlson, W. E. Moddeman, G. K. Schweitzer, W. E. Bull, and F. A. Grimm, *J. Chem. Phys.* **53,** 768 (1970).
115. M. O. Faltens, Ph.D. dissertation, University of California at Berkeley, UCRL–18706, February 1969.
116. T. Novakov, R. Stepic, and P. Janiujevic, unpublished results.
117. T. Novakov and J. M. Hollander, Lawrence Radiation Laboratory Report UCRL–18839, 1968.
118. J. Hedman, P.-F. Hedén, C. Nordling, and K. Siegbahn, *Phys. Lett.*, **29A,** 178 (1969).
119. C. Fadley, D. Shirley, A. Freeman, P. Bagus, and J. Mallow, *Phys. Rev. Lett.*, **23,** 1397 (1969).
120. C. S. Fadley and D. A. Shirley, *Phys. Rev. Lett.*, **21,** 980 (1968).
121. W. N. Delgass, *Catalysis Rev.*, **4,** 179 (1970).
122. L. A. Harris, *J. Appl. Phys.*, **39,** 1419 (1968).
123. W. T. Bordass and J. W. Linnett, *Nature*, **222,** 660 (1969).
124. W. N. Delgass and M. Boudart, *Catalysis Rev.*, **2,** 129 (1968).
125. References 7, p. 17.
126. C. Nordling, S. Hagström, and K. Siegbahn, *Z. Phys.*, **178,** 433 (1964); Ref. 7, p. 141.
127. J. W. Cooper and S. T. Manson, *Phys. Rev.*, **177,** 157 (1969).
128. J. W. Cooper, *Phys. Rev.*, **128,** 681 (1962).

129. S. T. Manson and J. W. Cooper, *Phys. Rev.*, **165,** 126 (1968).
130. J. Cooper and R. N. Zare, *J. Chem. Phys.*, **48,** 942 (1968).
131. M. O. Krause, *Phys. Rev.*, **177,** 151 (1969).
132. S. H. Lin, *Can. J. Phys.*, **46,** 2719 (1968).
133. J. Berkowitz, H. Ehrhardt, and T. Tekaat, *Z. Physik*, **200,** 69 (1967).
133a. J. W. McGowan, D. A. Vroom, and A. R. Comeaux, *J. Chem. Phys.*, **51,** 5626 (1969).
134. B. Schneider and R. S. Berry, *Phys. Rev.*, **182,** 141 (1969).
134a. J. C. Tully, R. S. Berry, and B. J. Dalton, *Phys. Rev.*, **176,** 95 (1968).
134b. T. A. Carlson, *Chem. Phys. Lett.*, **9,** 23 (1971).
135. P. Auger, *J. Phys. Rad.*, **6,** 205 (1925); *Ann. Phys.* **6,** 183 (1926).
136. E. H. S. Burhop, *The Auger Effect and Other Radiationless Transitions*, Cambridge University Press, Cambridge, 1952.
137. D. Stalherm, B. Cleff, H. Hillig, and W. Mehlhorn, *Z. Naturforsch.*, **24a,** 1728 (1969).
138. R. E. Weber and A. L. Johnson, *J. Appl. Phys.*, **40,** 314 (1969).
139. Editors of Analytical Chemistry, *Anal. Chem.*, **39** (#4), 26A (1967).
140. P. W. Palmberg and T. N. Rhodin, *J. Appl. Phys.*, **39,** 2425 (1968).
141. P. W. Palmberg, *J. Appl. Phys.*, **38,** 2137 (1967).
142. P. W. Palmberg, *Appl. Phys. Lett.*, **13,** 183 (1968).
143. R. E. Weber and W. T. Peria, *J. Appl. Phys.*, **38,** 4355 (1967).
144. L. A. Harris, *J. Appl. Phys.*, **39,** 1419 (1968).
145. L. A. Harris, *J. Appl. Phys.*, **39,** 1428 (1968).
146. L. A. Harris, *J. Appl. Phys.*, **39,** 4862 (1968).
147. L. A. Harris, *Anal. Chem.*, **40,** (#14) 24A (1968).
148. L. A. Harris, *Ind. Res.*, p. 53, February 1968.
149. L. I. Yin, I. Adler, and R. Lamothe, *Appl. Spectry.*, **23,** 41 (1969).
150. B. L. Moiseiwitsch and S. J. Smith, *Electron Impact Excitation of Atoms*, National Bureau of Standards NSRDS-NBS 25 (August 1968).

Chapter **VI**

MOLECULAR BEAMS IN CHEMISTRY*

John E. Jordan, Edward A. Mason, and the late Isadore Amdur

---

* This research was supported in part by the U.S. Office of Naval Research and by the National Science Foundation (MIT) and in part by the U.S. National Aeronautics and Space Administration (Brown Univ.).

1 INTRODUCTION

Molecular beams have been for many years a powerful tool for research into physical problems but have only fairly recently been applied to problems in chemistry. This is a rather curious situation because many of the pioneers in the development of molecular beams were physical chemists by training. Although this chapter emphasizes areas of interest primarily to chemists, the broader areas are discussed briefly to place applications to chemistry in proper context. The phrase "molecular beam" is usually used to denote a beam of either atoms or molecules. It is used in that general sense here; where discussion is limited to atoms the phrase atomic beam is used.

The basic molecular-beam technique was developed by Dunoyer [1] in 1911 to verify one of the fundamental postulates of the kinetic theory of gases, namely that molecules execute nearly rectilinear motions between collisions. He built an apparatus similar in principle to that shown in Fig. 6.1. The source chamber, S, was filled with vapor from heated metallic sodium, and the other two chambers were evacuated to the point where intermolecular collisions were infrequent. The temperature of the sodium

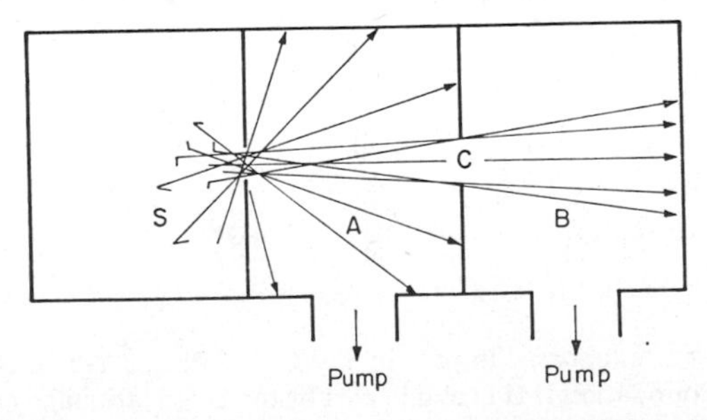

Fig. 6.1 Schematic apparatus for production of a molecular beam.

was controlled to ensure that the mean free path in the vapor was large with respect to the aperture between the two chambers S and A. Under these conditions the paths of the molecules became quite long, and the molecules could be collimated into rays or beams by installing a coaxial aperture C. Since chambers A and B were evacuated, the only intermolecular collisions that occurred were the relatively infrequent cases in which faster molecules overtook slower ones moving along the same path. Thus the beam in chamber B was essentially unidirectional and collision-free. Dunoyer found that the deposit of sodium on the cooled end of chamber B had the dimensions and form predicted by geometric optics, and further that objects placed in the path of the beam produced well-defined shadows. In addition, he performed the first molecular-beam scattering experiment by noting that the beam profile became diffuse when inert gas was introduced into chamber B.

In general, the molecular-beam technique is a means for producing isolated atoms or molecules within a narrow range of speed and solid angle. The basic strength of the method lies in the fact that the atoms or molecules are isolated from their surroundings; when any experiment is performed with the beam it can be analyzed in terms of an unperturbed single-particle interaction. It is difficult to generalize further because the degree of collimation, the range of particle velocities, and the technique of observing interaction of the beam all depend on the particular experiment being performed. There are, however, three general kinds of phenomena studied with molecular beams; they can be classified on the basis of how the molecular beam is made to interact. The classifications are single-particle interactions, many-particle interactions, and two-particle interactions. Examples of the types of experiments performed within these three categories are briefly indicated below.

Single-Particle Interactions

This area of molecular-beam research is a class of experiments in which the isolated atoms or molecules in the beam are introduced into an electromagnetic field. Observation is then made of the interaction of the individual atoms or the molecules with the field. The detailed manner in which the observation is made is determined by the particular experiment being performed. Ramsey [2] describes experiments of this type in detail.

Particles with resultant electric or magnetic moments can be deflected in inhomogeneous electric or magnetic fields. This fact has been used as the basis for a large number of extraordinary experiments, many of which were of great importance in the historical development of physics. One of the first of these was the Stern-Gerlach experiment [3], in which a beam of paramagnetic atoms, collimated from the effusive flow from a hot oven, was deflected in an inhomogeneous magnetic field into three distinct traces that could be correlated with the interaction of the field and the electron spins of the atoms. This experiment is an elegant proof of space quantization and of the need to include electron spin in accounting for the number of magnetic moments. Many similar experiments have been performed to determine atomic magnetic moments for a number of atoms and paramagnetic molecules. A molecule that is not paramagnetic, namely, that has no resultant electronic magnetic moment, can nonetheless be deflected in the inhomogeneous magnetic field if it has a nuclear magnetic moment, a rotational magnetic moment due to rotational angular velocity of the molecule, or a diamagnetic moment induced by an external magnetic field. All three of these magnetic moments have been determined in beam-deflection experiments. Another large and important area of research in physics that has evolved from beam-deflection experiments is referred to as radiofrequency spectroscopy. In the basic experimental arrrangement a beam passes through three successive regions, the first and last of which are inhomogeneous magnetic fields that produce, in molecules with the same nuclear magnetic moment, equal but opposite deflections. Radiofrequency energy is coupled into the middle region. If the frequency of the applied field is resonant with a state transition in the molecule, it will leave the field in a different state, usually with a change in effective nuclear magnetic moment. When the molecule traverses the second inhomogeneous field, the deflection it now receives does not exactly compensate for the deflection in the first field, so it misses the detector. The method then consists of observing resonances, that is, sharp intensity changes, as a function of the frequency of the exciting radiation. Thus spectra can be traced out that result from the coupling of nuclear moments with electronic motion in the molecule. Measurements of the structure of these spectra, called hyperfine structure, provide a wealth of information about nuclear properties, much of which has led to specific models of the nucleus.

Although the study of spectral fine structure is usually considered to be in the domain of optical spectroscopy, molecular-beam methods have been applied here as well. One of the classic discoveries of modern physics was that of the Lamb shift [4], where a technique similar to the molecular-beam method described above was the key to the experiment. Here resonant changes in intensity in a beam of metastable helium atoms were observed in a radiofrequency cavity when radiation of characteristic frequency induced intermultiplet transitions resulting in decay to the ground state. The observed multiplet splittings were in sharp disagreement with previous theory, which has now been modified to reflect the experimental observations.

Deflection of beams in inhomogeneous electric fields can also be induced [5], if the atoms or molecules in the beam have a permanent dipole moment. Electric deflection experiments have not as yet been exploited to the degree that magnetic deflection has, but the method has been used to measure atomic polarizabilities. An apparatus analogous to that used for magnetic resonance experiments can be used to study the interactions of electron spins with rotational states of polar molecules.

Inhomogeneous electric and magnetic fields can also be used to focus atoms with selected quantum states. Such state-selected beams are discussed at greater length below, but it can be mentioned here that this technique was instrumental in the development of the first maser [6]. The required population inversion (excess of atoms in the higher-energy state of a transition) was achieved by passing a beam of ammonia molecules through an inhomogeneous electric field. Those molecules in an upper inversion state were focused, while those in lower states were defocused; the beam then entered a resonant cavity where the downward transition was induced. This stimulated emission added power to the stimulating radiation, and so the apparatus served as an amplifier at the resonant frequency.

Many-Particle Interactions

This classification includes experiments in which beams of ions or neutral particles, usually at rather high energy, are directed at dense targets, either gas-phase or solid.

For many years investigations have been made of the range (that is, the depth of penetration) of fast particles in dense targets [7, 8]. The particles, alpha particles for example, are slowed down by successive collisions with atoms in the target, and the range therefore represents a gross effect of many collisions. Total stopping power is defined as the sum over all possible inelastic processes of the product of the inelastic cross section and the energy lost for each inelastic process. Elastic scattering is not usually considered because the kinetic energy loss in elastic events is small. In order to calculate the total stopping power all inelastic events must be considered. For beams of relatively low mass the interactions are principally those involving overlap of the orbital electron clouds, but for heavier and faster beams nuclear collisions must also be considered. In calculations of ionic stopping power of dense gases the possibility of electron capture and loss (charge exchange) must be accounted for because the ion can become a neutral particle during part of its trajectory [9]. A partial stopping power is that due to only one species in the beam; it is preferable, when possible, to investigate this rather than the total stopping power. It still is a dense-target experiment; although only one kind of beam particle is monitored, the effects of multiple collisions are observed.

The experimental procedure used in measuring the partial stopping power of a fast neutral beam is as follows [10]. The beam is passed through a long, gas-filled cell placed in a magnetic field transverse to the axis of the beam. The field is strong enough to deflect any charged particles from the beam, so that the only particles that can emerge from the stopping chamber are those that have remained neutral throughout their trajectory. The emerging neutral particles are then ionized, and their energy distribution is determined with an electrostatic analyzer. Thus measurement of the average energy loss and the target thickness (the product of the density of the stopping gas and the path length) permits a calculation of the stopping power. Partial ionic stopping powers can be measured in a similar apparatus [11]. In this case the magnetic field confines the ions to an approximately circular path through the stopping chamber, and the entrance and exit slits are arranged so that only ions of a given charge are transmitted. Thus any ion suffering a charge transfer collision is excluded. The energy distribution of the transmitted beam is determined, as before, with an electrostatic analyzer. Essentially all the work being done in this field is at high energy, upwards of at least 10 keV.

Another similar experiment is the measurement of the energy-loss spectra of fast beams after they have passed through single crystals [12–14]. The ions tend to be channeled between crystal planes as they pass through the target, and the energy loss spectra show structure that is attributed to transverse oscillation of the particles in the channels. With the aid of appropriate theory the detailed structure of the spectra can be used to map out the stopping power in the channels and to determine the interatomic potential of the beam particle with respect to the atoms in the channel. This work is also restricted to high energies.

Another area of research that can be included in the category of dense targets is the interaction of molecular beams with solid surfaces. The investigation of sputtering, the erosion of metal surfaces caused by ion impact, has been carried out for more than 100 years. A number of theories have been advanced [15] to account for the different sputtering yields observed for different materials, but the field is still largely an empirical one. This field has been rather inactive for some time, but in recent years renewed interest has developed.

Considerable work has been reported [16], and much more is currently in progress, on molecular beam-surface interactions where the beam energy is too low to damage the surface. Much of the practical information sought in these experiments involves energy transfer from the beam to the surface, and measurement of the thermal accommodation coefficient has been a popular activity for some time. The accommodation coefficient for momentum transfer or for energy transfer is measured by directing a molecular beam at a surface and determining the momentum or energy transfer. The surface in question can be mounted on a torsion balance whose deflection is a direct measure of the momentum transfer, or the kinetic energy of the beam can be determined (by time-of-flight methods) before and after striking the surface. The two extremes of energy transfer are permanent absorption with complete energy accommodation and completely elastic scattering. In the latter case it is possible to arrange the conditions so that diffraction can be observed. The necessary conditions are that the substrate be a single crystal whose lattice spacing has the proper relationship to the de Broglie wavelength of the beam atom and to the angle of incidence. Diffraction of atomic beams from crystals provided an early verification of the concept of the de Broglie matter wave [17].

Energy transfer between the two extremes is more common, and many experiments have been performed in which a beam is directed at a surface and the reflected particles are detected. Under some conditions diffuse reflection is observed, suggesting that the beam has been adsorbed and re-emitted at a later time. In other systems the beam is reflected into distinct directions, and the results can often be related to the interaction potential of the beam-lattice system and to the lattice dynamics of the crystal.

A serious problem in this type of experiment is the difficulty of characterizing the surface and in separating the effects of the lattice from those of surface adsorption. Gas-surface interaction is an area of research that has only recently attracted many workers; significant progress in this difficult field can be expected.

Two-Particle Interactions

This classification includes experiments in which atomic or molecular beams are passed through thin gas targets or allowed to interact with another beam. When the experimental conditions are established correctly, the experimental results can be interpreted in terms of two-body collision dynamics. In this sense the experiments probe single molecular collisions.

General statements about two-particle interaction experiments are not easy to make because they can be done in such a variety of ways and over such a wide range of energies. All such experiments require one or more sources that produce collimated beams of particles whose velocities are in a known range. Also required is a detector of some sort whose output response is a measure of the beam intensity or flux. The basic experiment (which has many variations) is to establish single collisions of the particles in the beam with other particles and then to determine the changes in energy and momentum of both colliding partners. The dynamics of the collisions can be interpreted with this information, and fundamental details of atomic and molecular interactions can be determined.

Two rather different kinds of experiment can be noted as representing the basic techniques. They can be called the crossed-beam technique [18–20] and the attenuation technique [21, 22]. In the crossed-beam method two independent collimated beams (usually one or both are velocity selected) are produced and permitted to intersect, as shown in Fig. 6.2. Both beams are so dilute that the probability that a particle will make more than one collision

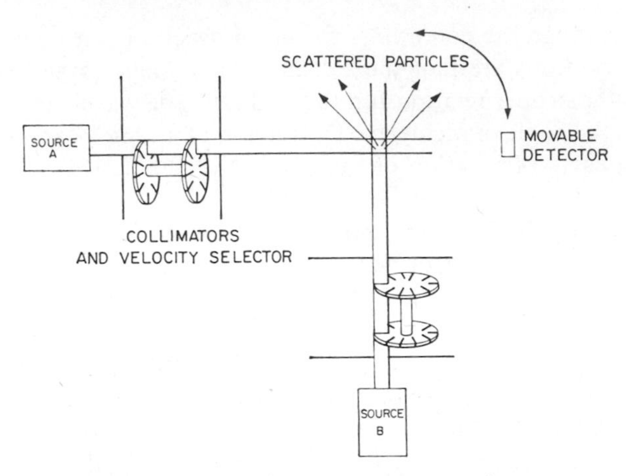

Fig. 6.2 Schematic crossed-beam experiment.

in the region of intersection is very small, and therefore any particles that are found to be scattered out of either beam result from single two-particle interactions. This method is used when the changes in momentum and energy are relatively large so that the scattering particles are well separated from the primary beams; the beam energies employed are accordingly relatively small.

The key to the experiment is the detector. It must have good resolution in order to determine the energy and momentum changes accurately, and it must be able to discriminate between the various chemical species that are involved in the experiment. In addition it should be quantitative, at least ideally. The latter criterion is difficult to achieve and is frequently disregarded when only qualitative information is sought. The requirement, however, can be met in several ways; for example, a mass spectrometer with a narrow input aperture has been used successfully. Other useful detectors are described in detail in a later section.

Providing that a suitable detector is selected, crossed-beam experiments can be performed which provide much fundamental information. For example, if beams of two species are allowed to interact and if a third species is found to have been produced, a chemical reaction has occurred. Careful investigation of the spatial distribution of the reaction product combined with the energy and momentum conservation laws permit estimates to be made of the energy transfer in the collision and shed some light on the mechanism of the reaction. Such studies can determine if a reaction occurs at large or at small separations of the reactants and whether some intermediate collision complex is formed. It is possible to produce beams with quite narrow distributions of velocity and to do an energy analysis (by a time-of-flight method, for example) of the reactive products. In addition, it is possible to produce beams in specific quantum states and with such detailed energy and spatial information that the reactive collisions can be extensively characterized.

The crossed-beam technique can also be employed to study nonreactive collisions. The basic technique is the same; the information sought is the energy and the spatial analysis of the scattered beam particles. If the collisions are elastic, that is, where there is no transfer of translational energy into any internal states of the colliding system, the experimental data may be analyzed to obtain information about the intermolecular potential. Where inelastic collisions occur the data can be interpreted to find out if the excitation can be assigned to a particular internal state (or group of states) and how the excitation energy can be transferred to other internal states.

The crossed-beam method is a powerful tool for the investigation of atomic and molecular collisions. Its particular strengths are flexibility and the lack of ambiguity in the interpretation of experiments. The only weakness of the method is that at the intersection of two molecular beams the intensities are so low that only a few percent of the beam particles collide, and the flux of scattered particles is therefore small and not easy to measure. It is difficult to devise a detector that gives a measure of absolute beam intensity, but it is especially difficult when the detector must be extremely sensitive at the same time. Consequently many crossed-beam experiments are performed with detectors without absolute calibrations. Nonetheless such experiments can be valuable, particularly in the study of reactive scattering. Even though it may be impossible to measure the absolute intensity of a particular reaction product, the fact that it is found to be scattered into a particular solid angle may provide much information about the details of the reactive collision.

The attenuation technique in molecular-beam research has a longer history and is still in wide use. It is useful over a large range of beam energy; it can employ conventional effusive oven beams, ion beams accelerated from an ion source, or high-velocity neutral beams produced by neutralizing fast ion beams. The interaction with the beam is achieved by allowing it to be attenuated as it passes through a thin layer of gas, as shown in Fig. 6.3. Two-particle interactions are assured by adjusting the gas pressure so that no beam

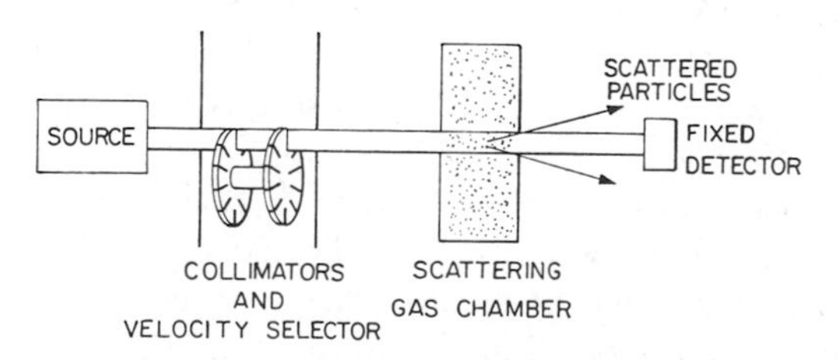

Fig. 6.3 Schematic attenuation experiment.

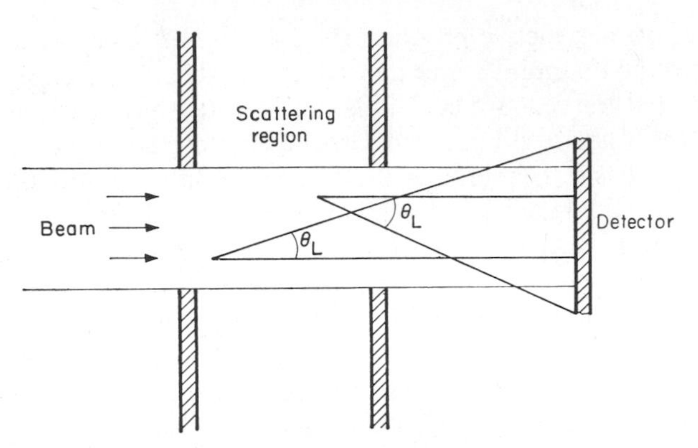

Fig 6.4 Detail of scattering region and detector geometry. The scattering gas is confined between two planes that establish the scattering region.

particle is likely to make more than one collision in traversing the layer of gas. Such a condition is established when the attenuation of the beam is linear with pressure. Even though single collisions predominate, the particle scattering density is much higher than it is in a crossed-beam experiment so that many more collisions take place and the requirements on the detector are less severe.

Both elastic and inelastic scattering can be studied in the same apparatus with only slight modification. Figure 6.3 is a schematic drawing of a typical apparatus for investigating elastic scattering. The beam attenuation resulting from scattering gas at density n is given by

$$I = I_0 \exp\,[-nS\,\Delta x(1 + \alpha)], \tag{6.1}$$

where I is the beam intensity after passing a distance Δx through the scattering gas, I_0 is the initial intensity, and S is the scattering cross section. The background term, $(1 + \alpha)$, accounts for the fact that a small part of the scattering gas escapes through the entrance and exit apertures and can scatter the beam in the regions outside the scattering box. The scattering cross section, S, is discussed in detail below, and it will be shown that measurements of S as a function of beam-particle velocity can by analyzed to obtain values of $V(r)$, the interaction potential for the colliding pair. Such an analysis is possible only when the scattering angle for each collision is known, so that a careful analysis of the geometry of the scattering region must be made. One possible experimental configuration is shown in Fig. 6.4. The cross section S is a function of the angle θ_L between the undeflected path of a particle and the deflected ray that causes that particle to strike an edge of the detector. As indicated in Fig. 6.4, every position within the scattering region is associated with a different value of θ_L and therefore a different value of S. Since a beam particle can be scattered at any point in the scattering region, it is clear that an experimental value of S actually represents an average over the finite dimensions of the beam, the scattering region, and the detector. The nature of the averaging process is discussed below. It should be noted that the dimensions in Fig. 6.4 are distorted and that the average scattering angles are generally small, on the order of 10^{-2} to 10^{-3} radians. This experiment makes selection of the detector relatively easy because (6.1) shows that only the intensity ratio is needed so that only a detector whose response is proportional to the beam intensity is satisfactory. Many such detectors are available.

As indicated above, inelastic scattering can be studied in a similar apparatus. For example, the cross section for ionization can be determined by measuring the initial beam intensity and the number of ionizing events. One arrangement involves collecting the slow ions with a transverse electric field and a collection electrode. Other possibilities for detecting inelastic events, such as observing light of characteristic wavelength in the scattering region, are conceptually straightforward but can be extremely difficult in practice.

Sometimes hybrid experiments using features from both crossed-beam and attenuation techniques are devised [23]. In these, the concept of a layer of scattering gas is employed, but the detector is moved to determine the intensity scattered out of the beam. These experiments are difficult to perform and to interpret, but when successful they provide much more information than the simpler attenuation experiment.

The category of two-particle interactions is introduced here only to provide a basis for the rest of this chapter. The first two categories were included in the discussion to provide an idea of the scope of molecular-beam research, but since they are generally of more interest to physicists or engineers than to chemists, they are not discussed further. The third category is of interest to physicists and chemists, but this chapter includes those topics that we feel would most appeal to chemists. The methods and techniques of molecular-beam research are common to both chemical and physical problems, so that the description of experimental details might be applicable to experiments in any of the three categories above; it is in the theoretical development that differences develop. Accordingly, the next section introduces the theoretical concepts needed for understanding two-particle experiments. The following sections describe some of the experimental methods used in molecular-beam research and discuss some of the significant results obtained from these experiments.

2 THEORETICAL BACKGROUND

The results of scattering experiments fall naturally into two categories: collisions that do not result in atomic rearrangements and collisions that do.

Since most chemistry involves rearrangement collisions, these are of greatest interest to the chemist. Unfortunately, they are also the most difficult to study, both experimentally and theoretically, and the bulk of our knowledge is presently limited to collisions that do not involve rearrangement. But even these collisions can supply considerable insight into chemical phenomena. In the first place the study of simple elastic collisions supplies the conceptual basis for the description of more complicated collisions. Secondly, elastic collisions give information on the interactions between atoms and small groups of atoms that can be applied to such chemically interesting phenomena as the conformation of large molecules and the degradation of the energy of "hot atoms." Thirdly, inelastic collisions that do not involve rearrangement but only changes in internal energy (rotational, vibrational, and electronic) can be considered prototypes of simple chemical reactions.

This section accordingly discusses elastic and inelastic collisions first, and then rearrangement collisions. All can be studied over a wide range of energies.

Elastic and Inelastic Collisions

For reasons made clear in the next section, experimental molecular-beam research can be divided into three energy regions. A similar division can also be made for theoretical reasons, depending on the relation of the de Broglie wavelength to molecular dimensions. When the wavelength is small compared to molecular dimensions, classical mechanics furnishes a good description of molecular collisions. This is the usual situation for high-energy beams. However, for thermal-energy beams the wavelength is often comparable to molecular dimensions, so that quantum effects are apparent. These take the form of wave-interference effects, which contribute structure to the observed scattering cross sections. Classical mechanics no longer furnishes a good description, but semiclassical approximations may be satisfactory. For intermediate-energy beams the situation is less clear; sometimes quantum effects are important, sometimes not, depending on particular experimental circumstances.

The principal aim of elastic scattering measurements is the determination of the potential energy of interaction between the particles. First a brief account of the mathematical description of elastic scattering is given, and then the scattering potential is related to other properties such as transport coefficients and the equation of state. Much of this carries over, in a general way, to the discussion of inelastic and rearrangement collisions. Each of the three energy regions is then discussed in turn, with special reference to the type of potential-energy information obtainable, and some indication of the application of this information to other problems.

Formalism and Definitions

The discussion of scattering is entirely in terms of the center-of-mass coordinate system, in which the collision is viewed as if an observer were located at the center of mass of the colliding system. The relation between the center of mass and the laboratory coordinate system depends only on conservation of momentum and energy, and is hence valid in both classical and quantum mechanics. Good accounts of the transformation between the coordinate systems are available in many books [24].

The angular distribution of particles scattered from a beam is described both classically and quantum-mechanically by a differential cross section $\sigma(\omega)$, defined by the statement that the number of particles scattered into solid angle $d\omega$ per unit time is $I\sigma(\omega)\,d\omega$, where I is the flux density of the incident beam in particles per unit area per unit time, so that $\sigma(\omega)$ has the dimensions of area. In molecular-beam scattering by gas targets there is usually axial symmetry of the beam and random orientation of target particles, in which case $\sigma(\omega)$ depends only on the polar deflection angle θ and not on the azimuth angle ϕ, and can be written as

$$\sigma(\omega)\,d\omega = \sigma(\theta)\sin\theta\,d\theta\,d\phi. \tag{6.2}$$

Often the angular distribution of scattered particles is not measured directly, but only the fraction of the beam that is scattered into all angles greater than θ_0, the angular aperture of the apparatus. This may be called the integrated cross section $S(\theta_0)$, and is given by

$$S(\theta_0) = 2\pi\int_{\theta_0}^{\pi}\sigma(\theta)\sin\theta\,d\theta. \tag{6.3}$$

The true total scattering cross section is obtained when $\theta_0 = 0$.

The maximum amount of direct information from a scattering experiment is thus obtained from the differential cross section as a function of angle and initial relative kinetic energy. The extension of this description to inelastic and simple rearrangement collisions is straightforward. A species A in internal energy state i collides with species B in internal state j with kinetic energy of relative motion ϵ; the collision produces species C in state k and D in state l, with kinetic energy of relative motion ϵ' and at angles θ, ϕ:

$$A(i) + B(j) = C(k) + D(l). \tag{6.4}$$

This will be described by a differential cross section $\sigma(k, l, i, j, \epsilon, \theta, \phi)$, just as for elastic collisions; the essential difference is that more quantities must be specified to describe the collision than were needed for the elastic case. If C and D are different from A and B, then there is said to be a chemical reaction if the speaker is a chemist, or a rearrangement collision if the speaker is a physicist. If C and D are the same as A and B, but states k and l are different from states i and j, then an inelastic collision is said to have occurred.

We record here for future reference the relation between the differential cross section $\sigma(\theta)$ and the potential $V(r)$, and defer comments on the extraction of $V(r)$ from $\sigma(\theta)$ (the inversion problem) to the discussions of scattering in the three different energy ranges.

In a classical description of the scattering of two particles with reduced mass μ and relative velocity $\vec{v}$ the collision trajectories are well-defined, and for a given kinetic energy of relative motion each value of the impact parameter b results in a definite value of the scattering angle, as shown in Fig. 6.5. Thus all the particles falling on the target area between b and $b + db$ will be scattered into angles between θ and $\theta + d\theta$; conservation of particles and the definition of $\sigma(\theta)$ then allows us to write $2\pi b\,|db| = 2\pi\sigma(\theta)\sin\theta\,|d\theta|$, from which the expression for $\sigma(\theta)$ is obtained:

$$\sigma(\theta) = \frac{b}{\sin\theta\,|d\theta/db|}\,. \tag{6.5}$$

The integrated cross section is then

$$S(\theta_0) = 2\pi\int_{b(\pi)}^{b(\theta_0)} b\,db = \pi b^2(\theta_0), \tag{6.6}$$

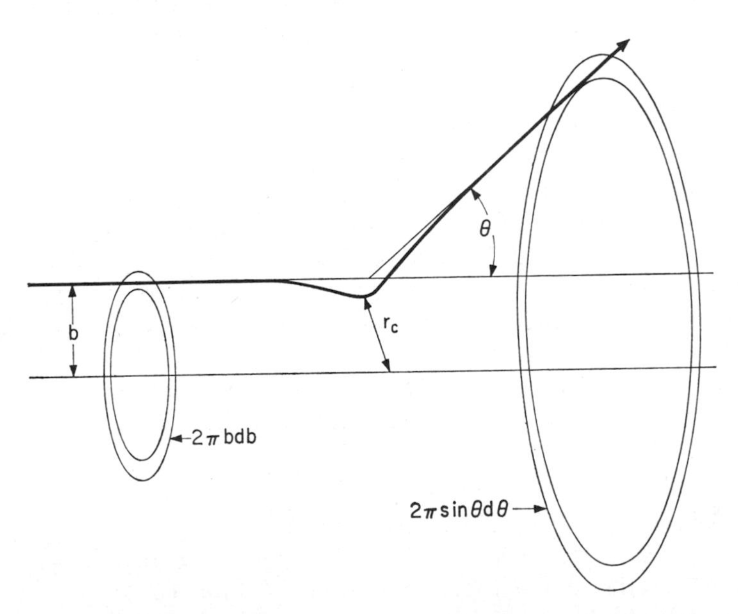

Fig. 6.5 Classical collision trajectory. All particles incident on the ring of area $2\pi b\,db$ appear after scattering on the ring of area $2\pi\sin\theta\,d\theta$.

since $b(\pi) = 0$ (head-on collision). The dependence of the cross section on $V(r)$ is implicitly contained in the relation between θ and b appearing in (6.5) and (6.6) [24],

$$\theta(b, \epsilon) = \pi - 2b \int_{r_c}^{\infty} \left[1 - \left(\frac{b}{r}\right)^2 - \frac{V(r)}{\epsilon}\right]^{-1/2} \frac{dr}{r^2}, \tag{6.7}$$

where r_c is the distance of closest approach, given by

$$1 - \left(\frac{b}{r_c}\right)^2 - \frac{V(r_c)}{\epsilon} = 0. \tag{6.8}$$

If the potential is not purely repulsive or attractive, several values of b may lead to scattering into the same θ. In such cases the expression for $\sigma(\theta)$ becomes a sum over the several values of b, but the expression for $S(\theta_0)$ involves both sums and differences [25].

Precise collision trajectories cannot be specified in quantum mechanics, and a wavelike description of scattering is necessary [24]. The total wave function consists of an incident wave plus a scattered wave; at distances far removed from the collision region the incident wave is a plane wave and the scattered wave is a spherical wave. The asymptotic form of the wave function is thus

$$\psi = \psi_{in} + \psi_{\text{scatt}} \sim e^{i\kappa z} + \frac{e^{i\kappa r}}{r} f(\theta), \tag{6.9}$$

where $\kappa = \mu v/\hbar$ is the wave number of relative motion and $f(\theta)$ is the scattering amplitude. The differential cross section is given by

$$\sigma(\theta) = |f(\theta)|^2. \tag{6.10}$$

In principle $f(\theta)$ must be found by solution of the Schrödinger equation. The usual procedure for elastic scattering is to expand the wave function into a series of partial waves, one for each value of the angular momentum quantum number, l. The incident and scattered partial waves then differ only in phase, and the problem is reduced to evaluation of the phase shift for each partial wave. The scattering amplitude is found to be

$$f(\theta) = (2i\kappa)^{-1} \sum_{l=0}^{\infty} (2l + 1)(e^{2i\delta_l} - 1)P_l(\cos \theta), \tag{6.11}$$

where δ_l is the phase shift of the lth partial wave and $P_l(\cos \theta)$ is a Legendre polynomial. The δ_l must usually be found by numerical integration of the radial Schrödinger equation, although useful approximations are sometimes available. The expression given in (6.3) for the integrated cross section still holds if the lower limit is permitted to extend to $\theta_0 = 0$, but it yields a simple result only for the total cross section,

$$S(0) = \frac{4\pi}{\kappa^2} \sum_{l=0}^{\infty} (2l + 1) \sin^2 \delta_l. \tag{6.12}$$

A semiclassical approximation is often useful for atomic and molecular scattering, in which the summations over phase shifts are replaced by integrations and the phase shifts are calculated by the Jeffreys-Wentzel-Kramers-Brillouin-Langer (JWKBL) approximation,

$$\delta_l \approx \delta(b) = \kappa \int_{r_c}^{\infty} \left[1 - \left(\frac{b}{r}\right)^2 - \frac{V(r)}{\epsilon}\right]^{1/2} dr - \kappa \int_{b}^{\infty} \left[1 - \left(\frac{b}{r}\right)^2\right]^{1/2} dr, \tag{6.13}$$

where the impact parameter is

$$b = \frac{l + \frac{1}{2}}{\kappa}. \tag{6.14}$$

The relation between the phase shifts and the classical deflection angle in this approximation is

$$\theta(b) = 2\frac{d\delta_l}{dl} = \frac{2}{\kappa}\frac{d\delta(b)}{db}. \tag{6.15}$$

Fairly extensive studies of the semiclassical approximation in atomic and molecular scattering are available [22, 26–28].

Relation to Other Phenomena

The complete differential cross section, if measured over a complete range of θ and a wide range of ϵ, contains a great deal of information. However, it is almost never fully measured. In practice, only partial information on the differential cross section is available—the ranges of angles and energies are limited by experimental difficulties. From such partial scattering information, therefore, only partial information on the potential can be extracted. In fact, to obtain any sensible answer at all it is usually necessary to inject some independent information on the potential, such as whether or not it is monotonic, and if so, whether it is attractive or repulsive, or indeed whether the scattering is caused by a single potential rather than by two or more potentials. Although in practice only partial information on $\sigma(\theta)$ is known, it is interesting to ask what phenomena other than scattering could be calculated if $\sigma(\theta)$ were known completely.

In the first place only binary collision phenomena could be calculated rigorously, since the scattering under discussion here involves collisions of only two molecules at a time. If only elastic collisions are possible, this means that only the two-body properties contained in the dilute-gas transport coefficients and the equation of state (viscosity, thermal conductivity, diffusion, thermal diffusion, and the second virial coefficient) can be obtained. If inelastic collisions are also possible, the absorption and dispersion of sound waves in gases and similar related phenomena can be calculated.

Information on rearrangement collisions extends the range to include calculations of bimolecular chemical rate coefficients in the gas phase. But strictly speaking, phenomena in dense gases, liquids, and solids would remain out of reach, as would chemical reactions that required three or more molecules to interact simultaneously. Of course, one always tries to make approximations that will extend the range of usefulness of strictly two-body information. Such approximations usually aim to build up many-body phenomena as collections of two-body phenomena: for example, the assumption of pair-wise additive interatomic potentials, of effective potentials such as appear in optical models, or the representation of a ternary collision as a transient orbiting binary pair struck by a third particle. Many of these approximations are astonishingly successful and are discussed in subsequent sections, but it is well to remember that they are approximations and are subject to failures.

This brings up a second limitation. A well-founded theory is required that connects the observable phenomena to $\sigma(\theta)$, and such a theory may not be simple. It is at this point that it is necessary to include the realm of general statistical mechanics, with its great fascination and many unsolved fundamental problems. For instance, if only elastic collisions occur, an excellent theory is available from which the viscosity, thermal conductivity, diffusion, and thermal diffusion coefficients of a gas can be calculated to any desired accuracy, once the differential cross section is given. But the calculation of the second virial coefficient from $\sigma(\theta)$ alone is a much more difficult task, for it requires knowledge of the bound two-body states. Moreover, if inelastic collisions can occur, the basic connecting theory for the transport coefficients becomes much more complex and involves some intractable features that are still not completely resolved. Finally, if rearrangement collisions can occur, then the theory is very complicated indeed, and only parts of it are as yet worked out.

As an illustration of these remarks, consider the viscosity of a dilute gas in which only elastic collisions occur. The Chapman-Enskog kinetic theory of gases gives the viscosity in terms of some integrations over the differential cross section [29, 30],

$$\frac{1}{\eta} = \frac{16}{5}\left(\frac{\pi}{mkT}\right)^{1/2} \int_0^\infty \gamma^7 e^{-\gamma^2} \left[\int_0^\pi (1 - \cos^2\theta)\sigma(\theta, \epsilon)\sin\theta\, d\theta\right] d\gamma, \quad (6.16)$$

where η is the viscosity, m the molecular mass, k the Boltzmann constant, and $\gamma^2 = \epsilon/kT$. Thus if $\sigma(\theta, \epsilon)$ is known for all values of θ and ϵ, η can be calculated for all temperatures. However, if $\sigma(\theta, \epsilon)$ is known only over a limited range of ϵ, η is reliably determined only over a limited range of T, corresponding roughly to the region of the maximum in the weighting function $\gamma^7 e^{-\gamma^2}\, d\gamma$ (that is, to $kT \approx \frac{1}{3}\epsilon$). Similar remarks apply to the other transport coefficients.

There are two interesting features about (6.16) and similar equations. The first is that there seems to be no need to determine the potential energy $V(r)$, for the desired property can be calculated directly from $\sigma(\theta, \epsilon)$. This appears to contradict the earlier remark about the principal aim of elastic scattering being the determination of $V(r)$. The second feature is that the calculation throws away most of the hard-won information contained in $\sigma(\theta, \epsilon)$ by integrating over all angles and all values of γ. This feature is aggravated if inelastic collisions are possible, for then the calculation further involves summation over all internal energy states [31]. Thus it appears that attempts to measure $\sigma(\theta, \epsilon)$ involve unnecessary complexity.

The rationalization of these features lies in the fact that scattering is studied in the hope of understanding enough about collisions and interactions to be able to make predictions and calculations on the basis of incomplete information. Furthermore, only fragmentary information on $\sigma(\theta, \epsilon)$ ever exists. Thus the aim is to proceed from partial information on $\sigma(\theta, \epsilon)$ to something deeper—the potential—from which useful calculations can be made. Such a program requires a well-founded theory and a careful analysis.

As an example, consider the calculation of transport properties from scattering experiments in which only small-angle scattering is studied. The weighting factor of $(1 - \cos^2 \theta)$ in (6.16) effectively discards small-angle deflections as far as any influence on the viscosity (as well as the other transport coefficients) is concerned. Thus one may get a superficial impression that small-angle scattering measurements are useless for determining transport coefficients, and this is true as far as only $\sigma(\theta, \epsilon)$ is concerned. The picture is quite different in terms of an analysis through the potential. In terms of internuclear separations small-angle scattering probes only the outer fringes of the potential for a given collision energy. However, at a lower energy even a head-on collision can penetrate only that far. Therefore the small-angle scattering at high energies probes that part of the potential that controls the large-angle scattering at much lower energies, and this is precisely what is needed to calculate the viscosity at temperatures corresponding to those low energies. For example, study of scattering at about 10^{-3} rad with a beam having kinetic energies in the range of 10^3 eV probes that part of the potential having a magnitude of about 1 eV. This in turn determines a property such as viscosity at temperatures of a few thousand degrees Kelvin.

In short, by analyzing scattering in terms of something more fundamental than the differential cross section the ability to correlate and predict can be greatly expanded. In the case of elastic scattering the potential is the more fundamental quantity, and such prediction ability is highly developed. For reactive scattering, on the other hand, as yet mostly only correlation theories or phenomenological theories similar to and modeled on those in nuclear physics are available—stripping reactions, complex formation, R-matrix theory, optical models, and so on.

The reader who wishes a comprehensive treatment of the relation of the intermolecular potential to transport properties and virial coefficients should refer to specialized monographs [29, 30, 32].

Rearrangement Collisions

Formalism and Definitions

The formalism used to describe elastic and inelastic two-body scattering may be used as a starting point for discussing rearrangement collisions. The general case is considered in which two isolated particles in given internal quantum states are converted by a single collision into two different isolated particles in specified internal quantum states. The treatment is therefore limited to reactions that are mechanistically bimolecular and kinetically second order. The origin of the limitation is the same as that previously discussed in connection with macroscopic transport and equilibrium properties of dilute gases.

The prototype reaction used earlier in connection with inelastic scattering is

$$A(i) + B(j) \rightleftharpoons C(k) + D(l). \tag{6.17}$$

$C(k)$ and $D(l)$ are now products of different chemical species than the reactants $A(i)$ and $B(j)$, and i, j, k, and l are the specified internal quantum states. The rate of reaction in the forward direction, Z_R, may be written as

$$Z_R = k_f c_{A(i)} c_{B(j)}, \tag{6.18}$$

where k_f is the rate constant and $c_{A(i)}$ and $c_{B(j)}$ are the volume concentrations of the reactants. On a molecular basis the reaction rate, which is simply the number of reactive collisions per unit time and unit volume, is proportional to the product of the number densities of the reactants and the average velocity with which they approach one another and, in addition, to the differential cross section for reaction $\sigma_R(k, l, i, j, p, \omega)$ which, as indicated, is dependent on the internal states of the four species as well as the initial relative momentum $\vec{p}$ and the solid angle ω of scattering. The reaction rate may therefore be written

$$Z_R = \int_0^{4\pi} \int_{-\infty}^{+\infty} \int_{-\infty}^{+\infty} c_{A(i)} c_{B(j)} \left| \frac{\vec{p}}{\mu} \right| \sigma_R(k, l, i, j, p, \omega) \times F(\vec{p}_{A(i)})\, d\vec{p}_{A(i)} F(\vec{p}_{B(j)})\, d\vec{p}_{B(j)}\, d\omega, \tag{6.19}$$

where the initial relative momentum corresponds to the initial kinetic energy of relative motion ϵ or initial relative velocity $\vec{v}$, μ is the reduced mass of $A(i)$ and $B(j)$, and $F(\vec{p}_{A(i)})$ and $F(\vec{p}_{B(j)})$ are the momentum distribution functions of the reactants. The differential cross section per unit solid angle for chemical reaction is similar in character to the differential cross section $\sigma(k, l, i, j, \epsilon, \theta, \phi)$ referred to earlier in connection with inelastic scattering.

An alternative and in some ways more useful method of deriving an expression for the reaction rate is through the impact parameter b defined earlier. It is shown in Fig. 6.5. The function $P(k, l, i, j, p, b, \phi)$ can be defined as the probability that a collision will result in reaction. Of those collisions involving impact parameters between b and $b + db$, the number per unit time and unit volume that result in reaction is the product of the density of reactants having specified differential ranges of momentum, their initial relative velocity and the reaction probability. Thus the total reactive collision frequency is

$$Z_R = \int_0^{c_{B(j)}} \int_0^{c_{A(i)}} \int_0^{2\pi} \int_0^{\infty} P(k, l, i, j, p, b, \phi) \left| \frac{p}{\mu} \right| b\, db\, d\phi\, dc_{A(i)}\, dc_{B(j)}. \quad (6.20)$$

Comparison of (6.19) and (6.20) establishes the following relationships:

$$P(k, l, i, j, p, b, \phi) b\, db\, d\phi = \sigma_R(k, l, i, j, p, \omega)\, d\omega$$

$$dc_{A(i)}\, dc_{B(j)} = c_{A(i)}\, F(\vec{p}_{A(i)})\, d\vec{p}_{A(i)}\, c_{B(j)}\, F(\vec{p}_{B(j)})\, d\vec{p}_{B(j)}\, . \quad (6.21)$$

In order to perform the integrations in (6.20) it is clear from (6.21) that the momentum distribution functions $F(\vec{p})$ must be known. It has been shown [33] that the equilibrium, or Maxwellian, distribution is a satisfactory approximation in almost all cases. Some additional transformations are required to relate the components of the momenta in (6.21) to components of the center-of-mass momenta and the relative momenta and to reduce the integrations to a single one over the relative velocity v. Details of these transformation procedures, in which it is customary to use velocities rather than the corresponding momenta, are found in texts on kinetic theory of gases or kinetics of chemical reactions [34, 35]. The result of this treatment of (6.20) leads to

$$Z_R = \left(\frac{\mu}{kT}\right)^{3/2} \left(\frac{2}{\pi}\right)^{1/2} c_{A(i)} c_{B(j)} \int_0^{\infty} \int_0^{2\pi} \int_0^{\infty} P(k, l, i, j, v, b, \phi) v^3 \times \exp\left(-\frac{\mu v^2}{2kT}\right) b\, db\, d\phi\, dv. \quad (6.22)$$

In analogy with the relations established above between the total scattering cross section and the differential cross section we express the total cross section for chemical reaction as

$$S_R(k, l, i, j, p) = \int_0^{4\pi} \sigma_R(k, l, i, j, p, \omega)\, d\omega = \int_0^{\infty} \int_0^{2\pi} P(k, l, i, j, p, b\ \phi) b\, db\, d\phi. \quad (6.23)$$

It follows from (6.18) through (6.23) that the temperature-dependent rate constant for the forward reaction between isolated $A(i)$ and $B(j)$ molecules is a weighted average, over all initial relative momenta, of the product of the total cross section for chemical reaction and the initial relative velocity. Thus

$$k_f(k, l, i, j, T) = \left(\frac{\mu}{kT}\right)^{3/2}\left(\frac{2}{\pi}\right)^{1/2}\int_0^\infty [S_R(k, l, i, j, v)v]v^2 \exp\left(-\frac{\mu v^2}{2kT}\right) dv. \quad (6.24)$$

Since the initial kinetic energy of relative motion, $\epsilon = p^2/2\mu = \mu v^2/2$, is also a convenient and frequently used variable, a useful third relation is

$$k_f(k, l, i, j, T) = \left(\frac{2}{kT}\right)^{3/2}\left(\frac{1}{2\pi}\right)^{1/2}\int_0^\infty \left[S_R(k, l, i, j, \epsilon)\left(\frac{2\epsilon}{\mu}\right)^{1/2}\right]\epsilon^{1/2} \exp\left(-\frac{\epsilon}{kT}\right) d\epsilon. \quad (6.25)$$

Since any function of p may also be written as a function of v or ϵ, the designations of the functional dependence of P and S_R in (6.22), (6.24), and (6.25) have been changed, when necessary, to be consistent with the selected variables of integration.

Although there has been significant recent progress in inelastic and reactive scattering in selecting beam particles in specific internal quantum states [22, 36–42], most beam studies of rearrangement collisions have involved reactants and products in a variety of quantum states. The difficulties of state selection are such, however, that much future work will almost certainly involve reactants and products in a wide variety of internal states. The formal procedure for removing the restriction that $A(i)$, $B(j)$, $C(k)$, and $D(l)$ represent isolated molecules each in a single quantum state is therefore of interest. When this restriction is removed, (6.17) is applicable to a mixture of reactants and products in all possible quantum states, and the relation for the rate constant for the forward direction becomes

$$k_f(T) = \left(\frac{2}{kT}\right)^{3/2}\left(\frac{1}{2\pi}\right)^{1/2}\int_0^\infty \left[S_R(\epsilon)\left(\frac{2\epsilon}{\mu}\right)^{1/2}\right]\epsilon^{1/2} \exp\left(-\frac{\epsilon}{kT}\right) d\epsilon, \quad (6.26)$$

where

$$S_R(\epsilon) = \sum_{i,j} x_{A(i)} x_{B(j)} \left[\sum_{k,l} S_R(k, l, i, j, p)\right]. \quad (6.27)$$

The fractions of $A(i)$ and $B(j)$ in given quantum states have been designated as $x_{A(i)}$ and $x_{B(j)}$. If these fractions are assumed to be equilibrium fractions at all times with respect to $x_{C(k)}$ and $x_{D(l)}$, then $k_f(T)$ in (6.26) is time-independent.

In general, the energy dependence of the total cross section for chemical reaction that appears in (6.26) will be determined by the dynamical motions

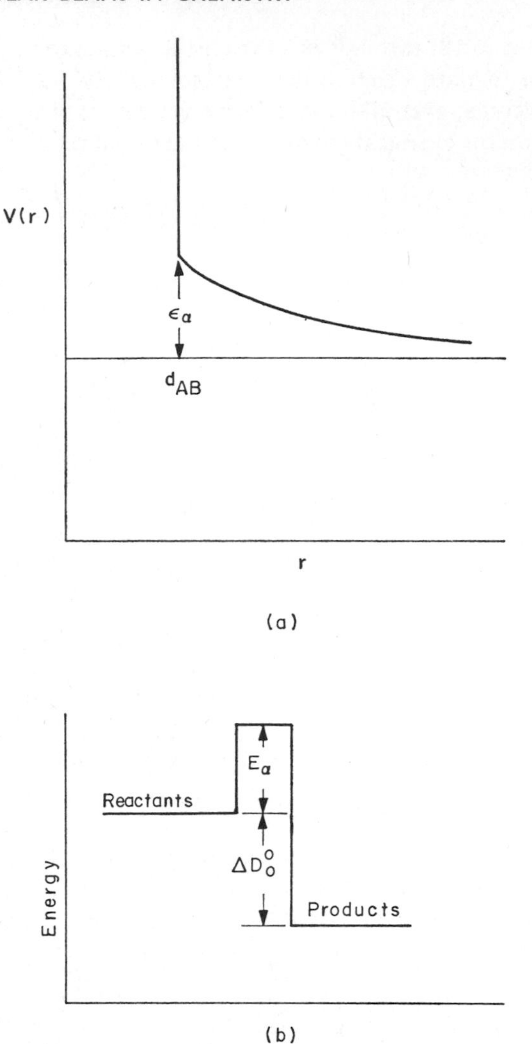

Fig. 6.6 (*a*) Variation of potential for repelling hard spheres that react when the energy of relative motion reaches a threshold value of ϵ_a. (*b*) Energy-level diagram showing the definition of E_a and $\Delta D_0{}^0$.

of the reactants and products on a multidimensional surface that represents the variation in potential energy of the reacting system as its components change their internuclear separations and spatial orientations. For molecular interactions that are physically realistic the problem of determining the energy dependence of $S_R(\epsilon)$ is formidable, and the likelihood of obtaining an analytical expression for $k_f(T)$ is very small. However, with a sufficiently simple model a qualitatively correct picture of the usual behavior of $S_R(\epsilon)$ and $k_f(T)$ can be obtained.

One model assumes that the reactants are structureless, repelling hard spheres that change into products when the kinetic energy of relative motion along the line of centers exceeds a threshold value ϵ_a. For this model, whose potential energy varies with internuclear separation as shown in Fig. 6.6, it has been shown [43] that

$$S_R(\epsilon) = \pi\, d_{AB}^2 \left[\frac{1-\epsilon_a}{\epsilon}\right] \quad \text{for} \quad \epsilon > \epsilon_a \qquad (6.28)$$

$$S_R(\epsilon) = 0 \quad \text{for} \quad \epsilon < \epsilon_a.$$

The quantity $\pi\, d_{AB}^2$ is the cross-sectional area of the sphere that envelops the centers of the reacting spheres when they are in contact. Integration of (6.26) with $S_R(\epsilon)$ given by (6.28) leads to

$$k_f(T) = \pi\, d_{AB}^2 \left(\frac{8kT}{\pi\mu}\right)^{1/2} \exp\left(-\frac{\epsilon_a}{kT}\right). \qquad (6.29)$$

Since ϵ_a is usually much greater than kT, the temperature dependence of $k_f(T)$ is largely determined by the exponential term in (6.29), and a differentiation produces the approximate relation

$$d \ln k_f(T)/dT \cong \frac{\epsilon_a}{kT^2}\,. \qquad (6.30)$$

Equation (6.30) describes the temperature dependence of the rate constant for a large number of bimolecular reactions for which a plot of $\ln k_f(T)$ versus $1/T$ gives a straight line whose slope, multiplied by k, is the so-called Arrhenius activation energy.

An addition to the notation is introduced here in order to conform to common usage. Most kineticists use molar energies rather than the molecular energies that have been used so far. The molar energy, E, is defined as $N\epsilon$, where N is Avogadro's constant. Both energies will be used interchangeably.

Collision Dynamics

In understanding microscopic features of reactive scattering it is often helpful to supplement the preceding theoretical treatment with a discussion of the conservation laws of energy and momentum [44–47]. If E is the initial kinetic energy of relative motion of the reactants, Z the associated internal energy (rotational, vibrational, or electronic), and $\Delta D_0{}^0$ the difference in dissociation energy of products measured from zero-point levels, as shown in Fig. 6.6*b*, then the final kinetic energy of relative motion of the products E' and the corresponding internal energy Z' are given by

$$E' + Z' = E + Z - \Delta D_0{}^0. \qquad (6.31)$$

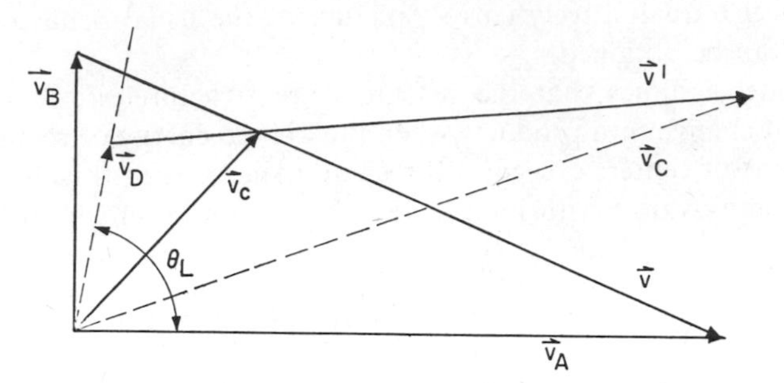

Fig. 6.7 Vector diagram of initial and final velocities of a collision.

In view of the fact that in most reactive scattering experiments neither reactants nor products are likely to be in specifically selected internal quantum states, the simplified designations A, B, C, and D will be used rather than $A(i)$, $B(j)$, $C(k)$, and $D(l)$ for reactants and products. Thus the center-of-mass velocity vector $\vec{v}_c$, also known as the centroid, is defined in terms of the masses and the initial velocity vectors of the reactants by the relation

$$\vec{v}_c = \frac{m_A\vec{v}_A + m_B\vec{v}_B}{m}, \tag{6.32}$$

where m represents the total mass of the reactants, $(m_A + m_B)$, which is of course also equal to that of the products. The total kinetic energy of the system is the sum of the kinetic energy of the reactants, which can be shown to be

$$\tfrac{1}{2}m_Av_A{}^2 + \tfrac{1}{2}m_Bv_B{}^2 = \tfrac{1}{2}(m_A + m_B)v_c{}^2 + \tfrac{1}{2}\mu v^2. \tag{6.33}$$

The first term on the right in (6.33) is the kinetic energy of the total mass moving at the velocity of the center of mass. This energy can be shown to be constant and has therefore been omitted from both sides of (6.31). If $\vec{v}_A$ and $\vec{v}_B$ are plotted with a common origin, as shown in Fig. 6.7, the initial relative velocity, $\vec{v} = \vec{v}_A - \vec{v}_B$ intersects $\vec{v}_c$ at a point determined by the relations

$$\vec{v}_A - \vec{v}_c = \frac{m_B\vec{v}}{m} \tag{6.34}$$

and

$$\vec{v}_B - \vec{v}_c = -\frac{m_A\vec{v}}{m}. \tag{6.35}$$

Equations (6.34) and (6.35) state that at large internuclear separations an observer stationed on the moving center of mass sees the reactants approach with velocities that are inversely proportional to their total mass and parallel to the initial relative velocity. Similarly the recoil velocities with which the products recede from the moving center of mass are given by the relations

$$\vec{v}_C - \vec{v}_c = \frac{m_D \vec{v}'}{m} \tag{6.36}$$

and

$$\vec{v}_D - \vec{v}_c = -\frac{m_C \vec{v}'}{m}, \tag{6.37}$$

where the final relative velocity $\vec{v}' = \vec{v}_C - \vec{v}_D$ may take any direction in space and therefore may be considered as pivoting around the end of $\vec{v}_c$. The vector diagram in Fig. 6.7 is a graphical representation of the relations of (6.34) through (6.37) for a binary collision involving mutually perpendicular initial velocity vectors. For the case illustrated, θ_L is the angle in the laboratory system, measured from $\vec{v}_A$ and in the plane of $\vec{v}_A$ and $\vec{v}_B$, into which the product D is scattered with velocity $\vec{v}_D$. When the collision between A and B is elastic, Z' and Z are equal, $\Delta D_0{}^0$ is zero, and E and E' are identical, or equivalently, $|\vec{v}'| = |\vec{v}|$. For a reactive collision, however, $|\vec{v}'|$ is restricted to $(2E'/\mu')^{1/2}$, where μ' is $m_C m_D/m$, and E' is subject to the condition specified by (6.31), in which neither $Z' - Z$ nor $\Delta D_0{}^0$ will in general be equal to zero.

It is interesting to see how the conservation relations of (6.33) through (6.37) and the vector diagram in Fig. 6.7 can be used to correlate or even to predict certain experimental features of reactive scattering. For example, a special case can be considered where $m_C \ll m_D$ so that $m_C/m \ll 1$, and also where E' is not much larger than E, so that $|\vec{v}'|$ is approximately equal to $|\vec{v}|$. Equation (6.37) then states that the velocity vector $\vec{v}_D$ may be confined to a small cone around $\vec{v}_c$ so that the product D of mass m_D will be found near the centroid. The imposed special conditions therefore result in an increased flux or intensity of D around a small range of θ near the centroid. However, because the intensity of D is so sharply peaked, it will be difficult to secure adequate angular resolution for accurate measurement of the differential scattering cross sections in this angular region. At the other extreme, if $|m_C\vec{v}'/m|$ is sufficiently larger than $|\vec{v}_C|$ the product D with velocity $\vec{v}_D$ may be found at any laboratory angle. Herschbach and coworkers [45, 46] have given a detailed analysis of the collision dynamics on which the conclusions for these special cases are based.

The discussion of collision dynamics thus far has illustrated the type of microscopic information that may be obtained by invoking the conservation laws for total energy and linear momentum. In general the total angular momentum is also a collisional invariant, and additional microscopic information can be obtained by requiring that this particular quantity also be conserved. The initial angular momentum of the system is a vector sum of the

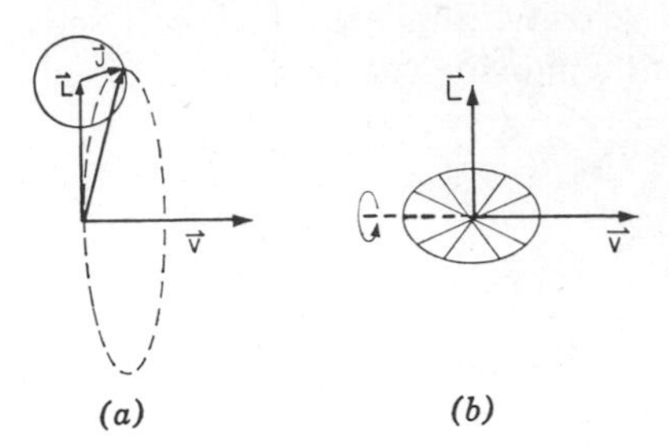

Fig. 6.8 (*a*) Orientation of initial angular momentum vectors. (*b*) Distribution of recoil vectors.

orbital angular momentum $\vec{L} = \mu\vec{v}b$ and the rotational angular momenta $\vec{J}_A$ and $\vec{J}_B$ of the reactants. If $\vec{J} = \vec{J}_A + \vec{J}_B$ and $\vec{J}' = \vec{J}_C + \vec{J}_D$, the conservation law for total angular momentum may be written as

$$\vec{L}' + \vec{J}' = \vec{L} + \vec{J}, \tag{6.38}$$

where $\vec{L}' = \mu'\vec{v}'b'$. For the special case of elastic scattering in a central potential $\vec{J}' = \vec{J}$ so that $\vec{L}' = \vec{L}$. Although the initial orbital angular momentum $\vec{L}$ must be perpendicular to the initial relative velocity $\vec{v}$, the initial rotational angular momentum $\vec{J}$ may have any spatial orientation as shown in Fig. 6.8*a*, which also shows the possible spatial orientation of the initial total angular momentum vector. Frequently $\vec{L} \gg \vec{J}$ and $\vec{L}' \gg \vec{J}'$, and under these conditions the final velocity vector $\vec{v}'$, which must be perpendicular to $\vec{L}'$, will also be almost perpendicular to $\vec{L}$. However, even if the final relative velocity vectors are assumed to be uniformly distributed about the initial orbital angular momentum, an anisotropy in the spatial distribution of the product D can be expected. This conclusion can be visualized by referring to Fig. 6.8*b* in which the final relative velocity vectors, for a particular $\vec{v}$ and $\vec{L}$, are represented as uniformly spaced radii in the circle perpendicular to $\vec{L}$. For a selected value of $\vec{v}$ it is possible to have all values of $\vec{L}$ that are uniformly distributed in directions perpendicular to $\vec{v}$, so that the spatial distribution of $\vec{v}'$ vectors is obtained by rotating Fig. 6.8*b* about $\vec{v}$ as shown. In the sphere generated by the rotation of the circle containing the final relative velocity vectors these vectors will be denser along the $\vec{v}$ axis than near the plane produced by the rotation of L [45].

If μ/μ' is greatly different from unity, a marked change may occur in the orbital angular momentum and, according to (6.38), a corresponding change in the rotational angular momentum. For example, if $\mu/\mu' \gg 1$ and if $\vec{L}$ is large, the products must be in highly excited rotational states since $\vec{J}' \gg \vec{L}'$, and the orientation of the final rotational angular momentum vectors will be approximately parallel to the initial total angular momentum [45, 48].

The discussion of the dynamics of collisions has shown some of the microscopic features of reactive scattering that are applicable, in certain special cases, to any bimolecular reaction that can be experimentally investigated by the method of crossed molecular beams. By combining the material in this section with the theoretical background for rearrangement collisions in the preceding section and with the theoretical background for elastic and inelastic collisions, we may proceed to analyze many of the microscopic features of specific bimolecular reactions. Limitations of space prevent a discussion of all the experiments that are relevant to this chapter. A selected group of reactions is used to illustrate those features of rearrangement collisions of particular interest and significance.

3 EXPERIMENTAL METHODS

Molecular-beam experiments involve a large variety of types of apparatus and techniques. These reflect both the numerous experiments that have been performed as well as the extensive ingenuity of the experimenters. A complete description of experimental methods must of necessity be long and detailed. In this section no effort is made to attain such completeness; instead some typical examples of the various experimental methods are given. The basic configurations of beam experiments were discussed in the introduction and are not repeated here. All experiments, however, require three basic elements; one or more sources to produce a collimated beam (or beams) with a known energy distribution, a detector to measure the beam intensity, and a region where two-body interactions can occur. These elements of beam experiment are discussed separately in this section. The experimental techniques also separate naturally into three categories according to the energy range of the beam particles. Methods of production, collimation, and detection are generally different in the three regimes, here rather arbitrarily defined as low energy (less than about 0.5 eV), high energy (above about 25 eV), and intermediate energy (between 0.5 and 25 eV). The following discussion briefly describes some of the general techniques useful in each range.

Low-Energy Beams

Production

Modern apparatus employed in research with thermal energy beams utilize sources operating on the same principle as that used in the very early experiments described in the introduction. Most apparatus consist of several individually pumped chambers, the first of which contains an apertured, temperature-controlled oven filled with permanent gas or with vapor from condensed material. Succeeding chambers provide additional collimating apertures and space for various operations with the beam. The design of an oven source, particularly one that is to operate at a high temperature, depends strongly on the material that is to constitute the beam. Other than an obvious criterion with respect to melting point, the principal requirement considered when selecting the material to fabricate the oven is that it be chemically inert. Some means to prevent spattering and condensation at the exit slit must be provided; either the slit can be heated separately or an interior baffle can be installed. Operation at very high temperatures may require the use of radiation shields to protect nearby elements of the apparatus. Some commonly used substances that constitute the beam (e.g., the alkali metals), present handling problems. They are usually sealed into glass ampules, and some means is provided for opening the ampules under vacuum and loading the ovens. Ovens for producing beams of permanent gases provide fewer design problems, but the demands placed upon the pumping system are usually much greater because all the gas emerging from the oven must be pumped away since there is no condensation on the walls to add to the effective pumping speed.

Various methods have been adopted to make beams emerging from the oven more directional. In all these procedures the oven exit is constructed as a canal [49] rather than a thin aperture; in order to increase beam intensity without loss of directionality many parallel canals are often used. Stacks of hypodermic needles [50], glass capillaries [51], and stacks of alternate flat and corrugated metal foils [51] have all been used successfully. Recently Laval nozzles have been used as exit slits [18]; as is discussed later, these nozzles provide large intensities and, in addition, considerably narrower velocity distributions than conventional source apertures.

Many other special thermal beam sources have been constructed. It is often necessary to form a beam of atoms from species that are stable only as molecules. In such cases some form of electrical discharge is produced in the oven, the most common type being the microwave arc. Beams of hydrogen atoms [52] and the halogens [53] have been obtained this way. Use of a Wood's tube discharge to produce atomic hydrogen has also been extensively reported [54]. Hydrogen atoms have also been obtained from a thermal dissociater [55], the most usual form of which is a tungsten tube with an exit slit that can be electrically heated to about 2500°K. Such thermal dissociaters have been used mostly for hydrogen. Many experiments require beams of atoms or molecules in excited states [56]. Sometimes this is achieved with state selectors, described below; at other times excitation is provided by electron impact or by optical excitation to a higher state that decays into a metastable state.

The molecules emerging from an effusive oven have a velocity distribution that can be written in terms of the Maxwellian distribution at the temperature of the oven. The energy of the particles in the beam can be varied to some extent, therefore, by adjusting the temperature of the oven. Only a limited temperature range is available, however; at low temperatures the vapor pressure of the material forming the beam becomes too low to provide a beam of useful intensity, and at high temperatures the beam atoms may be internally excited, usually to an unknown degree. A more useful method of varying the beam energy is to select from the beam only those particles in a narrow range of velocity. Not only does this permit the selection of a range of velocity any place within the distribution, but it always simplifies the interpretation of an experiment when the velocity spread is small. The first velocity selector was that of Lammert [57], who used it to verify that the density distribution of an effusive beam is indeed Maxwellian, as had been predicted. He passed a beam through slots in two successive disks that rotated at a speed such that only those atoms in a chosen range could pass. This fundamental concept has been used in the design of a number of velocity selectors.

The first basic design to receive widespread use was that of a rotating cylinder containing helical grooves [58–60]. This selector has high transmission and completely eliminates the velocity sidebands that are transmitted by the simpler selector containing two slotted disks. The cylinders are, however, difficult to fabricate and in general have a relatively large mass that poses problems in achieving high angular speeds. A more recent development has been the design of selectors employing multiple slotted disks [61–65]. These selectors are easier to make than grooved cylinders and can be made with a low moment of inertia. By adjusting the interdisk spacing and the relative displacement of the slots, the selector can be made to pass a beam of selected velocity range free of sidebands. An additional advantage of the slotted-disk configuration is that it can be used with beams of condensable vapors as well as with permanent gases because most of those atoms not transmitted are removed by collision with the face of a disk rather than with the sidewall of a groove.

At least three published discussions [63–65] deal with the optimum relative position and the angular displacement of the slotted disks in order to eliminate sidebands completely while minimizing the total rotating mass. The details are not repeated here because, in general, they involve specific features of each selector, for example whether or not the disks have varying thickness.

The various velocity selectors now provide, in fairly routine manner, beams of adequate intensity with energy spreads of a few percent. A recent innovation has been the development of techniques for selecting beams with atoms in specified quantum states. The basic techniques of quantum-state selection are those that have long been used in molecular-beam, high-frequency resonance experiments such as those described by Ramsey [2]. If a molecule has a permanent magnetic or an electric moment, interaction with an external inhomogeneous magnetic or an electric field is possible, and the amount of deflection of the beam and its direction in the external field is a function of the angular momentum quantum number and its projection on the local field direction. Pauly and Toennies [22, 26] discuss a number of experimental arrangements for such deflections and alignments, from the traditional Stern-Gerlach arrangement, which deflects the beam particles without focusing, to multipole electrode assemblies that can be used to focus molecules in different (j, m) states. The application of state selection to specific experiments is discussed later.

Detection

Methods of detecting molecular beams have been the subject of extensive research. The original method, that of condensation, is clearly inefficient and awkward because it requires a relatively long time to condense enough material to observe, and obviously provides difficulties for permanent gases and materials with low melting points. The search for a faster, more sensitive detector for thermal beams has been under way for many years and has produced a number of successful devices. An early example is the chemical detector [67, 68], which detects chemically active species whose chemical reaction with the target material at impact causes a color change. A more recent development [69] is a variation of the condensation detector in which the beam contains a radioactive species. Modern counting techniques are such that a measurable deposit is obtained much faster than by the simple condensation method. Quite recently two more investigations have been reported in the continuing search for useful beam detectors. Brooks and Herschbach [70] have applied modern electronic techniques to the Kingdon gauge [71] first proposed in 1923. This gauge is a diode operating with space-charge-limited electron emission. A beam particle entering the diode is ionized by electron impact, and the resulting positive ion partially neutralizes the space charge producing a large increase in plate current. Johnston and King [72] report the use of field ionization in the construction of a beam detector. Here the large electric fields near the point of a field emitter tip can quantitatively ionize atoms; the ions are then measured routinely.

Although many ingenious devices have been developed for use as detectors for thermal beams, only three are extensively used at present. These are the Pirani, surface ionization, and electron-bombardment detectors. The Pirani gauge [51] is generally used to detect beams of the permanent gases. The beam passes into an enclosure through a long narrow orifice where it scatters at the walls, equilibrates, and eventually effuses back through the orifice. The steady-state pressure rise measured with a Wheatstone bridge, as with a conventional hot-wire pressure gauge, is proportional to the beam intensity. The pressure rise is small, and for satisfactory performance it is necessary to measure the difference in response between two identical cells, one of which can respond to fluctuations in the background pressure but which is not exposed to the beam. It is possible to increase the sensitivity of the detector by making the entrance orifice longer and narrower, but only at a cost of increasing the equilibration time.

The Pirani gauge cannot be used where the material constituting the beam will condense on the walls. For many of such materials, surface ionization [73–75] can be used to produce an effective detector. In this method the beam particles are directed onto a heated wire from which they quickly evaporate. In general if the work function of the wire substrate is greater than the ionization potential of the beam particle, the probability is large that the particles will leave as positive ions. The ion current is therefore proportional to the beam intensity. The surface ionizer is effective for detecting only atoms with low ionization potentials but is nonetheless an extremely useful device. It has the advantages of short time constant, high ionization efficiency, and simplicity. The most common wire used in surface ionizers is tungsten; the tungsten must be oxidized to increase its work function in order to use it to detect the two lightest alkali metals, sodium and lithium. The alkali halides dissociate at the hot wire and also may be detected. Of critical importance to the study of reactive scattering, to be discussed below, was the discovery that both potassium and KBr (for example) are surface-ionized with a tungsten wire while platinum wire ionizes only potassium. Simultaneous use of both wires therefore permits the determination of potassium and KBr in a mixed beam.

A number of species, notably the halogen atoms, may be detected by the formation of negative atoms at a hot wire [51]. In this case it is the relationship of the electron affinity and the metal work function that determines whether or not an atom can be surface-ionized.

The third type of detector commonly employed is the electron-bombardment detector, sometimes called the universal detector. In essence it consists of a means for ionizing some fraction of the incident beam, discriminating between these ions and those due to the inevitable background gas, and collecting the primary ions. In principle, such detectors could be used with beams of any material, and considerable effort in various laboratories [76–86] has been devoted to their development.

The large variety of successful designs reflects the different constraints put on the detection system by various experiments. In some cases the ionizer is followed by a mass spectrometer so that the intensity of a particular species can be monitored in the presence of other material. In other cases the technique of phase-sensitive detection can be employed [87], as when an ion current must be observed in the presence of a much larger background current arising from the ambient gas. In this technique the signal that is to be extracted from the extraneous background is modulated, mechanically or otherwise, at a fixed frequency. An amplifier tuned to this frequency will reject all signals except the desired one.

Beams of metastable atoms are relatively easy to detect because they will eject an electron from a metal surface if the excitation energy of the metastable is greater than the work function of the metal [56]. This energy criterion is met by metastables of the rare gases and mercury that eject electrons from surfaces such as gold with efficiencies of about 0.2. In order to make absolute intensity measurements the ejection efficiency must of course be accurately determined. Detection of metastables other than those of the rare gases and mercury is usually done by an absorption method [88], in which light of a characteristic wavelength is absorbed, exciting the metastable to a higher level from which it can decay by emission of light.

Scattering

The final aspect of a molecular-beam experiment is the design of the interaction region, which can take many forms depending on the specific experiment being performed. As indicated in the introduction, one method of permitting particles in a beam to interact with other particles is to pass the beam through a thin layer of gas. The gas is contained in a small differentially pumped vessel, and the pressure is adjusted until the molecules in the beam are likely, on the average, to suffer no more than one collision in traversing the gas layer. Such conditions are established when the beam attenuation is linear with pressure. The presence of apertures in the vessel for beam entrance and exit means that there is a steady flow of gas out of the chamber and that there will be some scattering of the beam outside the chamber and in the finite length of the beam apertures. It is possible to account for this background scattering; the calculation is relatively simple if the mean thermal velocity of the scattering gas is less than the beam velocity, but is more difficult if the two velocities are comparable. Recently an experimental determination was made of the background scattering of the thermal beam [89]. A scattering chamber with adjustable scattering path length was designed. Two measurements of a scattering cross section were made with different, known, scattering path lengths. The two measurements were compared by a difference method such that the background correction was eliminated. In most cases it is not necessary to resort to such methods because it is usually possible to design a scattering chamber so that less than 10% of the scattering occurs outside the chamber.

The measurement of the scattering-gas density in a chamber is a matter of considerable importance. The elastic scattering cross section of atoms with thermal velocities is large (relative to fast atoms) so that the layer of scattering gas must be relatively dilute. The pressures that must be used fall generally into a region around 0.1 mtorr, where accurate measurements are difficult.

The McLeod gauge is in common use, but for this pressure it tends to be large and bulky and therefore to some extent fragile. The pumping effect of the cold trap recently discovered [90] also can contribute serious error in this pressure range unless it is accounted for or suppressed [91–93]. Another gauge that has recently been widely employed is the diaphragm gauge. This gauge is composed of two chambers, one at a known pressure (usually zero), separated from the other by a thin diaphragm. As pressure moves the diaphragm, its displacement is picked up with a capacitative sensor. These gauges require no cold trap, are relatively free of contaminants, can be readily calibrated, and the calibration is independent of gas species. Ionization gauges are sometimes used to measure the scattering-gas pressure but their use is fraught with difficulties. They tend to be unstable, are difficult to calibrate, and must be calibrated individually for each gas used. Increasing the required scattering-gas pressure by shortening the length of the chamber has only limited effectiveness since the decrease in path length only increases the importance of the background corrections.

The geometry of the scattering region must be tractable in order to interpret a scattering experiment. Tractability implies not only that the dimensions and the locations of the apertures defining the beam and the target be accurately known but that the exact dimensions of the region in which the interaction takes place also be known. The shape and the dimensions of this region, here called the scattering volume, are necessary for interpreting a scattering experiment because those particles in a beam scattered into any arbitrary element of solid angle can originate from any point within the scattering volume, and any experiment involves an average over the finite dimensions of the beam [94]. Definition of the scattering volume for an attenuation experiment with fixed axial detector is usually straightforward. If, however, the experiment measures the differential cross section at varying relative angle, the scattering volume can become not only rather complicated geometrically; it must be differently defined for each scattering angle. Many ingenious scattering chambers have been designed to reduce the difficulties that arise from this problem [23, 95, 96].

The other common arrangement for scattering a beam is to cross it with another beam or jet of molecules. The density of scattering centers in the crossed beam is determined by measurement of beam intensity in the same way as is done with the primary beam so that no pressure gauge is necessary. If a crossed-beam experiment is performed at various relative orientations, the definition of the scattering volume becomes complicated, just as for the measurement of the differential cross section discussed above. In a few crossed-beam experiments it is unnecessary to make quantitative intensity measurements so that variation of the scattering volume with angle is of no real consequence.

High-Energy Beams

High-energy beams are formed by accelerating ions to the desired energy and collimating either mechanically as with thermal beams or with electric or magnetic focusing fields. Fast neutral beams are produced by neutralizing the accelerated ion beam in a gas-filled region where charge transfer can occur. Most of the following discussion concerns the details of ion-beam experiments because clearly it is necessary to form an ion beam for both ion-beam and neutral-beam experiments. The few special features of neutral-beam experiments are presented after the discussion of ion beams.

Production

The development of ion sources has a history dating back more than 50 years. Most sources used in molecular-beam research produce the ions in a discharge, usually either dc or radiofrequency. A large number of different designs has been reported [97] which reflect the different requirements of various experiments. The two principal input considerations are power and gas-flow efficiency; the output performance criteria are beam divergence, energy spread, total current, and internal state of the ions. It is generally not possible to optimize all these parameters, and a given experiment usually indicates which conditions are most important.

Collimation and focusing of the ion beam usually must be considered as parts of the design of the ion source. The principal problem involved in focusing intense ion beams is that of space charge dispersion; the usual solution is a combination of keeping the source-to-detector distance short and of using some form of focusing electrodes. The theory of ion optics in the presence of space charge [98, 99] is only partially developed, so that most ion lenses that are used are operated with parameters chosen experimentally by performance tests. The difficulty of lens design is increased by the fact that the lenses must function in the presence of fringe fields (electric or magnetic) of the source that are difficult to estimate in advance and that the ions are usually extracted from a plasma boundary whose shape and position are not stable.

Energy determination and selection are also considerations in the design of beam collimators. The energy distribution of ions extracted from most sources is not broad, as in thermal-beam ovens. Generally speaking, there will be three species in the flux issuing from an ion source; thermal atoms, high-energy neutral particles, and high-energy ions, each of which will have different energy distributions. The thermal-energy atoms appear because the ion source is a gas-filled chamber with an exit slit, and it therefore acts as a thermal-beam oven. Since the energy distribution of these atoms is nearly Maxwellian and is peaked at an energy much lower than the high-energy particles or ions, the effect of this thermal beam can be eliminated by energy discrimination. The high-energy neutral component of the beam flux arises from charge-transfer collisions of the ion beam with atoms of the source gas during extraction of the beam. Since it is never certain where, in the acceleration history of the ion, the charge transfer event occurred, the energy distribution of this neutral beam is never known, and it can be expected to depend strongly on the detailed design of each source. This neutral beam can be removed, when desired, by a slight deflection of the ion beam with an electric or a magnetic field. The energy distribution of the ion beam can be quite narrow, although the energy resolution achieved in electron guns [95] can never be reached. The ions are extracted from a plasma, a region in which the pressure is high enough to ensure that collisions take place; and in addition, the ions are extracted from a finite region of the plasma in which the potential distribution may be nonisotropic. A typical energy spread is a few electron volts in a source producing a beam of a few hundred electron volts and upward.

The determination of the energy distribution of high-energy beams must be performed on the ion beam rather than on the neutral component. Rotating disk selectors are usually of little use at high energies because the rotor speeds required are much too high to be practicable. However, measuring the ion current transmitted as a function of an applied retarding potential will determine the energy spread of the beam. Although conceptually straightforward, retarding potential analysis involves a number of difficult experimental problems [100]. It is common to use electric or magnetic fields, separately or in combination, as velocity selectors [101, 102]. Since an electric field can be used to select an almost arbitrarily small range of energies, and magnetic fields can be used for selection of momentum, selection of both energy and momentum automatically selects a specific mass. Magnetic selection involves use of a sector-field deflector with entrance and exit slits. The size of the slits controls the resolution, and clearly high resolution is obtained only at the expense of decreased intensity. Energy analysis with electric fields can be accomplished in a variety of ways. In the cylindrical electrostatic analyzer [103] a radial, inverse first-power electric field is established in the annulus between two cylindrical electrodes. In such a field, charged particles follow circular paths. If two particles with the same energy enter the analyzer with divergent angles, they will recross the axis at a point 127 degrees with respect to the entrance point.

A similar device is the spherical analyzer where the electrostatic field is established between concentric spheres, a configuration that provides focusing in two dimensions. The spherical analyzer [104, 105], especially the 180-degree deflection version, has been extensively used for energy analysis of electrons [106] but is only slowly being applied to the analysis of ions. An electric field applied between parallel plates can also be employed as an energy selector [107]. A beam of charged particles can be injected at 45 degrees through a slit in the lower plate; the particles will follow parabolic trajectories between the plates, and those of a given energy will pass through exit slits some distance down from the entrance. A large variety of devices are available, many commercially, which perform velocity selection by a time-of-flight method [102, 108]. The most commonly used technique is to accelerate a group of ions to a given energy and then to observe a short pulse of these ions as it propagates down a drift tube. Ions with different velocities (which in this context means ions with different mass) arrive at the end of the drift tube at different times, and a measurement of current as a function of time will show a spectrum of velocities present in the ion pulse. This technique is more useful for measuring a distribution of velocities than for selecting ions of a given velocity. Other time-of-flight methods are used regularly for velocity selection, however, such as radiofrequency spectrometers, which can take many forms. One example is the Bennett spectrometer [109, 110], in which two sets of fine mesh grids normal to the axis of ionic motion are alternately spaced. Radiofrequency accelerating voltages of opposite polarity are applied to each set of electrodes. If an ion with the proper velocity enters the system with the proper phase with respect to the radiofrequency field, it will be accelerated during its flight between each electrode, while all other ions will lose synchronization with the field. A retarding potential grid placed as the last electrode can permit the passage of only those resonant ions that have received the maximum acceleration.

There is clearly a wide variety of techniques available for velocity selection or analysis of high-energy ion beams. Which method is used in a given experiment is frequently determined by intensity considerations. High-resolution magnetic mass spectrometers require long path lengths and narrow slits, and so inevitably provide low intensities. The radiofrequency spectrometer described above has quite a short length and no slits at all, but the fact that it does not transmit those ions with the resonant velocity whose entrance phase is wrong, again results in poor beam intensities. For this reason the energy analysis is sometimes combined with beam detection, by using some form of mass spectrometer directly as a beam detector [84]. This is particularly attractive for experiments with neutral beams because it permits the neutralization of the beam at a short distance from the ion source, thereby minimizing losses in intensity caused by space-charge effects. Although this advantage is partly offset by the inefficiency of reionization in the spectrometer source, the method is still attractive. The method essentially amounts to performing an experiment in the presence of species that are unwanted or that have a wide variety of velocities by observing only the particles of interest. Frequently this is possible, but in some cases the presence of unwanted species affects the interpretation of the experiment; in such cases the only recourse is to place the selector before the scattering chamber. Finally, it might be noted that some mass spectrometers, notably the radiofrequency quadrupole spectrometers [111, 112], perform their role well but completely destroy the beam collimation in the process of mass analysis. Such spectrometers are only useful as part of the detection system.

As indicated earlier, beams of neutral particles can readily be formed by passing the ion beam through a layer of the parent gas where a substantial fraction of the ion beam can be neutralized by resonant charge exchange [113]. The neutralization can be made to occur in a differentially pumped chamber with beam entrance and exit slits or in a jet of gas that crosses the ion beam. If the density of the charge-exchange gas is large, the ion beam can be almost entirely neutralized; at such high densities, however, the neutral beam that is formed is significantly attenuated by scattering. The usual procedure, therefore, is to adjust the charge-exchange gas density until the neutral-beam flux is optimized; the density so selected is such that the neutral-beam intensity is about half the ion-beam intensity [114]. The energy exchange in resonant charge transfer is very small, on the order of the ratio of the mass of the electron to that of the ion, so that velocity selection performed on the ion beam is essentially unchanged for the neutral beam that passes through the usual arrangement of axial collimating slits. The charge-transfer process also acts somewhat as a filter for impurities in the beam. The impurities will in general have different ionization potentials than the charge-exchange gas; in that case the nonresonant charge-exchange cross sections are not only small, but the energy defect requires that the neutral particle in a nonresonant collision be deflected. It has been experimentally verified that passing an ion beam through a nonresonant medium results in low intensities, and if the energy difference is large enough (e.g., Ar^+ into He) no axial neutral beam is produced even though a substantial part of the ion beam is neutralized [114]. Another experimental verification of this phenomenon was obtained when a beam of helium ions, mass analyzed to show several impurity peaks, was neutralized in helium gas. The resulting helium beam was mass-analyzed, and no impurity peaks were found [115].

Detection

The detection of fast ion beams is straightforward and usually provides no problems. Either the ion current is measured in a Faraday cup assembly or the ions are counted. The latter procedure has become more common recently as the elaborate electronic equipment developed in nuclear physics has become available. Direct measurement of ion currents is usually made only when the current levels are relatively high. Noise levels in electrometers are higher than in multipliers, and the electrometer time constant is longer so that it is difficult to follow transient phenomena. If either method is used, it is necessary to assure that secondary electrons emitted at the collector electrode are not permitted to escape since this would result in a false measurement. The collecting electrode can be made long and deep, or fitted with internal fins to assure that the secondary electrons do not escape, or a grid system can be placed in front of the collector to repel the secondaries back into the collector [95].

The detection of neutral particles is more difficult but a number of techniques have been developed over the years. One powerful technique is the calorimetric measurement of the total beam energy [21]. Thus if the energy of the individual particles is known, the flux can be calculated. Calorimetric detectors have taken many forms; they usually consist of a mask, slightly larger than the beam, that is attached to a thermocouple, a thermopile, or a bolometer. In order to have high sensitivity they are constructed with very fine wires and are therefore rather fragile and tend to have long time constants. At least two types of thermal detector that are mechanically sturdy but still have reasonably high responsivity and short time constant have been reported. The Harris thermopile [116] has the disadvantage, for some experiments, that the detector sensitivity is not uniform over its surface. The thermistor bolometer [117] can be built with a time constant of only a few milliseconds, and it is customarily operated in an ac mode by modulating the beam with a chopper wheel. This is done primarily to take advantage of the simplification of ac electronics rather than to distinguish the beam from the background gas. One of the greatest advantages of calorimetric detectors is that they automatically discriminate against the background gas. Interfering signals caused by the presence of background gas are usually negligible because, although the background gas is present in concentrations much greater than that of the beam, the detector does not respond to it because its energy is so much lower than the beam energy.

Calorimetric detectors are also used to detect beams of atoms that were formed from diatomic molecules [118]. Energies of several electron volts are released at the detector surface when the atoms recombine, and this energy can be detected since its effect is to heat the detector. Since the atomic recombination energy is known, measurement of the detector signal can be converted into a value of the flux.

Other neutral-beam detectors that have been devised provide a means for producing a number of charged particles proportional to the beam flux. The universal detector described in connection with thermal energy beams can be used for detection of fast beams, but its efficiency decreases with increasing beam energy because the ionization probability is proportional to the transit time in the ionizer. Its use for fast beams is therefore rare. More common is the use of electron multipliers [51], in which secondary electrons, released on impact of the fast beam on the cathode, are caused to strike a series of subsequent electrodes where each impact produces more secondary electrons. It is possible to obtain gains in such instruments of 10^6 to 10^7; that is, that many electrons are produced at the anode for each electron released at the cathode. The most popular instrument currently in use is the Bendix magnetic multiplier [119], which is less sensitive than most multipliers to surface contaminants. A recent development, also from Bendix [120], is the channeltron that takes the form of a narrow tube whose inner surface is a high-resistance, high-secondary-emission material. A potential is applied along the axis of the tube, and the fast beam is directed obliquely into the end of the tube. Secondary electrons released on impact propagate down the tube making zigzag paths and releasing more electrons at each wall collision. This simple device produces very high gains and can be made extremely compact. Arrays of channeltrons can be assembled to probe the beam intensity.

Multipliers can be operated either as current-measuring devices or as particle counters. In the current mode the multiplier can be used only for measurements of the relative beam intensity; determination of the absolute intensity from the measured current requires that both the overall gain and the secondary emission coefficient of the cathode be known, and neither of these can be obtained with accuracy. Operation in the counting mode is therefore more common because each incoming particle generates a pulse that can be counted accurately, so long as the secondary emission coefficient is greater than unity. Operation in either mode is affected at high intensities by saturation effects; for example, in the current mode when the current carried by the multiplied electrons becomes comparable to the current in the dynode strip, the final stages of amplification are in effect shorted out. Discrimination against background gas signal is usually done by maintaining the multiplier in a separate, differentially pumped chamber where the pressure can be reduced to as low as 10^{-9} torr. Mounting the multiplier in such a separate low-pressure chamber also reduces dynode contamination and subsequent gain deterioration. Discrimination against background gas can also be achieved by the traditional method of modulating the beam and switching the multiplier output into two scaling channels in phase with the beam modulation [121]. When one is operating in the counting mode, background discrimination can be accomplished by pulse-height analysis. The low-energy background particles in general create much lower pulse heights than those associated with the faster beam particles, and a discriminator can be placed before the scalar to reject all pulses below a certain height. It should be pointed out that multipliers are most useful at the higher beam energies where they have high gains and relatively good background discrimination.

Other secondary-electron detectors have been built that employ a positively biased electrode to collect the secondary electrons [122]. This detector has all the disadvantages of the multiplier without the advantage of producing high gains, and so it has not been widely used. Other similar detectors have been devised [122] in which the beam particles, or secondary electrons produced by the beam, are directed onto a phosphor in order to release photons that are detected with a photomultiplier. These devices have been used only to detect beams with very large energy.

Scattering

The design of scattering chambers and the measurement of the pressure of the scattering gas are in general simpler for fast molecular beams than for thermal energy beams. The principal reason for this is that the scattering cross sections for fast atoms are relatively small. Thus the gas pressure needed in typical fast-beam attenuation experiments is of the order of 10 mtorr, whereas for a thermal-beam experiment it might be a factor of 100 or more lower. In fast-beam experiments a variety of well understood pressure gauges are available, such as null-diaphragm, Pirani, and McLeod gauges. At these higher pressures the error in McLeod-gauge measurements caused by pumping of the cold trap is frequently negligible. McLeod gauges are usually the standard to which other gauges are referred, but such a calibration can in turn be checked against an accurate expansion system so that possible errors in the measurement of the pressure can be reduced to negligible levels. The smaller scattering cross sections for fast beams also result in less beam attenuation in the ambient gas in the apparatus, and so experiments with fast beams present less severe constraints on the pumping system. As noted earlier, the correction that must be applied to account for the attenuation of the beam in the background gas is much simpler to calculate for fast beams than for slow beams because it can be assumed that the scattering molecules are stationary. This assumption eliminates the complex problem of including a velocity distribution of the target atoms.

Intermediate-Energy Beams

Research with beams of intermediate energy, those with kinetic energies between about 0.5 and 25 eV, have been less fully developed than either the high-energy or the low-energy beams. For the most part this is a result of the much greater difficulty of producing beams in this energy range that are sufficiently stable (or reproducible), sufficiently monoenergetic to be useful for quantitative studies, and which in addition have adequate intensity. In large measure the underdevelopment of intermediate-energy beams must be attributed to the lack of interest in the past on the part of research workers in this energy range. Many methods of producing beams in this energy range have been investigated, but at this time the experimental situation is far from that for high- or low-energy beams.

Production

Most effort has gone into the development of nozzle beams. These beams have high intensities and can be manipulated to produce particles with energies between about 0.1 to 10 eV. The method substitutes for the Knudsen effusive source of thermal beams a supersonic jet produced by hydrodynamic expansion of a gas or suitable gas mixture through an appropriately designed nozzle. The technical difficulties are many and great, but the prospect of producing beams of high intensity in a hitherto inaccessible energy range has spurred a large number of investigators to enter this field and to continue in it. One group of the many avid workers in this field has been Anderson, Andres, and Fenn, whose review articles [124, 125] give a very detailed picture of the present state of the development of nozzle beams.

A recent approach [126, 127] to the study of two-body collisions of heavy particles in the energy range from a few tenths of an electron volt to about 100 eV involves the simultaneous use of two fast beams traveling in the same direction along a common axis. The technique is referred to as the use of merging beams, superimposed beams, overtaking beams, or confluent beams. The method has several advantages. Two fast beams, each with laboratory energy of the order of 1 or 2 keV and a modest energy spread, say 1.5 eV, can be merged to produce interaction energies as small as 0.25 eV or less, with a spread in the interaction energy of the order of 0.01 eV. The high energy of the fast beams minimizes loss of beam intensity associated with space-charge divergence of ions extracted from sources. The detection problem is also comparatively easy since, in addition to detectors that can be used for both slow and fast beams, other detectors uniquely suited to fast beams can also be used. These include thermal detectors and secondary-electron emission detectors with or without particle multiplier features. Particle-counting detection may also be used. It is worth mentioning that if a single source is used, only ion-neutral interactions can be studied by the merging-beam technique. This is achieved by partial charge transfer of the original ion beam to produce a mixed ion-neutral beam and then retarding the ion-beam component to achieve the desired difference in energy and accordingly, the desired small interaction energy. If two sources are used, the ions from each source can be used to obtain neutral beams by charge transfer, and after removal of ions that have not been neutralized, interactions between the two neutral beams may be investigated. A review of the rather small amount of research completed thus far with merging beams of heavy particles has been published [128]. That article describes the method, explains the theory, indicates the type of collision processes for which the technique is suitable, and cites references to published results.

Recently a new method was described for producing beams of potassium with energies in the approximate range of 0.25 to 50 eV [129]. This method, which had also been independently investigated at Freiburg [130], uses an ion source such as a uno-plasmatron to focus a beam of ions, usually Ar^+ or Xe^+, with energies of 5 to 30 keV, on a potassium surface. Neutral potassium atoms are sputtered from the bombarded mass of potassium metal to form a beam that has an approximate cosine-squared distribution and is peaked in a forward direction that is determined by the angle at which the ions impinge upon the potassium surface. By use of velocity selectors, collimated beams of small energy spread and good intensity are obtained. The method has reached the point where it can be used to obtain total scattering cross sections that appear to be of the right magnitude and energy dependence. It appears, therefore, that the technique is well on its way to becoming a useful tool for producing beams in the intermediate-energy range. It is interesting to observe that in principle, and probably in practice, the sputtering source can be used to produce beams of intermediate energy from a wide variety of materials, conductors as well as nonconductors. In the case of nonconductors the charge that would build up as the ions impinge on the substrate (and which, if not eliminated, would eventually repel the fast ions from the plasmatron source) could be prevented from forming simply if there is a source of electrons close to the substrate. Although these electrons could be swept across the surface of the nonconductor, it appears that all that is required is that they be close to the surface that would otherwise build up a charge. Additional experimental details of this cathode-sputtering method have been recently published [131].

The use of a shock tube as a high-temperature thermal source for the production of a molecular beam has been under investigation for some time [132]. Since it is possible to raise the temperature of a shocked gas to several thousand degrees, a source of this type usually produces a beam with an energy of a few tenths of an electron volt. Although it is possible to increase this energy by producing stronger shocks, this procedure introduces difficulties that are associated with production of a mixed beam that may contain one or more of the following species: ground-state particles; dissociated atoms or free radicals; excited and metastable atoms or molecules of the original gas or its dissociation products; and ions. However, an interesting application of shock tubes has been made that produces beams with an energy of about 1 eV and minimizes the difficulties introduced by extremely strong shocks [133–135]. The method uses a conventional shock-tube source whose shocked gas, heated to only several thousand degrees or less, is then hydrodynamically expanded to produce a nozzle beam similar to that described above.

The use of a shock tube as a primary source or as a source for a nozzle beam has an inherent disadvantage in that it is a pulsed experiment. This disadvantage not only restricts the time available for actual useful experiments but also greatly increases the difficulty of obtaining reproducible results.

Nevertheless the inclusion in this review of the technique of using shock tube sources for producing molecular beams is an interesting example of the ingenuity that has gone into the development of sources for molecular beams in the intermediate range of energy.

It was pointed out earlier that multipole fields can be used for state selection and focusing of molecules with permanent magnetic or electric moments. However, this method is applicable only for molecules in states that gain energy with applied external field. Another apparatus [136] has been designed and built that is applicable to situations in which there is either a gain or a loss in energy. The apparatus, which may be regarded as a linear accelerator for neutral molecules, operates as follows.

In its present form it uses 700 stages of dipole accelerators, and for molecules with a large dipole moment, such as LiF, there is a gain of about 3 meV at each stage so that at the completion of the acceleration a beam with a kinetic energy of about 2 eV is obtained. At one end of the apparatus dipolar molecules drift into an electric field produced by a pair of carefully aligned electrodes and are accelerated into the gap between the electrodes. At this moment the field is turned off, and the dipoles drift toward the next pair of electrodes. The field is then turned on again, and the dipolar molecules are further accelerated. The process is repeated until an intermediate-energy neutral beam emerges from the last pair of electrodes. At this time the apparatus is complete, and the difficult task of aligning the 700 pairs of electrodes, which extend over a length of more than 30 ft, is in progress. A brief description of the apparatus is contained in a recent review article [137].

Another example of the specialized apparatus developed for the intermediate-energy range is the scheme to form a pulsed beam of intermediate energy by mechanical acceleration. The procedure involves accelerating a gas sample contained in a hollow projectile by firing it from a rifle or light gas gun. When the projectile reaches the desired velocity, the gas sample is released, which expands downstream with a velocity that is the sum of its average thermal velocity and the projectile velocity. Calculations [138] indicate that the projectile can be accelerated without substantial heating of the gas inside, and because the gas sample can be loaded at a very high pressure, very large intensities can be expected. Research performed thus far has not produced a molecular-beam system, but it has been shown that it is possible to accelerate gas-loaded projectiles to high velocities, open them in flight, and observe the subsequent expanding gas cloud as it moves downrange [139]. The problems that remain before this device can be developed into a useful molecular-beam system are many and difficult. It is not yet certain that it will ultimately be a usable tool for molecular-beam research.

It is apparent that the intermediate-energy range experiments present difficulties, primarily in the matter of formation of the beam. Procedures for collimation, detection, and provision for scattering the beam can be taken directly from the techniques already established in high- or low-energy experiments. Because so much effort has been expended merely to develop new methods of beam formation, only a few experiments performed with intermediate-energy beams have been reported.

4 RESULTS

Molecular-beam research has been carried out for over 50 years. In that time an immense number of experiments have been performed and the results reported. To give even a representative sampling of this work would be a formidable task; this section presents only a few examples to illustrate the kind of information that can be obtained from elastic, inelastic, and reactive scattering. The elastic scattering section is separated, as before, into three regimes on the basis of energy.

Elastic Scattering

High-Energy Scattering

The most important property of high-energy molecular beams is their ability to probe the intermolecular potential at close distances of molecular approach, thus supplying information that is almost unattainable by other means. The range of beam energies is roughly 25 to 10^4 eV; at such energies the scattering is very strongly peaked in the forward direction, and the preferred procedure is therefore to measure the integrated cross section $S(\theta_0, \epsilon)$ rather than $\sigma(\theta, \epsilon)$, using a small value of θ_0.

Two detailed reviews of high-energy molecular scattering have appeared [21, 140] that pay particular attention to the technical problems involved in the inversion of the scattering data to obtain intermolecular potentials. In one respect the inversion problem is simple because classical mechanics furnishes an adequate description of the scattering and because the potentials are usually satisfactorily represented as single monotonic repulsions in the region probed by the scattering. For such a monotonic repulsion all scattering angles are monotonic, and (6.7) can be transformed to a power series in the energy

$$\theta(b, \epsilon) = \sum_n a_n(b)\epsilon^{-n}. \tag{6.39}$$

For large energies and small angles the series in (6.39) can be truncated after one or two terms for sufficient accuracy. If an algebraic form for the potential is assumed, the coefficients $a_n(b)$ can be evaluated and the series can be inverted to obtain the impact parameter, and then through (6.6), the cross section $S(\theta_0, \epsilon)$ corresponding to the angular aperture θ_0. As an example, for the inverse power potential, $V(r) = Kr^{-s}$, the first term in the series gives

$$S(\theta_0, \epsilon) = \pi\left(\frac{KC}{\epsilon\theta_0}\right)^{2/s} \tag{6.40}$$

where

$$C = \frac{\sqrt{\pi}\,\Gamma[(s/2) + (1/2)]}{\Gamma(s/2)}. \tag{6.41}$$

When necessary, higher terms in (6.39) can be evaluated, resulting in additional terms of powers of θ_0 in (6.40). Equation (6.39) has also been evaluated for the exponential and the screened coulomb potentials [114]. As indicated in the introduction, the experimental value of $S(\theta_0, \epsilon)$ represents an average over the dimensions of the beam and the detector because an attenuation experiment involves scattering over a range of angles. Considerable care is needed to take account of such instrumental effects. When this has been done properly the potential parameters can be deduced by comparing (6.40) with the experimentally determined cross sections.

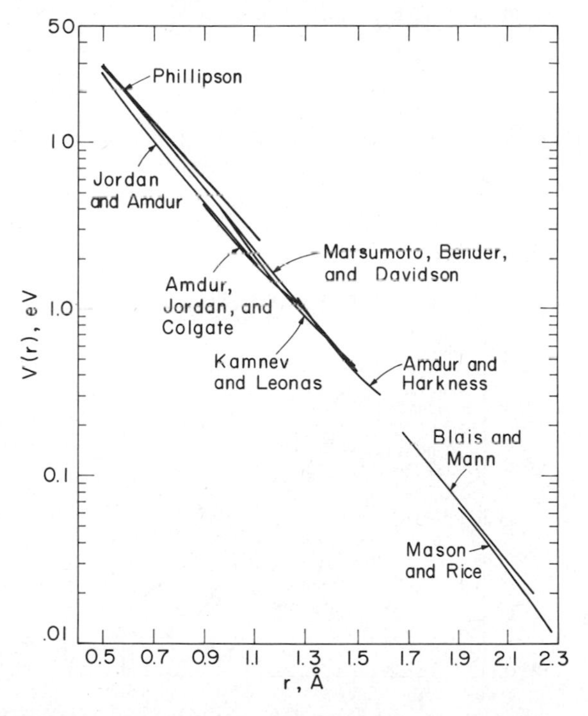

Fig. 6.9 Helium–helium potential energy determined from four different high-energy scattering experiments, from high-temperature thermal conductivity, from viscosity and virial coefficients, and from two *ab initio* quantum mechanical calculations.

The two reviews [21, 140] cover most of the work through 1964, which emanated largely from MIT; several subsequent papers have by now appeared [114, 141–144]. Recently Leonas and coworkers in Moscow have initiated a similar research program [145–150], and their results provide valuable corroboration and extension of the MIT work. Results published to date from both MIT and Moscow include the five noble gases and their 10 possible unlike pairs, plus the systems He → (H_2, D_2, N_2, CH_4, CF_4), Ar → (H_2, N_2, O_2, CO, CH_4), H → (He, H_2, N_2, O_2), N → (N_2, O_2), O → (N_2, O_2), N_2 → (N_2, O_2), H_2 → H_2, and O_2 → O_2. A number of other systems have been reported in conference proceedings [151, 152] and reviews [21], but have not yet been published in detail.

An impression of the general level of consistency and reliability of potentials obtained from high-energy elastic scattering can be gained from Figs. 6.9 and 6.10, where various results for He–He and Ar–Ar are compared.

The He–He system is probably the single system most thoroughly investigated, both experimentally and theoretically. The results of four separate high-energy scattering investigations [114, 145, 153, 154] using four different apparatuses are shown in Fig. 6.9; the consistency is quite good. Perhaps

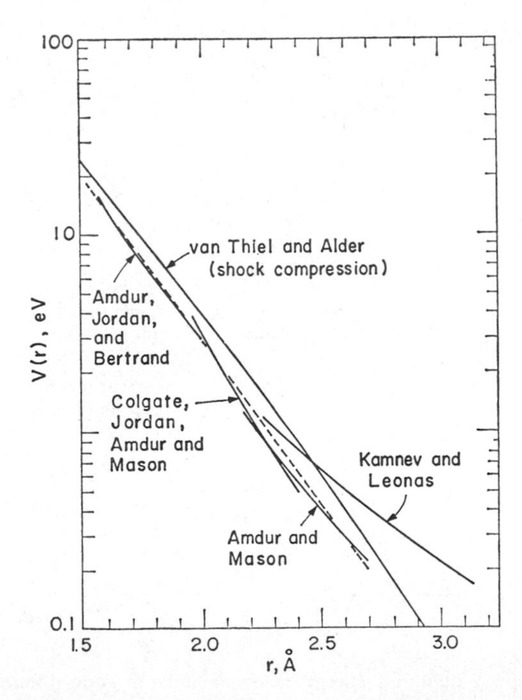

Fig. 6.10 Argon–argon potential energy determined from four different high-energy scattering experiments and from shock compression measurements on liquid argon. The dashed curve is a single exponential representation of the three MIT results shown.

more important is the agreement with potentials obtained by entirely different experimental methods. Figure 6.9 also shows potentials obtained from analysis of high-temperature thermal conductivity data [155] and from analysis of viscosity and second virial coefficient data [156]. Although these do not overlap the scattering potentials, they are close enough to show that there is a high degree of consistency. Also shown in the figure are the results of two accurate *ab initio* quantum-mechanical calculations [157, 158]. The only empirical information used in these calculations are the numerical values of Planck's constant and the charge and mass of the electron. Here the agreement is a little poorer, but within the estimated uncertainties of the calculations and the experiments.

The Ar–Ar results are shown in Fig. 6.10. Again four different scattering investigations are shown [143, 145, 159, 160]. The overall consistency is quite good, although the Moscow results seem high at their largest separation distances. No accurate quantum-mechanical calculations are available for a system with such a large number of electrons, nor are enough accurate high-temperature gas data available for argon to permit the additional comparisons shown for helium. However, a potential valid at small separations has been deduced for argon from the results of shock compression of the liquid [161, 162], and this is shown in Fig. 6.10. The agreement is remarkably good.

The validation of the absolute accuracy of the high-energy scattering potentials is so important that it is worth mentioning comparisons with the few other available accurate quantum-mechanical calculations. The most accurate other calculation is probably the recent one on H–He by Fischer and Kemmey [163]. These are in excellent agreement with the MIT scattering results [164], but the Moscow results [150] are inexplicably low. Results of more difficult calculations are available for the more complicated systems He–H_2 [165] and He–Ar [166], and the agreement with the experiments [142–145] is satisfactory.

Although a number of potentials have been determined by high-energy scattering, the amount of direct information available seems meager indeed in comparison with the variety and complexity of most chemical systems. It is therefore important to see whether it is possible to predict potentials for unknown complicated systems from those for simpler known systems. Only in this way is there any hope that molecular-beam results will be of much chemical use in the near future.

The simplest type of prediction is the calculation of a third noble-gas potential, from two known noble-gas potentials. An example would be the prediction of the He–Ar potential from measured He–He and Ar–Ar potentials. It happens that a geometric mean combination rule works surprisingly well,

$$V_{12} = (V_{11}V_{22})^{1/2}. \tag{6.42}$$

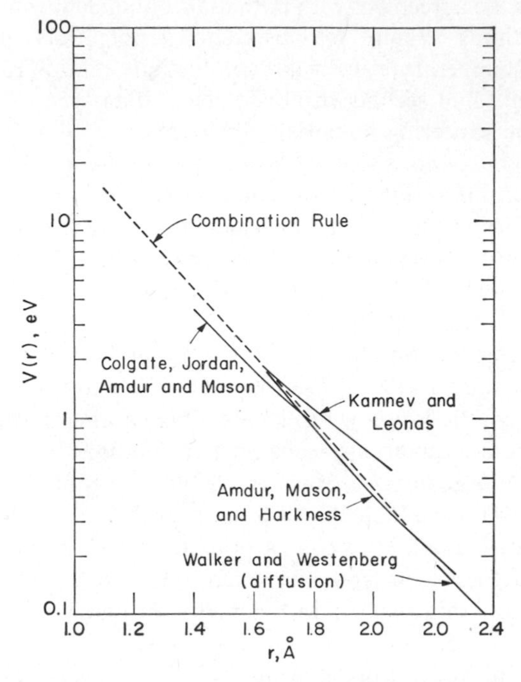

Fig. 6.11 Helium–argon potential energy predicted from the scattering potentials for He–He and Ar–Ar (Figs. 6.9 and 6.10) by geometric-mean combination rule, compared with direct measurements.

The basis for such a relation is discussed in a recent review [167]. The theoretical reasons are not very strong, and the relation should be regarded as mainly empirical. It has been tested by direct experiment a number of times, however; the most extensive tests have been in fact for He–Ar, and these results are shown in Fig. 6.11. The dashed curve shows the combination-rule prediction based on the results for He–He and Ar–Ar shown in Figs. 6.9 and 6.10; the solid curves represent three direct scattering measurements [143, 145, 168] plus one potential based on high-temperature diffusion measurements [169]. The agreement is very good.

The prediction procedure for interatomic potentials becomes much more complicated if free atoms having incomplete electron shells are involved. Here one must deal not with a single potential for each pair of atoms, but with multiple potentials corresponding to all possible molecular states involved. For instance, two ground-state hydrogen atoms can interact in either of two ways, depending on whether their electron spins are parallel or antiparallel, and two ground-state nitrogen atoms can interact in four different ways. In such cases it is necessary to invoke more elaborate molecular quantum mechanics in order to generate combination rules. This has been done with fair success, but the results are not of much direct interest yet in chemical applications, where it is seldom necessary to consider the interaction between two free valence-unsaturated atoms. At any rate, a recent review is available [167].

Of perhaps more chemical interest are interactions between an atom and a molecule, between two molecules, or between separate parts of large molecules. The same theoretical ideas used in the treatment of free valence-unsaturated atoms lead to the idea that each atom can be considered a separate source of potential and that the potential between two molecules can be approximated as the sum of all the atom-atom potentials acting between the two molecules [167, 170]. Empirical evidence for this idea comes from some of the beam-scattering results. For instance, the interaction between an argon atom and a CH_4 molecule was found to be better described as the interaction of the argon with the peripheral hydrogen atoms than as the interaction

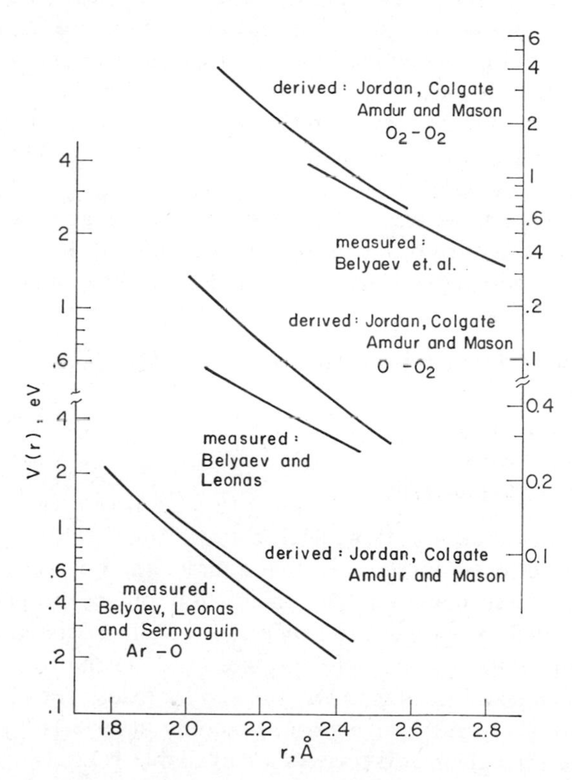

Fig. 6.12 Comparison of derived Ar–O, O–O_2, and O_2–O_2 potential energies with direct measurements.

with a single central potential [141]. Such an approximation opens up immense possibilities, for the geometric rule of (6.42) should apply to the effective atom-atom potentials if there is no correlation between their electron spins, as there is not if one or both of the atoms is bound as part of a molecule. Thus a few good beam measurements on relatively simple systems would be a sufficient basis for the calculation of interactions between rather complex molecules of chemical interest.

The subject has not yet progressed that far, but a promising beginning has been made. As an illustration, the Ar–O_2 and Ar–Ar measurements can be used to predict the Ar–O, O–O_2, and O_2–O_2 potentials, all of which have been measured independently. Details can be found elsewhere [21, 144], but the general procedure is as follows. The Ar–O_2 potential, averaged over all molecular orientations, is obtained from the scattering of a high-energy argon beam in O_2 gas; from this the effective Ar–O potential is deduced. Combination of the Ar–O potential with the measured Ar–Ar potential according to (6.42) then yields an effective O · · · O potential corresponding to no correlation of electron spins; this may be built up into O–O_2 and O_2–O_2 potentials, which can be averaged over all molecular orientations for comparison with directly-measured beam results. In Fig. 6.12 the predicted potentials [144] for Ar–O, O–O_2, and O_2–O_2 are compared with the measured [148, 149, 151] ones. The agreement is fair for O–O_2 and remarkably good for Ar–O and O_2–O_2.

Thus the prospects at present seem good for the application of high-energy beam potentials to problems of chemical interest. Five areas in which information on short-range potentials would have direct application are as follows.

1. High-temperature gas properties.
2. Vibrational relaxation.
3. Radiation damage.
4. Hot-atom chemistry.
5. Conformational analysis.

A few comments on each of these areas follow.

The previous discussion suggested that a direct and obvious application of potential-energy information is the calculation of gas properties at high temperatures. Such properties are needed for a full understanding of many phenomena in upper-atmosphere physics and chemistry, combustion, detonation, high-speed gas dynamics, and similar fields. The first application of beam potentials to such calculations was only in 1958 [171]. Much work has been done since then, and reviews are available [172].

The interchange of vibrational and translational energy requires energetic collisions, even at room temperatures, and potentials at close distances of approach are involved [173]. Such vibrational relaxation is important in determining the absorption and dispersion of sound in gases, as well as the fate of the energy stored in vibrationally "hot" molecules.

Radiation damage frequently causes the formation or deposition of gas atoms in crystal lattices. The rate at which these atoms can escape is controlled in part by the short-range forces between them and the atoms of the lattice. A similar situation arises in connection with the problem of the behavior of solutions of gases in solids [174, 175].

Hot-atom chemistry obviously involves high-energy collisions in a direct way [176, 177]. A central problem, for example, is the determination of the energy losses of hot atoms by collisions with inert moderator gas. The potential energies involved in such collisions are in the range in which molecular-beam information would be appropriate.

The geometric conformation of a complicated molecule is of fundamental importance to its chemical and physical behavior [178]. The conformation is dependent on the potential energy of interaction of nearby nonbonded atoms, and in many cases the short-range repulsions are important. Here information from molecular-beam experiments would be most helpful, for the usual extrapolations of potentials into the short-range region are notoriously unreliable. The first attempt to use molecular-beam information in conformational studies was restricted to the calculation of barriers to internal rotation and energy differences of rotational isomers for small molecules [179, 180]. For larger molecules, such as polymers, polypeptides, and proteins, many nonbonded interactions are involved and computers are necessary, but the basic potential energy information is still required. Conformation calculations have become an active research subject in recent years [181, 182], but little use has yet been made of relevant molecular-beam results.

Low-Energy Scattering

Thermal energy scattering experiments probe the intermolecular potential at large distances of molecular approach, the beam energies being generally much less than 0.5 eV. At these distances the potential energy has both attractive and repulsive components and is not monotonic as at close distances of approach; instead it is attractive at long range and repulsive at shorter range, with an attractive well in between. The measured cross sections for such potentials do not show the smooth behavior in energy or angle that occurs with high-energy beams, but tend to have more structure. Such structure may also depend strongly on the fact that the de Broglie wavelength of the relative motion is not negligible compared to molecular size. Of course a measured curve with structural features contains more information than a smoothly varying one, and so rather detailed potential information is often obtainable from thermal-energy scattering measurements.

Two excellent reviews on the determination of intermolecular potentials from thermal-energy molecular beams have recently appeared [22, 183], and this discussion is therefore confined to some remarks about the major features of the observed cross sections, the type of information on the potential that is obtained from those features, and some applications of the results.

Because the peaking of the scattered intensity in the forward direction is much less pronounced at low energies than at high, it is possible to make measurements of the total scattering cross section $S(0)$. It is possible to construct an apparatus with aperture θ_0 near enough to zero so that most measurements of integrated cross sections are therefore essentially equal to total cross section. Total cross sections probe the outermost fringes of the potential field and always require quantum mechanics for their interpretation. The reason is that a long-range force that nominally extends to infinity always gives an infinite $S(0)$ in a classical description, for a slight deflection is produced by any encounter, no matter how distant. The uncertainty principle of quantum mechanics limits the size of an observable deflection and leads to a finite value of $S(0)$ when the proper quantum treatment is carried out.

A convenient criterion for determining when quantum mechanics must be used can be written in terms of the scattering angle. When this angle is smaller than a critical angle, which may be written as [140]

$$\theta_{\text{crit}} \sim \frac{h}{2\mu v r_c}, \tag{6.43}$$

the scattering must be analyzed with quantum-mechanical methods.

For a potential of the form

$$V(r) = \frac{-C_s}{r^s}, \tag{6.44}$$

in which C_s is a constant, the total cross section is

$$S(0) = p(s)\left(\frac{C_s}{\hbar v}\right)^{2/(s-1)}, \tag{6.45}$$

where $p(s)$ is a constant whose value depends on s. For many systems the longest-range component of the potential is the London dispersion energy corresponding to $s = 6$. A large number of London C_6 constants have been measured in this way; Bernstein and Muckerman [183] give a comprehensive tabulation. Measurements of $\sigma(\theta, \epsilon)$ at small angles also give essentially the same information.

The total cross section is determined by the distance r_c at which a grazing collision occurs that just produces "zero" deflection within the meaning of the uncertainty principle. Another class of collisions also produces zero deflection. In these collisions the distances of closest approach are less than r_c. The trajectory is warped from a straight line by both attractive and repulsive forces, but their effects compensate each other, and no net deflection is suffered. Such trajectories are called glory trajectories after an analogous optical phenomenon [26, 184]. Waves following the glory trajectories have different phase shifts than those following the grazing trajectories; the resulting wave interference produces undulations of $S(0)$ versus v about a smooth curve. These undulations have been analyzed in considerable detail [183]. Their spacing essentially gives the area of the attractive well of the potential, and their total number gives the number of bound states that the well can accommodate; information from these glory undulations of $S(0)$ thus complements the information from the absolute magnitude of the average $S(0)$. Most of the important features of this type of scattering can be understood without detailed mathematical analysis through the use of optical analogies [185].

If the beam energy is somewhat larger than the depth of the potential-energy well, a large intensity of scattered particles will appear near some angle θ_r [25, 26]. This is called rainbow scattering by analogy with the well-known optical phenomenon. It occurs whenever there is a minimum in the deflection angle as a function of impact parameter, since trajectories from many different impact parameters around the minimum all emerge at nearly the same angle. In classical terms, $\sigma(\theta, \epsilon)$ becomes infinite whenever $d\theta/db = 0$, as (6.5) clearly shows, but the rainbow scattering is always substantially modified by wave interference (quantum) effects. These not only modify the shape of the rainbow, but cause it to break up into an interference pattern consisting of the primary rainbow plus a set of weaker peaks known as supernumerary rainbows [186]. An example of rainbow scattering is given in the next section.

The angle θ_r at which the primary rainbow occurs depends primarily on the well depth and the kinetic energy and only secondarily on the detailed shape of the potential well. The spacing of the supernumerary rainbows depends primarily on the ratio of the de Broglie wavelength to the distance at which the potential well has its minimum, and only secondarily on the shape of the well.

Other interference phenomena also occur in thermal-energy scattering, and are discussed in various reviews [22, 26, 28, 167, 183]. The important point is that thermal-energy beam scattering gives information on the part of the potential around the attractive well, and outward. In some ways this is less useful information from a chemical point of view than information on the short-range potential, for most of the interesting chemistry involves rather energetic collisions. Direct applications would probably include only gas properties and conformational analysis. Nevertheless, much of our chemical intuition about molecular interactions is based to some extent on what we know about the potential energy near the attractive well, so the indirect contribution is not to be slighted, even though much of this knowledge in the past came from studies of bulk properties, before the emergence of thermal-beam scattering as a research tool.

Intermediate-Energy Scattering

The energy range from about 0.5 to 25 eV involves extreme experimental difficulty, and what little knowledge we have comes mainly from interpolation between high-energy and thermal-energy results. This is most unfortunate from a chemical point of view, for much interesting chemistry occurs in this energy range. Most of the experimental effort to date has gone into the development of nozzle beams, merging beams, or sputtered beams, all of which are discussed in the section on experimental methods. So far almost no results of chemical interest have emerged, for the experimental techniques are still under development. No great theoretical surprises are anticipated even if the techniques are eventually successful, and the major contribution is expected to be in the form of quantitative information.

Inelastic Scattering

The performance and the interpretation of inelastic scattering measurements are more difficult and less highly developed than in the case of elastic scattering. This is unfortunate, although understandable, because inelastic collisions would be expected to be more characteristic of chemical reactions than are elastic collisions. At present only a few comments are offered on the available studies of rotational, vibrational, and electronic excitation by molecular beams. The nature of the discussion is different from that for elastic scattering because no particular aspect of inelastic scattering is generally applicable to all experiments. Furthermore, there has not as yet been any significant effort to extend the results to chemical problems.

Experimentally, the inelastic process most readily detected is ionization. Although ionization of neutrals by neutrals has been studied over a long period, the center-of-mass energy in such studies has generally been in the kilovolt range. The result of these high-energy ionization experiments has been reviewed in a number of articles and books [113, 187]. A few years ago, however, some careful measurements were reported in which parallel-plate collectors were used for slow charge collection, which showed that neutral beams of N_2 or O_2 could ionize N_2 or O_2 at center-of-mass energies not far removed from their ionization energies [188, 189]. In fact, extrapolation of these measurements indicates that the threshold for the ionization process is the same as that for ionization of these molecules by electrons. A similar result has been obtained more recently for the Ar–Ar system with a cylindrical ion collector with appropriate grids for suppressing secondary electron emission [190]. In this case the measurements indicate a threshold, in center-of-mass energy units, of about 15 eV, which is very close to the ionization potential of argon. Similar results have been cited for other systems in recent conference reports and in private communications.

Inelastic scattering that produces changes in rotational and vibrational energy has also been reported recently. The use of an inhomogeneous electric or magnetic field for selecting and focusing polar molecules has been used to measure cross sections associated with changes in the (j, m) states of LiF [191]. By using time-of-flight analysis to determine energy losses of velocity-selected beams of potassium atoms, the rotational deexcitation of ortho deuterium according to

$$K + o\text{-}D_2(j = 2) \rightarrow K + o\text{-}D_2(j = 0)$$

has been observed [192]. Larger changes in rotational energy have also been reported [193] wherein a K–CO_2 crossed-beam system was used and rotational excitation was detected corresponding to J changes between 8 and 24. These experiments also confirmed the presence of intermediate complexes produced by "sticky" collisions, as reported earlier [194] on the basis of a difference between the observed angular distribution of potassium atoms and that which would be expected for purely elastic scattering. By proper choice of the projectile and target and the range of energy, it was found possible to excite either rotation or vibration [195]. For example, Ar^+ ions scattered in H_2 produced only rotational excitation; O^+ ions in O_2, only vibrational excitation. The excitation was detected by careful retardation measurements of the primary ions.

Observation of vibrational excitation in H_2, without evidence for rotational excitation, has also been observed in scattering experiments with Li^+ ions having energies between 10 and 50 eV [196]. Time-of-flight techniques were used to determine energy loss, and excitation of several vibrational levels of H_2 was observed. Similar time-of-flight techniques are currently being used [197, 198] in the scattering of beams of K^+, Na^+, and Cs^+ into several different target gases, H_2, D_2, He^4, and He^3 at center-of-mass energies as high as 35 eV. When helium was used as the target gas, no inelastic scattering was observed, but in the case of H_2 and D_2 vibrational excitation, dissociation, and electronic excitation of the dissociated atoms could be inferred from the energy-loss spectrum.

A rather different approach was used [199] to obtain indirect evidence of rotational excitation. In a study of the scattering of beams of helium atoms with energies up to 2 keV by hydrogen isotopes, the total scattering cross sections for H_2 and D_2 were found to be identical and in good agreement with elastic cross sections computed from the theoretical He–H_2 potential [165]. The He–HD cross section was somewhat larger than expected, and application of perturbation theory [200] showed that the enhancement was consistent with rotational excitation of HD.

Finally, a particularly interesting example of the transfer of internal energy in molecular collisions is found in a chemiluminescence study [201]. Beams of potassium atoms and bromine were crossed to produce vibrationally excited $KBr^{\ddagger}$ according to the reaction

$$K + Br_2 \rightarrow KBr^{\ddagger} + Br.$$

The vibrationally excited $KBr^{\ddagger}$ was then reacted with sodium in a triple-beam arrangement to produce electronically excited K^* according to the rearrangement reaction

$$KBr^{\ddagger} + Na \rightarrow K^* + NaBr.$$

As the electronically excited K^* decayed to the ground state, radiation was emitted and detected by a photomultiplier.

Reactive Scattering

The results of reactive scattering need not be separated into energy regimes because most experiments are done at thermal energies or slightly higher. Many reactive scattering experiments have been performed since the original one of Taylor and Datz [73]. Several review articles that discuss such experiments are available [18, 19, 202, 203]. In general, differential cross sections for reactive systems can be expected to show structure, both because a chemical reaction has occurred and because of the effects of scattering by a nonmonotonic potential. Such structure is indeed observable experimentally under the right circumstances. For example, it was pointed out above that the discontinuity in the differential cross section at the rainbow angle as predicted on the basis of classical arguments is actually not a sharp discontinuity but a rather broadened peak. In addition, it was noted that the rainbow angle depends on the relative velocity, and therefore the effects of rainbow scattering are generally not discernible without velocity selection, since each initial relative velocity is associated with a different θ_r and the fine structure in the differential cross section will be washed out by the overlapping rainbow angles if the spread of initial relative velocities is too broad. This is just a sample of the rather specialized experimental conditions that must be met in order to obtain useful information concerning reactive scattering. Almost every experiment requires its own specific arrangement, but most such details are excluded in the following discussion. The results of four particular crossed-beam experiments are discussed, primarily in terms of the interpretation of the measurements; the experimental details can be obtained from the original papers or from one of the indicated review articles. The first example is a rather special case that illustrates many of the essential features of interpreting reactive scattering experiments. The other three that are discussed are specific reactions illustrating different types of mechanisms by which reactive scattering processes may occur. Although semantic differences exist in the characterization of these three types, there is general agreement concerning the features by which the mechanisms can be identified. The categories used here appear elsewhere under either identical or closely related designations [19, 204, 205]. The mechanisms are discussed in terms of the prototype reaction

$$A + BC \rightarrow AB + C$$

with appropriate specific illustrative examples.

Special Example

The first example is

$$\mathrm{K + HBr \rightarrow H + KBr}.$$

This system has several of the special features previously discussed, such as a small value of $\Delta D_0{}^0$ and a ratio of μ/μ' that is much larger than unity. The reaction was the subject of the first significant experiment on reactive scattering with crossed thermal beams. It was originally studied by Taylor and Datz [73], and the results of these first measurements were later reinterpreted by Datz, Herschbach, and Taylor [46]. Collimated beams of potassium effusing from an oven at temperatures between 541 and 837°K intersected at 90-degree noncollimated cones of HBr, which effused from a source kept between 373 and 400°K. Surface-ionization detectors were used to monitor the fluxes of potassium and KBr in a manner that made it possible to distinguish between them, since both potassium and KBr are surface-ionized by tungsten, whereas a platinum surface ionizes potassium but is essentially insensitive to KBr [73–75]. By taking the difference between the positive ion currents produced by the two types of detectors, the flux of product KBr was measured as a function of angle with respect to the potassium beam in the laboratory plane; that is, the plane defined by the direction of the potassium beam and the normal to the source from which the HBr effused. Because $\mu/\mu' \gg 1$ and $\Delta D_0{}^0$ is small (about 3.8 kcal/mole), the KBr product is expected to be distributed in a fairly small cone near the centroid. However, (6.32) shows that there will be a different centroid for each pair of velocities in the reactant beams. Neither beam was velocity-selected in this experiment, so that the product KBr was actually distributed over a range of centroids. The experiment could be analyzed only by an elaborate calculation of the distribution of KBr, which required the use of a number of relevant experimental details as well as an assumption about the energy-dependence of the reaction cross section. The threshold energy for reaction, E_a, was carried as a parameter in this calculation, and the best agreement between the experimental and calculated distributions of KBr was found for $E_a \simeq 3$ kcal/mole.

Beck, Greene, and Ross [48], in a reinvestigation of this reaction, used a velocity selector to obtain a very nearly monochromatic beam of potassium. The introduction of velocity selection has several advantages, among which is the partial elimination of the spread of centroids and a more precise determination of the threshold for reaction. Furthermore, a great deal of additional microscopic information can be obtained that is inaccessible when velocity selection is not performed.

The velocity selection of the potassium beam was such that if v_A were the maximum in a narrow band of selected velocities, the spread in velocity at half height would be $0.084v_A$. The HBr beam was not velocity selected but effused from a thermal source at T_B. The initial average kinetic energy of the system was taken as $\bar{\epsilon} = \mu(\bar{v}_A^2 + \bar{v}_B^2)/2$, where $\bar{v}_B = (8kT_B/\pi m_B)^{1/2}$. The distribution of the KBr product flux was measured and from this, after a rather involved analysis of the spatial distribution of the reactant HBr, the total cross section for reaction S_R was obtained. It was found that the threshold energy E_a was quite low, less than 0.4 kcal/mole, and that for values of $\bar{E}$

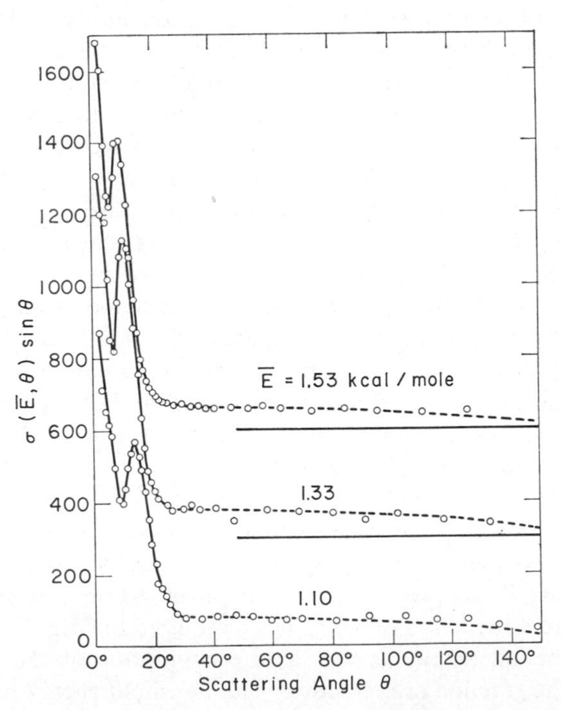

Fig. 6.13 Elastic scattering of the system K + Kr.

larger than this threshold the total cross section for reaction $S_R(E)$ was essentially independent of the initial kinetic energy of relative motion. The discrepancy between this low value of E_a and that reported in the earlier experiment simply reflects the improvement possible with velocity selection of the beam.

In the same study it was possible to obtain information about the reaction probability. The reaction probability that was determined was a special case of $P(k, l, i, j, p, b, \phi)$ first introduced in (6.20). In this experiment the probability was not completely specified since there was no state selection, no velocity selection of the HBr beam, and the flux was measured only as a function of θ. Accordingly, the reaction probability was designated as $P(\bar{E}, \theta)$, and the method used to obtain it was as follows.

The differential elastic cross section $\sigma(\bar{E}, \theta)$ was measured as a function of θ. Measurements were also made of $\sigma(\bar{E}, \theta)$ as a function of θ for the nonreactive analog of K + HBr, namely, K + Kr [206]. The experimental determination of $\sigma(\bar{E}, \theta)$ as a function of angle for a given value of the initial average kinetic energy of relative motion is straightforward. It requires measurement of the initial axial intensity of the beams of potassium atoms and the scattered flux of potassium atoms at selected laboratory angles associated with corresponding values of θ. The ratio of the off-axis flux to the axial intensity is the differential cross section for a given value of $\bar{E}$ and θ. For K + Kr the results for three different values of $\bar{E}$ are shown in Fig. 6.13 and the results for K + HBr for five different values of $\bar{E}$ are shown in Fig. 6.14.

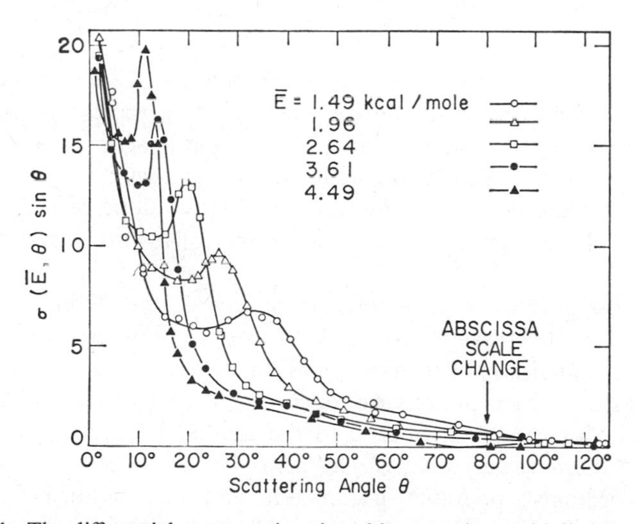

Fig. 6.14 The differential cross section, in arbitrary units, multiplied by sin θ versus the relative scattering angle θ for the system K + HBr at five different values of the kinetic energy of relative motion.

Although the ordinates in both figures are in arbitrary units, it is apparent that the scattering patterns of the two systems show qualitative similarities in that each curve shows a sharp decrease at very small angles and a well-defined peak. The peak is caused by rainbow scattering, as discussed previously. It was pointed out earlier that structure in the differential cross section is generally not observable without velocity selection. That it is so clearly observable in these two systems in which only the potassium beam was velocity-selected is in part the result of suppressing the spread in the initial relative velocity. This was accomplished by having the Kr and HBr, both of which are heavier than potassium, effuse from sources that were lower in temperature than the potassium source. Thus the value of $\bar{v}_B$ for krypton or HBr was lower than the rather well-defined selected velocity v_A for potassium. Furthermore, an intersection angle of 90° as opposed to a larger angle further reduced the spread in v, although this effect was of secondary importance. The influence of the angle of intersection on the spread in the initial relative velocity has been analyzed by Pauly and Toennies [22].

It can be assumed in analyzing the results shown in the figures that the molecular interactions were a combination of central finite repulsive and attractive forces for which the two-dimensional trajectory could be represented as shown in Fig. 6.5. The potential-energy function associated with the trajectory was the so-called exp-6 potential, which can be written as

$$V\left(\frac{r}{r_m}\right) = \frac{\epsilon_m}{1 - 6/\alpha}\left\{\frac{6}{\alpha} \exp\left[\alpha\left(1 - \frac{r}{r_m}\right)\right] - \left(\frac{r_m}{r}\right)^6\right\}, \tag{6.46}$$

where r_m is the position of the potential minimum and ϵ_m is the magnitude of its depth and α is a measure of the steepness of the exponential repulsion. Inversion procedures have been developed that permit the parameters ϵ_m and r_m in (6.46) to be evaluated (at least for the nonreactive system). The procedures utilize the position of θ_r and the negative slope of $\sigma(\bar{E}, \theta) \sin \theta$ versus θ; details have been described by a number of investigators [22, 202, 207, 208].

The two most striking qualitative features about the differential cross section for the reactive system in Fig. 6.14 are the similarity to the nonreactive system at angles smaller than θ_r and the distinct difference at large θ, where $\sigma(\bar{E}, \theta) \sin \theta$ is seen to decrease much more rapidly than for K – Kr. The latter behavior can be attributed to the chemical reaction. In general, scattering near the relatively small angles near θ_r is not affected by chemical reaction. Accordingly, potential parameters can be obtained from the differential cross sections in the region near θ_r by the same inversion procedure used for the nonreactive system. (The significance of the shift of the rainbow scattering to larger angles for the case of K – HBr means only that the potential well for this system is deeper than that for K — Kr.) The potential so determined can be used to calculate hypothetical elastic differential cross sections at larger angles where, in reality, reactive scattering has a large effect on the scattered flux of potassium. Thus if $[\sigma(\bar{E}, \theta)]_{el}$ is the elastic differential cross section calculated on the assumption that no reaction has occurred, $\sigma(\bar{E}, \theta)$ will be smaller than $[\sigma(\bar{E}, \theta)]_{el}$ since some of the potassium-beam atoms are lost because of chemical reaction. The reaction probability $P(\bar{E}, \theta)$ for a given value of $\bar{E}$ is therefore defined as

$$P(\bar{E}, \theta) = \frac{[\sigma(\bar{E}, \theta)]_{el} - \sigma(\bar{E}, \theta)}{[\sigma(\bar{E}, \theta)]_{el}}. \tag{6.47}$$

Beck, Greene, and Ross [48] have shown that it is possible to extrapolate these reaction probabilities to obtain a value of the threshold energy; that is, the value of $\bar{E}$ for which $P(\bar{E}, \theta)$ is zero. Their procedure involves extrapolation with respect to $\bar{E}$ at a fixed value of the potential that corresponds to a constant value of the distance of closest approach r_c. Their extrapolation method yielded a value of 0.15 kcal/mole for E_a. As previously indicated, their direct measurements based on detecting the onset of an appearance of a KBr signal gave an upper limit of 0.4 kcal/mole.

For a given value of $\bar{E}$ a threshold angle, θ_t, can be determined by noting the angle at which (6.47) indicates that $P(\bar{E}, \theta)$ is zero. Although θ_t is found to vary with $\bar{E}$, the threshold impact parameter, b_t, associated with θ_t is essentially invariant [48]. Thus the total reaction cross section for a given value of $\bar{E}$ is simply given by the analog of the general relation in (6.23) namely,

$$S_R(\bar{E}) = 2\pi \int_0^{b_t} P(\bar{E}, b) b \, db. \tag{6.48}$$

The values of $S_R(E)$ calculated as described above were in reasonable agreement with those obtained by direct measurement [202, 209]. Another independent calculation of $S_R(\bar{E})$ has been reported [210], in which a potential surface is constructed, and formal scattering theory is applied to the calculation of $S_R(\bar{E})$. Again, reasonable agreement is obtained with the experimental value of $S_R(\bar{E})$.

The use of elastic scattering in reactive systems involves approximations and assumptions. Yet the procedure is of special interest since in principle, and to a large extent in practice, it provides a nearly independent method for obtaining $S_R(\bar{E})$ with very little recourse to direct measurement of the yield of reaction product.

In concluding the discussion of the reaction of potassium and HBr it is appropriate to show the type of microscopic information that can be obtained by selecting a suitable type of system, using both elastic and reactive scattering results, and including velocity selection. The most significant results are summarized in Table 6.1, which is taken from the paper of Beck, Greene, and Ross [48]. In making their calculations they used a value of −4.2 kcal/mole for $\Delta D_0{}^0$ instead of the more recent value of −3.8 kcal/mole. This difference in $\Delta D_0{}^0$ will produce no important changes in the entries in the table, particularly in view of the assumptions and approximations that were

Table 6.1 Distribution of Energy in the Products of the Reaction Potassium + HBr

$\bar{E}$, kcal/mole		1.49			4.49	
Angular momentum of HBr, $j = 2$, g-cm²/mole-sec		0.00155			0.00155	
Rotational energy of HBr, $j = 2$, kcal/mole		0.15			0.15	
Potential at distance of closest approach, kcal/mole	0.15	0.6	1.2	0.15	0.6	1.2
Impact parameter b, Å	3.72	2.92	1.67	3.84	3.53	3.21
$\mu v b$, nearly equal to angular momentum of KBr, g-cm²/mole-sec	0.068	0.053	0.030	0.121	0.111	0.101
Rotational energy of KBr, kcal/mole	2.52	1.55	0.51	8.07	6.83	5.65
Rotational quantum number	104			186		
Sum of final relative kinetic energy and vibrational energy of KBr, kcal/mole	3.17	4.14	5.18	0.59	1.83	2.42
Maximum vibrational state of KBr	5	6	8	1	3	4

involved. For the sake of clarity and for emphasis of the special features of the reaction of potassium and HBr these assumptions and approximations are repeated.

The rotational angular momentum $\vec{J}_B$ of HBr is so small that it may be neglected in comparison with the initial angular momentum $\vec{L}$ of the system. The final orbital angular momentum $\vec{L}'$ of the system is also very small compared to $\vec{L}$. This follows from the fact that μ' is very much less than μ, that $\vec{v}'$ is not likely to be very much different than $\vec{v}$ because of the small value of ΔD_0^0, and that b' and b would be expected to have approximately the same magnitudes [18, 48]. Thus since $\vec{L} \gg \vec{J}$ and $\vec{L} \gg \vec{L}'$, the total angular momentum conservation condition given by (6.38) reduces to

$$\vec{L} \cong \vec{J}'. \tag{6.49}$$

The rotational angular momentum of the KBr is therefore very nearly equal to the initial orbital angular momentum of the system. In arriving at the entries in Table 6.1, HBr and KBr were treated as rigid rotors and harmonic oscillators. The main conclusions to be drawn from the rather detailed discussion of the reaction of potassium and HBr are that in this particular type of reaction the products that are formed will be in highly excited internal states, generally rotational states, and that the laboratory scattering angles at which the products would be expected to appear may be closely predicted without detailed knowledge of the dynamics of the collision [19].

The special case of K + HBr was discussed in detail because that system contains so many interesting features. In the next sections three reaction mechanisms that have been investigated with molecular-beam technique are discussed. The treatment is less detailed than the present section and is intended to show the kinds of general conclusions which can be reached in chemical-reaction studies with beams.

Rebound Reactions

These reactions typically have relatively small total reaction cross sections, about 50 Å^2 or less, and in the center-of-mass system they have quite anisotropic angular distributions of products. Figure 6.15 is a schematic representation of the collision trajectories for rebound reactions in the laboratory and in the center-of-mass systems for beams of A and BC that intersect at 90°, as indicated in Fig. 6.15*a*. In the laboratory system AB should appear at an angle close to 90° with respect to the direction of the beam of A atoms. As shown in Fig. 6.15*b* an observer traveling with the center of mass would see the A and BC beams approaching from his left and right to collide at his point of observation. After the collision he would observe that the AB product would return roughly in the direction from which the A atoms approached.

A typical example of this type of reaction is

$$K + CH_3I \rightarrow KI + CH_3,$$

which was first investigated by Herschbach and coworkers [47, 211]. Although velocity selection was not used in their experiments, it was observed that the distribution of KI product was strongly peaked at an angle of 83° as measured from the direction of the beam of potassium atoms. The experimental data

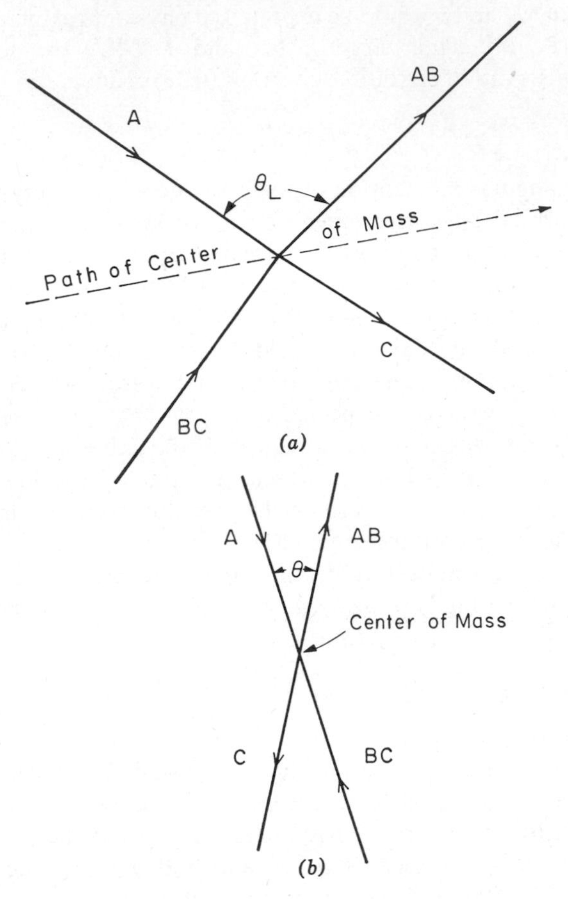

Fig. 6.15 Idealized trajectories of reactants and products for a rebound mechanism. (*a*) Laboratory system; (*b*) center-of-mass system.

are shown in Fig. 6.16. It was originally concluded that the total reaction cross section was small, of the order of 10 Å², and that most of the liberated energy, about 90%, appeared in the internal degrees of freedom of the products rather than in their average kinetic energy of relative motion. In a reanalysis of the experimental results it was later shown that the average kinetic energy of relative motion of the products was considerably larger than originally estimated and that the internal energy of the products was correspondingly smaller [212]. The reaction was later reinvestigated by Ross and coworkers [213] with velocity-selected beams of potassium, and the total reaction cross section, although still rather small, was found to be about 30 Å². Other recent investigations of this reaction using state selection of the

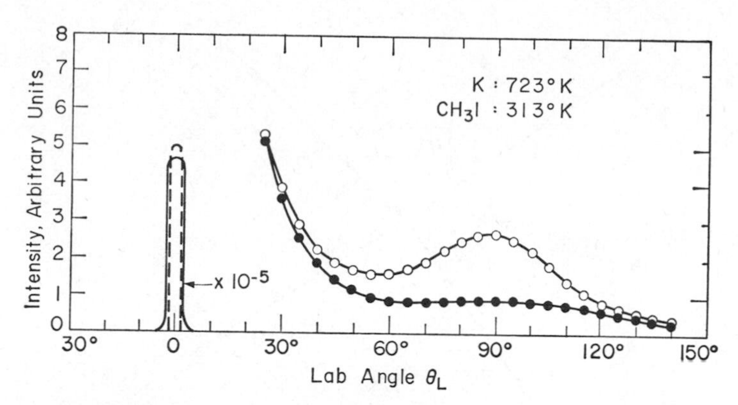

Fig. 6.16 The scattering of K + CH_3I. The curves show the intensity measurements from two detectors. The solid circles indicate the intensity of K, and the open circles the intensity of K + KI. The main potassium beam, shown at angle 0°, is attenuated about seven percent by the CH_3I crossed beam.

CH_3I have shown that there is a small but definite steric factor associated with the spatial orientation of the potassium atom relative to the CH_3I molecule as they approach each other prior to reaction [36–38]. It was possible to effect partial spatial orientation of the CH_3I molecules with strong inhomogeneous fields with several different multipole electrode arrangements. Both velocity-selected and nonvelocity-selected beams of potassium atoms were crossed with oriented CH_3I molecules and with nonoriented molecules. The analysis of the results permits a comparison of the total reaction cross sections for the two cases in which the potassium atoms approach the CH_3I from the iodine end or from the CH_3 end. The ratio was about 1.5 and, as might have been anticipated on physical grounds, the larger cross section was found for the approach of potassium toward the iodine end of the CH_3I. The experimental results are in accord with a theoretical study of this steric effect [38].

Stripping Reactions

Typically these reactions have large total reaction cross sections, of the order of 200 Å². As in the case of the rebound reactions the angular distribution of products is anisotropic. However, in stripping reactions, instead of a backward peaking of the molecular product in the center-of-mass system, there is strong forward peaking. This is shown in the schematic trajectories in Fig. 6.17 for 90° intersection of beams of A atoms and BC molecules. It can be seen that in both the laboratory and the center-of-mass systems the product AB continues to travel in much the same direction as the original A atoms.

Stripping reactions are often referred to as proceeding by a "harpooning" mechanism, a description first suggested by Polanyi [214] in connection with

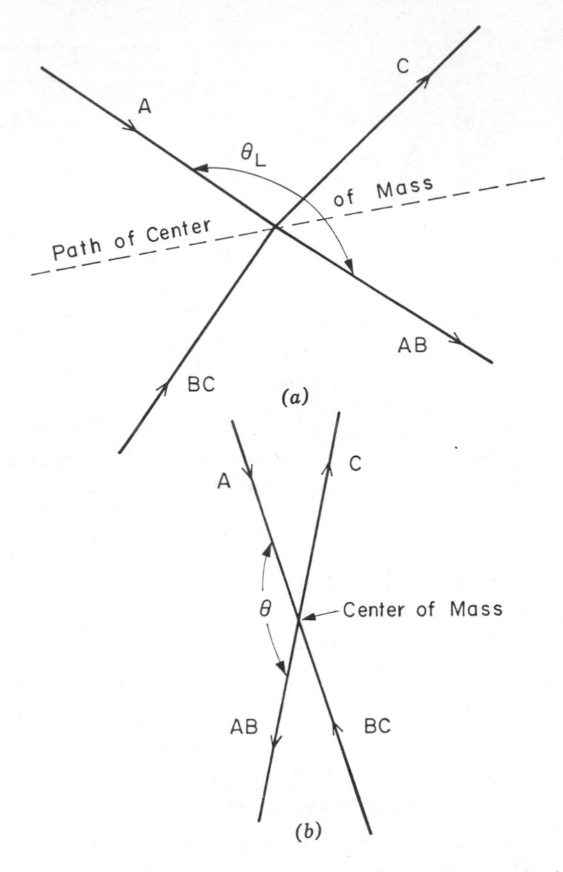

Fig. 6.17 Idealized trajectories of reactants and products for a stripping mechanism. (*a*) Laboratory system; (*b*) center-of-mass system.

his classic studies of reactions between alkali atoms and halogen molecules in dilute flames. Harpooning was pictured in terms of an atom A "tossing out its valence electron to hook atom B which it hauls in with the coulomb force of attraction." The spectrum of stripping reactions is quite broad and is a function of the initial kinetic energy of relative motion of the reactants and the magnitude of the total reaction cross section. A somewhat extreme type is referred to as spectator-stripping, which may occur when the kinetic energy of the relative motion of the reactants is high. In this case there is little interaction during the impulsive collision between A and the atom C in the BC molecule. Thus C plays the role of a "spectator" to the interaction of A and B, a situation that is likely to occur only if the impact parameter associated with the reaction is large. The products in this type of stripping reaction

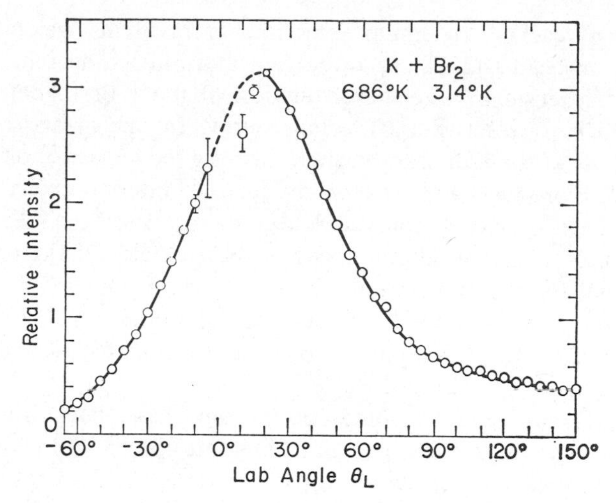

Fig. 6.18 The scattering of K + Br_2. The curve shows the distribution of KBr product in the laboratory system.

separate in a time so short that there is no time for transfer of much momentum to *C*.

The application of the concept of stripping in two-body collisions resulting in chemical reaction was first applied [215–217] as a result of mass-spectrometric studies of ion-molecule reactions. The first experimental studies of stripping reactions in neutral systems were reported simultaneously by two independent groups, one of which used a velocity selected cesium beam that reacted with Br_2 [218], the other a nonvelocity-selected beam of potassium atoms also reacting with Br_2 [219].

For the present discussion the example chosen to illustrate stripping reactions is

$$K + Br_2 \rightarrow KBr + Br.$$

The results of the first study of this reaction [219] are shown in Fig. 6.18, where the spatial distribution, in laboratory coordinates, of KBr product is shown. This was determined as usual with two detectors, and the points shown in the figure were obtained by subtraction. It was clear immediately that the KBr distribution showed a strong maximum near 0° with respect to the direction of the beam of potassium atoms and that the reaction did indeed correspond to the scheme shown in Fig. 6.17. A total reaction cross section of about 200 Å² was deduced. In a subsequent study [220] under more refined experimental conditions the first results were confirmed, and it was concluded from the angular distribution of the products that most of the energy available from the reaction appeared as internal excitation of KBr. These conclusions had also been reached in similar experiments involving velocity-selected potassium atom beams [221]. An important and interesting refinement was added to the reaction by direct examination of the KBr to determine the extent to which it was rich in internal energy. In one arrangement [222] velocity analysis of the KBr showed that most the the 46 kcal/mole of energy available to the products appeared in the form of internal excitation of the KBr. Another arrangement employed deflection analysis of the KBr molecules as they traversed a strong inhomogeneous electric field [223]. It was found that of the total angular momentum available to the products, approximately equal amounts were distributed between the rotational angular momentum of the KBr and the orbital angular momentum associated with the relative motion of the products KBr and Br.

Although rebound and stripping mechanisms have been illustrated by examples that have been chosen to best illustrate the special characteristics of these mechanisms, there exists a group of reactions whose features represent a transition between the two types of mechanisms. In general these reactions will have moderate total reaction cross sections, of the order of 100 Å^2 or less, and the distribution of molecular product in the center-of-mass system will be in angular regions between the backward and forward directions. Among the reactions that show these transition characteristics are K → SF_6, Cs + HBr, and Na + $SnCl_4$.

Collision Complexes

Both rebound and stripping reactions are characterized by having extremely short collision times. This is a direct conclusion from the observed anisotropic peaking, which implies that the incident particles "remember" their initial direction of approach. If the impact parameter associated with reaction is small, the molecular product immediately rebounds in the center-of-mass system; if the corresponding impact parameter is large, the atomic reactant strips an atom from the reactant molecule, and the product molecule immediately continues in the forward direction. In order that such directed motions of the product molecules occur, the reaction must proceed in a time that is short compared to the time of rotation of the molecular adduct formed by the reactants when they are sufficiently close to react. If there were a rotation of even half a cycle, the directional selectivity of the products would not take place. For these reactions there is therefore an upper limit of less than 10^{-12} sec on the reaction time, and therefore the *A-B-C* system at the time of collision is not in any sense a reactive complex; its lifetime is not greater than a vibrational period.

There are, however, a group of reactions in which the reactants do form a long-lived complex that can exist for a number of rotational and vibrational periods before dissociating into products or, in the event reaction does not occur, go back into reactants. Those systems that do react exhibit two interesting experimental features. First, two maxima, in the laboratory angle system, appear in the measured intensity of molecular product. Of the two peaks one, the smaller, is observed near 0°, near the direction of the beam atoms, while the other is found near 90°, close to the direction of the reactant BC beam. In the center-of-mass system these product distributions correspond to approximately equal distributions near 0 and 180°. Secondly, the total reaction cross sections are quite large, of the order of 200 Å².

For this discussion the reaction

$$Cs + RbCl \rightarrow CsCl + Rb$$

is used. It was investigated along with others of the same type by Herschbach and coworkers [224]. A collimated but nonvelocity-selected beam of cesium atoms from a source at 473°K intersected at 90° a similar but higher intensity beam of RbCl from a source at 950°K. In this experiment a Pt–W alloy that has been pretreated with oxygen serves as a surface-ionization detector that ionizes both cesium and CsCl with almost 100 percent efficiency. After pretreatment of the same alloy with methane, the detector ionizes essentially none of the incident alkali halide molecules. The detection system was therefore capable of the same kind of discrimination as that involved in the use of pure tungsten and platinum detectors in the investigation of the reaction of potassium and HBr.

The experimentally observed angular distribution of CsCl as a function of the laboratory angle θ_L is shown in Fig. 6.19. The two maxima, one near 0°,

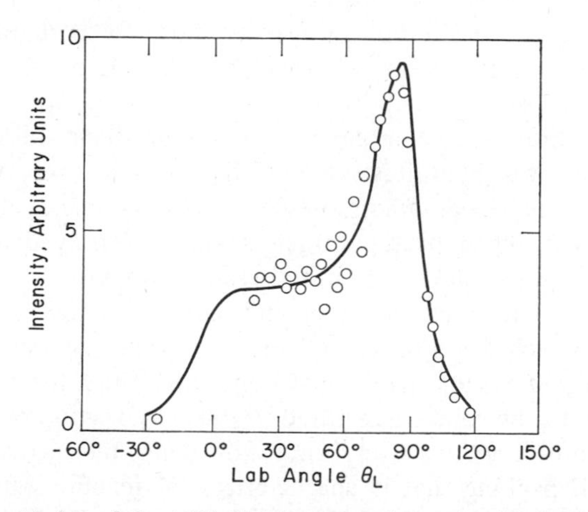

Fig. 6.19 The scattering of Cs + RbCl. The curve shows the distribution of CsCl in the laboratory system.

the other near 90°, stand out clearly. In the center-of-mass coordinate system the two maxima are of equal magnitude. As expected, the total reaction cross section was large, between 150 and 200 Å².

A specific reason determines whether a reaction will proceed by direct or impulsive mechanisms or by mechanisms that involve long-lived complexes. With the exception of RbCl the molecular reactants AB in all the reactions discussed above have covalent bonds. RbCl and the other alkali halide molecules, however, are ion pairs [225] so that the schematic equation for reactions that proceed through long-lived complexes is more informative if written as showing the ion pairs present during the reaction:

$$A + B^-C^+ \rightarrow A^+B^- + C.$$

The facts that the bond strengths of B^-C^+ and A^+B^- are nearly equal and that diatomic alkali molecule ions are relatively stable make it probable that a complex $(AC)^+B^-$ will be formed [226, 227]. The likelihood of the existence of such complexes is also strengthened by an electronic structure calculation for the K–Na–Cl system [228], which shows a basin in the potential energy surface, whose depth is about 13.5 kcal/mole lower than the energy of the products.

The symmetrical distribution of molecular products in angular regions in the center of mass that are concentrated close to 0 and 180° may be visualized in terms of Fig. 6.8 in the section on dynamics of collisions. The main features of the explanation have been mentioned above but are repeated here. The total initial angular momentum will consist for the most part of orbital angular momentum $\vec{L}$, particularly since the large total reaction cross sections imply that these reactions involve initial impact parameters that are quite large. If a long-lived complex is formed, this orbital angular momentum will be converted into rotational angular momentum of the three-atom complex, which is likely to be approximately linear. When, after random intervals, decomposition of the complex occurs, the products will fly off in directions perpendicular to $\vec{L}$, as illustrated in Fig. 6.8*b*. Each vector $\vec{L}$ can thus be associated with a plane containing these vectors $\vec{v}'$ uniformly distributed over 2π radians. Since the impact parameters are uniformly distributed about directions that are parallel to the initial relative velocity, all directions of $\vec{L}$ in a plane normal to $\vec{v}$ are possible, and a plane of the vectors $\vec{v}'$ can be associated with each direction of $\vec{L}$. Thus the overall spatial distribution of $\vec{v}'$ can be visualized by rotating the disk containing the $\vec{v}'$ for a given direction of $\vec{L}$ around $\vec{v}$. In the sphere generated in this way vectors of $\vec{v}'$ will be concentrated in both directions along the $\vec{v}$ axis. This concentration is precisely the 0 and 180° peaking that is characteristic of reactions that proceed by mechanisms involving long-lived complexes. A more detailed description of the dynamical features of reactions that proceed through long-lived complexes can be found in the original paper of Herschbach and coworkers [224] and in the review article of Toennies [19].

The existence of long-lived complexes may also be observed in nonreactive systems. Kinsey and co-workers [229] have given a preliminary report of a number of such systems that were studied by using velocity-selected beams of potassium with SO_2, CO_2, and NO, and velocity-selected beams of cesium with SO_2, CO_2, and NO_2. The angular distributions of potassium and cesium in the laboratory system showed definite evidence of long-lived complexes which, except in the case of NO_2, decomposed back into reactants. The authors concluded that the cross section for the formation of the complex was very small for NO, intermediate for CO_2, and large (in excess of 100 Å^2) for SO_2 and NO_2. A comprehensive paper has also been written [230].

Ion-Molecule Reactions

Molecular-beam techniques have also been used to investigate bimolecular reactions in which one of the reactants is a charged particle, a positive or a negative ion. These reactions fall into the same two broad experimental classes as those discussed earlier; attenuation [231–233] and crossed-beam experiments [20, 234]. The ion sources used are similar in many cases to mass-spectrometer sources, and indeed many ion-molecule experiments have been performed in suitably modified commercial mass spectrometers. More recently, special apparatus types have been constructed that employ appropriate focusing and decelerating electrodes to produce reasonably large intensities of ionic reactants in the approximate energy range of 1 to 100 eV [20].

As a class, ion-molecule reactions that are exothermic have no appreciable barrier to reaction. It is usually assumed that this is the result of the charge-induced dipole attraction that accelerates the ion and the neutral reactants toward each other. This energy, $-\alpha e^2/2r^4$, where α is the polarizability of the neutral reactant, e is the electronic charge, and r is the internuclear separation of the reactants, in the absence of repulsion resulting from electron cloud overlap, may be sufficient to overcome a barrier of several kcal/mole. There are several excellent review articles and a monograph in which the dynamics and detailed mechanisms of ion-molecule reactions are discussed at length and in which extensive bibliographies are included [187, 235–239].

The general characteristics of ion-molecule reactions are similar to those of neutral-neutral reactions. Many of the reactions proceed by direct or impulsive collisions, indicating that the ion and the neutral particle are within the reaction distance for a very short period, of the order of a vibrational period, and accordingly the reactions show rebound, stripping, and spectator-stripping features discussed previously. Of particular interest is the fact that in a number of reactions the actual mechanism is a function of the kinetic energy of relative motion of the reactants. An example is the reaction of Ar^+ and D_2 where at sufficiently high energies the nature of the products changes. This particular reaction is used to illustrate the salient features of ion-molecule reactions that proceed by impulsive collisions. At relatively low energies the reaction may be written as

$$Ar^+ + D_2 \rightarrow ArD^+ + D.$$

In the limit of very low energy if the reactants start essentially at rest, they will accelerate toward each other as the result of the attractive ion-induced dipole forces and will tend to collide in a head-on collision. The products will then rebound and the ArD^+ will go back in the direction from which the Ar^+ came. Under these conditions the reaction will have the characteristics associated with a rebound mechanism. As the initial kinetic energy of relative motion is increased somewhat and becomes comparable or slightly less than the interaction energy, the reactants will have a tendency to spiral with respect to each other, and the products may come off in several directions depending on the angle of spiralling. As the energy continues to increase the spiralling decreases, and the products will acquire a tendency to appear in a narrow, forwardly directed cone as in spectator reactions. At still higher energies the attractive ion-induced dipole interaction is unimportant, and the mechanism becomes for the most part that of spectator stripping as originally proposed by Henglein and coworkers [231]. It has been shown that even in this high-energy region there is also an important component of rebound reaction which may result from those Ar-D-D collisions that are essentially head-on. As the energy increases still further the total reaction cross section for ArD^+ formation becomes extremely small, and as the internal energy in ArD^+ tends to exceed the dissociation limit, fragmentation results so that a different set of products is observed [232] according to

$$Ar^+ + D_2 \rightarrow Ar^+ + 2D$$

or

$$Ar^+ + D_2 \rightarrow Ar + D^+ + D.$$

This brief description of the reaction between $Ar^+ + D_2$ is a summary of a comprehensive account given by Wolfgang and co-workers [240].

The discussion of the reaction of $Ar^+ + D_2$ shows that the features of direct, impulsive neutral-neutral bimolecular reactions may also be seen in ion-molecule reactions. Although long-lived collision complexes in ion-molecule reactions have actually been detected in mass spectrometric studies [241], complexes that exist for only a few rotational periods (about 10^{-10} sec) are not detectable by this method. However, the techniques that have shown the existence of short-lived complexes in neutral-neutral reactions have also shown that they may be present in ion-molecule systems. The most recent, and probably the best documented reaction [242], is

$$O_2^+ + D_2 \rightarrow DO_2^+ + D.$$

The authors concluded that the long-lived collision complex mechanism is dominant at lower kinetic energies of relative motion, but that the mechanism changes to one of direct, impulsive interaction as this energy increases. They conclude that there are two significant reasons why the reaction proceeds through the long-lived collision complex mechanism at the lower energies. First, the reaction is endothermic by 1.9 eV. Second, the $D_2O_2^+$ intermediate lies in a potential well whose depth, excluding activation barriers, is 2.6 eV lower than that of the reactants. Since the threshold for dissociation from the complex into products is therefore 4.5 eV, the lifetime of the complex, at least for low values of the initial kinetic energy of relative motion, would be expected to be reasonably long.

Charge Transfer

A type of rearrangement collision that is usually of more interest to physicists than to chemists involves a process known as charge transfer. It is convenient to distinguish between systems of like atoms that undergo symmetric resonant charge transfer according to

$$A^+ + A = A + A^+,$$

and systems of unlike atoms that undergo asymmetric nonresonant charge transfer according to

$$A^+ + B = A + B^+ + \Delta E.$$

The quantity ΔE is referred to as the energy defect, and when all particles are in their ground states, ΔE is the difference in ionization potentials of A and B. When ΔE is small or zero, the process is referred to as being near-resonant. Charge transfer was mentioned earlier in this chapter as a method for producing fast neutral beams, usually by symmetric resonant charge transfer of fast positive ions in neutral atoms of their own species. An example of this type of charge exchange is

$$He^+ + He = He + He^+.$$

When used to produce a fast neutral beam, the incident He^+ was a fast ion, the target He, a slow atom. After charge transfer the He was a fast neutral, and the He^+, a slow ion. As an example of nonresonant charge transfer that involves only atoms or atomic ions one might use

$$He^+ + Ar \rightarrow He + Ar^+ + 8.825 \text{ eV},$$

where 8.825 eV is the difference between the first ionization potentials of helium and argon, both in their ground states.

Since most of the detailed experimental and theoretical features of charge transfer are not within the scope of this chapter, only the most general features are presented. For symmetrical resonant charge transfer the cross sections are large at low kinetic energies of relative motion and decrease with increasing energy. The cross sections for nonresonant charge transfer, on the other hand, increase with increasing initial kinetic energy of relative motion and usually have relatively small values at lower energies, particularly if ΔE is of the order of several electron volts or higher. A comprehensive discussion of both experimental and theoretical aspects of charge transfer may be found in a number of excellent reviews [113, 243].

Other classes of charge-transfer reactions that are of interest are those in which the transfer is accompanied by dissociation as in

$$A^+ + BC = A + B^+ + C.$$

Specific instances of dissociative charge transfer have been studied by many investigators, among them Lindholm [244], Gustafson and Lindholm [245], and Giese and Maier [246], who have investigated the system

$$X^+ + \mathrm{CO} = X + \mathrm{C}^+ + \mathrm{O}$$

where X^+ represents either He^+, Ne^+, or Ar^+. This type of charge transfer may be regarded as a special category of ion-molecule reactions.

Finally, a useful type of charge transfer has been reported [247, 248] that uses highly polyatomic molecules such as butane or benzene as charge-transfer gases. The existence of a very large number of internal electronic states makes it highly probable that the positive ion will find an energy level in the molecule that will permit efficient charge transfer under near-resonant conditions. This may be referred to as accidental resonance. This type of charge transfer, which is frequently accompanied by dissociation of the target molecule, has been used to neutralize positive ions under conditions in which symmetrical resonant charge transfer would be difficult [247].

5 CONCLUSION

In this chapter an attempt has been made to relate the broad field of molecular-beam research to chemical problems. Although the treatment has been highly selective, topics have been treated in which the contributions or interest of chemists have been particularly significant.

Since there is no doubt that the contributions and interest will continue, it would seem relevant to try to predict the direction in which such molecular-beam research will go. Advances over a broad front are expected. Potential energies for complex molecular systems will be determined, and the anisotropies in such potentials will be studied in much more detail. More studies of inelastic scattering can be expected, particularly rotational and vibrational excitation. These studies, in which extensive use will be made of state selection, should provide valuable information for a clearer understanding of the microscopic features of reactive scattering. In the area of reactive scattering future studies will be less confined to reactions involving alkali atoms and will therefore probe reactions with threshold energies considerably larger than those encountered in the past. Finally it would appear that the relatively small amount of ion-molecule research by molecular-beam methods will be greatly increased.

If these predictions are reasonable, this chapter will require frequent revision.

References

1. L. Dunoyer, *Compt. Rend.*, **152,** 592 (1911).
2. N. F. Ramsey, *Molecular Beams*, Oxford University Press, London and New York, 1956.
3. W. Gerlach and O. Stern, *Ann. Physik*, **74,** 673 (1924).
4. W. E. Lamb, Jr. and R. C. Retherford, *Phys. Rev.*, **79,** 549 (1950).
5. B. Bederson and E. J. Robinson, *Adv. Chem. Phys.*, **10,** 1 (1966).
6. J. P. Gordon, H. J. Zeiger, and C. H. Townes, *Phys. Rev.*, **99,** 1264 (1955).
7. H. Bethe and J. Ashkin, *Experimental Nuclear Physics*, E. Segre, Ed., Wiley, New York, 1953, Chapter 2.
8. A. Dalgarno, *Atomic and Molecular Processes*, D. R. Bates, Ed., Academic, New York, 1962, Chapter 15.
9. A. Dalgarno and G. W. Griffing, *Proc. Roy. Soc.*, **232A,** 423 (1955).
10. S. K. Allison, J. Cuevas, and M. Garcia-Munoz, *Phys. Rev.*, **127,** 792 (1962).
11. M. N. Huberman, *Phys. Rev.*, **127,** 799 (1962).
12. S. Datz, T. S. Noggle, and C. D. Moak, *Phys. Rev. Lett.*, **15,** 254 (1965).
13. S. Datz, T. S. Noggle, and C. D. Moak, *Nucl. Inst. Methods*, **38,** 221 (1965).
14. H. O. Lutz, S. Datz, C. D. Moak, and T. S. Noggle, *Phys. Rev. Lett.*, **17,** 285 (1966).
15. H. S. W. Massey and E. H. S. Burhop, *Electronic and Ionic Impact Phenomena*, Clarendon, Oxford, 1952.
16. *Fundamentals of Gas-Surface Interactions*, H. Saltsburg, J. N. Smith, Jr., and M. Rogers, Eds., Academic, New York, 1967.
17. I. Estermann, R. Frisch, and O. Stern, *Z. Physik*, **73,** 348 (1931).
18. D. Herschbach, *Adv. Chem. Phys.*, **10,** 319 (1966).
19. J. P. Toennies, *Ber. Bunsenges. Phys. Chem.*, **72,** 927 (1968).
20. Z. Herman, J. D. Kerstetter, T. L. Rose, and R. Wolfgang, *Rev. Sci. Instr.*, **40,** 538 (1969).
21. I. Amdur and J. E. Jordan, *Adv. Chem. Phys.*, **10,** 29 (1966).

22. H. Pauly and J. P. Toennies, *Advances in Atomic and Molecular Physics*, D. R. Bates and I. Estermann, Eds., Academic, New York, 1965.
23. D. C. Lorents and W. Aberth, *Phys. Rev.*, **139,** A1017 (1965).
24. E. W. McDaniel, *Collision Phenomena in Ionized Gases*, Wiley, New York, 1964.
25. E. A. Mason, *J. Chem. Phys.*, **26,** 667 (1957).
26. K. W. Ford and J. A. Wheeler, *Ann. Phys.* (*N.Y.*), **7,** 259, 287 (1959).
27. R. J. Munn, E. A. Mason, and F. J. Smith, *J. Chem. Phys.*, **41,** 3978 (1964).
28. R. B. Bernstein, *Adv. Chem. Phys.*, **10,** 75 (1966).
29. S. Chapman and T. G. Cowling, *The Mathematical Theory of Non-Uniform Gases*, 2nd ed., Cambridge University Press, New York and London, 1952.
30. J. O. Hirschfelder, C. F. Curtiss, and R. B. Bird, *Molecular Theory of Gases and Liquids*, 2nd printing, Wiley, New York, 1964.
31. C. S. Wang Chang, G. E. Uhlenbeck, and J. de Boer, *Studies Stat. Mech.*, **2,** 243 (1964).
32. E. A. Mason and T. H. Spurling, *The Virial Equation of State*, Pergamon, London, 1969.
33. I. Prigogine and E. Xhrouet, *Physica*, **15,** 913 (1949).
34. R. D. Present, *Kinetic Theory of Gases*, McGraw-Hill, New York, 1958.
35. I. Amdur and G. G. Hammes, *Chemical Kinetics*, McGraw-Hill, New York, 1966.
36. K. H. Kramer and R. B. Bernstein, *J. Chem. Phys.*, **42,** 767 (1965).
37. P. R. Brooks and E. M. Jones, *J. Chem. Phys.*, **45,** 3449 (1966).
38. R. J. Beuhler, Jr., R. B. Bernstein, and K. H. Kramer, *J. Amer. Chem. Soc.*, **88,** 5331 (1966).
39. T. G. Waech, K. H. Kramer, and R. B. Bernstein, *J. Chem. Phys.*, **48,** 3978 (1968).
40. R. J. Beuhler, Jr. and R. B. Bernstein, *Chem. Phys. Lett.*, **2,** 166 (1968).
41. P. R. Brooks, *J. Chem. Phys.* **50,** 5031 (1969).
42. P. R. Brooks, E. M. Jones, and K. Smith, *J. Chem. Phys.*, **51,** 3073 (1969).
43. R. D. Present, *Proc. Nat. Acad. Sci.*, **41,** 415 (1955).
44. D. R. Herschbach, *J. Chem. Phys.*, **33,** 1870 (1960).
45. D. R. Herschbach, *Vortex*, Calif. Sec., ACS **22,** No. 8, October, 1961.
46. S. Datz, D. R. Herschbach, and E. H. Taylor, *J. Chem. Phys.*, **35,** 1549 (1961).
47. D. R. Herschbach, *Disc. Faraday Soc.*, **33,** 149 (1962).
48. D. Beck, E. F. Greene, and J. Ross, *J. Chem. Phys.*, **37,** 2895 (1962).
49. L. Davis, Jr., D. E. Nagle, and J. R. Zacharias, *Phys. Rev.*, **76,** 1068 (1949).
50. H. H. Stroke, Ph.D. dissertation, Physics Department, M.I.T., Cambridge, 1954 (unpublished).
51. J. G. King and J. R. Zacharias, *Adv. Electron. Elect. Phys.*, **8,** 1 (1956).
52. D. E. Nagle, R. E. Julian, and J. R. Zacharias, *Phys. Rev.*, **72,** 971 (1947).
53. J. G. King and V. Jaccarino, *Phys. Rev.*, **94,** 1610 (1954).
54. J. M. B. Kellogg, I. I. Rabi, and J. R. Zacharias, *Phys. Rev.*, **50,** 472 (1936).
55. W. L. Fite and R. T. Brackman, *Phys. Rev.*, **112,** 1141 (1958).
56. E. E. Muschlitz, Jr., *Adv. Chem. Phys.*, **10,** 171 (1966).
57. B. Lammert, *Z. Physik*, **56,** 244 (1929).

58. J. G. Dash and H. S. Sommers, Jr., *Rev. Sci. Instr.*, **24,** 91 (1953).
59. R. C. Miller and P. Kusch, *Phys. Rev.*, **99,** 1314 (1955).
60. E. F. Greene, R. W. Roberts, and J. Ross, *J. Chem. Phys.*, **32,** 940 (1960).
61. H. G. Bennewitz and W. Paul, *Z. Physik*, **139,** 489 (1954).
62. H. G. Bennewitz, W. Paul, and Ch. Schlier, *Z. Physik*, **141,** 6 (1955).
63. H. U. Hostettler and R. B. Bernstein, *Rev. Sci. Instr.*, **31,** 872 (1960).
64. S. M. Trujillo, P. K. Rol, and E. W. Rothe, *Rev. Sci. Instr.*, **33,** 841 (1962).
65. J. L. Kinsey, *Rev. Sci. Instr.*, **37,** 61 (1966).
66. H. Pauly and J. P. Toennies, *Methods of Experimental Physics*, Vol. 7A; B. Bederson and W. L. Fite, Eds., Academic, New York, 1968, Chapter 3.1.
67. T. H. Johnson, *J. Franklin Inst.*, **207,** 629 (1929).
68. O. E. Kurt and J. E. Phipps, *Phys. Rev.*, **34,** 1357 (1929).
69. A. Lemonuck and F. M. Pipkin, *Phys. Rev.*, **95,** 1356L (1954).
70. P. R. Brooks and D. R. Herschbach, *Rev. Sci. Instr.*, **35,** 1528 (1964).
71. K. H. Kingdon, *Phys. Rev.*, **21,** 408 (1923).
72. W. D. Johnston and J. G. King, *Rev. Sci. Instr.*, **37,** 475 (1966).
73. E. H. Taylor and S. Datz, *J. Chem. Phys.*, **23,** 1711 (1955).
74. S. Datz and E. H. Taylor, *J. Chem. Phys.*, **25,** 389 (1956).
75. S. Datz and E. H. Taylor, *J. Chem. Phys.*, **25,** 395 (1956).
76. G. Wessel and H. Lew, *Phys. Rev.*, **92,** 641 (1953).
77. G. Fricke, *Z. Physik*, **141,** 166 (1955).
78. F. Bernhard, *Z. Angew. Phys.*, **9,** 68 (1957).
79. H. Heil, *Z. Physik*, **120,** 212 (1943).
80. W. Paul, *Z. Physik*, **124,** 244 (1948).
81. H. Friedman, *Z. Physik*, **156,** 598 (1959).
82. M. Von Ardenne, *Phys. Zeit.*, **43,** 91 (1942).
83. W. E. Quinn, A. Pery, J. M. Baker, H. R. Lewis, N. F. Ramsey, and J. T. LaTourette, *Rev. Sci. Instr.*, **29,** 935 (1958).
84. R. Weiss, *Rev. Sci. Instr.*, **32,** 397 (1961).
85. G. O. Brink, *Rev. Sci. Instr.*, **37,** 857 (1966).
86. O. F. Hagena and A. K. Varma, *Rev. Sci. Instr.*, **39,** 47 (1968).
87. R. H. Dicke, *Rev. Sci. Instr.*, **17,** 268 (1946).
88. A. V. Phelps and J. L. Pack, *Rev. Sci. Instr.*, **26,** 45 (1955).
89. K. Kodera, I. Kusunoki, and A. Sakiyama, *Sixth International Conference on the Physics of Electronic and Atomic Collisions, Book of Abstracts*, 1969, p. 169.
90. H. Ishii and K. Nakayama, *Transactions of the Eighth National Vacuum Symposium*, Vol. I, Pergamon, Oxford, 1962, 519.
91. E. Rothe, *J. Vac. Sci. Tech.*, **1,** 66 (1964).
92. C. Meinke and G. Reich, *Vacuum*, **13,** 579 (1963).
93. A. deVries and P. Rol, *Vacuum*, **15,** 135 (1965).
94. P. Kusch, *J. Chem. Phys.*, **40,** 1 (1964).
95. C. E. Kuyatt, *Methods of Experimental Physics*, Vol. 7A; B. Bederson and W. L. Fite, Eds., Academic, New York, 1968, Chapter 1.1.
96. M. E. Rudd and T. Jorgenson, Jr., *Phys. Rev.*, **131,** 666 (1963).
97. M. Hoyaux and I. Dujardin, *Nucleonics*, **4,** 12 (1949).

98. H. F. Ivey, *Adv. Electron. Elect. Phys.*, **6,** 137 (1954).
99. J. R. Pierce, *Theory and Design of Electron Beams*, Van Nostrand, New York, 1954.
100. J. A. Simpson, *Rev. Sci. Instr.*, **32,** 1283 (1961).
101. K. T. Bainbridge, *Experimental Nuclear Physics*, E. Segre, Ed., Wiley, New York, 1953, p. 559.
102. L. Kerwin, *Adv. Electron. Elect. Phys.*, **8,** 187 (1956).
103. P. Marmet and L. Kerwin, *Can. J. Phys.*, **38,** 787 (1960).
104. E. M. Purcell, *Phys. Rev.*, **54,** 818 (1938).
105. F. R. Paolini and G. C. Theodoridis, *Rev. Sci. Instr.*, **38,** 579 (1967).
106. C. E. Kuyatt and J. A. Simpson, *Rev. Sci. Instr.*, **38,** 103 (1967).
107. G. A. Harrower, *Rev. Sci. Instr.*, **26,** 850 (1955).
108. J. A. Alcalay and E. L. Knuth, *Rev. Sci. Instr.*, **40,** 438 (1969).
109. W. H. Bennett, *J. Appl. Phys.*, **21,** 143 (1950).
110. R. L. F. Boyd and D. Morris, *Proc. Phys. Soc. London*, **A68,** 1 (1955).
111. W. Paul and M. Raether, *Z. Physik*, **140,** 262 (1955).
112. W. Paul, H. P. Reinhardt, and U. von Zahn, *Z. Physik*, **152,** 143 (1958).
113. J. B. Hasted, *Physics of Atomic Collisions*, Butterworth, London, 1964.
114. J. E. Jordan and I. Amdur, *J. Chem. Phys.*, **46,** 165 (1967).
115. N. Peterson, private communication, 1969.
116. L. Harris, *J. Opt. Soc. Amer.*, **36,** 597 (1946).
117. Barnes Engineering Company, Stamford, Conn.
118. M. A. D. Fluendy, *Rev. Sci. Instr.*, **35,** 1606 (1964).
119. G. W. Goodrich and W. C. Wiley, *Rev. Sci. Instr.*, **32,** 846 (1961).
120. G. W. Goodrich and W. C. Wiley, *Rev. Sci. Instr.*, **33,** 761 (1962).
121. J. Draper, *Methods of Experimental Physics*, Vol. 4A; V. W. Hughes and H. L. Schultz, Eds., Academic, New York, 1967, Chapter 2.1.2.
122. H. W. Berry, *Phys. Rev.*, **75,** 913 (1962).
123. N. R. Daly, *Rev. Sci. Instr.*, **31,** 264 (1960).
124. J. B. Anderson, R. P. Andres, and J. B. Fenn, *Advances in Atomic and Molecular Physics*, D. R. Bates and I. Esterman, Eds., Academic, New York, 1965.
125. J. B. Anderson, R. P. Andres, and J. B. Fenn, *Adv. Chem. Phys.*, **10,** 275 (1966).
126. S. M. Trujillo, R. H. Neynaber, and E. W. Rothe, *Rev. Sci. Instr.*, **37,** 1655 (1966).
127. V. A. Belyaev, B. G. Brezhnev, and E. M. Erastov, *Zh. Eksper. Teor. Fiz.-Pisma Redact.*, **3,** 321 (1966); English trans. *JETP Lett.*, **3,** 207 (1966).
128. R. H. Neynaber, *Methods of Experimental Physics*, Vol. 7A; B. Bederson and W. L. Fite, Eds., Academic, New York, 1968, Chapter 4.3.
129. J. Politiek, J. Los, J. J. M. Schipper, and A. P. M. Baede, *Entropie*, **18,** 82 (1967).
130. E. Hulpke and Ch. Schiler, *Z. Physik*, **207,** 294 (1967).
131. J. Politiek, P. K. Rol, J. Los, and P. G. Ikelaar, *Rev. Sci. Instr.*, **39,** 1147 (1968).
132. G. T. Skinner, *Phys. Fluids*, **4,** 1172 (1961).

133. G. T. Skinner and B. H. Fetz, *Rarefied Gas Dynamics*; J. H. Leeuw, Ed., Academic, New York, 1965, p. 536.
134. G. T. Skinner and J. Moyzis, *Phys. Fluids*, **8,** 452 (1965).
135. R. A. Oman, A. Bogan, C. H. Weiser, C. H. Li, and V. S. Calia, *Grumman Res. Dept. Rept.*, RE-166, 1963.
136. D. Auerbach, E. E. A. Bromberg, and L. Wharton, *J. Chem. Phys.*, **45,** 2160 (1966).
137. R. Wolfgang, *Sci. Amer.*, **219,** 44 (1968).
138. L. Sodickson, J. Carpenter, and G. Davidson, Rept. No. AFCRL-65-337, American Science and Engineering, Cambridge, Mass., 1965.
139. J. E. Jordan and O. Shepard, Rept. No. AFCRL-69-0095, American Science and Engineering, Cambridge, Mass., 1969.
140. E. A. Mason and J. T. Vanderslice, *Atomic and Molecular Processes*, D. R. Bates, Ed., Academic, New York, 1962, Chapter 17.
141. E. A. Mason and I. Amdur, *J. Chem. Phys.*, **41,** 2695 (1964).
142. I. Amdur and A. L. Smith, *J. Chem. Phys.*, **48,** 565 (1968).
143. S. O. Colgate, J. E. Jordan, I. Amdur, and E. A. Mason, *J. Chem. Phys.*, **51,** 968 (1969).
144. J. E. Jordan, S. O. Colgate, I. Amdur, and E. A. Mason, *J. Chem. Phys.*, **52,** 1143 (1970).
145. A. B. Kamnev and V. B. Leonas, *Dokl. Akad. Nauk SSSR*, **162,** 798 (1965), [*Soviet Phys.-Dokl.*, **10,** 529 (1965)].
146. A. B. Kamnev and V. B. Leonas, *Dokl. Akad. Nauk SSSR*, **165,** 1273 (1965) [*Soviet Phys.-Dokl.*, **10,** 1202 (1966)].
147. Yu. N. Belyaev and V. B. Leonas, *Zh. Tekh. Fiz.*, **36,** 353 (1966) [*Soviet Phys.-Tech. Phys.*, **11,** 257 (1966)].
148. Yu. N. Belyaev and V. B. Leonas, *Dokl. Akad. Nauk SSSR*, **170,** 1039 (1966) [*Soviet Phys.-Dokl.*, **11,** 866 (1967)].
149. Yu. N. Belyaev and V. B. Leonas, *Zh. Eksper. Teor. Fiz. Pisma, Redact.* **4,** 134 (1966) [*Soviet Phys.-JETP Lett.*, **4,** 92 (1966)].
150. Yu. N. Belyaev and V. B. Leonas, *Dokl. Akad. Nauk SSSR*, **173,** 306 (1967) [*Soviet Phys.-Dokl.*, **12,** 233 (1967)].
151. Yu. N. Belyaev, V. B. Leonas, and A. V. Sermyaguin, *Abstracts of Papers, Fifth International Conference on the Physics of Electronic and Atomic Collisions* Nauka, Leningrad, 1967, p. 643.
152. Yu. N. Belyaev, N. V. Kamyshov, V. B. Leonas, and A. V. Sermyaguin, *Entropie*, **30,** 173 (1969).
153. I. Amdur and A. L. Harkness, *J. Chem. Phys.*, **22,** 664 (1954).
154. I. Amdur, J. E. Jordan, and S. O. Colgate, *J. Chem. Phys.*, **34,** 1525 (1961).
155. N. C. Blais and J. B. Mann, *J. Chem. Phys.*, **32,** 1459 (1960).
156. E. A. Mason and W. E. Rice, *J. Chem. Phys.*, **22,** 522 (1954).
157. P. E. Phillipson, *Phys. Rev.*, **125,** 1981 (1962).
158. G. M. Matsumoto, C. F. Bender, and E. R. Davidson, *J. Chem. Phys.*, **46,** 402 (1967).
159. I. Amdur and E. A. Mason, *J. Chem. Phys.*, **22,** 670 (1954).

160. I. Amdur, J. E. Jordan, and R. R. Bertrand, *Atomic Collision Processes*, M. R. C. McDowell, Ed., North-Holland, Amsterdam, 1964, p. 934.
161. R. N. Keeler, M. van Thiel, and B. J. Alder, *Physica*, **31,** 1437 (1965).
162. M. van Thiel and B. J. Alder, *J. Chem. Phys.*, **44,** 1056 (1966).
163. C. R. Fischer and P. J. Kemmey, *J. Chem. Phys.* **53,** 50 (1970).
164. I. Amdur and E. A. Mason, *J. Chem. Phys.*, **25,** 630 (1956).
165. M. Krauss and F. H. Mies, *J. Chem. Phys.*, **42,** 2703 (1965).
166. R. L. Matcha and R. K. Nesbet, *Phys. Rev.*, **160,** 72 (1967).
167. E. A. Mason and L. Monchick, *Adv. Chem. Phys.*, **12,** 329 (1967).
168. I. Amdur, E. A. Mason, and A. L. Harkness, *J. Chem. Phys.*, **22,** 1071 (1954).
169. R. E. Walker and A. A. Westenberg, *J. Chem. Phys.*, **31,** 519 (1959).
170. H. Margenau and N. R. Kestner, *Theory of Intermolecular Forces*, Pergamon, London, 1969, Chapter 7.
171. I. Amdur and E. A. Mason, *Phys. Fluids*, **1,** 370 (1958).
172. E. A. Mason, *Kinetic Processes in Gases and Plasmas*, A. R. Hochstim, Ed., Academic, New York, 1969, Chapter 3.
173. K. F. Herzfeld and T. A. Litovitz, *Absorption and Dispersion of Ultrasonic Waves*, Academic, New York, 1959.
174. D. R. Olander, *J. Chem. Phys.*, **43,** 779 (1965).
175. D. R. Olander, *J. Chem. Phys.*, **43,** 785 (1965).
176. R. M. Martin and J. E. Willard, *J. Chem. Phys.*, **40,** 3007 (1964).
177. R. Wolfgang, *Ann. Rev. Phys. Chem.*, **16,** 15 (1965).
178. J. F. Williams, P. J. Stang, and P. von R. Schleyer, *Ann. Rev. Phys. Chem.*, **19,** 531 (1968).
179. E. A. Mason and M. M. Kreevoy, *J. Amer. Chem. Soc.*, **77,** 5808 (1955).
180. M. M. Kreevoy and E. A. Mason, *J. Amer. Chem. Soc.*, **79,** 4851 (1957).
181. G. N. Ramachandran and V. Sasisekharan, *Adv. Protein Chem.*, **23,** 283 (1968).
182. P. J. Flory, *Statistical Mechanics of Chain Molecules*, Interscience, New York, 1969, Chapter 5.
183. R. B. Bernstein and J. T. Muckerman, *Adv. Chem. Phys.*, **12,** 389 (1967).
184. M. Minnaert, *Light and Colour in the Open Air*, Dover, New York, 1954.
185. P. Kong, E. A. Mason, and R. J. Munn, *Amer. J. Phys.*, **38,** 294 (1970).
186. E. A. Mason and L. Monchick, *J. Chem. Phys.*, **41,** 2221 (1964).
187. C. F. Barnett and H. Gilbody, *Methods of Experimental Physics*, Vol. 7A; B. Bederson and W. L. Fite, Eds., Academic, New York, 1968, Chapter 4.2.
188. N. G. Utterback and G. H. Miller, *Rev. Sci. Instr.*, **32,** 1101 (1961).
189. N. G. Utterback and G. H. Miller, *Phys. Rev.*, **124,** 1477 (1961).
190. R. H. Hammond, J. M. S. Henis, E. F. Greene, and J. Ross, *Sixth International Conference on the Physics of Electronic and Atomic Collisions, Book of Abstracts*, 1969, p. 408.
191. J. P. Toennies, *Z. Physik*, **182,** 257 (1965).
192. A. R. Blythe, A. E. Grosser, and R. B. Bernstein, *J. Chem. Phys.*, **41,** 1917 (1964).
193. D. Beck and H. Förster, *Sixth International Conference on the Physics of Electronic and Atomic Collisions, Book of Abstracts*, 1969, p. 634.

194. D. O. Ham and J. L. Kinsey, *J. Chem. Phys.*, **48,** 939 (1968).
195. T. Moran, private communication.
196. J. Schlötter and J. P. Toennies, *Z. Physik*, **214,** 472 (1968).
197. P. F. Dittner and S. Datz, *Sixth International Conference on the Physics of Electronic and Atomic Collisions, Book of Abstracts*, 1969, p. 469.
198. P. F. Dittner and S. Datz, *J. Chem. Phys.*, **49,** 1969 (1968).
199. M. C. Fowler, J. E. Jordan, and I. Amdur, *Sixth International Conference on the Physics of Electronic and Atomic Collisions, Book of Abstracts*, 1969, p. 516.
200. R. J. Cross and R. G. Gordon, *J. Chem. Phys.*, **45,** 3571 (1966).
201. M. C. Moulton and D. R. Herschbach, *J. Chem. Phys.*, **44,** 3010 (1966).
202. E. F. Greene, A. L. Moursound, and J. Ross, *Adv. Chem. Phys.*, **10,** 135 (1966).
203. "Molecular Dynamics of the Chemical Reactions of Gases," *Disc. Faraday Soc.*, **44,** 1967.
204. J. I. Steinfeld and J. L. Kinsey, *Progress in Reaction Kinetics*, Vol. 5, Pergamon, 1969.
205. K. J. Laidler, *Theories of Chemical Reaction*, McGraw-Hill, New York, 1969, pp. 189–190.
206. D. Beck, *J. Chem. Phys.*, **37,** 2884 (1962).
207. J. R. Luoma and C. R. Mueller, *J. Chem. Phys.*, **46,** 680 (1967).
208. R. E. Olson and C. R. Mueller, *J. Chem. Phys.*, **46,** 3810 (1967).
209. J. R. Airey, E. F. Greene, K. Kodera, G. P. Reck, and J. Ross, *J. Chem. Phys.*, **46,** 3287 (1967).
210. R. J. Suplinskas and J. Ross, *J. Chem. Phys.*, **47,** 321 (1967).
211. D. R. Herschbach, G. H. Kwei, and J. A. Norris, *J. Chem. Phys.*, **34,** 1842 (1961).
212. E. A. Entemann and D. R. Herschbach, *Disc. Faraday Soc.*, **44,** 289 (1967).
213. J. R. Airey, E. F. Greene, G. P. Reck, and J. Ross, *J. Chem. Phys.*, **46,** 3295 (1967).
214. M. Polanyi, *Atomic Reactions*, Williams and Nargate, London, 1932.
215. A. Henglein and G. A. Muccini, *Z. Naturforsch.*, **17a,** 452 (1962).
216. A. Henglein and G. A. Muccini, *Z. Naturforsch.*, **18a,** 753 (1963).
217. A. Henglein, K. Lacmann, and B. Knoll, *J. Chem. Phys.*, **43,** 1048 (1965).
218. S. Datz and R. E. Minturn, *J. Chem. Phys.*, **41,** 1153 (1964).
219. K. R. Wilson, G. H. Kwei, J. A. Norris, R. R. Herm, J. H. Birely, and D. R. Herschbach, *J. Chem. Phys.*, **41,** 1154 (1964).
220. J. H. Birely, R. R. Herm, K. R. Wilson, and D. R. Herschbach, *J. Chem. Phys.*, **47,** 993 (1967).
221. R. E. Minturn, S. Datz, and R. L. Becker, *J. Chem. Phys.*, **44,** 1149 (1966).
222. A. E. Grosser and R. B. Bernstein, *J. Chem. Phys.*, **43,** 1140 (1965).
223. R. R. Herm and D. R. Herschbach, *J. Chem. Phys.*, **43,** 2139 (1965).
224. W. R. Miller, S. A. Safron, and D. R. Herschbach, *Disc. Faraday Soc.*, **44,** 108 (1967).
225. E. Rittner, *J. Chem. Phys.*, **19,** 1030 (1951).
226. R. F. Barrow and A. J. Merer, *Ann. Rept. Prog. Chem.*, **59,** 121 (1962).
227. Y. T. Lee and B. H. Mahan, *J. Chem. Phys.*, **42,** 2893 (1965).

228. A. C. Roach and M. S. Childs, *Mol. Phys.*, **14,** 1 (1968).
229. D. O. Ham, J. L. Kinsey, and F. S. Klein, *Disc. Faraday Soc.*, **44,** 174 (1967).
230. D. O. Ham and J. L. Kinsey, *J. Chem. Phys.* **53,** 285 (1970).
231. A. Henglein, K. Lacmann, G. Jacobs, *Ber. Bunsenges. Phys. Chem.*, **69,** 279 (1965).
232. R. L. Champion, L. D. Doverspike, and T. L. Bailey, *J. Chem. Phys.*, **45,** 4377 (1966).
233. W. R. Gentry, E. A. Gislason, B. H. Mahan, and C. W. Tsao, *J. Chem. Phys.*, **49,** 3058 (1968).
234. B. R. Turner, M. A. Fineman, and R. F. Stebbings, *J. Chem. Phys.*, **42,** 4088 (1965).
235. B. H. Mahan, *Acc. Chem. Res.*, **1,** 217 (1968).
236. R. Wolfgang, *Acc. Chem. Res.*, **2,** 248 (1969).
237. E. W. McDaniel, *et al.*, *Ion-Molecule Reactions*, Wiley, New York, 1970.
238. W. R. Gentry, E. A. Gislason, Y. T. Lee, B. H. Mahan, and C. W. Tsao, *Disc. Faraday Soc.*, **44,** 137 (1967).
239. E. W. McDaniel, *Methods of Experimental Physics*, Vol. 7A; B. Bederson and W. L. Fite, Eds., Academic, New York, 1968, Chapter 4.1.
240. Z. Harman, J. Kerstetter, T. Rose, and R. Wolfgang, *Disc. Faraday Soc.*, **44,** 123 (1967).
241. R. F. Pottie and W. H. Hamill, *J. Phys. Chem.*, **63,** 877 (1959).
242. E. A. Gislason, B. H. Mahan, C. W. Tsao, and A. S. Werner, *J. Chem. Phys.*, **50,** 5418 (1969).
243. R. F. Stebbings, *Adv. Chem. Phys.*, **10,** 195 (1966).
244. E. Lindholm, *Arkiv Fysik*, **8,** 433 (1954).
245. E. Gustafson and E. Lindholm, *Arkiv Fysik*, **18,** 219 (1960).
246. C. F. Giese and W. B. Maier II, *J. Chem. Phys.*, **39,** 197 (1963).
247. V. L. Talroze, private communication.
248. W. A. Chupka and E. Lindholm, *Arkiv Fysik*, **25,** 349 (1963).

Chapter **VII**

NEUTRON ACTIVATION ANALYSIS

Vincent P. Guinn

---

1 INTRODUCTION

Neutron activation analysis may be defined as a nuclear method of quantitative elemental analysis. The method determines elements irrespective of their physical or chemical states, and provides no information regarding the valence states or chemical forms of the elements detected. Because its current form is very sophisticated and highly instrumented, the method is often regarded as a fairly new analytical method. However, its history actually starts in the mid-1930s. In that period the first few activation analysis measurements were carried out (with a cyclotron), using charged particles for charged-particle activation analysis. With the development of the nuclear reactor, during the 1940s, neutrons—generated in great numbers by the reactor—were rapidly shown to be a more widely useful bombarding particle, and the basic neutron activation analysis (NAA) method was developed. This form of the method was developed extensively in the 1950s, and continues to be refined and extended. It is now the most widely used form of the activation analysis method. During the 1950s it was also shown that relatively modest electron- and charged-particle accelerators could provide more modest, but still quite useful, fluxes of slow and fast neutrons for NAA work. The use of such moderate-cost accelerator neutron sources has greatly increased the use of NAA in industrial analytical laboratories. Quite recently the possibility of using the newly available californium-252, spontaneous-fission, isotopic-neutron sources for moderate-flux NAA with thermal neutrons has aroused considerable interest.

When the purely instrumental (nondestructive) form of the neutron activation analysis method, based upon γ-ray spectrometry, can be employed, it can be a very fast method. Samples can then in many cases be analyzed in a matter of minutes per sample. If the highest of the available fluxes of neutrons are used, the method is extremely sensitive for a large number of elements. The method then provides the most sensitive method known at present, for the quantitative determination of the majority of the elements of the periodic system. This is not true, of course, if relatively low fluxes are used. The dynamic range of the method is large, extending rigorously from the limit of detection on up to the pure element itself (100% concentration).

As a result of the great penetrability of neutrons, in most matrices, the range of practical sample sizes usable in NAA work is very large, extending from microscopic samples on up to even 100 g in some cases. Since the absolute limits of detection of the various elements, under prescribed irradiation and counting conditions, increase only moderately with increasing sample size, throughout this range of sample sizes, the method is especially powerful for the detection of low concentrations. Thus for a typical element, detectable down to 0.001 μg [by NAA in a nuclear reactor for one hour, at a thermal-neutron flux of 10^{13} neutrons/(cm^2)(sec)], the concentration limit of detection for that element is 0.1% in a 1-μg sample, 1 ppm in a 1-mg sample, and 0.001 ppm in a 1-g sample.

Although charged-particle activation analysis and photonuclear activation analysis both have significant applications, neither is so generally applicable, sensitive, available, or as fully developed as NAA; hence this chapter is devoted entirely to the subject of neutron activation analysis.

2 THEORY OF NEUTRON ACTIVATION ANALYSIS

If one considers a particular stable nuclide (nucleus of a particular number of protons, Z, and a mass number, A) of an element that is capable of undergoing a particular reaction with an incident neutron to form a radionuclide product, when bombarded with neutrons, one can readily show that:

$$\frac{dN^*}{dt} = N\phi\sigma, \tag{7.1}$$

in which dN^*/dt is the rate of formation (in nuclei/sec) of these radioactive product nuclei, N is the number of "target" stable nuclei of that type in the sample, ϕ is the average flux of neutrons (in neutrons/(cm^2)(sec)) to which the sample is exposed, and σ is the nuclear reaction cross-section (in cm^2/nucleus) for that particular neutron reaction, at the neutron energy employed. (*Note.* Tables of nuclear reaction cross-sections usually list σ values in "barns" (b), where 1 b = 10^{-24} cm^2/nucleus.)

The radioactive product nuclei formed decay according to the first-order radioactive decay equation, both during and after the irradiation:

$$-\frac{dN^*}{dt} = \lambda N^* = \frac{0.693N^*}{T}, \tag{7.2}$$

in which λ is the decay constant of that radionuclide (in sec^{-1}) and is equal to $0.693/T$, where T is the half-life of the radionuclide in seconds.

Thus during a steady-flux irradiation the net rate of formation of product radioactive nuclei of a given type is simply

$$\frac{dN^*}{dt} = N\phi\sigma - \lambda N^*. \tag{7.3}$$

When this equation is integrated to obtain an expression for N^* present after some irradiation time, t_i, one obtains the expression

$$N^* = \frac{N\phi\sigma S}{\lambda}, \tag{7.4}$$

in which S, called the "saturation" term, is simply

$$S = 1 - e^{-0.693t_i/T}. \tag{7.5}$$

Since at the end of the irradiation period λN^* is equal to $-dN^*/dt$ [see (7.2)], one can finally write the standard activation analysis equation:

$$A_0 = \left(-\frac{dN^*}{dt}\right)_0 = N\phi\sigma S. \tag{7.6}$$

The quantity A_0, then, is the initial activity in disintegrations per second (dps) of a particular radionuclide, formed by a particular neutron reaction—just at the end of the irradiation.

Irradiation Time

The saturation term, S, is a dimensionless quantity, ranging in value only from 0 (at $t_i = 0$) to 1 (at $t_i = \infty$, or $t_i/T \gg 1$). It asymptotically approaches a value of 1 with increasing t_i/T. Thus at t_i/T values of 0, 1, 2, 3, 4, 5, and so on, S has values of 0, $\frac{1}{2}$, $\frac{3}{4}$, $\frac{7}{8}$, $\frac{15}{16}$, $\frac{31}{32}$, and so on. Each induced activity follows exactly the same curve of S versus t_i/T. It can also be represented as a straight line if log $(1 - S)$ is plotted versus t_i/T. Because of the rapid asymptotic approach of S to 1, one seldom, in practice, irradiates a sample for a period of time (t_i) longer than one or a few half-lives of the induced activity of interest. At very long irradiation times (relative to T) a steady state is approached (or "saturation" condition) in which the rate of formation of new nuclei of that particular type, $dN^*/dt = N\phi\sigma$, is equal to the rate of decay of such nuclei already formed, $-dN^*/dt = \lambda N^*$. Thus the activity level at saturation, A_0 (sat'n), is equal to $N\phi\sigma$. Because of the saturation term the activity levels of short-lived induced activities are enhanced, relative to longer-lived activities, by the use of a short irradiation period—followed quickly by counting. Similarly the activity levels of longer-lived induced activities are enhanced by use of a longer irradiation period, followed by an appreciable delay (for decay of interfering short-lived activities) prior to counting. Optimum choices of irradiation and decay times are important in minimizing interferences.

Radioactive Decay

Typically, the activated sample is counted at some decay time, t, after the end of the irradiation (t_0). Each induced activity decays according to (7.2), the basic first-order radioactive-decay equation. In integrated form (7.2) becomes

$$A_t = A_0 e^{-0.693t/T}, \tag{7.7}$$

or

$$\log A_t = \log A_0 - \frac{0.301t}{T}, \tag{7.8}$$

or

$$\log \frac{A_t}{A_0} = -\frac{0.301t}{T}, \tag{7.9}$$

in which log refers to the common logarithm (base 10) of the quantity. It is seen that for all radionuclides the log (A_t/A_0) versus t/T relationship is the same straight line. The relationship between the activities at any two decay times (one not necessarily being t_0) is given by the same equations if t is replaced by Δt, the difference between the two decay times.

One seldom, in practice, determines the actual disintegration rate of an induced activity, but rather one measures its counting rate under a given set of counting conditions. For a given disintegration rate of a particular radionuclide in a sample the counting rate of that radionuclide depends upon the type of counter used, the size of the counter, the decay scheme of the radionuclide, the sample-to-counter geometry (solid angle subtended by the counter), self-absorption or scattering within the sample, and, in some situations, also on other smaller factors. It is instructive to consider an illustrative case; for example, that of the ^{52}V γ-ray photopeak (total-absorption peak) counting rate of a sample in which the ^{52}V disintegration rate, at the time of counting, is 1000 dis/sec. From the very useful Table of Isotopes by Lederer and others [10], the decay scheme of 3.75-min ^{52}V is seen to be a simple one: in every one of its disintegrations a 2.47-MeV (E_{max}) β^- particle is emitted, promptly followed by a 1.434-MeV γ-ray photon. (*Note.* The common nuclear unit of energy is the electron volt (eV); multiples are 1 keV = 10^3 eV, 1 MeV = 10^6 eV). If the sample is small ($\leq$1 cm^3) and of low density ($\leq$2 g/cm^3), absorption or Compton scattering of these relatively high-energy γ-ray photons within the sample itself will be small or negligible. If the sample is in a small polyethylene vial, placed on top of a 1.5-cm-thick polystyrene β^- particle absorber, which in turn rests on top of an aluminum-encased thallium-activated sodium iodide scintillation crystal detector, the crystal being 3 in. in diameter and 3 in. high, the distance from the center of the sample to the top surface of the NaI(Tl) crystal will be about 2 cm. From Heath's excellent and highly useful *Scintillation Spectrometry Gamma-Ray Spectrum Catalogue* [8] one finds that the overall detection efficiency for such a crystal, for this sample-to-detector distance, for 1.434-MeV γ-rays, is 10.8% (26.5% geometry; 40.7% detection efficiency for those 1.434-MeV γ-rays that strike the crystal). Thus the total ^{52}V γ-ray counting rate of the sample in this case would be 1000 × 0.108, or 108 counts/sec; that is, 6480 counts/min. However, in most instances one ignores the counts other than the characterizing photopeak counts. From Heath's Catalogue [8] one finds that the photofraction (peak-to-total ratio) for 1.434-MeV γ-rays under these conditions of crystal size and geometry is

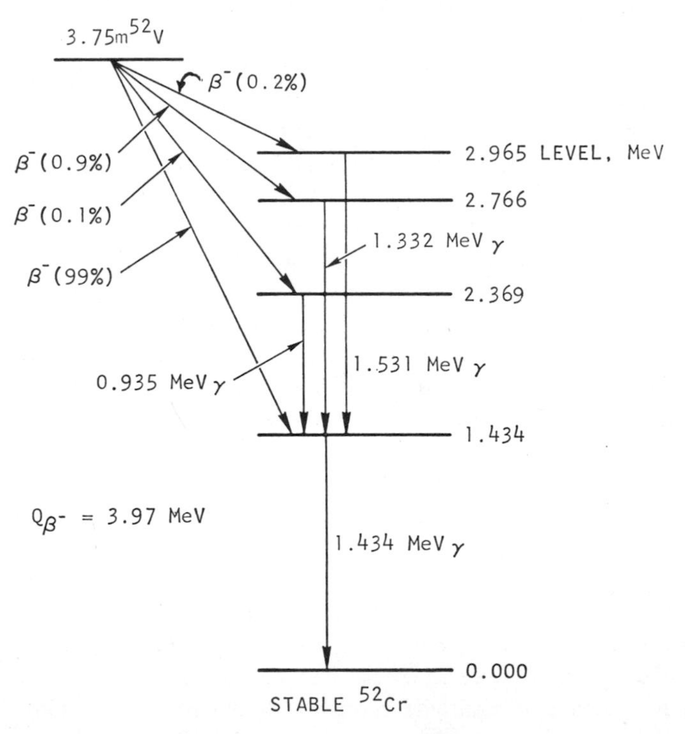

Fig. 7.1 Decay scheme of ^{52}V.

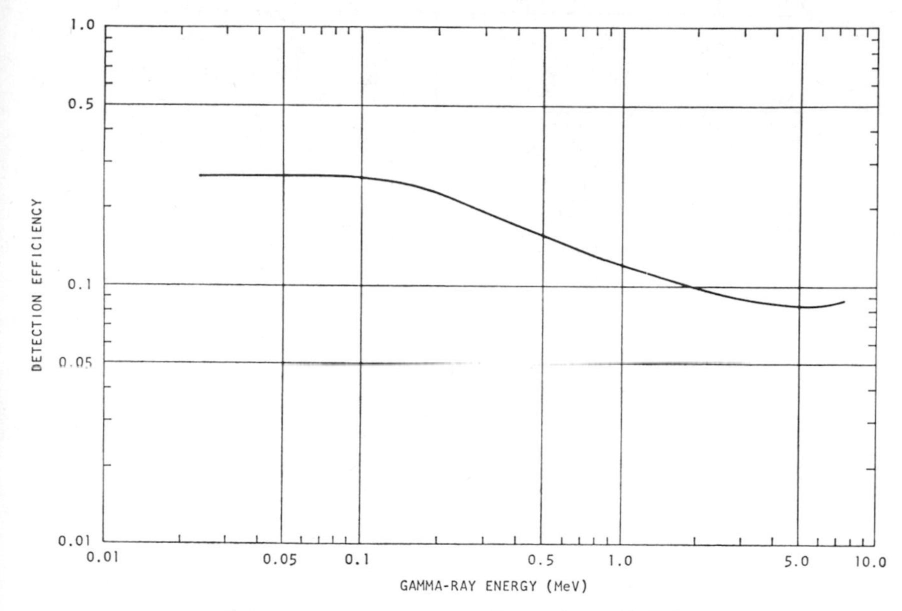

Fig. 7.2 Counting efficiency versus γ-ray energy (3×3-in. NaI(Tl) detector, 2-cm distance). From Heath's Catalogue [8].

0.334 (i.e., 33.4%). Thus the ^{52}V net photopeak counting rate in the 1.434-MeV region of the samples' observed pulse-height spectrum would be 6480×0.334, or 2164 counts/min.

In Fig. 7.1 the decay scheme of 3.75-min ^{52}V, taken from the Table of Isotopes [10] is shown. Figure 7.2, taken from Heath's Catalogue [8], shows the relationship between counting efficiency and γ-ray energy for a 3×3-in. solid NaI(Tl) crystal, and a sample-to-detector distance of 2 cm. Figure 7.3, also taken from Heath's Catalogue [8], shows the relationship between photofraction and γ-ray energy for a 3×3-in. solid NaI(Tl) crystal—in this case for a sample-to-detector distance of 3 cm. (However, the photofraction is not very sensitive to variations in the sample-to-detector distance.)

The Comparator Technique

In practical activation analysis work one seldom uses (7.6). There are a number of reasons for not using it directly, to compute the desired analytical value, N (or, really, w):

1. It is difficult and often not very accurate to convert an observed radionuclide counting rate to the corresponding disintegration rate (due to often inadequate knowledge of the decay scheme and of the counting efficiency).

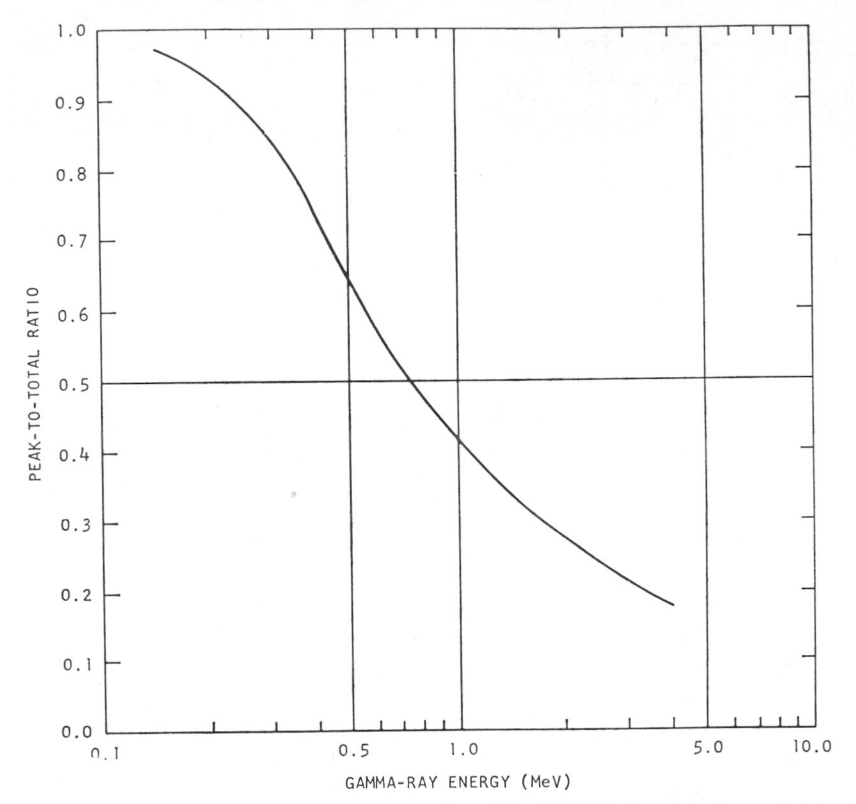

Fig. 7.3 Photofraction versus γ-ray energy (3 × 3-in. NaI(Tl) detector, 3-cm distance). From Heath's Catalogue [8].

2. It is also difficult and often not very accurate to determine the true average value of the neutron flux (of a particular energy or energy range), ϕ, at the sample irradiation location.

3. Many of the reaction cross sections (σ) found in the literature are known only approximately.

One usually wishes to determine the amount of a particular element in a sample rather than the amount of the particular stable nuclide of that element that could form the radionuclide detected. Thus one must also know the natural fractional isotopic abundance (a) of the particular stable nuclide that is activated (amongst the various stable nuclides of that element) and the chemical atomic weight (AW) of the element. The weight, w (in grams), of the element present is related to N and other quantities by the equation

$$w = \frac{N \times AW}{N_A \times a}, \tag{7.10}$$

where N_A is Avogadro's number (6.023×10^{23}).

In order to avoid dependence upon these various values one normally instead employs a comparator technique, which is both simpler and more accurate than the "absolute" method. If a sample or a series of samples is to be analyzed for one or several elements of particular interest, one prepares standard samples of these elements, activates them with neutrons at the same time as the unknowns—under exactly the same conditions—and counts them in exactly the same way that the unknowns are counted. Then correcting all of the appropriate counting rates of a particular induced activity to either t_0, or to any other specified decay time, t [by means of (7.7)], one can replace the disintegration-rate values (A_0) of (7.6) by the corresponding counting-rate values (A_0') divided by ϵ, the fractional counting efficiency; replace the N of (7.6) by its equivalent, from (7.10); and divide the modified (7.6) for the unknown sample by that for the known standard of that element. Whatever their exact, true, values may be, one at least knows that the values for ϵ, N_A, a, AW, ϕ, σ, and S are exactly the same for unknown and standard. Thus in dividing one equation by the other, these terms all cancel out, leaving the very simple final equation that is used in practice:

$$\frac{A_0'(\mathrm{unk})}{A_0'(\mathrm{std})} = \frac{w(\mathrm{unk})}{w(\mathrm{std})}, \tag{7.11}$$

or for any specified decay time,

$$\frac{A_t'(\mathrm{unk})}{A_t'(\mathrm{std})} = \frac{w(\mathrm{unk})}{w(\mathrm{std})}. \tag{7.12}$$

If it is not possible to irradiate the unknown sample and the standard at exactly the same neutron flux, a correction factor, correcting both counting rates to what they would be at the same flux, must be applied before using (7.12). The relative counting rates of flux-monitor samples can be used to make such corrections. A similar normalization is necessary if it is not possible to count them both at exactly the same counting geometry. Such corrections, where needed, are obvious, and can be made quite accurately.

3 NEUTRON SOURCES AND TYPES OF NEUTRON REACTIONS

Of the approximately 4000 papers published to date on the subject of activation analysis and its applications, the great majority pertain to high-flux, thermal-neutron activation of samples in research-type nuclear reactors. A smaller number involve fast-neutron activation, usually with accelerator-produced fast neutrons. Even smaller numbers pertain to activation with high-energy photons or energetic charged particles—both accelerator-produced. There are both practical and historical reasons for the wider usage of reactor thermal-neutron activation analysis.

Thermal-Neutron and Fast-Neutron Activation Analysis with Research-Type Nuclear Reactors

Research Reactors

The most potent practical source of thermal neutrons (i.e., neutrons in kinetic-energy equilibrium with the surrounding temperature; at 20°C this corresponds to an average neutron kinetic energy of about 0.025 eV) is the nuclear reactor. Modern research reactors especially suitable for NAA work are safe, relatively simple, and only moderately expensive ($150,000 to $350,000, depending largely upon the power level, and hence neutron flux, required). They are usually of the pool type, in which a lattice of fuel elements, each containing some ^{235}U, is located in a deep pool of distilled, deionized water (15 to 25ft deep). The water serves as the principal moderator and as coolant and radiation shield. The reactor core contains slightly more than enough ^{235}U to make a critical mass, under the conditions specified (i.e., geometry, reflector, moderator). When several boron-loaded control rods are inserted into the core, the thermal-neutron ^{235}U fission chain reaction is stopped by the appreciable absorption of slow neutrons by the boron; that is, the system is then subcritical and the reactor is shut off. The thermal-neutron absorption cross section of ordinary boron is very high (762 b); that of pure ^{10}B is higher yet (4017 b).

To turn the reactor on, the control rods are withdrawn sufficiently to make the system slightly supercritical. The ^{235}U fission chain reaction (initiated by a small isotopic or sealed-tube accelerator neutron source) then rapidly accelerates, and hence the power level and the neutron flux increase rapidly. When the power level (hence neutron flux) has reached the desired or prescribed level (typically, in a few minutes), the reactor operator reinserts the control rods just far enough to maintain criticality and steady operation at that power level. Modern pool-type reactors of various kinds and costs can be operated at maximum steady power levels of from 10 to 1000 kW (1 Mw), providing useful thermal-neutron fluxes, at locations in or near the core where samples can be placed for activation, in the range of 10^{11} to 10^{13} neutrons/(cm^2)(sec). Some research reactors can even be operated at power levels as high as 10 Mw, giving thermal-neutron fluxes of about 10^{14} neutrons/(cm^2)(sec).

When a thermal neutron is captured by a ^{235}U nucleus and fission of the $(^{236}U)^*$ excited compound nucleus results, two intermediate atomic-number radionuclides are formed, and two or three energetic neutrons are emitted. The most probable initial energy of these fission neutrons is about 1 MeV, with rapidly decreasing numbers of neutrons being emitted with higher energies. A few even have initial energies up to about 25 MeV. By collisions with moderator nuclei, such as hydrogen, beryllium, or carbon, these fast neutrons are slowed down to thermal velocities (some being lost by escape or capture). Thus especially within the core of a reactor there are appreciable fluxes of fast neutrons, as well as high thermal-neutron fluxes. The fast-neutron flux decreases by a factor of approximately 10 for each 3-MeV increase in energy beyond about 2 MeV. Outside the core the thermal/fast-flux ratio is higher, but both fluxes are lower than within the core. For example, Table 7.1 gives the thermal-neutron flux and the fluxes of neutrons above

Table 7.1 Neutron Energy-Flux Distribution in the 250-kw TRIGA Mark I Nuclear Reactor

Irradiation Position	*Neutron Flux* (*in* neutrons/(cm²)(sec)				
	Thermal	>10 keV	>1.35 MeV	>3.68 MeV	>6.1 MeV
D ring	5.3×10^{12}	1.1×10^{13}	2.3×10^{12}	4.1×10^{11}	6.2×10^{10}
F ring (rabbit terminus)	2.5×10^{12}	5.3×10^{12}	7.5×10^{11}	1.2×10^{11}	1.9×10^{10}
Rotary specimen rack	1.8×10^{12}	1.5×10^{12}	1.8×10^{11}	2.5×10^{10}	4.0×10^{9}

various measured energies at three different locations in one of the three TRIGA reactors at the author's former laboratory—a 250-kW TRIGA Mark I reactor. A photograph of this reactor, looking down at the core through about 16 ft of water, is shown in Fig. 7.4. A cutaway scale diagram of this reactor, showing the 40-tube rotary specimen rack in the annular graphite reflector, just outside the core, is shown in Fig. 7.5.

The TRIGA type of nuclear reactor includes a rotary specimen rack (shown in Fig. 7.5), which is especially useful in neutron activation analysis work, since up to 40 samples can be exposed simultaneously in it to exactly the same average neutron flux. Each sample, which is in a small polyethylene vial, is placed inside a larger polyethylene tube, lowered into one of the 40 rotary-rack tubes, the rack is advanced one position, the next sample lowered into place, and so on, until all the samples (anywhere from 1 to 40) are in place. The loading of the rack only takes about 1 min or less per tube. The rack motor drive is then started (1 rpm) and the reactor is brought up to power. After the samples have been irradiated for the desired period of time (usually 30 min, 1 hr, or a few hours), the reactor is shut down, the rack rotation is stopped, and the samples are brought up, one at a time, each being

Fig. 7.4 TRIGA Mark I nuclear reactor.

briefly monitored, for safety, while still in the in-out tube—just as it arrives at the top of the pool. If the rack were not rotated during the irradiation, there would be small but significant differences in the neutron fluxes at the 40 different tube locations. These are averaged out by the rack rotation. Thus one standard sample is all that is needed for up to 39 unknown samples—a considerable advantage. There is some vertical flux gradient in the sample tubes, but this can be measured, to provide a calibration curve. Also, if desired, up to six small polyvials, stacked vertically, can be placed in each large polyethylene tube, giving a total sample capacity of 240 samples (in half-dram polyvials). The rotary rack is used wherever the main induced activity of interest is moderately long-lived, that is, where the half-life of the radionuclide is of the order of an hour, several hours, days, or longer. For such elements a moderately long irradiation is helpful for increased sensitivity (because of the saturation term of (7.6)), and it is more economical to irradiate many samples at the same time. Because of the longer half-lives of interest in such cases the activated samples, after a suitable decay period (to allow shorter-lived interfering activities to decay out) can be counted one at a time without excessive loss of activity by decay. In the rotary-rack positions of the TRIGA Mark I reactor the thermal-neutron flux (at 250 kW

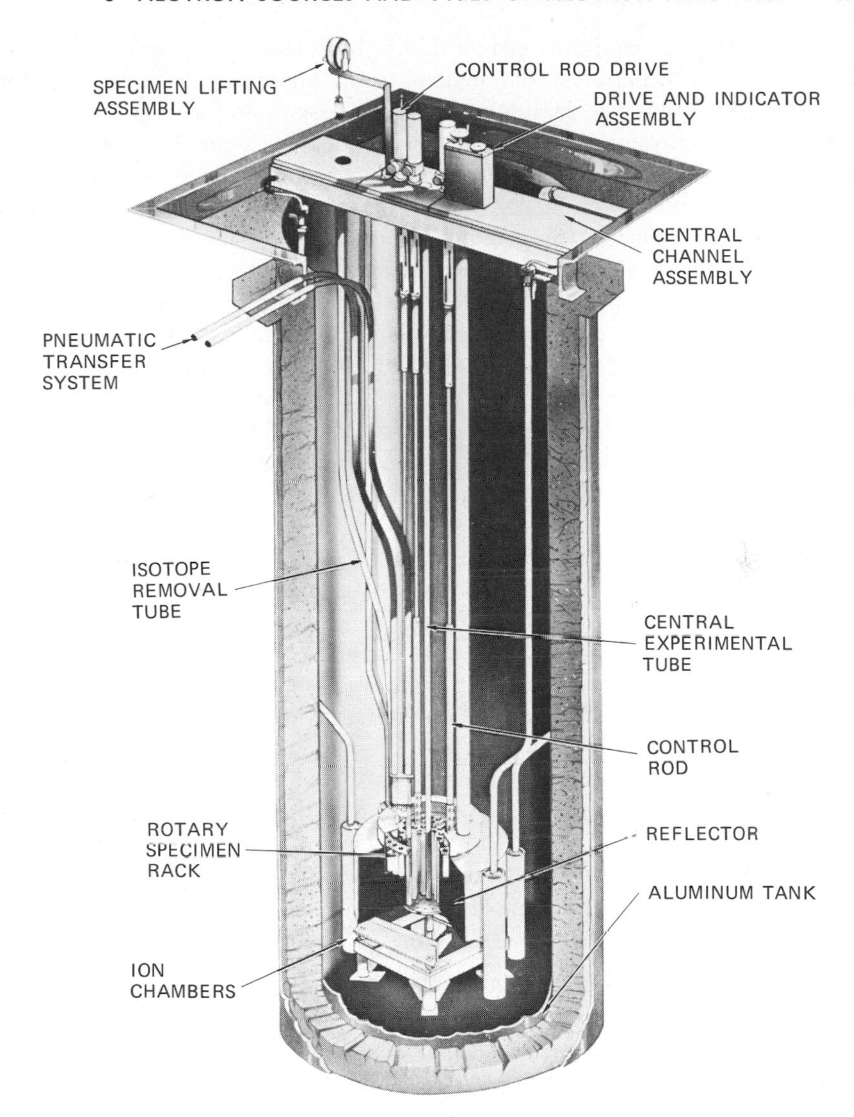

Fig. 7.5 Cutaway diagram of TRIGA Mark I reactor.

operation) is 1.8×10^{12} neutrons/(cm^2)(sec). Especially in irradiations of 30 min or longer one normally transfers the activated sample to a fresh polyvial before counting. This eliminates contributions from activated impurities in the polyethylene (such as sodium, aluminum, chlorine, titanium, and vanadium), and also from ^{41}Ar formed from argon in the air that is present in the vial or dissolved in the sample.

If one wishes to determine elements that form short-lived activities (half-lives of seconds or minutes), one instead uses a pneumatic transfer tube and activates and counts one sample at a time. The sample vial is placed inside a sturdier polyethylene "rabbit" for irradiation. With the TRIGA Mark I reactor the normal pneumatic tube has a transfer time of about 2.5 sec (between adjacent counting room and the reactor core), and terminates in a vacant fuel-element position in the outermost ring (F ring) of fuel elements. In this position the thermal-neutron flux (at 250 kW operation) is 2.5×10^{12} neutrons/(cm^2)(sec). The desired irradiation time can be preset on a timer, and the irradiated sample then is automatically returned to the counting room (where it was originally inserted), monitored automatically for safety, and the sample vial is then removed from the rabbit, ready for counting on the multichannel γ-ray spectrometer. A precise timer is actuated as the rabbit leaves the reactor, and it is stopped automatically when the counting is started, thus giving an accurate value for the decay time prior to counting.

The TRIGA reactor can also be pulsed to extremely high neutron fluxes [10^{16} to 10^{17} neutrons/(cm^2)(sec)]. The pulse durations are in the range of 5 to 30 msec. When the TRIGA Mark I reactor is pulsed to a peak-power level of about 1000 MW, it is found [19] that the amount of a given radionuclide activity generated in the pulse is $70/T$ times that which could be produced, even at saturation, with the reactor operated at its normal power level of 250-kW (T being the half-life of the induced activity, in seconds). This relationship applies to both thermal-neutron and fast-neutron reaction products. Thus for products with half-lives of 0.1, 1, and 10 sec the pulse enhancement values are about 700, 70, and 7, respectively. The pulsing technique is thus of real use for very short-lived induced activities. A special 0.5-sec transfer-time pneumatic tube without an outer rabbit is used in the author's laboratory for such reactor-pulsing NAA work.

Reactor Activation with Thermal Neutrons

When exposed to a flux of thermal neutrons, elements undergo the (n, γ) reaction; that is, when a nucleus captures a thermal neutron, the resulting excited compound nucleus promptly ($\sim 10^{-12}$ sec) drops down to the ground state of the product nucleus, emitting one or more "prompt" "capture γ-ray" photons to get rid of the excess excitation energy. The product nucleus is a nucleus of the same element, but now one atomic mass number higher (since the mass number of the neutron is 1). For example, if an ordinary stable ^{19}F nucleus captures a thermal neutron, a nucleus of ^{20}F is formed:

$$^{19}\mathrm{F} + {}^{1}n \longrightarrow ({}^{20}\mathrm{F})^{*} \xrightarrow{\text{prompt}} {}^{20}\mathrm{F} + \gamma$$

or more simply written,

$$^{19}\mathrm{F}(n, \gamma)^{20}\mathrm{F}.$$

If the product nucleus is a radioactive nucleus of that element, as is the case of ^{20}F, its later decay can be detected and can thus be of use for NAA (^{20}F decays with a half-life of 11.56 sec, emitting a β^- particle and a 1.63-MeV γ-ray photon in each disintegration). If the (n, γ) reaction only forms another stable nuclide of that element, the product is of no use in conventional NAA work (although prompt γ-rays can be utilized in a different form of activation analysis that is not discussed here). For example, the $^{16}O(n, \gamma)^{17}O$ reaction only produces another stable nuclide of oxygen, ^{17}O. It should be remarked that even with a fairly long irradiation at a thermal-neutron flux as high as 10^{13}, the elemental composition of a typical sample is not significantly altered. For an element with an (n, γ) cross section of 1 b, for example, 1 hr at a 10^{13} flux results in only $4 \times 10^{-6}\%$ of the amount of that element present in the sample undergoing reaction.

Reactor Activation with Fast Neutrons

For a few elements, activation by reactor fast neutrons (fission-spectrum neutrons) is more sensitive than activation by reactor thermal neutrons, and/or forms a product that is more readily detected (perhaps a γ emitter instead of a pure β^- emitter), or has a more convenient half-life relative to interfering activities, or is more easily radiochemically separated. In such situations one can employ one or more cadmium- or boron-encased sample vials for activation in the rotary rack (or, for short-lived activities one can use a cadmium or boron liner in the pneumatic-tube terminus and employ a slightly smaller rabbit. With even a 1-mm annulus of cadmium or boron (ordinary or enriched in ^{10}B) the incident thermal neutrons are almost completely absorbed, whereas the incident reactor-fast neutrons pass through, into the sample, undisturbed. The thermal-neutron absorption cross sections of Cd, B, and ^{10}B are, respectively, 2537, 762, and 4017 b. Epithermal and "resonance" neutrons are not appreciably absorbed by the cadmium or boron, however, and still cause some (n, γ) activation of various elements in the sample. The ratio of the amount of (n, γ) activation of an element with no cadmium around it, to that with a cadmium annulus, is called the "cadmium ratio" at that reactor position. The cadmium ratio varies somewhat from element to element, because of the different shapes of the (n, γ) cross section, versus neutron-energy curves for the different elements. Their differences in the epithermal-neutron and resonance-capture neutron regions, especially, affect the cadmium ratios. With a 1-mm annulus of cadmium the attenuation of purely thermal neutrons is over 100,000-fold, but cadmium ratios in the range of 10 to 30, for example, are usually found, experimentally, in the pneumatic-tube (F-ring) position of the author's TRIGA Mark I reactor. It is also possible to use an ordinary, or somewhat ^{235}U-enriched, annulus of uranium instead of cadmium or boron. The ^{235}U absorbs thermal neutrons quite well ($\sigma(n, \gamma) = 101$ b, $\sigma(f) = 582$ b), and, in addition, generates more fission-spectrum neutrons. Similarly, a lithium deuteride annulus (ordinary, or enriched in ^{6}Li) can be used. Lithium-6 is a strong absorber of thermal neutrons (σ for ordinary Li = 70.4 b; for ^{6}Li = 950 b). When a ^{6}Li nucleus absorbs a thermal neutron, the $^{6}Li(n, t)^{4}He$ reaction results. The resulting energetic tritons, striking nearby deuterons, can cause the $^{2}H(t, n)^{4}He$ reaction, producing 14 to 15-MeV neutrons.

Some elements that are often, or sometimes, determined with reactor fast neutrons, instead of with reactor thermal neutrons, are the elements nitrogen, oxygen, fluorine, silicon, phosphorus, chromium, and iron. Knowledge of the possible fast-neutron products from samples is also important when one is conducting thermal-neutron activation analyses, since, as shown in Table 7.1, the samples are typically also exposed to appreciable fast-neutron fluxes. For example, if one is analyzing samples for low levels of phosphorus (via the $^{31}P(n, \gamma)^{32}P$ reaction), one must ascertain that the samples do not contain sufficient sulfur and/or chlorine to produce relatively significant amounts of ^{32}P via the fast-neutron reactions, $^{32}S(n, p)^{32}P$ and $^{35}Cl(n, \alpha)^{32}P$. This requires, in this case, activations with and without boron or cadmium around the samples and a separate measurement of chlorine, via the $^{37}Cl(n, \gamma)^{38}Cl$ reaction. From the corresponding counting rates of ^{32}P from standard samples of P, S, and Cl, and of ^{38}Cl from standard samples of Cl, the amounts of ^{32}P generated in samples arising, respectively, from their contents of P, S, and Cl can readily be calculated. Fortunately, this multiplicity situation does not arise or is not serious for very many elements, and even then it involves only an appreciable correction in those situations in which the desired (n, γ) reaction has a rather low cross section and/or involves a stable nuclide of rather low abundance (relative to the one or two possible fast-neutron interfering reactions from the elements of atomic number one or two units larger), or where the ratio of interfering element(s) to the sought-for element is very high. A few other common difficult analyses of this type should, however, be mentioned:

1. Thermal-NAA of aluminum (or aluminum-rich matrices) for trace levels of sodium (since ^{24}Na is formed by both the $^{27}Al(n, \alpha)^{24}Na$ and $^{23}Na(n, \gamma)^{24}Na$ reactions).
2. Thermal-NAA of silicon (or silicon-rich matrices such as quartz, silicones, glass, or siliceous rocks) for trace levels of aluminum (since ^{28}Al is formed by both the $^{28}Si(n, p)^{28}Al$ and $^{27}Al(n, \gamma)^{28}Al$ reactions; high levels of phosphorus can also interfere, via the $^{31}P(n, \alpha)^{28}Al$ reaction).
3. Thermal-NAA of chromium (or chromium-rich matrices) for trace levels of vanadium (since ^{52}V is formed by both the $^{52}Cr(n, p)^{52}V$ and $^{51}V(n, \gamma)^{52}V$ reactions).

Measurements analogous to those described above for the phosphorus, silicon, chlorine case, however, can provide accurate results—unless the interfering elements are present at relatively very high levels.

Fast-Neutron (and Thermal-Neutron) Activation Analysis with Accelerator-Produced Neutrons

Isotopic neutron sources, such as Pu–Be and Am–Be sources, are not discussed here, since the maximum fluxes available with them are only around 10^5 to 10^6 neutrons/(cm^2)(sec)—too low to be of much practical use. Recently, however, spontaneous-fission ^{252}Cf neutron sources (2.646-year half-life) have become available, and should find practical uses, since 1 mg of ^{252}Cf can provide a thermal-neutron flux of about 10^8 neutrons/(cm^2)(sec).

Accelerator Sources of 14-MeV Neutrons

A commonly-used, relatively low-cost source of 14-MeV neutrons is a small Cockcroft-Walton deuteron accelerator. A typical commercial neutron generator of this type is shown in Fig. 7.6. These cost approximately $20,000, and with a fresh tritium target can produce about 2×10^{11} neutrons per second (isotropic). At positions in which one or a few samples of appreciable size (1 to 5 cm^3) can be placed, this output rate corresponds to an average 14-MeV neutron flux within the sample of about 10^9 neutrons/(cm^2)(sec).

Typically these machines operate as follows: (1) deuterium gas from a small cylinder is fed through a palladium leak into an ionizing chamber; (2) there

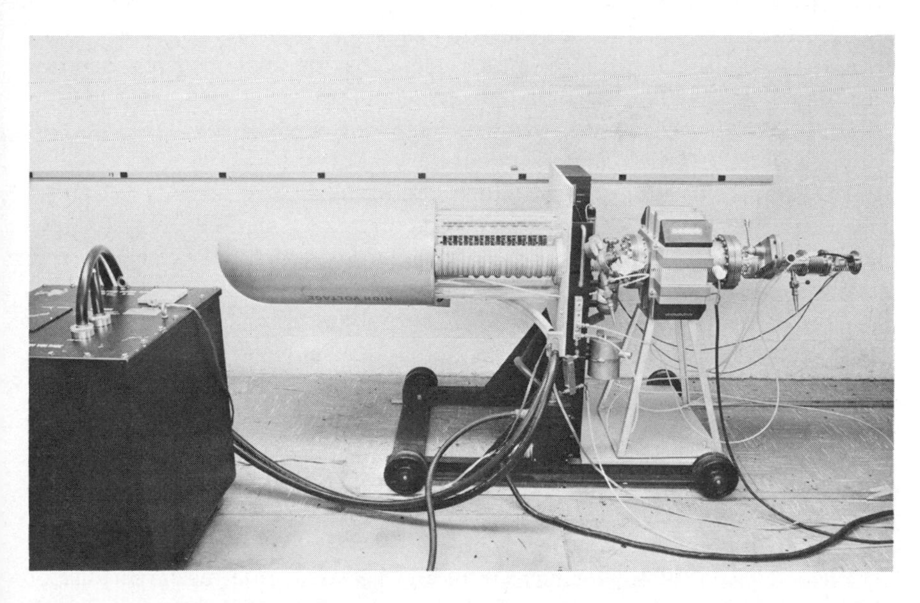

Fig. 7.6 Typical Cockcroft-Walton neutron generator (Courtesy of Texas Nuclear Division, Nuclear-Chicago Corp.).

it is ionized, forming D^+ and D_2^+ ions—either with a radiofrequency source or with a Penning source; (3) the ions emerge through a slit and are accelerated through a potential of 100 to 200 kV down an evacuated "drifttube"; (4) the collimated beam of accelerated D^+ and D_2^+ ions, usually defocused so that they cover an area of 1 to 10 cm^2 (to avoid melting a hole in the target), impinges upon a water-cooled, metal-tritide target (typically, about 1 in. in diameter); and (5) ~ 14-MeV neutrons are thus produced in the target, via the 17.6 MeV exoergic $^3H(d, n)^4He$ reaction, in which the neutron receives 14.0 MeV of the 17.6 MeV released. Because of the conservation of momentum, with 150 keV deuterons, the neutrons emitted in the forward direction range in energy (assuming a thick target) up to 14.9 MeV; those in the backward direction down to 13.4 MeV; those at right angles, 14.0 MeV.

The entire system is maintained at a reasonably good vacuum ($\sim 10^{-5}$ mm Hg) by means of a Vac-Ion, molecular, oil-diffusion, or mercury-diffusion pump, backed up by a mechanical vacuum pump. For NAA work these accelerators are usually operated steadily, but the neutron beam can also be pulsed, if desired, by either pulsing the ion source, magnetically or electrostatically deflecting the beam on or off the target, or both. Various pulse durations and repetition rates can be selected. At its maximum steady neutron output rate, such a machine requires approximately 7 ft of concrete (or equivalent) shielding, in all directions, in order to reduce the fast-neutron flux outside the shielding to a value less than the tolerance dose rate for fast neutrons (~10 neutrons/(cm^2)(sec) for 40-hr/week exposures). This also reduces the subsidiary thermal-neutron and the γ-ray levels to less than tolerance. Additional distance and/or additional thermal- and fast-neutron shielding (besides the usual lead γ-ray shielding) around the scintillation counter is also needed, to prevent increased counter background and activation of the sodium and iodine in the NaI crystal.

The major limitation of such moderate-flux neutron sources is the target lifetime problem. The tritium targets used consist of a metal backing (copper or other metal) about 10 mils thick, on one face of which has been deposited a thin layer (one or a few microns thick) of titanium or zirconium—which has then been tritiated up to a level of 1 to 10 curies of tritium per square inch. Even when the impinging beam of 150-keV deuterons is spread over an area of several square centimeters, a 1 to 2 mA beam will result in a rapidly declining neutron output rate from the water-cooled target. Usually, at such full-power operation the output rate will drop in half in an hour or less, continuing to decline with further use. A number of methods can be employed to extend the period of time before a target has ceased to generate enough neutrons per second to be of much practical use and a fresh target has to be installed. These methods include use of a larger-area, rotating target; use of thicker, higher curie-per-square-inch targets; intermittent or continuous target replenishment with accelerated tritons; and use of refrigerated target coolants. These methods each help, but each adds to the cost of the system and poses some complications. The changing of a target requires only a few minutes, but because of the time needed for subsequent pumping to a good vacuum, a total time of about 30 to 60 min usually elapses before operation with the new target can be started. If the very maximum neutron flux is not needed, it is often more desirable to start operation with a fresh target at a reduced deuteron beam current (perhaps 0.5 mA) and then gradually increase the beam current so as to maintain a constant neutron output rate. Operating in this fashion, an output rate of perhaps 5×10^{10} neutrons/sec can be maintained for quite a few hours of on-target operation—until the beam current has been increased to the maximum possible value with the accelerator being used (typically, about 2 mA). If a neutron-flux monitor counter (of almost any type) is used, the gradual beam-current increase can be performed automatically, by a servomechanism, if desired.

A sealed-tube 14-MeV neutron generator is also available, that is guaranteed to yield 1×10^{11} neutrons/sec for at least 100 hours of on-target operation. This type eliminates problems of pumping, target decline, target changing, deuterium input, palladium leaks, and ion sources. It utilizes a deuterium-tritium source inside the tube, as well as a metal tritide target. When its output drops, the whole tube is simply replaced with a new one.

Just as with reactor installations, a pneumatic transfer tube is employed for work with short-lived induced activities (half-lives of seconds to a few minutes). Typically, transfer times in the range of 1 or 2 sec are employed. Because of the difficulty in maintaining highly constant, reproducible, neutron fluxes with such generators (compared with a nuclear reactor), it is useful to employ a dual pneumatic-tube system—one for sample, one for standard or flux monitor. Such a dual transfer system is used in the author's former laboratory for routine oxygen determinations, with sample and standard both spinning during the irradiation period (oxygen is determined via the 7.14-sec half-life ^{16}N formed by the $^{16}O(n, p)^{16}N$ reaction, using 14-MeV neutrons). The sample, and a standard of known oxygen content (and of the same volume and shape), are simultaneously sent into position in front of the target, the beam is turned onto the target, and then, after a preset irradiation period, the activated sample and the standard are sent back to their respective (cross-calibrated) counters and counting is started simultaneously on both and for exactly the same preset length of time. Sample and standard are thus processed identically, and therefore their respective oxygen contents are in exactly the same ratio as their net ^{16}N counts (corrected to the same counting efficiency). This presupposes that the neutron flux is the same for both sample and standard. This is true—for a perfectly symmetrical system, with perfect centering of the beam and a completely uniform target. Since these ideal conditions may not be met exactly, however, the system is checked with each new target (or more frequently), using identical samples in the two tubes to provide a small correction factor if the two positions are exposed to slightly different neutron fluxes. An alternative is to employ a dual-axis (rather than a single-axis) rotator for sample and standard. For elements whose fast-neutron products are longer-lived, one can activate a number of samples simultaneously (usually for 5 to 10 min) in a spinning sample rack.

Activation Analysis with Accelerator-Produced 14-MeV Neutrons

Using a source of 14-MeV neutrons, quite a few elements can be effectively activated by means of (n, n'), $(n, 2n)$, (n, p), or (n, α) reactions. The (n, n') type of reaction is termed a "fast-neutron inelastic scattering" reaction. If the product is a nuclear isomer (with the same Z and same A as the target nuclide) with an adequately long half-life, it can be of use for activation analysis. An example of this type of fast-neutron reaction is the

$$^{89}\mathrm{Y}(n, n')^{89m}\mathrm{Y}$$

reaction, forming 16.1-sec ^{89m}Y. All (n, n') reactions are endoergic, but they are usually so only to the extent of about a megaelectron volt or less. They of course have no Coulomb barrier. The $(n, 2n)$ reactions result in a nuclide with the same Z, but one unit lower in A. An example of interest in activation analysis is the ^{63}Cu$(n, 2n)^{62}$Cu reaction, forming 9.76-min ^{62}Cu. All such reactions are also endoergic, usually to the extent of 6 to 12-MeV, and have no Coulomb barrier. The (n, p) reactions result in a product nuclide of the same A, but one unit lower in Z. An example of interest is the ^{16}O$(n, p)^{16}$N reaction, forming 7.14-sec ^{16}N. Some (n, p) reactions are endoergic (and some are exoergic). They do exhibit a Coulomb barrier. The (n, α) reactions form a product nuclide with an A three units lower than the target nuclide, and a Z two units lower. An example of this type of fast-neutron reaction is the ^{31}P$(n, \alpha)^{28}$Al reaction, forming 2.31-min ^{28}Al. Such reactions are in some cases endoergic, and in other cases, exoergic—and they also exhibit a Coulomb barrier. All of these reactions can be produced in many elements, not only by 14-MeV neutrons, but also by reactor fission-spectrum neutrons that have energies greater than the thresholds of the particular reactions.

The small Cockcroft-Walton deuteron accelerator can also be used to generate 2.8-MeV neutrons, via the ^{2}H$(d, n)^{3}$He reaction. In this case one bombards a metal deuteride (rather than metal tritide) target with deuterons. The advantages of employing 2.8-MeV neutrons, instead of 14-MeV neutrons, are (1) the target lifetime is longer since the deuteron beam is replacing deuterium lost from the target, and (2) use of 2.8-MeV neutrons prevents the occurrence of possibly interfering $(n, 2n)$ reactions and some high-Coulomb-barrier or high-threshold (n, p) and (n, α) reactions—particularly with high-Z elements (e.g., if one wishes to determine fluorine in the presence of oxygen via the formation of ^{16}N, 2.8-MeV neutrons are helpful since they are above the 1.6-MeV threshold of the $^{19}F(n, \alpha)^{16}N$ reaction but below the 10.2-MeV threshold of the $^{16}O(n, p)^{16}N$ reaction; 14-MeV neutrons produce both of these reactions). The principal disadvantage of employing 2.8-MeV neutrons is that, at a 150-kV accelerating potential the (d, d) yield of 2.8-MeV neutrons is only about $\frac{1}{100}$ of the (d, t) yield of 14-MeV neutrons.

Activation Analysis with Accelerator-Produced Thermal Neutrons

The 14-MeV neutron generator can also be used to provide moderate thermal-neutron fluxes by surrounding the target assembly with a tank of moderator (such as water, oil, or paraffin) into which sample tubes are inserted. At a position about 2 cm in front of the tritium target the 14-MeV neutron flux, with a fresh, high-curie target, may be about 10^9 neutrons/(cm²)(sec). With a moderator present the 14-MeV neutron flux at this position is only slightly lower, and the maximum thermal-neutron flux is a few centimeters further out, and has a maximum value of perhaps 10^8 neutrons/(cm²)(sec). The thermal-neutron flux of course also declines rapidly, with full-power operation. However, a thermal-neutron flux of about 5×10^7 neutrons/(cm²)(sec) can be maintained for a number of hours of total on-target operation by starting a fresh target with a reduced-beam current, then gradually raising the beam current with continued usage, so as to maintain a constant neutron output rate. A moderator tank (cubic, cylindrical, or spherical) with maximum dimensions of about 2 ft is sufficient. A pneumatic tube can be inserted into the moderator tank for work with short-lived induced activities. For longer-lived activities several cross-calibrated stationary tubes can be fixed in the moderator, at a constant distance from the center of the target, or a spinning multisample rack can be installed in the moderator. Because of the target deterioration problem it is not very practical to employ irradiation periods of longer than 5 or 10 min.

Where interferences from fast-neutron reactions are appreciable, it is sometimes useful to employ (d, d) 2.8-MeV neutrons instead of (d, t) 14-MeV neutrons. The 2.8-MeV neutrons are more readily thermalized and do not induce $(n, 2n)$ reactions or many high-Coulomb-barrier, or high-threshold, (n, p) and (n, α) reactions, particularly in high-Z elements. However, as noted above, the yield of (d, d) 2.8-MeV neutrons is typically only about $\frac{1}{100}$ of the yield of (d, t) 14-MeV neutrons.

Lower-output accelerator 14-MeV neutron sources, both of the continuously pumped and the sealed-tube types, can be obtained for about half the cost of the 2×10^{11} neutrons/sec machines, but these give output rates about

10 times lower, namely, around 1 or 2 × 10^{10} neutrons/sec. The lower-output machines usually operate at 100 kV (instead of 150 or 200 kV) and at a beam current of only 0.5 to 1 mA (instead of 2 mA), hence the lower cost and lower output rate.

A somewhat more expensive accelerator (a 2-MeV Van de Graaff deuteron accelerator costing about $40,000) can be used to provide very constant thermal-neutron fluxes of about 5 × 10^8 neutrons/(cm^2)(sec) (at a deuteron beam current of about 150 μA), if a moderator is used. Here the target employed is one of beryllium (water-cooled), and the nuclear reaction in the target is the $^{9}Be(d, n)^{10}B$ reaction. This target does not deteriorate appreciably with time, and it produces neutrons with energies in the 4 to 6-MeV range. These can be employed unmoderated, to induce (n, p) and (n, α) reactions in many low-atomic-number elements, and (n, n') reactions in some others. The energy is not high enough to induce $(n, 2n)$ reactions in any element except beryllium (and ^{2}H).

If more expensive accelerators (higher-energy Cockcroft-Walton and Van de Graaff positive-ion accelerators, higher-energy Van de Graaff electron accelerators, electron linear accelerators, and cyclotrons) are used, thermal- and fast-neutron fluxes in the range of about 10^8 to 10^{11} neutrons/(cm^2)(sec) can be produced. With proper choices of particle, particle energy, and target, neutrons of various selected energy ranges can be produced. This can be advantageous for various particular problems—often allowing one to use neutrons high enough in energy to activate the element of interest but below the threshold energy of one or more possible interfering reactions of other elements present. However, the available neutron fluxes are still modest compared with those available in a modern research-type nuclear reactor, and the cost is as high as or higher than such a reactor.

Sensitivities attainable for a number of elements, with 14-MeV neutron fluxes of 10^9 neutrons/(cm^2)(sec), and thermal-neutron fluxes of 10^8 neutrons/(cm^2)(sec), are presented in a later section.

4 FORMS OF THE NEUTRON ACTIVATION ANALYSIS METHOD

There are two forms of the activation analysis method: (1) the purely instrumental (nondestructive) form and (2) the radiochemical-separation (destructive) form. It should be remarked that in long irradiations at very high fluxes, radiolysis effects caused by the γ-radiation, fast neutrons, recoils, and radioactive-decay radiations can cause significant chemical and physical changes in samples, especially in organic and some inorganic materials. Thus in such cases the term, "nondestructive" must be used with some reservations.

The Instrumental Form

As implied by the name, this form of the method is entirely instrumental in character, involving no chemical operations. If one has the necessary equipment to utilize this technique, it is the preferred form wherever it is applicable, since it minimizes the use of a chemist's time. The instrumental technique involves only (1) activation of the sample, and (2) γ-ray spectrometry of the activated sample.

Decay Schemes

The instrumental method requires that the induced activity (or activities) of interest emit γ-rays, characteristic X-rays, or positrons in an appreciable fraction of its disintegrations. Positron emitters are detected via the 0.511-MeV "annihilation" radiation produced when emitted positrons have slowed down (usually within the sample and its container, or in adjacent matter) to the point that the positron can be annihilated by reaction with an electron. The reaction is: $\beta^+ + e^- \rightarrow 2\gamma$. From the masses of the electrons (both e^+ and e^- have the same mass) and the necessity of conservation of momentum, one should obtain from each annihilation event two 0.511-MeV γ-ray photons going in opposite directions, and such is found to be the case.

Characteristic X-rays can result from radioactive decay via electron-capture (EC) or internal-conversion (IC) mechanisms. Positron emission and/or electron capture are modes of radioactive decay that are common amongst neutron-poor (proton-rich) nuclei. If the decay energy (Q) available is more than 1.022 MeV, both EC and β^+ emission can occur, and there are many examples amongst the neutron-poor radionuclides in which a certain fraction of the disintegrations occur via EC and a different fraction occur via β^+ emission. The probability of EC decay increases with increasing atomic number. In both β^+ emission and EC decay the product of the decay is a nucleus of the same mass number, A, but one atomic number unit lower (e.g., ${}^{13}_{7}N \rightarrow {}^{13}_{6}C + \beta^+$ and ${}^{55}_{26}Fe + e^- \rightarrow {}^{55}_{25}Mn$). Neutron-poor nuclides are often the products of $(n, 2n)$ reactions.

Electron-capture decay involves the conversion, within the nucleus, of a proton to a neutron, by interaction with an orbital electron of that atom, usually a K-shell electron. This process thus leaves a vacancy in the K (or other) electron shell, which promptly results in an orbital-electron cascade to fill the vacancy, thereby producing characteristic X-rays of the daughter element. The K X-ray photon energies of practical interest range from about 10 keV (for zinc, $Z = 30$) to about 114 keV (for uranium, $Z = 92$), increasing with increasing Z.

With decay via internal conversion some fraction of the possible expected number of γ-ray photons instead interact with an orbital electron of the atoms whose nuclei are disintegrating, ejecting the orbital electron, with the complete disappearance of the expected γ-ray photon. This is then a "nuclear photoelectric-effect" process. If the γ-ray energy available is greater than the binding energy of a *K*-electron in the atom, ejection of a *K*-electron is favored (otherwise, ejection of an *L* electron occurs). The energy of the ejected monoenergetic electron is equal to the γ-ray energy minus the binding energy of the orbital electron. Again, as in the case of electron-capture decay, the filling of the resulting vacancy in the *K* shell (or *L* shell) by an orbital-electron cascade results in the emission of characteristic X-rays by the atom. The probability of internal conversion of γ-ray photons increases with increasing *Z* and with decreasing γ-ray energy. For any particular γ-ray-emitting radionuclide the percentage extent of IC has a definite value—anywhere from almost 0 to almost 100%. For example, the 0.0586-MeV γ-rays of 10.47-min ^{60m}Co are 97.9% internally converted, only 2.1% being emitted as unconverted γ-ray photons.

Neutron-rich radionuclides (the type produced in most instances of thermal-neutron capture (n, γ) reactions and of fast-neutron (n, p) and (n, α) reactions) in most cases decay by β^- emission. The product of such decays is thus always a nucleus of the same mass number, but of one atomic number unit higher (e.g., $^{32}_{15}\mathrm{P} \rightarrow {}^{32}_{16}\mathrm{S} + \beta^-$, and an antineutrino.)

Some radionuclides decay by isomeric transition (IT), in which a metastable nuclear isomer decays by simple γ-ray emission. Here the product has the same *A* and *Z* as the nuclear isomer (e.g., $^{60m}\mathrm{Co} \rightarrow {}^{60}\mathrm{Co} + \gamma$). In a number of instances nuclear isomers are formed by (n, γ) and (n, n') reactions.

Whether the primary mode of decay is β^- emission, β^+ emission, or electron capture, the product may be formed in the ground state or in an excited state (depending upon the decay scheme of the radionuclide)—or some of each. If a certain percentage of the disintegrations result in the formation of an excited nucleus, each of these will promptly fall to the ground state, with the emission of one or more γ-ray photons of characteristic discrete energies.

An excellent compilation of decay schemes is the Table of Isotopes compiled by Lederer, Hollander, and Perlman [10].

The Scintillation Counter

The typical counting arrangement employed in γ-ray spectrometry is described below. The detector most commonly employed is a solid cylindrical 3-in. diameter × 3-in. high thallium-activated sodium iodide single crystal, coupled to a photomultiplier tube (PMT)—the whole assembly made light-tight. The scintillation counter is typically placed inside a 4-in.-thick lead shield (to reduce the "background" counting rate) of reasonably large inside dimensions (perhaps 18 by 18 by 24 in. high). The reason for using such a large shield is to minimize the number of γ-ray photons from the sample that are backscattered from the shield back to the NaI(Tl) crystal. Because such photons undergo an approximately 180° Compton scattering, their energies are related to the energies of the original γ-ray photons by the equation:

$$E_{BS\gamma} = \frac{E_\gamma}{(1 + 2E_\gamma/0.511)} \quad \text{MeV.} \tag{7.13}$$

Therefore a sample emitting γ-ray photons of just one particular energy will exhibit a pulse-height spectrum with an undesired small peak at the backscattered energy, in addition to the desired total-absorption peak ("photopeak"). Using (7.13), it can be calculated that the maximum backscattered-photon energy possible (i.e., from very-high-energy γ-rays) is 0.255 MeV. For E_γ values of 0.1, 0.3, 0.5, 1.0, 2.0, and 3.0 MeV, the corresponding $E_{BS\gamma}$ values are 0.079, 0.138, 0.169, 0.204, 0.226, and 0.236 MeV, respectively. Samples emitting γ-rays of several energies thus can exhibit several backscatter peaks of various sizes and at various energies, or a broad composite backscatter peak (due to overlapping of the various individual backscatter peaks). Increasing the internal dimensions of the shield decreases the magnitude of the backscatter peaks but does not eliminate them entirely.

The lead shield is usually lined with a thin (0.030-in.) sheet of cadmium and a thin (0.015-in.) sheet of copper (in the order, Pb–Cd–Cu) to absorb lead X-rays generated in the shield by interaction with sample γ-rays. The lead X-rays are efficiently absorbed by the cadmium, and the resulting cadmium X-rays are efficiently absorbed by the copper. The resulting copper X-rays are low in intensity and have an energy of only about 8 keV. Without these liners an undesired photopeak would appear in the sample pulse-height spectrum at about 75 keV.

When making γ-ray spectrometry measurements one wishes to eliminate or to reduce the smearing of the pulse-height spectrum that can be caused by sample β particles reaching the NaI(Tl) crystal. To accomplish this one typically places a 1.5-cm-thick plastic absorber between the sample and the top of the detector crystal. The sample itself, if of appreciable size, will stop or slow down a large fraction of its β particles. Then the sample vial, the plastic absorber, and the aluminum casing of the NaI(Tl) crystal combine to stop virtually all the remaining ones from reaching the crystal—without producing an appreciable number of Bremsstrahlung photons (since low-Z materials are used).

When a γ-ray photon strikes a NaI(Tl) crystal, four possible things can occur: (1) it can pass through the crystal with no interaction; (2) it can give part of its energy to an electron by a Compton-scattering event; (3) it can lose all of its energy to an electron by a photoelectric-absorption event;

or (4) it can lose all of its energy by formation of an e^-/e^+ pair (a "pair-production" event). Pair production can only occur if the γ-ray energy is greater than 1.022 MeV (i.e., the energy equivalent of two electron rest masses). In NaI(Tl) photoelectric absorption is the predominant mode of interaction for energies up to about 0.3 MeV, Compton scattering between 0.3 and about 6 MeV, and pair production above 6 MeV. For a given sample and a given counting geometry the overall counting rate and the fraction of the pulses in the photopeak both increase with increasing crystal thickness. With the thicker crystals many of the γ-ray photons that first interact by Compton scattering are absorbed in a second (or third) interaction by photoelectric absorption. Since the two (or three) events occur within the resolving time of the detector, the output pulse size is the same as if the absorption had occurred in a single event. Hence with larger crystals, the peak-to-total ratio (or photofraction) of counts is increased, producing a more desirable pulse-height spectrum, since the photopeak pulses identify a particular radionuclide (via its characteristic γ-ray energy or energies), whereas the smear of Compton pulses do not aid in the identification. With a 3 × 3-in. crystal the photofractions for γ-rays of 0.1, 0.3, 0.5, 1, 2, and 4-MeV energies are about 0.99 (i.e., 99%), 0.84, 0.63, 0.41, 0.27, and 0.18, respectively [8]. The corresponding overall detection efficiencies (geometry factor excluded) are approximately 100, 72, 59, 46, 37, and 33%, respectively [8]. For a 2-cm sample-to-detector distance, with a 3 × 3-in. detector, the geometry factor is 0.265 (i.e., 26.5%).

As the result of conservation of momentum, the maximum energy that a γ-ray photon can lose by Compton interaction is less than the full photon energy and is given by the equation:

$$E_{CE} = \frac{E_\gamma}{(1 + 0.511/2E_\gamma)} \quad \text{MeV}. \tag{7.14}$$

This produces the "Compton edge" in the pulse-height spectrum. Because of the relatively poor resolution of the NaI(Tl) detector, the region of the spectrum between a Compton edge and the corresponding photopeak is partially filled in to form a "Compton trough."

As mentioned above, the most common NaI(Tl) detector is a solid cylindrical 3 × 3-in. crystal. However, for the detection and measurement of low-energy γ-rays ($\leq$0.5 MeV) in the presence of considerable interference from high-energy γ-rays, a thinner crystal is preferable. Conversely, for optimum detection of high-energy γ-rays ($\geq$2 MeV) an even thicker crystal is desirable. One limitation on one's choice is that of cost. The approximate costs of canned 1 × 1-in., 2 × 2, 3 × 3, and 5 × 5 NaI(Tl) crystals are 50, 250, 750, and $2000, respectively. Also, larger crystals have higher background counting rates and generally have somewhat poorer resolution.

As referred to above, sodium iodide detectors have, in general, rather poor energy resolution. That is, the photopeaks obtained are quite broad even though the result of interactions with monoenergetic γ-rays. For example, a particular 3 × 3-in. detector might show percentage resolutions (percentage full width at half maximum of the photopeak) of about 15% at (0.1 MeV), 11% (0.3 MeV), 9% (0.5 MeV), 7% (1 MeV), 5.5% (2 MeV), and 5% (3 MeV) [8]. The %R, of a particular NaI(Tl) detector varies roughly according to the equation: $\%R = a/(b + E_\gamma^{1/2})$, where a and b are constants. As mentioned above, this poor resolution accounts for the rounding of the pulse-height spectrum at the Compton edge and the presence of pulses in the spectrum between the Compton edge and the photopeak, as well as for the appreciable width of the photopeaks.

When a γ-ray photon interacts with an electron in the NaI(Tl) crystal, the struck electron slows to a stop in about a millimeter or less, giving up its kinetic energy to the crystal. Approximately 10% of this energy appears as Tl^+ fluorescence light, as a burst of ~3 eV photons, the other 90% degrading to heat. The photons produced are emitted isotropically, but because of the efficiency of the diffuse white reflector coating (MgO or Al_2O_3) on the outer surface of the crystal (except at the face adjacent to the photomultiplier tube), a high percentage of them strike the photocathode of the PMT essentially instantaneously. The photocathode (typically a thin layer of Cs–Sb) has about a 10% photoelectric efficiency, so about 10% of the incident 3-eV photons eject a photoelectron into the vacuum of the PMT. In the tube the photoelectrons are accelerated by an electric potential and focused onto the first "dynode" (coated with a material with a high secondary-emission coefficient, such as Cs–Sb or Ag–Mg). The accelerating potentials employed are such that about three electrons are ejected from the dynode surface by each impinging electron. The process is repeated down the whole series of dynodes (usually 10), so that a large number of electrons appear as an output pulse, compared with the smaller initiating number that were produced at the photocathode. For example, the absorption of a single 1-MeV γ-ray photon by the crystal produces some 33,000 light photons, which in turn generate about 3300 photoelectrons, which can result in an output pulse of the order of 10^8 electrons. It is important to note that the size of the output pulse is directly proportional to the amount of energy absorbed by the crystal in the interaction event.

Gamma-Ray Spectrometry

WITH NaI(Tl) DETECTORS

The complete γ-ray spectrometer consists of (1) the scintillation detector in its shield, (2) a pulse preamplifier and a linear amplifier, (3) a pulse-height analyzer, and (4) one or more data-readout modes or displays. A typical modern transistorized multichannel pulse-height analyzer is shown in Fig. 7.7. Such analyzers operate on an amplitude-to-time conversion principle, sorting the incoming amplified electrical pulses into 100, 128, 200, 256, 400, 512 (or a larger number) of "channels," or pulse-height sizes. With NaI(Tl) detectors a 200-channel spectrum is often sufficient, so the observed pulse-height spectrum consists of 200 data points. While the radioactive sample is being counted (usually for just one or a few minutes), one observes the pulse-height spectrum accumulating on the face of the oscilloscope. When the counting is finished, the static 200-point spectrum (counts-per-channel versus channel number) is observed on the face of the oscilloscope. The pulse-height data can then be plotted out on a precision X-Y or strip-chart recorder, or they can be printed out in digital form on paper tape or an electric typewriter, or they can be rapidly transferred onto punched-paper tape, punched cards, or magnetic tape. After the data from one sample have been recorded, the magnetic-core memory of the analyzer can be erased in preparation for the counting of the next sample.

Because of the small but finite time required by the analyzer to measure the size of each pulse (the larger the pulse, the longer the time needed to

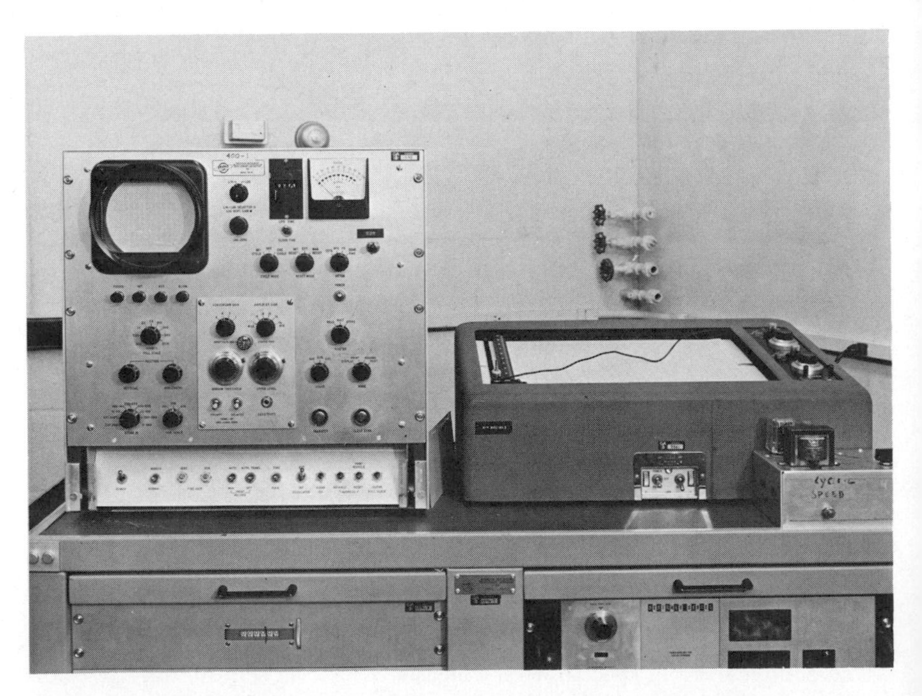

Fig. 7.7 Typical multichannel pulse-height analyzer (Courtesy of RIDL Division, Nuclear-Chicago Corp.).

measure its size), an appreciable dead-time loss results if the counting rates are high (typically, negligible at counting rates of less than about 100,000 counts/min). Counting periods can be preset for any desired "clock time" or "live time," as desired. The "dead-time" meter on the analyzer indicates the percentage dead time while the sample is being counted. Typically, counting rates in excess of about 100,000 counts/min are to be avoided, since they can lead to undesirable gain-shift problems unless an electronic gain stabilizer is employed. When a sample or series of samples is to be counted, the gain of the analyzer system is first carefully adjusted to the exact value desired, (e.g., 3.75, 7.50, or 15.0 keV/channel). This is done by adjusting the gain setting so that the photopeaks of one or more long-lived standard radionuclides fall exactly in the appropriate channels. Common calibration radionuclides used for this purpose are 270-day ^{57}Co (0.136 MeV), 2.62-yr ^{22}Na (0.511 MeV from β^+, and 1.275 MeV), 30.0-yr ^{137}Cs (0.662 MeV), and 5.263-yr ^{60}Co (1.173 MeV and 1.332 MeV, in cascade). Thus if a gain setting of exactly 7.50 keV/channel is desired (200 channels thus covering the spectrum from 0 to 1.5 MeV), the gain should be adjusted so that the 0.662-MeV photopeak of ^{137}Cs centers in channel 88. A typical multichannel pulse-height analyzer with one or two readout devices costs about $10,000.

An illustrative NaI(Tl) pulse-height spectrum, that of 3.75-min ^{52}V, is shown in Fig. 7.8. This radionuclide emits a β^- particle and a 1.434-MeV γ-ray photon in each disintegration. Principal features of its NaI(Tl) (3 × 3-in.) pulse-height spectrum are thus the Compton-continuum region (0 to 1.215 MeV), the Compton edge (at 1.215 MeV), and the photopeak (at 1.434 MeV). Smaller features are a small peak at 0.511 MeV (the result of β^+ annihilation photons, generated in the lead shield by the 1.434-MeV γ-rays, striking the crystal), and a small backscatter peak at 0.217 MeV. Higher-energy γ-rays also produce single- and double-escape peaks (at 0.511 and 1.022 MeV less than the γ-ray photon energy). In addition, with β^+ emitters and with high-energy γ-ray emitters a small peak is sometimes observable at 0.681 MeV (0.511 MeV plus a 0.170-MeV backscattered photon from the other 0.511-MeV annihilation photon); and with radionuclides that decay via a γ-ray cascade, small "sum" peaks are observed (e.g., the ^{60}Co 1.173-MeV and 1.332-MeV cascade γ-ray photons result in major photopeaks at these two energies and a smaller peak at their sum, 2.505 MeV, the result of coincident γ-ray photons striking the crystal simultaneously). With very-low-energy γ-rays (particularly those with energies of less than 100 keV), a satellite "iodine escape peak" is observed at about 29 keV less than the photopeak—because of the escape, in some of the absorption events, of an iodine *K* X-ray photon (29 keV) from the NaI(Tl) crystal (which can occur fairly frequently with low-energy γ-rays because of their very slight penetration into the crystal).

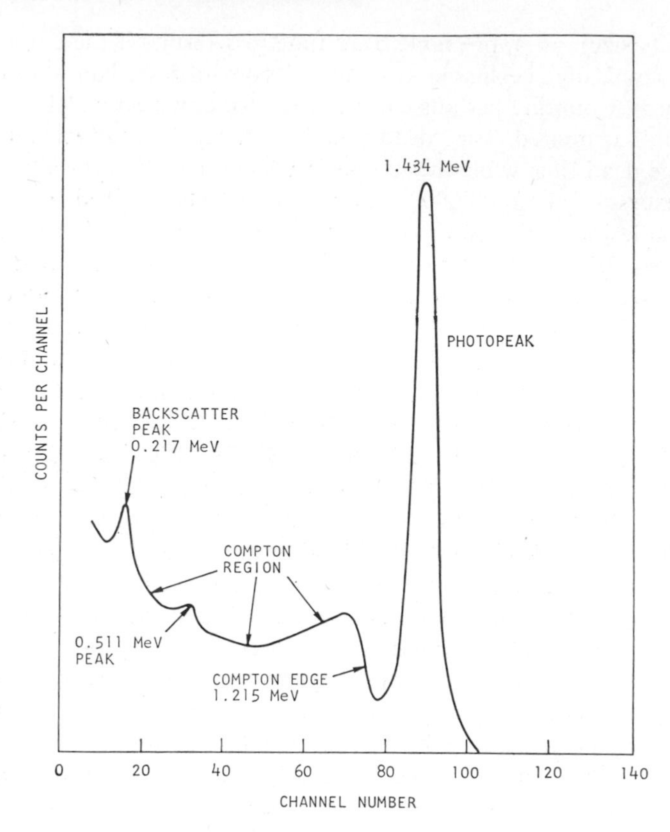

Fig. 7.8 Illustrative γ-ray NaI(Tl) pulse-height spectrum (3.75-min ^{52}V).

In some cases, when improved geometrical counting efficiency is desired, a well-type NaI(Tl) crystal is used. In such cases the sample vial is placed down in the cylindrical well of the crystal, thereby increasing the geometry from a typical 25% value (sample on top of a β absorber, on top of solid crystal) to a value of about 90%. For a given set of external crystal dimensions (e.g., 3 × 3-in.), the effective NaI(Tl) path length of the well-type crystal is much less than that of the corresponding solid crystal, so the interaction efficiency and the photofraction are both poorer for the well-type crystal, at least for γ-ray energies above about 0.5 MeV. With well-type crystals cascade (coincident) sum peaks are more pronounced. A special case is that of a β^+ emitter, where the 1.022-MeV peak may be about the same size as the 0.511-MeV peak, whereas with a solid crystal the 0.511-MeV peak is essentially the only one observed.

When one wishes to preferentially detect a radionuclide that emits γ-rays in cascade, in the presence of considerable γ-radiation of higher energy, one can sometimes utilize a well-type crystal and a sum peak, as mentioned above —or one can employ a two-crystal coincidence technique with the sample placed in between the two crystals. If one detector is set on a group of pulse-height channels that encompasses one of the cascade γ-ray energies, the other group of channels is set to encompass the other cascade γ-ray energy, and the circuitry is set to record only coincident pulses from the two detectors, detection of that activity is enhanced relative to the detection of the interfering activities. For maximum efficiency (1) the two detectors should be large in area, compared to the size of the sample, (2) they should be thick enough to give good total-absorption efficiencies for the two γ-ray energies involved, and (3) they should be as close together as possible. For selective detection of β^+ emitters both groups of channels are set to encompass the same photopeak (0.511 MeV). Similarly a β^-/γ-coincidence technique can be used effectively in some cases—using a β^- detector (organic crystal, gas-proportional, plastic phosphor, or thin silicon semiconductor detector) and a γ-detector (NaI(Tl)) in coincidence.

WITH SEMICONDUCTOR DETECTORS

More recently, increasing usage is being made of lithium-drifted germanium (Ge(Li)) semiconductor detectors, instead of, or in addition to, NaI(Tl) detectors, for γ-ray spectrometry measurements. These are markedly superior in resolution to NaI(Tl) detectors, percent resolutions (percentage full-width at half-maximum (FWHM) of the total-absorption peak—photopeak) being typically about 15-fold better than for NaI(Tl). However, because of the relatively shallow effective-depth ones thus far available (maximum of about 2 cm), they are not very efficient (compared with a 3-in.-thick NaI(Tl) detector) for γ-rays of energies greater than a few tenths of a megaelectron volt. Also, they must be maintained at liquid-nitrogen temperature at all times; they are considerably more expensive; and (for maximum resolution) they require a more elaborate pulse-height analyzer—one with a few thousand channels (e.g., 4096), rather than just a few hundred. Small Ge(Li) detectors with a depletion layer of just a few millimeters (rather than 2 cm), are not so expensive and are very useful in the X-ray region ($\sim$10 to 100 keV). Larger ones, with large areas and deeper depletion zones, are available, with volumes approaching 100 cm^3, but they cost around \$10,000. For maximum resolution a special field-effect transistor (FET) preamplifier should be used with them. Whereas photopeaks in, for example, the region around 1 MeV, that differ in energy by only about 50 keV or less, are not resolved from one another by a NaI(Tl) detector, they are completely resolved by a Ge(Li) detector even if their energies differ by as little as 5 keV. An illustrative Ge(Li)

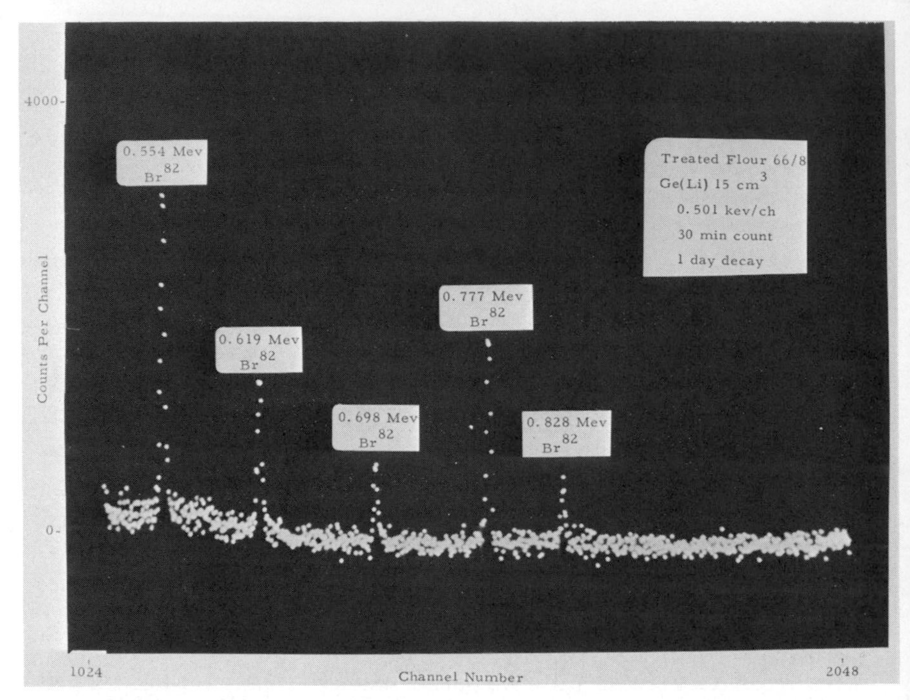

Fig. 7.9 Illustrative γ-ray Ge(Li) pulse-height spectrum (35.34-hr ^{82}Br).

pulse-height spectrum is shown in Fig. 7.9: a portion of the pulse-height spectrum of 35.34-hr ^{82}Br, showing the complete resolution of its five major γ-rays.

In the lower-energy part of the X-ray region a similar lithium-drifted silicon [Si(Li)] semiconductor detector is also very good (high resolution, good efficiency). This type of detector does not need to be maintained at liquid-nitrogen temperature except when in use (for maximum resolution).

Calculations

In instrumental activation analysis work one typically uses a particular γ-ray spectrometer to count a series of activated samples and standards (all activated under identical conditions) under exactly the same conditions but at different decay times after t_0 (the end of the irradiation) and perhaps for different lengths of time. In order to make proper calculations one usually computes the net photopeak area (above the Compton continuum, using a linear baseline approximation) counting rate of the photopeak of interest. One then corrects all of these counting rates to t_0, or to any other desired reference time, using (7.7). To make the decay corrections one takes the calculated counting rate (net photopeak counts divided by live counting time) to be the same as the true net photopeak counting rate at the midpoint of the counting period (clock time). This is a fairly accurate assumption so long as the percentage dead time during the counting period is small ($<10\%$) and the sample is not counted for a period of time longer than the half-life of the radionuclide of interest; otherwise corrections must be applied. Since one knows the exact amount of the element in question in the standard sample, he can then readily compute the amount in each unknown sample, using (7.11) or (7.12). If desired, a computer program can be used to process the data, making all of the necessary calculations from the raw data and printing out an answer (in micrograms or parts per million) along with its standard deviation (based upon its counting statistics) for each sample. Some multichannel pulse-height analyzers have the capability of automatically adding together the counts in any group of channels selected, such as those included within a photopeak.

In many cases a person can analyze one or more samples for a number of elements of interest at the same time, instead of one element at a time. Also, if one employs carefully standardized irradiation and counting conditions, he can develop a collection of sensitivity and γ-ray spectral information for each radionuclide formed from the various elements. Then he can analyze samples instrumentally for all elements that show up, even though some may not have been anticipated. Well-developed computer programs are available (1) to search the raw pulse-height digital data for all statistically-significant photopeaks present; (2) from these photopeak energies, the counter resolution, and other criteria (such as irradiation time and flux, decay time, and counting conditions) determine which radionuclides might possibly be contributing significantly in each peak; (3) set up a matrix and solve by weighted least-squares the combination of pure radionuclide reference spectra that best fit the data from the sample, (4) calculate and print out the amount (in micrograms) or concentration (in parts per million) of each element detected, along with its standard deviation (from counting statistics only). If desired, another computer program may also be used to calculate a firm upper limit to the possible amount or concentration of each of many other elements that do not show up in the spectra. In this fashion either actual values or upper limits can be obtained for up to about 70 elements in a sample. A weighted least-squares computer program is also available for resolving a single photopeak (such as the 0.511-MeV β^+ annihilation peak) into its possible contributors of different half-lives. A number of these computer techniques are used regularly in the author's laboratory, with a great saving in time and effort.

The pulse-height spectrum of a mixture of radionuclides is, when measured properly, simply the sum of the spectra of its various components. Most multichannel pulse-height analyzers include the possibility of subtracting one spectrum from another. This feature can be of use for (1) subtracting out the counts resulting from background radiation, (2) subtracting out the spectra of one or more interfering activities, or (3) subtracting the sample spectrum at a longer decay time from that obtained at a shorter decay time in order to obtain a better representation of the short-lived species present. Such operations are termed "spectrum stripping," and may be carried out by "live" counting in subtract mode, or by subtraction of spectra stored in a second group of channels in the analyzer memory, or on punched-paper tape, or on magnetic tape, using a variable multiplier. By being operated on a time-sharing basis one pulse-height analyzer can simultaneously handle the input from two or four detectors, each going into a different group of channels. Most multichannel analyzers can also be operated in a "multiscaler" mode, that is, with all the incoming pulses being stored in the first channel for a preset period of time, then advancing regularly, in succession, to the following channels. In this way a gross decay curve is readily obtained fully automatically.

The Radiochemical-Separation Form

Where (1) the induced activity of interest emits only β^- particles in all or almost all of its disintegrations or (2) the interferences from other γ-emitting induced activities prevent the purely instrumental detection of the activity of interest, one can or must resort to a post-irradiation radiochemical-separation procedure. After the activity of interest has been separted from all other induced activities, it can be counted on a β-sensitive counter, or if it is a γ-emitter, on a γ-ray detector.

Procedure

After activation, the sample is dissolved and chemically equilibrated with a relatively large amount (accurately known, and typically about 10 mg) of the element (or elements) of interest, as a carrier. Frequently the equilibration is accomplished by alternate oxidation and reduction in acidic solution. If high levels of high specific-activity interfering nuclides of other elements are present, one usually dilutes them with similar amounts (which, however, do not need to be known accurately) of "hold-back" carriers of those elements. Then the element of interest is separated out and purified by any suitable separation procedures (e.g., precipitation, volatilization, solvent extraction, electrodeposition, ion-exchange), and is then counted in an appropriate fashion. The amount of carrier element recovered is measured quantitatively, by any suitable means (e.g., gravimetrically, volumetrically, by atomic absorption, by reactivation), so that the results can be normalized to 100% recovery. It should be noted that whenever a nucleus undergoes a nuclear reaction, there is a recoil energy imparted to the product nucleus in the prompt deexcitation of the compound nucleus—sufficient to eject it from the molecule it was in. The energetic recoil nucleus slows down in the medium and undergoes various chemical reactions. As a result, most of the radionuclides formed by activation end up in a variety of chemical forms, mostly different from the original chemical forms, and many of these are virtually carrier-free. This must be kept in mind when carrying out the later chemical equilibration with added carrier. Recoils can also lead to losses of induced activity in the halogens and other sometimes volatile elements, and to losses by adsorption on the wall of the polyvials.

Freedom from Reagent Blank

Since any elemental contamination of the sample after activation cannot affect the results, no reagent-blank correction is necessary. For example, if one wished to analyze a sample for mercury and if the sample contained only 0.01 μg Hg, the reagents used for the radiochemical separation of the thermal-neutron induced ^{197}Hg activity, after activation, could even contain thousands of times more mercury than the sample itself, without affecting the accuracy of the result. This is because only the mercury originally present in the sample has been made radioactive and because the mercury present as impurity in the reagents is negligible compared to the 10 mg of mercury added as carrier.

Freedom from Loss Errors

Because of the rapid equilibration of the induced activity of interest with a gross amount of carrier element, losses caused by adsorption, coprecipitation, and other processes become of negligible importance. Without such carrier dilution, of course, such losses could be very serious—when one is determining microgram and sub-microgram amounts of an element. After equilibration with the carrier has been established, even sizable losses do not affect the accuracy of the result. This fact permits one to employ fairly rapid, nonquantitative, separations since the recovery of the carrier is measured and the results are normalized to complete recovery. In practice, of course, one still strives for fairly good recoveries—in the range of 50 to 100%—since the higher the recovery, the better the counting statistics.

Beta Counting

When the separated activity must be β-counted, or is best counted in this fashion, some precautions must be observed. One of these precautions differs appreciably from that involved in γ-ray spectrometry; namely, the effect of self-absorption. Because of the high penetrability of γ-rays (at least of energies above a few tenths of a megaelectron volt) in most materials (except those of very high density and high atomic number), activated samples with volumes of the order of 1 cm^3 or less can be counted by γ-ray spectrometry with negligible difficulty from self-scattering or self-absorption of the γ-rays within the sample itself. This is not true for β particles. First, the β^- particles emitted by a particular radionuclide are not monoenergetic (whereas the γ-rays emitted by a particular radionuclide are), but instead exhibit an energy spectrum ranging from zero to a maximum value (E_{max}) that is equal to the energy release in the β^- transition. The average and most-probable β^- energies are typically around $\frac{1}{3}$ of the E_{max} (in each β^- emission event the total energy release, E_{max}, is distributed between the readily detected β^- particle and the simultaneously emitted, virtually undetectable antineutrino, $\tilde{\nu}$). Tables, such as the Table of Isotopes [10] and the General Electric Chart of the Nuclides [6] list the E_{max} values of the β^- particles emitted by the various radionuclides.

Secondly, β particles are nowhere near as penetrating as γ-ray photons of comparable energy, and unlike γ-rays they exhibit a maximum range in a given medium, the range depending upon the energy. For β particles the range, R (in g/cm² of aluminum), is given by the equations (E in MeV):

$$R = 0.407E^{1.38} \quad \text{for} \quad E < 0.8 \text{ MeV}, \tag{7.15}$$

and

$$R = 0.542E - 0.133 \quad \text{for} \quad E > 0.8 \text{ MeV}. \tag{7.16}$$

Thus, the ranges in aluminum of 0.5, 1, 2, 3, and 4-MeV β^- particles are 0.16, 0.41, 0.95, 1.49, and 2.03 g/cm², respectively. Since the density of aluminum is 2.70 g/cm³, these ranges correspond to only 0.058, 0.15, 0.35, 0.55, and 0.75 cm, respectively. The range in most other materials is quite close to the range in aluminum, in units of grams per centimeter squared.

The relatively small penetration of β^- particles requires that there be very little material between the radioactive sample to be counted and the sensitive region of the detector. It also requires, when the β^- counting rates of two or more samples are to be accurately compared, not only that the overall counting geometry be exactly the same but also that the sample chemical composition and thickness (expressed in grams per centimeter squared) be almost exactly the same (or all at least "infinitely thick"). For an "infinitely-thin" sample self-absorption of the β^- particles within the sample itself is zero (or at least negligible). With a sample of a thickness less than the range of the maximum-energy β^- particles emitted by the radionuclide, the self-absorption of the β^- particles is appreciable. The fraction of the β^- particles generated within the sample that can get out of the sample decreases as the sample thickness is increased. An "infinitely-thick" sample is one whose thickness (in grams per centimeter squared) is greater than the range of the β^- particles. For a sample of a given specific activity (β^- dis/min of a particular E_{max} per gram of sample), counted at a fixed geometry and detector efficiency, the β^- counting rate is constant for a series of samples that are all infinitely thick or thicker, even if they are of different thicknesses.

In β^- counting one usually employs a counter that has a high-detection efficiency for charged particles, but a low-detection efficiency for γ-radiation. Suitable counters are Geiger-Mueller and gas-proportional counters; organic crystal or plastic phosphor scintillation detectors; Cerenkov detectors; and thin silicon semiconductor detectors.

For β^- counting following radiochemical separation of the activity of interest, it is usually desirable to verify that the activity being measured is, in fact, the one that it is supposed to be. This is done in two ways: (1) by counting the sample at two or more decay times, appropriately spaced, to see if the observed half-life agrees with that of the radionuclide in question, and (2) by checking that the β^- E_{max} value is the same as that of the radionuclide in question. The energy check is usually done by counting the sample with and without an aluminum absorber of a suitable thickness placed between the sample and the detector. The same measurements are also made on the reference sample of that radionuclide—a sample of the same chemical composition and thickness (in grams per centimeter squared). An aluminum absorber that decreases the counting rate by a factor of around two or three is quite suitable. The fraction, net counts per minute with absorber divided by net counts per minute without absorber, should be closely the same for sample and standard. If the half-life and β^--energy measurements do not indicate that the sample activity is quite pure and that of the radionuclide in question, the sample must be purified further. It is also wise to check for the presence of any γ-emitting impurity in the sample by counting it also on the γ-ray spectrometer.

Positrons emitted by radionuclides are also emitted with a spectrum of energies (rather than as monoenergetic particles), and they exhibit an average or most-probable energy of around $\frac{1}{3}$ of the E_{max}. Their energy-range dependence is essentially the same as for β^- particles (i.e., they are also related by (7.15) and (7.16). However, since positrons readily undergo an annihilation reaction with electrons once they have slowed down sufficiently, β^+ emitters are usually counted by γ-ray spectrometry instead of by β-particle counting. Also, if the separated activity of interest is an X-ray or a γ-emitter, it can often best be counted by γ-ray spectrometry or simple γ-ray counting.

In some cases a sample matrix can become so highly radioactive that it is difficult or even hazardous to handle during the radiochemical-separation steps. If the activity of interest is relatively long-lived and if most of the interfering activities are relatively short-lived (as is often the case), one can simply postpone the handling of the sample until the interferences have decayed away to a large extent. If this is not possible, one must carry out the radiochemical separations under "semi hot-lab" conditions (i.e., using some protective lead shielding and long-handled tongs, remote-control pipets, or other such devices). In most activation analysis work the actual samples handled are not dangerously radioactive. In practice, one estimates in advance, wherever possible, the probable resulting radioactivity level of a sample to be activated, then monitors it before handling it. All operations are carried out to ensure that the chemist does not receive a whole-body exposure of more than 100 milliroentgens (mr) per week—a generally accepted tolerance level. Cumulative doses are checked regularly with film badges and pocket dosimeters.

In some instances, especially where the induced activity of interest is rather short-lived (seconds to a few hours), one may prefer to give up the advantage of freedom from reagent blank and instead employ a preirradiation chemical separation. If so, one must use very pure reagents, check every step of the procedure, and run a reagent blank—if one is attempting to determine very small amounts or very low concentrations. An example of this procedure is the preirradiation separation of vanadium in blood-serum samples, used in the author's laboratory. The half-life of the thermal-neutron activated vanadium, ^{52}V, is only 3.75 min. In this case the reagent blank is typically only about 0.001 μg V.

5 SENSITIVITY OF NEUTRON ACTIVATION ANALYSIS FOR VARIOUS ELEMENTS

The discussion here is limited to the interference-free limits of detection for various elements when exposed to (1) the high (10^{13} neutrons/(cm^2)(sec)) thermal-neutron flux of a $\sim$1 MW research reactor, and (2) the moderate (10^9 neutrons/(cm^2)(sec)) 14-MeV flux of a small neutron generator.

High-Flux Thermal-Neutron Activation Analysis

It can be seen from (7.5), (7.6), and (7.10) that the amount of a particular radioactivity induced per unit mass of a particular element, when exposed to a flux of thermal neutrons, is (1) directly proportional to the thermal-neutron flux, ϕ; to the fractional isotopic abundance, a; and to the isotopic thermal-neutron capture cross section, σ; (2) inversely proportional to the chemical atomic weight (AW) of the element; and (3) dependent in a more complex manner upon the irradiation time, t_i, and the half-life of the radionuclide, T (both involved in the saturation term, S). The attainable counting rate, from a given disintegration rate of a particular radionuclide, depends upon (1) the decay scheme of the radionuclide, (2) the counting geometry, and (3) the efficiency of the detector.

In order to compile a useful table of limits of detection one therefore selects or defines values for each of these variables and then makes arbitrary, but reasonable, assumptions regarding the minimum counting rate that is still measurable with some accuracy. This has been done in somewhat different ways by various authors, but the one developed in the author's laboratory by J. D. Buchanan [3] is used here. Using his same assumptions, but calculating limits of detection for a thermal-neutron flux of 10^{13} instead of his 1.8×10^{12} neutrons/(cm^2)(sec) and making a few minor changes, one obtains the sensitivity values shown in Table 7.2 (listed in the order of increasing atomic number). The values shown are calculated detection limits (in the absence of interferences) for a 1-hr (maximum) irradiation at 10^{13} flux, with radiochemical separation and β^- counting assumed for the β^- detection limits (allowing for decay and yield), and γ-ray spectrometry, using the photopeak counting rate of the γ-ray photopeak listed, for the γ-detection limits. A 3×3-in. solid NaI(Tl) detector is assumed, with a mean sample-to-crystal distance of 2 cm.

In the case of β^- counting the defined minimum detectable net counting rate is taken as 100 counts/min if the half-life is less than 1 hr, and 10 counts/min if it is greater than 1 hr. For the γ-ray spectrometry the defined minimum detectable photopeak counting rate is taken as 1000 counts/min (for $T \leq 1$ min), 100 counts/min (for $T > 1$ min but < 1 hr), and 10 counts/min (for $T \geq 1$ hr). The decay schemes, half-lives, and γ-ray energies used or cited are taken from the Table of Isotopes [10], and the photopeak counting efficiencies are taken from Heath's Catalogue [8]. Although the sensitivities listed in Table 7.2 are calculated ones, many of them have been checked experimentally in various laboratories, including the author's. The experimental and calculated values in most cases agree at least within a factor of about 2. In some cases the experimental sensitivity is severalfold better than the calculated sensitivity because of resonance absorption of neutrons, in addition to absorption of thermal neutrons. Only rarely is there a discrepancy as large as a factor of 10. In most instances where no γ limit is shown, the radionuclide is a pure β^- emitter. In most instances where no β^- limit is shown either the half-life is too short (<1 min), or the radionuclide decays by electron capture or isomeric transition. In a few cases detection via a radioactive daughter is listed (e.g., Pt via the ^{199}Au daughter of ^{199}Pt, and Th via the ^{233}Pa daughter of ^{233}Th). For many elements not all of the (n, γ) products are listed in the Table 7.2. Just the one, two, or three (in a few cases four or five) predominant ones are given (i.e., the ones of most analytical usefulness). (*Note*: 20 of the elements of the periodic system are monoisotopic in nature, hence form only a single (n, γ) product—except for a few elements, such as cobalt, which form an isomeric state of the product radionuclide also. The remaining stable elements exist in nature in the form of from 2 to

Table 7.2 Limits of Detection at a Thermal-Neutron Flux of 10^{13} neutrons/(cm^2)(sec) (1-hr maximum irradiation time)

				Limit of detection, μg	
Element	*Radionuclide*	*Half-life*	*γ-ray energy,* MeV	*β^- counting*	*γ-ray spectrometry*
F	^{20}F	11.56 sec	1.63	—	0.1
Ne	^{23}Ne	37.6 sec	0.439	—	0.5
Na	^{24}Na	14.96 hr	1.369	0.0004	0.0005
Mg	^{27}Mg	9.46 min	0.842	0.04	0.03
Al	^{28}Al	2.31 min	1.780	0.02	0.0009
Si	^{31}Si	2.62 hr	1.26	0.009	30.
P	^{32}P	14.28 days	—	0.02	—
S	^{35}S	87.9 days	—	2.	—
	^{37}S	5.07 min	3.09	10.	20.
Cl	^{38}Cl	37.29 min	1.60	0.002	0.01
Ar	^{41}Ar	1.83 hr	1.293	—	0.004
K	^{42}K	12.36 hr	1.524	0.002	0.04
Ca	^{49}Ca	8.8 min	3.10	0.09	0.2
Sc	^{46m}Sc	19.5 sec	0.142	—	0.0003
	^{46}Sc	83.9 days	0.889	0.002	0.005
Ti	^{51}Ti	5.79 min	0.320	0.07	0.01
V	^{52}V	3.75 min	1.434	0.0004	0.00008
Cr	^{51}Cr	27.8 days	0.320	—	0.1
Mn	^{56}Mn	2.576 hr	0.847	0.000004	0.000007
Fe	^{59}Fe	45.6 days	1.095	9.	20.
Co	^{60m}Co	10.47 min	0.059	—	0.0002
	^{60}Co	5.263 yr	1.173	0.04	0.07
Ni	^{65}Ni	2.564 hr	1.481	0.004	0.04
Cu	^{64}Cu	12.80 hr	(0.511)β^+	0.0002	0.0006
	^{66}Cu	5.10 min	1.039	0.002	0.005
Zn	^{65}Zn	245. days	1.115	0.9	3.
	^{69m}Zn	13.8 hr	0.439	0.02	0.01
Ga	^{70}Ga	21.1 min	1.040	0.0004	0.1
	^{72}Ga	14.12 hr	0.835	0.0002	0.0004
Ge	^{75}Ge	82. min	0.265	0.0004	0.002
As	^{76}As	26.4 hr	0.559	0.0002	0.0006
Se	^{75}Se	120.4 days	0.265	—	0.1
	^{77m}Se	17.5 sec	0.161	—	0.04
	^{81}Se	18.6 min	—	0.001	—
	^{81m}Se	56.8 min	0.103	—	0.009
Br	^{80}Br	17.6 min	0.618	0.00005	0.005
	^{80m}Br	4.38 hr	0.037	0.0001	0.0003
	^{82}Br	35.34 hr	0.554	0.0009	0.0008
Kr	^{81m}Kr	13.sec	0.190	—	0.003
	^{85m}Kr	4.4 hr	0.150	—	0.001

Table 7.2 continued

Element	Radionuclide	Half-life	γ-ray energy, MeV	Limit of detection, μg β⁻ counting	γ-ray spectrotrmey
Rb	^{86}Rb	18.66 days	1.078	0.04	0.6
	^{86m}Rb	1.02 min	0.56	—	0.06
	^{88}Rb	17.8 min	1.863	0.009	0.09
Sr	^{87m}Sr	2.83 hr	0.388	0.0009	0.0006
Y	^{90}Y	64.0 hr	—	0.002	—
Zr	^{97}Zr	17.0 hr	0.747	0.2	0.3
Nb	^{94m}Nb	6.29 min	0.871	—	0.2
Mo	^{99}Mo	66.7 hr	0.181	0.02	0.1
	^{101}Mo	14.6 min	0.191	0.02	0.03
Ru	^{103}Ru	39.5 days	0.497	0.09	0.09
	^{105}Ru	4.44 hr	0.726	0.002	0.002
Rh	^{104m}Rh	4.41 min	0.051	—	0.00003
	^{104}Rh	43. sec	0.56	—	0.002
Pd	^{109m}Pd	4.69 min	0.188	—	0.004
	^{109}Pd	13.47 hr	0.088	0.0002	0.0008
Ag	^{108}Ag	2.42 min	0.632	0.0009	0.002
	^{110}Ag	24.4 sec	0.658	—	0.00006
Cd	^{111m}Cd	48.6 min	0.150	—	0.03
	^{115}Cd	53.5 hr	0.53	0.009	0.03
In	$^{116ml}In$	54.0 min	1.293	0.000005	0.00001
Sn	^{123m}Sn	39.5 min	0.160	0.07	0.02
	^{125m}Sn	9.5 min	0.325	0.05	0.03
Sb	^{122m}Sb	4.2 min	0.061	—	0.004
	^{122}Sb	2.80 days	0.564	0.0009	0.001
Te	^{131}Te	24.8 min	0.150	0.005	0.004
I	^{128}I	24.99 min	0.441	0.0002	0.0004
Xe	^{135}Xe	9.14 hr	0.250	—	0.01
Cs	^{134m}Cs	2.895 hr	0.128	—	0.00009
	^{134}Cs	2.046 yr	0.605	0.05	0.04
Ba	^{139}Ba	82.9 min	0.166	0.005	0.0002
La	^{140}La	40.22 hr	1.596	0.0004	0.0008
Ce	^{141}Ce	32.5 days	0.145	0.2	0.03
	^{143}Ce	33. hr	0.293	0.02	0.04
Pr	^{142}Pr	19.2 hr	1.57	0.0001	0.006
Nd	^{147}Nd	11.06 days	0.533	0.02	0.04
Sm	^{153}Sm	46.8 hr	0.103	0.00009	0.00006
Eu	$^{152ml}Eu$	9.3 hr	0.963	0.0000009	0.00001
Gd	^{159}Gd	18.0 hr	0.363	0.001	0.007
Tb	^{160}Tb	72.1 days	0.966	0.005	0.1

Table 7.2 *continued*

Element	Radionuclide	Half-life	γ-ray energy, MeV	Limit of detection, μg: β⁻ counting	Limit of detection, μg: γ-ray spectrometry
Dy	^{165}Dy	139.2 min	0.361	0.0000004	0.000004
	$^{165ml}Dy$	1.26 min	0.108	—	0.000002
Ho	^{166}Ho	26.9 hr	0.081	0.00004	0.00004
Er	^{171}Er	7.52 hr	0.308	0.0005	0.0009
Tm	^{170}Tm	134. days	0.084	0.002	0.03
Yb	^{169}Yb	31.8 days	0.198	—	0.004
	^{175}Yb	101. hr	0.396	0.0004	0.003
	^{177}Yb	1.9 hr	0.151	0.0002	0.002
Lu	^{176m}Lu	3.69 hr	0.088	0.000009	0.00004
	^{177}Lu	6.74 days	0.208	0.0001	0.0005
Hf	^{175}Hf	70. days	0.343	—	0.04
	^{178m}Hf	4.3 sec	0.326	—	0.00005
	^{179m}Hf	18.6 sec	0.217	—	0.004
	^{180m}Hf	5.5 hr	0.444	—	0.00005
	^{181}Hf	42.5 days	0.482	0.02	0.03
Ta	^{182m}Ta	16.5 min	0.172	—	0.02
	^{182}Ta	115.1 days	1.122	0.009	0.03
W	^{187}W	23.9 hr	0.686	0.0002	0.0004
Re	^{186}Re	88.9 hr	0.137	0.0002	0.0003
	^{188}Re	16.7 hr	0.155	0.00004	0.0001
Os	^{191}Os	15.0 days	0.129	0.02	0.09
	^{193}Os	31.5 hr	0.28	0.004	0.02
Ir	$^{192ml}Ir$	1.42 min	0.058	—	0.002
	^{192}Ir	74.2 days	0.317	0.0005	0.0004
	^{194}Ir	17.4 hr	0.328	0.00002	0.00005
Pt	^{199}Pt	31. min	0.197	0.004	0.01
	^{199}Au	3.15 days	0.158	0.02	0.02
Au	^{198}Au	2.697 days	0.412	0.00005	0.00004
Hg	^{197}Hg	65. hr	0.077	—	0.006
	^{197m}Hg	24. hr	0.134	—	0.002
	^{199m}Hg	43. min	0.158	—	0.3
	^{203}Hg	46.9 days	0.279	0.08	0.05
	^{205}Hg	5.5 min	0.205	0.06	0.04
Tl	^{204}Tl	3.81 yr	—	0.5	—
	^{206}Tl	4.19 min	—	0.04	—
Pb	^{207m}Pb	0.80 sec	0.570	—	0.04
	^{209}Pb	3.30 hr	—	2.	—
Bi	^{210}Bi	5.013 days	—	0.05	—
Th	^{233}Pa	27.0 days	0.31	0.007	0.01
U	^{239}Np	2.346 days	0.106	0.0007	0.0004

as many as 10 stable nuclides of various abundances. These elements, in general, can each form a number of different (n, γ) radionuclide products). With some elements the fast-neutron flux in the reactor can also generate quite significant activities via (n, n'), $(n, 2n)$, (n, p), or (n, α) reactions. These have been discussed in an earlier section and are not included in Table 7.2.

If one takes the best β^- sensitivity for each element listed in Table 7.2, the values for the 67 elements range from as low as 4×10^{-7} μg (dysprosium) to as high as 9 μg (iron), with a median of about 0.002 μg. Similarly, if one takes the best γ sensitivity for each element, the values for the 71 elements range from as low as 2×10^{-6} μg (dysprosium) to as high as 30 μg (silicon), with a median of about 0.002 μg. If the best sensitivity (either β^- or γ) is taken for each element, the values for the 75 elements range from as low as 4×10^{-7} μg (dysprosium) to as high as 9 μg (iron), with a median of about 0.001 μg. Thus a typical element can be detected down to about one nanogram, or 0.001 ppm in a 1-g sample.

It may be remarked that, if needed, the sensitivities shown in Table 7.2 can all be improved further by employing an even higher neutron flux, if it is available. In addition, the sensitivities for those elements that form rather long-lived activities (half-lives in excess of a few hours) can all be improved by using an irradiation time longer than 1 hr, and also by counting for a longer-than-usual period of time.

Moderate-Flux 14-MeV Neutron Activation Analysis

With presently available 14-MeV neutron generators the maximum 14-MeV-neutron flux to which samples of appreciable size (1 to 5 cm^3) can be exposed for any length of time is about 10^9 neutrons/(cm^2)(sec). The 14-MeV neutron limits of detection for the 41 elements shown in Table 7.3 are experimentally determined values, measured in the author's laboratory. They are for a maximum irradiation time of 5 min (in view of the target-decline problem), at a 14-MeV neutron flux of 10^9 neutrons/(cm^2)(sec). The limits of detection in this case are defined as the amount of the element, under these conditions, that will give 100 photopeak counts in a counting period of 10 min or less, again assuming a 3×3-in. solid NaI(Tl) detector and a mean sample-to-crystal distance of 2 cm. They are listed in the order of increasing atomic number. In this experimental study the rare-earth and inert-gas elements were not studied, nor were some of the more unusual elements (such as ruthenium, rhodium, and osmium). The limits of detection are seen to range from as low as 1 μg (rubidium, strontium, and barium) to as high as 300 μg (nickel), with a median of about 20 μg—under the stated conditions.

Such 14-MeV neutron generators can, with a moderator, provide thermal-neutron fluxes of about 10^8 neutrons/(cm^2)(sec). Thus the thermal-neutron

Table 7.3 Limits of Detection at a 14-MeV Neutron Flux of 10^9 Neutrons/(cm^2)(sec) (5 min Maximum Irradiation Time)

Element	*Reaction*	*Product*	*Half-Life*	*γ-ray energy,* MeV	*Limit of detection,* μg
N	$^{14}N(n, 2n)$	^{13}N	9.96 min	$(0.511)\beta^+$	90
O	$^{16}O(n, p)$	^{16}N	7.14 sec	6.13	30
F	$^{19}F(n, p)$	^{19}O	29.1 sec	0.197	20
Na	$^{23}Na(n, p)$	^{23}Ne	37.6 sec	0.439	20
Mg	$^{24}Mg(n, p)$	^{24}Na	14.96 hr	1.369	80
Al	$^{27}Al(n, p)$	^{27}Mg	9.46 min	0.842	6
Si	$^{28}Si(n, p)$	^{28}Al	2.31 min	1.780	2
P	$^{31}P(n, \alpha)$	^{28}Al	2.31 min	1.780	8
K	$^{39}K(n, 2n)$	^{38}K	7.71 min	$(0.511)\beta^+$	90
Sc	$^{45}Sc(n, 2n)$	^{44}Sc	3.92 hr	$(0.511)\beta^+$	20
Ti	$^{46}Ti(n, 2n)$	^{45}Ti	3.09 hr	$(0.511)\beta^+$	90
V	$^{51}V(n, p)$	^{51}Ti	5.79 min	0.320	7
Cr	$^{52}Cr(n, p)$	^{52}V	3.75 min	1.434	10
Mn	$^{55}Mn(n, \alpha)$	^{52}V	3.75 min	1.434	40
Fe	$^{56}Fe(n, p)$	^{56}Mn	2.576 hr	0.847	30
Co	$^{59}Co(n, \alpha)$	^{56}Mn	2.576 hr	0.847	50
Ni	$^{60}Ni(n, p)$	^{60m}Co	10.47 min	0.059	300
Cu	$^{63}Cu(n, 2n)$	^{62}Cu	9.76 min	$(0.511)\beta^+$	9
Zn	$^{64}Zn(n, 2n)$	^{63}Zn	38.4 min	$(0.511)\beta^+$	30
Ga	$^{69}Ga(n, 2n)$	^{68}Ga	68.3 min	1.078	20
Ge	$^{76}Ge(n, 2n)$	^{75m}Ge	48. sec	0.139	5
As	$^{75}As(n, p)$	^{75m}Ge	48. sec	0.139	4
Se	$^{78}Se(n, 2n)$	^{77m}Se	17.5 sec	0.161	20
Br	$^{79}Br(n, n')$	^{79m}Br	4.8 sec	0.21	60
Rb	$^{85}Rb(n, 2n)$	^{84m}Rb	20. min	0.250	1
Sr	$^{88}Sr(n, 2n)$	^{87m}Sr	2.83 hr	0.388	1
Zr	$^{90}Zr(n, 2n)$	^{89m}Zr	4.18 min	0.588	4
Mo	$^{92}Mo(n, 2n)$	^{91}Mo	15.49 min	$(0.511)\beta^+$	30
Pd	$^{110}Pd(n, 2n)$	^{109m}Pd	4.69 min	0.188	4
In	$^{113}In(n, 2n)$	^{112m}In	20.7 min	0.156	30
Sb	$^{121}Sb(n, 2n)$	^{120}Sb	15.89 min	$(0.511)\beta^+$	7
Te	$^{130}Te(n, 2n)$	^{129}Te	68.7 min	0.455	60
I	$^{127}I(n, 2n)$	^{126}I	12.8 days	0.386	80
Ba	$^{138}Ba(n, 2n)$	^{137m}Ba	2.554 min	0.662	1
Ce	$^{140}Ce(n, 2n)$	^{139m}Ce	54. sec	0.746	9
Hf	$^{180}Hf(n, 2n)$	^{179m}Hf	18.6 sec	0.217	80
Ta	$^{181}Ta(n, 2n)$	^{180m}Ta	8.15 hr	0.093	20
W	$^{186}W(n, 2n)$	^{185m}W	1.62 min	0.131	20
Pt	$^{198}Pt(n, 2n)$	^{197m}Pt	78. min	0.346	200
Hg	$^{200}Hg(n, 2n)$	^{199m}Hg	43. min	0.158	20
Pb	$^{208}Pb(n, 2n)$	^{207m}Pb	0.80 sec	0.570	100

limits of detection would all be 100,000 times higher than the reactor 10^{13} flux limits shown in Table 7.2, if a 1-hr irradiation at 10^8 flux were employed. Because of the target lifetime problem, however, it is more reasonable to consider a maximum irradiation time of about 5 min. Hence for those elements of Table 7.2 that form an (n, γ) product with a half-life of the order of one hour or longer, the 10^8-flux thermal-neutron limits are about 10^6 times those of Table 7.2. For those with half-lives of a few minutes or less the factor is 10^5.

6 SOURCES OF ERROR, ACCURACY, AND PRECISION

Some of the potential sources of error in the activation analysis method have already been mentioned in previous sections, so this discussion is short, and limited mostly to factors involved in thermal-neutron activation analysis. In general, if care is taken and if good counting statistics are obtained, absolute accuracies of the order of $\pm 2\%$ of the value are attainable in NAA work. It is very difficult, except by extreme care and measurement replication and averaging, to attain accuracies of the order of $\pm 0.1\%$. If one becomes careless, the accuracies can even be as poor as $\pm 10\%$.

Geometrical Factors

When a sample analyzed by NAA is compared with a standard sample of the element being determined, it is assumed that both were exposed to exactly the same neutron flux for the same period of time and were counted in exactly the same way. If any of these assumptions are not true to a high degree of accuracy, a source of error has been introduced. If deviations from these assumptions are known to have occurred, accurate corrections can usually be made. For example, differences in neutron flux at the various positions at which samples and standards are placed can be measured by means of flux-monitor samples placed at these locations (often, thin gold foil for thermal-NAA, thin copper foil for 14-MeV-NAA). Differences in counting efficiency between counters and for different sample-to-detector distances can be determined by calibration measurements. It is usually necessary to make the sample shape and volume exactly the same as that of the standard in order to avoid difficultly-correctable effects of flux gradients. For most work, carefully-prepared aqueous standard solutions can be used (some exceptions are noted below). Such solutions are prepared at accurately known concentrations—high enough to give very good counting statistics under the irradiation and counting conditions employed. Some such reference solutions (e.g., gold and mercury solutions) are unstable in storage and need to be made up fresh.

Thermal-Neutron Self-Shielding Errors

Since some elements have very large atomic cross sections (10 barns and higher) for the absorption of thermal neutrons, it is possible unless due care is taken to introduce errors in the thermal-NAA results caused by "self-shielding" within the sample. For the comparator method to be accurate, either (1) sample and standard must be exposed, throughout the volume of each, to the same average neutron flux, or (2) suitable corrections must be applied if this is not the case.

With aqueous standard samples of even several cubic centimeters volume, self-shielding is quite negligible, since oxygen has a very small (n, γ) cross section (<0.0002 b), and that of hydrogen is also fairly small (0.33 barn). With 1 cm^3 unknown samples self-shielding may or may not be appreciable, depending on the overall thermal-neutron capture cross section of the matrix. For a matrix that is essentially a pure element, the extent of self-shielding will depend upon the shape and the volume of the sample (increasing with increasing sample volume and symmetry), and will, for a given volume, increase with increasing density and atomic thermal-neutron capture cross section and decrease with increasing atomic weight. With low-Z, low-density, elements with low thermal-neutron capture cross section (such as Al: $\sigma(n, \gamma)$ of 0.23 b, K: $\sigma(n, \gamma)$ of 2.1 b), self-shielding is negligible even for a 1-cm^3 sample. However, for germanium, which has a modest $\sigma(n, \gamma)$, 2.4 b, but a higher density (5.35 g/cm^3), self-shielding in a 1-cm^3 sample can be as large as several percent ($\sim$5%). Some elements that can contribute to serious self-shielding, if present in samples even at fairly low levels, are lithium, boron, rhodium, silver, cadmium, indium, neodymium, samarium, europium, gadolinium, dysprosium, holmium, erbium, thulium, lutetium, hafnium, rhenium, iridium, gold, and mercury. These each have atomic thermal-neutron capture cross sections of $\geq$50 b. The one with the largest capture cross section is samarium: $\sigma(n, \gamma) = 5828$ b. All of these capture thermal neutrons by (n, γ) reaction, except lithium and boron, which exhibit large thermal-neutron cross sections for (n, α) reactions: $Li^6(n, \alpha)H^3$ and $B^{10}(n, \alpha)Li^7$. Fortunately, most of these high cross-section elements do not commonly occur in samples at levels of 0.1% or higher, and all but lithium and boron readily reveal their presence upon activation with thermal neutrons, by forming γ-emitting radionuclide products, thereby calling to the analyst's attention the possibility that self-shielding may be significant.

In order to avoid errors resulting from thermal-neutron self-shielding, when analyzing samples of appreciable thermal-neutron capture cross section, one must do one of three things: (1) use a sample much smaller than 1 cm^3, (2) prepare the standard in a matrix of the same capture cross section (also, of course, matching the shape and volume of the sample), or (3) measure the extent of self-shielding in the sample and make a correction. The third alternative can be employed by measuring the attenuation of a thermal-neutron beam caused by the sample or by placing a tiny bead or rod of a suitable detector element at the center of the sample. With careful calibrations fairly accurate corrections for self-shielding can then be made. For a sample of a given volume, self-shielding can also be reduced by spreading the sample out in a thin layer.

With fast neutrons the neutron absorption and scattering within a sample of the order of 1 cm^3 volume are quite small, since none of the cross sections are anywhere near as large as many of the thermal-neutron capture cross sections.

Gamma-Ray Self-Absorption Errors

When one employs γ-ray spectrometry for comparing the net photopeak counting rate of a particular radionuclide with that of a standard sample of the element in question, it is assumed that either the absorption and the Compton scattering of these particular γ-ray photons are negligible in sample and standard, or very closely the same in each. With low-to-medium-density samples ($d \leq 3$ g/cm^3) of about 1 cm^3 volume or less, and γ-rays of energies above about 0.3 MeV, self-absorption of the γ-ray photons within the sample is quite negligible. However, even in such samples Compton scattering within the sample can be appreciable, and its extent will depend upon the matrix. Since photons that are Compton-scattered within the sample have reduced energies, they cannot possibly contribute to the total-absorption peak observed with the detector. Hence errors are possible unless suitable steps are taken. As an example, the percentages of the photons that undergo Compton scattering within a 1-cm^3 sample are very roughly 5, 11, and 28%, in H_2O, Al, and Fe, respectively, for 0.5-MeV photons; 3, 8, and 21%, respectively, for 1-MeV photons; and 2, 4, and 12%, respectively, for 3-MeV photons. With high-density, high-Z matrices, both self-absorption and self-scattering can be quite large. To avoid errors in the results caused by these effects, one either (1) reduces the sample size to such a point that the effects are negligible, (2) prepares the standard in a matrix that matches the sample in these respects, or (3) makes suitable corrections for these effects.

Errors from Recoil and Other Effects

The corrections needed if an (n, γ) product of interest is also formed appreciably in the sample by an (n, p) fast-neutron reaction on the element one unit above it in Z, and/or by an (n, α) fast-neutron reaction on the element two units above it in Z, have already been discussed.

Recoil protons can, in a very few cases, introduce a different kind of possible error. Fast neutrons impinging upon a sample containing an appreciable amount of hydrogen can produce many "knock-on" (or "recoil") protons, which in turn can generate some induced activity, particularly via (p, n) or (p, γ) reactions. For example, in analyzing samples that contain hydrogen and either carbon or oxygen (or both) for low levels of nitrogen, via the $^{14}N(n, 2n)^{13}N$ reaction, one must apply a correction for the small amount of ^{13}N formed by the $^{13}C(p, n)^{13}N$ and $^{16}O(p, \alpha)^{13}N$ recoil-proton reactions. (Note: This effect can also be put to use, as has been done in the determination of ^{18}O in water or organic materials, using the $^{18}O(p, n)^{18}F$ recoil-proton reaction. A somewhat related technique is that of using recoil tritons to determine oxygen, via the $^{16}O(t, n)^{18}F$ reaction—generating the recoil tritons in place by mixing the sample with finely powdered LiF and exposing this to a high thermal-neutron flux; the recoil tritons ($\sim$2.7 MeV energy) are produced by the high thermal-neutron cross-section $^{6}Li(n, t)^{4}He$ reaction.)

Since ^{235}U is readily fissioned by thermal neutrons, and ^{238}U and ^{232}Th by fast neutrons, the various fission-product radionuclides thus formed, if the sample contains appreciable amounts of either or both of these elements, must not be confused with the (n, γ) products of the elements being determined. (*Note.* In some cases the fission products of uranium and thorium are used for their activation-analysis determination. Also, uranium and thorium can be determined via the very short-lived delayed-neutron fission products which they form—unusual radionuclides with half-lives in the range of seconds to about one minute, which decay by neutron emission. The resulting neutrons can be detected by neutron-sensitive counters, such as BF_3 or ^{3}He counters.) A more unusual possible type of error is that illustrated by the thermal-neutron determination of very low levels of phosphorus in high-silicon matrices. Here Si forms 2.62-hr ^{31}Si, via the $^{30}Si(n, \gamma)^{31}Si$ reaction. This decays by β^- emission to form stable ^{31}P. Thus prolonged irradiation will gradually form a small, but increasing, amount of phosphorus in the sample.

The Effect of Counting Statistics upon Precision

Because radioactive decay is a statistically random process, the counts obtained when one employs a detector to measure and record a fraction of the emitted particles or photons also exhibit statistical fluctuations. With low-activity samples the statistical uncertainty of the final analytical results is usually mainly caused by the effect of the counting statistics. A very brief discussion of counting statistics follows, discussing only the principal factors and cases.

If a radioactive sample is counted on some kind of detector and produces N counts during the counting period, t, the standard deviation (σ) of N is simply $\pm N^{1/2}$. The meaning of this σ is that if the same sample were counted in exactly the same way time and time again (assuming the decline in disintegration rate of the sample, due to decay, is negligible during the whole course of the measurements), 68.3% of the observed values for N would fall within the limits of the average (or "true") value, $\bar{N}$, $\pm 1\sigma$, about 15.85% would be less than $\bar{N} - 1\sigma$, and about 15.85% would be greater than $\bar{N} + 1\sigma$. It is relevant to note that counts (N) of 100, 10,000, and 1,000,000 have σ's of ± 10, ± 100, and ± 1000 counts, respectively, or % σ's of ± 10, ± 1, and ± 0.1, respectively.

Since measurements of samples, standards, and background are often carried out for different counting periods, the results of each are usually expressed as counts per unit time, for example, counts per minute (cpm). Because the counting period can usually be measured very accurately, its contribution to the statistical uncertainty of the answer is generally negligible, and the value for t is thus treated as an absolute value. Hence if σ_N is $\pm N^{1/2}$, then $\sigma_{N/t}$ is $\pm N^{1/2}/t$.

When even a pure radionuclide is counted, one automatically is acquiring counts resulting from the overall "background" radiation at the same time. Thus the observed N is really N_{s+b}; that is, $N_s + N_b$, where the subscripts, s and b, refer to sample and background, respectively. In order to obtain the desired N_s one must subtract a separately determined N_b from the observed N_{s+b}. If the background counting rate, B, is measured during a counting period, t_b, the standard deviation of the value obtained for the background counting rate, σ_B, is simply $\pm N_b^{1/2}/t_b$. If the sample is counted for a period of time, t, the standard deviation of the gross counting rate, σ_{S+B}, is $\pm N_{s+b}^{1/2}/t$, where S denotes the net sample counting rate.

When one adds two or more quantities together (in the same units), each with its own standard deviation, or subtracts one number from another, the standard deviation of the result is equal to the square root of the sum of the squares of the individual standard deviations. That is, in addition (with n being the number of terms added):

$$\sigma_{\text{sum}} = \pm \left[\sum_{1}^{n} \sigma_i^2 \right]^{1/2}, \tag{7.17}$$

and in subtraction,

$$\sigma_{\text{diff}} = \pm [\sigma_1^2 + \sigma_2^2]^{1/2}. \tag{7.18}$$

Therefore the standard deviation of the net sample counting rate, σ_S, is given by the equations,

$$\sigma_S = \pm [\sigma_{S+B}^2 + \sigma_B^2]^{1/2}, \tag{7.19}$$

or

$$\sigma_S = \pm \left[\frac{N_{s+b}}{t^2} + \frac{N_b}{t_b^2} \right]^{1/2}, \tag{7.20}$$

or

$$\sigma_S = \pm \left[\frac{S + B}{t} + \frac{B}{t_b} \right]^{1/2}. \tag{7.21}$$

For the special case where $t = t_b$,

$$\sigma_S = \pm\left[\frac{S + 2B}{t}\right]^{1/2}. \tag{7.22}$$

In γ-ray spectrometry measurements one usually determines a particular net photopeak counting rate by subtracting the counts in the underlying Compton and background continuum from the gross counts in the analyzer channels that define the peak; he then divides by the live counting time. A linear base line is usually assumed, going from the channel located at the trough of the Compton-edge dip to the left (lower-energy) side of the photopeak to an arbitrarily selected channel on the descending right (higher-energy) side of the photopeak. The area of this base is then $n(N_L + N_R)/2$, where in this case N_L and N_R are the numbers of counts in these two channels, and n is the number of analyzer channels included in the photopeak. The precision of the value obtained for the Compton and background region is determined by the σ's of these two channels, σ_{N_L} and σ_{N_R}. The standard deviation of the Compton and background region is thus $\pm n[N_L + N_R]^{1/2}/2$. The standard deviation of the gross photopeak counts, N_{gpp}, is simply $\pm N_{\mathrm{gpp}}^{1/2}$, where N_{gpp} is the sum of the observed counts in the channels included in the photopeak. The desired net photopeak counts, N_{npp}, are obtained by subtracting the base counts from the gross counts, that is,

$$N_{\mathrm{npp}} = N_{\mathrm{gpp}} - \frac{n(N_L + N_R)}{2}. \tag{7.23}$$

The standard deviation of N_{npp} is equal to the square root of the sum of the squares of the standard deviations of the two quantities:

$$\sigma_{N_{\mathrm{npp}}} = \pm\left[N_{\mathrm{gpp}} + \frac{n^2(N_L + N_R)}{4}\right]^{1/2}. \tag{7.24}$$

Then N_{npp} and $\sigma_{N\mathrm{npp}}$ can each be converted to the corresponding counting-rate values by dividing each by the live counting time.

Where the net counting rate of a sample is compared with the net counting rate of a standard (same radionuclide, corrected to the same decay time, same counting conditions), the standard deviation of the final answer will depend upon the standard deviations of the two quantities. However, in the comparator method one divides the sample net by the standard net counts per minute, and then multiplies this quotient by the micrograms of the element in the standard in order to obtain the micrograms of the element in the sample. In the multiplication of two or more quantities, each having its own standard deviation, or in dividing one quantity by another, the fractional or percentage standard deviation of the result is equal to the square root of the sum of the squares of the individual fractional or percentage standard deviations (with n being the number of terms multiplied together):

$$\%\sigma_{\text{product}} = \pm \left[\sum_{1}^{n} (\%\sigma_i)^2 \right]^{1/2}, \tag{7.25}$$

and

$$\%\sigma_{\text{quotient}} = \pm[(\%\sigma_1)^2 + (\%\sigma_2)^2]^{1/2}. \tag{7.26}$$

In many cases, of course, the $\%\sigma$ of the standard is negligible, compared with the $\%\sigma$ of the sample (if the sample is of very low activity and the standard is of high activity). It is assumed that the amount of the element present in the standard is known very accurately, so that it can be taken as an absolute value (i.e., one of negligible uncertainty).

It should be noted that the σ obtained from the counting of a sample merely means that the counting statistics indicate that the true value probably (to a 68.3% confidence level) falls within $\pm 1\sigma$ of the value found—if all other sources of error are negligible, compared with the uncertainty introduced by the counting statistics. In other words, the true uncertainty, expressed as a standard deviation, is at least that large and could be larger if other sources of variation are significant. One can multiply the σ obtained by various factors to express the result to greater confidence levels (CL); that is, 90% CL = value $\pm$ 1.645σ, 95% CL = value $\pm 1.960\sigma$, 99% CL = value $\pm 2.576\sigma$, and 99.9% CL = value $\pm 3.291\sigma$.

Where one repeats the analysis of a given sample a number of times, the uncertainty, $\sigma_{\bar{R}}$, of the mean value, $\bar{R}$—including all sources of variation (not just the counting statistics)—can be determined by taking the square root of the sum of the squares of the individual deviations from the mean value and dividing this by the square root of $(n - 1)$, where n is in this case the number of individual determinations.

$$\sigma_{\bar{R}} = \pm \left[\sum_{1}^{n} (\bar{R} - R_i)^2 \right]^{1/2} \Big/ (n - 1)^{1/2}. \tag{7.27}$$

7 APPLICATIONS OF NEUTRON ACTIVATION ANALYSIS IN CHEMISTRY

With a suitably high neutron flux, neutron activation analysis provides a method capable of quantitatively determining almost every element in the periodic system from levels ranging all the way from gross levels down to submicrogram amounts. For approximately half of these elements, even subnanogram amounts can be determined. Because of this ultra-sensitivity, the wide dynamic range, and the wide range of sample sizes that can be accommodated, the NAA method has found or is finding extensive use in almost every branch of science, industry, and medicine. In those cases in which the method can be applied in its purely instrumental form the speed of analysis and the nondestructive feature can also be of real value in its applications to various fields.

It is beyond the scope of this condensed treatment of the subject of neutron activation analysis to review the large number of applications of the method that have been made to date in various fields. In particular, the high-flux (reactor) thermal-NAA method has found many applications. Extensive use of this method is being made in such fields as chemistry (organic, inorganic, polymer), physics (very pure materials and semiconductors), metallurgy, industry (petroleum, chemical, mining, plastics, metals, rubber, solvents, paper, lumber, cement), biology, medicine, geochemistry, oceanography, archaeology, and criminalistics (scientific crime investigation). Since neutron activation analysis really determines specific stable nuclides of a polyisotopic element, forming different products from each, the method can also be used in studies of isotopic fractionation and in stable-isotope tracer studies.

One special chemical application of the NAA method should be particularly noted: the nondestructive analysis of very small samples of pure compounds, for the establishment of empirical formulas. This is an important application to chemists engaged in the synthesis of new compounds, for example, organic and inorganic compounds, and coordination compounds. Often, such syntheses result in only a few milligrams of high-purity material whose chemical formula is thought to be known, but must be verified. Milligram amounts of organic compounds containing in their molecules one or more of the elements, boron, nitrogen, oxygen, fluorine, silicon, phosphorus, sulfur, chlorine, bromine, or iodine can be analyzed nondestructively, precisely, and accurately for one or many of these elements via activation with either thermal or fast neutrons (or both), followed by γ-ray spectrometry. Similarly, milligram amounts of inorganic, metallo-organic, and coordination compounds can be analyzed nondestructively for these same elements and essentially all metallic and metalloid elements.

An extremely valuable bibliographical reference source on activation analysis and its applications is the National Bureau of Standards publication: Activation Analysis: A Bibliography [13]. Part 1 gives complete identification of some 3500 publications in this field (papers, books, reports, and so on) that have appeared up to mid-1968. All forms of activation analysis are included, but the great majority of the references listed are concerned with high-flux thermal-NAA, since this is the most generally powerful and most extensively developed and applied form of the method. In Part 2 the bibliography is excellently cross-indexed according to (1) authors, (2) elements determined, (3) matrices analyzed, and (4) techniques used. The NBS plans to publish supplements of this bibliography annually, so as to regularly update it.

For more complete discussions of many of the topics mentioned briefly in this chapter the reader is referred to the books by Bowen and Gibbons [2], Lyon [14], Lenihan and Thomson [11], Taylor [16], Koch [9], and Albert [1]. Biomedical applications are discussed in detail in the book edited by Comar [4]. Forensic (crime investigation) applications are presented in detail in the Conference Proceedings edited by Guinn [7]. The Proceedings of the 1961 and 1965 International Conferences on Modern Trends in Activation Analysis, held at Texas A&M University, and edited by Wainerdi and Gibbons, [17, 18] contain a large number of excellent papers on the subject of activation analysis and its applications. Similarly, the proceedings of the corresponding 1968 conference, held at the National Bureau of Standards, edited by DeVoe [5], is an extremely useful reference. An excellent compilation of all possible reactor thermal-neutron and fission-spectrum neutron reactions that form γ-emitting products, and their yields, has been prepared by Lukens [12]. The General Electric Chart of the Nuclides [6] is a very convenient and useful source of information. Valuable information on radiochemical-separation procedures is contained in the National Research Council Radiochemistry Monograph series [15]. The very useful Scintillation Spectrometry Gamma-Ray Spectrum Catalogue prepared by Heath [8] has already been cited, as has the invaluable Table of Iostopes prepared by Lederer, Hollander, and Perlman [10].

References

1. P. Albert, *l'Analyse par Radioactivation*, Gauthier-Villars & Cie, Paris, 1964 (in French).
2. H. J. M. Bowen and D. Gibbons, *Radioactivation Analysis*, Oxford University Press, London, 1963.
3. J. D. Buchanan, *Atompraxis*, **8,** 272, 1962 (also available as General Atomic Report GA-2662, San Diego).
4. D. Comar, Ed., *l'Analyse par Radioactivation et ses Applications aux Sciences Biologiques*, Presses Universitaires de France, Paris, 1964 (mostly in English).
5. J. R. DeVoe, Ed., *Modern Trends in Activation Analysis*, Vols 1 and 2, National Bureau of Standards Special Publication 312, Superintendent of Documents, U.S. Government Printing Office, Washington, D.C., 1969.
6. General Electric Company, *Chart of the Nuclides*, General Electric Co., San Jose, Calif.
7. V. P. Guinn, Ed., *Proc. First Intern. Conf. Forensic Activation Analysis*, General Atomic, San Diego, 1966.

8. R. L. Heath, *Scintillation Spectrometry Gamma-Ray Spectrum Catalogue*, 2nd ed., 2 vols, IDO-16880-1 and -2, Office of Technical Services, U.S. Dept. of Commerce, Washington, D.C., 1964.
9. R. C. Koch, *Activation Analysis Handbook*, Academic, New York, 1960.
10. C. M. Lederer, J. M. Hollander, and I. Perlman, Table of Isotopes, 6th Ed., Wiley, New York, 1967.
11. J. M. A. Lenihan and S. J. Thomson, Eds., *Activation Analysis: Principles and Applications*, Academic (London), 1965.
12. H. R. Lukens, *Estimated Photopeak Specific Activities in Reactor Irradiations*, General Atomic Report GA-5073, General Atomic, San Diego, 1964.
13. G. J. Lutz, R. J. Boreni, R. S. Maddock, and W. W. Meinke, Eds., *Activation Analysis: a Bibliography*, Parts 1 and 2, NBS Technical Note 467, Superintendent of Documents, U.S. Government Printing Office, Washington, D.C., 1968.
14. W. S. Lyon, Jr., Ed., *Guide to Activation Analysis*, Van Nostrand, Princeton, N.J., 1964.
15. National Research Council, *Radiochemistry Monographs* (series of monographs by different authors in the NRC Nuclear Science series, National Research Council, Washington, D.C.).
16. D. Taylor, *Neutron Irradiation and Activation Analysis*, Van Nostrand, Princeton, N. J., 1964.
17. R. E. Wainerdi and D. Gibbons, Eds., *Proc. 1961 Intern. Conf. Modern Trends in Activation Analysis*, Texas A&M University, College Station, Tex., 1961.
18. R. E. Wainerdi and D. Gibbons, Eds., *Proc. 1965 Intern. Conf. Modern Trends in Activation Analysis*, Texas A&M University, College Station, Tex., 1965.
19. H. P. Yule and V. P. Guinn, in *Radiochemical Methods of Analysis*, Vol. 2, International Atomic Energy Agency, Vienna, 1965, p. 111.

Chapter **VIII**

POSITRON ANNIHILATION

Joseph A. Merrigan, James H. Green, and Shu-Jen Tao

1 INTRODUCTION

Study of the interactions of antimatter with matter has progressed from intriguing verifications of quantum dynamic predictions to the applications of antimatter as a scientific tool. The most thoroughly characterized antiparticle is the antielectron, that is, the positron. The existence of the positron was publicly predicted by Dirac [1] in 1930. The positron was discovered [2–5] in cosmic-ray showers in 1933. Annihilation of the positron with an electron with resultant conversion of the mass of the two particles to electromagnetic energy was found soon after discovery of the positron [6–9]. Experimental studies of the positron and its interactions with matter were minimal until the late 1940s, when scintillation counters became available. Some of the first studies involved measurement of the anticolinearity of the photons emitted when the positron was annihilated [10–12]. The existence of positronium [13], a relatively stable bound state of the positron with an electron was found in 1949 [14–17]. During the 1950s considerable study was made of the fundamental characteristics of positrons and positronium. Since the recent advent of commercially available fast electronic devices that make it feasible to measure annihilation rates to a high degree of accuracy, the positron is gaining its place as a very sensitive microprobe of the chemical-physical properties of materials. The chemistry and physics of positrons and positronium have been the subject of several reviews [18–28] and two books [29, 30].

2 FUNDAMENTAL CONSIDERATIONS

The Positron

One of the most scientifically significant predictions made from Dirac's relativistic quantum mechanics of the electron [1, 31] was the existence of the positron. The relativistic wave equation for the total energy of the electron, E, has two solutions: $E = \pm(p^2c^2 + m^2c^4)^{1/2}$, where p is the momentum of the electron, m is the rest mass, and c is the velocity of light. In the absence of external fields electrons may have energies of $\pm mc^2$ to $\pm\infty$. It is assumed that all negative energy states are normally occupied by electrons. When an electron is excited to a positive energy state a "hole" is left in the midst of the occupied negative energy states, and this appears as an electron of positive charge. The positron may also be thought of as a positive electron. It has the same mass as an electron but the opposite charge [32–34].

Positrons are found in cosmic radiation and as decay products of many radioactive nuclides. One of the most well-known conversions of electromagnetic energy to mass is pair production. Pair production is conversion of a high-energy photon to a positron and an electron by interaction with the electric field of a nucleus. The energetic photon may be produced by nuclear disintegration or as a consequence of high-energy particle interactions with matter. For pair production to occur the photon must have an energy at least as great as the mass equivalent of a positron and electron, $2mc^2$. This energy is 1.02 MeV. Any excess energy is carried by the positron and the electron. Positrons born by nuclear decay generally have an energy below 2 MeV, while those produced by pair production may have energies in the hundreds of megaelectron volts, depending upon the source of high-energy photons or particles. The stopping power of a material for an energetic positron is similar to that for a negatron. However, because of the opposite electrical charge the positron may have a range slightly greater than a β particle of the same energy [35, 36].

Conversely to light-energy conversion to matter, a positron may annihilate with an electron and produce electromagnetic radiation. When a positron collides with an electron, the electron has a high propensity for making a radiative transition to the unoccupied negative energy level. This transition results predominantly [37–39] in the conversion of the mass of the electron and the positron to electromagnetic radiation of total energy $E_\gamma = 2mc^2 + E_+ + E_-$, where m is the rest mass of the electron or positron, c is the velocity of light, and E_+ and E_- are the kinetic energies of the positron and the electron, respectively. The electromagnetic energy may consist of one photon if there is sufficient interaction of the positron with closely bound electrons (mostly K-shell) [40, 41], but two quantum emission is by far the most probable process. Because of conservation-of-momentum requirements, the two photons are emitted in nearly opposite directions. Each photon has an energy of 0.51 MeV (mc^2). Three-photon annihilation also occurs but to a much lesser extent than the two-photon process. The three photons have a summed energy of 1.02 MeV with the energy shared in accordance with their relative directions. The energy spectrum of photons for three- and two-quanta annihilation is shown in Fig. 8.1 [42].

Distinct selection rules govern the number of quanta emitted in a collision of a free positron with an electron [43]. If the particles meet with their spins antiparallel, that is, a singlet ^1S interaction, an even number of annihilation photons is allowed. Because the probability of multiple-photon annihilation decreases greatly with increasing multiples, two-photon emission is always observed. Two-photon emission is forbidden for collisions where the particles meet with their spins parallel, that is, triplet ^3S interactions. In this case three annihilation quanta are emitted. The relative number of two- and three-photon annihilations depends on the triplet and singlet interactions and the rate of annihilation from each state. The singlet state has total angular momentum $J = 0$; hence the z-component of angular momentum, $\mathbf{m}$, is 0.

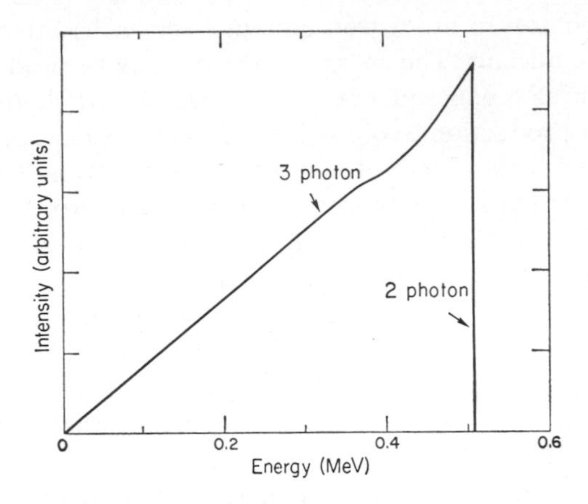

Fig. 8.1 Energy spectrum of photons for two and three quanta annihilation of positrons. Reproduced by permission of the American Institute of Physics, from Ref. 42.

The triplet state has $J = 1$ and $\mathbf{m} = 0, \pm 1$. Thus the statistical ratio of singlet to triplet interactions is 1:3. The ratio of the singlet to triplet annihilation rates [29,42] is approximately 1115:1. Combining these factors the ratio of probabilities of two-photon (singlet) to three-photon (triplet) annihilations resulting from *free* positron interactions with electrons is 1115:3 or 372:1.

A free positron may interact on the average with more or less than one electron depending on the electron density of the medium. The free annihilation life-time is generally between 1 and 5×10^{-10} sec in condensed phases. For the hypothetical case of unpolarized electrons and positrons annihilating with nonrelativistic velocities, Dirac [1] calculated the cross section for two photon production: $\sigma_s = \pi r_0{}^2 c/v$, where r_0 is the classic electron radius (e^2/mc^2), c is the velocity of light, and v is the relative velocity of the positron and the electron. The rate of free annihilation depends on σ_s and the number of electrons in the area. If n is the number of electrons per cm^3, the rate of annihilation, λ_s, is given by $\lambda_s = \sigma_s nv = \pi r_0{}^2 cn = 4.5 \times 10^9\ dZ/A\ \text{sec}^{-1}$, where d, Z, and A are the density, the atomic number, and the atomic weight of the medium, respectively. Thus free annihilation of positrons is a function of electron density.

Because of the overlapping effects of the coulombic attraction of electrons and positrons, nuclear repulsion, the Pauli exclusion principle and other field and space effects [20, 21] it is difficult to calculate the effect of electron density. Equations involving the radius of the unit electron sphere [44, 45], r_s, have been found that describe the free annihilation rates in metals relatively well. This parameter is the radius, in units of Bohr radius, of the sphere whose volume is that occupied by each valence electron in the metal. It is proportional to the cube root of the inverse of the density of valence electrons: $r_s = 1.384 (A/dZ)^{1/3}$. One such equation [20] is given by $\lambda = 12(r_s^{-3/2} + 0.659r_s^{-3/4})^2 \times 10^9$ sec^{-1}. Obviously, the greater the electron density the greater the annihilation rate, hence the shorter the lifetime* of the free positron.

Besides undergoing free annihilation, positrons may form a relatively stable "atom" by uniting with an electron. This bound system, called positronium, may have a lifetime as long as 1.4×10^{-7} sec. Thus not all positrons have a characteristic free annihilation lifetime. This may result in more than one positron annihilation rate in a given material, typically a free and one or more bound-state annihilation rates are observed. Bound states of positrons are discussed in the next section.

Positronium

Characteristics

Positronium is a metastable combination of a positron and an electron. If a high-energy positron falls within a certain energy boundary in its thermalization process (around 10 eV), it may strip an electron from the environment and form a bound electron-positron pair, positronium. One can envision this entity as an electron-positron pair rotating around their center of mass, their electrostatic and centrifugal forces balanced. In this circumstance the positron and the electron lose their identity, and the pair is a very light, neutral-free radical. Although positronium (Ps) certainly has unique physical and chemical properties, several analogies can be drawn between it and hydrogen. These analogies are convenient for becoming acquainted with Ps on terms that are familiar to most scientists.

Positronium exists in the singlet and triplet states with particle spins antiparallel and parallel, respectively. The singlet state is called *para* positronium (p-Ps) and has a lifetime $1.25n^3 \times 10^{-10}$ sec in the absence of external effects. The triplet state is *ortho* positronium (o-Ps) with a lifetime of $1.4n^3 \times 10^{-7}$ sec when only self-annihilation occurs. Since Ps is born in the ^1S state or de-excites from the ^3S to the ^1S state before annihilation, the lifetimes are observed as when $n = 1$, where n is the total quantum number. The reduced mass of Ps is one-half the mass of an electron, $m/2$, while that of the hydrogen atom is m. Hence in quantum-mechanical comparisons of hydrogen and Ps atoms the factor of $\frac{1}{2}$ or 2 is repeatedly encountered. The gross structure of Ps states is like that of hydrogen except for the different reduced mass effects. The ionization potential is 0.5×13.6 eV $= 6.8$ eV. The binding energy is the same as the ionization potential in this two particle system. The energy

* Lifetime will refer to average lifetime for convenience.

levels of Ps are one-half those in hydrogen, and the Lyman α line has a wavelength of 2430 Å. The intraparticle distance is twice the Bohr radius in hydrogen, 1.06 Å.

Because of the statistical nature of Ps formation three times more *o*-Ps ($J = 1$; $\mathbf{m} = 0, \pm 1$) is formed than *p*-Ps ($J = 0$; $\mathbf{m} = 0$). The positron in Ps experiences electron effects predominantly from its partner, which has a specific spin orientation. If all of the positrons born in a system formed Ps which only underwent self-annihilation, 75% would be triplet and would decay by three-photon emission. The other 25% would be singlet and would emit two annihilation photons. The ratio of the two-photon to the three-photon annihilations would be 0.33. This is very different from the ratio of 372 for annihilation of positrons from the unbound state; hence one criterion for determining if Ps is formed is by measuring the $2\gamma/3\gamma$ ratio. If it is below 372 some Ps is formed. Measurement of a 1.4×10^{-7} lifetime for positron decay that is independent of the density of electrons in the environment (pressure in gases) also indicates formation of Ps. A third method of indicating Ps formation is the measurement of the energy spectrum for the annihilation photons. As shown in Fig. 8.1, photons of less than 0.51 MeV are emitted when three photon (triplet) annihilation occurs. Thus if more 0.20- to 0.40-MeV photons and fewer 0.51-MeV photons are observed relative to the free positron annihilation spectrum, Ps is present. In reality, positrons annihilate from both the free and bound states (Ps) and the resultant characteristic spectra are mixtures that are separated into the respective components.

The fine structure of Ps has been studied thoroughly [46–49]. The fine structure constant, α, has been calculated and measured to an agreement of five significant figures [50, 51]. The second-order Zeeman splitting of the ground state of positronium is illustrated [50] in Fig. 8.2. The sublevels $\mathbf{m} = \pm 1$ of the triplet state ($J = 1$) are unaffected by a magnetic field. However, the sublevel $\mathbf{m} = 0$ can mix with the singlet state where $J = 0$ and $\mathbf{m} = 0$. This gives rise to a magnetic quenching of the triplet state; that is, if a magnetic field is applied to triplet Ps, as much as one-third of it can be mixed with the singlet state and decay by two quantum annihilation with a shorter lifetime.

Because of its unique structure, Ps has been one of the favorite systems for theoretical study. As early as 1934 the electron-positron pair and isomers of it were suggested to be present in stellar atmospheres and inside stars [52]. The most probable excited levels of Ps have been predicted to radiate to the ground level before annihilation [25, 53]. The stability of several polyelectron systems has been calculated [54, 55]; for example, $e^+e^-e^+e^-$ is stable [56] by 0.95 eV and $e^+e^-e^+$ by about 0.2 eV. The positronium negative ion, $e^-e^+e^-$, has a binding energy of 0.3 eV, and the three-photon annihilation should have a

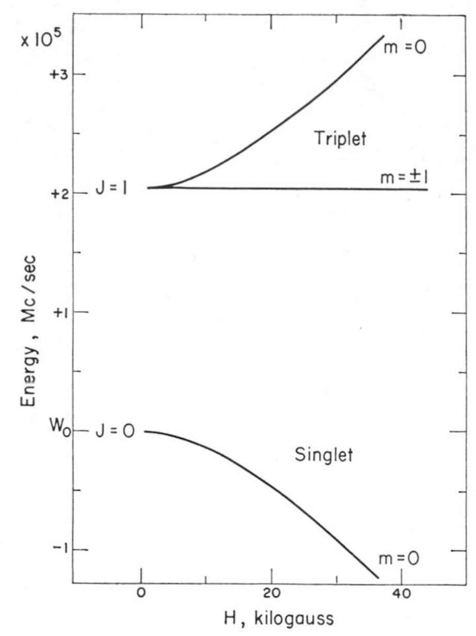

Fig. 8.2 Positronium levels in a magnetic field. Reproduced by permission of The American Institute of Physics, from Ref. 50.

lifetime [57] of 5.66×10^{-7} sec. However, formation of such systems is highly improbable in most materials. Striking hypotheses have been made that the relativistic state of Ps (where the energy states are $E_n \simeq 137nmc^2$) might be a fundamental particle in the makeup of all matter [58, 59].

Formation

In a typical laboratory situation a radioactive nuclide is used as a source of positrons. The positron is born when the radioactive nuclide decays to a nuclide of lower energy. The positron is emitted into the environment of the source with all or part of the energy involved in the nuclear decay. This energy may be a few hundred kiloelectron volts. To form Ps the positron must possess a certain threshold energy but not so much that the Ps will disintegrate upon its next collision. The positron must have enough energy to extract an electron from its immediate environment, but energy is regained from the binding of the electron. The threshold energy that a positron must possess in gases is the ionization potential (expressed in electron volts) minus the Ps binding energy; that is, $I_A - 6.8$ eV. If the positron had a higher energy than I_A and if Ps formation occurred, the Ps would be born with an energy greater than its binding energy and would break up. Thus the energy range in which Ps formation may occur is I_A to $(I_A - 6.8)$ eV. If an excitation level, E, of

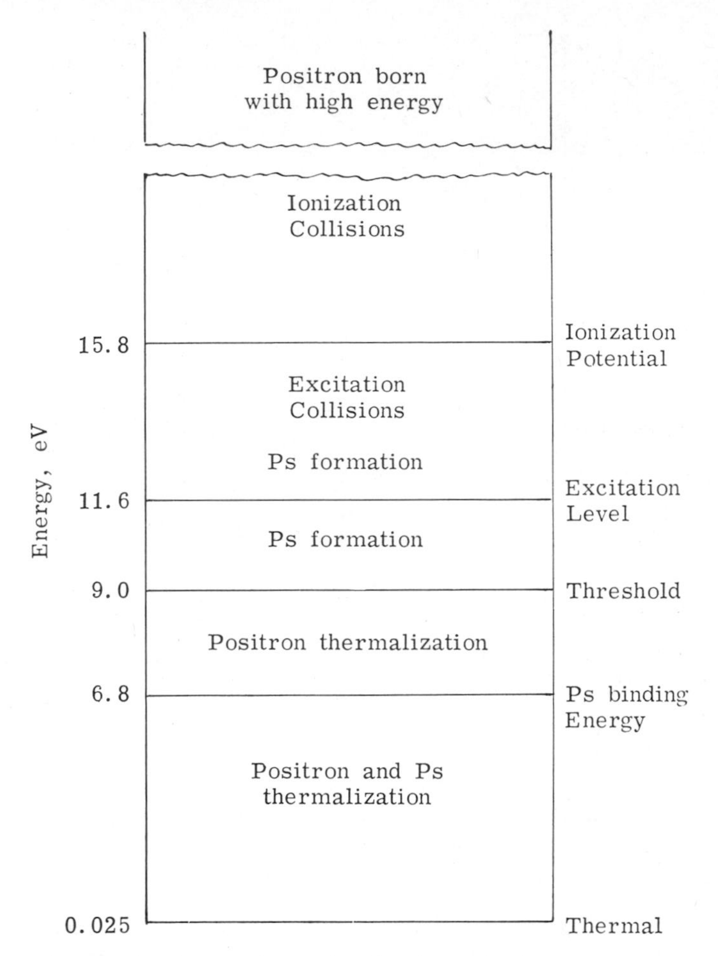

Fig. 8.3 Ore diagram for positronium formation in gaseous argon.

the gas exists in this range, excitation collisions compete with Ps formation between I_A and E. If we assume that a high-energy positron has equal probability of deexciting to any given level below I_A, the fraction of positrons that may form Ps should be between $[I_A - (I_A - 6.8)]/I_A$ and $[E - (I_A - 6.8)]/E$. This analysis is attributable to Ore [60]; an "Ore diagram" of Ps formation in argon gas is illustrated in Fig. 8.3. The maximum *intensity* of Ps, that is, the fraction of positrons forming Ps, is $(15.8 - 9.0)/15.8 = 0.43$. The minimum intensity is $(11.6 - 9.0)/11.6 = 0.23$.

This method of calculating an expected intensity of Ps is a good starting point but does not always agree with experimental findings. Perhaps it is because of insufficient knowledge of the excitation levels in the medium or an improper assumption of a uniform positron energy distribution in the region of formation. However, some success has been found by using this approach in gases [25, 61], and it has been modified to explain Ps formation in solids [20] and liquids [62, 63]. In an ionic crystal the ionization potential and the threshold energy are considered as in the gaseous model, but the excitation level is the bottom of the conduction band. The effect of the dielectric constant of the substance may also affect the threshold energy by its effect on the ionization potential of Ps (binding energy). If Ps is considered somewhat analogous to an exciton, the higher the dielectric constant the weaker would be the bonding between the positron and an electron.

The ionization potentials, the ion association, and the solvation effects of liquids are not always known accurately. It is very difficult to predict Ps intensities under these circumstances. In general, Ps formation in liquids depends on the ionization potential and the threshold energy as above, but the rates of Ps formation in a collision and the positron slowing-down rate in the Ps formation energy range are important [64]. The intensity of Ps is given by $I = (6.8/I_A)[\lambda_p/(\lambda_p + \lambda_s)]$, where λ_p is the average Ps formation rate and λ_s is the average slowing-down rate for removal of the positron from the Ps formation range. The maximum formation fraction is $6.8/I_A$ and would be higher the lower the ionization potential of the liquid. Chemical reactions may oxidize Ps to Ps^+, that is, the positron, and cause an experimental indication that there is a lower-than-predicted intensity of Ps. These reactions are dependent on the ionization potentials and the bond strengths of the liquid [63, 64].

Obviously, Ps formation does not occur simultaneously with positron emission, but only after the positron has slowed to an energy in the vicinity of 10 eV. The positron thermalizes according to the electron density in the surroundings. In metals it reaches thermal energies, 0.025 eV, in approximately 10^{-11} sec [65, 66]. In ionic or amorphous solids and in liquids the thermalization time is about 10^{-10} sec [67]. In gases, because of the much lower density, it may require 10^{-7} sec to thermalize [68]. Positronium is born with up to 6.8-eV energy and thermalizes at about the same rate as a positron. Thus in condensed phases Ps spends most of its lifetime at thermal energies, while in the gas phase it may be an epithermal "hot atom" for most of its existence.

Quenching and Enhancement

Quenching refers to any means by which the Ps lifetime is shortened from its self-annihilation lifetime. Quenching occurs in all practical systems. Hence the *o*-Ps lifetime, which is 1.4×10^{-7} sec in free space, may be as short as a few tenths of a nanosecond in condensed phases. The most noticeable quenching is of *o*-Ps because of its long lifetime relative to *p*-Ps. Indeed, quenching of *o*-Ps by conversion to *p*-Ps is often cited. There are three main types of quenching: conversion, pick-off, and chemical reaction.

Conversion quenching is the process of shortening the normal lifetime of *o*-Ps by a spin reversal of the electron relative to the positron. If *o*-Ps collides with an atom or a molecule containing one or more unpaired electrons it may be converted to *p*-Ps. Of course *p*-Ps may be converted to *o*-Ps, but the annihilation rate of *p*-Ps is much greater than that of *o*-Ps, so relatively few *para*-to-*ortho* conversions can occur. Conversion may be a result of a spin-flip process [69], that is, electron exchange, but the only essential requirement is that the collision involve a paramagnetic system [70]. The effect of conversion quenching is to yield an *o*-Ps lifetime less than 140 nsec, the minimum being 0.5 nsec [71] because of statistical averaging of *para*-to-*ortho* and *ortho*-to-*para* conversions. Typical molecules that cause conversion quenching are NO, NO_2, and O_2. Most of the transition-metal cations and complexes have unpaired electrons and cause conversion.

Pick-off is the annihilation of the bound positron with an "outside" electron during a collision [72]. When the wave function of the positron in Ps overlaps sufficiently with an electron in the surrounding medium, the pair can annihilate. In *o*-Ps the positron is bound to an electron with parallel spin; however, in a collision with another atom or molecule the positron may interact with an electron of antiparallel spin and annihilate during the collision with a rate approaching that of *p*-Ps. Hence the positron in positronium "picks off" an electron from the surroundings. This phenomenon reduces the lifetime of *o*-Ps from 140 to a few nanoseconds or less in the condensed phase, depending upon the molar density of the medium. The *o*-Ps lifetime is generally shorter the higher is the molar density. Pick-off quenching becomes less prevalent as the free volume in a system increases; thus temperature and phase changes that result in a higher free volume will result in a lower rate of pick-off and a relatively longer *o*-Ps lifetime [73]. Pick-off quenching occurs to some extent in all substances, but it is most prevalent in the solids and the liquids that disallow conversion quenching because of few unpaired electrons.

Chemical reaction of Ps may oxidize it to the free positron, bring the Ps into close proximity to electrons by bond formation, or reduce it to Ps^- ($e^-e^+e^-$), which is stable by only 0.3 eV. This causes the lifetime of *o*-Ps to be shortened toward the free annihilation lifetime. There are four types of chemical reactions of Ps [64]: (1) oxidation by electron transfer; (2) oxidation by compound formation; (3) oxidation by double decomposition; and (4) reduction. The energetic conditions required for these reactions are discussed in Section 6.

Perhaps it is appropriate at this point to summarize the interactions of positrons and positronium in a common environment. A schematic diagram of the various possible fates of a positron in water [24, 74, 75] is presented in

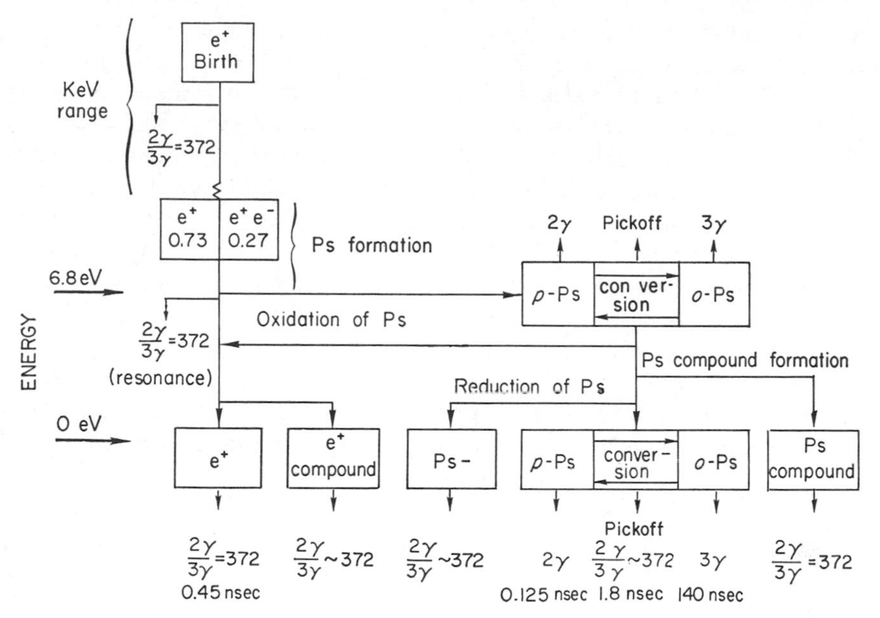

Fig. 8.4 Positron and positronium decay in water.

Fig. 8.4. The high-energy positron may annihilate from the free state while above thermal energies, at thermal energies, or in possible chemical combination. The free annihilation lifetime is 4.5×10^{-10} sec. About 27% of the positrons form positronium. Three-fourths of this is in the triplet state (*o*-Ps) and the other one-fourth is in the singlet state (*p*-Ps). Pick-off may occur from Ps formation energies down to thermal. This results in a 1.8×10^{-9} sec lifetime of *o*-Ps and a lifetime of less than 1.25×10^{-10} sec for *p*-Ps. Self-annihilation of *p*-Ps and *o*-Ps yields 1.25×10^{-10} sec and 140×10^{-9} sec lifetimes, respectively. If paramagnetic species are present, conversion quenching causes the averaging of *p*- and *o*-Ps annihilation rates to yield a lifetime of 0.5×10^{-9} sec. If strong enough oxidizing or reducing agents or other solutes that may form positronium compounds are present, the lifetime of *o*-Ps is shortened to near that of free annihilation; 4.5×10^{-10} sec. The lifetime of the positron in *p*-Ps is possibly lengthened slightly by oxidation, but in no case should it be longer than the free-annihilation lifetime. The ratio of two-photon to three-photon annihilations is about 372 in all cases except self-annihilation of *p*-Ps and *o*-Ps, where the annihilation is essentially 100% by two-photon and three-photon emission, respectively.

Besides the main types of Ps quenching that involve the chemical and physical nature of the environment, outside magnetic forces can be applied that cause magnetic quenching [76]. By the presence of a magnetic field the

m = 0 state of triplet Ps (Fig. 8.2) is mixed during part of its lifetime with the **m** = 0 state of singlet Ps. This causes an equality between the singlet and one-third of the triplet states. Hence one-third of the *o*-Ps can decay by two photon emission at the same rate as *p*-Ps with the magnetic field absorbing the unbalanced angular momentum. A radiofrequency signal corresponding to the magnetic substate splitting in the magnetic field enhances this type of quenching [50]. When full magnetic quenching occurs, the number of three photon annihilations arising from triplet Ps decay is diminished by one-third, and the relative amount of long lived *o*-Ps is decreased by the same fraction.

Another type of outside force that may be applied to a system to affect Ps formation and decay is an electrostatic field. With the aid of an electric field it is possible to accelerate positrons with low kinetic energy up to the Ps-formation energy range (Ore gap). Thus positrons that did not fall into the Ore gap in their thermalization process may still form Ps. This results in an abnormally high fraction of positrons forming Ps [50]. In gases such as A, Ne, N_2, and O_2 nearly 80% formation of Ps can be produced [77–79]. In more complex molecules such as C_2H_6, CH_4, CO_2, and freon electric fields have little effect, presumably because of energy transfer away from the accelerating positron by excitation collisions with the gaseous molecules. Although it is conceptually easy to visualize a Ps intensity enhancement by an electric field, there are cases in condensed phases in which the opposite occurs [80]. The electric field may cause quenching of Ps [81, 20]. In this case it is believed [82] that the field induces diffusion out of the Ore gap, which decreases Ps formation. This effect is more important at small than at large fields.

3 EXPERIMENTAL TECHNIQUES

Positron Sources and General Consideration

The most commonly used source of positrons is sodium-22. This radioactive nuclide ($t_{1/2}$ 2.6 yr) can be obtained from most radioisotope suppliers as sodium chloride in carrier-free aqueous solution. The predominant decay of sodium-22 to neon-22 produces a positron of 542-keV maximum energy, a 1.28-MeV γ-ray, and a neutrino. Because of energy sharing between the positron and the neutrino, the average positron is born with a kinetic energy of 120 keV [83]. The γ-ray is emitted within 10^{-12} sec of the positron. Experimentally, it is treated as though it were emitted simultaneously. Advantageous characteristics of sodium-22 include (1) its long half-life, which allows for a relatively constant strength of the source over a prolonged period, (2) the simultaneous emission of a γ-ray and a positron, which is important in measuring the lifetime of the positron, and (3) the separation in energy of the γ-ray (1.28 MeV) and the annihilation photons ($\leq$0.51 MeV), which provides for easy discrimination between them. Copper-64 is often used as a source of positrons in studies of the angular correlation between annihilation photons. It is also used in studies of annihilation spectra where the Compton scatter from the sodium-22, 1.28-MeV γ-ray would interfere. Copper-64 emits a 1.34-MeV γ-ray, but only one for every 20 positrons. It has a short half-life (12.9 hr) and can be obtained as a thin copper disk that may be irradiated with thermal neutrons to replenish the source.

A typical sodium-22 source is prepared by evaporating a drop of the carrier-free solution onto aluminum or Mica about 10 μ thick. The holding material is folded or sealed in a leakproof "sandwich," which can be inserted into the systems under study. The Mica or aluminum holder should be as thin as structurally possible to ensure that a maximum fraction of positrons interact with the sample rather than the holder. On the other hand, the holders should be strong enough to withstand handling and any corrosive action of the sample or the sodium-22 chloride. Typically 10 to 50% of the positrons are stopped by the holder. Since no long-lived Ps is formed in these materials, measurement of positron lifetimes in the sample is not hampered. The observed intensity of Ps (percent of positrons that form Ps) in a sample must be corrected for the fraction of positrons that decay in the holder. This may be accomplished by depositing carrier-free sodium-22 between two pieces of a suitable solid sample, measuring the intensity of long-lived Ps and comparing it with the intensity measured when the source in the holder is placed between two identical pieces of the sample. This comparison yields the fraction of positrons annihilating in the source holder, and all measured intensities may be corrected appropriately.

Ultrathin film materials such as Mylar may be used to encase sodium-22 when the sample yields a high intensity of long-lived Ps. These materials should not be used when studying the lifetime spectra in materials that yield a very low intensity of fairly short-lived Ps. Most organic materials used as thin films yield a Ps lifetime of 1 to 3 nsec and perhaps 30% intensity. If 10% of the positrons annihilated in the holder, a 3% intensity of Ps would result from the holder. This may interfere with proper data reduction.

The ideal source activity is dictated by the type of investigation and apparatus used. In angular correlation measurements the source may be in the millicurie to curie region if localized radiation damage is not appreciable. In lifetime measurements the radioactivity is a function of the maximum-allowable random coincidences between 1.28-MeV γ-rays and 0.51-MeV annihilation photons arising from a separate nuclear decay event (background that may interfere with data reduction). The appropriate source activity can be calculated from the equation:

$$\text{activity} = \frac{\text{random rate}}{\text{time range studied} \times \text{real rate.}}$$

If the time range is 100 nsec and the maximum desired ratio of random to real events is 0.01, a source activity of 10^5 counts/sec may be used, that is, 2.7 μCi. If the source activity were increased, the influence of random events would increase. Hence license-free quantities of sodium-22 are enough for most studies.

The macroscopic irradiation damage to a sample caused by the positron source is negligible in most lifetime measurement cases. In general, the lifetime of Ps will not be affected by the small amounts of radiation products that build up during the collection of data. Prolonged exposure or a very high source activity may have a significant effect owing to production of free radicals or other products that quench Ps [84]. The localized radiation-induced reactions that occur in the vicinity of the slowing-down positron should be considered with regard to interpretation of results in many systems. Because of the deposition of considerable energy in a localized region over a very short time, local concentrations of free electrons, ions, and radicals around the positron may contribute to the lifetime or intensity of positronium formed. This effect may be most evident in semiconductors and photoconductive materials in which electrons that are excited to the conduction band by the energetic positron have a lifetime comparable to or longer than that of Ps.

The thickness of material under study should be chosen to allow absorption of all positrons that escape the source holder. The maximum penetration of positrons in a sample can be calculated in the same manner as for negative β-particles. However, because of its opposite charge, the positron has a range in some materials up to 80% greater than an electron of the same energy. The stopping power for electrons in metals is about 25% greater than for positrons. In liquids such as toluene, benzene, and water it is about 75% greater [85]. A thickness of 200 mg/cm^2 is adequate for stopping positrons emitted from sodium-22 in condensed materials. The maximum range in gases such as argon and nitrogen is about 150 cm atm [86].

The physical nature of a sample and the way it is prepared may affect experimental results. The lifetime of Ps in certain solid powders can be altered by changing the particle size. This results from diffusion of Ps to the surface where it annihilates with a rate dependent upon the gaseous environment [87]. Hence if there is a significant internal (defects and dislocations) or external surface, annihilation of positrons may not be a function of the solid lattice alone. This can be advantageous in some cases in which the nature of defects or their concentrations is under study. Extreme care must be taken to ensure purity in samples. The presence of a foreign substance can affect the experimental results, particularly if it is paramagnetic in a diamagnetic sample. Very small amounts of impurities perhaps included during sample preparation can cause large variations in results. Nonreproducible heat treatments or formation of oxide layers on metal surfaces may also cause nonreproducible results.

Positronium is extremely sensitive to unpaired spin density and is quenched by paramagnetic materials. Oxygen is a very effective Ps quencher in liquids [88, 89]. Hence degassing is necessary to obtain unambiguous data. Degassing by repeatedly freezing the sample, evacuating, and thawing is generally sufficient. Oxygen is an effective quencher in gases also [90], which makes it necessary to obtain gases of high purity for gaseous studies. Although oxygen has been stressed as one of the main offenders, caution must be exercised to remove all unwanted paramagnetic species, highly oxidative agents, radicals, and radical-scavenging materials from all solid, liquid, and gaseous samples.

Energy Spectra of Annihilation Photons

Observation of Positronium

Some of the earliest research on the positronium atom was performed by studying the annihilation spectra of positrons in a gas as the pressure was varied and a quenching gas such as NO or NO_2 was added [14–16]. The annihilation photons are detected on a sodium iodide scintillation crystal that is optically coupled to a photomultiplier. The scintillation and the photomultiplier output are proportional to the energy deposited in the crystal by the photons. The annihilation photons may deposit all of their energy by photoelectric absorption, or a part of it by Compton scattering. Thus, the electrical pulse height is an indication of the energy of the photons. The scintillation detector is connected to a multichannel pulse-height analyzer, which stores the number of pulses versus the pulse height. A single-channel analyzer and a scaler may be used in lieu of the multichannel analyzer, but it requires more time and more frequent standardizations. Several million annihilation photons are detected, and the resultant data are displayed as the counting rate versus energy of the detected event.

As illustrated in Fig. 8.1, the three-quanta annihilation of *o*-Ps yields a continuous-energy spectrum of photons (γ-rays) up to 0.51 MeV. Decay of *p*-Ps produces two 0.51-MeV γ-rays, and the energy spectrum is a sharp line. This difference in the energy spectra arising from two- and three-quanta annihilation is used to determine the relative amount of *o*-Ps in a sample. Since free annihilation yields 99.7% (372/373) and *p*-Ps yields essentially 100% two quanta annihilation, the detection of a relatively large number of γ-rays of less than 0.51 MeV indicates that triplet Ps is present and undergoing three-quanta annihilation.

The relationship of energy spectra of annihilation photons with and without significant triplet Ps decay is shown in Fig. 8.5. The sharp peak at 0.51 MeV is produced by γ-rays undergoing photoelectric absorption. The plateau at lower energies is due to Compton scatter. The dashed curve may represent positron decay in a gas mixture such as argon and a small amount of NO, which converts *o*-Ps to *p*-Ps. Hence the spectrum is from about 100% two-quanta annihilation. The solid line represents the spectrum in the same gas but without the quencher. Some *o*-Ps decays by three-photon annihilation causing more counts in the low-energy region at the expense of the peak value. The "peak-to-valley" ratio is less the more *o*-Ps present. The count

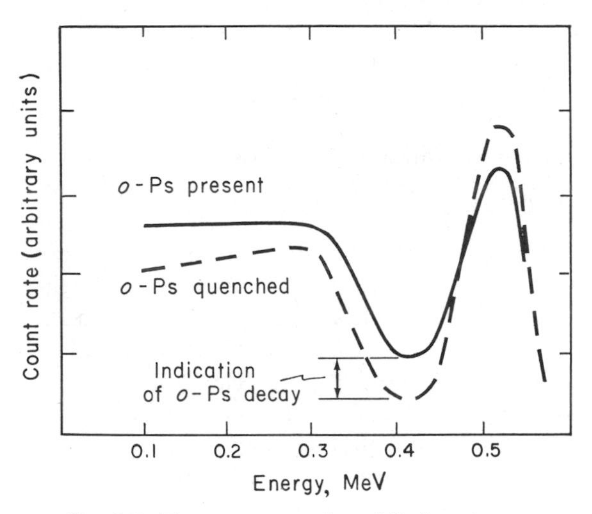

Fig. 8.5 Energy spectra of annihilation photons.

rate in the valley region from about 350 to 450 keV as compared with the peak count rate at 511 keV gives a measure of the relative importance of annihilations from triplet and singlet states. This method has been used in gas studies [91, 92] where *o*-Ps is prevalent and in solids [93] where "positronium-like" systems yield a significant fraction of three-quanta triplet annihilation.

In practice, the annihilation spectrum of positrons decaying in a condensed sample is compared with the spectrum obtained in a reference material. Aluminum is a common reference material because no significant fraction of long-lived Ps is formed in it. The spectrum obtained in aluminum is assumed to be caused by free annihilation, where 99.7% of the annihilations are by the emission of 0.51-MeV γ-rays. The ratio of the fraction of three-quanta events $\mathbf{F}_3$, to the fraction of two-quanta events $\mathbf{F}_2$, is $\frac{1}{372}$. The fraction $\mathbf{F}_3$ in a material under investigation can be derived from the equation

$$\frac{R\mathbf{v}}{{}^0R\mathbf{v}} = \frac{1 + (\mathbf{F}_3/\mathbf{F}_2)(\epsilon_3/\epsilon_2)(R\mathbf{p}/R\mathbf{v}^*)}{1 + {}^0(\mathbf{F}_3/\mathbf{F}_2)(\epsilon_3/\epsilon_2)(R\mathbf{p}/R\mathbf{v}^*)},$$

where $R\mathbf{v}$ and $R\mathbf{v}^*$ are the count rates in the valley region with and without three-quanta annihilation, respectively; $R\mathbf{p}$ is the peak count rate at 511 keV; ϵ_3 and ϵ_2 are the probabilities for detection of the three- and two-quanta annihilation events, respectively; and the superscript "0" refers to the reference material aluminum. The ratio $R\mathbf{p}/R\mathbf{v}^*$ is obtained from aluminum, ϵ_3/ϵ_2 is calculated from the spectral shape for three-photon annihilation (Fig. 8.1), and the detection efficiency of the scintillation detector in the peak and valley regions as determined by using γ-ray standards with energies in these regions. By the above assumption for aluminum, $^0(\mathbf{F}_3/\mathbf{F}_2)$ is 1/372. Thus by measuring $R\mathbf{v}$ in the sample and $^0R\mathbf{v}$ in the reference aluminum the ratio $\mathbf{F}_3/\mathbf{F}_2$ in the sample is obtained. From this the fraction of three-quanta events can be calculated.

Annihilation spectra are convenient for studying quenching rates and intensities of Ps in gases [92]. If $R_v{}^p$ is the count rate in the valley region at pressure p and $R_v{}^0$ is the rate when the gas contains a complete quencher, $R_v' = (R_v{}^p - R_v{}^0)$ is a measure of the intensity of o-Ps decaying by three photon emission in the unquenched gas. If N positrons are emitted into the unquenched gas per second, g forms Ps, $\frac{3}{4}gN$ forms o-Ps, and a fraction $\mathbf{n}$ of these is detected in the valley region, the detection rate would be $\frac{3}{4}gN\mathbf{n}$/sec. When there is quenching at a rate Λ_q, the fraction of o-Ps annihilating by three-photon emission is $\lambda_0/(\lambda_0 + \Lambda_q)$, and the count rate is $R_v' = \frac{3}{4}gN\mathbf{n}\,\lambda_0/(\lambda_0 + \Lambda_q)$. If Λ_q is a function of pressure, that is, $\Lambda_q = \lambda_q p$, then $R_v' = \mathbf{n}A(1 + p\lambda_q/\lambda_0)^{-1}$, where $A = \frac{3}{4}gN$, which can be held constant. Thus

$$\frac{1}{R_v'} = \frac{(\mathbf{n}A)^{-1}p\lambda_q}{\lambda_0} + (\mathbf{n}A)^{-1}.$$

If the detector is placed about 2 ft from the source, $\mathbf{n}$ will be independent of p, and a graph of $1/R_v'$ versus p will be linear with an intercept—λ_0/λ_q on the p-axis and $1/\mathbf{n}A$ on the $1/R_v'$-axis. In this manner the quenching constant λ_q and g (from $1/\mathbf{n}A$) can be obtained. A reference gas where g is known can be helpful in obtaining $\mathbf{n}$.

A single plot of the valley-to-peak ratio versus concentration of quencher can yield valuable information about relative effectiveness of quenchers. An analogous plot can be used to determine the effect of magnetic or electric fields on formation or quenching of Ps. The main use of the energy spectra of annihilation photons is in determining the intensity of Ps formation and the changes in it because of a change in a controlled parameter.

Measurement of Electron-Positron Pair Momentum

The high resolution of semiconductor detectors, particularly lithium-drifted germanium, has made it possible to measure the Doppler broadening of the annihilation photoelectric peak caused by the distribution of electrons in the environment. Because the photoelectric peak at 511 keV may have a full width at half maximum (FWHM) as low as 1.55 keV for monoenergetic γ-rays, the spreading of the annihilation photon peak caused by the motion of the electron-positron pair with respect to the detector can be observed [94]. If the annihilating pair is moving with a momentum $\mathbf{p}$ and if one of the two quanta is emitted in the direction of $\mathbf{p}$ and is detected, the detector receives a photon of energy $h\nu$. If the momentum were in the opposite direction, the detected photon would have energy $h\nu'$. Conservation of energy and momentum requires that $2mc^2 = h\nu + h\nu'$, and

$$\mathbf{p} = \frac{h\nu}{c} - \frac{h\nu'}{c},$$

where m is the rest mass of the electron and c is the velocity of light. The pair momentum can be expressed in terms of the rest mass energy $h\nu_0$ (511 keV), which is observed when $\mathbf{p} = 0$; $\mathbf{p} = 2(h\nu - h\nu_0)/c$.

If the positron is thermalized, the energy distribution of annihilation photons is the same as the momentum distribution of electrons in the direction of photon emission. The electron kinetic energy is

$$E = \frac{\mathbf{p}^2}{2m} = \frac{2(h\nu - h\nu_0)^2}{mc^2}.$$

An annihilation energy spectrum shift of 1 keV corresponds to an electron energy 3.9 eV.

Experimentally, a high-resolution detector, a drift-free linear preamplifier and amplifier, and a multichannel analyzer are needed. These can be obtained commercially. The resolution of the system is measured by detecting the energy spectrum from a radioactive nuclide that produces a γ-ray in the 511-keV region. Strontium-85 emits a γ-ray of 514 keV. Because of the finite resolution of the spectrometer, photons of the same energy are counted as a distribution. This causes a distortion of the positron annihilation spectrum. Removal of this distortion is accomplished by stripping out the spectrometer response function with a computer program [94]. An example of the "unstripped" Doppler broadening of the annihilation photoelectric peak in silver is shown [94] in Fig. 8.6 along with the associated response function.

After the energy spectrum is corrected for the response function of the spectrometer, the momentum distribution of the electrons with which the positrons annihilated may be found. The electron-momentum space density function, $\rho(\mathbf{p})$, is related to the pulse-height distribution of the counting rate, $I(n)$, by

$$I(n) = \text{const} \iint \rho(\mathbf{p})\, d\mathbf{p}_y\, d\mathbf{p}_z,$$

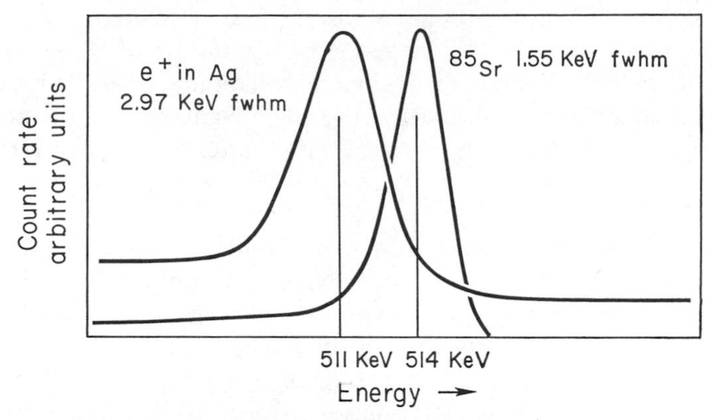

Fig. 8.6 Annihilation photon energy broadening caused by e^+e^- momentum. Reproduced by permission of The American Institute of Physics, from Ref. 94.

where n is the channel number, and the x-axis is chosen parallel to the direction of annihilation photon emission. Assuming that distribution in momentum space is isotropic, the integral is

$$I(n) = \text{const} \times 2\pi \int_{\mathbf{p}_x}^{\infty} \mathbf{p}\rho(\mathbf{p})\, d\mathbf{p}.$$

The differential of this equation yields

$$\rho(\mathbf{p}) = \frac{dn/d\mathbf{p}_x}{\text{const} \times 2\pi\mathbf{p}_x} \frac{dI(n)}{dn}.$$

Hence the pulse-height distribution can be differentiated numerically to convert it to an electron-momentum space density function. This distribution for silver is plotted [94] in Fig. 8.7. The open points derived by this method are in agreement with the same distribution measured by angular correlation methods (solid line in Fig. 8.7) [95].

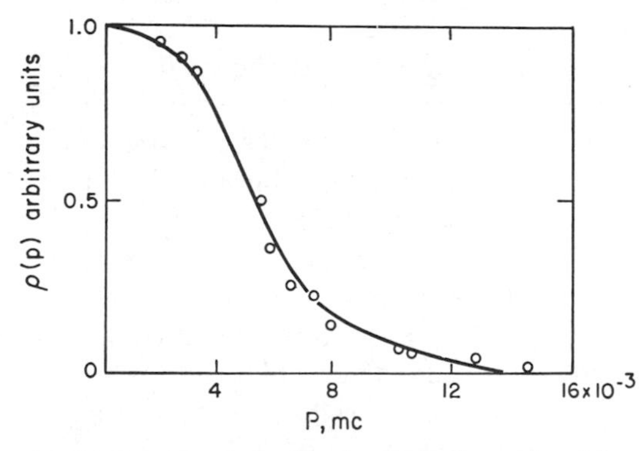

Fig. 8.7 Electron-momentum space density in silver. Reproduced by permission of The American Institute of Physics, from Ref. 94.

The detector energy resolution limits this method of determining electron-momentum distributions. At this stage the momentum resolution attainable is not equal to that attained in angular correlation studies. However, it offers several advantages [96], particularly the advantage that the positron source strength may be relatively low and the radiation damage effects reduced.

Lifetime Technique

The essentially simultaneous emission of a 1.28-MeV γ-ray with the positron during sodium-22 decay provides a convenient signal of the birth of a positron. Annihilation of the positron is signaled by the annihilation γ-rays. The lifetime of the positron can be measured by establishing the time delay between these two events. Some of the first measurements involved the method of delayed coincidences in which a start pulse generated by detection of the 1.28 MeV γ-ray was delayed for various times and routed to a fast coincidence circuit that received a stop pulse arising from detection of a 0.51-MeV γ-ray. The plot of coincidence rate versus time delay was analyzed to yield the average positron lifetime [16]. Although this method was improved with development of fast circuitry [97], accurate results were hard to obtain.

Perhaps the most significant technological advancement that allowed relatively easy determination of positron lifetimes was the invention of the time-to-pulse-height converter (TPHC) [98, 99]. The start 1.28-MeV signal triggers the TPHC. When the stop 0.51-MeV signal arrives, the TPHC puts out an electrical pulse proportional in height to the time between the start and stop signals. This analog signal is stored in a multichannel analyzer (MCA) according to its height. The higher the pulse, the higher the channel number in which it is stored. Hence the channel number is an indication of the time between start and stop, that is, the lifetime of the positron. Summation of about 10^6 pulses in the MCA yields a lifetime spectrum that can be analyzed to determine the average lifetime associated with one or more modes of positron decay.

A typical positron lifetime spectrum is shown in Fig. 8.8. The spectrum exhibits two components; the shorter is associated with free annihilation and p-Ps decay, while the longer is from o-Ps. The lifetimes are calculated from the inverse of the slopes of best-fit straight lines through the data points after background is subtracted. The short lifetime is calculated after the longer lifetime is subtracted from the spectrum. Zero time is determined by the centroid of a prompt peak derived from Compton scatter detections of simultaneous 1.33- and 1.17-MeV γ-rays from cobalt-60 with all spectrometer settings identical with those used in determining the positron lifetime spectrum. The counts, Y, in any channel on the right side of the spectrum can be defined by the equation

$$Y = Ae^{-\alpha t} + Be^{-\beta t} + \text{background},$$

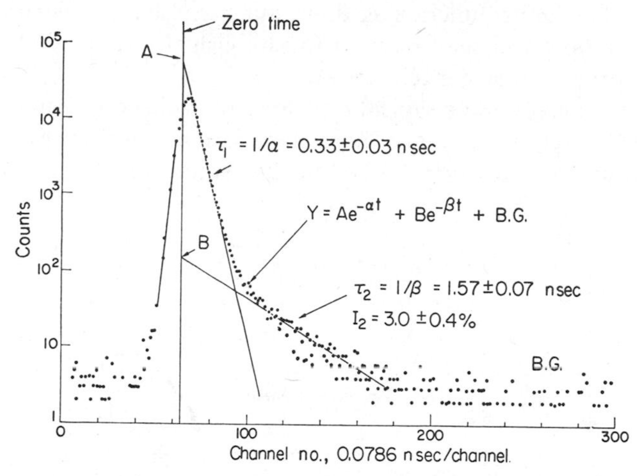

Fig. 8.8 Typical positron decay spectrum.

where A and B are the extrapolated zero intercepts of the lifetime components with decay rates α and β, respectively, and t is time. The short lifetime τ_1 is $1/\alpha$ and the longer lifetime τ_2 is $1/\beta$. The fraction of positrons decaying by the longer lifetime is given by the integral of $Be^{-\beta t}$ from zero to infinite time divided by the total counts under the background subtracted spectrum, that is, the area under the long-lived component divided by the area under the entire curve. The intensity of long-lived *o*-Ps is given by this fraction times 100.

The timing resolution is defined by the full width at half maximum (FWHM) of the prompt peak and the apparent half-life associated with its slope. This should be obtained with complete spectrometer settings identical with those used in positron lifetime determinations. The present state of the art allows measurement of the lifetime of an individual positron to within 0.3 nsec (FWHM). By spectrum-slope analysis the average lifetime can be measured to within a few hundredths of a nanosecond. Complete systems of detectors and electronic components, including excellent time-to-pulse-height converters, necessary to achieve this resolution are available from several manufacturers.* Components from various manufacturers that meet the AEC-NIM requirements can be interchanged at will. There are several methods of constructing a timing spectrometer, all of which must allow for precise measurement of time between two signals *and* establishment of the identity of the signals (1.28 MeV or 0.51 MeV) by energy discrimination.

* Manufacturers of fast timing components include Chronetics Inc., Mt. Vernon, N.Y.; E. G. and G., Inc., Boston, Mass.; LeCroy Research Systems Corp., West Nyack, N.Y.; Nanosecond Systems, Inc., Fairfield, Conn.; and ORTEC, Oak Ridge, Tenn.

Obviously, the timing function is done quickly. Energy selection may be either fast or slow and may be used to establish the validity of the timing signal by a proper coincidence network.

A schematic diagram of a typical fast-slow coincidence system is illustrated in Fig. 8.9. Detections of γ-rays by a scintillation detector result in anode signals that are routed to the TPHC for measurement of the time between

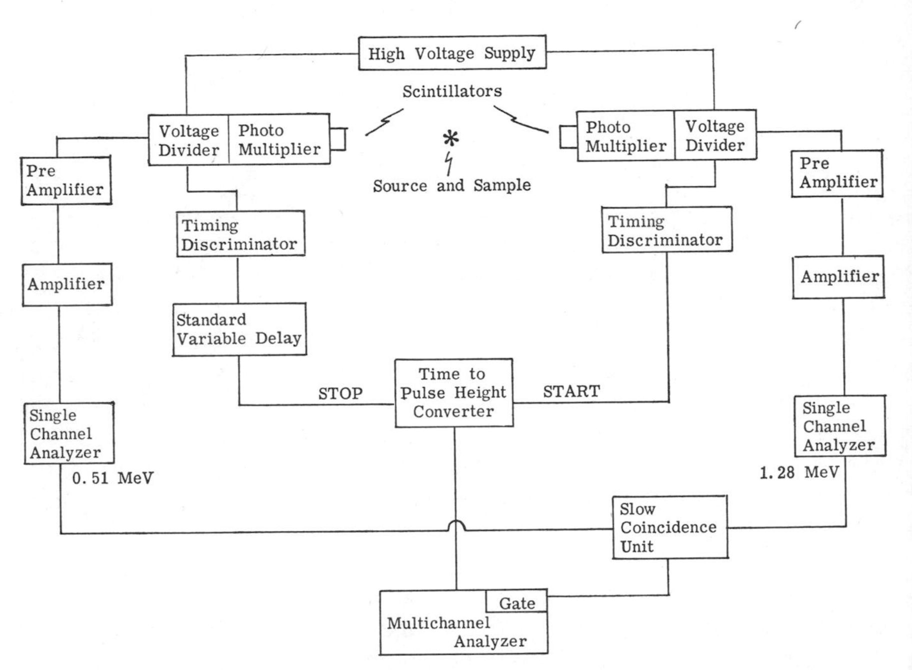

Fig. 8.9 Fast-slow coincidence system for determining positron lifetime spectra.

detections. For the TPHC to produce an output the start and stop signals must arrive within the time range selected, perhaps 100 nsec. This is considered a fast coincidence. The output from the TPHC is delayed within the component to coincide with the opening of the gate at the MCA. Dynode signals are amplified and routed to single-channel analyzers (SCA) that select the pulses arising from 1.28- and 0.51-MeV γ-rays on the start and stop sides, respectively. If the energy requirements are met, a coincidence unit triggers a gate that allows the MCA to accept the signal from the TPHC. This energy selection may be relatively slow—1 to 5 μsec.

A fast-fast coincidence system involves energy selection by SCA's that operate in a few nanoseconds. The fast SCAs trigger a fast coincidence unit, which gates the TPHC. Proper delay is allowed in the line between the detectors and the TPHC for energy selection before the timing signals arrive. If the energy requirements are met and the timing pulses arrive within the time range selected, the TPHC will send a signal to the MCA. A variation of this system is when the fast SCAs gate the timing discriminators.

A third type of timing spectrometer employing a TPHC is a fast crossover coincidence system [100]. This type requires only one signal from the detector. Energy selection and crossover timing may be achieved simultaneously in a fast crossover SCA, which then triggers the TPHC. When the energy requirements are met and the start and stop pulses arrive at the TPHC within the selected time range, the TPHC sends a signal to the MCA. This system has the advantage that the complexity and cost are reduced. However, because the fast crossover method of timing is relatively new, the technique has not been exploited widely.

The requirements of the various components in a timing spectrometer are treated in several publications [101–105]. Major factors that affect the time resolution are the scintillation lifetimes of scintillators used to detect the γ-rays, the rise time of the photomultiplier (PM) output signals, and the method of time pick-off. The timing jitter due to introduction of noise in the timing signals is a minor factor in currently available PMs, timing discriminators, and time-to-pulse-height converters. The available linearity of the TPHC output with time between start and stop signals is better than 99.9% integral and 98% differential. Multichannel analyzers of equal or better linearity are available from various nuclear instrumentation manufacturers. The machine-time resolution available is less than 10 psec. It is measured by splitting a PM signal with a Tee connector and by feeding the pulses to the start and stop timing discriminators and the TPHC (Fig. 8.9).

The scintillator used for γ-ray detection should have a very fast scintillation decay time and a high efficiency for light output. The shorter the decay time, the faster the γ-ray detection is transmitted to the PM. If the decay time is longer than the PM rise-time capabilities, the output rise time will be dictated by the scintillator. A higher efficiency of light output provides for a more uniform PM pulse shape, hence less time jitter. The time resolution decreases (becomes better) approximately as $1/\sqrt{N}$, where N is the number of photoelectrons produced by the γ-ray interaction with the scintillator. Plastic NATON-136, NE-111, and PILOT-B scintillators are used extensively because of their short decay times. Sodium iodide is used much less because of its long decay time, but it has the advantage that γ-ray energy resolution is greater than in the low-atomic-weight plastics. Recently NaI was used to achieve subnanosecond timing resolution [106]. This was possible because of its high output efficiency and the use of a newly developed bialkali photocathode PM (RCA-8575). Timing resolution is limited somewhat by the scintillators available. However, liquid scintillators have been found with much shorter decay times [107], and various chemicals can be added to shorten them [108].

The scintillator is optically coupled to the face of a fast-rise-time PM with a suitable grease such as Dow-Corning Optical Coupling Compound. Air bubbles or dirt in this coupling may cause poor timing resolution. The smaller the scintillator, the better is the timing resolution, at the expense of efficiency of γ-ray detection. The time spreading caused by photoelectron travel to the first dynode in the PM is least in the center of the PM face. A one-inch right cylinder is a good compromise size. The scintillator should be wrapped or painted with a reflective material, such as PILOT white paint, to improve light output at the photocathode. This yields a higher PM output and better timing resolution. The scintillator is held in place by wrapping it and the PM with electrical tape.

Photomultiplier tubes having output rise times of less than 1 nsec are available from some of the major manufacturers. Until scintillators become available that can deliver light to the PM more rapidly, a rise time of about 2 nsec is adequate. The PM must be designed to minimize the variation in transit times of photoelectrons caused by different path lengths or initial energy and angle of secondary electrons. The amperex XP-1021, RCA-70045, and RCA-8575 PMs can be used to achieve resolution times of less than 0.3 nsec FWHM. The PMs are powered by up to 3000 V dc through voltage dividers designed not to degrade the rise time of the output. BNC couplers are present on most commercial voltage dividers to connect high-voltage supply, dynode, and anode outputs. Selection of pulses arising from 1.28- and 0.51-MeV γ-rays may be performed on the dynode output while time pick-off is achieved on the anode pulse (Fig. 8.9).

Although the rise of the timing signal may be around 2 nsec, timing pick-off from that signal may be to within 10 psec. The derivation of a time signal from the PM output may be achieved by leading edge, constant fraction of pulse height, or zero crossover timing. These are illustrated in Fig. 8.10. In leading-edge timing the rise of the signal above a preset threshold on a timing discriminator (Fig. 8.9) triggers the time mark. In constant fraction of pulse-height timing [109] the PM signal is split into two, and one is inverted and delayed behind the other by a time approximately equal to the pulse rise time. The other is attenuated by a factor F that determines the fraction of pulse height. The two pulses are then added and the time mark taken when the added pulse crosses zero. In zero crossover timing a doubly differentiated bipolar signal is derived from the PM, and the time of base-line crossover is marked by circuits sensing zero amplitude after an initial threshold trigger.

Leading-edge timing is most prevalent because it was the first method to have low jitter. When the PM pulse reaches the height necessary to trigger a

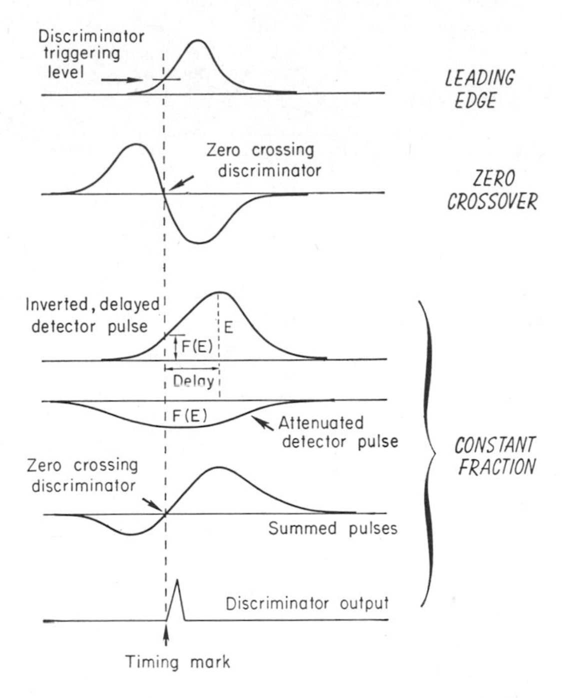

Fig. 8.10 Methods of time pick-off from detector pulses.

timing discriminator, the discriminator sends a standardized, highly reproducible output to the TPHC. Thus the time mark is the point in time at which the PM signal causes the discriminator to trigger. The time mark may vary with respect to a given pulse because of noise in the pulse or the discriminating circuit. This is termed jitter (Fig. 8.11). The time mark may vary between separate signals because of differences in the signal heights. This is termed "walk" (Fig. 8.11), and is a major source of timing inaccuracy. Walk may be decreased by taking a narrower "window" of start and stop pulses with the SCAs. Wider windows will yield faster data collection but poorer resolution. A 20% window is common, that is, from pulses corresponding to 1.28 to 1.03 MeV or 0.51 to 0.41 MeV dissipations of energy in the scintillators. Walk can be minimized also by proper selection of the leading-edge triggering height relative to the PM-output pulse shape and amplitude. The optimum resolution for a timing spectrometer occurs at a triggering point on the PM pulse where the charge C collected on the PM anode is a certain fraction of the total charge R collected on the anode from the scintillation event. More simply, it is optimum where the voltage of the triggering level is a certain fraction of the peak voltage. The C/R fraction is around 0.1 for

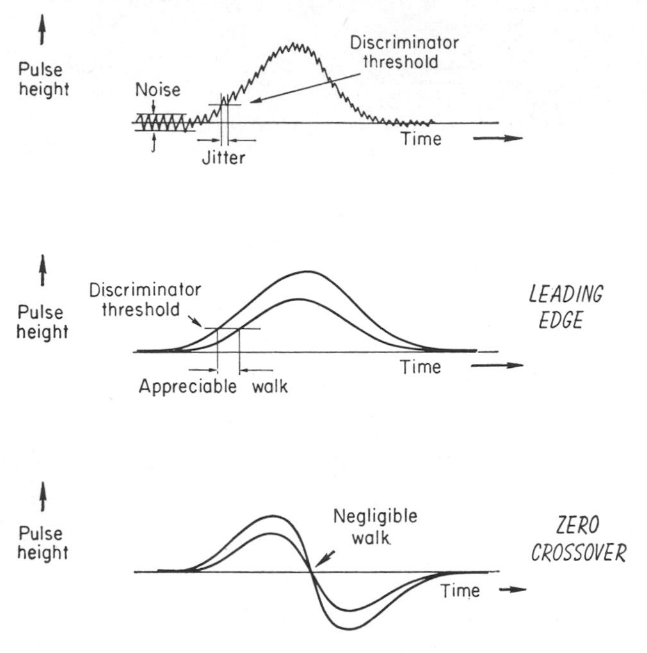

Fig. 8.11 Factors that influence time pick-off.

most systems or about 20 to 25% of the pulse height for approximately symmetric pulses. The optimum value of timing resolution as a function of C/R should be determined for individual systems by setting the SCAs to accept the 1.28- and 0.51-MeV pulses, and adjusting the triggering level on the timing discriminator to obtain optimum time resolution. If the triggering level is fixed, an attenuator before discriminator may be used to adjust the pulse height relative to the triggering level.

Constant fraction of pulse height [109] overcomes the problem of walk while retaining the low-jitter character of leading-edge timing. As illustrated in Fig. 8.10, the time mark is taken at a constant C/R ratio by splitting the PM pulse, inverting one signal, delaying it behind the other attenuated signal, adding them together, and detecting when the added pulses cross the base line. Since the inverted pulse varies proportionately in height with that of the noninverted, when they are summed, the zero crossover exhibits very little walk. Zero crossover timing using doubly differentiated pulses offers the same advantage. These types of timing have a wide dynamic range. Leading-edge timing may yield a walk of 0.1 nsec over a 20% range (window), whereas the latter methods will yield less walk over a 95% range. This allows acceptance of a greater number of detections with equal time resolution. Thus all start pulses can be accepted if they exceed in height those arising from 0.51-MeV photons, and stop pulses can be accepted from 0.51-MeV down to the noise in the detector due to backscatter and other spurious effects.

Another means of diminishing the walk caused by pulse-height variations is to build a pulse-height compensator into the system. A compensator sends to the TPHC output correcting signals whose amplitudes are functions of the pulse heights from both detectors [101, 110]. Another device that may be beneficial in some spectrometers is a pile-up rejector [101, 111]. When two pulses appear in a detector within about 1 μsec, the time and amplitude information may be distorted because of lack of electronic recovery between the pulses. This causes a broadening in the lower region of the time spectrum. This type of pulse pile-up is less critical at lower source strengths and in the fast-fast coincidence systems where most components recover in about 10 nsec, that is, operate at 100 MHz or greater.

Even though fast electronics are very sophisticated and should be handled with special care, their use and maintenance is not prohibitive. The advent of modular instrumentation makes it easy to send a defective unit back to the manufacturer for repair. However, some basic information regarding connecting the components of the timing spectrometer is valuable. Connections should be made with coaxial cable that has an impedance matching the input and outputs of the components. Most timing components have 50-Ω impedance connections, and RG 58A/U or equivalent 50-Ω cables should be used. The connecting cables should be as short as practicable. In determining the proper cable length for delaying a pulse, an approximation of 1 ft/nsec may be used. If the impedances of component inputs and outputs are not equal, they should be adjusted to match by proper terminators (resistors to ground) that can be bought or put together with a Tee connector. Short connecting cables and impedance matching reduces reflection and ringing of pulses.

Most components are designed to have very little temperature and voltage dependence. It is still advisable to have thermostatic control in the laboratory, and a voltage regulator should be used to power the equipment.

Standardization and optimization of the timing spectrometer is fairly easy. An oscilloscope is very useful. The single channel analyzers (Fig. 8.9) are set to accept the desired "window" of start and stop pulses, respectively. The timing discriminators are set to the optimum C/R by observing the narrowest time-spectrum distribution with respect to discriminator settings. The linearity of the system and the calibration of time per channel on the MCA is established by inserting standard delays before the stop side of the TPHC. If the spectrum is shifted 100 channels per 5-nsec delay and if this ratio is constant over the time range studied, the time calibration is 0.05 nsec/channel. Good linearity is necessary. The resolution of the system is found by inserting a ^{60}Co source between the detectors and monitoring the Compton interactions of the simultaneous 1.17- and 1.33-MeV γ-rays. This prompt peak also establishes zero time if the settings are not changed before collecting a positron annihilation spectrum. The background or noise-spectrum response is obtained by shielding the detectors from one another, placing separate ^{22}Na sources next to each detector, and collecting timing pulses from random coincidences. This spectrum should be flat with a low deviation in order to facilitate data reduction.

Two- and Three-Photon Coincidences

Qualitative and quantitative indications of positronium formation in a sample can be obtained by comparing the count rate of two- or three-photon coincidences with the rate in a substance in which Ps is formed in a known amount. The 2γ coincidence rate of 0.51-MeV photons is highest in substances such as aluminum where no *o*-Ps is formed, that is, when the $2\gamma/3\gamma$ ratio is 372. If *o*-Ps is formed, the relative 2γ coincidence rate is less than in aluminum. Conversely, the 3γ coincidence rate is least in aluminum and greater when a significant intensity of the positrons decay from *o*-Ps by 3γ annihilation.

The equipment is basically the same as that used to obtain annihilation energy spectra. In 2γ coincidence counting two scintillation detectors with associated power supplies, amplifiers if necessary, and single-channel analyzers are connected to a coincidence unit and scaler. The decay of a singlet electron-positron pair presents two 0.51-MeV photons in nearly opposite directions that are detected in coincidence by detectors collinear with the sample and the ^{22}Na source. The detectors are shielded from extraneous detections and the SCAs are set to accept only pulses arising from 0.51-MeV γ-rays.

Three-photon coincidence measurements are made by adding a third detection system to the apparatus described above and requiring a triple coincidence. Three-photon decay occurs when a triplet positron-electron pair annihilates, as when *o*-Ps decays. Energy and momentum conservation does not require a unique directional correlation between the three photons, only that they are emitted in the same plane and no more than two in the same half-plane. Hence the detectors and the sample must be placed in coplanarity. A typical arrangement is shown in Fig. 8.12. When the angles between the detectors are 120°, the SCAs should be set to accept pulses arising from γ-rays of $\frac{2}{3}mc^2$ or 340 keV energy. If the angles differ from 120°, the SCA settings should be adjusted appropriately (Fig. 8.1).

Fast, slow, or fast-slow coincidence systems may be used. In a fast coincidence system (Fig. 8.12) energy selection is made on each detector pulse, and if the detected γ-rays meet the energy requirements, the SCA signals arrive at the coincidence unit simultaneously and the event is registered on

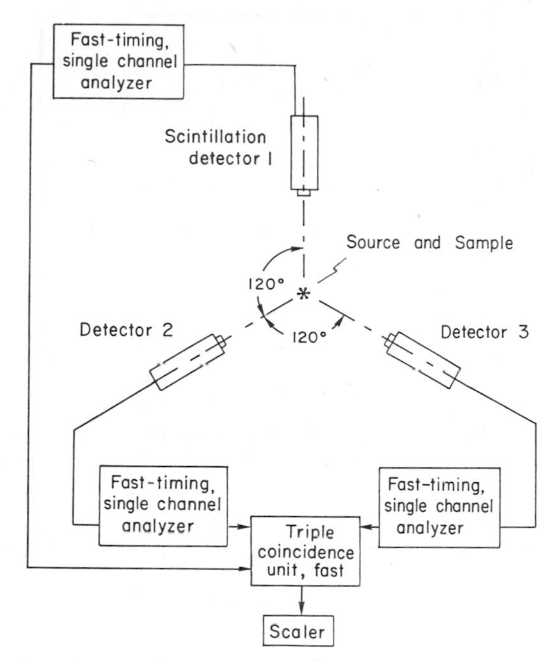

Fig. 8.12 Triple-coincidence spectrometer.

the scaler. By using fast scintillators and PMs, commercially available fast SCAs (particularly zero-crossover types), and a fast coincidence unit, resolving times of less than 10 nsec can be obtained. The shorter the resolving time, the lower the spurious counting rate and the higher the source strength usable. A fast-slow system employs three timing discriminators that trigger a fast triple coincidence unit. Energy selection is done by relatively slow SCAs. The SCAs and the fast coincidence unit are connected to a quadruple coincidence unit and scaler. A slow system can be constructed from three ordinary NaI single-channel analysis systems. The SCA systems are connected to a coincidence unit that triggers a scaler. The random coincidence rate is relatively high in the slow system, and the lower resolving time requires a weaker source strength.

The source-to-detector distance should be adjusted to achieve equal counting rates on each detector. A typical distance is less than 20 cm. To facilitate comparison of counting rates between samples, the thickness (>200 mg/cm^2) and size should be identical. Differential γ-ray absorption effects in solids can be reduced by preparing the sample, reference material, and source in a five-layer sandwich [112]. The reference rate is obtained by mounting the reference material next to the source with the sample on the

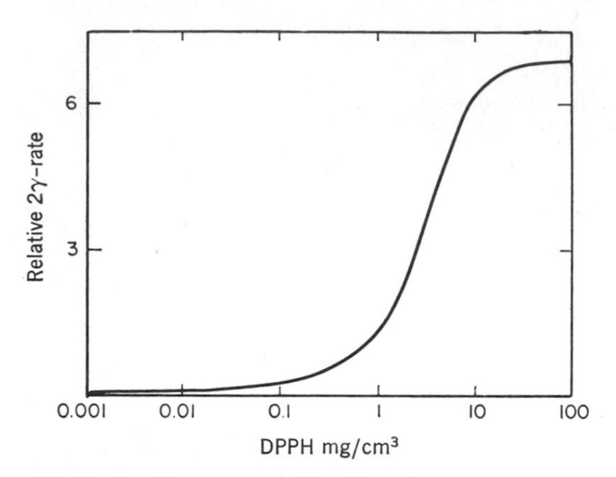

Fig. 8.13 Two-photon counting rate as a function of concentration of DPPH in benzene. Reproduced by permission of The American Institute of Physics, from Ref. 114.

outside. The sample rate is obtained by mounting the sample next to the source and the reference material on the outside. Comparisons in gases and liquids should be made under similar γ-ray absorption conditions or as a function of a parameter, such as gas pressure or chemical concentration, respectively [113]. The random coincidence rate is found by tilting one detector out of coplanarity in 3γ or colinearity in 2γ counting while maintaining the same source-to-detector distance. The counting rate in a reference material should be measured frequently in case electrical components drift. Because of the stringent requirements, 3γ counting rates are perhaps two or three cpm, and long collection times are required. Source strengths up to a millicurie may be used if the resolving time of the system is low and the response of the SCAs is not affected by high pulse rates.

An example [114] of information obtainable by 2γ coincidence counting is shown in Fig. 8.13. The relative counting rate in benzene solvent increases as the concentration of the free radical, diphenyl picryl hydrazyl (DPPH), increases. This indicates that *o*-Ps formed in benzene is quenched by DPPH with a resultant increase in singlet two-photon emission. The 3γ coincidence rate in the same system should decrease as the DPPH concentration increases. Similar types of information concerning the Ps quenching rate coefficients of gases [113] and the relative intensities of Ps in insulators and other materials [112] can be obtained.

Angular Correlation

Annihilation of a singlet positron-electron pair produces two 0.51-MeV γ-rays. Conservation of momentum requires that they are emitted in opposite directions. However, the angle between the γ-rays is not exactly 180°

because of the momentum associated with the pair. An elementary vector diagram shows that the sine of the angle of perpendicular deviation from collinearity, θ, is the ratio of the pair momentum, $P_{\perp}$, to the momentum of an annihilation photon, mc. For the small angles involved, $\sin \theta = \theta$. Hence $\theta = P_{\perp}/mc$, where θ is measured in radians. Measurement of the angular distribution of annihilation photons can yield considerable information about the electronic environment of the positron. Qualitative and semi-quantitative indications of Ps formation can also be obtained.

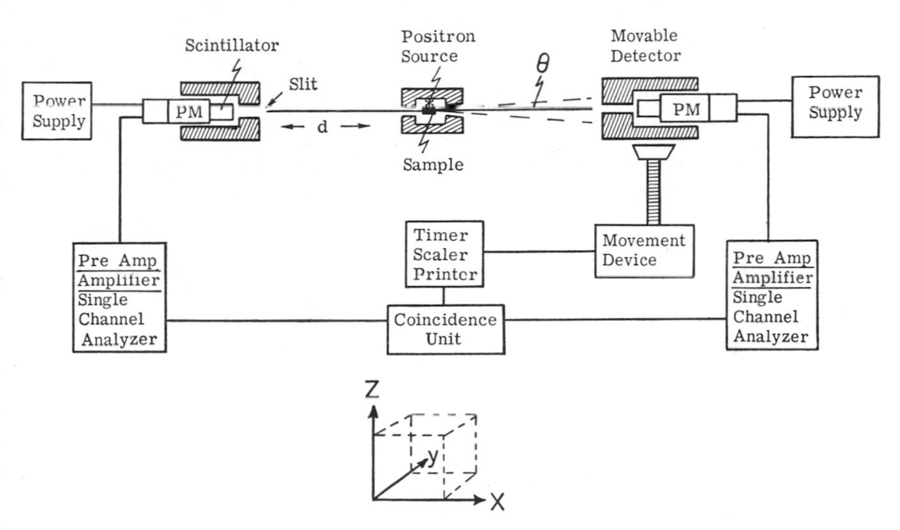

Fig. 8.14 Angular correlation spectrometer.

The apparatus used for angular correlation studies is a collimated 2γ-coincidence system with accessories necessary to move one detector with respect to the positron source and the opposite detector. As illustrated in Fig. 8.14, two γ-ray spectrometers set to accept 0.51-MeV γ-rays are connected to a coincidence unit and a scaler. The count is printed out at a prescribed time, and an electromechanical device moves one detector to another angle θ with respect to the source and the second detector. This process may be performed manually or it may be automated to sweep the spectrum of coincidence count rate versus θ. The design of the movement device may take many forms, including a heavy-duty Mössbauer driver with associated movement-sensing devices and multichannel analyzer. This part of the system is generally left to the ingenuity of the researcher [115] and need not be expensive or elaborate. The detectors should be shielded and the coincidence resolution time should be short to maintain low background. Sodium iodide scintillation detectors are advantageous because of their high efficiency and energy resolution, but plastic scintillators are often used because they are available in unusual shapes and are less expensive.

The most common geometrical arrangement is shown in Fig. 8.14. The Z and X axes are as illustrated, and the Y axis is in and out of the plane of the figure. Slit collimators are placed coplanar with the X-Y plane, and one collimated detector is moved through a spectrum of angles, θ, perpendicular to the X-Y plane. This arrangement detects a slice of the cone of annihilation photons surrounding collinearity. In order to simplify data interpretations, the slit and the scintillator should be long enough to extend through the entire cone. The angle associated with the maximum momentum of the positron-electron pair is seldom as great as 15×10^{-3} rad. Then for a hypothetical point source and distance, d, between the sample and the detectors the intersections of a 15×10^{-3} rad cone with the X-Y plane are $Y = \pm d(15 \times 10^{-3})$. Then Y should equal at least $2d(15 \times 10^{-3}) = 0.03d$. If d is a meter, the detector and slit length should be about 3 cm. A slit that is too short causes a loss in efficiency for photon pairs with large angular deviation from 180°; corrections to experimental data may be required.

The narrower the slit width, the better the resolution. If the slit width were 1 mm it would subtend an angle of 10^{-3} rad in the case above. The geometrical resolution function of the system with slit widths equal would be an isosceles triangle centered at 180° with a base of 2×10^{-3} rad. In practice, a point source is not used, and the resolution is a function of the slit and the thickness of the sample viewed [116]. The base of the resolution function is 2 times the sum of the slit width and the sample thickness, divided by the distance d [117], that is, if the thickness is 1 mm the base is 4×10^{-3} radian in the case above. The actual resolution function can be measured by moving the detectors close enough to the source that angular divergence of the γ-rays is negligible, plotting the coincidence counting rate as a function of θ or distance one detector is moved, and extrapolating the function to the geometry used in obtaining the angular distribution spectra.

With the geometry described the coincidence counting rate as a function of vertical detector displacement, Z, depends only on the Z component of momentum of the center of mass of the annihilating pair. The counting rate at a given angle θ is proportional to the probability that the momentum vectors, $\mathbf{P}$, in momentum space are equal to $\theta\, mc$. If the electrons in the sample are assumed to behave as a Fermi gas, the momentum distribution, $N(\mathbf{P})$, in the approximation of zero temperature is given by $N(\mathbf{P}) = \mathbf{P}^2$ for $\mathbf{P}^2 < \mathbf{P}_m{}^2$ and $N(\mathbf{P}) = 0$ for $\mathbf{P}^2 > \mathbf{P}_m{}^2$, where $\mathbf{P}_m$ is the maximum momentum. Since a positron is thermalized before annihilation in most condensed systems, the momentum measured is that associated with the electron. Assuming that the probability of annihilation is independent of the electron momentum, the distribution of Z momentum [118], $N_z(\mathbf{P}_z)$ is $(\mathbf{P}_m{}^2 - \mathbf{P}_z{}^2)$ for $\mathbf{P}_z{}^2 < \mathbf{P}_m{}^2$

and 0 for $\mathbf{P}_z^2 > \mathbf{P}_m^2$. Then for settings with θ greater than $\mathbf{P}_m/mc$ the counting rate is 0. Where θ is less than $\mathbf{P}_m/mc$ the rate varies as $(\mathbf{P}_m^2 - \mathbf{P}_z^2)$ and at $\theta = 0$, $\mathbf{P}_z = 0$, and the counting rate is highest. The relation between $N_z(\mathbf{P}_z)$ and $N(\mathbf{P})$ may be visualized by assuming that the occupied electronic states are uniformly distributed in a sphere of radius $\mathbf{P}_m$ in momentum space. The spectrometer views slices of this sphere that lie perpendicular to the $\mathbf{P}_z$ axis. The volume of the slice and the number of electronic states within depend on the angle θ and are proportional to $\mathbf{P}_m^2 - \mathbf{P}_z^2$. Thus an inverted

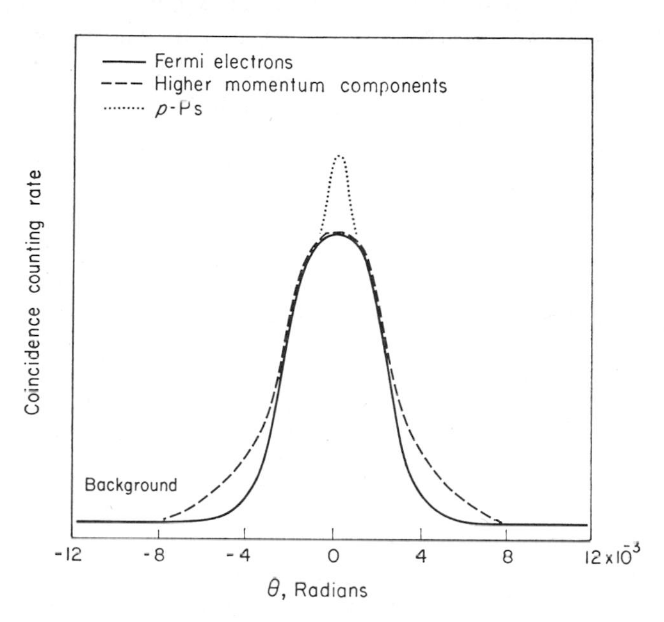

Fig. 8.15 Typical angular correlation curves.

parabola would be expected with $\mathbf{P}_m$ at the angle where the count rate decreases to 0.

The types of spectra obtained are illustrated in Fig. 8.15. The solid line represents the expected angular correlation curve from the above discussion. If the background were subtracted, the count rate would be 0 at about 4×10^{-3} rad. Thus the Fermi energy would be

$$E = \mathbf{P}_m^2/2m = \theta_m^2(mc)^2/2m = 4.08 \text{ eV}.$$

Parabolic angular correlation curves or curves consisting predominantly of a parabolic component are common in metals [95] where a free electron gas is approximated. Bell-shaped spectra delineated by the dashed curve arise when positrons annihilate with higher-momentum electrons. This may be caused by annihilation with core electrons or with electrons that are confined by the periodic potential of the lattice such that they have an unusually high zero-point motion. Bell-shaped curves may consist of a parabolic portion and a tail portion that approaches an exponential function. They are observed in noble metals, copper, silver, and in many ionic crystals and liquids.

The narrow peak (dotted curve) in Fig. 8.15 represents the angular spreading of annihilation photons that arise from thermalized *p*-Ps decay. If *p*-Ps has an energy of 0.025 eV with two electronic masses, by the equation above; $\theta^2 = 2m(2)(0.025\text{ eV})/(2mc)^2 = 0.025\text{ eV}/0.51\text{ MeV} = 4.90 \times 10^{-8}$, then $\theta = 0.22$ mrad. Very narrow slit widths and long source-to-detector

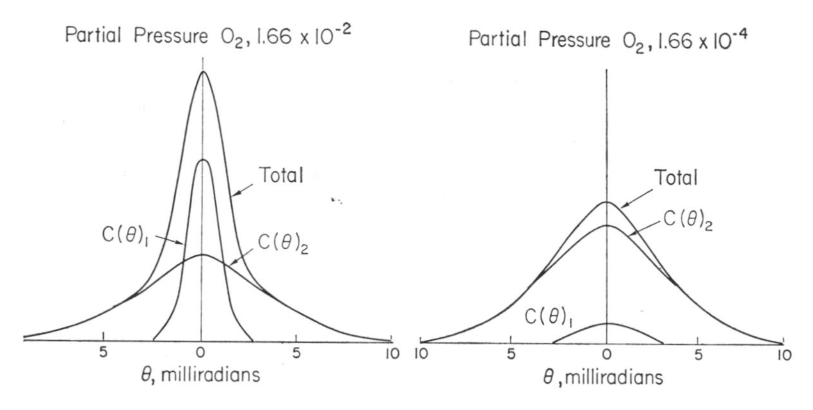

Fig. 8.16 Angular correlation in A—O_2 mixtures at 120-atmosphere total pressure. Reprinted by permission of *Soviet Physics—JETP*, from Ref. 119.

distances are required to achieve less than 0.2-mrad resolution. Hence in many systems the width of the peak is a function of the resolution, but the "peakedness" of the spectrum (above the theoretical parabolic curve defined by the curvature at higher angles) is an indication of positronium formation. The change in Ps formation as a function of temperature, electric field, or other parameters may be found by comparing the angular correlation spectra before and after changing the parameter. Such changes allow isolation of various mechanisms of *o*-Ps quenching that may not be identified by simple shortening or decreasing intensity of the long-lifetime component in positron lifetime spectra. For example, quenching by conversion to yield *p*-Ps will cause a narrow angular distribution (higher peak), while quenching by chemical oxidation to yield a positron will broaden the angular distribution. An example [119] of the change in the angular distribution in an argon–oxygen mixture as a function of partial pressure of O_2 is presented in Fig. 8.16.

There are various ways of treating the angular correlation spectra to obtain desired information. A subtraction of one spectrum from another yields a qualitative and semiquantitative indication of the change in angular distribution as a function of a parameter change. The spectra may be decomposed into a broad and a narrow component [62, 119] to study the effects of *o*-Ps quenching or formation in samples whose compositions are varied in a systematic manner. A common method of displaying angular correlation data obtained from metals in more meaningful fashion is to plot the momentum space density $\rho(\mathbf{P})$ and the momentum distribution, $N(\mathbf{P})$, which are calculated from the experimental spectra.

As illustrated earlier, if $\mathbf{P}_z$ is the z component of the momentum of the positron-electron pair, the coincidence counting rate $C(z)$ is proportional to the total number of pairs with $\mathbf{P}_z = mc(Z/d)$, where Z is the distance the movable detector is displaced from a line extending through the stationary detector and source, d is the distance from source to detectors, m is the mass of an electron, and c is the velocity of light. Then

$$C(z) = \text{constant} \times \iint \rho(\mathbf{P})\, d\mathbf{P}_x\, d\mathbf{P}_y,$$

where $\rho(\mathbf{P})$ is the momentum space density of annihilating pairs, and the integral is over the plane of $\mathbf{P}_z$ corresponding to the z displacement of the detector. If we assume that the momentum space distribution is isotropic, the integral is

$$C(z) = \text{const} \times 2\pi \int_{\mathbf{P}_z}^{\infty} \mathbf{P}\rho(\mathbf{P})\, d\mathbf{P}.$$

Differentiating with respect to $\mathbf{P}_z$ yields

$$\rho(\mathbf{P}) = \text{const}\,\frac{dC}{dz}\,\frac{1}{z}$$

and

$$N(\mathbf{P}) = \text{const}\,\frac{dC}{dz}\,z = 4\pi\mathbf{P}^2\rho(\mathbf{P}).$$

The relative change of $\rho(\mathbf{P})$ and $N(\mathbf{P})$ as a function of momentum $\mathbf{P}$ in sodium [95] is shown in Fig. 8.17. The values of $\rho(\mathbf{P})$ and $N(\mathbf{P})$ are obtained by numerically differentiating the spectrum at given angles θ. That is, $(C_2 - C_1)/(Z_2 - Z_1)$ is the derivative at $z = (Z_1 + Z_2)/2$, where C_1 and C_2 are the observed counting rates at Z_1 and Z_2, respectively. In the sample illustrated, the theoretically calculated Fermi energy corresponds closely to the maximum decrease in $\rho(\mathbf{P})$ and $N(\mathbf{P})$. The constant and quadratic nature of $\rho(\mathbf{P})$ and $N(\mathbf{P})$, respectively, up to the Fermi cutoff indicates that the free-electron theory is applicable in this case.

The linear-slit angular correlation spectrometer has been used to illustrate the angular correlation technique. It is most widely used, but other designs

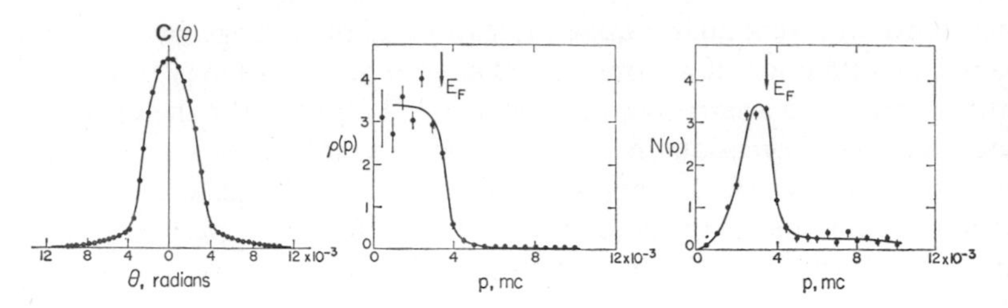

Fig. 8.17 Angular distribution curve in sodium with the derived momentum distribution N(p) and momentum space density ρ(p) in relative units. Reproduced by permission of The National Research Council of Canada, from Ref. 95.

employing circular [120] and point collimators [121, 122] have been successful. The circular collimators accept various conical shells of annihilation photons allowing fast data collection, while point slits yield low counting rates but very high resolution. Point slits are useful in determining the angular distribution for various orientations of a sample and in resolving the nature of the Fermi surface.

Multiparameter Methods

The techniques for studying positron annihilation that were described above can be combined in various manners to obtain results that are applicable to distinct decay mechanisms. For example, an angular correlation system may be interconnected with a lifetime spectrometer in a way that will allow determination of the lifetime of positrons decaying with electrons of a prescribed energy. Conversely, the angular distribution of annihilation photons may be studied as a function of the positron lifetime. A high-resolution Ge(Li) γ-ray spectrometer in conjunction with a lifetime spectrometer will yield the momentum distribution as a function of the positron lifetime. Thus measured lifetimes can be attributed more easily to decay of *p*-Ps, quenching of *o*-Ps by appropriate mechanisms, or annihilation of the free positron with electrons.

Abbreviated schematic diagrams of three "hybrid" systems appear in Fig. 8.18. The scintillators and electronics in each part of the systems should conform to the requirements outlined in the previous sections. Features such as the side energy-selection channels on the lifetime spectrometer part of the diagrams are not included for reasons of clarity in presenting the main functions of the overall systems.

The system presented in Fig. 8.18*a* is a combined angular correlation and lifetime spectrometer that yields positron lifetime spectra as a function of the angular correlation of the annihilation photons. The 1.28-MeV start photon emitted during positron emission is registered by detector 1. Detector 2

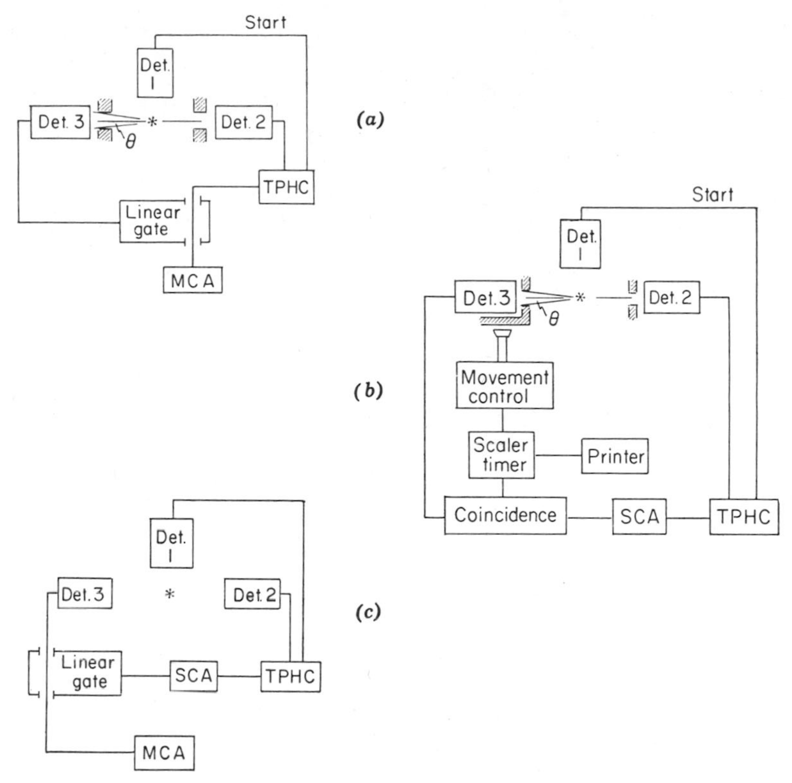

Fig. 8.18 Multiparameter analyses systems: positron lifetime spectra as a function of angular correlation (*a*), angular correlation spectra as a function of positron lifetime (*b*), and Doppler broadening of annihilation photons as a function of positron lifetime (*c*).

registers an associated 0.51-MeV stop photon in coincidence with detector 1 within the time range on the time-to-pulse-height converter. The TPHC produces an output that can be stored in the multichannel analyzer only if the linear gate has been triggered. The linear gate is opened by a signal from detector 3, which arises from detection of the other 0.51-MeV annihilation photon emitted at a proper angle with respect to the first 0.51-MeV γ-ray to allow detection through the collimator. Thus there is a triple coincidence allowing a timing pulse to be stored in the MCA with the constraint that the decaying positron-electron pair should have a momentum consistent with the angle, θ, between detector 3 and a line through the source and detector 2. The lifetime spectra of positrons decaying with high or low momentum (large or small θ, respectively) can be separated by the proper setting of the angular correlation part of the system. Because of the constraints involved, the time necessary to collect a lifetime spectrum is several days.

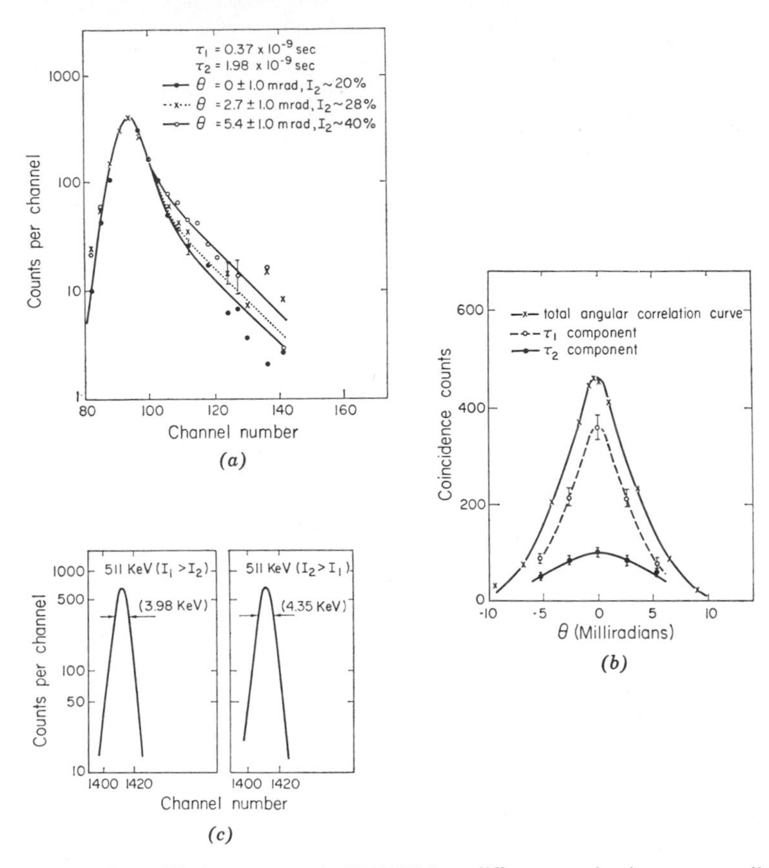

Fig. 8.19 Positron lifetime spectra in NATON at different angles between annihilation photons (*a*). Angular distributions of annihilation photons corresponding to different lifetimes (*b*). Annihilation photon energy broadening corresponding to different positron lifetimes (*c*). Reproduced with permission of the Physical Society, from Ref. 123.

A set of positron lifetime spectra at different angles between annihilation photons in plastic scintillator material NATON is shown [123] in Fig. 8.19*a*. The longer lifetime τ_2 was more prevalent for positrons annihilating with higher energy electrons, as shown by the intensity I_2 increasing from 20 to 40% as θ was increased from 0 to 5.4 mrad. This indicates that the τ_2 component arises from formation of a bound state of the positron, possibly *o*-Ps, which annihilates by picking off an electron from the environment. The momentum associated with pick-off reflects the momentum distribution of available electrons and shows a fairly broad spectrum.

The system illustrated in Fig. 8.18*b* is a combined angular correlation and lifetime spectrometer connected to yield angular distributions as a function of positron lifetime. Detectors 1 and 2 signal 1.28- and 0.51-MeV photons, respectively. If they are coincident within the range of the TPHC an output pulse is produced proportional in height to the lifetime of the positron. A single-channel analyzer is set to trigger on pulses corresponding to the lifetimes selected for study. Detector 3 registers the other 0.51-MeV annihilation photon. If the signal from detector 3 and the SCA are in coincidence, a count is registered on the scaler. Detector 3 is moved through a spectrum of angles θ with respect to the source and detector 2, resulting in an angular correlation spectrum with the constraint that it arises from positrons with a certain range of lifetimes. For example, if the SCA is set to trigger on pulses that correspond to channels 110 to 140 in Fig. 8.19*a*, a relatively wide angular distribution would be found. If it were set to accept pulses corresponding to channels 80 to 110, a narrower distribution would result. The angular correlation curve in NATON is resolved into short- and long-lifetime components [123] in Fig. 8.19*b*. The full width at half maximum is greater in the long-lifetime case.

A system of increasing utility is illustrated in Fig. 8.18*c*. A high-resolution Ge(Li) detector is used to detect momentum distributions of annihilating positron-electron pairs. It is gated by the selected output of a lifetime spectrometer such that only detections corresponding to positrons decaying with lifetimes under study are stored in the MCA. Detectors 1 and 2 register 1.28- and 0.51-MeV γ-rays, respectively. The time between the detections is reflected in the output of the TPHC. The SCA is set to trigger on the pulses arising from selected positron lifetimes. The output of the SCA opens the linear gate and allows the pulse arising from detection (by detector 3) of the other 0.51-MeV γ-ray to be stored in the MCA. Although there is a triple coincidence requirement, the data-collection rate is much greater than in systems (a) and (b) because narrow slit widths and long distances between source and detectors are not required. Increasing timing capabilities of high resolution, semiconductor detectors may allow them to be used for detecting the 0.51-MeV stop photon required for timing while simultaneously yielding momentum information. Thus only two detectors (1 and 3) may be required, with the output of detector 3 split such that it goes both through the linear gate and to the TPHC.

The energy spectrum of the annihilation photopeaks for positrons in NATON is illustrated [123] in Fig. 8.19*c*. When the SCA was set to accept most of the long-lifetime component (channels 110 through 140, Fig. 8.18*a*), the spectrum was broader than when the short-lifetime region was selected. Hence the Doppler broadening of the annihilation photons by the energy associated with the electron in the positron-electron pair was greater for decay by the longer lifetime.

As illustrated in Fig. 8.1, self-annihilation of *o*-Ps is accompanied by 3γ emission with a wide energy distribution. Self-annihilation of *p*-Ps yields 2γ emission with a narrow energy distribution of 0.51-MeV. Theoretically it is possible to separate p and o-Ps annihilations by γ-ray energy discrimination and assign lifetimes and other properties accordingly. In practice, this method can be applied only to positron annihilation in media that do not quench the lifetime of o-Ps substantially, such as in the rare gases. In condensed media most of the o-Ps decays by pick-off or chemical mechanisms, resulting in predominantly 2γ annihilation.

Because of the lack of high-energy resolution in the plastic scintillators used in lifetime measurements, this technique has not been used extensively. However, NaI scintillators have much greater energy resolution and sufficient timing resolution for use in certain work with gases. If the energy-selection window is set to accept γ-rays below the 0.51-MeV peak, the o-Ps decay component is more intense. Conversely, with the energy selection at 0.51 MeV, the decay of p-Ps and free positrons is more evident. Thus one can assign lifetimes to the proper annihilation mechanisms.

The advent of large-volume semiconductor detectors with high-energy resolution and efficiency makes studies of this type more convenient and meaningful. If a large multichannel analyzer with multiparameter capacity is used, the positron lifetime and energy of the annihilation γ-ray may be recorded simultaneously. In addition to the energy information, the time resolution of each lifetime spectrum at a specified annihilation photon energy is improved owing to less timing walk contributed by the energy spread of γ-rays.

4 METHODS OF DATA ANALYSIS

The methods of treating experimental data were explained in general terms in the last section. The complexity and wide use of lifetime and angular correlation spectra dictate further treatment.

Lifetime Technique

Positrons are born with high energies, and they annihilate via several mechanisms during or after thermalization. The resultant annihilation-lifetime spectrum can be quite complicated, with several lifetime components. Considering the three simplest species, free positrons, o-Ps and p-Ps, the changes in population of these states can be described by

$$\frac{dN_f(t)}{dt} = -f_o(t)N_f(t) - f_p(t)N_f(t) - \lambda_f(t)N_f(t),$$

$$\frac{dN_o(t)}{dt} = f_o(t)N_f(t) - \lambda_o(t)N_o(t),$$

$$\frac{dN_p(t)}{dt} = f_p(t)N_f(t) - \lambda_p(t)N_p(t).$$

Here N represents the numbers of positrons in different states, f the formation rates, λ the annihilation rates, t the time after the positron is born, and $_f$, $_o$, $_p$ denote free positron, o-Ps and p-Ps, respectively. In most of the cases the process of Ps formation occurs in a very short time compared with other processes. Hence for the time region observable we can rewrite the above formulas as

$$\frac{dN_f(t)}{dt} = -\lambda_f(t)N_f(t),$$

$$\frac{dN_o(t)}{dt} = -\lambda_o(t)N_o(t),$$

$$\frac{dN_p(t)}{d_t} = -\lambda_p(t)N_p(t).$$

In gaseous media these annihilation rates are sometimes time-dependent. However, if the thermalization process is fast compared with the annihilation rates, as in the condensed phases, or if the annihilation rates do not vary significantly within the energy region concerned, the annihilation rates are not significantly time-dependent. In these cases the resultant annihilation lifetime spectra will appear as three-component multi-exponentials,

$$Y(t) = \sum_{j=1}^{3} a_{2j-1} \exp(-a_{2j}t).$$

In condensed media the p-Ps annihilation component is often indistinguishable from the free annihilation component because of the short lifetime of free positrons. Thus in most cases the annihilation lifetime spectra will appear as only two-component multiexponentials,

$$Y(t) = \sum_{j=1}^{2} a_{2j-1} \exp(-a_{2j}t),$$

as shown in Fig. 8.8. The long-lifetime component is attributed to the annihilation of a relatively stabilized positron, such as o-Ps, and the short-lifetime component is attributed to the annihilation of p-Ps and free positrons. However, in some cases two long-lifetime components are observed and the annihilation lifetime spectrum appears as a 3-component multiexponential.

The method of least squares may be used to obtain the values of the parameters a_{2j} in the formula above. A set of data $y_i(t)$, the counts in a given channel, i, where $i = 1, 2, 3, \ldots, n$, is to be fitted to a curve

$$Y_i(a) = \sum_{j} a_{2j-1} \exp(-a_{2j}t_i) \qquad j = 1, 2 \quad \text{or} \quad 1, 2, 3.$$

Using the criterion of least squares, the expression

$$Q = \sum_{i=1}^{n} W_i(y_i - Y_i[a])^2$$

is minimized, where W_i is a statistical weighting factor, for example,

$$W_i = \frac{1}{Y_i(a) + \text{background}},$$

which helps correct for the greater error associated with the lower values of $Y_i(a)$. The usual procedure in linearizing and solving these exponential equations is to estimate a set of initial values of a, say a_k, and write

$$Y_i(a_{k+1}) = Y_i(a_k + \Delta a_k)$$

and

$$Q_k = \sum_{i=1}^{n} W_i(y_i - Y_i[a_k + \Delta a_k])^2.$$

Now the problem is to find the values of the set Δa_k that will give a minimum of Q_k.

If $Y_i(a_k + \Delta a_k)$ is substituted by its first-order Taylor's series expansion, the equation

$$Q_k^* = \sum_{i=1}^{n} W_i(y_i - Y_i[a_k] - \sum_j Z_{ij}[a_{jk}]\,\Delta a_{jk})^2$$

is obtained, where

$$\sum_j Z_{ij}(a_{jk})\,\Delta a_{jk} = \sum_j \exp(-a_{2j,k}t_i)\,\Delta a_{2j-1,k} - \sum_j t_i \exp(-a_{2j,k}t_i) a_{2j-1,k}\,\Delta a_{2j,k}.$$

To obtain the values of Δa_k that make Q_k a minimum, Q_k is differentiated with respect to Δa_k, and a set of simultaneous linear equations with Δa_k as variables is obtained and solved to give Δa_k. Then the new values $a_{k+1} = a_k + \Delta a_k$ are tested to find out whether they are really best-fit ones. If not, the process is repeated, and a_k is replaced with a_{k+1}. The iterative process is repeated until a satisfactory result is obtained, and the values of a_{2j-1} and a_{2j} converge at the least error. The values of the lifetimes associated with the exponential curves are then calculated from the inverse of the slopes of the lines, that is, $1/a_{2j}$, and the fraction of positrons (intensity) annihilating with a given lifetime is calculated from the ratio of the area under the respective exponential to the total area under the curve.

In many cases it is difficult to decide whether the spectrum is better represented by a single, a double, or a triple exponential curve fit. In these cases it is helpful to consider the weighted sum of the squared deviations, sometimes called the chi-square test, to choose the proper representation. Since counting experiments are governed by Poisson statistics, the error of any data point is the square root of the number of counts. A reasonable requirement of the derived expression to the decay spectrum is that it reproduce the experimental data to within the experimental error. If

$$Y_i(a) = Y_i(a) \pm \sqrt{Y_i(a)}$$

and

$$W_i = \frac{1}{Y_i(a)},$$

then for n data points,

$$Q = \sum_{i=1}^{n} W_i(y_i - Y_i(a))^2 = \sum_{i=1}^{n} \frac{1}{Y_i(a)} (y_i - (Y_i(a) \pm \sqrt{Y_i(a)})^2 = n.$$

A check on the reasonableness of a given fit to the data is to compare the weighted sum of the squared deviations with the number of points chosen for analysis. In many cases it will enable one to decide the number of components in a spectrum.

More detailed information about the various methods used to fit the experimental points is available [81, 124–131]. The calculations are very laborious if a computer is not available. In general, the data from the multichannel analyzer are fed into a computer and are reduced to lifetimes and intensities with associated errors according to the program used. Many researchers in this field will gladly share their knowledge of computer programming if they are asked.

Angular Correlation

The experimentally determined angular-distribution curve is the true angular distribution superimposed on the finite angular resolution of the apparatus. This must be corrected before interpretation. The background-subtracted angular distribution—$C'(\theta_0)$, coincidence count rate C as a function of angle θ—can be related to the true angular distribution, $C(\theta)$, by the following integral equation,

$$C'(\theta_0) = \int_{-\infty}^{\infty} C(\theta)R(\theta_0 - \theta)\, d\theta,$$

where $R(\theta)$ is the angular resolution function, θ_0 is the angle before correction, and θ is the angle after correction.

If the angular distribution of $C'(\theta_0)$ is broad and the resolution function $R(\theta_0 - \theta)$ is very narrow, the resolution function can be treated as a delta function, that is, $R(\theta_0 - \theta) = \delta(\theta_0 - \theta)$. Substitution yields $C'(\theta_0) = C(\theta_0)$. Therefore no correction is needed. This is the case for positron annihilation in solids of high density where the angular distributions are broad and resolution functions narrow.

Angular distributions in liquids are often rather narrow. Therefore the equation above must be solved for $C(\theta)$ to give the desired correction. An iterative numerical technique may be used. Each iteration generates an approximation solution $C_n(\theta_0)$, where n is the iteration number. This approximate solution is obtained from that of the preceding iteration by the relation

$$C_n(\theta_0) = C_{n-1}(\theta_0) + C'(\theta_0) - \frac{\int_{-\infty}^{\infty} R(\theta_0 - \theta)C_{n-1}(\theta)\, d\theta}{\int_{-\infty}^{\infty} R(\theta)\, d\theta}.$$

For the zeroth approximation, that is, the initial estimate of $C(\theta_0)$, $C_0(\theta_0) = C'(\theta_0)$. The iteration is continued until some arbitrary degree of convergence is attained (e.g., $C_n(\theta_0) \simeq C_{n-1}(\theta_0)$).

The resolution function can be calculated in terms of displacement along the Z axis (Fig. 8.14) and converted to an angular function. For two slits of widths s (i.e., s_z) at a distance d on either side of an infinitely thin source of annihilation γ-rays, the function is an isosceles triangle with base width $2s$. The relative value of the height of intersection of the isosceles triangle with unit height centered at z_0 when evaluated at z is

$$\begin{aligned} f(z, z_0) &= 1 - (z - z_0)/s \qquad \text{for} \quad z - z_0 < s \\ &= 0 \qquad \text{for} \quad z - z_0 > s. \end{aligned}$$

In practice a finite thickness of source is viewed through the slits. If the intensity of annihilation radiation is uniformly distributed through the thickness, th, of the sample viewed, the resolution function is the sum of isosceles triangles for each infinitely thin source within the thickness, that is,

$$R(z) = \int_{-th}^{th} f(z, z_0)\, dz_0 = \int_{-th}^{th} (1 - (z - z_0)/s)\, dz_0.$$

In this case the base of $R(z)$ is the sum of two times the thickness and two times the slit width. A geometrical plot readily reveals the same.

The assumption of a uniformly distributed source is not realistic. The intensity of the source is a function of the penetration depth of the positrons. The transmission curve for a continuous positron spectrum is nearly exponential but falls to zero at the maximum range R_m^+. Hence the intensity of the positron annihilation at a depth z in a sample being bombarded by positrons is approximately

$$I = I_0 \exp(-\mu z),$$

where I_0 is the intensity of the source at the highest level viewed (Fig. 8.14), I is the intensity at a depth z, and μ is the absorption coefficient of positrons, about $4.5/R_m^+$, which is chosen [35] to bring I to $0.01I_0$ for $z = R_m^+$. The triangular resolution functions must be weighted by $\exp(-\mu(th - z_0)/2)$ as z_0 varies from $-th$ to th. The total resolution function is

$$R(z) = \int_{-th}^{th} [f(z, z_0) \exp(-\mu(th - z_0))/2] \, dz_0.$$

Substituting $k = -\mu/2$ and integrating for five conditions,

$$R_1(z) = [\exp(-k(z + s - th)) - (1 - zk - sk - thk) \exp(2kth)]/k^2 s, \quad \text{for} \quad -(th + s) \leq z \leq -th.$$

$$R_2(z) = [\exp(-k(z + s - th)) - 2\exp(k(th - z)) + (sk - zk - thk + 1)\exp(2kd)]/k^2 s, \quad \text{for} \quad -th \leq z \leq th - s.$$

$$R_3(z) = [(thk + 1 - sk + zk) - 2\exp(k(th - z)) + (sk - zk - thk + 1) \exp(2thk)]/k^2 s, \quad \text{for} \quad th - s \leq z \leq s - th.$$

$$R_4(z) = [(thk + 1 - sk - zk) + \exp(k(th - z + s)) - 2\exp(k(th - z))]/k^2 s, \quad \text{for} \quad s - th \leq z \leq th.$$

$$R_5(z) = [\exp(k(th - z + s)) - (sk - zk + thk + 1)]/k^2 s, \quad \text{for} \quad th \leq z \leq th + s.$$

The resolution function in terms of displacement of one slit detector, $R(z)$, can be transformed to a function with respect to angle θ, $R(\theta)$, by dividing the displacement z by the distance from source to detector d (Fig. 8.14). The effect of the exponential absorption of positrons on $R(\theta)$ is shown in Fig. 8.20. In this case [132] the positrons were absorbed in liquid methane, where $\theta = z/d$ in milliradians and $s = th = 0.60$ mrad.

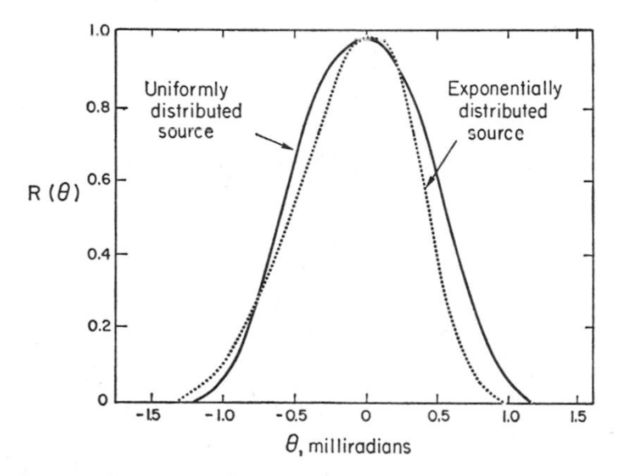

Fig. 8.20 Angular correlation resolution functions assuming different source distributions. From Ref. 132.

After the resolution function has been ascertained and the experimental data corrected accordingly, the angular correlation spectrum is reduced to more meaningful terms. As illustrated in the experimental techniques section, the momentum space density of annihilating pairs, $\rho(\mathbf{P})$, is defined by [95]

$$\rho(\mathbf{P}) = \text{const}\,\frac{dC}{dz}\frac{1}{z}$$

and the momentum distribution, $N(\mathbf{P})$, is

$$N(\mathbf{P}) = \text{const}\,\frac{dC}{dz}\,z = 4\pi\mathbf{P}^2\rho(\mathbf{P}).$$

The slope of the angular distribution curve, dC/dz, may be taken from the difference in adjacent measurements of C, that is,

$$\left(\frac{dC}{dz}\right)_{(z_n+z_n+1)/2} = \frac{C(z_n) - C(z_{n-1})}{z_n - z_{n-1}}.$$

When the counting rate C is low, a better method of slope determination should be used. For instance, a five-point least-square fit of a parabola can be valuable.

$$\left(\frac{dC}{dz}\right)_{z_n} = \frac{2C(z_{n+2}) + C(z_{n+1}) - C(z_{n-1}) - 2C(z_{n-2})}{10}$$

After the values of dC/dz are found at various values of z the relative distribution of $\rho(\mathbf{P})$ and $N(\mathbf{P})$ can be plotted versus z. Since $\sin\theta = \theta = \mathbf{P}_z/mc$ for small values of θ, by substituting z/d for θ, it is obvious that $\mathbf{P}_z$ is proportional to z and to θ. Hence in the previous equations involving a constant multiplier, const, $\mathbf{P}_z$ or θ may be substituted for z. The momentum distribution is commonly plotted as $N(\mathbf{P})$ versus $\mathbf{P}$, that is, $\theta(mc)$, as shown in Fig. 8.17 [95].

5 POSITRON AND POSITRONIUM DECAY IN MATTER

Gases

The existence of positronium was shown by experiments with positrons annihilating in the rare gases, beginning with 1949. Many experiments have been conducted in a variety of gases to measure the Ps formation fraction (the fraction of all positrons which form Ps) and the annihilation processes as a function of pressure and application of magnetic and electric fields. Extensive reviews of this work exist [24–29].

Ever since Ps was proposed to exist there have been speculations on the existence of compounds with atoms and molecules. Theoretical estimates of bond energies have been made (subject to enormous limitations in mathematical handling) for compounds such as PsH, PsH^+, and PsCl. Some early suspicions were voiced [17] that there was experimental evidence for compound formation between Ps and chlorine or Freon. This evidence was the observation of a large "positron attachment coefficient" for the gas molecule. Positronium was assigned [12] a chemical symbol, Ps, and in recent years a considerable body of experimental results shows evidence for positron and positronium compound formation in gases. These studies will certainly be considerably extended, and it is likely that more precise information will be obtained with positron-gas systems than with liquids and solids.

In general, the fate of a positron in gases is dependent upon the pressure and the composition of the sample. Narrow angular distributions of annihilation photons [133] and high lifetime quenching probabilities are exhibited by gases with unpaired electrons [90], such as O_2 and NO, which may cause conversion of *o*-Ps to *p*-Ps. Table 8.1 shows the resolved lifetimes of positrons

Table 8.1 Positron Lifetimes in Freon-12 at Various Gas Pressures

Pressure at 20°C (atm)	τ_1 (10^{-9} sec)	τ_2 (10^{-9} sec)
5.3[a]	nd[b]	100
4.2	nd[b]	114
3.1	nd[b]	117
2.0	2.8 ± 0.3	124
1.2	4.5 ± 0.5	135

Reproduced by permission of the American Institute of Physics, from Ref. 134.
[a] Density of Freon at 5.3 atm is 0.026 g/cm³.
[b] nd: not determined, too short.

annihilating in Freon 12 at various gas pressures [134]. The results indicate a small change in the long lifetime (from a little less than the theoretical lifetime of *o*-Ps to 100 nsec) as the pressure increases but a substantial decrease in the short component. This component includes the "free" positron annihilation, and the decrease is interpreted as an enhanced annihilation probability caused by the formation of the collision complex ("compound") $e^+CCl_2F_2$.

Similar and more extensive studies [135] show that the actual annihilation rates of positrons in several gases are considerably higher than the calculated Dirac free annihilation rates (p. 504). The ratios are argon, 3; methane, 20; ethane, 50; propane, 193; *n*-butane, 623; *iso*-butane, 555; and carbon tetrachloride, 702. The explanation of these enormously higher annihilation rates is that bound positron-molecule ions are formed, and increased positron-electron interaction leads to faster annihilation.

A phenomenon that has led to greater insights into the mechanism of positron annihilation is that of delayed-annihilation lifetimes. These lifetimes are exhibited in gases such as argon by positrons that fail to form positronium and whose energies fall below the threshold for formation of positronium. A theoretical analysis [68] predicted these delayed lifetimes, and later experiments [136] proved their existence in argon and argon–nitrogen mixtures. Figure 8.21, taken from [136], shows a "shoulder" in the

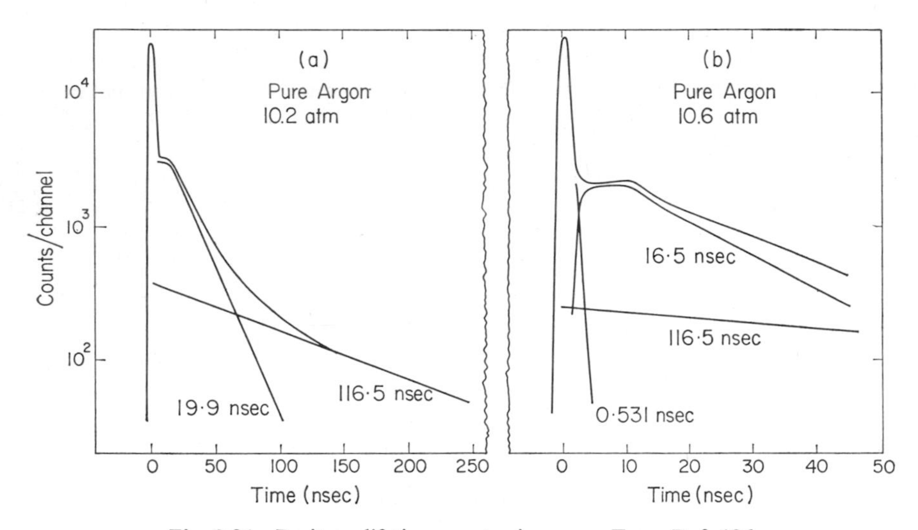

Fig. 8.21 Positron lifetime spectra in argon. From Ref. 136.

lifetime distribution of positrons in argon that is due to the time-dependence of the annihilation rate of positrons with energies below the threshold for Ps formation. Similar results were obtained by Osmon [137], and increasing electric fields have been demonstrated [138] to increase this time-delay in annihilation. The same phenomenon occurs in helium gas at 4.2°K but not at room temperature [139]. New computer methods of data reduction and higher-resolution spectrometers are revealing the fine details of the positron annihilation lifetime spectra.

The study of these fine details in positron annihilation in gases depends primarily on equipment not only with a time resolution of about 0.5 nsec but also a range up to 400 nsec. In experiments with argon–chlorine mixtures Tao [140] used two time-to-amplitude converters. One had a resolution of 0.5 nsec (full width at half maximum) and a useful range of 25 nsec; the other had a resolution of 1.2 nsec and a useful range of 360 nsec. Two relatively short components appeared in the annihilation lifetime spectrum of positrons in pure chlorine and argon–chlorine mixtures, but a shoulder appeared in the second component. A longer third component was seen in the mixtures but only when the partial pressure of chlorine was very small. This was presumably the *o*-Ps component. When $\lambda(t)$, the free annihilation rate at a given time during the positron thermalization process, was plotted versus time in the A–Cl_2 mixtures, four peaks were found. Two of these were seen in similar plots for argon and two for pure chlorine. Estimates of the positron energies at these resonance annihilation peaks were 0.9 and 0.7 eV for argon and 0.5 and 0.3 eV for chlorine.

Inspection of the positron lifetime spectra in A–Cl_2 mixtures shows an additional shoulder at short times near the peak. This second shoulder appears when the chlorine partial pressure is low enough for the third component, *o*-Ps, to be seen. The sum of the intensities of the additional shoulder and the third component remains practically constant over a wide range of intensities. Thus *o*-Ps annihilation is responsible for both phenomena. The *o*-Ps is considerably influenced by the presence of chlorine, and a strong annihilation resonance occurs at 0.5 eV. This is likely caused by the reaction

$$Ps + Cl_2 \rightarrow PsCl + Cl,$$

which requires a threshold positronium energy of about 0.5 eV [64]. The annihilation lifetime of PsCl has been calculated to be very short [141, 142], and a strong resonance interaction at 0.5 eV to form PsCl—followed by a resonance at 0.3 eV (a peak in the annihilation rate) caused by annihilation from the compound is a reasonable expectation. These phenomena are not seen in nitrogen–chlorine mixtures presumably because thermalization of Ps occurs faster, and the Ps escapes the resonance interaction region much faster than in argon–chlorine mixtures.

Liquids

There is a substantial literature on Ps in organic liquids, solutions in nonpolar solvents, condensed gases, and aqueous solutions that has been reviewed by several authors [25, 29, 30]. In this section a few examples are taken to illustrate the advances and the chemical interest in studies of Ps in liquids.

Organic Liquids and Solutions

Very extensive studies were made [143–147] to unravel the constitutive and substituent effects of organic molecules on *o*-Ps formation and annihilation. However, two factors prevented general agreement on the value of the results; there was insufficient agreement between results from different laboratories, and only after 1965 was the need for thorough degassing of the liquids realized [148]. (Paramagnetic dissolved oxygen is an efficient quencher

Table 8.2 Partial *o*-Ps Quenching Cross Sections

		$(\sigma v)_{A_v}$ (10^{-14} cm^3/sec)
CH_3		0.855
CH_2		0.971
CH		0.924
C		0.705
C_6H_{11}	Monosubstituted cyclohexanes	5.515
C_6H_{10}	Disubstituted cyclohexanes	5.373
C_5H_9	Monosubstituted cyclopentanes	4.469
C_6H_5	Monosubstituted benzenes	4.679
C_6H_4	Disubstituted benzenes	4.844
—CH=CH—		1.906
CH_2=CH—		1.667
CH_2=C(CH_3)—		2.635
—CH=C(CH_3)—		3.076
H		0.129
OH		0.749
—O—	Ethers	0.836
CHO	Aldehydes	1.673
COOH	Acids	1.839
C=O	Ketones	1.786
COO—	Esters	1.939
F		0.084
Cl		1.122
Br		1.872
NH_2		1.624
$PO_3\equiv$	Phosphites	3.382
$PO_4\equiv$	Phosphates	3.754
SH	Thiols	2.330

Reproduced by permission of the American Institute of Physics, from Ref. 149.

of *o*-Ps.) A recent and very extensive series of experiments was reported [149] on the correlation of *o*-Ps annihilation parameters with structural electronic properties of organic liquids. Results for 193 deoxygenated pure compounds were reported, and the *o*-Ps lifetime and quenching cross section for various families of compounds were compared. In *n*-alkanes, normal primary alcohols, 1-chloro-substituted *n*-alkanes, and others, τ_2 (lifetime of *o*-Ps) decreases as the molecule increases in length and becomes nearly constant for higher members of the series. However, $(\sigma v)_{A_v}$, the cross section for quenching *o*-Ps, is a linear function of the number of carbon atoms in the molecule. The quenching cross section is equal to $(M/\rho N_0)(1/\tau_2)$, where M is the molecular weight, ρ the density, N_0, Avogadro's number, and τ_2 is the *o*-Ps lifetime.

Partial quenching cross sections attributable to various groups on molecules were calculated, and the group characteristic values are given in Table 8.2. From these values, empirical annihilation cross sections were calculated and compared with the observed values for about 150 compounds. The percentage error was never more than two standard deviations. For example, the average error of the calculated cross sections for 75 hydrocarbons was 0.95%.

Annihilation parameters depend on the isomeric form of molecules, as shown in Table 8.3 [149]. Other isomeric systems show similar trends. The

Table 8.3 Annihilation Properties of Isomeric Pentanes

Compound	τ_2 (10^{-9} sec)	$(\sigma v)_{Av}$ (10^{-14} cm^3/sec)
Pentane	4.12 ± 0.11	4.69 ± 0.13
2-Methylbutane	4.45 ± 0.07	4.36 ± 0.07
2,2-Dimethylpropane	5.03 ± 0.07	4.08 ± 0.06

Reproduced by permission of the American Institute of Physics, from Ref. 149.

straight-chain molecule has the shortest lifetime and the largest quenching cross section.

The intensity of *o*-Ps varies regularly with the substituent in halogenated propanes [147, 149], as shown in Table 8.4. The value of I_2 depends greatly on the functional group, not on a property of the molecule as a whole.

Radicals and radical scavengers have a considerable effect on positron annihilation [88, 103, 150]. Both iodine in heptane and DPPH in benzene reduce τ_2 without much change in the intensity. DPPH, a free radical, is effective at a concentration about 100 times less than that of iodine. Carbon tetrachloride, which may supply chlorine by localized radiation damage at the positron, reduces I_2 in benzene from 35 to 8% at 0.12 mole % concentration without altering τ_2.

The annihilation rates of *o*-Ps in solutions of iodine in degassed diethyl ether, ethyl alcohol, *n*-pentane, *n*-hexane, *n*-heptane, and others and in the pure liquids over temperature ranges of about 50°C have been studied to

Table 8.4 Intensity of *o*-Ps in Organic Compounds

Compounds	*Intensity* (I_2)
Hydrocarbons	30 to 40%
Alcohols	18 to 22%
Ethers	23 to 26%
Aldehydes, ketones, and esters	12 to 16%
Monofluoro compounds	18 to 24%
Monochloro compounds	10 to 12%
Monobromo compounds	4 to 6%
Monoiodo compounds	1 to 2%
Acids	Increases with length 12.6 to 36.7%

Reproduced by permission of the American Institute of Physics, from Ref. 149.

delineate the mechanism of positron annihilation [151, 152]. That positronium atoms might be solvated was suggested earlier [29], and quenching of *o*-Ps by dissolved oxygen was shown [148] to be a diffusion-controlled process. The *o*-Ps annihilation rates in the pure liquids increased with increasing density as expected for pick-off annihilation. When iodine was dissolved, the annihilation rates depended inversely on the density and I_2 remained constant. Therefore one can write:

$$\frac{-d(o\text{-Ps})}{dt} = (\lambda_0 + k_q[C])(o\text{-Ps}).$$

where λ_0 is the annihilation rate in the pure liquid and k_q is the rate constant for chemical quenching by iodine of concentration [C]. The Debye relation for the rate of diffusion of a reactant in a solution shows that the rate is inversely dependent on the viscosity of the medium. The rate constant k_q was inversely dependent on the viscosity of the iodine solutions and the lifetime quenching rate, $d(o\text{-Ps})/dt$, depended on the temperature, as predicted by the Arrhenius equation. The activation energies calculated for chemical quenching from the Arrhenius equation were similar in magnitude to the activation energies for diffusion.

Hence the positronium and the iodine in solvent cages likely diffuse together through the solvent until they meet and undergo a reaction such as

$$\text{Ps} + \text{I}_2 \rightarrow \text{PsI}_2 \rightleftarrows e^+\text{I}_2^- \rightleftarrows \text{PsI} + \text{I}.$$

The positron would annihilate in the electron cloud of the iodine at a rate described by the above equation.

Aqueous Solutions

Ortho-positronium formed in pure water annihilates at a rate determined by pick-off annihilation, that is, by interaction with the molecular electrons of the medium. If dissolved substances are present, the *o*-Ps lifetime and the formation fraction (I_2) will be modified by two principal processes: (1) triplet-to-singlet conversion resulting from interaction with molecules or ions which have a bulk paramagnetism caused by the presence of unpaired electrons and (2) chemical interaction between positrons and the solute (giving decreased formation of Ps) or between Ps and the solute, which usually causes an increase in the pick-off annihilation rate.

A large number of salts quench the *o*-Ps lifetime when dissolved in water at high enough concentrations [71, 153]. The narrowing of 2γ angular correlation distributions when cations such as Mn^{+2} and Co^{+2} are added to water also indicates that conversion quenching occurs [62, 154]. This is particularly true for salts containing paramagnetic cations. The degree of paramagnetism should affect the probability of *o*-Ps quenching by conversion to *p*-Ps. The chlorides are convenient to use because Cl^- has little effect on *o*-Ps in water. Quenching cross sections are defined by $\sigma = \gamma/vN$, where γ is the conversion quenching rate, v is the average velocity of Ps, and N is the number of atoms per cm^3. The cross section can be expressed as a function of molarity and can be compared for different salts at the molarity that yields equal quenching [71]. Quenching cross sections vary considerably with the nature of the cation but cannot be correlated directly with the number of unpaired electrons in the ion, even when the cross sections are calculated [29] in terms of thermodynamic ion activities instead of molarities as shown in Table 8.5. Antimony, which is usually considered diamagnetic, quenches similarly to paramagnetic ions. The chemical state of the ion in solution and

Table 8.5 Ps Conversion Cross Section σ of Paramagnetic Ions in Aqueous Solution

Ion	*Ionic Activity* a_+	$\sigma \times 10^{18}$ cm^2	*Unpaired Electrons*
Cr^{3+} ($CrCl_3$)	0.00364	33.2	3
Mn^{2+} ($MnCl_2$)	0.0627	1.92	5
($MnSO_4$)	0.0138	8.76	5
Fe^{3+} ($FeCl_3$)	0.0064	18.9	5
Fe^{2+} ($FeCl_2$)	0.0434	2.79	4
Co^{2+} ($CoCl_2$)	0.0432	2.8	3
Ni^{2+} ($NiCl_2$)	0.0399	3.03	2
Cu^{2+} ($CuCl_2$)	0.024	5.04	1

other factors, such as diffusion of solvated Ps, must be taken into account to describe this phenomenon.

Lifetime measurements on *o*-Ps in various solutions of diamagnetic salts, especially nitrates [153], show that the anion can also affect the fate of the positron. As the nitrate ion concentration increases, the intensity of the long-lifetime component decreases while the lifetime remains practically unchanged. These positrons are lost from the τ_2 component in the lifetime curves and appear in the short-lifetime component τ_1. The three-photon coincidence counting rate in water is decreased considerably when a nitrate salt is added. This quenching of *o*-Ps may be caused by oxidation of Ps by the nitrate [155]. Other anions with less oxidative potential, such as $SO_4^=$ and Cl^-, do not exhibit the effect. However, the redox potentials of the ions, which are normally obtained under equilibrium chemical conditions, do not correlate directly with Ps quenching. This is caused by the nonequilibrium nature of Ps reactions and other novel features of positronium [74, 75, 156]. The kinetics and energetics of aqueous solution reactions of Ps and possible tunneling of electrons [25] through the potential barrier between positronium and acceptor are not fully understood.

Some extensive studies of the chemical reactions of Ps in acids have been reported recently [157]. Four oxyacid-water systems were studied over the widest practical range of compositions, limited in the cases of HNO_3 and $HClO_4$, but extending fully into the P_2O_5 and SO_3 regions in the case of H_3PO_4 and H_2SO_4. Values of τ_1, τ_2, and I_2 from positron lifetime spectra were obtained. In the oxyacid series τ_1 remains constant, τ_2 changes only slightly, but there are significant changes in I_2 as shown in Fig. 8.22. The regions

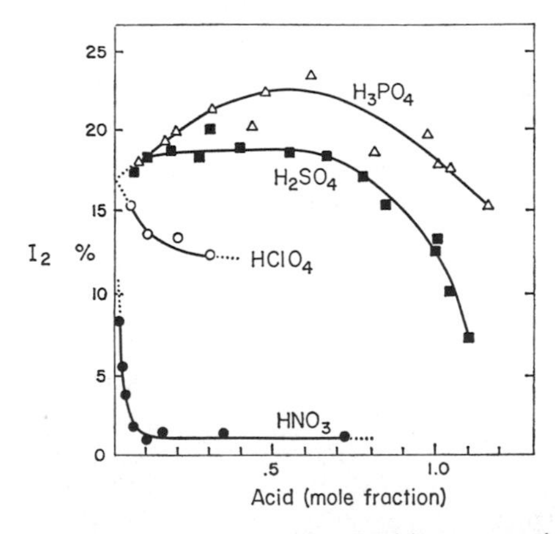

Fig. 8.22 Intensity of long-lived Ps as a function of acid concentration in water. From Ref. 157.

above 1 m.f. in this figure are the P_2O_5 and SO_3 regions. The curves of intensity, I_2, versus mole fraction of oxyacid extrapolate to about 17% at 0 mole fraction, instead of 23% observed in pure water. This should not be caused by a change in the positronium formation fraction. Oxidation of Ps by electron transfer is probable by

$$Ps + H^+(aq) \rightarrow Ps^+ + H(aq).$$

If the excitation energy is negligible, the threshold for this reaction is $I_{ps} - I_H$, where I_H is the ionization potential of H(aq) and I_{ps} is the ionization potential of Ps. (See Section 6.) The reduction in Ps intensity below that for pure water is 6%, about $\frac{1}{4}$ of 23%, indicating that $\frac{1}{4}$ of the Ps atoms are oxidized in this fashion. Evidently I_{ps} is much greater than I_H, and the rate of the oxidation reaction is very fast.

Figure 8.22 shows that the intensity of τ_2 is nearly constant from acid mole fractions 0.3 to 0.7. For $HClO_4$, I_2 becomes constant at about 0.5 mole fraction. The values of I_2 are 22, 18.5, 12, and 1% for H_3PO_4, H_2SO_4, $HClO_4$, and HNO_3, respectively. At these concentrations the acids are in their first dissociated form and probably react with Ps as follows:

$$Ps + H_2PO_4^- \rightarrow PsO + H_2PO_3^-$$

$$Ps + HSO_4^- \rightarrow PsO + HSO_3^-$$

$$Ps + ClO_4^- \rightarrow PsO + ClO_3^-$$

$$Ps + NO_3^- \rightarrow PsO + NO_2^-.$$

These are double decomposition reactions that assume the formation of the compound PsO to explain the experimental evidence.

The threshold energy possessed by Ps before these reactions can occur [64] is $D_{AB} - D_{PsO}$, where D_{AB} is the dissociation energy of AB (H_2PO_3–O^-, HSO_3–O^-, and others), and D_{PsO} is the dissociation energy of the compound PsA. Values of D_{AB} for $H_2PO_4^-$, HSO_4^-, and NO_3^- are known (see Table 8.9 in the next section) and the value of D_{PsO} can be obtained from the extrapolation plot of D_{AB} versus the values of I_2 given above. As I_2 approaches zero the extrapolation gives 2.2 eV. This is the bond energy of PsO if activation energy for the reaction is ignored. From the same graph the bond energy of $(ClO_3–O)^-$ has been estimated to be 3.3 eV.

Similar arguments indicate bond energies of the following Ps compounds with a maximum estimated error of 1 eV: PsOH (<1.5 eV), PsF (2.9 eV), PsCl (2.0 eV). There is much fascinating chemical information that may be gleaned from positronium studies. However, most of the current studies are directed toward defining Ps reactions rather than using Ps as an indicator of the liquid chemical environment.

Solids

Organic Compounds

Positrons in organic solids generally exhibit two lifetime components. The lifetime and the intensity of the longer component are usually less than in the melted substance. The lifetime of the long component is commonly less than 2 nsec (with the exception of certain polymeric materials), whereas in organic liquids it is in the region of 1 to 4 nsec. The short-lifetime component is about 0.33 nsec compared to approximately 0.40 nsec in liquids. Values of lifetimes in various solids are shown in Table 8.6 along with some

Table 8.6 Positron Annihilation in Selected Organic Materials

Material	*Phase*	τ_1 (nsec)	τ_2 (nsec)	I_2 (%)	*Reference*
Benzene	Solid	0.33	1.35	31	158
	Liquid	0.38	2.94	38	158
Naphthalene	Solid[a]	0.31	0.94	20	159
	Solid	0.3–0.4	1.14	9	160
	Liquid	0.3–0.4	2.68	29	160
Anthracene	Solid[a]	0.32	—	0	158
	Solid	0.37	—	0	161
	Liquid	0.37	2.25	35–40	161
Iodobenzene	Solid	0.30	0.87	4	162
3-Nitropropionic acid	Solid	0.28	1.04	3.4	159
DPPH[b]	Solid	0.34	1.35	2.9	159
Hemoglobin	Solid	0.36	1.60	8.4	159
Hemin	Solid	0.35	1.47	2.2	159
Fused pure cane sugar	Solid	0.34	1.02	16	159

[a] Zone-refined single crystal.
[b] 2,2-Diphenyl-1-picrylhydrazyl.

liquid comparisons at temperatures above the melting point.

The largest changes of τ_1, τ_2, and I_2 as a function of temperature occur at the solid-liquid phase transition. The annihilation rates (the inverse of the lifetimes) and the intensity of the long component in benzene [158] as a function of temperature are illustrated in Fig. 8.23. The positron annihilation characteristics are different in benzene of various purities but in all cases there is a large change in the fate of the positron at the solid-liquid phase transition. The effects of minor impurities on decay rates is significant in both phases. However, air increases the annihilation rates in liquid but not solid benzene.

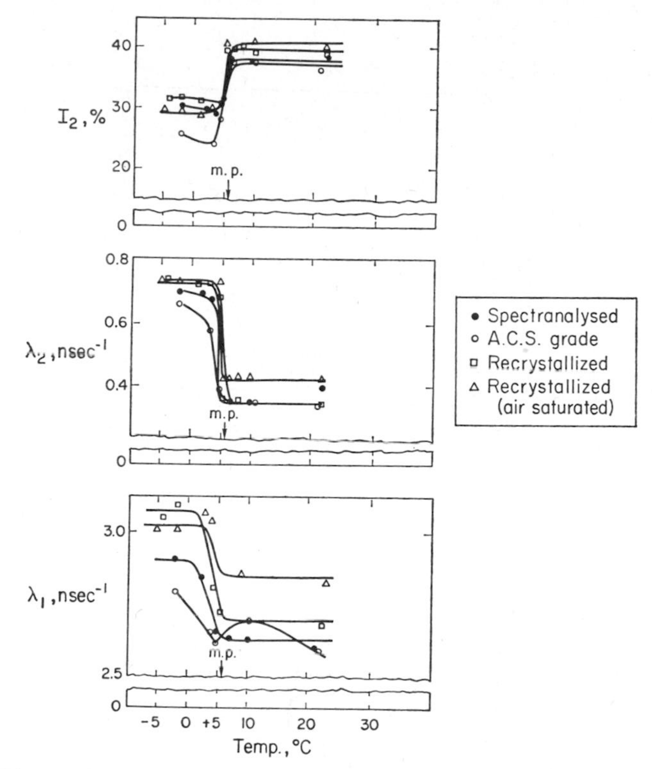

Fig. 8.23 Changes in positron decay as a function of temperature in benzene. From Ref. 158.

Naphthalene has been studied by both the lifetime and the angular correlation techniques. The lifetime of the long-lived component, τ_2, as a function of temperature in shown [160] in Fig. 8.24*a*. There is an abrupt change in the lifetime at the melting point, approximately 81°C. The intensity of the long component increases from 9 to 29% as the melting point is exceeded. The angular distributions of annihilation γ-rays [163] at 28 and 86°C are shown in Fig. 8.24*b*. When the curves are normalized to equal areas the count at the peak yields a measure of how the angular distributions vary. These "peak rates" as a function of temperature are shown in Fig. 8.24*c*. The intensity of the narrow angular distribution component increases markedly at the solid-to-liquid phase transition. This is ascribed to the decay of positronium, which is more prevalent in the liquid phase.

The change in positron annihilation properties at phase transitions may be used to study more subtle transitions than melting points. The lifetime of the long component in the lifetime spectrum in cholesteryl myristate [164]

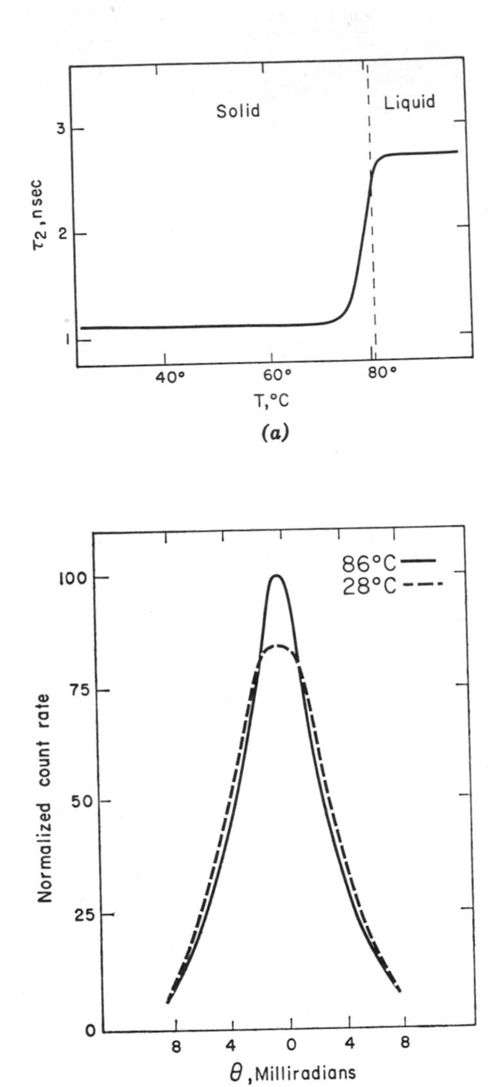

Fig. 8.24 Positron decay characteristics in naphthalene as a function of temperature: (a) lifetime τ_2, (*b*) angular distribution, (*c*) angular correlation peak rates. Reproduced by permission of The American Institute of Physics, from Refs. 160 and 163.

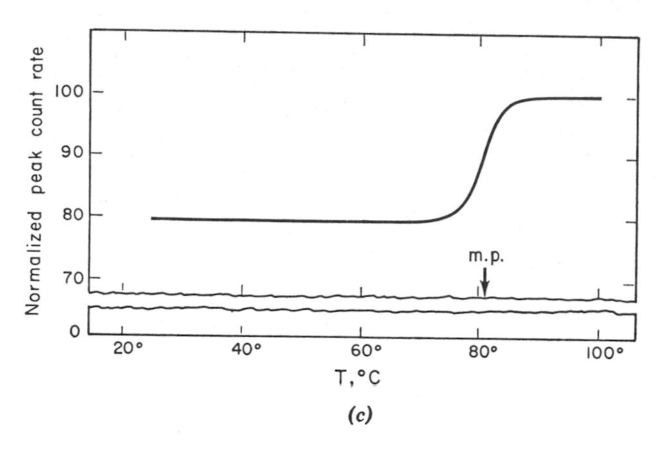

(c)

Fig. 8.24 *Continued*

increases markedly at the solid-to-smectic state transition, 71°C, and again at the smectic-to-cholesteric transition, 81°C. In solid cyclohexane there is a significant change in the long lifetime at about 110°K which may be caused by a solid-solid transition from monoclinic to cubic structure [165].

Positron annihilation in very pure, zone-refined anthracene yields no long lifetime. However, a lifetime of about 1 nsec and 3% intensity appears in anthracene of less-pure-grade [158] and neutron-irradiated samples. A long lifetime of 35–40% intensity appears when the sample is melted [161]. Hence the positron is sensitive to disorder due to radiation damage, chemical impurities, and the phase. It seems that positrons are highly sensitive to the degree of order or density of the organic substance. A direct relation of positron lifetime to the free volume in molecular materials has been found [166].

Inorganic Compounds

Most inorganic solids may be classified into three groups with respect to their effect on the positron lifetime. One group would contain strong oxidizing compounds and conversion-quenching agents that do not yield a long lifetime; for example, $KMnO_4$, $FeSO_4 \cdot 7H_2O$, and $K_2S_2O_8$. The second group would contain mild oxidizing compounds, weak conversion agents, many anhydrous salts, and slightly hydrated salts that produce a long lifetime component but with low intensity; for example, $MnCl_2 \cdot 4H_2O$, $NaBO_3 \cdot 4H_2O$, and Na_2SO_4. The third group would include many ordinary salts with water of hydration that yield a long-lifetime component with intensity greater than 10%, for example, $ZnSO_4 \cdot 7H_2O$, $Na_2SO_4 \cdot 10H_2O$, and H_3BO_3. These classifications necessarily have to be fairly qualitative because the intensity of the long component varies with the physical structure and impurity content in many compounds. Particularly, the amorphous state may cause a higher intensity than the crystalline state in many oxides.

The positron lifetimes are often close together and sometimes inseparable. A short lifetime below 0.4 nsec is generally ascribed to free annihilation and possibly some *p*-Ps. An intermediate lifetime of 0.4 to 0.7 nsec may result from a stabilized positron or *o*-Ps. Lifetimes above 0.7 nsec are called long-lifetime components. Table 8.7 illustrates the lifetimes and intensities in several inorganic solids [159] and a specific family, the alkali halides [167]. In the alkali halides τ_1 is separated into short and intermediate lifetimes $\tau_1{}^0$ and τ_1', respectively. Within each series of alkali halides the lifetime, τ_1', decreases with increasing molecular density. Comparing the variations of lifetimes as a function of cation with those as a function of anion, the anion is shown to have a greater influence on the positron lifetimes. The anion also has a greater effect on the 2γ angular distributions [168]. Three-quantum counting rates [169] do not yield any clear and convincing correlation with properties of the constituents.

The importance of physical structure to the fate of positrons is demonstrated in many inorganic oxides [170, 171]. In crystalline quartz (silicon oxide), positrons have a short lifetime and a broad 2γ angular distribution; however, a long lifetime and a narrow angular component appear in fused quartz [69, 172]. A very long positron lifetime of about 50 nsec exists in some SiO_2 powders, possibly caused by *o*-Ps annihilation in cavities [173] or in the gas at the surface of the crystals [87]. The intensity of Ps depends on the size of the particles [174]; by studying the intensity as a function of particle size, a Ps diffusion constant of 6×10^{-5} cm^2/sec has been measured [87].

Positron annihilation characteristics in Al_2O_3 are dependent on the crystallinity of the material, the gas adsorbed on the surface of powders, and the particle size [175]. The adsorption of NO_2 reduces the narrow component of the 2γ angular distribution compared with the distribution in vacuum. Hydrogen, N_2, and A cause no change, while O_2 increases the narrow component caused by conversion of *o*-Ps to *p*-Ps. A long lifetime exists with an intensity depending on the particle size. The diffusion constant of *o*-Ps has been measured at 2.6×10^{-4} cm^2/sec with an average diffusion distance of 140 Å before pick-off annihilation [87].

Positron lifetimes of about 1 nsec and a few percent intensity exist in the majority of bulk inorganic compounds studied. Typical examples are the halides of group IB in the periodic table and many metal hydrides [176, 177]. Semiconductors also may be included in this category [178]. The electron density that the positron experiences is substantially due to the valence band and those electrons in the conduction band that were excited by the localized radiation damage associated with the slowing-down positron. Silicon and

Table 8.7 Positron Annihilation Lifetimes in Various Inorganic Compounds

Sample[a]	τ_1 (nsec)	τ_2 (nsec)	I_2 (%)
$KMnO_4$	0.27	—	0
$CuSO_4 \cdot 5H_2O$	0.29	—	0
$FeSO_4 \cdot 7H_2O$	0.30	—	0
Pb acetate	0.35	—	0
$K_2S_2O_8$	0.30	—	0
$Na_2S_2O_3 \cdot 6H_2O$	0.29	1.11	0.6
$NaBO_3 \cdot 4H_2O$	0.31	1.90	0.5
$MnCl_2 \cdot 4H_2O$	0.31	1.15	1.9
Na_2SO_4	0.37	1.52	2.2
$Na_3PO_4 \cdot 12H_2O$	0.39	0.89	27
$Na_2SO_4 \cdot 10H_2O$	0.31	0.88	50
$ZnSO_4 \cdot 7H_2O$	0.29	0.65	61

Sample[b]	τ_1^0	τ_1' (nsec)	τ_2 (nsec)	I_2 (%)
LiF	~0.2	0.40	2.7	0.8
NaF	↓	0.50	2.5	1.4
KF		0.62	2.7	0.4
CsF		0.76	—	0
LiCl		0.42	—	0
NaCl		0.50	2.0	0.9
KCl		0.61	—	0
RbCl		0.63	—	0
CsCl		0.64	—	0
LiBr		0.46	3.6	1.0
NaBr		0.56	3.6	0.9
KBr		0.67	—	0
CsBr		0.68	—	0
NaI		0.51	3.3	1.2
KI		0.64	—	0
RbI		0.64	3.9	0.2
CsI		0.64	3.9	0.8

[a] From Ref. 159.
[b] From Ref. 167.

germanium with p-type and n-type dopants yield the same lifetime spectra as the undoped materials [179]. Neutron irradiation, which introduces lattice defects, causes changes in the lifetime spectra.

The present evidence strongly indicates that Ps is formed in inorganic solids. In compounds that are strong oxidizers or conversion quenching agents the lifetime of o-Ps may be reduced so much that no long lifetime is observable. In other compounds of lower quenching power the high-electron density may be sufficient to modify the lifetime by pick-off such that the longest lifetime is approximately 0.6 nsec. Both pick-off and oxidation quenching yield a broad 2γ angular distribution. Therefore as quenching by these mechanisms increases and as the lifetime becomes shorter, the narrow angular distribution component is reduced. Defects and imperfections in the solid apparently allow added space for Ps to exist and a longer lifetime component is found. This property allows Ps to be used as a tool in studying F-centers [180] and other types of defects in crystals [181].

Polymers

Positron annihilation in organic polymers has been studied by many researchers using various techniques. The more common polymers have received the most attention (e.g., polytetrafluoroethylene (Teflon), polyethylene, polystyrene and polypropylene). The positron lifetime spectra

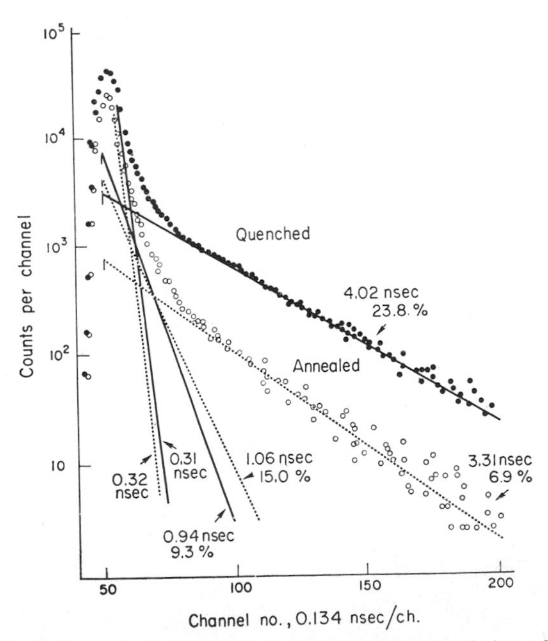

Fig. 8.25 Positron lifetime spectra in Teflon. Reproduced by permission of The Institute of Physics and the Physical Society, from Ref. 182.

exhibit the usual short component, τ_1, caused by p-Ps and free annihilation, along with one or two longer components with relatively high intensity. If the resolution of the lifetime spectrometer is sufficient, many spectra can be resolved into three components, τ_1, τ_2, and τ_3 [182, 183]. If a two-component fit is used, the longer lifetime consists of the combination of τ_2 and τ_3. Angular distribution curves for 2γ annihilation are narrower and more bell-shaped than in Fermi metals [184]. The long lifetimes and narrow angular distributions are attributed to the decay of positronium.

The dependence of positron characteristics on physical structure is manifested in Teflon. The positron lifetime spectra in quenched (44% crystallinity) and annealed (76% crystallinity) samples [182] are shown in Fig. 8.25. Three component fits to the spectra from three samples each of starting material, quenched, and annealed Teflon yield the average results shown in Table 8.8. The three lifetimes do not change as a function of crystallinity but

Table 8.8 Positron Annihilation Lifetimes and Intensities in Polymers[a]

Sample	τ_1 (nsec)	τ_2 (nsec)	I_2 (%)	τ_3 (nsec)	I_3 (%)
Teflon					
Quenched, 44% crys.	0.32	0.96	9.2	3.95	22.0
Starting material	0.32	0.93	11.4	3.89	16.0
Annealed, 76% crys.	0.32	1.03	14.3	3.69	8.0
Polyethylene	0.29	0.70	9.0	2.70	21.5
Polypropylene	0.34	0.84	4.4	2.50	21.4
Paraffin	0.33	0.90	5.5	3.20	19.0
Nylon-6	0.34	0.68	5.5	1.71	15.6

[a] Reproduced by permission of the Institute of Physics and the Physical Society, from Ref. 182.

the intensities I_2 and I_3 vary. As the crystallinity increases from the quenched to the untreated to the annealed samples, the intensity of τ_2 increases. Hence the third component with the longest lifetime is related to annihilation of o-Ps with amorphous Teflon while the second component is related to o-Ps interactions in crystalline Teflon. The lifetime is shorter in the crystalline material than in the relatively less ordered amorphous part. Radiation damage by neutrons or γ-rays has the effect of reducing I_3 and increasing the ratio $I_2/(I_2 + I_3)$. This means that radiation damage increases the crystallinity of Teflon [183]. Gamma radiation releases free radicals, such as fluorine atoms, which increase in concentration with radiation dose. This eventually causes a quenching of the long lifetimes [185, 186].

The crystallinity of Teflon also affects the 2γ angular correlations. When samples are quenched and annealed similarly to the above examples the curves are more peaked in the quenched (more amorphous) case. Typical spectra are shown in Fig. 8.26 along with the difference between them [187]. This difference in peaks is an indication of more Ps formed in the less crystalline sample. The narrow component in the 2γ angular correlation results

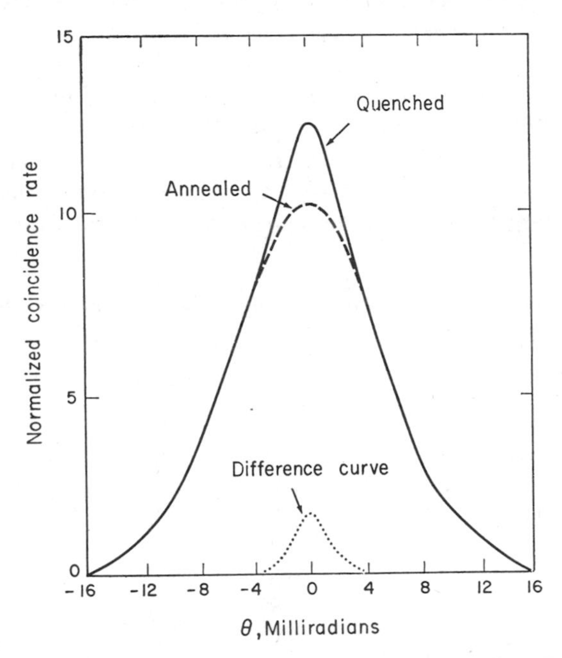

Fig. 8.26 Angular distribution of annihilation photons in Teflon. Reproduced by permission of The National Research Council of Canada, from Ref. 187.

from decay of a long-lifetime component of Ps. Application of a magnetic field causes an increase in the intensity of the narrow component as part of the *o*-Ps is converted to *p*-Ps [184]. Magnetic quenching of *o*-Ps is also shown by 3γ coincidence counting [188]. A static electric field quenches the formation of Ps [81].

Positron annihilation characteristics in Teflon change considerably as the temperature is changed. In general, the value of the longest lifetime increases as the temperature is increased with certain specific exceptions. At 1.25°K the lifetime is longer than at 4.2°K but increases from that point up to room temperature. The anomalous decrease in lifetime in raising the temperature from 1.25 to 4.2°K has been suggested to indicate a transition in the Teflon [189]. There is a sharp increase in the lifetime and a drop in its intensity near the triclinic-hexagonal lattice transition at about 20°C. This is illustrated [190] in Fig. 8.27. Another discontinuity in the lifetime versus temperature curve is found [191] near the temperature of crystalline melting, ~300°C. The increase in lifetime with temperature is probably caused by an increase in the disorder or free volume within the material as the temperature increases.

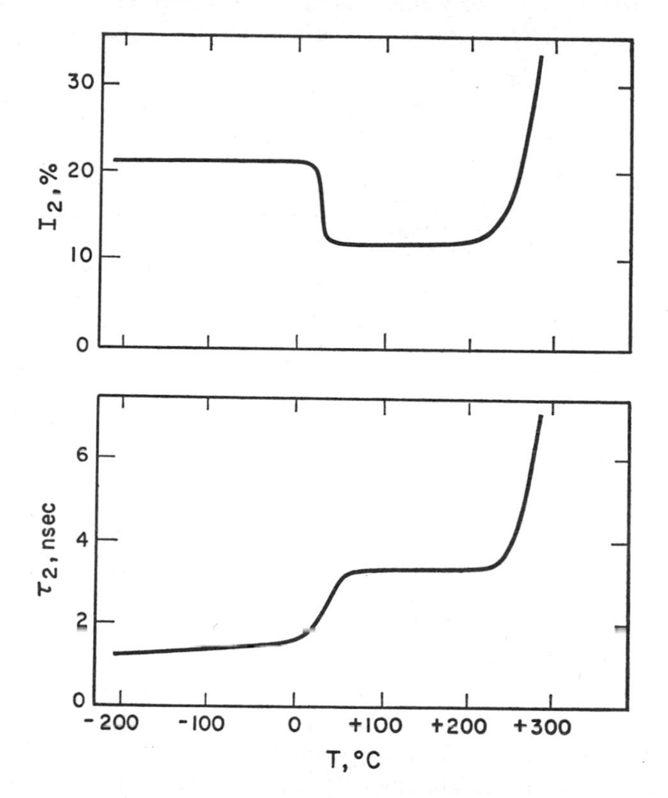

Fig. 8.27 Positron decay characteristics in Teflon as a function of temperature. Reproduced by permission of The American Institute of Physics, from Ref. 190.

The characteristics of positron annihilation in polyethylene, polystyrene, and similar polymers are affected by temperature, radiation damage, and bulk properties in much the same way as in Teflon. References to studies in these materials are the same as listed in the discussion of Teflon. Changes in positron lifetime as a function of temperature in polystyrene, Lucite, and Marlex 50 are presented in [192]. A compilation of lifetime data [193] in many polymers, including polyacrylonitrile, polyvinyl chloride, isotactic and atactic polypropylene, polystyrene, and polyethylene, indicates that pick-off quenching of *o*-Ps is dependent on the preparation and orientation of the polymers. There is also a qualitative relationship between the lifetime and the molar cohesive energy density. The higher this density, the shorter is the lifetime, that is, the pick-off quenching is greater.

The average molecular weight of polypropylene and polyethylene glycol has an effect on the positron lifetime spectra [194]. A three-component fit to the spectra in polypropylene yields lifetimes in the regions of 0.3, 1, and 2 nsec. All three lifetimes decrease as the average molecular weight of the

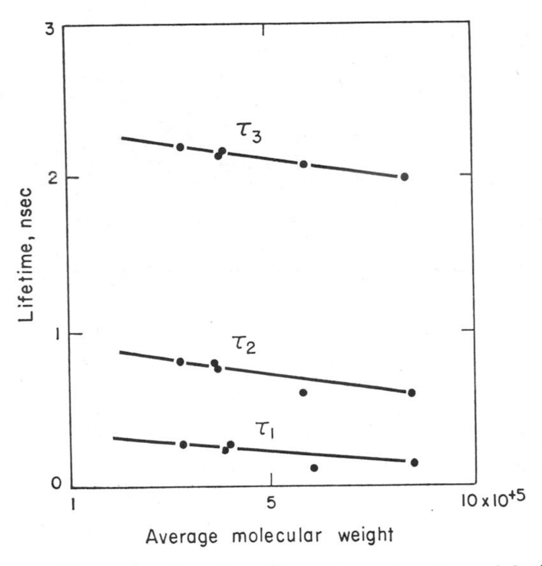

Fig. 8.28 Positron lifetimes as a function of average molecular weight in polypropylene.

polymer increases (Fig. 8.28). The intensity of the middle lifetime component increases from 10 to about 25% as the average molecular weight increases over the range illustrated. The intensity of the longest component remains constant at 20% within experimental error. In polyethylene glycol the lifetime spectra are best separated into two components. The longer component is about 2.3 nsec for average molecular weights from 200 to 600, but it dips to 1.9 nsec at 1000 and remains in this region up to a weight of 9000.

Metals

Positron annihilation in metals has received extensive study. The interaction of positrons with the electrons in metals yields information about the band structure and the Fermi distribution of electrons. Because the electron density is high and the free electrons are easily sampled by a positron, the positron lifetime is very short. Meaningful lifetime studies can be done only with high-resolution equipment. Angular correlation of 2γ annihilation quanta is a very popular technique used in studies of metals. Angular distributions in over 30 metals are illustrated in [95, 195].

Angular distribution curves can be analyzed to yield the momentum space density, $\rho(\mathbf{P})$, and the momentum distribution, $N(\mathbf{P})$, of the electrons with which the positrons annihilate. This is explained in the section on angular correlation and shown in Fig. 8.17. Theoretically, both functions should drop to zero above the Fermi energy in metals. This phenomenon is not observed strictly, but to various degrees according to the type of metal investigated. The function most sensitive to a change in the metal is $N(\mathbf{P})$. Metals including calcium, strontium, indium, and the group IA elements yield a narrow peak, a sharp cutoff near the Fermi energy E_f, and low values of $N(\mathbf{P})$ above E_f, in the plot of $N(\mathbf{P})$ versus $\mathbf{P}$. A second category of metals including barium, mercury, iron, and the group IB elements gives a steep rise in $N(\mathbf{P})$ on the low momentum side of E_f, a peak near E_f, and a very slow decline as the momentum increases. A third type of momentum distribution that is broad and nearly symmetrical with a peak slightly less than E_f is characteristic of beryllium, boron, aluminum, carbon, and antimony. Most other metals may be placed in these three categories. The momentum distributions [95] for sodium, copper, and aluminum are illustrated in Fig. 8.29.

Positrons annihilate mainly with conduction electrons in metals with simple electronic configurations, such as the alkalis. In these cases only a small portion of the positrons annihilate with the high momentum core electrons in the inner shells of the atoms. Hence an almost ideal momentum distribution corresponding to a Fermi free-electron gas is obtained. In metals of more complicated electronic structure the momentum distribution does not follow the free-electron theory as closely but contains contributions resulting from the lattice. This has a spreading effect on $N(\mathbf{P})$, as shown by aluminum in Fig. 8.29. For metals of even more complicated electronic configuration,

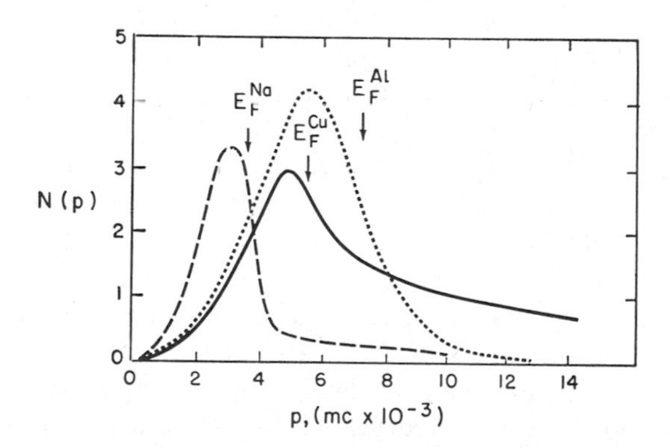

Fig. 8.29 Electron momentum distributions in Na, — — —; Cu, ———; and Al, - - - - -. Reproduced by permission of The National Research Council of Canada, from Ref. 95.

that is, the transition metals, the contributions of core electrons [196, 197] and excluded volume [20] in the lattice become important. This causes an increase in the higher-momentum components of the angular distribution curves, for example, copper (Fig. 8.29). Electron-hole excitation during positron interactions with the metal [198], incomplete thermalization of the positron [65], and an abnormally high effective mass of the positron [199] caused by association with electrons may contribute partially to the higher-momentum components.

Angular correlation of 2γ positron annihilations obviously can be used to study the dynamic nature of electrons in metals. It can locate the Fermi energy approximately by the peak in the momentum distribution (Fig. 8.29) and by the width of the angular correlation curve. There is a smooth relationship of Fermi energy to the width of the angular correlation curve at half maximum in the alkali and alkaline earth metals [117]. As the Fermi energy increases, the half-width increases. Besides measurement of the gross electronic nature of the metals, the Fermi surfaces can be studied by orienting single crystals with respect to the spectrometer. Point slits or very-high-resolution long slits are necessary in this work. Beryllium [200], monocrystals of bismuth and zinc [201], and hexagonal monocrystals of holmium, erbium, and yttrium [202] yield anisotropic orientation effects on the angular distribution. Sodium, lithium [203], and magnesium [200] show negligible anisotropy and are considered to have a nearly spherical Fermi surface. The Fermi surface in copper is not spherical, and a large change in the 2γ angular distribution occurs with a change in orientation of single crystals [204]. Figure 8.30 shows the angular distributions taken when: (a) the direction of radiation is parallel to the [111] face and the momentum examined is parallel to the [$1\bar{1}0$] face; (b) the direction of radiation is parallel to the [$1\bar{1}0$] face and the momentum examined is parallel to the [111] face. The side views of the Fermi surface are shown schematically above the correlation curves. The correspondence of changes in the correlation curves to various "necks" in the Fermi surface is illustrated by dashed lines.

Angular correlation curves in metals as a function of temperature indicate the relative population of the conduction-band electrons in the liquid and solid states. In mercury [205] the curves can be separated into a narrow and a broad component attributable to positron annihilation with conduction and core electrons, respectively. Conduction electron annihilations increase from about 25 to 40 percent when the solid is melted. This trend is also observed in gallium, zinc, cadmium, indium [206], and bismuth and bismuth–mercury alloys [207]. Other metals, such as lead, tin, and aluminum, yield small changes in angular distribution curves with temperature, but no more than expected from changes in lattice volume. A third classification of metals including sodium, lithium, and selenium shows very small temperature

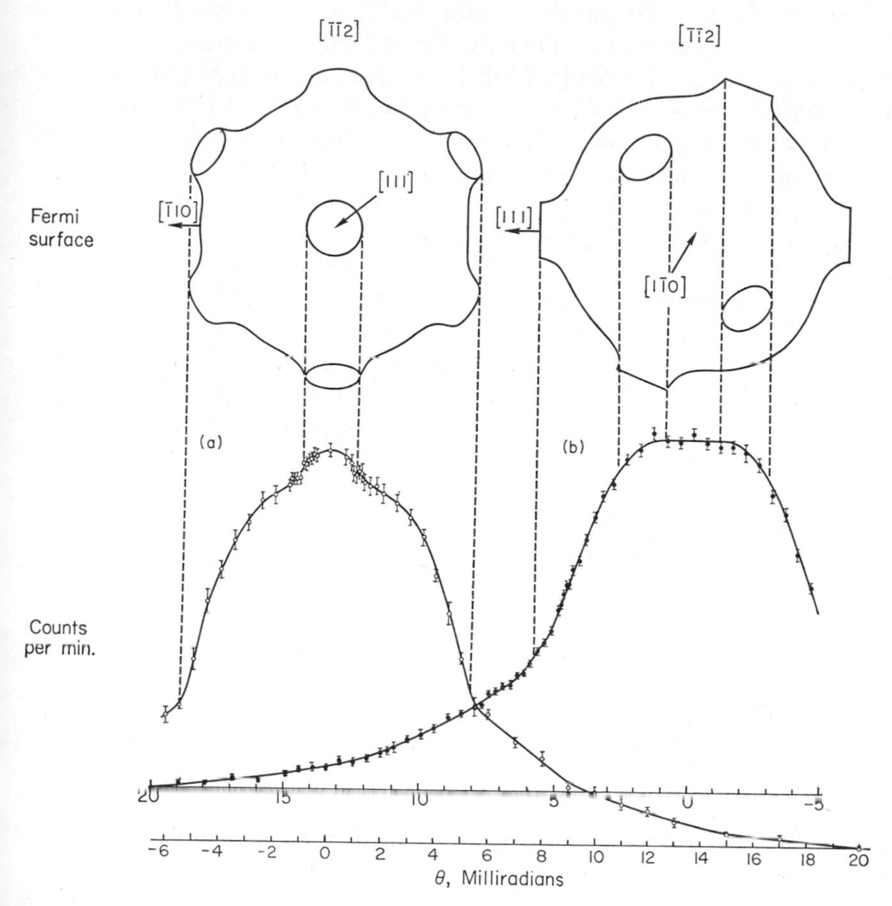

Fig. 8.30 Angular distributions of annihilation photons as a function of orientation of a copper single crystal. Reproduced by permission of The Physical Society of Japan, from Ref. 204.

effects [208]. The major changes in angular distributions at the melting point may be attributed to a decrease in the mean free path of electrons caused by heating the metal [209].

The angular correlation technique may be used to study lattice defects, such as deformation, stacking faults, dislocations, and impurity centers. Effects of magnetic fields [210] and the polarization character of the electronic bands in ferromagnets [211, 212] may be studied. A survey of this field appears in [204].

Positron lifetimes in metals are very short, and considerable care is necessary to prevent impurity effects caused by oxide layers or source or sample preparation [213]. Annihilation in pure metals is characterized by a single exponential decay with a low intensity, longer-lifetime component appearing as an artifact of sample preparation [214]. The annihilation rate increases with increased conduction electron density but for equal densities the rate is lower in semiconductors silicon and germanium that in the transition metals. Doping of the semiconductors makes little difference. In the group IIIA elements, selenium, yttrium, and the lanthanide series, the positron lifetime is nearly the same in all the trivalent metals but in europium and ytterbium, which are normally divalent, the lifetime is longer [215]. This is probably due to a greater radius of the unit electron sphere, that is, more free volume, in the divalent cases. The radius of the unit electron sphere, r_s, is 1.384 $(A/dz)^{1/3}$, where A, d, and z are the atomic weight, density, and valence, respectively. In general [111], the annihilation rate is very high (short lifetime) for low values of r_s, and decreases to the positronium spin averaged rate, 0.5 nsec, at higher values of r_s where the electron density is lower.

The positron lifetime is dependent on the density of electrons it can sample. In metals with body- and face-centered cubic structure, the lifetime increases as the lattice spacing increases. It is proposed [216] that in metals with large lattice spacings the annihilation is primarily with conduction band electrons but slightly enhanced by pick-off with core electrons. As the spacing decreases, the pick-off probability increases with concomitant shorter lifetimes and a large, high-momentum component in angular correlation curves.

Relatively long positron lifetimes appear in aluminum that is plastically deformed [217–219]. This appears to arise from positrons trapped in dislocations or vacancies [217, 220] where the electron density is reduced. Trapping of positrons in these defects also narrows the angular correlation curves [221] and Doppler-broadened annihilation lineshape [222]. This phenomenon may have very useful applications for measuring defect concentrations.

6 THEORETICAL INTERPRETATIONS

The previous section illustrated that positrons and positronium are sensitive to the electronic environment. In samples where the electron density is low at the free or bound (Ps) positron, the annihilation rate is low. The positron lifetime is from 0.1 to 3 nsec in most solids, 0.3 to 5 nsec in the majority of liquids, and up to 140 nsec in gases at low pressures. This property of the positron to sample the density of electrons that it "sees" makes it valuable for studying the internal state of a material. For example, changes in sample structure arising from variants such as temperature and radiation dose cause a pronounced change in the positron lifetime. Subtle phase transitions, small quantities of impurities, defects in crystals, disorder in solid lattices, and degrees of certain physical states such as crystallinity cause variations in the number of electrons the positron can sample; hence its lifetime is affected. If the positron stops in a defect site in a solid in which the electronic density is relatively low, it will have a longer lifetime than in the more ordered portion of the sample.

The positron lifetime is short when electrons in the environment are relatively free. For example, it is shortest in metals where an electron gas may be visualized. The fact that the momentum distribution of electrons in metals can be observed by 2γ angular correlation indicates that the conduction-band electrons are much more available to the positron than the more tightly bound core electrons. Hence in solid materials that are insulators or semiconductors the lifetime is longer than in metals of comparable macroscopic density. Fewer free electrons are available, and the positron must annihilate with a bound electron or an electron excited to the conduction band by the slowing-down positron. Doping (n- or p-typing) of semiconductors causes no change in the positron lifetime even though n-typing should increase the number of conduction-band electrons. This may be caused by still too few conduction electrons for the positron to "see," or the localized radiation damage from the positron may create an overriding high local concentration of conduction electrons in both doped and undoped cases.

Positronium is very sensitive to the paramagnetism of the environment. The bound positron in triplet positronium samples "outside" electrons for availability for pick-off, chemical reaction, or converting the triplet Ps to singlet. If unpaired electrons are available conversion occurs readily. Hence in gases such as the rare gases the Ps lifetime is relatively long but as soon as a small amount of nitric oxide is added the Ps is quenched. The cross section for quenching is greatest in paramagnetic and highly oxidative materials. Thus positrons may be very efficient sensors of certain types of impurities [126] in otherwise innocuous gases.

Positronium is sensitive to unpaired electrons and highly reactive species in liquids also. Concentrations of radical scavengers, free radicals, and oxidizing agents higher than about 10^{-4} to 10^{-3} M cause a pronounced quenching of any long-lived Ps. If water, which has 55 moles in a liter, is considered, the Ps is sensitive to these reactive species in quantities of about 2 to 20 mole ppm. The sensitivity of positronium to impurities is dependent on the reactivity with the impurity, the rate at which Ps annihilates in the solvent, and the Ps mobility. If it annihilates very rapidly with the solvent, its chances of sampling an impurity are reduced.

By virtue of the high energy with which a positron is born, it may impart localized properties to the sample. This radiation damage in the immediate environment of the positron may be negligible in cases such as metals in which many conduction-band electrons exist, but it should not be overlooked in samples in which the damage may persist during the lifetime of the positron. There is also a certain radiation and hot-atom chemistry associated with the positron and Ps, respectively. As noted in certain gases, there are "resonance" annihilations at specific energies of the positron. The interactions of positrons may be analyzed from either a chemical or a physical point of view, but both views must be taken into account in data interpretation. If the positron forms positronium, the chemistry becomes rather important.

Positronium, which consists of an electron and a positron, can be treated as a free radical with a structure something like a hydrogen radical. However, because of its low mass and the particular origin of the positron, positronium has quite special features.

1. Positronium is usually formed with a kinetic energy well above thermal energies, and it may react with the medium before it is thermalized, that is, it may react in the epithermal or "hot" region and undergo certain reactions that are not feasible for thermalized positronium.

2. Positronium is very much lighter than any other atomic or molecular species. Before a collisional reaction between positronium and a molecule, the total kinetic energy of the system is contributed mainly by the positronium, and after reaction the energy distribution depends on the masses of the products. If these masses are high, that is, a positronium compound is formed, the total kinetic energy will be distributed fairly evenly between the products; but if one of the products is a positron, it will have almost all the kinetic energy. This means that only the kinetic energy possessed by positronium has to be considered in discussing the energy required for a positronium reaction.

3. The bond energies of positronium compounds, in general, should be lower than for the corresponding hydrogen compounds because of the small mass and high mobility of positronium. Consequently positronium reactions are energetically less favorable than the corresponding hydrogen radical reactions under the same conditions. The activation energy for a positronium reaction is expected to be fairly low.

4. A chemical reaction of positronium never reaches equilibrium. The short lifetime, low concentration, and, more important, continuous thermalization obviously preclude the establishment of such an equilibrium. Consequently, no chemical or physical parameter derived from equilibrium measurements, (e.g., oxidation-reduction potentials) can be used in positronium chemistry without limitation. Obviously, positronium chemical kinetics cannot be treated as in classical chemistry, when an approach to thermal equilibrium is assumed.

5. Thermalized positronium will usually be solvated or caged in the medium.

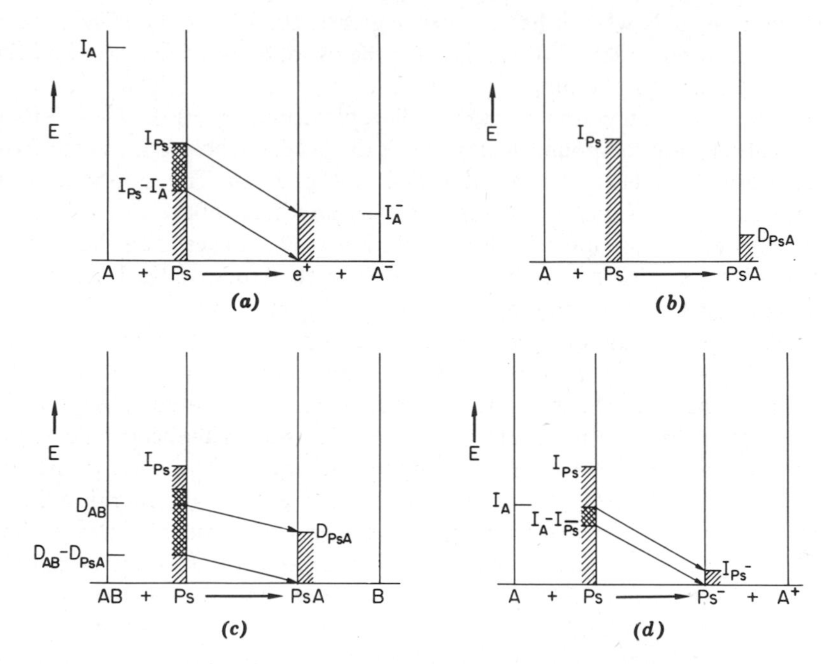

Fig. 8.31 Energetics of chemical reactions of positronium. *I* is the ionization potential and *D* is the bond energy of the subscripted molecules. (*a*) oxidation by electron transfer, (*b*) oxidation by compound formation, (*c*) oxidation by double decomposition, (*d*) reduction. Reproduced by permission of The Chemical Society, from Ref. 64.

However, because of its small mass, its wave function, particularly the positron part, may spread over a long distance and may exhibit some long-range effect.

Chemical reaction of Ps may oxidize it to the free positron, bring the Ps into close proximity to electrons by bond formation, or reduce it to Ps^- ($e^-e^+e^-$), which is stable by only 0.3 eV. This causes the lifetime of *o*-Ps to be shortened toward the free-annihilation lifetime. There are four types of chemical reactions of Ps [64]: (1) oxidation by electron transfer, (2) oxidation by compound formation, (3) oxidation by double decomposition, and (4) reduction. The energetic conditions required for these reactions are illustrated in Fig. 8.31.

Oxidation by electron transfer may be represented by:

$$\mathrm{Ps} + A \longrightarrow e^+ + A^-,$$

where A is a neutral or charged atom or molecule. As illustrated in Fig. 8.31*a*, if the ionization potential of A^-, I_{A^-}, is greater than the ionization potential of Ps, I_{Ps}, the Ps will be oxidized. If I_{A^-} is less than I_{Ps}, Ps may be oxidized to e^+ only when it has an energy greater than $I_{Ps} - I_{A^-}$. The greater the oxidation potential of A, the greater the probability of reaction and the more o-Ps quenching occurs.

Oxidation by compound formation is represented by $Ps + A \longrightarrow PsA$. The condition for compound formation is $D_{PsA} > 0$, where D_{PsA} is the Ps-A bond energy. The energetics are depicted in Fig. 8.31*b*. The annihilation of the positron in Ps is greatly enhanced by the proximity of the electron cloud in A. This may cause quenching to a lifetime even shorter than the normal free-annihilation lifetime. Interaction of Ps with strong radical scavengers such as O_2 or halogens may result in this type of reaction.

Positronium oxidation by double decomposition is represented by $Ps + AB \longrightarrow PsA + B$, where AB is a compound or ion, and PsA may be neutral or charged. The energetics of reaction are illustrated in Fig. 8.31*c*. The energy of Ps must be greater than the difference in dissociation energies of the reactant and the product compounds, $D_{AB} - D_{PsA}$. Obviously, if D_{PsA} is greater than D_{AB}, oxidation will occur at thermal energies. More practically, if D_{AB} is greater than D_{PsA}, the energy of Ps must be greater than $D_{AB} - D_{PsA}$ and the quenching effect of the reaction will be only to vary the observed Ps intensity. If $D_{AB} - D_{PsA}$ is greater than I_{Ps}, no reaction will occur because Ps would not be energetically stable above its binding energy. An example of double-decomposition chemical quenching is the $Ps + Cl_2 \longrightarrow PsCl + Cl$ reaction which was indicated in Ps studies in Cl_2–Ar mixtures.

Reduction of Ps to Ps^- ($e^-e^+e^-$), which has an ionization potential of 0.3 eV, results in an annihilation rate approaching free annihilation in the medium. Because the spin of the added electron may be either parallel or antiparallel with respect to the positron, nearly the same rate of decay considerations hold in reduction chemical quenching as in free annihilation. As illustrated in Fig. 8.31*d*, the condition for Ps^- formation is that $I_A - I_{Ps^-} < I_{Ps}$. This condition exists in many metals, particularly those with relatively low-conduction band levels.

When the chemical-reaction approach is used to interpret experimental data in liquids of early and recent research, some interesting and convincing features are observed. Table 8.9 lists values of the ionization potential, I_{AB}, and bond strength, D_{AB}, for selected molecules and ions and also the bond strength of hydrogen compounds. These values can be used to estimate the energetic conditions for reaction [157]. The observed values [143, 144, 145, 147] of I_2 for halogenated benzene and propane are as follows: ϕF, 24%; ϕCl, 14%; ϕBr, 6%; ϕI, 4%; PrCl, 16%; PrBr, 10% and PrI, 4%. These results cannot be interpreted as caused by a variation in the molecule as a whole and point to a direct interaction of Ps at the carbon-halogen bond.

The carbon–halogen bond strength, D_{AB}, and the value of D_{PsA}, depend almost entirely on the halogen, and so the threshold $D_{AB} - D_{PsA}$ depends

Table 8.9 Ionization Potentials and Bond Strengths of Selected Compounds and Ions

Compound or Ion	I_{AB}[a] (eV)	*Bond*	D_{AB}[b] (eV)	*Bond*	D_{HA}[b] (eV)	I_2 (%)
H_2O	12.6	H—OH	4.8	—	—	23.0[c]
H_2O_2	12.1	HO—OH	1.5	H—OOH	3.9	~4[c]
NH_3	11.0	H—NH_2	4.0	—	—	~25[c]
Cl_2	11.3	Cl—Cl	2.5	H—Cl	4.5	0[d]
I_2	9.4	I—I	1.6	H—I	3.1	0[e]
O_2	12.1	O—O	5.2	H—O	4.4	40.0[f]
HF	17.0	H—F	5.9	—	—	~17[c]
HCl	12.9	H—Cl	4.7	—	—	~23[c]
H_2PO_4	—	O—PO_3H_2—	6.8	H—O	4.4	~23[c]
HSO_4	—	O—SO_3H—	4.5	H—O	4.4	~18[c]
ClO_4	—	O—ClO_3	—	H—O	4.4	~12[c]
NO_3	—	O—NO_2—	2.3	H—O	4.4	~1[c]
C_3H_7Cl	10.7	Cl—C_3H_7	3.5	H—Cl	4.7	16[g]
C_3H_7I	9.41	I—C_3H_7	2.4	H—I	3.1	4[g]

From Ref. 157.
[a] Reference 223.
[b] These are the best values of the thermochemical bond energy terms listed by Cottrell [224].
[c] Reference 157.
[d] Reference 140.
[e] Reference 159.
[f] Reference 90.
[g] Reference 147.

on the halogen. In the interaction between Ps and halogen hydrocarbon the most probable reaction is

$$\text{Ps} + \text{RX} \longrightarrow \text{PsX} + \text{R}$$

because the bond between Ps and halogen is stronger than either the Ps–H or the Ps–RX bond. For this reaction D_{PsX} should be generally less than D_{RX}, and the threshold D_{RX}–D_{PsX} is positive; the quenching consequently will affect only I_2. The quenching effect of the halogen atom in a halogenated hydrocarbon will be in the order F < Cl < Br < I, because the order of the threshold energies is

$$(D_{RF}-D_{PsF}) > (D_{RCl}-D_{PsCl}) > (D_{RBr}-D_{PsBr}) > (D_{RI}-D_{PsI}).$$

Another example is the quenching due to oxyacids (Fig. 8.22), where the same relationship is found between quenching and threshold energy.

In very dense samples with resctricted free space for existence of positronium, a more physical approach to data interpretation is usually taken. For example, it has been suggested [225, 226] that long-lived Ps can exist in liquid helium only by virtue of its existing in a cavity. If no cavity existed, the Ps should have a short lifetime because of the high density of the medium.

In solids where there is very limited room for existence of free positronium [20], long positron lifetimes are often rationalized as "virtual" positronium. In repeating unit cells of a lattice the stabilized positron may exist in the "free" volume between atomic particles. The rate of pick-off annihilation of positronium trapped in one of these sites depends on the overlap of its wave function with the electronic wave function of the lattice [21, 73]. The Dirac pick-off annihilation rate is $\pi r_0^2 cn$; where r_0 is the electron radius, c is the velocity of light, and n is the number of electrons/cm^3. This may be restated as

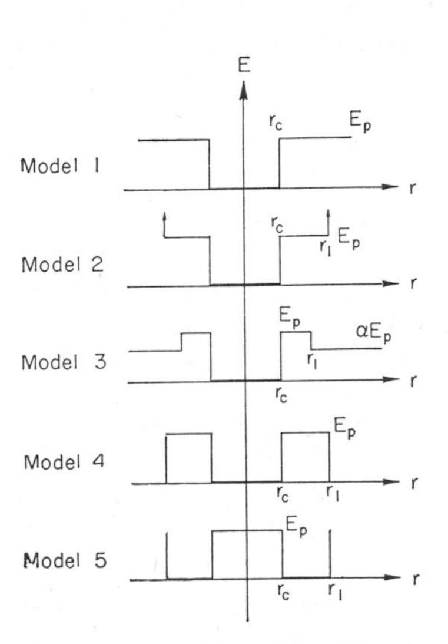

Fig. 8.32 Models of cavities in matter.

$$\lambda_{po} = \pi r_0^2 c[\Psi_-(r_+)]^2,$$

where the squared term is the density of the electron at the position of the positron. If a simplified case of positronium in a square well is considered, such as model 1 in Fig. 8.32, the pick-off annihilation rate, λ_{po}, may be mathematically defined by

$$\lambda_{po} = \pi r_0^2 c(\rho_e)G,$$

where ρ_e is the average electron density in the potential barrier and G is the fraction of the Ps wave function penetrating from the well into the barrier. The greater the overlap of the Ps wave function with the electronic wave functions in the barrier, the greater is the annihilation rate.

Elementary quantum mechanics [227] shows that for a particle to be stable in a cavity, that is, its energy is less than the potential barrier, the radius of the cavity with a potential barrier, E_p, must adhere to the following equation:

$$r_c^2 \leq \frac{\pi^2 \hbar^2}{2ME_p}.$$

The radius is defined by r_c, $\hbar$ is Planck's constant divided by 2π, M is the mass of the particle, and E_p is the energy of the barrier. If we substitute two times the mass of an electron (m) for the mass of Ps, the radius is

$$r_c \leq \frac{\pi}{2}\left(\frac{\hbar^2}{mE_p}\right)^{1/2}.$$

If the potential of the barrier is expressed in units of momentum, that is,

$$k_p = \left(\frac{mE_p}{\hbar^2}\right)^{1/2}$$

then the equation becomes

$$k_p r_c \leq \frac{\pi}{2}.$$

For $k_p r_c$ less than $\pi/2$ the potential barrier is not great enough to contain Ps and it will not be localized. When $k_p r_c$ is greater than $\pi/2$, the barrier may contain the Ps to an extent depending on the actual shape of the potential well.

At present, theoretical treatments of the extent that Ps may exist in a cavity are difficult because E_p and r_c are unknown in practical systems. The value of the product of the two terms expressed in momentum units times r_c, that is, $k_p r_c$ determines to some extent how much of the Ps wave function is in and out of the cavity. G was defined earlier as the fraction of the Ps wave function penetrating into the potential barrier. This represents the wave function out of the cavity which would cause pick-off annihilation with the surrounding medium. The fraction inside the cavity is $1 - G$. The ratio of the fraction of the Ps wave function in and out of the cavity, $F = (1 - G)/G$, yields an indication of the probability for pick-off reactions. The lower the value of F, the greater the pick-off annihilation rate should be.

Substituting F into the earlier equation: $\lambda_{\mathrm{po}} = \pi r_0^2 c(\rho_e)G$, an expression,

$$\lambda_{\mathrm{po}} = \frac{\pi r_0^2 c(\rho_e)}{1 + F},$$

is obtained. A rough approximation of ρ_e is the average electron density of the medium. With this assumption the value of λ_{po} can be calculated for values of F. The theoretical variation of F as a function of $k_p r_c$ is shown in Fig. 8.33. For values of $k_p r_c$ less than $\pi/2$, F is zero, that is, the Ps will not be localized in the cavity and λ_{po} should be high. For larger values of $k_p r_c$, F increases and the Ps is stabilized in the cavity with a resultant decrease in the pick-off annihilation rate. Even though $k_p r_c$ is unknown in a given system, pick-off annihilation rates can be measured and F calculated from the last equation. The values of F and $k_p r_c$ derived from the annihilation rate of the long component in positron-lifetime spectra in various substances are shown in Table 8.10. Obviously these values are uncertain estimates, particularly in the cases of liquids and gases, where polarization, diffusion, and kinetic factors are involved.

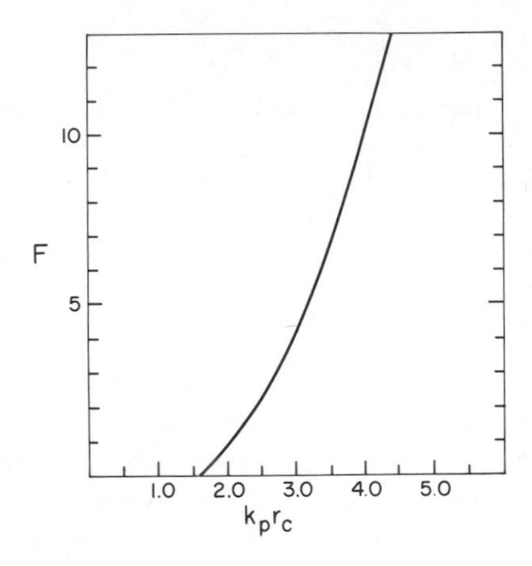

Fig. 8.33 Variation of F with $k_p r_c$ for the square well model.

Table 8.10 The Values of F and $k_p r_c$ for Positronium Annihilation in Various Substances

Material	F	$k_p r_c$
Polyvinyl chloride	5.38	3.26
Polystyrene	4.16	3.02
Polyethylene	4.25	3.03
Polypropylene (AMW 200,000)	4.21	3.03
Polypropylene (AMW 1,000,000)	3.33	2.82
Teflon, low-temperature	5.14	3.22
Teflon, high-temperature	13.6	4.5
Fused quartz	6.3	3.4
Quartz	<0.2	<1.7
Naphthalene, solid	1.3	2.2
Naphthalene, liquid	5.5	3.3
Benzene, solid	1.8	2.35
Benzene, liquid	5.7	3.27
Water	3.45	2.85
Nitric acid	2.4	2.55
Sulfuric acid	12.8	4.4
Argon	18.4	5.0
Nitrogen	19.0	5.05

In liquid media the size of the cavity is determined principally by the presence of the Ps atom. A potential barrier, such as in model 3, Fig. 8.32, is an approximation of the system, where the higher immediate barrier is due to association of molecules or electrons with the Ps atom. The barrier in the region beyond the immediate barrier cannot be described accurately and is represented by an average that is smaller than the immediate barrier. This increases the penetration of the Ps wave function and decreases F for small values of $(r_1 - r_c)/r_c$, where r_1 is the radius to the end of the immediate barrier. That is, the wave function may extend through thin immediate potential barriers, resulting in a decrease in F. The change in F as a function of $(r_1 - r_c)/r_c$ is shown schematically in Fig. 8.34 for models 2, 3, and 4 (Fig. 8.32).

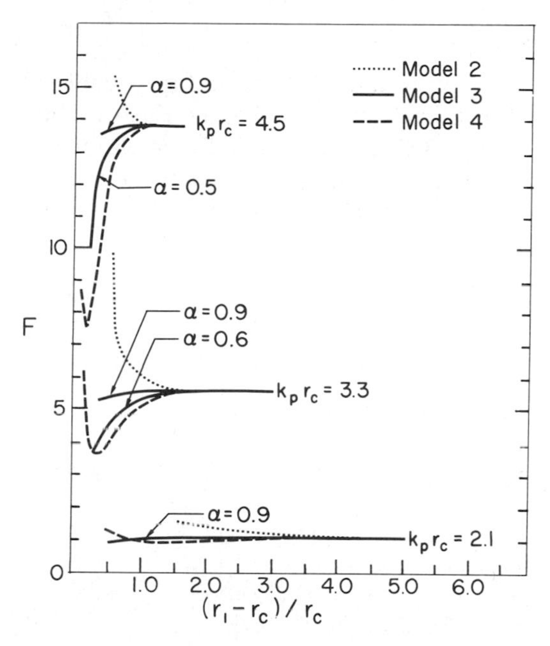

Fig. 8.34 Variation of F with cavity and barrier size.

In solids the size of the cavity is determined primarily by the structure, or because of the rigid lattice the Ps may occupy just the interstitial sites. Model 4 (Fig. 8.32) illustrates an approximation of the repeating energy barriers in solids. Tunnelling from one site to a nearby site may occur if the potential barriers are thin enough. The value of F has a minimum (maximum λ_{po}) with respect to $(r_1 - r_c)/r_c$ for this model. However, if the barrier is too thin, the volume of the barrier relative to the unit cell radius, r_1, is small enough that F increases. In solids with a regular lattice that allows only low values of

$k_p r_c$ the annihilation rates are high. Impurities or radiation damage may cause defect sites with a higher $k_p r_c$ value. Hence the Ps may tunnel to the defect sites where it is stabilized. This results in an abnormally long lifetime.

Positrons and positronium exhibit many unusual features that are directly related to the environment. Unfortunately, there is as yet no *general* explanation that relates the positron decay characteristics to a given property or set of properties in *all* materials. The positron used as a scientific probe into the microstructure of matter is still in an exploratory stage. Its present use lies mainly in specific problems where the interactions of the positron are well-known and where they vary as a function of a given parameter. However, indications are strong that future applications of the positron will yield valuable scientific information.

Acknowledgment

We thank T. M. Kelly and S. Y. Chuang for valuable discussions. J. H. Green and S. J. Tao acknowledge the Atomic Energy Commission and the United States Air Force Office of Scientific Research for support of some of the research reviewed in this chapter. We also thank the many authors who consented to have parts of their work reproduced in this review.

References

1. P. A. M. Dirac, *Proc. Cambridge Phil. Soc.*, **26,** 361 (1930).
2. C. D. Anderson, *Phys. Rev.*, **41,** 405 (1932).
3. C. D. Anderson, *Phys. Rev.*, **43,** 491 (1933).
4. C. D. Anderson, *Phys. Rev.*, **44,** 406 (1933).
5. P. M. S. Blackett and G. P. S. Occhianlini, *Proc. Roy. Soc. (London)*, **A139,** 699 (1933).
6. J. Thibaud, *Compt. Rend.*, **197,** 1629 (1933).
7. J. Thibaud, *Phys. Rev.*, **45,** 781 (1934).
8. F. Joliot, *Compt. Rend.*, **197,** 1622 (1933).
9. F. Joliot, *Compt. Rend.*, **198,** 81 (1934).
10. S. DeBenedetti, C. E. Cowan, and W. R. Konneker, *Phys. Rev.*, **76,** 440 (1949).
11. J. W. M. DuMond, D. A. Lind, and B. B. Watson, *Phys. Rev.*, **75,** 1226 (1949).
12. S. DeBenedetti, C. E. Cowan, W. R. Konneker, and H. Primakoff, *Phys. Rev.*, **77,** 205 (1950).
13. A. E. Ruark, *Phys. Rev.*, **68,** 278 (1945).
14. J. W. Shearer and M. Deutsch, *Phys. Rev.*, **76,** 462 (1949).
15. J. W. Shearer and M. Deutsch, *Phys. Rev.*, **76,** 462 (1949).
16. M. Deutsch, *Phys. Rev.*, **82,** 455 (1951).
17. M. Deutsch, *Phys. Rev.*, **83,** 866 (1951).
18. M. Deutsch, *Progr. Nucl. Phys.*, **3,** 131 (1953).

19. S. DeBenedetti and H. C. Corben, *Ann. Rev. Nucl. Sci.*, **4,** 191 (1954).
20. R. A. Ferrell, *Rev. Mod. Phys.*, **28,** 308 (1956).
21. P. R. Wallace, "Positron Annihilation in Solids and Liquids," in F. Seitz and D. Turnbull, Eds., *Solid State Physics*, Vol. 10, Academic, New York, 1960.
22. F. F. Hegmann, *Endeavour*, **20,** 225 (1961).
23. J. H. Green, *Proc. Roy. Aus. Chem. Inst.*, January 1965, pp. 7–11.
24. J. H. Green, *Endeavour*, **25,** 16 (1966).
25. V. I. Goldanski, *Atomic Energy Review*, Vol. VI, IAEA, Vienna (1968), p. 1.
26. B. G. Hogg, G. M. Laidlaw, V. I. Goldanskii, and V. P. Shantarovich, *Atomic Energy Review*, Vol. VI, IAEA, Vienna (1968), p. 149.
27. B. V. Thosar, *Proc. 56th Ind. Sci. Congress*, Part 2, Presidential Address (1969), p. 1.
28. P. A. Fraser, "Positrons and Positronium in Gases," in D. R. Bates and I. Estermann, Eds., *Advances in Atomic and Molecular Physics*, Vol. 4, Academic, New York, 1968.
29. J. Green and J. Lee, *Positronium Chemistry*, Academic, New York, 1964.
30. A. T. Stewart and L. O. Roellig, *Positron Annihilation*, Academic, New York, 1967.
31. W. Heitler, *The Quantum Theory of Radiation*, 3rd ed., Oxford University Press, London, 1954.
32. J. Thibaud, *Compt. Rend.*, **197,** 447 (1933).
33. A. H. Spees and C. T. Zahn, *Phys. Rev.*, **58,** 861 (1940).
34. L. A. Page, P. Stehle, and S. B. Gunat, *Bull. Amer. Phys. Soc.*, **28,** No. 1, 48 (1953).
35. P. S. Takhar, *Phys. Lett.*, **23,** 219 (1966).
36. P. S. Takhar, *Brit. J. Appl. Phys.*, **18,** 246 (1967).
37. F. Perrin, *Compt. Rend.*, **197,** 1302 (1933).
38. J. Brunings, *Physica*, **1,** 996 (1934).
39. R. D. Present and S. C. Chen, *Phys. Rev.*, **85,** 447 (1952).
40. W. R. Johnson, D. J. Buss, and C. O. Carroll, *Phys. Rev.*, **135,** A1232 (1964).
41. H. Mazaki, M. Nishi, and S. Shimizu, *Phys. Rev.*, **171,** 408 (1968).
42. A. Ore and J. L. Powell, *Phys. Rev.*, **75,** 1696 (1949).
43. C. N. Yang, *Phys. Rev.*, **77,** 242 (1950).
44. J. P. Carbotte and H. L. Arora, *Can. J. Phys.*, **45,** 387 (1967).
45. F. Seitz, *The Modern Theory of Solids*, McGraw-Hill, New York, 1940, p. 340.
46. J. Pirenne, *Arch. Sci. Phys. Nat.*, **29,** 121, 207 (1947).
47. V. B. Berestetzki, *Z. Exp. Teor. Fiz. SSSR*, **19,** 673, 1130 (1949).
48. R. A. Ferrell, *Phys. Rev.*, **84,** 858 (1951).
49. R. Karplus and A. Klein, *Phys. Rev.*, **87,** 848 (1952).
50. M. Deutsch and S. C. Brown, *Phys. Rev.*, **85,** 1047 (1952).
51. R. Weinstein, M. Deutsch, and S. Brown, *Phys. Rev.*, **98,** 223 (1955).
52. S. Mohorovičic, *Astron. Nachr.*, **253,** 94 (1934).
53. A. Ore, *Univ. i Bergen Arbok. Naturvitenskap. Rekke*, No. 12 (1949).
54. J. A. Wheeler, *Ann. N.Y. Acad. Sci.*, **48,** 219 (1946).
55. E. A. Hylleraas, *Phys. Rev.*, **71,** 491 (1947).
56. R. R. Sharma, *Phys. Rev.*, **171,** 36 (1968).

57. G. Ferrante, *Phys. Rev.*, **170,** 76 (1968).
58. P. F. Browne, *Nature*, **193,** 1019 (1962).
59. P. F. Browne, *Nature*, **211,** 810 (1966).
60. A. Ore, *Univ. i Bergen Arbok. Naturoitenskap. Rekke*, No. 9 (1949).
61. T. A. Pond, *Phys. Rev.*, **85,** 489 (1952).
62. G. Trumpy, *Phys. Rev.*, **118,** 668 (1960).
63. C. R. Hatcher, *J. Chem. Phys.*, **35,** 2266 (1961).
64. S. J. Tao and J. H. Green, *J. Chem. Soc.*, **A,** 408 (1968).
65. J. P. Carbotte and H. L. Arora, *Can. J. Phys.*, **45,** 387 (1967).
66. G. E. Lee-Whiting, *Phys. Rev.*, **97,** 1157 (1955).
67. P. R. Wallace, *Phys. Rev.*, **100,** 738 (1955).
68. S. J. Tao, J. H. Green, and G. J. Celitans, *Proc. Phys. Soc.* (*London*), **81,** 1091 (1963).
69. R. E. Bell and R. L. Graham, *Phys. Rev.*, **90,** 644 (1953).
70. R. A. Ferrell, *Phys. Rev.*, **110,** 1355 (1958).
71. R. E. Green and R. E. Bell, *Can. J. Phys.*, **36,** 1684 (1958).
72. R. L. Garwin, *Phys. Rev.*, **91,** 1571 (1953).
73. W. Brandt, S. Berko, and W. W. Walker, *Phys. Rev.*, **120,** 1289 (1960).
74. S. J. Tao, *Phys. Rev. Lett.*, **14,** 935 (1965).
75. J. D. McGervey, H. Horstmann, and S. deBenedetti, *Phys. Rev.*, **124,** 113 (1961).
76. M. Deutsch and E. Dulit, *Phys. Rev.*, **84,** 601 (1951).
77. W. B. Teutsch and V. W. Hughes, *Phys. Rev.*, **103,** 1266 (1956).
78. S. Marder, V. W. Hughes, C. S. Wu, and W. Bennett, *Phys. Rev.*, **103,** 1258 (1956).
79. F. E. Obenshain and L. A. Page, *Phys. Rev.*, **125,** 573 (1962).
80. A. Bisi, F. Bisi, A. Fasana, and L. Zappa, *Phys. Rev.*, **122,** 1709 (1961).
81. A. Bisi, A. Fasana, and L. Zappa, *Phys. Rev.*, **124,** 1487 (1961).
82. W. Brandt and H. Feibus, *Phys. Rev.*, **174,** 454 (1968).
83. R. E. Leamer, *Phys. Rev.*, **96,** 1607 (1954).
84. S. J. Tao and J. H. Green, *Proc. Phys. Soc.*, **82,** 1002 (1963).
85. P. S. Takhar, *Phys. Lett.*, **24A** (1967).
86. C. J. Celitans and J. H. Green, *Proc. Phys. Soc.*, **82,** 1002 (1963).
87. W. Brandt and R. Paulin, *Phys. Rev. Lett.*, **21,** 193 (1968).
88. J. Lee and G. J. Celitans, *J. Chem. Phys.*, **44,** 2506 (1966).
89. C. K. Majumdar and M. G. Rhide, *Phys. Rev.*, **169,** 295 (1968).
90. G. J. Celitans, S. J. Tao, and J. H. Green, *Proc. Phys. Soc.* (*London*), **83,** 833 (1964).
91. V. W. Hughes, S. Marder, and C. S. Wu, *Phys. Rev.*, **98,** 1840 (1955).
92. F. F. Heyman, P. E. Osmon, J. J. Veit, and W. F. Williams, *Proc. Phys. Soc.*, **78,** 1038 (1961).
93. C. Bussolati and L. Zappa, *Phys. Rev.*, **136,** A657 (1964).
94. H. P. Hotz, J. M. Mathiesen, and J. P. Hurley, *Phys. Rev.*, **170,** 351 (1968).
95. A. T. Stewart, *Can. J. Phys.*, **35,** 168 (1957).
96. H. R. Reddy and R. A. Carrigan, Jr., *Nuovo Cimento*, **66B,** 105 (1970).
97. R. E. Bell, *Ann. Rev. Nucl. Sci.*, **4,** 93 (1954).

98. R. E. Green and R. E. Bell, *Nucl. Instr.*, **3,** 127 (1958).
99. G. C. Neilson and D. B. James, *Rev. Sci. Instr.*, **26,** 1018 (1955).
100. C. W. Williams, private communication, ORTEC, Oak Ridge, Tennessee, December (1967).
101. A. Schwarzschild, *Nucl. Instr. Methods*, **21,** 1 (1963).
102. A. Ogata, S. J. Tao, and J. H. Green, *Nucl. Instr. Methods*, **60,** 141 (1968).
103. G. Present, A. Schwarzschild, I. Spirn, and N. Wotherspoon, *Nucl. Instr. Methods*, **31,** 71 (1964).
104. A. Z. Schwarzschild and E. K. Warburton, *Ann. Rev. Nucl. Sci.*, **18,** 265 (1968).
105. R. E. Bell, *α, β, and γ-Ray Spectroscopy*, Vol. II, 905, K. Siegbahn, ed., North-Holland, Amsterdam, 1965.
106. F. Lynch, *IEEE Trans. Nucl. Sci.*, **NS-13,** No. 3, 140 (1966).
107. F. Lynch, *IEEE Trans. Nucl. Sci.*, **NS-15,** No. 15 (1968).
108. R. M. Lambrecht, T. M. Kelly, and J. A. Merrigan, *J. Phys. Chem.*, **74,** 2222 (1970).
109. D. A. Gedcke and W. J. McDonald, *Nucl. Instr. Methods*, **58,** 253 (1968).
110. C. W. Shinners, *Bull. Amer. Phys. Soc.*, **13,** 1406 (1968).
111. R. E. Bell and M. Jorgensen, *Can. J. Phys.*, **38,** 652 (1960).
112. M. Bertolaccini, C. Bussolati and L. Zappa, *Phys., Rev.* **139,** A696 (1965).
113. G. J. Celitans and J. H. Green, *Proc. Phys. Soc.*, **83,** 823 (1964).
114. T. A. Pond, *Phys. Rev.*, **93,** 478 (1954).
115. P. Colombino, I. Degregori, L. Mayrone, L. Trossi, and S. DeBenedetti, *Nuovo Cimento*, **18,** 632 (1960).
116. O. E. Mogensen, *Nucl. Instr. Methods*, **84,** 293 (1970).
117. R. E. Green and A. T. Stewart, *Phys. Rev.*, **98,** 486 (1955).
118. G. Lang and S. DeBenedetti, *Phys. Rev.*, **108,** 914 (1957).
119. A. D. Mokrushin and V. I. Goldanskii, *Sov. Phys. JETP.*, **26,** 314 (1968); *Z. Eksp. Teor. Fiz.*, **53,** 478 (1967).
120. W. E. Millett and R. Castillo-Bahena, *Phys. Rev.*, **108,** 257 (1957).
121. P. Colombino, B. Fiscella, and L. Trossi, *Nuovo Cimento*, **27,** 589 (1963).
122. K. Fujiwara and O. Sueoka, *J. Phys. Soc. Japan*, **21,** 1947 (1966).
123. F. H. H. Hsu and C. S. Wu, *Phys. Rev. Lett.*, **18,** 889 (1967).
124. H. O. Hartley, *Biometrika*, **51,** 347 (1964).
125. S. J. Tao, *IEEE Trans. Nucl. Sci.*, **15,** 431 (1968).
126. S. J. Tao and J. Bell, Reference 27, p. 393.
127. D. W. Marquardt, *J. Soc. Indust. Appl. Math.*, **11,** 421 (1963).
128. H. O. Hartley, *Technometrics*, **3,** 269 (1961).
129. D. G. Gardner and J. C. Gardner, NAS-NS, 3107, 33 (1962).
130. M. Bertolaccini, A. Bisi, and L. Zappa, *Nuovo Cimento*, **46B,** 237 (1966).
131. D. A. L. Paul and P. C. Strangeby, Reference 27, p. 417.
132. S. Y. Chuang, Thesis, University of Manitoba, 1968.
133. M. Heinberg and L. A. Page, *Phys. Rev.*, **107,** 1589 (1957).
134. J. H. Green and S. J. Tao, *J. Chem. Phys.*, **39,** 3160 (1963).
135. D. A. L. Paul and L. Saint-Pierre, *Phys. Rev. Lett.*, **11,** 493 (1963).
136. S. J. Tao, J. Bell, and J. H. Green, *Proc. Phys. Soc.* (*London*), **83,** 453 (1964).
137. P. E. Osmon, *Phys. Rev.*, **138,** B216 (1965).

138. W. R. Falk and G. Jones, *Can. J. Phys.*, **42,** 1751 (1964).
139. T. M. Kelly and L. R. Roellig, Reference 27, p. 387.
140. S. J. Tao, *Phys. Rev. Lett.*, **14,** 935 (1965).
141. L. Simons, *Phys. Rev.*, **90,** 165 (1953).
142. R. A. Ferrell, *Phys. Rev.*, **103,** 1266 (1956).
143. C. R. Hatcher and W. E. Millett, *Phys. Rev.*, **112,** 1924 (1958).
144. C. R. Hatcher, W. E. Millett, and L. Brown, *Phys. Rev.*, **111,** 12 (1958).
145. C. R. Hatcher, T. W. Falconer, and W. E. Millett, *J. Chem. Phys.*, **32,** 28 (1960).
146. B. G. Hogg, T. R. Sutherland, D. A. L. Paul, and J. W. Hodgins, *J. Chem. Phys.*, **25,** 1082 (1956).
147. D. P. Kerr and B. G. Hogg, *J. Chem. Phys.*, **36,** 2109 (1962).
148. J. Lee and G. J. Celitans, *J. Chem. Phys.*, **42,** 437 (1965).
149. P. R. Gray, C. F. Cook, and C. P. Sturn, *J. Chem. Phys.*, **48,** 1145 (1968).
150. S. Berko and A. J. Zuchelli, *Phys. Rev.*, **102,** 724 (1956).
151. T. M. Kelly and S. J. Tao, *Bull. Amer. Phys. Soc.*, **13,** 103 (1968).
152. S. J. Tao, *J. Chem. Phys.*, **52,** 752 (1970).
153. R. E. Green and R. E. Bell, *Can. J. Phys.*, **35,** 398 (1957).
154. R. L. deZafra, *Phys. Rev.*, **113,** 1547 (1959).
155. J. D. McGervey and S. DeBenedetti, *Phys. Rev.*, **114,** 495 (1959).
156. J. E. Jackson and J. D. McGervey, *J. Chem. Phys.*, **38,** 300 (1963).
157. S. J. Tao and J. H. Green, *J. Phys. Chem.*, **73,** 882 (1969).
158. S. J. Tao and J. Lee, *Bull. Amer. Phys. Soc.*, **12,** 534 (1967).
159. S. J. Tao, New England Institute, Ridgefield, Conn., unpublished data.
160. H. S. Landes, S. Berko, and A. J. Zuchelli, *Phys. Rev.*, **103,** 828 (1956).
161. C. Cottini, G. Fabri, E. Gatti, and E. Germagnoli, *J. Phys. Chem. Solids*, **17,** 65 (1960).
162. E. Germagnoli, G. Poletti, and G. Randone, *Phys. Rev.*, **141,** 419 (1966).
163. R. L. deZafra and W. T. Joyner, *Phys. Rev.*, **112,** 19 (1958).
164. G. D. Cole and W. W. Walker, *J. Chem. Phys.*, **62,** 1692 (1965).
165. H. C. Clark and B. G. Hogg, *J. Chem. Phys.*, **37,** 1898 (1962).
166. R. G. Lagu, V. G. Kulkarni, B. V. Thosar, F. A. Sc., and G. Chandra, *Proc. Indian Acad. Sci.*, **69,** 48 (1969).
167. A. Bisi, A. Fiorentini, and L. Zappa, *Phys. Rev.*, **134,** A328 (1964).
168. A. T. Stewart and N. K. Pope, *Phys. Rev* , **120,** 2033 (1960).
169. A. Bisi, C. Bussolati, S. Cova, and L. Zappa, *Phys. Rev.*, **141,** 348 (1966).
170. P. Sen and A. P. Parto, *Nuovo Cimento*, **64B**, 324 (1969).
171. A. D. Tsyganov, A. Z. Varisov, A. D. Mokrushin, and E. P. Prokop'ev, *Soviet Phys.–Solid State*, **11,** 1679 (1970).
172. R. E. Bell and R. L. Graham, *Phys. Rev.*, **87,** 236 (1952).
173. R. Paulin and G. Ambrosimo, *Compt. Rend.*, **263,** 207 (1966).
174. J. Lovseth, *Phisica Norvegica* **1,** 145 (1963).
175. R. Paulin and G. Ambrosimo, *J. Phys.*, **26,** 263 (1968).
176. S. Cova, A. Dupasquier, and M. Manfredi, *Nuovo Cimento*, **47,** 263 (1967).
177. A. Gainotti, C. Ghezzi, M. Manfredi, and L. Zecchima, *Nuovo Cimento*, **56B,** 47 (1968).

178. R. Fieschi, A. Gainotti, C. Ghezzi, and M. Manfredi, *Phys. Rev.*, **175,** 383 (1968).
179. G. Fabri, G. Poletti, and G. Randone, *Phys. Rev.*, **151,** 356 (1966).
180. D. Herlach and F. Heinrich, *Phys. Lett.*, **31A,** 47 (1970).
181. G. Coussot, *Phys. Lett.*, **30A,** 138 (1969).
182. S. J. Tao and J. H. Green, *Proc. Phys. Soc.*, **85,** 463 (1965).
183. S. J. Tao and J. H. Green, *Brit. J. Appl. Phys.*, **16,** 981 (1965).
184. L. A. Page and M. Heinberg, *Phys. Rev.*, **102,** 1545 (1956).
185. G. Chandra, V. G. Kulkarni, R. G. Lagu, A. V. Patankar, and B. V. Thosar, *Phys. Lett.*, **16,** 40 (1965).
186. M. Kahn, D. Carswell, and J. Bell, *Brit. J. Appl. Phys.*, **1,** 1833 (1968).
187. I. K. MacKenzie and B. T. A. McKee, *Can. J. Phys.*, **44,** 435 (1966).
188. V. L. Telegdi, J. L. Sens, D. D. Yovanovitch, and S. D. Warshaw, *Phys. Rev.*, **104,** 867 (1956).
189. T. M. Kelly, K. F. Canter, and L. O. Roellig, *Phys. Lett.*, **18,** 115 (1965).
190. G. Fabri, E. Germagnoli, and G. Randome, *Phys. Rev.*, **130,** 204 (1963).
191. W. Brandt and I. Spirn, *Phys. Rev.*, **142,** 231 (1966).
192. B. C. Groseclose and G. D. Loper, *Phys. Rev.*, **137,** A939 (1965).
193. M. N. G. A. Khan, thesis, University of New South Wales, Australia, 1968.
194. S. J. Tao and S. Y. Chuang, to be published.
195. G. Lang, S. DeBenedetti, and R. Smaluchowski, *Phys. Rev.*, **99,** 596 (1955).
196. S. Berko and J. S. Plaskett, *Phys. Rev.*, **112,** 1877 (1958).
197. J. A. Arias-Limonta and P. G. Varlashkin, *Phys. Rev. B*, **1,** 142 (1970).
198. V. L. Sedov, *Fizikatverdogo Tela*, **9,** 1957 (1967).
199. A. T. Stewart, J. B. Shand, and S. M. Kim, *Proc. Phys. Soc.*, **88,** 1001 (1966).
200. S. Berko, *Phys. Rev.*, **128,** 2166 (1962).
201. I. Ya. Dekhtyar and V. S. Mikhalenkov, *Dokl. Akad. Nauk S.S.S.R.*, **133,** 60 (1960); **140,** 1293 (1961).
202. Z. Frait and D. Fraitova, *Cs. Cas. Fys.*, **A18,** 1947 (1966).
203. J. J. Donaghy, A. T. Stewart, D. M. Rockmore, and J. H. Kusmiss, *Proceedings of 11th International Low-Temperature Meeting*, Plenum, New York, 1965, p. 835.
204. K. Fujwara and O. Sueoko, *J. Phys. Soc. Japan*, **21,** 1947 (1966).
205. D. R. Gustafson, A. R. Mackintosh, and D. J. Zaffarano, *Phys. Rev.*, **130,** 1455 (1963).
206. I. K. MacKenzie, G. F. O. Langstroth, B. T. A. McKee, and C. G. White, *Can. J. Phys.*, **42,** 1837 (1964).
207. R. N. West, R. E. Borland, J. R. A. Casper, and N. E. Cusack, *Proc. Phys. Soc.*, **92,** 195 (1967).
208. I. Ya. Dekhtyar, *Czech. J. Phys.*, **B18,** 1509 (1968).
209. A. T. Stewart, J. H. Kusmiss, and R. H. March, *Phys. Rev.*, **132,** 495 (1963).
210. D. R. Gustafson and G. T. Barnes, *Phys. Rev. Lett.*, **18,** 3 (1967).
211. S. Berko and J. Zuckerman, *Phys. Rev. Lett.*, **13,** 339a (1964).
212. I. Ya. Dekhtyar, V. S. Mikhalenkov, and S. G. Sakharova, *Dokl. Akad. Nauk S.S.S.R.*, **174,** 803 (1967).
213. H. W. Kugel, E. G. Funk, and J. W. Mihelich, *Phys. Lett.*, **20,** 364 (1966).

214. H. Weisberg and S. Berko, *Phys. Rev.*, **154,** 249 (1967).
215. J. L. Rodda and M. G. Stewart, *Phys. Rev.*, **131,** 255 (1963).
216. G. Jones and J. B. Warren, *Can. J. Phys.*, **39,** 1517 (1961).
217. P. Hautojarvi, A. Tamminen, and P. Jauko, *Phys. Rev. Lett.*, **24,** 459 (1970).
218. B. McKee, H. Bird, and I. MacKenzie, *Bull. Amer. Phys. Soc.*, **12,** 687 (1967).
219. J. Grosskreutz and W. Millett, *Phys. Lett.*, **28A,** 621 (1969).
220. D. Connors and R. West, *Phys. Lett.*, **30A,** 24 (1969).
221. S. Berko and J. Erskine, *Phys. Rev. Lett.*, **19,** 307 (1967).
222. I. K. MacKenzie, *Phys. Lett.*, **30A,** 115 (1969).
223. F. H. Field and J. L. Franklin, *Electron Impact Phenomena*, Academic, New York, 1957.
224. T. L. Cottrell, *The Strengths of Chemical Bonds*, Butterworth, London (1958).
225. R. A. Ferrell, *Phys. Rev.*, **108,** 167 (1957).
226. L. O. Roellig and T. M. Kelly, *Phys. Rev. Lett.*, **18,** 387 (1967).
227. L. I. Schiff, *Quantum Mechanics*, McGraw-Hill, N.Y., 1955.

Chapter **IX**

THE MEASUREMENT OF RADIOACTIVITY

Herbert M. Clark

---

1 INTRODUCTION

The methods for measuring radioactivity are usually described in terms of the kind of radiation detected and the type of detection system used. In some cases it is appropriate to specify also the physical form and method of preparation of the radioactive source. For some radionuclides there may be several approximately equivalent choices of methods, whereas for others one method may be far superior to others. Before a method can be selected, the measurement problem must be defined in terms of the radiation and half-life of the radioisotope or radioisotopes to be used, the level of radioactivity expected in the assay samples, the number of measurements anticipated, and the acceptable level of error in the data. One method may be preferable in terms of routine processing of a large number of samples, another in terms of sensitivity, another in terms of efficiency, another in terms of cost, and so on. The method of choice is not necessarily a relatively new one. Thus the detector used by Becquerel when he discovered radioactivity in 1896 is still the best detector for certain types of measurement and new applications continue to be reported.

Two types of measurement are involved in the use of radioactivity, those that provide the desired experimental data and those needed to monitor the laboratory operations from the point of view of safety. In either case, an understanding of the methods requires consideration of the types of measurable radiation emitted in the various decay processes and the ways in which the radiations interact with matter. For this reason the chapter contains brief accounts of such topics as well as sections dealing with instrumentation and techniques for the measurement of radioactivity. In addition, self-decomposition of labeled compounds, radiological safety, and sources of error are covered briefly.

Somewhat greater coverage has been given to those nuclear radiation detectors and measurement methods that have become popular relatively recently, are still being investigated and improved, and seem to have the potential for many new applications. Certain methods that are rather restricted in application, for example, calorimetry, are not covered at all.

Literature sources of detailed information are given for the reader wishing to pursue a particular topic in depth. In the case of relatively well-established methods and techniques, the references are mainly to review articles, monographs, treatises, and so on, with an additional few examples of recent journal articles. The review articles, monographs, and books generally contain many references and are intended to provide entry into the literature. The recent journal references have been chosen in many instances to indicate current trends and to identify some of the journals which should be consulted for information on the "state-of-the-art" for some of the newer methods. Applications and new measurement techniques will also be found in many other journals dealing with geology, nuclear physics, biochemistry, medical science, radiochemistry, and so on. Reference 11 is less restrictive in coverage than its title suggests and is a source of reference material on many of the topics covered in this chapter. For topics such as liquid scintillation counting and the characteristics and uses of semiconductor detectors, the body of literature that has been generated in the last ten years and is currently being generated is simply overwhelming and cannot be listed in depth in this brief survey.

2 RADIOACTIVE DECAY

Rate Law and Units of Activity

Radioactive decay is a first order process for which the rate equation for a single radionuclide is

$$A = \left| \frac{dN}{dt} \right| = \lambda N, \tag{9.1}$$

where A is the activity or disintegration rate, λ is the disintegration constant, and N is the number of atoms of radionuclide. The integrated form is

$$A = A_0 e^{-\lambda t} = A_0 e^{-0.693t/t_{1/2}}, \tag{9.2}$$

where A_0 is the initial activity at the beginning of the decay period t and $t_{1/2}$ is the half-life of the radionuclide. Equations for the growth and decay of genetically related radionuclides and for activation of a target by nuclear reaction are given in a number of textbooks [1–4].

Activity is expressed in terms of the curie, where 1 Ci is 3.700×10^{10} disintegrations per second (dps). In tracer experiments the activity used is commonly in the range of microcuries to millicuries. Sources employed for counting are often in the range of nanocuries to a few microcuries, although low-level sources may be encountered at the subpicocurie level.

Specific activity, the activity per unit quantity of radioactive sample, is expressed in a variety of ways, by dps per unit weight or volume or, more commonly, in units such as μCi or mCi per ml, per gram, or per millimole (mM). The last is preferable for labeled compounds.

Modes of Decay

Radioisotope Data

A radioisotope is characterized by its half-life, the type of transition involved when it decays, and the type and energy of the radiation emitted. Such information is essential for the recognition and understanding of the problems associated with the measurement of radioisotopes. An extensive compilation of radioisotope data and decay schemes (energy level diagrams showing the various transitions) is contained in *Nuclear Data Sheets* [5], and in the table prepared by Lederer, Hollander, and Perlman [6]. New and revised data and decay schemes are published in *Nuclear Data* [7]. Martin and Blichert-Toft [8] have published a table of decay schemes and decay characteristics for 105 radioactive nuclides of special importance in nuclear medicine, health physics, and industrial applications.

Handy charts of nuclidic data are also available [9, 10]. For a list of nuclear data compilations the reader should consult Reference 11.

Alpha Decay [1–3]

For each α-transition a group of monoenergetic α-particles (^{4}He nuclei) of kinetic energy $E\alpha$ is emitted. Alpha spectra are, therefore, line spectra with each line corresponding to an α-group. The transition energy includes the recoil energy, $4E_\alpha/A$-4, given to the residual nucleus in the decay of an α-emitter of mass number A.

Alpha emission may or may not be accompanied by γ-emission. Competitive decay by an alternate route to a different product atom by way of electron capture or negatron emission may also occur. These processes are described below.

Beta Decay [1–3]

There are three modes of β-decay: negatron emission or β^--decay, positron emission or β^+-decay, and electron capture (EC) or ϵ-decay. (β^- and β^+ designate, respectively, negative and positive electrons originating in a nucleus; otherwise e^- and e^+ are used.)

In β^--decay the atomic number increases by one unit. Examples of pure β^--emitting radioisotopes of elements of particular interest are ^{3}H, ^{14}C, ^{32}P, ^{35}S, ^{36}Cl, and ^{45}Ca. Most β^--emitters are also γ-emitters. Two examples of β^--γ emitters are ^{60}Co and ^{137}Cs-^{137m}Ba.

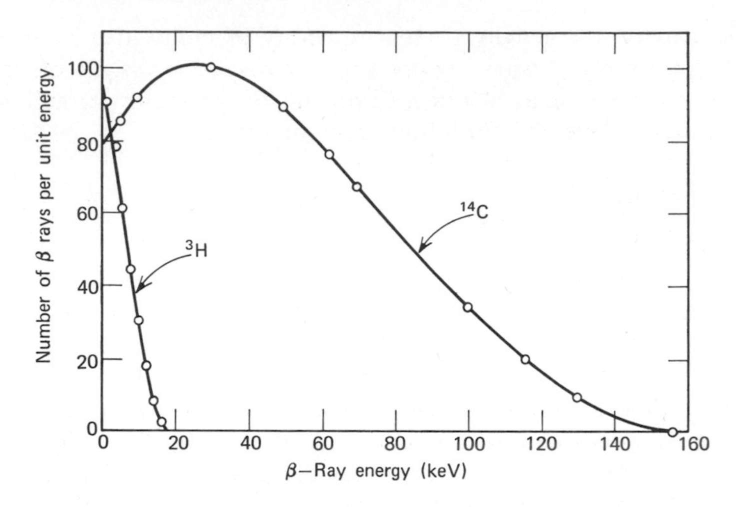

Fig. 9.1 Theoretical β-ray spectra of ^3H and ^{14}C. (From E. F. Polic, in G. J. Hine, Ed., *Instrumentation in Nuclear Medicine*, Vol. 1, Academic Press, New York, 1967.)

β^--ray spectra are continuous as shown in Fig. 9.1 for ^3H and ^{14}C. A given β^--ray shares the transition energy with an antineutrino emitted when neutron to proton conversion occurs in the nucleus. Few have the energy E_{max}, the upper limit of the spectrum corresponding to the transition energy. Spectral shapes vary depending upon E_{max}, the atomic number of the β^--emitter (or β^+-emitter), and the nuclear spin and parity changes for the transition [12]. The average energy is 30–40% of E_{max} for many β^--emitters [13]. (See also references in Section 9.)

When β^+-decay occurs, there is a decrease in atomic number by one unit. Because positron emission cannot occur unless 1.022 MeV (twice the rest mass of an electron) is available, it is usually accompanied by electron capture. Electron capture leads to the same end-product but has a much lower threshold energy. Although not shown in decay schemes, a β^+-emitter is a source of 0.511 MeV photons as discussed in Section 3.

As for negatrons, positrons are emitted with a continuous distribution of energy having an upper limit, E_{max}, equal to the transition energy. There are relatively fewer low-energy positrons, however, and E_{av} tends to be somewhat higher [13] than for negatrons. The difference arises from the interaction between the Coulomb potential of the nucleus and the ejected β-ray.

Electron capture is a process in which the unstable nucleus captures an orbital electron in the same atom to bring about the conversion of a proton to a neutron with the emission of a neutrino. The threshold energy is the binding energy of the captured electron, for example, a *K*-electron. The ion formed, for example, a *K*-ion, is a source of characteristic X-rays of the product element.

Since neutrinos are nondetectable for practical purposes, the detectable radiation emitted by a radionuclide decaying by pure electron capture consists of characteristic X-rays and Auger electrons (both from the daughter atom with discrete energies) and possibly internal bremsstrahlung (X-radiation with a continuous spectrum emitted when the nuclear charge charges). For example, ^{55}Fe decays by electron capture only and is a source of 5.95 keV *K* X-rays of manganese and 0.63 and 5.19 keV Auger electrons from manganese. By contrast, ^{54}Mn undergoes EC-γ decay resulting in the emission of a readily detectable 0.83 MeV γ-photon from ^{54}Cr in addition to X-rays and Auger electrons from ^{54}Cr.

Gamma Decay [1–3]

There are three processes by which a nucleus can undergo de-excitation without a resultant change in atomic number. The most common process is γ-emission. Gamma rays may be emitted promptly following α-decay or one of the modes of β-decay or their emission may be delayed. If the excited state has a measurable half-life, it is called metastable and is designated by an *m* after the mass number. Deexcitation from a metastable state is referred to as isomeric transition (IT). The 0.662 MeV photon associated with ^{137}Cs is emitted when the 2.55-min ^{137m}Ba daughter undergoes isomeric transition to the ground state. An example of a radionuclide decaying entirely by IT is 120-day ^{123m}Te.

Gamma rays are monoenergetic and γ ray spectra consist of discrete lines.

A nucleus may lose excitation energy without γ-ray emission by two alternate processes. When the process is internal conversion, the transition energy is transferred to an atomic electron, which is ejected from one of the shells of the atom with a kinetic energy equal to the difference in energy between the two nuclear levels minus the binding energy of the emitted electron, the conversion electron (*ce*). Thus for a given transition there will usually be several groups of monoenergetic conversion electrons differing in energy according to the binding energy for the shell from which they are ejected. Characteristic X-rays and Auger electrons are also emitted. Internal conversion becomes more probable with decreasing transition energy. The relative probability of conversion-electron emission to γ-ray emission is called the conversion coefficient. Because the intensity of a γ-ray decreases as the conversion coefficient increases and the detection system used to count conversion electrons is quite different from that used for γ-rays, internal conversion can complicate the measurement of a radionuclide.

Another decay mode which is an alternative to γ-ray emission is electron-positron pair emission. The pair, $e^{\pm}$, shares that portion of the transition energy that is in excess of 1.022 MeV. This process is relatively rare and will not be considered further.

3 INTERACTION OF NUCLEAR RADIATION WITH MATTER

Mechanisms of Interaction

Alpha Particles [1–3, 14, 15]

Because of their positive charge, α-particles passing through matter undergo Coulombic interaction with the electrons of atoms along the path. They very rarely interact with nuclei. During a primary interaction event, the α-particle transfers some of its energy to the atom or molecule as it causes either electronic excitation or ionization or both. Ionization is the formation of an ion pair, that is, a negative electron plus a positive ion, in gases or a negative electron-hole pair in solids. In gases an electron may receive up to a few keV of kinetic energy and can itself produce ionization.

As an alpha particle (or a β-ray) slows down while moving through a gas, the specific ionization, the number of ion pairs produced per unit track length, increases and reaches a maximum very near the end of the track. In air, w, the average energy lost by an α-particle per ion pair formed, is about 35 eV. This value is not the energy required to form an ion pair, but is obtained by dividing the total energy that an ionizing particle loses by the total number of ion pairs produced and, therefore, includes energy loss by excitation. As an example, a 7.0 MeV α-particle would produce, in air, a total of about 2×10^5 ion pairs along its track, but over half of the energy of the α-particle would be expended in causing excitation.

Beta Rays [1–3, 14, 15]

Negatrons and positrons, like α-particles, interact with atomic electrons to cause electronic excitation and ion-pair formation. Because of their smaller mass, their velocities are much greater than those of α-particles and they are scattered very easily.

Positive electrons differ from negative electrons by suffering an annihilation reaction which usually occurs at the end of the range where positronium [16], $e^{+} \cdot e^{-}$, forms and exists momentarily before conversion to two 0.511 MeV annihilation photons. These photons are emitted in opposite directions consistent with the conservation of linear momentum.

When a negatron or a positron is deflected in the Coulomb field of an atomic nucleus, it can lose energy by emitting electromagnetic radiation. This radiation, called bremsstrahlung (slowing-down or braking radiation) is in the X-ray spectral region. Its spectrum is continuous having an approximately exponential decrease in intensity with increasing energy from zero energy to the upper limit, which is E_{max} of the beta ray. (Because the intensity of bremsstrahlung varies inversely with the square of the mass of the moving charged particle, bremsstrahlung production is negligible for particles heavier than electrons.)

The linear rate of energy loss by bremsstrahlung production for a β-ray passing through matter is approximately proportional to the square of the atomic number of the absorbing material. For a given energy-absorbing medium the rate of energy loss increases with increasing β-ray energy. Below 100 keV it is negligible. The fraction of the β-ray energy radiated as bremsstrahlung is approximately $Z \cdot E_{max}/3000$ for passage through matter with atomic number Z. Thus for β-rays with $E_{max} = 1$ MeV interacting with aluminum about 0.4 percent of the energy appears as bremsstrahlung.

Bremsstrahlen of low intensity are also emitted when β-rays interact with the field of the nucleus from which they are emitted. These are called internal bremsstrahlen [17] to distinguish them from the more intense (external) bremsstrahlen produced along the β-ray track. Internal or "inner" bremsstrahlen are also emitted in electron capture.

High-speed β-rays can also interact with matter to cause the emission of electromagnetic radiation known as Cerenkov radiation. This low-intensity radiation occurs mainly in the ultraviolet spectral region but extends into the visible region where it can be observed as a bluish-white light. It is emitted when a β-ray passes through a transparent dielectric medium such as water or glass with a velocity v exceeding that of light in the medium, that is, c/n, where c is the velocity of light in vacuum and n is the refractive index of the medium. When v exceeds c/n, polarization of the atoms of the medium leads to the emission of electromagnetic waves in phase at an angle θ relative to the particle track. The radiation is emitted in the forward direction in a conical surface having its angle given by $\cos\theta = c/nv = 1/n\beta$. There is, then, a threshold velocity for charged particles below which Cerenkov radiation cannot be emitted. It is $\beta = 1/n$, corresponding to $\theta = 0$. For water, where $n = 1.332$, the threshold velocity is $0.7508c$, which leads to a minimum energy of 260 keV for β-rays.

Cerenkov radiation involves interactions of swift β-rays with atoms somewhat more distant than those involved in ion-pair and bremsstrahlung production. Although the fraction of the β-ray energy radiated as bremsstrahlung is much greater than that radiated as Cerenkov radiation, the intensity of the latter is much greater in the visible and ultraviolet regions. For a detailed discussion of Cerenkov radiation the reader should consult the monograph by Jelley [18].

Gamma Rays and X-rays [1–3, 14, 19, 20]

For our purposes, there are three important ways in which γ-rays or γ-photons can interact with matter to cause ionization and excitation. Interactions with atomic nuclei, including the special case of recoil-free resonance absorption (Mössbauer effect), will not be considered here.

In the photoelectric effect the γ-ray of energy E_γ disappears in a single event in which a bound electron is ejected from an atom with a kinetic energy equal to $(E_\gamma - E_B)$, where E_B is the binding energy of the electron. The photoelectron loses its energy by ion-pair production and excitation. The original positive ion formed is a source of characteristic X-rays and of Auger electrons. The X-rays are also absorbed by the photoelectric effect and in turn produce lower energy characteristic X-rays until, barring escape of radiation at some stage in the chain of events, the incident photon energy is degraded to thermal energy. The probability of occurrence of the photoelectric effect varies approximately as Z^5 for the absorber atoms and as $1/E_\gamma^{7/2}$.

By contrast, a γ-ray undergoing a Compton interaction is scattered by and transfers some of its energy to a free or loosely bound electron. After scattering, the photon energy, E_γ' in MeV, is given by

$$E_\gamma' = \frac{E_\gamma}{1 + \dfrac{E_\gamma(1 - \cos\theta)}{0.511}}, \tag{9.3}$$

where θ is the angle through which the photon is scattered. E_γ' is a minimum for $\theta = 180°$. The Compton electron has energy $(E_\gamma - E_\gamma')$ which it expends by causing excitation and ion-pair formation.

The probability of the Compton effect is approximately proportional to Z for the absorber and $1/E_\gamma$. Eventually the γ-ray energy is degraded by successive Compton interactions to a value where the photon undergoes the terminal photoelectric effect.

For γ-rays having energy in excess of 1.022 MeV, pair production is possible. In this type of interaction the γ-ray interacts with the electric field of a nucleus and is converted to a pair (a positive electron plus a negative electron). The nucleus participates in the conservation of momentum and energy. Gamma-ray energy in excess of 1.022 MeV is shared by the pair as kinetic energy which is used for ion-pair production and excitation.

The probability of interaction by pair production increases with E_γ starting at zero below $E_\gamma = 1.022$ MeV and is approximately proportional to Z^2 for the absorber.

When the positron member of the pair is annihilated, the resulting 0.511 MeV photons undergo Compton scattering and photoelectric absorption.

Absorption, Self-Absorption and Scattering of Nuclear Radiation [1–3, 14]

Alpha Particles

The range of α-particles in air expressed in centimeters is given in Fig. 9.2. The linear range may be converted to "density-thickness" range by multiplying by the density of air, for example, 1.293 mg/cm² (for dry air at 760 torr and 0°C). When expressed in mg/cm² the range is a relatively slowly varying function of the atomic number of the absorber. As an approximation, the range R_Z in an elemental absorber (with $Z > 10$) can be calculated from the range in air R_a by the equation [2]

$$\frac{R_Z}{R_a} = 0.90 + 0.0275Z + (0.06 - 0.0086Z) \log \frac{E_\alpha}{4}, \qquad (9.4)$$

where R_Z and R_a are in mg/cm² and E_α is in MeV.

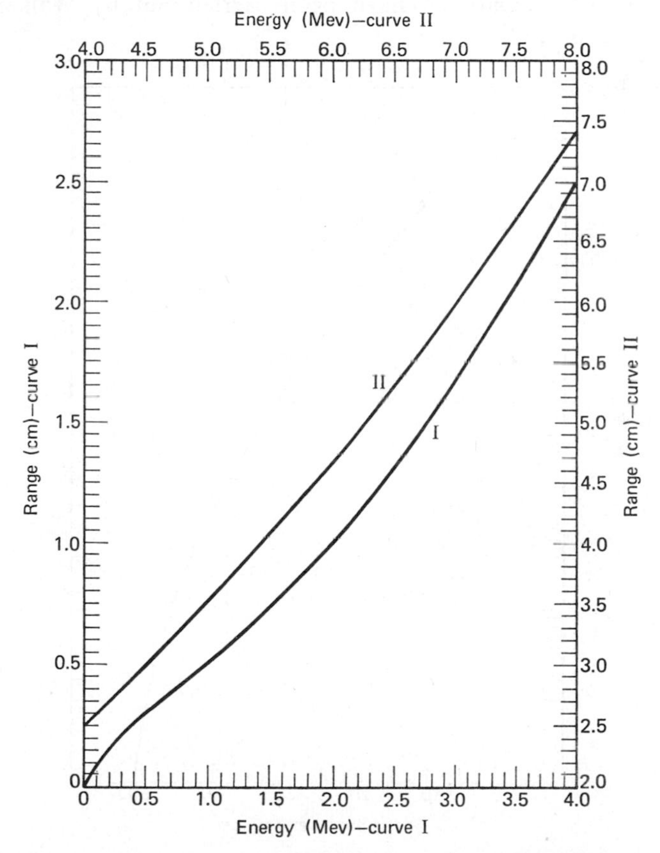

Fig. 9.2 Range-energy relation for alpha-particles in air (15°C, 760 torr) [14].

Although the range-energy curve shown in Fig. 9.2 is not linear, a rather useful rule of thumb is that for most α-emitters (E_α between 5 and 8 MeV) the range in air is approximately equal numerically to the energy in MeV.

In the preceding discussion it was assumed that the α-particles escape from the matter that constitutes the α-emitting source. Because of the short ranges of α-particles, however, they are easily self-absorbed, that is, absorbed within the source. For example, 5 MeV α-particles emitted at a depth greater than about 8 μm below the top of a solid source having a density of 5 gm/cm^3 would not escape. Solid sources for α-measurement must, therefore, be exceedingly thin.

Scattering of nuclear radiation is a potential source of error in radioactivity measurements. As shown by Rutherford, however, α-particles are not readily scattered. Scattering from metallic source mounts is the order of 2–4% and is readily taken into account. Relatively recent studies of the backscattering of α-particles have been carried out by Walker [21] and Hutchinson et al. [22].

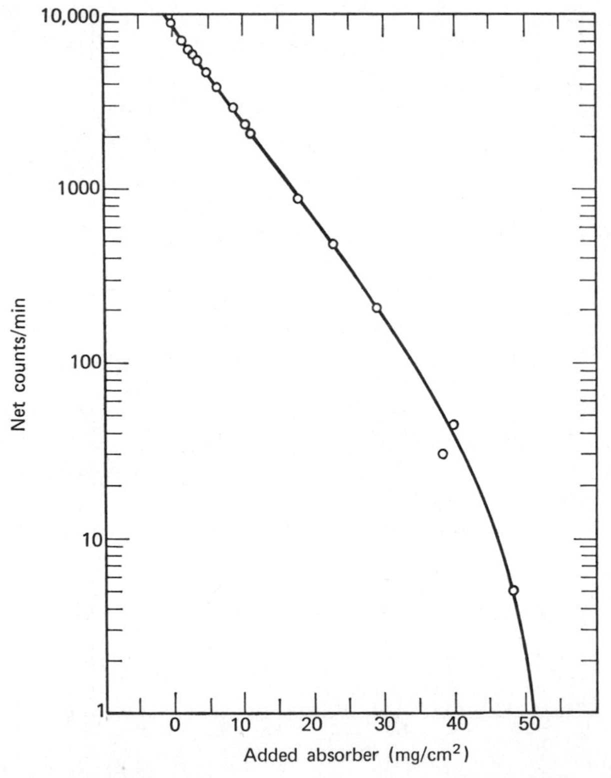

Fig. 9.3 Aluminum absorption curve for ^{45}Ca.

Beta Rays

Although β^--rays and β^+-rays have a finite range, their absorption is approximately exponential for absorber thickness much less than the range as shown in Fig. 9.3. In the exponential region

$$R(\text{cpm}) = R_0(\text{cpm})e^{-\mu_m x}, \tag{9.5}$$

where R and R_0 are the counting rates of a source measured with and without an absorber of thickness x (g/cm²), respectively, and μ_m is the mass absorption coefficient in cm²/g. The value of μ_m varies with E_{max} and is relatively insensitive to Z of the absorber. It can be estimated by the relationship [1]

$$(\mu_m) = \frac{17}{E_{max}^{1.14}}. \tag{9.6}$$

By contrast, the transmission of monoenergetic electrons through absorbers is essentially a linear function of absorber thickness in g/cm².

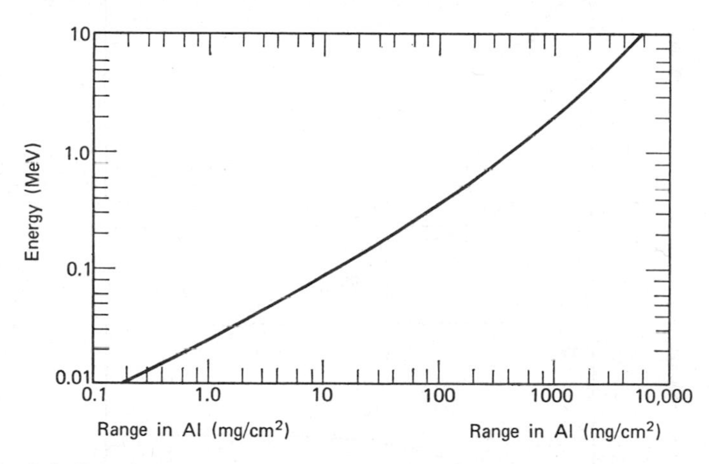

Fig. 9.4 Range-energy relation for β-particles and electrons in aluminum [2].

The experimentally determined relationship between E_{max} and range is nonlinear. Graphical representation of the range-energy relation for β-rays in aluminum is shown in Fig. 9.4. Various empirical equations have been devised to fit regions of the curve. Two of these are as follows:

$$\text{range (mg/cm}^2) = 412E_{max}^n \tag{9.7}$$

where $n = 1.265 - 0.0954 \ln E_{max}$. This equation [23] is for E_{max} from 0.01 to about 3 MeV. A simpler equation [1] for the range 0.5 to 3 MeV is

$$\text{range (g/cm}^2) = 0.52E_{max} - 0.09. \tag{9.8}$$

From (9.8) one can extract the rule of thumb that the range of β-rays in g/cm^2 is approximately equal, numerically, to one-half E_{max} in MeV.

The extent of self-absorption of β-radiation in a solid source can be estimated by assuming the absorption to be an exponential function of absorber thickness [24, 25]. For a source of thickness x (mg/cm^2) the counting rate R is related to the rate R_0 which the source would have in the absence of self-absorption by

$$R = \frac{R_0}{\mu_m x}(1 - e^{-\mu_m x}). \tag{9.9}$$

Here, the absorption coefficient in cm^2/mg is for the source material. In practice a set of sources containing the same total activity but varying

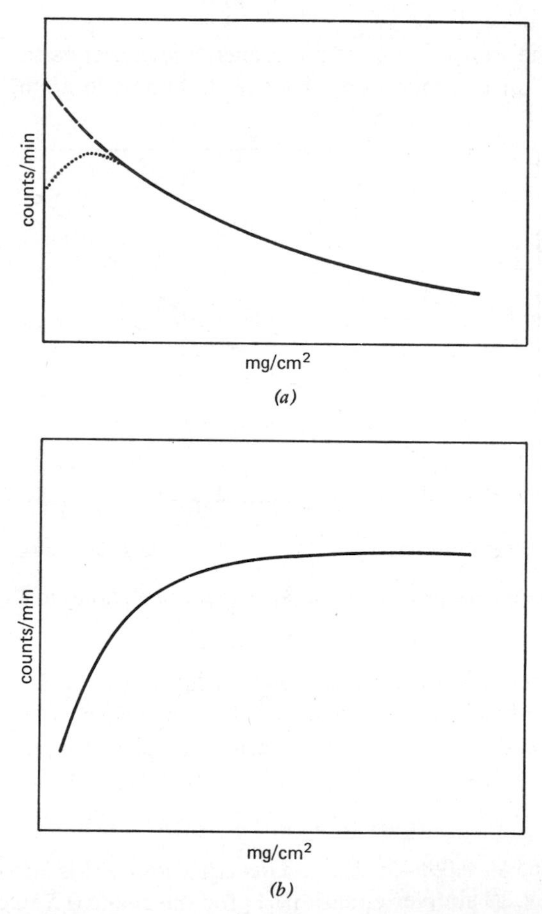

Fig. 9.5 (*a*) Counting rate as a function of sample thickness, constant total activity, (*b*) Counting rate as a function of sample thickness, constant specific activity.

thickness is prepared, the observed rates are plotted (Fig. 9.5*a*) as a function of source thickness, and the curve is extrapolated back to zero source thickness if R_0 is needed.

When the counting rates of sources having constant specific activity, for example, dpm/mg, but varying thickness are measured, a plot of the type shown in Fig. 9.5*b* is obtained. At saturation of "infinite" thickness the counting rate becomes independent of the thickness and depends only on the specific activity because beta rays emitted below a characteristic depth do not escape from the source. Saturation thickness is about 75% of the range in the source material.

Self-absorption is a potential problem whenever a radioactivity measurement involves β-counting of solid sources, but it is a major problem for weak or "soft" β-emitters. Self-absorption losses for tritium ($E_{max} = 0.0185$ MeV) are so great that solid sources are not practical for tracer experiments. For a β-emitter such as ^{14}C having E_{max} of 0.15 MeV, (9.6, 9.9) predict $R/R_0 = 0.53$ for a source 10 mg/cm in thickness. In the case of ^{32}P ($E_{max} = 1.709$ MeV) a source of the same thickness would have an estimated $R/R_0 = 0.95$.

Two types of scattering of β-rays complicate β-activity measurement. Self-scattering can be observed when the counting rates of very thin sources containing the same amount of radionuclide but different thickness are compared. With increasing source thickness the counting rate increases at first because of the increased scattering in the forward direction until self-absorption becomes the dominant effect. The dotted line in Fig. 9.5*a* represents self-scattering.

Backscattering is the reflection of β-rays from the source backing. Beta rays leaving the source in a direction which would take them away from the detector are scattered back toward the detector. The effect is to increase the counting rate. The magnitude of the effect depends upon the thickness and the atomic number of the backing [26, 27]. As shown in Fig. 9.6*a*, backscattering increases with increasing thickness of backing until a saturation thickness is reached at about 20% of the range for the particular backing material. The increase in backscattering with atomic number of the backing for saturation thickness is shown in Fig. 9.6*b*. For a thin source, saturation backscattering from a high Z backing material can almost double the number of β-rays leaving the source in the forward direction. As one might expect, the backscattered radiation is degraded in energy.

Gamma Rays and X-Rays

Gamma rays and X-rays are absorbed exponentially and have no finite range. The attenuation of γ-rays with incident intensity I_0 by an absorber of thickness x g/cm² is given by

$$I = I_0 e^{-\mu_m x}. \tag{9.10}$$

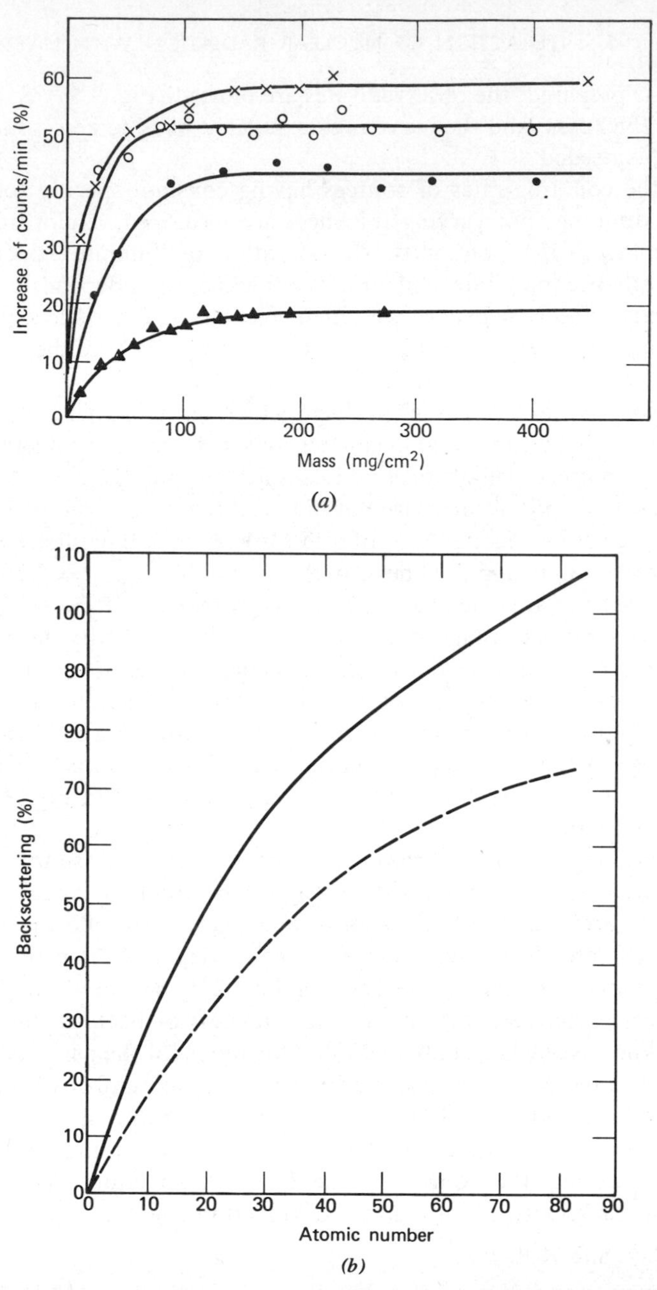

Fig. 9.6 (*a*) Backscattering for ^{32}P with various thicknesses of lead ×, copper ●, silver ○, aluminum ▲ [26]. (*b*) Saturation backscattering of beta radiation as a function of the atomic number of the backscattering material. The solid line represents data for the source directly on the surface of the backscatterer and the broken line represents data for the source on a thin film separating it from the backscatterer [90].

The mass attenuation coefficient μ_m (cm^2/g) includes the contributions from the photoelectric effect, the Compton effect, and pair production. Values for μ_m for various absorbing materials over a range of energies can be found in a number of publications [1, 20, 28–30]. (See Fig. 9.7). Equation 9.10 is often written [2, 4] in terms of the half-thickness or half-value layer, $0.693/\mu_m$, the thickness of absorber required to reduce the intensity of a beam of γ-rays by 50%. Very roughly, the half-thickness for lead in g/cm^2 is about 10 times the γ-ray energy in MeV and the linear half-thickness in cm of lead is about 10% less than the γ-ray energy in MeV.

Attenuation coefficients are measured in "good geometry" experiments, that is, experiments in which collimation of the incident and transmitted beam prevents Compton-scattered photons from reaching the detector. In the case of a shield, however, the scattered radiation that escapes from the shield is not removed by a collimator and must be considered. The buildup of scattered radiation can be taken into account by introducing a buildup factor B in the equation to give

$$I = BI_0e^{-\mu_m x}. \tag{9.11}$$

The buildup factor is a function of the γ-ray energy, and both the atomic number and the thickness of the attenuating material. The thickness is usually expressed as the number of mean free paths, $\mu_m x$. For $E_\gamma = 0.5$ MeV and $\mu_m x$ values from 1 to 19, B varies from 2.52 to 77.6 for water and from 1 24 to 2.27 for lead [28–30].

In order to calculate the energy absorbed, the mass absorption coefficient rather than the mass attenuation coefficient must be used. It is less than μ_m because the attenuation by scattering is no longer included [1, 20, 28].

Self-absorption (self-attenuation) in typical sources for γ-counting is generally much less than it is for β-sources and is certainly negligible relative to that for α-sources. It can, however amount to a few percent [31] and must be considered for accurate work.

For characteristic X-rays and for low-energy γ-rays self-absorption losses can be significant. For example, μ_m for the 14.4 keV γ-rays of ^{57}Fe in copper (a common matrix for Mössbauer sources) is 85 cm^2/g, whereas for a 500 keV photon μ_m is 0.081 cm^2/g.

Compton scattering of γ-rays can create problems in the measurement of γ-activity, especially in the case of γ-ray spectroscopy. According to (9.3) the energy of the scattered radiation for a given E_γ for the incident radiation varies from a maximum equal to the incident photon energy (for $\theta = 0°$) to a minimum of less than 0.255 MeV (for $\theta = 180°$). For E_γ above 0.25 MeV, E'_γ decreases more rapidly with increasing θ for small scattering angles the higher E_γ is, but it always approaches the same limiting value of about 0.2 MeV for scattering angles above about $\theta = 120°$. The spectrum of the

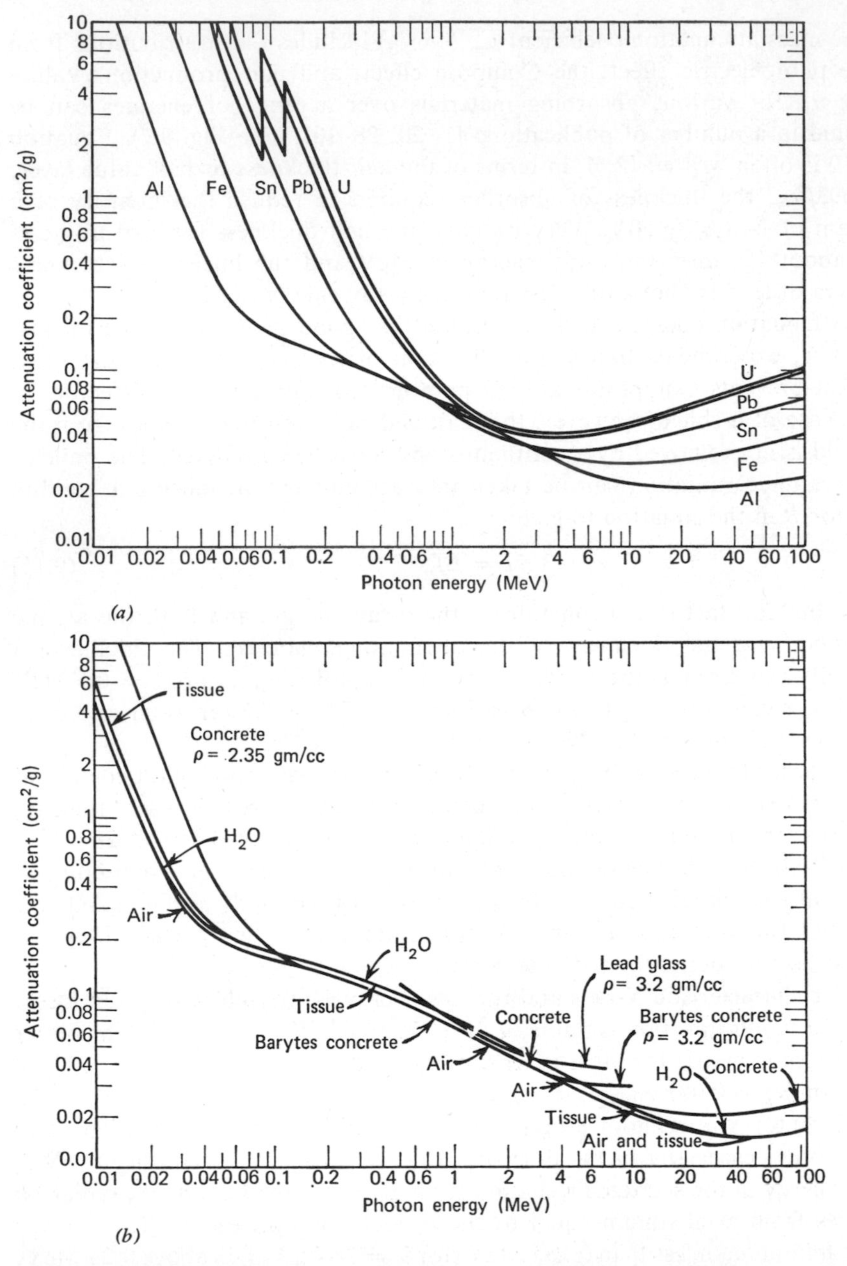

Fig. 9.7 (*a*) Mass attenuation coefficients for various elements, (*b*) Mass attenuation coefficients for various materials [28].

backscattered radiation, that is, $\theta = 120\text{–}180°$, is, therefore, peaked at about 0.2 MeV for E_γ in excess of this value.

4 SELF-DECOMPOSITION OF LABELED COMPOUNDS [32–40]

Mechanisms

Self-absorption can not only create a measurement problem, as pointed out in the preceding section, but it can also invalidate a tracer experiment, for example, by causing self-decomposition of labeled organic tracer compounds before they are put to use. Self-decomposition, in the broadest sense, includes normal thermodynamic instability, which is independent of labeling. In some cases the effect is negligible and presents no problem. In other cases, however, the compound may be essentially destroyed while in storage.

There are three mechanisms of self-decomposition which result from radioactive decay. Except for the case of isomeric transition, radioactive decay is accompanied by a change in atomic number and produces a compound of different chemical composition. Depending upon the recoil energy, the product atom (usually a stable nuclide) may or may not remain bound in the parent molecule. Unless the molecule is multiply labeled, this self-destruction, often called a primary, internal effect, does not change the radiochemical purity (the fraction of the radioisotope present in a specified chemical form). The amount of chemical impurity formed in a given storage time can be calculated from the specific activity, for example, millicuries per mmole, mCi/mM, and the half-life of the radionuclide. This type of decomposition is normally of minor importance.

A second effect, also a primary effect, but external, arises from self-absorption of the radiation in the pure substance. The ionization and excitation caused by the self-irradiation lead to chemical changes such as oxidation, condensation, and polymerization. Since it is the labeled material that undergoes chemical change in this case, the radiochemical purity is affected. In this case the percentage decomposition depends not only on the specific activity, the half-life, and the storage time, but also on the radiation sensitivity of the compound, the average energy of the radiation, and the fraction of the energy self-absorbed. The radiation sensitivity varies from one compound to another, being smallest for aromatic compounds. It is expressed in terms of a radiation destruction coefficient or "G-value" where $G(-M)$ is the number of molecules undergoing permanent damage (chemical change) for each 100 eV absorbed in the sample.

The third effect, considered to be a secondary effect, is again a consequence of absorption of the radiation, but by other substances present with the labeled compound. Examples are hydroxylic solvents, adsorbed water, and dissolved oxygen. Free radicals and other reactive species formed in these substances react with the labeled compound. The reactive species formed from water include H, OH, HO_2, H_2O_2, and H_2O^-. In addition, the energy absorbed may be transferred to the labeled compound. The effects are difficult to separate from those resulting from direct absorption of radiation.

$G(-M)$ values for the combined self-irradiation effects are commonly between 1 and 10, but they range from a few hundredths to a few thousand for cases in which chain reactions are most likely involved. They depend on the structure of the compound and on many factors such as impurities present initially and produced by self-irradiation, the physical state of the sample, the nature of a solvent, temperature, and so forth. For an example of the determination of impurities in ^{14}C-labeled hydrocarbons by gas-liquid radiochromatography, the reader is referred to the work of Muhs, Bastin, and Gordon [41].

In the absence of a measured value, $G(-M)$ may be taken as 3, based on the assumptions that the damage is caused by ionization and that the average energy expended by β-rays per ion pair formed in the compound is about 34 eV. By using this value of $G(-M)$, an estimate of the theoretical percentage of self-decomposition can be made. Although self-decomposition is a first-order process, the usual linear approximation for an exponential can be used in the calculation when the self-decomposition amounts to only a few percent. For example, if a compound has 10 mCi/mM as ^{14}C ($E_{max} = 0.156$ MeV, $E_{av} = 49.2$ keV), then assuming that $G(-M) = 3$ and that all the β-ray energy is absorbed in the sample, there would be about 1.47×10^3 molecules damaged per disintegration or 1.72×10^{19} per year corresponding to 2.8% per year. In the same period only 0.012% of the ^{14}C atoms would have decayed for any initial specific activity.

Storage Precautions

Self-decomposition is potentially a major problem for organic compounds labeled with sulfur-35, carbon-14, or tritium. Tritiated compounds are especially susceptible to radiation damage because self-absorption of the β-radiation approaches 100%. The magnitude of self-decomposition may be negligible, that is, less than 1% a year, or it may be almost complete in a few months. Self-decomposition data have been tabulated in several publications [32–40] which also contain detailed discussion of the phenomenon and ways to minimize it.

Briefly, the precautions that can be taken to control self-decomposition during storage are (1) storage of the labeled compound in solution containing a protective radiation-resistant aromatic solvent (preferably benzene), (2) addition of a free radical scavenger such as benzyl alcohol to the solution, (3) dispersion of a solid into thin layers or even adsorption on filter paper to reduce the extent of self-absorption, (4) storage at a reduced temperature (but above the freezing point, in the case of a solution) to reduce secondary radiation effects and thermal decomposition, (5) storage in a vacuum or under an inert atmosphere after removal of oxygen (and moisture, if no solvent or a nonaqueous solvent is used), and (6) conversion to a less radiation-sensitive derivative. Although these precautions are generally applicable, they may not all be equally suitable for a particular compound. The purity should be checked at least every three months and the specific activity should not be higher than necessary.

5 RADIOLOGICAL SAFETY [11, 29, 30, 42–49]

External and Internal Exposure

In order to limit the scope of coverage on this topic, it will be assumed that the activity which a radionuclide user might handle is in the range of nanocuries to millicuries.

External exposure, that is, exposure to radiation from a source external to the body, is generally not a major problem in the case of α-emitters because α-particles are stopped by a few centimeters of air and by the outer layer of about 70 μm of dead skin tissue. Beta rays, however, can travel several meters in air, and millimeters to centimeters in tissue depending upon their energy. They can, therefore, cause skin burns and can deposit energy at some depth within the body and thereby cause biological damage. The lens of the eye is particularly radiosensitive. Gamma rays and X-rays (including bremsstrahlung) can reach and damage the internal organs of the body. Alpha-emitters and β-emitters are very likely, of course, to be γ-emitters as well.

Internal exposure can arise only if the user has introduced the radionuclide into his body by some means such as inhalation, ingestion, absorption through the skin, or contamination of an open wound.

In many ways the hazards of internal exposure to the three types of radiation are reversed in importance, at least qualitatively, to those for external exposure. The hazards of each radioisotope must now be evaluated not only in terms of the type and energy of the radiation, but also in terms of its chemistry and the metabolism of the element. In general, α-emitters are potentially the most dangerous because α-particles have a relatively high specific ionization which makes them very destructive within a small volume of tissue. Radium-226, for example, is an α-emitter with a relatively high specific activity (mCi/mM) and a simple chemical behavior which makes it consistently deposit in bone, where it has a relatively long biological half-life. By contrast, the chemical behavior of plutonium, for example, ^{239}Pu, enables it to deposit in the liver or in bone or in both regions depending upon the extent to which it is in a colloidal or a soluble form in the blood stream, and to deposit in the lungs and lymph nodes, if inhaled.

Beta-emitters having a long radioactivity half-life and a long biological half-life are potential hazards in terms of internal exposure. Thirty-year strontium-90 with its 64 hr yttrium daughter is a well-known example. Precautions for handling tritium are discussed by Evans [37] and for carbon-14 by Catch [33] and Raaen, Ropp, and Raaen [38].

Because the radiation from a γ-emitter is likely to be absorbed at some distance from its point of deposit or even escape from the body, it is less likely to cause intense local damage. This would not be true, however, for an emitter of low energy γ-rays or X-rays.

Units of Dose

Both the units of dose and the maximum permissible dose levels have been changed several times since the radiation standards were first set in the late 1920s and early 1930s. According to current usage, the roentgen, R, which is a measure of the ionization produced in air, is the unit of exposure. Initially it was used as a unit of dose for X-rays and γ-rays and later, briefly, as a unit of exposure dose. An exposure of 1R produces in air 1 e.s.u. of charge in 0.001293 gram of air or 2.58×10^{-4} coulomb in 1 kilogram of air. Exposure rate is then expressed in roentgens per hour (R/hr). For work at the tracer level, milliroentgens per hour (mR/hr) is a more convenient unit.

As distinct from exposure, absorbed dose, D, is a measure of the radiation energy actually absorbed by a system, for example, a person working in a radiation area, and can be correlated with the effects of radiation. The unit of absorbed case is the rad, which is equal to 100 ergs absorbed per gram of absorber. It is approximately equal to the absorbed dose resulting from exposure to 1R of X-radiation in medical diagnosis. For X-rays and γ-rays the absorbed dose in air is approximately equal to the exposure, since 1R = 0.877 rad [50].

Because the effect of ionizing radiation on the body depends on factors in addition to the absorbed dose, an additional quantity, the dose equivalent, DE, has been defined. It takes into account variables such as the energy locally imparted along the track of a charged particle and any nonuniformity in the distribution of an internal source of radiation. The former is the linear energy transfer (LET) for the radiation in a given medium. In terms of ion pairs per millimeter or keV per micrometer, α-particles have a high LET compared with that for β-rays of the same energy.

The dose equivalent is obtained by multiplying the absorbed dose by modifying factors such as QF, the quality factor (which depends on LET) and DF, the distribution factor. Values for QF are similar to those for RBE (relative biological effectiveness), which are now restricted to use in radiobiology. The unit of dose equivalent is the rem. It differs from the absorbed dose to the extent that the modifying factors differ from unity.

For X-rays, γ-rays, and β-rays, QF is unity for all values of LET. For α-particles, QF varies from 1 to 20 corresponding to a variation of the LET in water from less than 3.5 to 175 keV/μm.

Maximum Permissible Dose (MPD)

The maximum permissible dose level has decreased by an order of magnitude since the value was first set at 0.2R per day in 1934. As established by the International Commission on Radiological Protection, ICRP, and the National Council on Radiation Protection and Measurements, NCRP, for an individual over age 18 working in a controlled area, that is, for an occupational worker, the maximum permissible accumulated dose to the whole body from external exposure is $5(N - 18)$ rems, where N is the age of the worker. For a calendar quarter it may be 3 rems. The MPD from internal exposure is 5 rems/yr and 3 rems/quarter. Levels have also been set for specific organs and parts of the body [42–48]. Current trends in radiation protection have been summarized by Taylor [51].

Safe Handling of Radioactive Materials

Appropriate safety measures are those that reduce the exposure to radiation to the minimum. In tracer work this is generally far below the MPD levels and can often be kept very close to background.

Precautions to avoid internal exposure should be taken in work involving all levels of activity. This means "good housekeeping" procedures must be used to avoid the spread of contamination. It means no pipetting by mouth and no smoking, eating, or drinking in the laboratory area. Large glass dishes, stainless steel, or fiberglass trays can be used to prevent contamination of working areas such as bench tops and hoods. The potential hazard is obviously least at the submicrocurie level and greatest at the millicurie level and above. At the higher levels and especially for high specific activity material, small losses of powders and liquids encountered in normal operations such as transfer and evaporation can lead to microcurie amounts dispersed as airborne material or as skin contamination. A hood or a glove box is required to control airborne contamination. At the high millicurie range the exhaust air from the hood should be filtered if the handling results in the release of particulate matter which would result in discharge of air with contamination exceeding the maximum permissible levels in air established by law. Wang, Adams, and Bear [52] have briefly reviewed some of the design aspects of radioisotope laboratories. (See also Reference 11.)

Rubber or plastic gloves should be worn whenever the laboratory operation is likely to lead to contamination of the hands, for example, a separation involving a separatory funnel. Routine use of gloves is seldom necessary, except perhaps to cover a skin abrasion. Gloves may, in fact, increase the chance of accidental spills because they may lead to awkwardness.

Precautions against external exposure are required for γ-emitters, β^+-emitters, and to a somewhat lesser extent for pure β^--emitters. For point sources the inverse square law provides the simplest means for reducing the exposure rate. At a distance of 3 cm, γ-emitters become a significant source of radiation at activity levels of about 100 μCi. Thus at 30 cm from an unshielded point source of 100 μCi of ^{137}Cs, the absorbed dose rate in tissue would be about 0.35 mrad/hr or somewhat over 10 times normal background. (On an hourly basis for a 40-hr week, the MPD is 2.5 mrem/hr). Within about 20%, the dose rate for point source γ-emitters having energies between 0.07 and 4 MeV can be estimated by the equation

$$\text{mrad/hr at 1 ft} = 6nEC, \tag{9.12}$$

where n is the number of photons of energy E(MeV) per disintegration and C is the source strength in mCi. The analogous expression [36] for the dose rate from a point source of a β-emitter is

$$\text{mrad/hr in tissue at 10 cm} = 3100C. \tag{9.13}$$

Equation 9.13 does not take into account (1) a minor dependence on β-ray energy, (2) backscattering, and (3) bremsstrahlung production. At greater distances the dose rate will be less than that calculated by the inverse square law because of absorption in air.

The combination of maximum convenient distance and minimum exposure time may not reduce the exposure to an acceptable level. Shielding of the source is then required. Glass storage vessels often provide adequate absorption of β-rays; however, the possible presence of penetrating radiation even in the case of pure β-emitters should not be overlooked. For β^+-rays annihilation photons are emitted from the vessel and for β^+-rays and β^--rays bremsstrahlen are also emitted from the vessel. For example, the bremsstrahlung exposure from 100 mCi of ^{32}P in aqueous solution in a glass bottle is approximately 1 mR/hr at 1 ft. To minimize bremsstrahlung production, additional shielding should be of low Z material, for example, glass, aluminum, or, preferably, plastic.

Lead, iron, and concrete are suitable for shielding against γ-radiation. Lead is the most effective and, therefore, provides the most compact shielding. In estimating shielding requirements for large sources using "good geometry" attenuation coefficients or half-thickness layers, the buildup factor (9.11) should be used to correct for Compton-scattered radiation emitted by the shield.

In setting up the equipment for an experiment involving radioactive materials and during the course of the experiment, area monitoring should be performed using instruments selected according to the type or types of radiation and the level of activity to be encountered. Portable, battery-operated instruments are commercially available for high- and low-level monitoring, including contamination monitoring. Line-operated monitors are also useful for contamination monitoring. Some commonly used instruments have been described by Grupton and Davis [53] and McCall and Wall [54].

Decontamination and Disposal

Removal of radioactive materials from working surfaces, glassware, and floor covering is basically a chemical problem and, by and large, this problem reduces to the selection of a solvent which will remove the contaminant without destroying the surface material. Depending on the chemical properties of the contaminant and the contaminated surface, one first tries the simplest solvent and then adds solubilizing agents such as soaps, detergents, complexing agents, chelating agents, or dilute acids or bases. Often the contaminant can be removed mechanically by means of a scouring powder.

If the contaminant is relatively short-lived and the contaminated items can be set aside for a sufficient number of half-lives, decontamination occurs automatically. In the case of a long-lived radionuclide such as 30 yr ^{137}Cs or 2.6 yr ^{22}Na, it is advisable that they be removed from glassware such as pipets with minimum delay because they undergo exchange-adsorption with the glass and become very difficult, if not impossible, to remove.

Waste disposal is controlled by state and federal regulations. Disposal as sewage requires a knowledge of dilution in the sanitary sewer system unless the waste is initially diluted to the maximum permissible concentration (MPC). In addition, caution is required because of the possibility that the diluted radioactive material may be reconcentrated in traps and other parts of the plumbing in the building.

Short-lived liquid waste can be stored until decayed to a negligible level and, therefore, should not be mixed with long-lived waste. Long-lived liquid waste not discharged into the sewer system can be converted to solid waste (after neutralization of any acid) by treatment with plaster of Paris. The resulting solid can be packaged in cans or drums for disposal.

Disposal by burial of solid radioactive wastes having relatively long half-lives must again conform with the applicable governmental regulations. Waste disposal and decontamination service is commercially available [40].

6 NUCLEAR RADIATION DETECTORS [11]

Ionization Chambers [55–60]

An ionization chamber (or ion chamber) consists of a vessel containing a gas and two electrodes which separate and collect the ion pairs produced by the passage of ionization radiation through the gas. The potential across the electrodes is adjusted to minimize recombination of the ion pairs without causing multiplication (gas amplification). Pulse ion chambers produce for each ionizing event an electronic pulse proportional to the number of electrons released by the ionizing radiation. Current ion chambers integrate the events and provide a dc current.

Gas-filled Proportional Counters [55–61]

These detectors are traditionally called proportional counters, although the term counter is best used to indicate a counting system, that is, a detector plus associated electronics. A gas-filled proportional counter, like an ion chamber, produces a signal by collecting ion pairs produced in a gas by absorption of energy from ionizing radiation. In this case, however, the electrons of the ion pairs produced by the ionizing radiation are given sufficient energy as they move to the anode so that they produce secondary ionization. Thus there is amplification of the original charge within the tube. The amplification factor (the number of secondary electrons produced per electron released by the ionizing radiation) increases with increased applied voltage and is typically 10^2–10^4.

The term proportional is used because the charge collected and, therefore, the output current pulse (which is converted to a voltage pulse at the pre-amplifier) is proportional to the charge produced by the ionizing radiation. It follows that the pulse is also proportional to the energy deposited in the tube by the radiation. This characteristic of a gas-filled proportional detector (and certain scintillation detectors and semiconductor detectors) makes possible the use of pulse height discrimination to distinguish between different types of ionizing radiation and the performing of energy analysis for radiation of a given type.

Geiger-Müller Tubes [55, 56, 58, 59, 61]

As for a proportional detector, gas amplification occurs in a Geiger-Müller (G-M) tube, but the proportionality is lost. A G-M tube produces larger pulses than does the proportional detector, but the pulses are essentially the same size for all ionizing events without regard to the type or energy of the incident radiation. Because the G-M tube requires low amplification, G-M counting systems are relatively simple.

Scintillation Detectors [56, 58, 59, 62–64]

A scintillation detector consists of a scintillator (phosphor) optically coupled to a photomultiplier tube which produces a current pulse when it receives light from the scintillator. In terms of chemical composition, scintillators can be placed in three categories: inorganic, organic, and noble gas. Fluorescence following excitation, and ionization within an inorganic ionic crystal by ionizing radiation, involves de-excitation of the crystal as a whole. In the case of organic scintillators, fluorescence occurs as the individual molecules undergo electronic de-excitation. For noble gases, light is emitted as excited atoms and ions de-excite.

Several types of scintillation detectors produce output voltage pulses which are proportional to the energy received by the scintillator.

Semiconductor Detectors [58, 59, 65–71]

Semiconductor detectors are often referred to as solid-state ionization chambers. The charge carriers produced by ionizing radiation are electron-hole pairs rather than ion pairs. An electrical signal is obtained by separating and collecting the electron-hole pairs, whose number is proportional to the energy received by the detector.

In a junction detector the ionizing radiation creates electron-hole pairs in the sensitive volume (the depletion or space charge region) between n^+-p or p^+-n-type layers. The depletion region in a surface-barrier detector is created when a p-type layer is formed on the surface of n-type material by oxidizing the surface. Greater depletion depths are attainable for surface barrier detectors.

Lithium-ion drifted p-i-n detectors provide much larger sensitive volumes, for example, 50 cm^3, for penetrating radiation. Lithium is diffused into p-type material to form a lithium-diffused n^+-layer. At a controlled temperature and with reverse bias (negative voltage applied to the p-type layer and positive voltage to the n-type layer) the lithium ions drift into the bulk p-type material where they compensate for the acceptor centers and thereby create an "intrinsic" region that constitutes the sensitive volume.

Junction and surface-barrier detectors are usually made from silicon and can be operated at room temperature. Similar detectors made from germanium must be cooled to reduce the leakage currents to acceptable levels. Lithium drifted silicon detectors Si(Li) can be operated at room temperature although leakage current is reduced by cooling. Ge(Li) detectors must be cooled to liquid nitrogen temperature not only for operation, but also for storage to prevent reduction in compensation that would result from diffusion at room temperature.

Semiconductor detectors are ideally suited for certain types of nuclear spectroscopy. Taking w, the average energy required to form an electron-hole pair, to be 3.65 eV for silicon and 2.95 eV for germanium, one can see that for a given amount of energy absorbed about ten times as many electron-hole pairs are formed as ion pairs formed in gases. This implies intrinsically better spectral resolution. (Note that w is greater than the energy gap because, as for the interaction with gases, the ionizing radiation loses a large fraction of its energy by causing electronic excitation.)

Photographic Emulsion [59, 72–76]

Ionizing radiation traversing a silver halide emulsion produces a latent image of its track. When the image is developed it provides a permanent record of the interaction. Depending upon the technique used, a photographic emulsion can be used as a single event detector (micro autoradiography) by recording the track of a single ionizing particle. It can also be used to locate the sites of radioactivity on a surface (gross or macro autoradiography), or it can be used to integrate many events over a period of time (dosimetry) by measurement of the change in optical density.

Thermoluminescent Detectors [46, 59, 77–80]

A thermoluminescent detector is a crystalline phosphor which emits light when it is heated after exposure to radiation. Some of the electrons that are released and some of the resulting lattice vacancies that are formed by ionizing radiation, are trapped in metastable states at lattice imperfections, for example, impurity sites. When the material is heated (usually in the range of 200–300°C), trapped electrons and holes receive sufficient energy to escape from the traps and recombine with the emission of light called thermoluminescence (TL). The light, which is in the visible spectral region and whose intensity increases with increasing integrated number of ionizing events, can be detected by a photomultiplier tube connected to an electrometer. One source of error in the use of TL phosphors arises from nonradiation-induced thermoluminescence induced mechanically. These detectors are used mainly for γ-ray dosimetry although they respond to β-rays.

Radiophotoluminescent Detectors [78]

When silver-activated phosphate glasses absorb ionizing radiation, stable photoluminescent centers are formed. Subsequent irradiation with ultraviolet light produces an orange fluorescence which can be measured with a photomultiplier tube. The signal from the photomultiplier is amplified and then used to operate a read-out device. The intensity of the radiophotoluminescence (RPL) increases with increased absorption of ionizing radiation.

Solid-State Track Detectors [81–85]

Charged particles impinging on the surface of an object such as a glass plate or a film of a high polymer produce trails of damage which can be developed (chemically etched) into tracks that can be individually examined or counted with an optical microscope or by spark counting [86, 87]. These detectors, also known as dielectric track detectors (DTD), respond to heavy charged particles, for example, α-particles and fission fragments, and protons and other ions from accelerators. Lexan and cellulose nitrate are two commonly used plastic detectors. Two of the criteria for track registration in use are the primary ionization criterion [88] and the restricted energy loss criterion [89].

7 NUCLEAR RADIATION MEASUREMENT SYSTEMS [11]

Introduction

It is important to introduce certain terms that are frequently used to characterize and compare detectors and detection systems.

Geometry Factor

The geometry factor (or simply the "geometry"), G, is the fraction of the radiation emitted by a source which starts out in the direction of the sensitive volume of the detector. In other words, it is the fraction of the solid angle about the source subtended by the sensitive volume of the detector, and is usually expressed as percent solid angle. The two special cases of 100 and 50% are designated as 4π and 2π, respectively.

The geometry factor depends on the dimensions of both source and detector and on the distance of separation. Formulas for calculating geometry factors are given by Burtt [27], Price [59], and Steinberg [90]. Computer-calculated geometry values are given by Williams [91], Ruffle [92], and Gardner and Verghese [93].

Detection Efficiency

The efficiency E (Y is also used) of a detection system is the fraction or percent of the disintegrations occurring in a source that is detected. It can vary with any of a number of factors [4, 27, 59, 90], for example, (1) geometry (2) intrinsic efficiency of the detector, that is, the fraction of the radiation reaching the sensitive volume that is actually detected; (3) resolving time of the detector; (4) absorption in a source cover, air, or detector window; (5) self-absorption; (6) backscattering, self-scattering, and scattering from shielding, detector mount, and so forth; and (7) fraction of detectable radiation emitted per disintegration.

In addition, E depends also on the way in which the signal from the detector is processed by the associated electronics, that is, on the instrument settings. For example, raising the level of the pulse height selector in a pulse counter will reduce the counting rate in proportion to the arrival rate of pulses with height less than the level. If such pulses include those produced by radiation from the source and are needed for counting, the efficiency will be reduced. Also, if the associated electronics, for example, an amplifier, is unable to resolve pulses being detected at a high rate, the efficiency will be reduced.

For a discussion of the efficiency of a γ-ray spectrometer, see Chapter X, Gamma-Ray Spectroscopy.

Figure of Merit

Various figures of merit have been suggested for comparing detectors, comparing counting systems, and for selecting the optimum operating conditions for a given counter. Two rather commonly used figures of merit are S^2/B and E^2/B, where S is the net counting rate of a given radioactive source, E is the detection efficiency in percent for the source, and B is the background rate determined with the same instrument settings used to obtain S or E. These evaluation criteria are of use for low-level samples where the net sample rate is less than the background rate. As pointed out in Section 11, they are obtained from consideration of the statistical aspects of radioactive decay. There is an optimum division of total time between sample measurement and background measurement that gives the minimum theoretical error in the net sample rate. For $S \ll B$, the predicted statistical error will be a minimum for a given total measurement time when the counter with the largest value of S^2/B is used or when a given counter is operated to give the largest value of S^2/B. Similarly, the counter with the largest S^2/B will require the shortest total counting time for a specified error for a given sample. Because $S = E \cdot A$, where A is the disintegration rate of the source, the figure of merit can also be written E^2/B. In this form it is independent of the sample counting rate.

Detection Sensitivity

Detection sensitivity is also called detector sensitivity because it is often treated as if a property of the detector alone. The term detection sensitivity is often used without being precisely defined. For example, it is sometimes used to mean the lowest level of activity that can be quantitatively measured with the detection system. A very high sensitivity means that a source having a very low activity can be detected. This would require a very low background (signal obtained with a blank). Low background does not guarantee high detection sensitivity, however, because the sensitivity also depends on efficiency. Sensitivity is often expressed in terms of a figure of merit, but not necessarily those discussed in the previous section. Thus for spectrometers spectral line (peak) detection sensitivity is important.

Resolving Time

An important characteristic of a nuclear radiation detection system using a pulse detector is its resolving time, which is the minimum time required between two successive pulses if they are to be detected as separate pulses. The resolving time, τ, of a detection system is commonly limited by the detector itself, but it is also dependent upon the level of pulse height selector. In the case of detectors with extremely short resolving times, for example, in the nanosecond range, the resolving time of the auxiliary electronics can be limiting.

One characteristic of a G-M tube is that it is a relatively slow detector. Resolving times are typically 100–600 μsec for a selfquenching tube. Consequently, appreciable counting rate errors can be encountered with G-M tubes because of coincidence losses. By comparison, τ may be a few microseconds to a few milliseconds for pulse ion chambers, depending upon whether the time constant of the amplifier is set to give a pulse for electron or positive ion collection, respectively; a few microseconds for proportional counters; 10 nanoseconds to a microsecond for scintillation detectors; and in the nanosecond range for semiconductor detectors.

The coincidence correction for a counting rate is given by

$$R_{\text{true}} - R_{\text{obs}} = \frac{\tau \cdot R_{\text{obs}}^2}{1 - \tau R_{\text{obs}}}, \tag{9.14}$$

where τ is assumed to be independent of counting rate. A simplified equation, obtained from (9.14) by standard series expansion, is

$$R_{\text{true}} - R_{\text{obs}} \cong \tau R_{\text{obs}}^2 \tag{9.15}$$

For $\tau = 300$ μsec and $R_{\text{obs}} = 12{,}000$ cpm, R_{obs} would be low by 768 cpm or 6.4%, but for $R_{\text{obs}} = 1200$ cpm, the correction is only 4 cpm.

The resolving time of a G-M counter can be determined by several methods which depend on measuring the difference between the expected and the observed counting rate at rates where the losses are significant. One method is to measure the counting rate of a series of sources of a long-lived material differing in strength and having accurately known relative strengths. A second is to follow the decay of a source of a short-lived nuclide having an accurately known half-life, and a third is the paired source method [59, 90, 94–95]. The third is the simplest and the most popular.

If the halves of a pair of point sources are counted separately and then together under essentially identical conditions of geometry, the observed counting rate, R_{12}, when the sources are counted together will be less than the sum of the separate rates, $R_1 + R_2$, because of coincidence losses. The resolving time can be calculated by means of the relationship

$$\tau = \frac{R_1 + R_2 - R_{12} - R_b}{R_{12}^2 - R_1^2 - R_2^2} \tag{9.16}$$

where R_b is the background rate. As an approximation, the denominator can be replaced by $2R_1R_2$. A coincidence correction curve can be plotted once τ is known. Actually, τ for a G-M tube is not strictly a constant for a given discriminator setting [96]. It depends slightly on the counting rate and the applied voltage (decreases with increased counting rate and increased voltage) and is difficult to measure with precision better than a standard deviation of about 10%. It is preferable to reduce the counting rate rather than add a large correction when the highest accuracy is required.

Pulse Counting Systems [58, 59, 97–101]

Pulse counting systems (or pulse counters or simply "counters") provide either a total count of the pulses from a pulse-type detector in a known period of time or an instantaneous average pulse rate. As used here it is understood that "pulse counting systems" do not have pulse height analysis capability but do have a pulse height selector, which may or may not be a readily accessible control. The components that may be used in a pulse counting system are shown in Fig. 9.8. The number of components and the

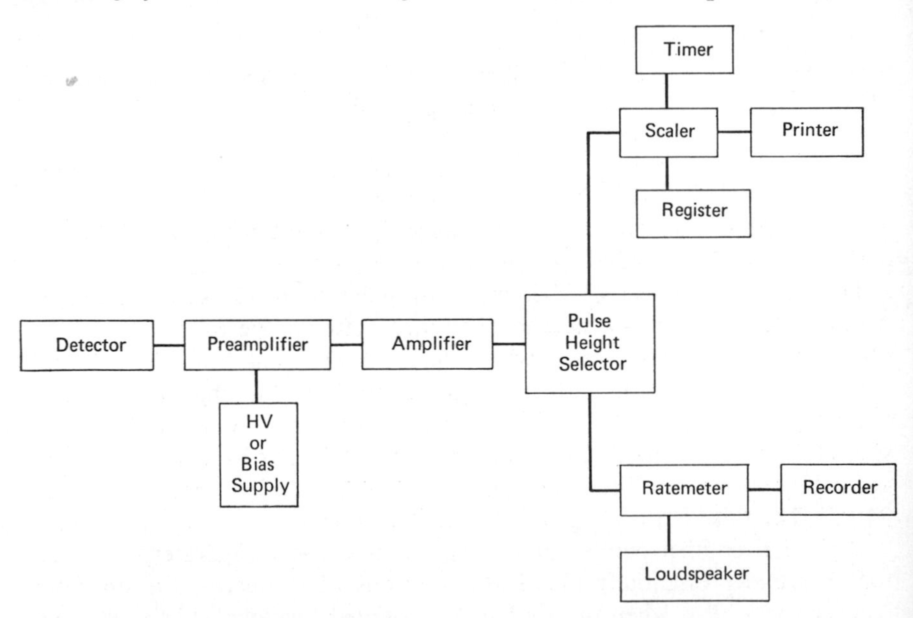

Fig. 9.8 Block diagram for a pulse counter.

specifications of each vary with the type of detector. For example, a preamplifier is seldom needed for a G-M tube. Counting rate meters integrate the pulses electronically and provide a direct panel-meter readout of the rate at which the ionizing events are detected. The dc output of the ratemeter can also be used to drive a strip chart recorder. Battery operated and line operated G-M tube ratemeter systems are particularly convenient for background monitoring and decontamination measurements in the laboratory. For these an audible signal, preferably as audio pulses, is a very useful feature.

Pulse counting systems are available as integral units or can be assembled from nuclear instrument modules (NIM) to provide the optimum electronic system for the detector and intended use of the counter. Although the NIM's are standardized with respect to physical dimensions, connectors, and power requirements, the units made by different manufacturers (or even the same manufacturer) are not necessarily compatible with respect to input and output pulse requirements, for example, polarity, width, and timing.

Ion Current Systems [58, 59, 97, 102]

The dc current from an integrating ion chamber can vary over many orders of magnitude, but it is often very small. For example, if 6000 dpm of ^{14}C, perhaps as an internal $^{14}CO_2$ source, occurred within an ion chamber, the maximum current, assuming no recombination, could be as high as 2×10^{-14} amp, depending on the detection efficiency. Similarly, the ion current from a 500 ml ion chamber containing air at 1 atmosphere pressure and located at a point where the exposure rate is 1 mR/hr could be about 4×10^{-14} amp.

Portable ion chamber systems used for area monitoring, for example, employ high gain dc amplifiers. When the highest sensitivity and accuracy are required, a vibrating-reed (dynamic capacitor) electrometer is employed for current measurement.

Spectrometers

There are many types of spectrometer for the measurement of the spectral distribution of nuclear radiation. The discussion will be limited to those which are based on pulse height analysis. For pulse height analysis to provide a meaningful spectrum, the height of the voltage pulses resulting from the detector signal must be proportional to the energy transferred to the detector by the incident radiation. In the limiting case in which (1) the incident radiation transfers all of its energy to the detector in the interaction time of perhaps 10^{-10} sec and (2) all of the energy received is used in producing a pulse, the pulse height will be proportional to the energy of the incident radiation. It is also essential that the preamplifier and linear amplifier retain the proportionality of the pulses reaching the pulse height analyzer.

For line spectra (α-, X-, γ-, and conversion electron spectra) the width of a line is expressed in terms of the full width at half maximum height (FWHM) in energy units, for example, keV. This width is 2.36 times greater than the standard deviation for a Gaussian-shaped line. The broadening of spectral lines beyond the natural line width can arise from (1) interactions such as absorption and scattering experienced by the radiation before reaching the detector, (2) fluctuations in the overall detector response, and (3) random noise fluctuations in the detector itself and random electronic noise fluctuations in the preamplifier. The total noise width is the square root of the sum of the squares of the separate noise widths.

The resolution of a spectrometer or a detector for a specified type of radiation at a specified energy is commonly taken to be the minimum attainable FWHM at that energy. For scintillation spectrometers the resolution is also specified in percent obtained by dividing the FWHM of a line by the energy corresponding to the center of the line.

With respect to the detector alone, the line width for monoenergetic radiation observed with ionization-type detectors is found to be less than that expected by assuming a statistical fluctuation in the production of ions and complete charge collection for each ionizing event. Thus in terms of w (ϵ is also used) the average energy expended to produce a current-carrying pair by an ionizing particle of energy E, the line width in keV is equal to 2.36 $\times$ $(FEw)^{1/2}$, where F is the Fano factor. The Fano factor is an intrinsic constant for each material and determines the ultimate detector resolution. For ionization chambers F can be less than 0.1, for pure silicon or germanium reported values fluctuate between 0.08 and 0.15. A limit of about 0.05 has been predicted for both. In order not to lose the high-resolution capabilities of semiconductor detectors, cooled low-noise field-effect-transistor (FET) preamplifiers are used.

Pulse height analysis is achieved by either a single channel analyzer (SCA) or a multichannel analyzer (MCA). A block diagram is shown in Fig. 9.9. Typically, a single channel analyzer [58, 59, 99, 103] consists of a variable lower level discriminator (LLD) (also called the base level) and a variable upper level discriminator (ULD) which can be set to remain a fixed number of volts above the lower discriminator. The difference between the two discriminator levels is the window and represents an energy difference ΔE. Pulses from a detector producing an output voltage pulse proportional to the energy absorbed from the incident radiation go first to a preamplifier (located very close to the detector) to minimize degradation of the signal in the cable before it reaches the linear amplifier which increases the pulse heights (e.g. from millivolts to volts) to meet the requirements of the pulse height analyzer. The linear amplifier is designed to produce output pulses which have the same relative heights as the input pulses. Pulses from the linear amplifier that are

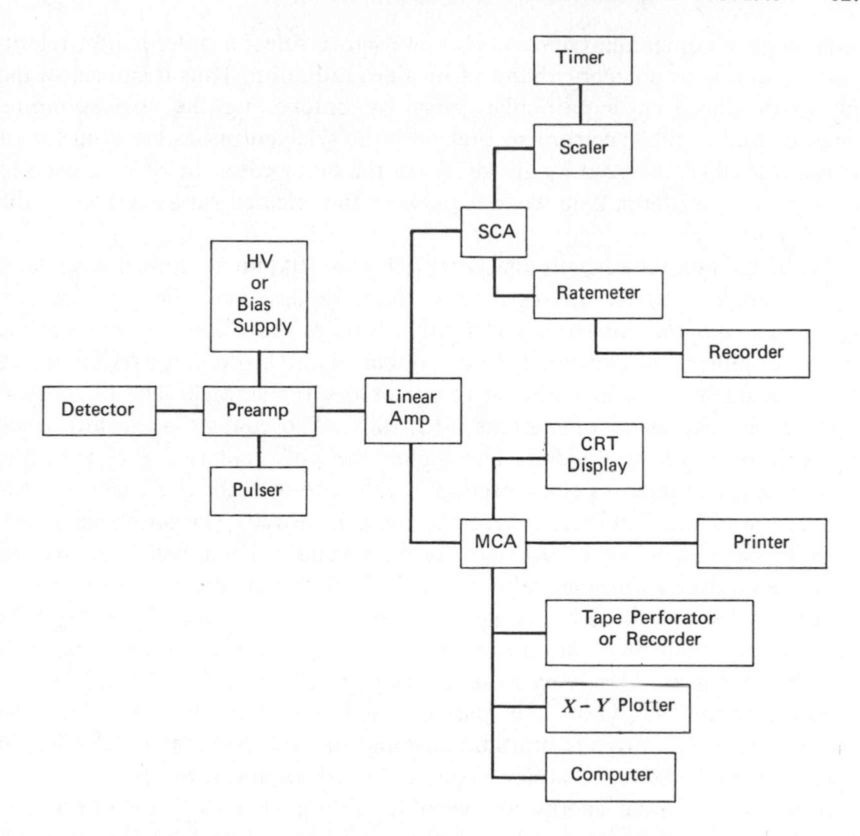

Fig. 9.9 Block diagram for SCA and MCA spectrometers.

less than the lower discriminator level of the SCA are rejected by it; pulses above the upper discriminator level and, therefore, above the lower one also are passed by both discriminators, but are rejected by an anticoincidence circuit. Only pulses that are above the base level and between the two discriminator levels, that is, fall in the window range ΔE, are routed to a scaler or ratemeter. The radiation spectrum is obtained by plotting counting rate versus the base setting. This may be done either point-by-point using a scaler or continuously by automatically sweeping over the base range while recording the output of a ratemeter on a strip chart recorder. The relationship between base setting (in volts) and energy of the radiation (in MeV) is established for the settings of potential applied to the detector and linear amplifier gain by means of sources of the same type of radiation having known energies in the range of interest.

A single channel analyzer can also be used to select a pulse height region corresponding to an energy band of incident radiation. Thus it can select the pulses produced by a particular γ-ray by bracketting the corresponding peak or line of the spectrum so that only the selected pulses are counted or perhaps further analyzed by an MCA. In the latter case, the SCA is used to turn on a coincidence gate when a pulse in the selected range arrives at the MCA input.

Multichannel pulse height analyzers [99, 103–105] are no longer a series of preset single channel analyzers as originally designed. They utilize an analogue-to-digital converter (ADC) which takes the voltage pulse from the linear amplifier and converts it to a number of clock oscillator pulses which are counted by a scaler and sent to a memory for storage. The memory is divided into channels, for example, 512, 1024, 4096, and so forth, into which the spectrum can be divided. The higher the pulse entering the ADC the greater is the number of clock oscillator pulses counted by the scaler and the higher the channel number where the pulse is stored. The spectrum is obtained as a plot of counts per channel versus channel number. This may be done manually or automatically by an *X-Y* plotter or by means of a computer. The counts in each channel may be displayed on a cathode ray tube (CRT), may be printed out, may be stored on paper tape or magnetic tape, or may be processed directly by a computer [106, 107].

Multichannel analyzers can also be used with two or more detectors simultaneously to provide multidimensional or multiparameter pulse height spectra as described in Chapter X on Gamma-Ray Spectroscopy.

Sources of known energy are used to obtain an energy calibration—a correlation between channel number and energy received by the detector. Procedures for calibration and evaluation of stability, linearity, and so forth, of MCA systems are given by Crouch and Heath [108]. The pulse generator shown in Fig. 9.9 is used to determine the resolution and linearity of various components of the spectrometer.

Perhaps a few comments on the relative merits of SCA and MCA spectrometers are in order. The MCA makes much more efficient use of the detector pulses than does the SCA. It is capable of accepting a wide range of pulse heights and sorting them out rapidly, whereas the SCA rejects all pulses that do not fall within the window. Use of the SCA is, therefore, impractical when it becomes necessary to subdivide the spectrum into very small energy increments (pulse height increments) in order to fully utilize the capabilities of a high-resolution detector. Another consequence of the inefficiency of the SCA is that even when high resolution is not required, the SCA is a relatively slow device. On the other hand, it is less expensive and is useful for routine checking of simple γ-ray spectra, for example, for radiochemical purity check, using a sodium iodide scintillation detector.

Suppliers of Nuclear Instruments, Radioisotopes, and Accessories

The names and addresses of manufacturers and suppliers of detectors, instrument modules, complete measurement systems, radioisotopes, labeled compounds, accessories, and so forth, are listed in buyers' guides such as those issued annually by *Analytical Chemistry*, *Nuclear News*, and *Science*.

8 MEASUREMENT OF ALPHA ACTIVITY [11, 109]

Source Preparation [58, 90, 110–116]

Solid sources are most commonly used for α-counting. In order to minimize self-absorption losses in α-counting and spectral distortion, for example, line broadening, in α-spectroscopy the sources must be very thin (a few $\mu g/cm^2$) and uniform in thickness. The methods used to prepare sources vary, of course, with the chemical element involved. They include electrodeposition, vacuum sublimation of a volatile compound, electrospraying, and evaporation of a solution to dryness. In each case an appropriate backing material such as a metal foil or a plastic film is required. The last method mentioned is the simplest, but it is generally the least satisfactory because the residue obtained by evaporation does not form a deposit of uniform thickness. Various organic spreading agents or suspending agents have been used successfully. These can be removed by ignition.

Liquid sources have also become of interest as major improvements have been made in the instrumentation for liquid scintillation counting.

Gaseous sources are used for measuring naturally occurring gaseous α-emitters such as radon-222.

Proportional Counting [58, 59]

Windowless gas-flow proportional counters have a high intrinsic efficiency for α-particles. The α-emitting source is placed inside the detector (commonly one with 2π geometry), which is then sealed and flushed with a counting gas such as a mixture of argon and methane or even pure methane. During operation the gas flows through the tube continuously at a slow rate from a cylinder of compressed gas and is maintained at a very slight positive pressure by means of a bubbler-type trap.

Very thin windows (<1 mg/cm^2) of a strong, opaque, conducting material such as aluminized or gold-coated Mylar permit the passage of α-particles, but absorption in the window reduces the detection efficiency. Such windows are not likely to be airtight so that continuous gas flow is required to exclude or remove air from the detector. (Oxygen, because of its relatively high electron affinity, has a deleterious effect on the performance of gas filled detectors.)

The α-particles from an α-β-emitting source such as one containing ^{210}Pb-^{210}Bi-^{210}Po can be counted separately from the β-rays by means of a proportional counter. For a fixed-pulse height selector level, the relatively large pulses from the α-particles are counted with a lower gas amplification factor (corresponding to a lower applied voltage) than those from the β-rays. Thus for α-counting, the operating potential is set at a lower value than for β-counting. Depending upon the particular detector the number of β-rays entering the detector per minute may be on the order of 10^6/min before pile-up produces spurious pulses comparable to those from α-particles. The α-background for such a counter is about 0.1 cpm.

Scintillation Counting

For routine α-counting, zinc sulfide (activated with silver or copper or manganese ions) is a convenient scintillator [58, 59, 64, 117]. Its luminescent efficiency is about 100%. Commonly a thin layer (perhaps 10 mg/cm^2) is attached by a very thin layer of cement to the window of a photomultiplier or to a transparent plastic support which is then optically coupled to the photomultiplier with a high viscosity silicone oil. Sometimes a very thin aluminum foil is placed over the zinc sulfide layer to make the assembly light tight and act as a reflector to increase the number of photons reaching the photomultipler cathode. The geometry can be close to 2π. Gaseous α-emitters can also be counted using zinc sulfide on the inner wall of a vessel containing the sample. Thick layers, for example, > 25 mg/cm^2 or so, are opaque to the fluorescent light. The pulses obtained from ZnS (Ag) are not suitable as a basis for α-spectroscopy.

Liquid scintillation counting (Section 9) of α-emitters is capable of essentially 100% efficiency and has been used for a number of nuclides [118-119].

Alpha-Ray Spectroscopy [58, 59, 64, 67, 68, 71, 119]

Surface-barrier semiconductor detectors are well suited for the detection of heavy charged particles such as alpha particles. Because of the limitations in detector area, however, they are not efficient for large sources. Pulse-height analysis by means of multichannel analyzer provides a spectrum of the type shown in Fig. 9.10.

Pulse-ion chambers and gas-filled proportional counters have also been used to measure the spectral distribution of α-particles. The former has somewhat better resolution than the latter, that is, perhaps 15 and 70 keV, respectively, at about 5 MeV. Surface-barrier detectors are capable of better than 12 keV resolution for 5.477 MeV α-particles from ^{241}Am. Scintillation spectrometers with NaI(Tl) or CsI(Tl) detectors have a resolution of about 100 keV at 5 MeV.

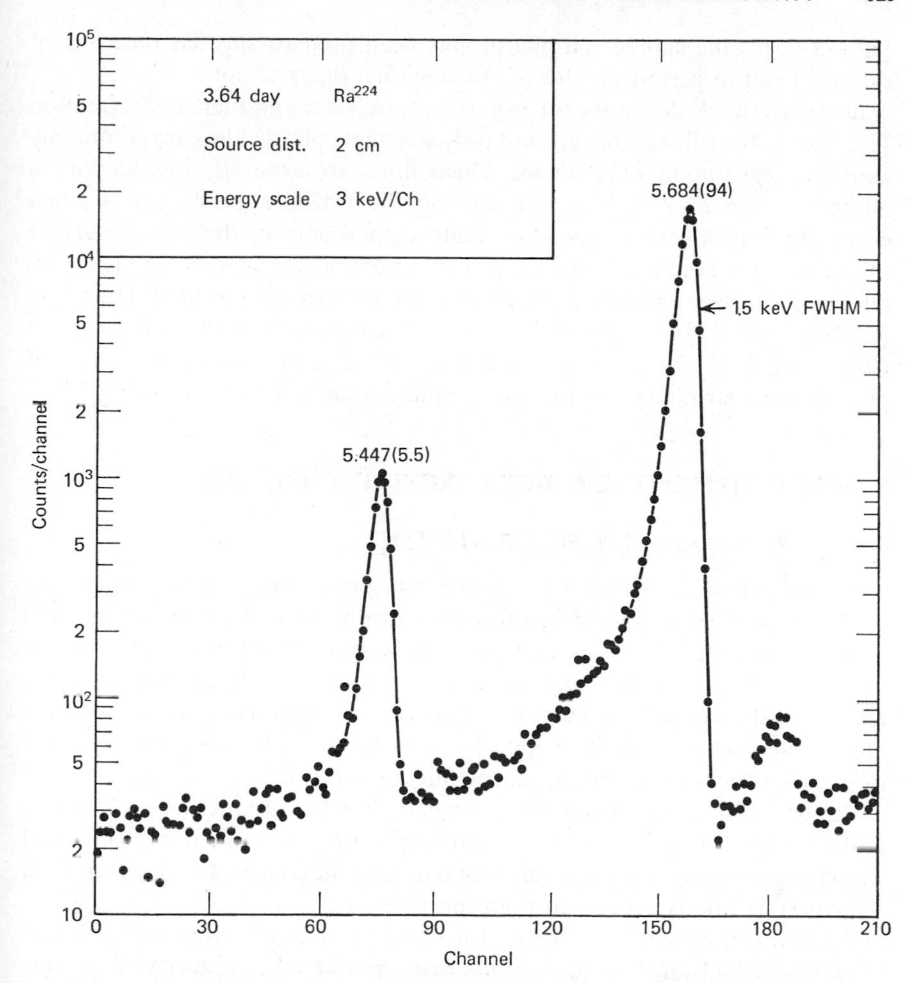

Fig. 9.10 Alpha-particle spectrum of ^{224}Ra obtained with a silicon surface-barrier detector. Energies are given in MeV with relative intensities in parentheses [115].

Alpha energy standards for calibration of spectrometers are listed in Reference 6.

Other Methods

Current ion chambers are suitable for the determination of low levels of α-emitters in air. The chambers are equipped with valves and can be connected to a standard gas-transfer apparatus for evacuation and filling.

Autoradiographic methods for locating α-emitters in tissue and on surfaces using nuclear track emulsions and stripping emulsions are well established

[59, 72–75]. Zinc sulfide scintillator has been used to shorten the time of exposure and to permit the use of conventional types of film.

Dielectric track detectors for α-particles are receiving increased attention [84, 120–124]. Cellulose nitrate and polycarbonate plastic films are commonly used for registration of α-tracks. These films are generally etched with a dilute solution of NaOH. After etching the tracks appear as pits when examined with a microscope. For high α-track density the etched surface of the detector becomes translucent. One of the advantages of DT detectors relative to nuclear emulsions is that they are not sensitive to light. Hamilton [84] has pointed out the potential use of DT detectors in the life sciences, and Cole et al. [121] have compared the use of cellulose nitrate with that of photographic stripping film for α autoradiography of bone containing ^{239}Pu.

9 MEASUREMENT OF BETA ACTIVITY [11, 109]

Source Preparation [58, 90, 110, 112, 113]

For β-emitters for which E_{max} of the beta radiation is greater than about 0.2 MeV, solid sources are often the easiest to prepare. This applies to most inorganic material. Aqueous solutions can be evaporated in planchets made of suitably inert material such as copper, stainless steel, aluminum, glass, and so forth. Sample spinners for improving the uniformity of the deposit are commercially available. Similarly, slurries of precipitates can be dried in planchets or the precipitate can be filtered and dried.

Powders can also be placed in planchets, but usually require a binder (a dilute solution of a cement in a volatile solvent) to prevent loss of material or a change in uniformity as a result of handling the source. For such sources, the counting rate is dependent upon particle size.

When it is necessary to compare the counting rates of samples of different thickness, self-absorption corrections will very likely be necessary. The same is true when a disintegration rate is to be calculated from a counting rate.

Liquids can be counted in planchets, but such sources are generally unsatisfactory. They usually require self-absorption correction, which itself may require a correction for solvent evaporation during the measurement. If a volatile solvent is present and if corrosive vapors are evolved, liquid sources are not suitable as internal sources in gas-filled detectors. The preparation of sources for liquid scintillation counting has been developed to a high degree and is discussed below as a key part of this counting method.

Gaseous sources are generally unpopular because of the handling, purification, and transfer problems. They are, however, advantageous when very high sensitivity is required, particularly for soft-β-emitters [33, 37, 38, 102, 112, 125].

Liquid Scintillation Counting [37, 38, 40, 62–64, 118, 126–132]

Liquid scintillation counting has become a very popular method for measuring β-activity and appears to have a high potential for new applications. Its popularity stems mainly from the fact that it is especially well suited for the measurement of low energy pure β-emitters such as tritium, carbon-14, and sulfur-35 with minimum sample preparation. It is also of use for higher energy β-emitters, for electron capture nuclides such as ^{55}Fe, and α-emitters as mentioned in Section 8. The method is readily instrumented for automatic counting of large numbers of samples and can be used to simultaneously measure more than one, for example, three, β-emitters in multiply labeled samples. The various techniques and procedures have been developed to a high degree and are described in great detail in the literature. Fortunately, there are a number of review papers such as [129, 130] and proceedings of conferences and symposia [126, 127, 131, 132] containing relevant papers. References 128, 129, and 131 will be found especially useful. Only a very brief account of the method can be given here.

The radioactive sample is added to a liquid scintillation solution ("cocktail") containing typically an organic solvent such as toluene, a primary scintillator solute such as PPO or butyl-PBD, a secondary scintillator such as POPOP or dimethyl POPOP, and possibly a secondary solvent or a solubilizing agent to increase the solubility of the radioactive sample. Concentrations used are usually in the range of 4–8 g/l for the primary solute with corresponding concentration of the secondary solute 0.05–0.10 g/l, or, about 80:1. Costs vary, depending upon the specific scintillator, the quantity purchased and the supplier, but primary scintillators are about 30 cents/g and secondary scintillators are about 2–5 dollars/g.

The secondary solute absorbs the light emitted by the primary scintillator and re-emits it at a longer wavelength more suitable for ejection of electrons from the photocathode of the photomultipliers. Because the β-rays are emitted and absorbed in the sensitive volume of detector, where absorption is mainly by the solvent which transfers the excitation energy to the primary scintillator, self-absorption losses, which adversely affect the use of solid samples of tritium and carbon-14, can be avoided. This is true if the sample dissolves in the scintillator solution.

In most liquid scintillation counters the sample, contained in a closed vial, is placed between two photomultipliers in a light-tight chamber. An example of a block diagram for a liquid scintillation counter is shown in Fig. 9.11. Because of the high gain required to obtain usable pulses from low-energy β-emitters, background pulses from electrons released thermally within the photomultiplier are a problem. This thermal noise can be reduced by cooling the photomultipliers. Furthermore, the use of two photomultipliers

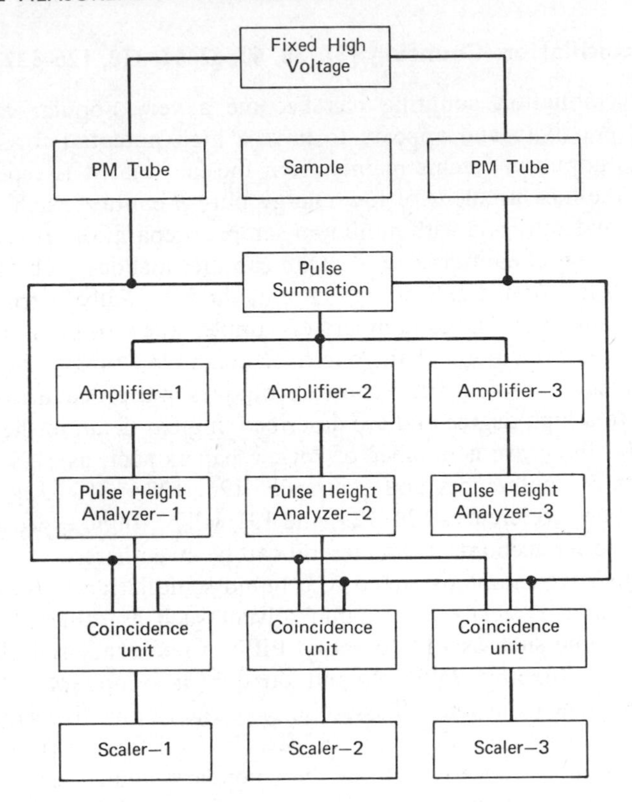

Fig. 9.11 Block diagram of a three-channel liquid scintillation counter with linear amplifiers and pulse summation [129].

makes it possible to reduce background electronically. A scintillation event in the sample vial that is seen by both photomultipliers should produce pulses which are in coincidence. Because thermal noise is random in each photomultiplier, the chance of random coincidence is low.

The random coincidence rate is given by $R_{\text{coinc}} = 2R_1R_2\tau$ in terms of the two individual rates and the coincidence resolving time. For modern photomultipliers, τ can be reduced to about 1 nanosecond and, therefore, the coincidence rate to a negligible value with resultant improvement is sensitivity. Only those pulses in coincidence, therefore, are counted after suitable single channel pulse height analysis. The pulse from one photomultiplier may be taken for analysis or the two coincident pulses may be summed and then analyzed. If it is not important to have an extremely low background, the two-photomultiplier type of scintillation counter can be operated at room temperature.

The variations in detailed procedures that can be used for liquid scintillation counting seem almost unlimited. In addition to having several scintillators to choose from one also has a choice of vials made from normal glass, low-background glass (low in ^{40}K), and a variety of plastics. The radioactive material does not have to be in solution. It can be present as a suspension, in a rigid or a thixotropic gel, in an emulsion, on the surface of filter paper (preferably glass fiber paper), or as a two liquid-phase system obtained by solvent extraction. Gases such as $^{14}CO_2$ can be dissolved directly in the liquid scintillation solution or absorbed first in a substance such as ethanol amine. Similarly ^3H_2O can be absorbed first in absolute methanol. Thus the labeled products from the oxygen-flask method of combustion can be readily counted. Finally, the method can be used for flow measurements.

In the case of aqueous solutions a polar secondary solvent such as methanol can be added to the toluene scintillator solution, but only up to about 20% for methanol before the lowering of the scintillation efficiency becomes a major problem. Instead, a different scintillator solution consisting, for example, of water-miscible 1,4-dioxane, naphtahalene (50 g/l, to improve the energy transfer from dioxane to the primary scintillator); PPO (7 g/l); and POPOP (0.05 g/l) may be used.

By using two or more channels, each having its width determined by an upper and a lower level discriminator, it is possible to measure more than one beta emitter, for example, ^3H and ^{14}C, at the same time and to correct for what is perhaps the greatest source of error in liquid scintillation counting, namely, lowering of the scintillation counting efficiency by quenching of the fluorescence. There are two major types of quenching: chemical quenching and color quenching. The first type lowers the detection efficiency by collisional removal of excitation energy from the scintillator molecules before they can undergo radiative deexcitation. Molecules containing strongly electronegative elements, for example, oxygen in air, water, ketones, and organic halides, are strong quenchers. The sample itself may also be a chemical quencher or a colored substance which absorbs the photons from the scintillator. In most cases colored impurities in the sample can be removed by adsorption.

A quench correction can be evaluated in a number of ways including the following:

1. Internal standard method. A known quantity of a standard sample, for example, ^{14}C labeled toluene of known μCi/mM, is added to the sample after a single or repetitive measurement and the efficiency is determined. This procedure requires removing each sample from the counter to add the internal standard and defeats the advantage of automatic sample measurement for a large number of samples.

2. Channels ratio method. Quenching shifts the distribution of pulse heights for a weak β-emitter as shown in Fig. 9.12. If the spectrum is divided into two regions by three discriminators as shown in Fig. 9.12, the ratio of counts in two channels, for example, $(L_1 - L_2)/(L_1 - L_3)$ for a given measurement varies with the degree of quenching as the spectrum is shifted and the total number of pulses is decreased. The quench correction

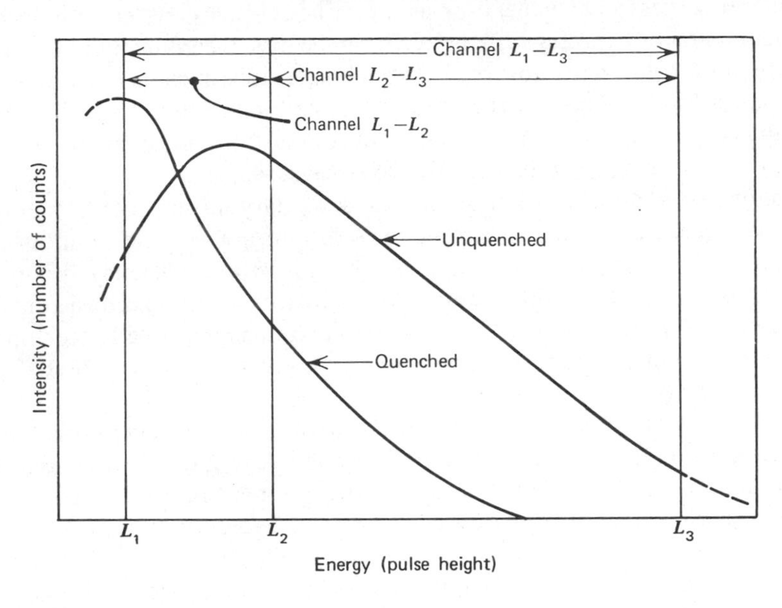

Fig. 9.12 Beta-ray spectra and spectrometer channels. Effect of quenching on spectrum. (From E. T. Bush, *Anal. Chem.* **35,** 1024 (1963).)

can be obtained for any sample from a plot of the channels ratio versus efficiency measured for a series of samples, each containing a known quantity of the radionuclide but a different amount of added quenching agent.

3. External standard method. An external source of γ-rays such as ^{226}Ra or ^{133}Ba can be used to introduce electrons (photoelectrons and Compton electrons) into the scintillation mixture after measurement of the counting rate of the internal sample. The efficiency is determined by relating the relative increase in counting rate to efficiency by means of a calibration curve. The method can be automated so that the external source, for example, a small pellet, is removed from its shield and positioned near the vial at the appropriate time in the counting schedule. An external channels ratio can also be used and has the advantage of minimizing the effect that might arise from variation in vial wall thickness.

In dual-isotope analysis involving two β-emitters having appreciably different values of E_{max}, the isotopes can be counted simultaneously by using separate channels to bracket the major portion of the pulse height spectrum of each isotope. The more sophisticated counters can also be programmed to automatically correct for quenching for each isotope by means of the external standard channels ratio method. If sufficient channels are available, three β-emitters such as ^{3}H, ^{14}C, and ^{32}P can be measured. Although it lengthens the sample preparation time, conversion of a labeled organic sample to CO_2 and H_2O, which are absorbed and transferred to separate scintillation solutions, removes quenching impurities and permits separate measurement of ^{14}C and ^{3}H when both are present.

Quite apart from quenching within the liquid, fluorescence and phosphorescence of the vials induced by ultraviolet light can cause interference and make it necessary to dark-adapt the samples prior to counting. The problem can be a major one in the case of tritium. Adsorption of the radioactive sample on the walls of the vial and self-absorption in suspended solid samples are additional potential sources of error.

The efficiency for carbon–14 can be as high as about 90%. For tritium, however, it is usually much less, for example, 30 60%.

The liquid scintillation counter used for assay can also be used for checking the laboratory for contamination. A small piece of paper (filter paper or paper towel) is used to remove a "wipe sample" from the floor or bench top, for example, and is then placed in a vial containing liquid scintillation solution.

Many off-line computer programs for simplifying data reduction for liquid scintillation counting have been published. Computer programs are available from the Packard Instrument Company, which maintains a computer program library.

Proportional Counting and G-M Counting [33, 37, 38, 58, 90, 125, 133]

For routine measurements of solid sources of β-emitters having E_{max} above about 0.25 MeV, simple counters with thin-window G-M tube are about the least expensive to use. Gas-filled, thin-window proportional counters permit counting at higher rates, are somewhat more stable and have the possibility, through pulse height selection, of adjustment of sensitivity and efficiency. Often a flow-type end-window counter can be used interchangeably in the GM or the proportional mode by merely changing the counting gas, for example, helium (99.05%) and isobutane (0.95%) for the G-M mode, and P-10, which is argon (90%) and methane (10%), for the proportional mode. In the proportional mode greater external amplification is required and a somewhat higher operating voltage may also be required. The organic constituent of the counting gas is used to quench the detector after each response to ionizing radiation.

When an α-β emitter is β-counted with a proportional counter, the counter is operated at a higher voltage than for α-counting. At the higher voltage the counting rate is the sum of the α- and the β-counting rates. The gross beta rate is determined by subtracting the α-counting rate. For accurate work, the α-counting rate should be determined at the operating potential for β-counting. The background rate for β-counting is determined mainly by the size of the detector and the amount of shielding.

Windowless, gas-flow proportional counters for internal source counting are useful for β-activity measurements of very thin sources, or infinitely thick sources of low-energy β-emitters. For high sensitivity and high efficiency counting of ^{14}C, for example, gaseous samples can be introduced into a proportional tube as CO_2 to which a counting gas is added. Gases such as H_2 and C_2H_6 can be used for ^{3}H, and H_2S for ^{35}S.

Cerenkov Counting [134–136]

Cerenkov counting with a liquid scintillation counter offers certain potential advantages over liquid scintillation counting. For example, sample preparation is simpler, chemical quenching is not a problem, relatively large sample volumes can be used, the sample can be recovered relatively easily for chemical analysis or other processing, and the cost of the organic scintillator is saved. There are limitations, of course. These are the lower limit of 260 keV for the β-ray energy in water, an efficiency less than for liquid scintillation counting, and color quenching. The directional nature of the radiation is a disadvantage in terms of the response of a typical liquid scintillation counter having two photomultipler tubes in the usual 180° arrangement and connected to a fast coincidence circuit.

The method is suitable for measurement of single β- or β-γ emitters. In most applications reported the radionuclide has been contained in an homogeneous aqueous solution, but heterogeneous systems involving plant tissue have also been used effectively [137]. For biological tissue [138] the minimum β-ray energy required is about 0.5 MeV. Self-absorption lowers the efficiency somewhat.

Water-soluble wavelength shifters can be used to increase the counting efficiency by perhaps a factor of two or so. They absorb the Cerenkov radiation in the ultraviolet region and emit light in the visible region. The ultraviolet radiation would otherwise be absorbed in the glass vial and the glass envelope of the photomultipler tube. In addition, the light from the wavelength shifter is emitted isotropically, as it is in a liquid scintillation solution. This reduces the directional effect of the Cerenkov radiation. Wavelength shifters are used in concentrations of 50–200 mg/l. Ross [139] recommends 4-methylumbelliferone for its effectiveness and relative insensitivity to pH.

Examples of counting efficiencies (with wavelength shifter present) reported by Parker and Elrick [136] are 4.7% for ^{36}Cl ($E_{max} = 0.71$ MeV), 50% for ^{32}P ($E_{max} = 1.71$ MeV), and 85% for ^{42}K ($E_{max} = 2.0$ and 3.5 MeV in the ratio of 1:4.6). The efficiency increases with E_{max} since the fraction of the β-ray spectrum above 0.260 MeV increases with E_{max}. For a radionuclide having E_{max} near the threshold for water, the efficiency can be increased markedly by increasing the refractive index of the solution [135]. By using α-bromonaphthahene with $n = 1.6582$ Ross [140] was able to measure ^{14}C ($E_{max} = 0.15$ MeV) by Cerenkov counting with an efficiency (12.3%) much higher than expected, apparently because of anomalous refractive dispersion.

Counting efficiency can also be enhanced by using plastic vials. There is improved transmission of the ultraviolet light and the light is diffused by the plastic. Background is also lower, because of the absence of ^{40}K, as for scintillation counting. A plastic vial can also act as the dielectric medium and lower the Cerenkov threshold below that for water.

The larger sample volume and the lower background (perhaps half that for liquid scintillation counting) for Cerenkov counting tend to compensate for the lower efficiency relative to scintillation counting.

The methods used for obtaining quench corrections for liquid scintillation counting can be used to obtain corrections for color quenching [141, 142]. When external standardization is used, the external source, for example, ^{226}Ra, must emit γ-rays of sufficient energy to produce Compton electrons with energy above the Cerenkov threshold. Because the Cerenkov radiation so produced is emitted in the walls of the vial as well as in the sample, the method is sensitive to variations in the position of the vial and in the wall thickness of the vials. This problem is alleviated in liquid scintillation counting by using a channels ratio for the external source.

Cerenkov counting is being used routinely in some laboratories and its use is likely to grow, particularly as liquid scintillation counters are modified to improve their response to Cerenkov radiation.

Other Methods of Measurement

Although current ion chambers designed for the measurement of the radioactivity of gases as internal sources are tedious to use and inconvenient for handling large numbers of samples, they are capable of very high efficiency, sensitivity, and accuracy for low-energy β-emitters. The detection efficiency can approach 100% for ^{14}C as CO_2 and perhaps 70% for ^{3}H as H_2. Sample preparation and measurement techniques have been reviewed in several publications [33, 37, 38, 102, 112, 125].

Ion chambers with thin windows are useful for area monitoring when β-emitters are used at the mCi level. For monitoring at the μCi level or for checking for low-level contamination a thin-window or thin-wall G-M probe used in conjunction with a ratemeter is a sensitive, rugged, and relatively inexpensive device.

Solid scintillation detectors have been used for β-counting in a number of ways. Tubing and cells fabricated from plastic scintillators have been used for flow measurements, for example, for effluents from a chromatograph. Well-type NaI(Tl) crystals, normally used for γ-counting, will detect the bremsstrahlung from β-emitters. Although the efficiency suffers because of the low bremsstrahlung intensity, the geometry factor can approach 4π and the sample can be counted as a liquid in a test-tube-type container. The method is useful for radionuclides such as ^{32}P and ^{90}Sr-^{90}Y which emit relatively energetic β-rays.

Autoradiographic techniques can be used for a variety of measurements that supply geographical information. Examples are (1) location of a biological cell where beta decay occurred by identifying β-ray tracks and following them to their origin, (2) locating areas of radioisotopically labeled compounds on chromatograms, and (3) tracer studies in semiconductor [143] research. Sensitivity for weak β-radiation can be improved by means of scintillation fluorography in which a thin film of plastic scintillator is placed between the sample, for example, a TLC plate or paper chromatogram, and the film. As a variation, a primary liquid scintillant can be applied directly to the sample. Autoradiographic methods for ^{3}H have been described by Evans [37] and for ^{14}C by Raaen, Ropp, and Raaen [38].

Determination of E_{max}

Absorption Methods [4, 23, 144–148]

Determination of E_{max} for beta radiation by measurement of the range in aluminum is a relatively simple method even though somewhat old-fashioned and not very sophisticated. Usually, the range is determined using an end-window proportional counter or a G-M tube and a graded set of aluminium absorbers. The counting rate is measured with an absorber in place between the source and the detector and located near the counter window. An absorptinn curve such as that shown in Fig. 9.3 is obtained for a pure β-emitter. The range can be estimated visually from the curve as the point where the gross rate reaches background or where the net rate drops to zero—where the curve becomes vertical on a log scale. From the range in mg/cm^2 of aluminum, the value of E_{max} can be obtained from a curve such as that in Fig. 9.4 or from (9.7 or 9.8). For β-γ emitters, for positron emitters (source of 0.511 MeV annihilation photons) and for high counting rate sources of pure β-emitters, especially those with high E_{max} that produce measurable bremsstrahlung, the absorption curve has a "tail" above background beyond the range of the β-rays. Since end-window tubes have relatively low intrinsic efficiency for γ-rays ($\sim 1\%$) compared to that for β-rays ($\sim 100\%$), the photon contribution to the counting rate is usually very small.

Methods for resolving complex absorption curves obtained when the source emits β-rays with different E_{max} values or when there are photon contributions and comparative methods (e.g., the Feather method of obtaining the range from an absorption curve) are described in a number of publications [1, 4, 23, 90].

In a recent paper Baltakmens [147] described an absorption method requiring only two measurements with absorbers made of different metals, for example, aluminum and copper, but having the same density-thickness. The determination of E_{max} using thick solid sources containing a low-level of activity by measurement of the half-thickness in aluminum has been studied by Rodriguez-Pasques and co-workers [148].

Beta Ray Spectroscopy [1, 58-59, 64, 67, 68, 71, 118, 149–154]

The most accurate E_{max} values are obtained by β-ray spectroscopy employing a magnetic type spectrometer. This is a specialized measurement carried out in relatively few laboratories. For many purposes a relatively simple spectrometer utilizing a gas proportional, organic scintillation, or semiconductor detector and a multichannel pulse height analyzer can be used to determine E_{max}. The β-emitting source must be very thin to avoid self-absorption and must be mounted on a very thin backing to avoid backscattering which distorts the spectrum. Similarly, scattering from a shield or support for the source will distort the spectrum.

If a plot is made of the spectrum, it will appear about as shown in Fig. 9.1 although the detailed shape, especially in the low energy region, will vary from one β-emitter to another. It will also show some distortion introduced by the spectrometer. Corrections are applied to the data for resolution of the spectrometer and for backscattering from the detector in the case of silicon semiconductor detectors.

An accurate value of E_{max} cannot be obtained from the high energy endpoint of a plot of the spectrum because the shape of the curve makes it difficult to extrapolate to an intercept on the energy axis and because the experimental points near E_{max} are subject to a relatively large statistical error. An accurate value of E_{max} can be obtained by means of a Kurie plot (also called a Fermi plot and a Fermi-Kurie plot) which, for a single β-emitter having an allowed spectrum, gives a straight line that can be readily extrapolated to the energy axis. A Kurie plot is based on the Fermi theory of beta decay and can be obtained by plotting energy (or channel number) versus a quantity which involves the square root of the number of counts $N(W)$ observed in an energy increment (channel). For example, the quantity plotted may be $[N(W)/\eta WF]^{1/2}$ where W and η are, respectively, the total energy (kinetic plus rest energy) and the momentum of the β-rays in rest mass units, m_0c^2, where $m_0c^2 = 0.511$ MeV. F is $F(Z, \eta)$, the Fermi function which depends on the momentum of the β-ray and on the atomic number Z of the product atom. A number of related quantities may be plotted, for example $[\eta N(W)/Wf]^{1/2}$, where $f = \eta^2F(Z, \eta)$. Both F and f have been tabulated. Extrapolation to the energy axis gives W_0, the transition energy, from which $E_{(max)}$ in MeV can be calculated.

For forbidden or nonallowed β-ray spectra the Kurie plot deviates from a straight line reflecting its dependence on the shape of the spectrum which in turn depends on the β-transition probability. The lower the probability, the more forbidden the transition is. By introducing a correction term, called a "shape factor," in the calculation of the square root quantity, the Kurie plot becomes essentially linear for forbidden transitions. An example (from a paper published by Johnson, Johnson, and Langer [155]) is shown in Fig. 9.13 for the case of ^{204}Tl, which has a unique first forbidden transition

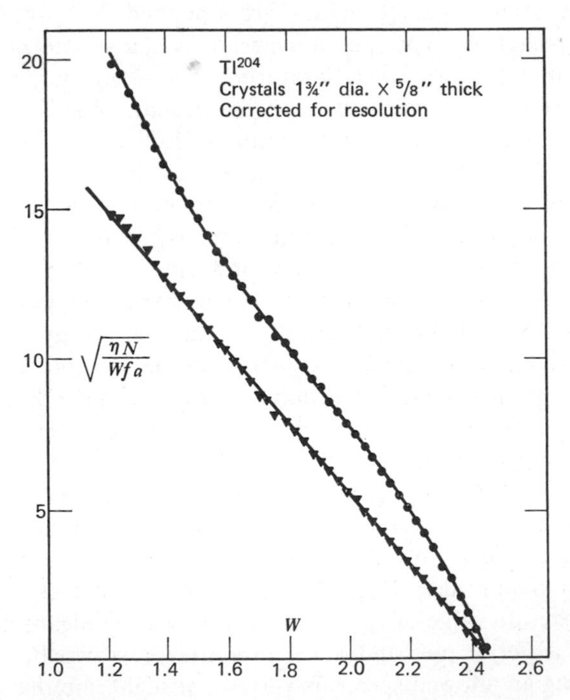

Fig. 9.13 Kurie plot for ^{204}Tl. Lower curve shows effect of shape factor a. (Adapted from Ref. 155.)

(a spin change of two and a parity change). When the appropriate shape factor, a, is introduced, the lower curve is obtained. For the upper curve $a = 1$. (The ^{204}Tl was placed between two plastic scintillation "crystals".)

When a source emits β-rays having different E_{max} values but comparable intensity, the Kurie plot is the sum of two or more straight lines. Under favorable conditions, that is, good data and only two or three β-transition with very different E_{max} values, the plot can be resolved or unfolded into the component lines by starting with the straight line portion corresponding to the highest energy β-ray.

Prevo and Cate [156] have described a two-point method of analysis of β-ray spectra to determine E_{max} without using a Kurie plot.

Calibration of a β-ray spectrometer can be achieved by using β-ray sources having a known E_{max} for example, ^{32}P, ^{90}Y, and sources emitting monoenergetic electrons, that is, conversion electrons. Examples of the latter type of source are ^{207}Bi, ^{109}Cd, ^{113}Sn, and ^{137}Cs-^{137m}Ba. A list of electron energy standards is given in Reference 6.

10 MEASUREMENT OF GAMMA ACTIVITY [11, 109]

Source Preparation [110]

The methods used to prepare solid sources of α- and β-emitters are equally applicable to γ-emitters. In general, however, it is not necessary to use a method especially intended to produce very thin sources of uniform thickness of a few mg/cm² (except for very low energy γ-ray or X-ray emitters as mentioned in Section 3).

Liquid sources contained in test tubes, capped vials, or Marinelli beakers can be counted with relatively little decrease in detection efficiency from absorption in the solvent or vessel wall. Flowing samples of γ-emitters can be measured through the container wall. Gaseous samples can also be measured through the container wall.

Scintillation Counting [58, 59, 63, 64, 90, 112, 133, 157]

Before crystals of NaI(Tl) became commercially available for scintillation detection of γ-rays, users of β-γ emitters generally used the more efficient methods for pulse counting of beta radiation, that is, G-M or proportional counting for routine measurements. The situation is now reversed and for many radioisotopes γ-counting is preferable.

Sodium iodide (Tl activated) crystals are most commonly used, although CsI(Tl) has some advantages. It has a higher average atomic number, is not so hygroscopic, and is less subject to thermal and mechanical shock. Its light output, however, is about 85% of that of NaI(Tl). Another scintillator,

$CaF_2(Eu)$, is less sensitive to high energy γ-rays but can be used for X-rays. Thin NaI(Tl) crystals are also sensitive to X-rays and relatively insensitive to γ-rays.

Cylindrical crystals of NaI(Tl) are available in a wide range of diameters and thicknesses. They have been made as large as 30 in. in diameter. The 3 × 3 in. size has been popular with users and for many purposes is considered as a "reference" size. Well-type crystals which allow the sample to be placed inside the crystal are especially convenient for chemists. The geometry can be close to 4π and liquid samples can simply be placed in a vial or a test tube without lengthy source preparation. A 2 × 2 in. or a 3 × 3 in. well-type crystal is suitable for samples having a volume of 1–5 ml. A crystal about 9 × 9 in. will hold 500 ml in the well [158]. Although the sample volume must be held within limits for accurate work, the volume effect is not an extremely sensitive one. A summary of techniques for using well-type NaI(Tl) scintillation detectors is given by Hine [157]. The method lends itself to the use of automatic sample changers.

Detection efficiency for NaI(Tl) crystals depends on the energy of the γ-rays, the size of the crystal, the decay scheme for the radionuclide, and the method of counting. The latter may involve merely a scaler with discriminator to reduce background or it may involve the use of a spectrometer in order to determine the counts under a specific γ-ray peak.

Large volumes of liquid, for example, 100–1000 ml, can also be γ-counted with a solid crystal by means of a Marinelli beaker (a beaker having an annular shape) that is placed over the crystal.

Other Methods of Measurement [58, 59, 133]

There are several γ-ray measurement devices that employ ion chambers. Because of their long-term stability, ion chambers used with vibrating-reed electrometers are well-suited for very accurate work with γ-emitters. For routine assay work (especially for relatively high-level sources) and for comparison of sources with standards, the re-entrant type [36, 40, 133] (cylindrical chamber with an axial well) having a large reproducible solid angle is particularly convenient.

Ion chambers find wide application in personnel monitoring. Perhaps the simplest device is the pen-size pocket dosimeter. It has a built-in, direct-reading quartz fiber electrometer which provides instantaneous evaluation of accumulated exposure within the full scale range, for example, 200 mR to 600 R.

Portable, battery-operated ion chambers equipped with dc amplifiers are suitable for measurement of gamma ray exposure rates over a range of many orders of magnitude, but unless filled to a pressure over 1 atm their low efficiency for γ-rays limits their accuracy at low exposure rates, for example, below 1 mR/hr.

Portable line-operated ratemeters equipped with G-M tubes are suitable for low level area monitoring and decontamination applications for γ-emitters. They are the same units used for β-ray monitoring.

The photographic film [76] has long served as a γ-ray detector for personnel monitoring. By means of absorbers they can be used to discriminate between β- and γ-radiation and they provide a reasonably permanent record. Despite the use of carefully standardized calibration and processing procedures, they are not particularly accurate (perhaps ± 30 mrem) and are not useful below an absorbed dose of about 50 mrem.

Thermoluminescent dosimetry (TLD) [77–80] has been developed to a stage of reliability, automation, and availability of the instrumentation so that it is generating considerable interest as a replacement for photographic film dosimetry for personnel monitoring and radiation therapy. TL phosphors include LiF, $Li_2B_4O_7$:Mn, CaF_2:Mn, natural CaF_2, $CaSO_4$:Mn, and Al_2O_3. As indicated, the commonly used activator impurity is manganese, although europium and samarium have also been used. The phosphors can be used as a powder, in the form of rods or ribbon made by extrusion, and imbedded in a substance such as Teflon. They are insensitive to exposure rate and have a wide useful exposure range, for example, from a few mR to 10^6 R, depending on the phosphor and quantity used. Their energy response is almost constant with γ-ray energy from 30 keV to 1 MeV for phosphors of low Z, for example, LiF, but is higher in the low energy range for higher Z materials since they have higher absorption coefficients for the photoelectric effect. The relationship between the logarithm of the response and the logarithm of the exposure is approximately linear over the useful range for some phosphors. The more sensitive phosphors such as $CaSO_4$:Mn, which will respond at the μR exposure level, show more rapid fading during the time between exposure and reading. For LiF, probably the most commonly used phosphor at the present, the fading is less than 5% in 12 weeks [80]. In terms of size, the dosimeter can be small, having dimensions of a few millimeters. A TL dosimeter does not provide a permanent record and a reading cannot be checked because the effect of ionizing radiation is removed in the reading process. It can be re-used after being annealed. Various sources of inaccuracy in TLD have been pointed out by Kartha [159]. Cameron, Suntharalingam, and Kenney [80] and Becker [160] compare TLD with photographic film dosimetry.

Glass radiophotoluminescent dosimeters [78] are similar to TL dosimeters in many ways. They can be remeasured, however, because the RPL centers are not destroyed during the measurement. The centers can be removed by annealing, thereby permitting re-use of the glass. As for photographic film and TL dosimeters, RPL dosimeters require calibration for each batch of radiation-sensitive material. Becker [160] has compared the advantages and disadvantages of the three types of dosimetry.

Gamma-Ray Spectroscopy [67, 68, 71, 161–166]

Before the development of scintillation detectors and pulse height analyzers, the measurement of γ-ray energy by absorption measurements [4, 144] with lead absorbers was an unsatisfactory method for obtaining an uncertain result. Spectrometers for measuring γ-ray energies indirectly by the deflection of electrons in a magnetic field to measure the energies of photoelectrons, and Compton electrons and pairs produced by the γ-rays have been in use for years, but they are not suitable for routine measurements. Introduction of γ-ray spectrometers with NaI(Tl) crystals having a good intrinsic efficiency and providing pulses with heights proportional to the energy over a wide range of γ-ray energy, revolutionized all γ-ray measurements which depend on measuring or selecting γ-ray energies. Well-type crystals mentioned earlier are less satisfactory than solid crystals for spectroscopy because of spectral distortion resulting from summing the energy given to the crystal by more than one γ-ray. This is particularly troublesome for radionuclides such as ^{60}Co that emit more than one γ-ray per disintegration.

Another major step forward has resulted from the use of semiconductors such as the lithium-drifted germanium type. They have much better resolution than sodium iodide detectors but, so far, their detection efficiency is less and they must be cooled, usually with liquid nitrogen. Radioisotopes suitable for energy calibration of γ-ray spectrometers have been listed by Black and Heath [167] and by Lederer et al. [6].

Semiconductor detectors, especially Si(Li) detectors, have added a new dimension to the measurement of X-rays and low energy γ-rays. They are capable of resolution better than 200 eV FWHM at 6 keV. A recent review on this topic has been published by Walter [168].

The application of liquid scintillator solution to the measurement of X-rays and γ-rays of energy less the 60 keV has been reviewed by Horrocks [118] in a very useful general review covering many other applications.

Gas-filled proportional tubes with very thin windows, for example, beryllium, can be used for measurement of X-ray or low energy γ-ray spectra by pulse height analysis. They are particularly well suited for specific energy ranges of photons because the gas mixture can be selected to optiminize the absorption of photons having an energy near the photoelectric absorption edge.

Details of the techniques of γ-ray spectroscopy are given in Chapter X on Gamma-Ray Spectroscopy.

11 ERRORS IN RADIOACTIVITY MEASUREMENTS [1, 2, 4, 95, 169–172]

Theoretical Counting Errors

Radioactivity measurements, like other physical measurements, are subject to systematic errors and they are subject to the usual random errors such as those associated with the preparation of replicative samples. In addition, there is a random variation which arises from the nature of the property being measured, that is, radioactivity. The rate of radioactive decay of a radionuclide is not a constant with time but fluctuates randomly about an average value. The associated theoretical minimum standard deviation σ which could be expected (with a 68.3% probability) from the radioactive decay itself can be readily calculated. One can, therefore, plan the radioactivity measurement part, that is, the counting, of an experiment with an acceptable confidence limit, for example, σ, 2σ, or 95%, in mind.

Because both the disintegration rate of a source and background rate can be described by the Poisson probability distribution function, the standard deviation in a single measured value of a counting rate R measured over a counting time t is controlled by the *total number of counts* N obtained in the measurement. Thus assuming that N is a good estimate of the mean value that would be obtained in a series of repetitive measurements, the theoretical standard deviation in N is

$$\sigma_N = (N)^{1/2} = (Rt)^{1/2} \qquad (9.17)$$

and in the case of R

$$\sigma_R = \sigma_N/t = (R/t)^{1/2}. \qquad (9.18)$$

The relative standard deviation, $1/(N)^{1/2}$, for a measured total number of counts or for a measured rate decreases with increasing total number of counts. Thus for $N = 100$ counts there is a 10% σ, but for $N = 10{,}000$ there is a 1% σ error. From (18) one can also see that for a source having a given rate, σ in the measured rate decreases with increasing counting time.

For most rate measurements a net rate R_s is calculated by subtracting a separately measured background rate R_b from the observed (gross) sample rate, R_{s+b}. Then, according to the rules for the calculation of errors propagated through a calculation involving the measured quantities R_{s+b} and R_b,

$$\sigma_{R_s} = (\sigma^2_{R_{s+b}} + \sigma^2_{R_b})^{1/2} = (R_{s+b}/t_s + R_b/t_b)^{1/2}. \qquad (9.19)$$

Equation 9.19 shows how the standard deviation in a net counting rate (based on a single sample measurement and a single background measurement) is related to the gross rate, the background rate, and the counting times. This equation and others that can be derived from it provide the basis for planning an experiment with respect to items such as the amount of radioisotope required, the possible need to take steps to reduce background, and the allocation of time for sample and background counting to attain the desired precision to be associated with the counting. When $R_s \gg R_b$ and R_b can be neglected, the relative or fractional standard deviation F_σ $(= \sigma_{R_s}/R_s)$ can be obtained from (18) and is simply equal to $1/(R_s t)^{1/2}$. Similarly, the value of R_s required for a given F_σ is $1/(F_\sigma^2 t)$.

When background cannot be neglected, the question of the allocation of time between sample measurement and background measurement becomes of interest. From (19) the minimum value of σ_{R_s} is obtained when the total available counting time T is divided between t_s and t_b according to the relationship

$$t_s/t_b = (R_{s+b}/R_b)^{1/2}. \tag{9.20}$$

The value of R_{s+b} can be estimated from a preliminary measurement of short duration for the sample and a typical value of R_b is usually known from previous measurements of background.

There are numerous applications of (20). Loevinger and Berman [170] give a plot for determining the optimum total number of counts for minimizing the combined counting time when the percent error is assigned in advance and the gross rate-to-background rate is known. England and Miller [173] present charts and equations for determining the measuring time or the number of counts required to control the precision of the net source rate when equal time is allocated to source and to background or equal numbers of counts are allocated optimally to background and to a number of sources. Goldin [174] calculates the optimum distribution of counting time for several sample counting periods and one background counting period.

An equation of interest in connection with low-level counting can be derived [175] from (20). If both the fractional standard deviation F_σ and the total time T are specified, then the minimum net counting rate consistent with these is

$$R_{s,\,\mathrm{min}} = \frac{1 + 2F_\sigma(TR_b)^{1/2}}{F_\sigma^2 T}, \tag{9.21}$$

where T is apportioned according to (20).

When the gross sample counting rate is very close to the background rate, equations such as (21) can be used to formulate a figure of merit (Section 7) that can be used as a criterion for selection or adjustment of counting systems to give the smallest F_σ for a given source for a given counting time or will require the shortest counting time for a specified F_σ for the source. In the limiting case for $R_s \ll R_b$ the counter with the highest "S^2/B" is preferable or, for a given counter, the control settings, for example, high voltage, discriminator level, and so forth, giving the maximum S^2/B is preferable. Here $S = R_s$ and $B = R_b$. The figure of merit when $R_s \gg R_s$ can be simply S (or $S^{1/2}$ as suggested by DeVolpi [176]).

The statistical aspects of low-level counting have been discussed in a number of publications [112, 175, 177, 178].

Observed Counting Errors

For a single measured value of the counting rate for a source, one can calculate a theoretical standard deviation taking into account only the statistical fluctuations in the decay process. The result would be indicative of how repeated measurements would scatter about a mean value, if there were no other cause for scatter. There may be other causes such as random timing errors, random spurious detector pulses, and so forth. To evaluate the total error it is necessary to make repeated measurements and calculate the overall standard deviation of n measured values of N (or R) by the usual root-mean-square calculation.

Standard statistical tests, for example, the χ^2 test, can be applied to the data to test the statistical behavior of the counting equipment and detect instability or nonrandom behavior. Similarly, Chauvenet's criterion can be used to reject "bad" data. For checking the long term stability of a counting system it is customary to measure the counting rate of a long-lived reference source.

When the error in R_b can contribute significantly to the error in the net sample rate and when many samples are counted or when very long counting times are required, it is generally necessary to measure R_b periodically and use an average value for calculation of the net sample rates. The number of required background measurements required (made, for example, with the source replaced by a blank planchet or by a vial containing liquid scintillation mixture, etc.) depends on the importance of σ_{R_b} and the magnitude and frequency of change in the average background rate R_b. One might, for example measure R_b, R_{s+b} three times for two sources and then R_b again. In general it is better to make duplicate or triplicate counting rate measurements for samples and for background and average the results rather than to make one measurement of long duration. There is then at least a chance of detecting an instrument malfunction, a drift in high voltage or amplifier gain, or a momentary power fluctuation that might occur during a measurement.

Additional sources of error in radioactivity measurements become apparent when counting sources are prepared in duplicate and triplicate. There are the usual aliquoting errors, possible losses associated with source preparation, variations in counting efficiency associated with source positioning (geometry), variations in thickness of solid sources or homogeneity of liquid scintillation sources, to mention but a few examples. In the design and execution of an experiment one attempts to reduce the overall error to the minimum arising from the radioactivity itself.

12 EFFICIENCY CALIBRATION OF INSTRUMENTS AND STANDARDIZATION OF RADIONUCLIDES

Efficiency Calibration [4, 27, 36, 90, 133, 179, 180]

The detection efficiency of a measurement system must be known if the absolute activity, that is, the disintegration rate, of a source is to be calculated from a measured counting rate or ion current. Even if all measurements involved in an experiment are on a relative basis so that only directly measured values need be compared, at least an estimate of the detection efficiency is needed in the overall planning of an experiment in order to determine the quantity of radionuclide needed. Furthermore, precautions must be taken in a comparative experiment to make certain that the detection efficiency remains constant or that sufficient information is available to normalize the data to a reference efficiency. One must, then, be aware of the factors that affect the detection efficiency. Finally, at least a good estimate of the detection efficiency is needed for instruments used for such radiological safety purposes as the determination of levels of contamination and levels of activity in radioactive waste.

In a general way, detection efficiency can be treated as a product of many of the following factors which take into account characteristics of the measurement system:

1 Geometry, which takes into account the size of the source, the size of the detector, and the source-to-detector distance.
2. Intrinsic efficiency (response) of the detector for the type and energy of radiation emitted by the source.
3. Rejection of signals from the detector by a pulse height selector.
4. Fraction of disintegrations providing the radiation being detected.
5. Self-absorption and self-scattering of the radiation within the source.
6. Absorption of the radiation in a source cover, air and detector window, or wall.
7. Backscattering and scattering from source holder, shield, and so forth.

The most direct way to determine the detection efficiency for a given radionuclide is to measure the counting rate for a source of the radionuclide having a known disintegration rate. The source must be prepared and measured in exactly the same way as the sources for which the calibration is to be applied. Standardized solutions of many radionuclides for the preparation of such calibration sources are available from the National Bureau of

Standards and from several commercial suppliers. (See buyer's guides mentioned in Section 7). These solutions are generally supplied in flame-sealed glass ampoules. Recommendations for the use of such standards are given in "Users' Guides for Radioactivity Standards" [180].

The activity of standardized solutions may be expressed in a variety of ways. For example, it may be in microcuries per milliliter or per gram of solution, in disintegrations per second per gram of solution, or in terms of the emission rate for a particular type of radiation (for which the energy may be specified). Once an ampoule of a standard solution has been opened, extreme care is required to avoid altering the specific activity. Obviously, loss of solvent should be prevented. Loss of the solute, not usually a problem in chemistry, can be serious in the case of radioactive standard solutions in which concentrations of the radioactive species may be exceedingly low, as 1.14×10^{-10} M in ^{32}P for a solution containing 1.0 μCi of ^{32}P in 1.0 ml. To reduce adsorption losses, the standardized solutions contain an inactive carrier (isotopic, if possible, and perhaps 10–100 μg/ml). If dilutions are required, the diluent solution (including the carrier) should be the same as that specified on the certificate for the standard solution.

The "accuracy" specified for a standard solution must be interpreted very carefully. In terms of precision measure, 95% confidence levels from a few tenths of a percent to five percent are reported, with 3–5% a common range. Unfortunately, the certified accuracy is not an infallible indication of the agreement to be expected between standard solutions for the same radionuclide from completely independent suppliers. The user of a standard should keep in mind the effects of radiochemical impurities and in particular the increase in importance of a long-lived radiochemical impurity with increased age of the standard.

For liquid scintillation counting, standardized solutions containing ^{3}H, ^{14}C, and ^{35}S, for example, as toluene-H^{3}, toluene-C^{14}, and ^{35}S in elemental sulfur dissolved in toluene, respectively, are available. Standardized water-H^{3} is also used for tritium. Precautions against loss by volatilization should be taken when standardized solutions are aliquoted. Less volatile *n*-hexadecane labeled with ^{3}H or ^{14}C is also available as a standard material. Sets of quenched standards for determining the effect of quenching on efficiency are offered by commercial suppliers.

Calibrated α-, β-, and γ-sources are also available. These may be essentially "point" sources in the case of β-emitters and γ-emitters and may be mounted on plastic film or metal foil. The activity may be expressed in terms of microcuries or in terms of emission rate. To the extent that such sources differ (area, backing, cover, etc.) from those to be made for the same radionuclide in the laboratory, various corrections for efficiency factors are required. These reduce the accuracy of the calibration.

When a standard solution or a reference source for the radionuclide of interest is not available, a semiquantitative calibration can sometimes be made using a standard solution or reference source of a radionuclide emitting radiation similar in type and energy. Appropriate corrections to the efficiency factors are required and differences in the decay scheme must be taken into account. Consider, for example, the emission rate for 0.010 μCi of each of the following [8]: ^{32}P, 2.22 × 10^4 β^--rays/min; ^{204}Tl, 2.16 × 10^4 β^--rays/min; ^{22}Na, 2.01 × 10^4 β^+-rays and 2.22 × 10^4 γ-rays/min, but for a source in plastic sufficiently thick to absorb the β^+-rays, 4.02 × 10^4 annihilation photons instead of the β^+-rays; ^{137}Cs-^{137m}Ba, 2.22 × 10^4 β^--rays plus 0.21 × 10^4 conversion electrons (mostly 0.624 meV), 1.88 × 10^4 γ-rays, and 1.5 × 10^3 barium K X-rays/min.

A long-lived reference source with radiation similar to that of the radionuclide of interest is useful for monitoring the stability and calibration of a counting system. Its activity can be compared to that of a short-lived standard at the time of calibration.

Standardization of Radionuclides [114, 133, 179, 181–187]

There are two types of radioactivity standard, absolute and relative. Absolute standards, formerly referred to as primary standards, are obtained without the use of other standards. Relative or reference standards are obtained by measurement with an instrument calibrated with an absolute standard and are generally less accurate.

Two methods are commonly used for making absolute radioactivity measurements. In the first, the source is counted with a detector for which the detection efficiency is known from a knowledge of the solid angle and the other efficiency factors. Ideally, the efficiency factors are all unity. If they are not, the method requires an accurate evaluation of all correction factors. In the case of an ideal 4π detector, each disintegration in the source would release ionizing radiation that would be totally absorbed in the sensitive volume of the detector and would produce a signal. The counting rate would then be identical with the disintegration rate. Examples of specific 4π detectors are $4\pi\beta$ gas-filled proportional or G-M detectors with the source at the center, cylindrical proportional counters with internal gaseous samples, liquid scintillation detectors, and plastic scintillation sandwich detectors with a β-emitting source at the center. All of these require some corrections and have limitations with respect to the useful energy range. For example, corrections for absorption of the radiation in the source and in the extremely thin support for the source are required for a $4\pi\beta$ counter. These counters have been constructed in various shapes, two of which are the face-to-face hemisphere type and the pillbox type shown in Fig. 9.14. Measurements made with internal gaseous samples require corrections for the wall-effect (absorption of radiation in the wall rather than the sensitive volume of the detector) and the end-effect (fraction of the gas volume at the ends of the tube beyond the region of the sensitive volume).

The second absolute method, the coincidence method, is suitable for a radionuclide emitting two types of radiation per disintegration, for example, β and γ. The two types of radiation must be detectable in separate counters

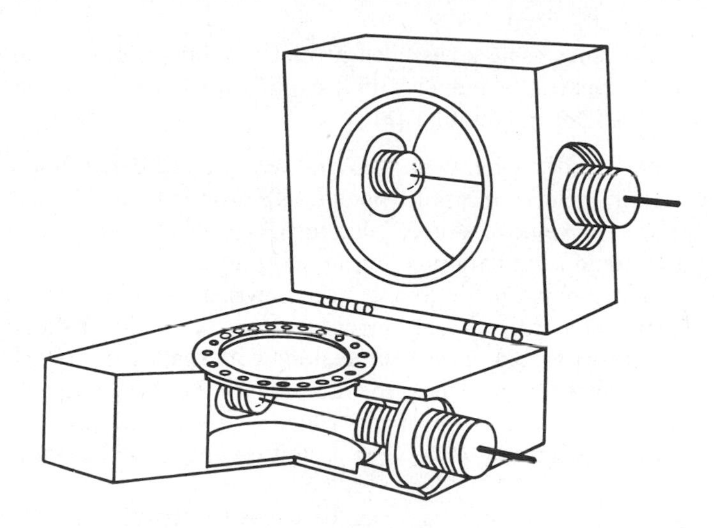

Fig. 9.14 $4\pi\beta$ proportional counter: pillbox type [133].

and they must be emitted together or at least within the resolving time of the coincidence circuit. As an example, a β-γ coincidence detection system might consists of a $4\pi\beta$ counter sandwiched between two large NaI(Tl) scintillation detectors for the γ-rays.

Taking as an example an ideal β-γ coincidence measurement with a β-detector and a "γ-only" detector, the two counting rates R_β and R_γ for a source having a disintegration rate A would be $R_\beta = AE_\beta$ and $R_\gamma = AE_\gamma$, where E_β and E_γ are the detection efficiencies of the two counters. The coincidence rate would be $R_{\beta\gamma} = AE_\beta E_\gamma$. By replacing E_β by R_β/A and E_γ by R_γ/A in the equation for $R_{\beta\gamma}$, the disintegration rate is given by $A = R_\beta R_\gamma/R_{\beta\gamma}$, which is independent of the unknown efficiencies of the detectors. For highest accuracy various correction terms are required for dead-time of the detectors, accidental coincidences, background, nonzero detection efficiency for the other type of radiation, and so on. A recent review of coincidence methods for the determination of absolute disintegration rates has been given by Remsberg [185]. The preferred method of standardization for β-γ emitters is $4\pi\beta$-γ counting. This method has been described in detail by Campion [188].

A $4\pi\beta$-γ anticoincidence spectrometry method has been described recently by Kawada, Yura, and Kimura [189].

The following methods are used for absolute α-counting:

1. Low geometry (to reduce errors from backscattering), defined solid angle counting with the source at one end of a vacuum chamber and the detector (a gas-filled proportional detector or a ZnS or plastic scintillation detector) at the other end.
2. α-γ Coincidence using a gas-filled proportional detector or scintillation detector for the α-particles and a NaI(Tl) scintillation detector for the γ-rays.
3. Liquid scintillation counting [118].

For β-emitters and β-γ emitters, $4\pi\beta$ and $4\pi\beta$-γ coincidence counting are generally used. For electron capture nuclides 4π proportional X-ray counting with absorbers to remove Auger electrons, 4π(X and e)-γ coincidence counting and liquid scintillation counting [118] are used.

An efficiency tracing method has been devised [190] to improve the accuracy for the standardization of low-energy pure β-emitters. A standardized β-γ emitter is incorporated in the source along with the low-energy β-emitter and is used to determine the detection efficiency for the low-energy β-emitter. Other methods include the use of calorimeters for α- and β-emitters and measurement of the loss-of-charge for β-emitters mounted on an insulated electrode.

Absolute counting of γ-emitters can be carried out with NaI(Tl) scintillation detectors for solid crystals and a defined solid angle. Absolute detection efficiencies for γ-rays for right circular cylindrical solid NaI(Tl) crystals have been calculated [90, 162] taking into account (1) the geometry for sources and crystals of various sizes and (2) the absorption coefficient as a function of γ-ray energy. Coincidence measurements (γ-γ) involving pulse-height selection (or analysis) can also be used for standardizing γ-emitters.

An excellent review of absolute measurement techniques is given by Allen [182] and many papers on details of the techniques are contained in Refs. 181 and 183.

Relative standardization requires the use of a stable counting system and very reproducible geometry. Typical procedures used to prepare relative standards are described in References 133 and 179.

13 LOW-LEVEL COUNTING [112, 191–197]

Low-level counting, also known as high-sensitivity counting, is the counting of sources having a very low disintegration rate. Usually this situation arises because the counting source is prepared from a sample having a very low specific activity, but it can also arise because the total amount of the sample is extremely small.

In practice, low activity samples have a gross counting rate (sample plus background) close to the background rate. The measurement problem is to obtain a value for the net sample rate R_s having an acceptable error. In increasing the sensitivity of a measurement method the objective is to make the net rate for a given sample as large as possible or to lower the minimum measurable net rate. Essentially, the goal is to obtain the highest figure of merit R_s^2/R_b (Sections 7 and 11). If the efficiency can be increased, say by a factor of two, the effect is equivalent to a factor of four in background reduction.

Methods for reducing the background counting rate include (1) removing all other samples of radioactive material from the counting area, (2) avoiding or reducing to a minimum all sources of naturally occurring radionuclides such as ^{40}K and the members of the uranium, thorium, and actinium series; fission products from fallout; and long-lived neutron-produced radionuclides in materials used to construct the detector and its shield and in materials used with the counting source such as glass for planchets, vials for γ-counting in a well-type scintillation counter or in a liquid scintillation counter, and (3) increasing the shielding of the detector itself. There is a limit to the background reduction attainable by the selection of materials for source containers, detector construction, and increasing the shielding because the materials cannot be obtained completely free of radioactivity. Studies [112, 191, 198, 199] have been made of low-level radioactive contamination in materials such as aluminum, steel, brass, and lead. Mercury is effective as a dense low-background shielding material. Graded shields consisting of layers of material of different atomic number increase the efficiency of shielding against cosmic radiation.

At a number of installations low-background counting rooms have been constructed, sometimes underground, sometimes using selected low-background materials of construction and several inches of steel plate with a layer of lead to line the walls.

A second approach to reducing the background counting rate involves the design and operation of the system itself. Electronic components can be selected for low-noise characteristics. In the case of photomultipliers, cooling and the use of two photomultipliers in anticoincidence to reduce thermal noise has been mentioned in connection with liquid scintillation counting. For a detector giving pulses with heights proportional to energy received, pulse height selection can often be used effectively to discriminate between pulses from the source and background pulses and thereby lower the background counting rate. A very effective approach for reducing background from radiation sources external to the detector also involves the use of two (or more) detectors in anticoincidence. A "guard" or "shield" detector (or detectors) partially or essentially completely surrounds the sample detector. The guard detector may be a gas-filled proportional or G-M detector, a large scintillation detector, or some other type. For background radiation originating outside the detector array and producing pulses in the guard detector and in the sample detector within a predetermined time increment, that is, in slow coincidence, the pulses are rejected by an anticoincidence circuit.

The monograph by Watt and Ramsden [112] contains detailed discussions of experimental methods and is a good source of references on low-level counting. A few additional examples of specific systems and techniques are given in References 119, 200–208.

Acknowledgments

The author wishes to thank J. M. Nielsen for his helpful comments and suggestions and D. S. Miller for helpful discussions on solid-state track detectors.

References

1. R. D. Evans, *The Atomic Nucleus*, McGraw-Hill, New York, 1955.
2. G. Friedlander,J. W. Kennedy, and J. M. Miller, *Nuclear and Radiochemistry*, 2nd ed., Wiley, New York, 1964.
3. I. Kaplan, *Nuclear Physics*, 2nd ed., Addison-Wesley, Reading, Mass. 1963.
4. R. T. Overman and H. M. Clark, *Radioisotope Techniques*, McGraw-Hill, New York, 1960.
5. "Nuclear Data Sheets," National Academy of Sciences—National Research Council (1959–1965), reprinted by Academic Press, New York, 1965.
6. C. M. Lederer, J. M. Hollander, and I. Perlman, *Table of Isotopes*, 6th ed., Wiley, New York, 1967.
7. *Nuclear Data*, A Journal Published in Two Sections: Section A, Nuclear Data Tables; Section B, Nuclear Data Sheets, Academic Press, New York.
8. M. J. Martin and P. J. Blichert-Toft, *Nuclear Data*, **A-8**, 1 (1970).
9. *Chart of the Nuclides*, Prepared by KAPL, Available from Educational Relations, General Electric Company, Schenectady, N.Y. 12305.
10. *Chart of the Nuclides*, Prepared by Battelle-Northwest, for the U.S. Atomic Energy Commission, Available for 30 cents as Y 3 AT 7:26N88, Superintendent of Documents, U.S. Government Printing Office, Washington, D.C. 20402.
11. "Source Material for Radiochemistry," Subcommittee on Radiochemistry. Committee on Nuclear Science, National Academy of Sciences—National Research Council, Nuclear Science Series Report No. 42 (1970 revision),

Available from Printing and Publishing Office, National Academy of Sciences, 2101 Constitution Ave., Washington, D.C. 20418.
12. J. H. Marshall, *Nucleonics*, **13,** (8), 34 (1955).
13. J. C. Widman, J. Mantel, N. H. Horwitz, and E. R. Prowsner, *Int. J. Appl. Radiat. Isotopes*, **19,** 1 (1968).
14. A. Bethe and J. Ashkin, "Passage of Radiations through Matter," in E. Segrè, Ed., *Experimental Nuclear Physics*, Vol. I., Wiley, New York, 1953.
15. G. Knop and W. Paul, "Interactions of Electrons and Alpha Particles with Matter," in K. Siegbahn, Ed., *Alpha-, Beta- and Gamma-Ray Spectroscopy* Vol. 1, North-Holland Publ. Co., Amsterdam, 1965.
16. M. Deutsch and S. Berko, "Positron Annihilation and Positronium," in K. Siegbahn, Ed., *Alpha-, Beta- and Gamma-Ray Spectroscopy*, Vol. 2, North-Holland Publ. Co., Amsterdam, 1965.
17. B. G. Pettiersson, "The Internal Bremsstrahlung," in K. Siegbahn, Ed., *Alpha-, Beta- and Gamma-Ray Spectroscopy*, Vol. 2, North-Holland Publ. Co., Amsterdam, 1965.
18. J. V. Jelley, *Cerenkov Radiation and Its Applications*, Pergamon Press, New York, 1958.
19. C. M. Davisson, "Interaction of γ-Radiation with Matter," in K. Siegbahn, Ed., *Alpha-, Beta- and Gamma-Ray Spectroscopy*, Vol. 1, North-Holland Publ. Co., Amsterdam, 1965.
20. R. D. Evans, "X-Ray and γ-Ray Interactions," in F. H. Attix and W. C. Roesch, Eds., *Radiation Dosimetry*, 2nd ed., Vol. 1, Academic Press, New York, 1968.
21. D. H. Walker, *Int. J. Appl. Radiat. Isotopes*, **16,** 183 (1965).
22. J. M. R. Hutchinson, C. R. Naas, D. H. Walker, and W. B. Mann, *Int. J. Appl. Radiat. Isotopes*, **19,** 517 (1968).
23. L. Katz and A. S. Penfold, *Revs. Mod. Phys.*, **24,** 28 (1952).
24. W. F. Libby, *Anal. Chem.*, **19,** 2 (1947).
25. E. K. Gora and F. C. Hickey, *Anal. Chem.*, **26,** 1158 (1954).
26. L. Yaffe and K. M. Justus, *J. Chem. Soc.*, **S,** 341 (1949).
27. B. P. Burtt, *Nucleonics*, **5** (2), 28 (1949).
28. J. Neufield, L. C. Emerson, F. J. Davis, and J. E. Turner, "The Passage of Heavy Charged Particles, Gamma Rays, and X-rays Through Matter," in K. Z. Morgan and J. E. Turner, Eds., *Principles of Radiation Protection*, Wiley, New York, 1967.
29. *Radiological Health Handbook*, rev. ed. (1970), U.S. Dept. of Health, Education and Welfare, Superintendent of Documents, U.S. Government Printing Office, Washington, D.C. 20402.
30. *Safe Handling of Radioactive Materials*, U.S. Natl. Bur. Std. Handbook 92 (1964), Superintendent of Documents, U.S. Government Printing Office, Washington, D.C. 20402.
31. R. D. Evans and R. O. Evans, *Rev. Mod. Phys.*, **20,** 305 (1948).
32. B. M. Tolbert, *Nucleonics*, **18** (8), 74 (1960).
33. J. R. Catch, *Carbon-14 Compounds*, Butterworths, Washington, 1961.
34. P. Rochlin, *Chem. Rev.*, **65,** 685 (1965).

35. R. J. Bayly and E. A. Evans, *J. Labelled Compds.*, **2**, 1 (1966); **3, S1**, 349 (1967). (Reprinted in Ref. 40.)
36. *The Radiochemical Manual*, 2nd ed., The Radiochemical Centre, Amersham, Bucks., England, 1966.
37. E. A. Evans, *Tritium and Its Compounds*, Van Nostrand-Reinhold, New York, 1966.
38. V. F. Raaen, G. A. Ropp, and H. P. Raaen, *Carbon-14* McGraw-Hill, New York, 1968.
39. L. E. Geller and N. Silberman, "Stability of Labeled Compounds After Storage," in E. D. Bransome, Jr., Ed., *The Current Status of Liquid Scintillation Counting*, Grune and Stratton, New York, 1970.
40. Y. Wang, Ed., *Handbook of Radioactive Nuclides*, The Chemical Rubber Co., Cleveland, O., 1969.
41. M. A. Muhs, E. L. Bastin, and B. E. Gordon, *Int. J. Appl. Radiat. Isotopes*, **16**, 537 (1965).
42. *Maximum Permissible Body Burdens and Maximum Permissible Concentrations of Radionuclides in Air and in Water for Occupational Exposure*, U.S. Natl. Bur. Std. Handbook 69 (1959), Superintendent of Documents, U.S. Government Printing Office, Washington, D.C. 20402.
43. There are many very useful reports and handbooks (e.g., Refs. 30 and 42) on the various aspects of radiological health issued by the U.S. National Bureau of Standards, the Federal Radiation Council, the National Council on Radiation Protection and Measurements, the International Commission on Radiological Protection, and the International Atomic Energy Agency. Titles of such publications are contained in References 44–49.
44. K. Z. Morgan, "Techniques of Personnel Monitoring and Radiation Surveying," in A. H. Snell, Ed., *Nuclear Instruments and Their Uses*, Wiley, New York, 1962.
45. H. Blatz, *Introduction to Radiological Health*, McGraw-Hill, New York, 1964.
46. K. Z. Morgan and J. E. Turner, Eds., *Principles of Radiation Protection*, Wiley, New York, 1967.
47. H. F. Henry, *Fundamentals of Radiation Protection*, Wiley-Interscience, New York, 1969.
48. A. Brodsky and F. J. Bradley, "Radiation Protection and Regulations," Part VIII in Y. Wang, Ed., *Handbook of Radioactive Nuclides*, The Chemical Rubber Co., Cleveland, O., 1969.
49. H. Kiefer, R. Maushart, and V. Mejdahl, "Radiation Protection Dosimetry," in F. H. Attix and E. Tochlin, Eds., *Radiation Dosimetry*, 2nd ed., Vol. III, Academic Press, New York, 1969.
50. M. F. Fair, "Radiation Quantities and Units," in K. Z. Morgan and J. E. Turner, Eds., *Principles of Radiation Protection*, Wiley, New York, 1967.
51. L. S. Taylor, *Health Phys.* **20**, 499 (1971).
52. C. H. Wang, R. A. Adams, and W. K. Bear, "Coordinated Design of Radioisotope Laboratories," in S. Rothchild, Ed., *Advances in Tracer Methodology*, Vol. 2, Plenum Press, New York, 1965.

53. E. D. Gupton and D. M. Davis, "Health Physics Instruments," in K. Z. Morgan and J. E. Turner, Eds., *Principles of Radiation Protection*, Wiley, New York, 1967.
54. R. C. McCall and J. A. Wall, "Radiation Safety Instruments," in G. J. Hine, Ed., *Instrumentation in Nuclear Medicine*, Vol. 1, Academic Press, New York, 1967.
55. D. H. Wilkinson, *Ionization Chambers and Counters*, Cambridge University Press, Cambridge, 1950.
56. J. Sharpe, *Nuclear Radiation Detectors*, Wiley, New York, 1955.
57. W. Franzen and L. W. Cochran, "Pulse Ionization Chambers and Proportional Counters," in A. H. Snell, Eds., *Nuclear Instruments and Their Uses*, Wiley, New York, 1962.
58. N. R. Johnson, E. Eichler, and G. Davis O'Kelley, *Nuclear Chemistry*, *Technique of Inorganic Chemistry*, Vol. II, Wiley-Interscience, New York, 1963.
59. W. J. Price, *Nuclear Radiation Detection*, 2nd ed., McGraw-Hill, New York, 1964.
60. S. C. Curran and H. W. Wilson, "Proportional Counters and Pulse-Ion Chambers," in K. Siegbahn, Ed., *Alpha-, Beta- and Gamma-Ray Spectroscopy*, Vol. 1, North-Holland Publ. Co., Amsterdam, 1965.
61. C. V. Robinson, "Geiger-Müller and Proportional Counters," in G. J. Hine, Ed., *Instrumentation in Nuclear Medicine*, Vol. 1, Academic Press, New York, 1967.
62. E. Schram and R. Lombaert, *Organic Scintillation Detectors*, Elsevier Publishing Co., New York, 1963.
63. R. B. Murray, "Scintillation Counters," in A. H. Snell, Ed., *Nuclear Instruments and Their Uses*, Wiley, New York, 1962.
64. J. B. Birks, *The Theory and Practice of Scintillation Counting*, Pergamon Press, New York, 1964.
65. J. M. Taylor, *Semiconductor Particle Detectors*, Butterworths, London, 1963.
66. F. S. Goulding, *Nucleonics*, **22** (5), 54 (1964).
67. W. M. Gibson, G. L. Miller, and P. F. Donovan, "Semiconductor Particle Spectrometers," in K. Siegbahn, Eds., *Alpha-, Beta- and Gamma-Ray Spectroscopy*, Vol. 1, North-Holland Publ. Co., Amsterdam, 1965.
68. G. Dearnaley and D. C. Northrop, *Semiconductor Counters for Nuclear Radiation*, 2nd ed., Wiley, New York, 1966.
69. A. J. Tavendale, *Ann. Rev. Nucl. Sci.*, **17**, 73 (1967).
70. G. Bertolini and A. Coche, Eds., *Semiconductor Detectors*, North-Holland Publ. Co., Amsterdam, 1968.
71. G. T. Ewan, "Semiconductor Spectrometers," in F. J. M. Farley, Ed., *Progress in Nuclear Techniques and Instrumentation*, Vol. 3, North-Holland Publ. Co., Amsterdam, 1968.
72. H. Yagoda, *Radioactive Measurements with Nuclear Emulsion*, Wiley, New York, 1949.
73. G. A. Boyd, *Autoradiography in Biology and Medicine*, Academic Press, New York, 1955.

74. R. H. Herz, *Nucleonics*, **9** (3), 24 (1951).
75. W. P. Norris and L. A. Woodruff, *Ann. Rev. Nucl. Sci.*, **5**, 297 (1955).
76. R. A. Dudley, "Dosimetry with Photographic Emulsions," in F. H. Attix and W. C. Roesch, Eds., *Radiation Dosimetry*, 2nd ed., Vol. II, Academic Press, New York, 1966.
77. F. H. Attix, Ed., *Proceedings of the International Conference on Luminescence Dosimetry*, Stanford University, June 21–23, 1965, USAEC Conf. 650637.
78. J. F. Fowler and F. H. Attix, "Solid State Integrating Dosimeters," in F. H. Attix and W. C. Roesch, Eds., *Radiation Dosimetry*, 2nd ed., Vol. II, Academic Press, New York, 1966.
79. J. A. Auxier, K. Becker, and E. M. Robinson, Eds., *Proceedings of the Second International Conference on Luminescence Dosimetry*, Gatlinburg, Tenn., Sept. 23–26, 1968, USAEC Conf. 680920.
80. J. R. Cameron, N. Suntharalingam, and G. N. Kenney, *Thermoluminescent Dosimetry*, University of Wisconsin Press, Madison, 1968.
81. R. L. Fleischer, P. B. Price, R. M. Walker and E. L. Hubbard, *Phys. Rev.*, **133,** A1443 (1964).
82. R. L. Fleischer, P. B. Price, and R. M. Walker, *Ann. Rev. Nucl. Sci.*, **15,** 1 (1965).
83. R. L. Fleischer, P. B. Price, and R. M. Walker, *Science*, **149,** 383 (1965).
84. E. I. Hamilton, *Int. J. Appl. Radiat. Isotopes*, **19,** 159 (1968).
85. P. B. Price and R. L. Fleischer, *Radiat., Eff.* **2,** 291 (1970).
86. N. L. Lark, *Nucl. Instrum. Methods*, **67,** 137 (1969).
87. H. A. Khan, *Radiat. Eff.*, **8,** 135 (1971).
88. R. L. Fleischer, P. B. Price, R. M. Walker, and E. L. Hubbard, *Phys. Rev.* **156,** 353 (1967).
89. E. V. Benton, *Radiat. Eff.*, **2,** 273 (1970).
90. E. P. Steinberg, "Counting Methods for the Assay of Radioactive Samples," in A. H. Snell, Ed., *Nuclear Instruments and Their Uses*, Vol. 1, Wiley, New York, 1962.
91. I. R. Williams, *Nucl. Instrum. Methods*, **44,** 160 (1966).
92. M. P. Ruffle, *Nucl. Instrum. Methods*, **52,** 354 (1967).
93. R. P. Gardner and K. Verghese, *Nucl. Instrum. Methods*, **93,** 163 (1971).
94. T. Kohman, *Anal. Chem.*, **21,** 352 (1949).
95. P. C. Stevenson, *Processing of Counting Data*, NAS-NS 3109 (1966), available from Natl. Technical Info. Service, U.S. Dept. of Commerce, Springfield, 22151.
96. T. Tojo, T. Nakajima, and M. Kondo, *Nucl. Instrum. Methods*, **89,** 39 (1970).
97. E. Fairstein, "Electrometers and Amplifiers," in A. H. Snell, Ed., *Nuclear Instruments and Their Uses*, Vol. 1, Wiley, New York, 1962.
98. C. C. H. Washtell and S. G. Hewitt, *Nucleonic Instrumentation*, George Newnes, London, 1965.
99. H. H. Chiang, *Basic Nuclear Electronics*, Wiley-Interscience, New York, 1969.
100. E. Fenyves and O. Haiman, *The Physical Principles of Nuclear Radiation. Measurements*, Academic Press, New York, 1969.

101. E. Kowalski, *Nuclear Electronics*, Springer-Verlag, Berlin, 1970.
102. B. M. Tolbert and W. E. Siri, "Determination of Radioactivity," in A. Weissberger, Ed., *Technique of Organic Chemistry*, 3rd. ed., Vol. I, Interscience New York, 1960.
103. R. L. Chase, *Nuclear Pulse Spectrometry*, McGraw-Hill, New York, 1961.
104. G. S. Stanford, *Nucl. Instrum. Methods*, **34,** 1 (1965).
105. F. S. Goulding, "Multi-channel Pulse-Amplitude Analyzers," in K. Siegbahn, Eds., *Alpha-, Beta- and Gamma-Spectroscopy*, Vol. 1, North-Holland Publ. Co., Amsterdam, 1965.
106. G. D. O'Kelley, Ed., *Applications of Computers to Nuclear and Radiochemistry*, NAS-NS 3107 (1963), Available from Natl. Technical Info. Service, U.S. Dept. of Commerce, Springfield, Va. 22151.
107. E. Der Mateosian, *Nucl. Instrum. Methods*, **73,** 77 (1969).
108. D. F. Crouch and R. L. Heath, "Evaluation and Calibration of Pulse-Height Analyzer Systems for Computer Data Processing," in Ref. 106.
109. An excellent source of state-of-the-art information on the measurement of radioactivity is the review on nucleonics published in the Analytical Reviews (Fundamentals) issue of *Analytical Chemistry*.
110. Information on source preparation is contained in two series of monographs, *Radiochemistry of the Elements* and *Radiochemical Techniques*. These are publications of the Subcommittee on Radiochemistry of the Committee on Nuclear Science of the NAS-NRC. The monographs are listed in Ref. 11 and are available from the National Technical Information Service, U.S. Department of Commerce, Springfield, Va. 22151.
111. L. Yaffe, *Ann. Rev. Nucl. Sci.*, **12,** 153 (1962).
112. D. E. Watt and D. Ramsden, *High Sensitivity Counting Techniques*, Pergamon Press, New York, 1964.
113. W. Parker and H. Slätis, "Sample and Window Technique," in K. Siegbahn, Ed., *Alpha-, Beta- and Gamma-Ray Spectroscopy*, Vol. 1, North-Holland Publ. Co., Amsterdam, 1965.
114. R. J. Brouns, *Absolute Measurement of Alpha Emission and Spontaneous Fission*, NAS-NS 3112 (1968), Available from Natl. Technical Info. Service, U.S. Dept. of Commerce, Springfield, Va. 22151.
115. R. A. Deal and R. N. Chanda, *Nucl. Instrum. Methods*, **69,** 89 (1969).
116. J. Leon and N. H. Steiger-Shafrir, *Nucl. Instrum. Methods*, **84,** 102 (1970).
117. R. D. Cherry, "Alpha Particle Detection Techniques Applicable to the Measurement of Samples from the Natural Radiation Environment," in J. A. S. Adams and W. M. Lowder, Eds., *The Natural Radiation Environment*, University of Chicago Press, 1964.
118. D. L. Horrocks, "Liquid Scintillator Solutions in Nuclear Physics and Nuclear Chemistry," in H. A. Elion and D. C. Stewart, Eds., *Progress in Nuclear Energy*, Series IX, *Analytical Chemistry*, Vol. 7, Pergamon Press, New York, 1966.
119. D. L. Horrocks, *Int. J. Appl. Radiat. Isotopes*, **17,** 441 (1966).
120. E. V. Benton, *A Study of Charged Particle Tracks in Cellulose Nitrate*, USNRDL-TR-68-14, Jan. 1968 (Uncl.).

121. A. Cole, D. J. Simmons, H. Cummins, F. J. Congel, and J. Kastner, *Health Phys.*, **19,** 55 (1970).
122. G. Somogyi, M. Varnogy, and L. Medveczky, *Radiat. Eff.*, **5,** 111 (1970).
123. J. Anno, "Use of Cellulose Nitrate for Alpha-Particle Detection" (Thesis, in French), U.S. Atomic Energy Comm. 1970, NP-18411 (on Microfiche).
124. G. Somogyi and D. S. Srivastava, *Int. J. Appl. Radiat. Isotopes*, **22,** 289 (1971).
125. R. F. Glascock, *Isotopic Gas Analysis for Biochemists*, Academic Press, New York, 1954.
126. C. G. Bell, Jr., and F. N. Hayes, Eds., *Liquid Scintillation Counting*, Pergamon Press, New York, 1958.
127. S. Rothchild, Ed., *Advances in Tracer Methodology*, Vol. 1, 1963; Vol. 2, 1965; Vol. 3, 1966; Vol. 4, 1968, Plenum Press, New York.
128. E. Rapkin, "Preparation of Samples for Liquid Scintillation Counting," in G. J. Hine, Ed., *Instrumentation in Nuclear Medicine*, Vol. 1, Academic Press, New York, 1967.
129. Y. Kobayashi and D. V. Maudsley, "Practical Aspects of Liquid Scintillation Counting," in D. Glick, Ed., *Methods of Biochemical Analysis*, Vol. 17, Interscience, New York, 1969.
130. J. H. Parmentier and F. E. L. Ten Haaf, *Int. J. Appl. Radiat. Isotopes*, **20,** 305 (1969).
131. E. D. Bransome, Jr., Ed., *The Current Status of Liquid Scintillation Counting*, Grune and Stratton, New York, 1970.
132. D. L. Horrocks and C. T. Peng, Eds. "Organic Scintillators and Liquid Scintillation Counting," *Proceedings of the International Conference on Organic Scintillators and Liquid Scintillation Counting*, San Francisco, July 7–10, 1970, Academic Press, New York, 1971.
133. *A Manual of Radioactivity Procedures*, U.S. Natl. Bureau of Standards Handbook 80 (1961), Superintendent of Documents, U.S. Government Printing Office, Washington, D.C. 20402.
134. E. H. Belcher, *Proc. Roy. Soc.*, **A216,** 90 (1953).
135. H. H. Ross, *Anal. Chem.*, **41,** 1260 (1969).
136. E. P. Parker and R. H. Elrick, "Cerenkov Counting as a Means of Assaying β-Emitting Radionuclides," in E. D. Bransome, Jr., Ed., *The Current Status of Liquid Scintillation Counting*, Grune and Stratton, New York, 1970.
137. A. Läuchli, "Application of Cerenkov Counting to Ion Transport Studies in Plants," in D. L. Horrocks and C. T. Peng, Eds., *Organic Scintillators and Liquid Scintillation Counting*, Academic Press, New York, 1971.
138. A. Läuchli, *Int. J. Appl. Radiat. Isotopes*, **20,** 265 (1969).
139. H. H. Ross, "Performance Parameters of Selected Waveshifting Compounds for Cerenkov Counting," in D. L. Horrocks and C. T. Peng, Eds., *Organic Scintillators and Liquid Scintillation Counting*, Academic Press, New York, 1971.
140. H. H. Ross, "Cerenkov Radiation: Photon Yield Application to ^{14}C Assay," in E. D. Bronsome, Jr., Ed., *The Current Status of Liquid Scintillation Counting*, Grune and Stratton, New York, 1970.

141. R. D. Stubbs and A. Jackson, *Int. J. Appl. Radiat. Isotopes*, **18,** 857 (1967).
142. A. T. B. Moir, *Int. J. Appl. Radiat. Isotopes*, **22,** 213 (1971).
143. J. F. Osborne, *Int. J. Appl. Radiat. Isotopes*, **18,** 829 (1967).
144. L. E. Glendenin, *Nucleonics*, **2,** (1), 12 (1948).
145. G. I. Gleason, J. D. Taylor, and D. L. Tabern, *Nucleonics*, **8** (5), 12 (1951).
146. J. H. Harley and N. Hallden, *Nucleonics*, **13** (1), 32 (1955).
147. T. Baltakmens, *Nucl. Instrum. Methods*, **82,** 264 (1970).
148. R. H. Rodriguez-Pasques, P. A. Mullen, G. A. George, and J. E. Harding, *Int. J. Appl. Radiat. Isotopes*, **18,** 835 (1967).
149. A. C. G. Mitchell, "Procedures for the Investigation of Disintegration Schemes—General Procedures," in K. Siegbahn, Ed., *Alpha-, Beta- and Gamma-Ray Spectroscopy*, Vol. 1, North-Holland Publ. Co., Amsterdam, 1965.
150. R. V. Lieshout, A. H. Wapstra, R. A. Ricci, and R. K. Girgis, "Scintillation Spectra Analysis," in K. Siegbahn, Ed., *Alpha-, Beta- and Gamma-Ray Spectroscopy*, Vol. 1, North-Holland Publ. Co., Amsterdam, 1965.
151. C. S. Wu, "The Shape of β-Spectra," in K. Siegbahn, Ed., *Alpha-, Beta- and Gamma-Ray Spectroscopy*, Vol. 2, North-Holland Publ. Co., Amsterdam, 1965.
152. E. J. Konopinski, *The Theory of Beta Radioactivity*, Oxford University Press, Oxford, 1966.
153. C. S. Wu and S. A. Moszkowski, *Beta Decay*, Wiley-Interscience, New York, 1966.
154. D. G. Gardner and W. W. Meinke, *Int. J. Appl. Radiat. Isotopes*, **3,** 232 (1958).
155. R. G. Johnson, O. E. Johnson, and L. M. Langer, *Phys. Rev.*, **102,** 1142 (1956).
156. C. T. Prevo and J. L. Cate, *Nucl. Instrum. Methods*, **55,** 173 (1967).
157. G. J. Hine, "γ-Ray Sample Counting," in G. J. Hine, Ed., *Instrumentation in Nuclear Medicine*, Vol. 1, Academic Press, New York, 1967.
158. R. W. Perkins, *Health Phys.*, **7,** 81 (1961).
159. M. Kartha, *Health Phys.*, **20,** 431 (1971).
160. K. Becker, *Health Phys.*, **12,** 955 (1966).
161. F. Adams and R. Dams, Eds., *Applied Gamma-Ray Spectrometry*, 2nd. ed. Pergamon Press, New York, 1970.
162. R. L. Heath, *Scintillation Spectrometry*, 2nd ed., Vols. 1 and 2, U.S. Atomic Energy Comm. IDO-16880 (1964).
163. F. S. Goulding, *Nucl. Instrum. Methods*, **43,** 1 (1966).
164. J. W. Mayer, *Nucl. Instrum. Methods*, **43,** 55 (1966).
165. J. M. Hollander, *Nucl. Instrum. Methods*, **43,** 65 (1966).
166. V. C. Rogers, *Anal. Chem.*, **42,** 807 (1970).
167. W. W. Black and R. L. Heath, *Nucl. Phys.*, **A90,** 650 (1967).
168. F. J. Walter, *IEEE Trans. Nuclear Science*, **NS-17** (3) (June 1970), Twelfth Scintillation and Semiconductor Counter Symposium (March, 1970).
169. A. A. Jarrett, *Statistical Methods Used in the Measurement of Radioactivity* U.S. Atomic Energy Comm. AECU-262 (MonP-126) (1946).
170. R. Loevinger and M. Berman, *Nucleonics*, **9** (1), 26 (1951).
171. A. H. Jaffey, *Nucleonics*, **18** (11), 180 (1960).

172. G. E. A. Wyld, "Statistical Confidence in Liquid Scintillation Counting," in E. D. Bransome, Jr., Ed., *The Current Status of Liquid Scintillation Counting*, Grune and Stratton, New York, 1970.
173. J. M. England and R. G. Miller, Jr., *Int. J. Appl. Radiat. Isotopes*, **20,** 1 (1969).
174. A. S. Goldin, *Nucl. Instrum. Methods*, **48,** 321 (1967).
175. J. L. Putman, *Int. J. Appl. Radiat. Isotopes*, **13,** 99, 222 (1962); **20,** 205 (1969).
176. A. DeVolpi, *Int. J. Appl. Radiat. Isotopes*, **22,** 103 (1971).
177. B. Altshuler and B. Pasternak, *Health Physics*, **9,** 293 (1963).
178. L. A. Currie, *Anal. Chem.*, **40,** 586 (1968).
179. *Preparation, Maintenance, and Application of Standards of Radioactivity*, U.S. Natl. Bureau of Standards Circular 594 (1958), Superintendent of Documents, U.S. Government Printing Office, Washington, D.C. 20402.
180. *Users' Guides for Radioactivity Standards*, Subcommittee on the Use of Radioactivity Standards, Committee on Nuclear Science, NAS-NRC, Washington, D.C. 20418.
181. *Metrology of Radionuclides*, Proceedings of a Symposium Organized by the IAEA, Vienna, 14–16 October 1959, IAEA, STI/PUB/6, Vienna (1960).
182. R. A. Allen, "Measurement of Source Strength," in K. Siegbahn, Ed., *Alpha-, Beta- and Gamma-Ray Spectroscopy*, Vol. 1, North-Holland Publ. Co., Amsterdam, 1965.
183. *Standardization of Radionuclides*, Proceedings of a Symposium on Standardization of Radionuclides Held by the IAEA in Vienna, 10–14 October 1966, IAEA, STI/PUB/139, Vienna (1967).
184. *Standards of Activity*, R. C. C. Review 4 (1967 rev. ed.), The Radiochemical Centre, Amersham, Bucks, England.
185. L. P. Remsberg, *Ann. Rev. Nucl. Sci.*, **17,** 347 (1967).
186. G. C. Lowenthal, *Int. J. Appl. Radiat. Isotopes*, **20,** 559 (1969).
187. *Radioactivity Calibration Standards*, Proceedings of a Seminar Held in Washington, D.C., November 1968, U.S. Natl. Bureau of Standards, Special Publication 331 (1970), U.S. Government Printing Office, Washington, D.C. 20402.
188. P. J. Campion, *Int. J. Appl. Radiat. Isotopes*, **4,** 232 (1959).
189. Y. Kawada, O. Yura, and M. Kimura, *Nucl. Instrum. Methods*, **78,** 77 (1970).
190. J. S. Merritt, J. G. V. Taylor, W. F. Merritt, and P. J. Campion, *Anal. Chem.*, **32,** 310 (1960).
191. J. R. DeVoe, *Radioactive Contamination of Materials Used in Scientific Research*, Nuclear Science Series Report No. 34, NAS-NRC, Washington, D.C. (1961).
192. T. T. Sugihara, *Low-level Radiochemical Separations*, NAS-NS 3103 (1961), Available from Natl. Technical Info. Service, U.S. Dept. of Commerce, Springfield, Va. 22151.
193. J. A. S. Adams and W. M. Lowder, Eds., *The Natural Radiation Environment*, University of Chicago Press, 1964.
194. W. V. Mayneord and C. R. Hill, "Natural and Man-made Background Radiation," in F. H. Attix and E. Tochlin, Eds., *Radiation Dosimetry*, 2nd ed., Vol. III, Academic Press, New York, 1969.

195. *Radioactivity*, National Bureau of Standards Handbook 86 (1963), Superintendent of Documents, U.S. Government Printing Office, Washington, D.C. 20402.
196. J. A. Cooper, *Nucl. Instrum. Methods*, **82,** 273 (1970).
197. B. S. Pasternak and N. H. Harley, *Nucl. Instrum. Methods*, **91,** 533 (1971).
198. R. I. Weller, "Low-Level Radioactive Contamination," in J. A. S. Adams and W. M. Lowder, Eds., *The Natural Radiation Environment*, The University of Chicago Press, 1964.
199. R. I. Weller, E. C. Anderson, and J. L. Barker, *Nature*, **206,** 1211 (1968).
200. F. J. Walter and R. R. Boshart, *Nucl. Instrum. Methods*, **42,** 1 (1966).
201. R. A. Allen, D. B. Smith, R. L. Otlet, and D. S. Rawson, *Nucl. Instrum. Methods*, **45,** 61 (1966).
202. P. R. B. Ward and P. Wurzel, *Int. J. Appl. Radiat. Isotopes*, **19,** 529 (1968).
203. A. S. Stenberg and I. U. Olsson, *Nucl. Instrum. Methods*, **61,** 125 (1968).
204. H. Perschke, M. B. A. Crespi, and G. B. Cook, *Int. J. Appl. Radiat. Isotopes.* **20,** 813 (1969).
205. W. Roedel, *Nucl. Instrum. Methods*, **83,** 88 (1970).
206. N. A. Wogman, *Nucl. Instrum. Methods*, **83,** 277 (1970).
207. S. R. Lewis and N. H. Shafrir, *Nucl. Instrum. Methods*, **93,** 317 (1971).
208. J. A. Cooper and R. W. Perkins, *Nucl. Instrum. Methods*, **94,** 29 (1971).

Chapter **X**

GAMMA-RAY SPECTROMETRY*

Julian M. Nielsen

1 INTRODUCTION

Gamma-ray spectrometry is the measurement of the energies and intensities of electromagnetic radiation of discrete wavelengths. This technique is used mainly in two study areas. The first of these is the determination of the energy level structures and their relationships in nuclei. The second area is the utilization of the technique as a means of isotopic or elemental analysis. To be best suited for this work the spectrometer should have high energy resolution and high efficiency; however, no present system combines both of these requirements to the degree desired. The nuclear spectroscopist must

* This paper is based on work performed under United States Atomic Energy Commission Contract AT(45-1)-1830.

resolve all of the gamma rays from an often complex spectrum and then measure their intensities to elucidate the nuclear structure. This leads him to place the greatest emphasis on the resolution, with the efficiency of high but lesser importance. On the other hand the radiochemist might not need complete resolution of the spectrum in order to select a portion for intensity measurement so that his requirements often emphasize high efficiency. This high efficiency also allows many analyses to be made in a given period of time with consequent lower cost. Nevertheless the resolving power determines the limit of the complexity of the spectra that can be examined so that for analysis of gamma rays in complex mixtures of radionuclides high resolution becomes increasingly important. However, even though the instrument requirements for the two main study areas are rapidly becoming more similar, the chapter will emphasize the instruments and techniques of the radiochemist for gamma-ray spectrometric analysis.

Rutherford and Andrade [1] made the first measurements of gamma-ray spectra using crystal diffraction. Although the resolution was poor, further development, particularly by DuMond [2], resulted in instruments of very high precision which played an important part in determining the gamma-ray energies and intensities resulting from nuclear transitions of radionuclides. In this method the crystal elastically scatters the photons with an angular distribution which is measured to determine the energies of the incident photons. In principle this is an absolute method. When germanium crystals are used a resolution of about 1% at 1 MeV was obtained with about 0.1% at 100 keV [3].

Another high resolution system makes use of the magnetic beta-ray spectrometer. In one mode of operation, internal conversion spectrometry, the internal conversion electrons ejected in a process competing with photon emission between nuclear energy levels are measured. In external conversion magnetic spectrometry the photoelectron spectrum produced when gamma rays strike a heavy element foil are analyzed in the beta spectrometer. These magnetic spectrometry techniques provide the highest resolutions obtainable in the energy range of a few tenths to a few MeV. The resolution obtained is about 0.3% at 1 MeV [4].

The high resolution spectrometer systems discussed above have only limited applicability for use in radiochemical analysis and will not be discussed in detail here. For further information see *Alpha, Beta- and Gamma-Ray Spectrometry*, edited by Siegbahn [5]. The efficiencies of these systems are on the order of one part in 10^7 to 10^{10} events, requiring intensely radioactive sources and long counting times. The instruments are essentially "single channel" devices which allow only a study of a small fraction of the energy interval at a given time. This has led to acceptance of lower resolution systems of higher efficiencies. Two of these, scintillation spectrometry and semiconductor spectrometry, have high efficiencies and multichannel capabilities for widespread usefulness in gamma-ray spectrometry. The scintillation detector utilizes the excitation resulting from charged particles passing through it to produce fluorescent light. The semiconductor detector utilizes the ionization produced by charged particle interactions.

2 INTERACTIONS OF GAMMA RAYS WITH DETECTORS

Gamma rays transfer all or part of their energy to a detector by three primary processes. These are the photoelectric (pe) effect, Compton (C), and electron-positron pair production (pp). If by one or a combination of these processes all of the energy of an incident photon is transferred to the detector, an output signal can be obtained which will be characteristic of that photon energy. The signal resulting from this total energy absorption is the basis for gamma ray spectrometry wherein a peak ("photopeak") is produced in a plot of number of signals (counts) versus the signal energy (photon energy). A typical spectrum obtained with a lithium drifted germanium [Ge(Li)] detector is shown in Fig. 10.1. Although total absorption of monoenergetic gamma rays by a crystal might be expected to result in a single line spectrum at the photon energy, a peak of essentially Gaussian shape results with finite width due to statistical variation in the detection process and non-proportional response of the detector system. The narrower the peak the better the resolution and generally the more useful the detector is for gamma

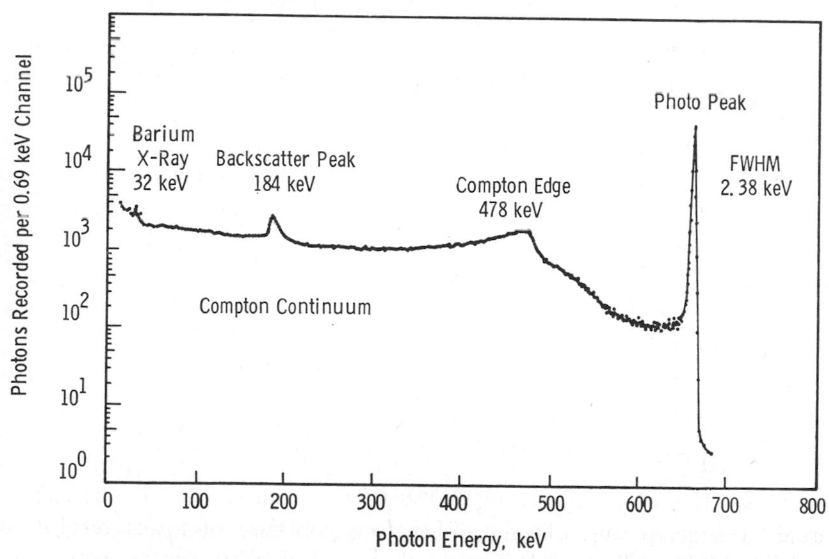

Fig. 10.1 Typical gamma-ray spectrum of ^{137}Cs showing components observed.

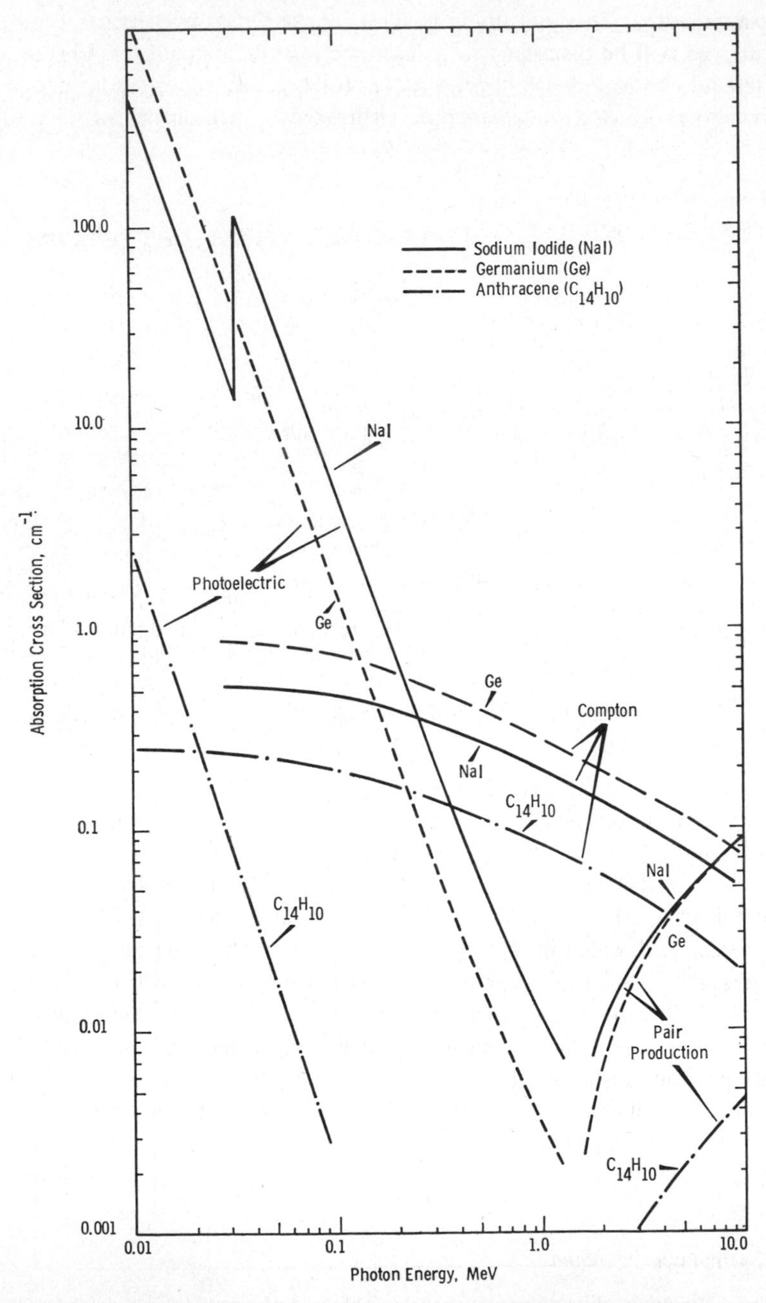

Fig. 10.2 Partial absorption coefficient curves for the three interaction processes in three types of detector materials.

ray spectrometry. This parameter varies markedly with detector type and material and will be discussed in a later section of this chapter. Figure 10.2 gives partial absorption coefficient curves for the three interaction processes in three types of detector material, anthracene, germanium, and sodium iodide.

The Photoelectric Process

The photoelectric process is the main process by which low energy gamma rays interact with a detector. Interaction efficiencies become very small in the region of about 1 MeV except for the elements of highest Z. The efficiency of this process decreases at first about as the $-\frac{7}{2}$ power of the photon energy changing to more nearly the -1 power with increasing energy. The efficiency for this process is approximately proportional to Z^5 so that high Z materials are chosen for gamma ray detectors if other factors permit.

In the photoelectric process the incident photon transfers all of its energy to an electron bound to an atom, ejecting the electron with kinetic energy equal to the incident photon energy less the binding energy of the electron in the atom. The process occurs only with a bound electron where the atom and the electron can conserve momentum. For photons with energies greater than the K shell binding energies of the detector material about 4 out of 5 photoelectric events occur with electrons in the K shell with most of those remaining occurring with the L shell electrons. Thus the photoelectric absorption curves have discontinuities at energies corresponding to the K, L, and so on, electron binding energies.

The ejected "photoelectron" generally expends its energy in ionization although some bremsstrahlung also results. The vacancy left in the orbital shell of the atom is then filled by an electron from an outer shell followed by the emission of X-rays or Auger electrons. If these X-rays or Auger electrons transfer all of their energy to the detector by single or multiple events the total process will occur in a time short relative to the time constant of the detector system and the event energy recorded will be the total energy of the incident photon. Should the X-ray (or less likely the Auger electron) escape absorption in the detector, the event recorded will be that of the incident photon less that of the energy carried by the escaping radiation. Such events result in an "escape peak" in the spectrum and add generally undesirable complexity to the spectra as well as reducing the number of events recorded in the photopeak. In practice effort is made to reduce or eliminate such losses.

The Compton Process

In the Compton effect the photon transfers only part of its energy to an electron. In this interaction the electron need not be bound in an atom as in the case of the photoelectric effect. However, in most detector materials the Compton interaction does take place with bound electrons although these are generally the outer electrons; hence Compton interactions do not result in K or L X-rays to the degree resulting from the photoelectric effect. The degraded photon from the Compton interaction is also deflected from its original direction and may escape unabsorbed from the detector. The resulting spectrum from this process has events recorded not in peaks such as with the photoelectric process, but as a continuum up to a Compton "cut-off" energy value. This value (Compton edge) represents those events wherein the incident photon has transferred the maximum amount of energy to the electron for this process. The energy transferred for a photon of given energy is a function of the scattering angle of the degraded photon, with the maximum energy transferring when the degraded photon is deflected back in the direction of the incident photon. These relationships are given by the equation (for a deviation of this equation see Ref. 6)

$$\frac{1}{E'_\gamma} = \frac{1}{E_\gamma} + \frac{1 - \cos\theta}{E_e} \tag{10.1}$$

where E'_γ = the energy of the degraded photon
E_γ = the energy of the incident photon
θ = the angle between the incident and the scattered photons
E_e = the rest mass energy (511 keV) of the electron

Thus for the condition of maximum energy transfer $\theta = 180°$ and $\cos\theta = -1$ we can calculate E'_γ for ^{137}Cs to be 184 keV. The Compton edge for ^{137}Cs should be at $662 - 184 = 478$ keV as can be verified by referring to Fig. 9.1. Under some conditions this Compton edge is more prominent than the photopeak and has been used for gamma-ray spectrometry. Also prominent on the Compton continuum in this spectrum of ^{137}Cs given in Fig. 10.1 is a peak at 184 keV. This results from absorption in the crystal of photons which have been degraded by transferring the maximum energy to material in the vicinity of the detector and are returned to ("backscattered" into) the detector. These are called backscattered photons and the resulting peak in the spectrum is called the backscatter peak. These peaks can complicate analysis of spectra by being mistaken for or by masking photopeaks.

The average energy lost for each Compton interaction varies from about 10% at 75 keV and 34% at 500 keV to 51% at 2 MeV. Thus each time a photon loses energy by a Compton interaction it is reduced in energy and then, on the average, loses a smaller fraction of energy during the next Compton interaction. With each reduction in energy, however, the probability of a photoelectric interaction becomes markedly more likely since the photoelectric process is the predominant one at low photon energies. For instance, the photoelectric process is more probable than the Compton process for germanium at photon energies below about 150 keV and below about 230 keV for sodium iodide. This would mean that for germanium on the average five Compton interactions would degrade the photon from 1 MeV to about 150 keV at which energy the interaction would most probably be of the photoelectric type. If this chain of interactions all takes place in the detector, total photon energy absorption is obtained and the resultant signal is indistinguishable (except for small differences in X-ray abundances) from a single photoelectric event of an incident photon of the original energy. By this means Compton events play an important part in gamma ray spectrometry. In order to enhance the role of this interaction large detectors are used within which multiple events are more likely without escape of a degraded photon.

The Compton process is a function of the electron density of a material and hence is proportional to Z. As noted earlier the photoelectric process is nearly proportional to Z^5. Therefore, all other factors being equal, a high Z material would be chosen for gamma ray spectrometry to enhance the photoelectric process over the Compton process.

The Pair Production Process

When the incident photon has an energy in excess of 1.02 MeV interaction with the detector by pair production is possible. In this process a positron-electron pair is created in the field of a nucleus or an electron which must be present to conserve momentum and energy. Since the creation of this pair requires energy equivalent to two electron rest masses (1.02 MeV) a photon of at least that energy is required. In gamma-ray spectrometry this process is of little significance except for photons of energy in excess of about 1.5 MeV. The positron and the electron transfer their kinetic energies to the detector by ionization or bremsstrahlung processes with little chance for escape. However, when the positron has nearly dissipated all of its kinetic energy it undergoes the process of annihilation with an electron, usually with the simultaneous emission of two 0.511 MeV photons at an angle of approximately 180° with respect to each other. Thus gamma ray spectrometry of high energy photons is always complicated by the secondary production of these 0.511 MeV photons. If these two photons completely transfer their energies to the detector by photoelectric or multiple Compton-photoelectric processes such that all of the incident photon energy is dissipated in the detector, a signal is recorded which is indistinguishable from a single photoelectric process on the original incident photon. This process is, then, also useful for gamma ray spectrometry. However, if either or both of these annihilation photons escape unabsorbed from the detector, "escape" or "double escape"

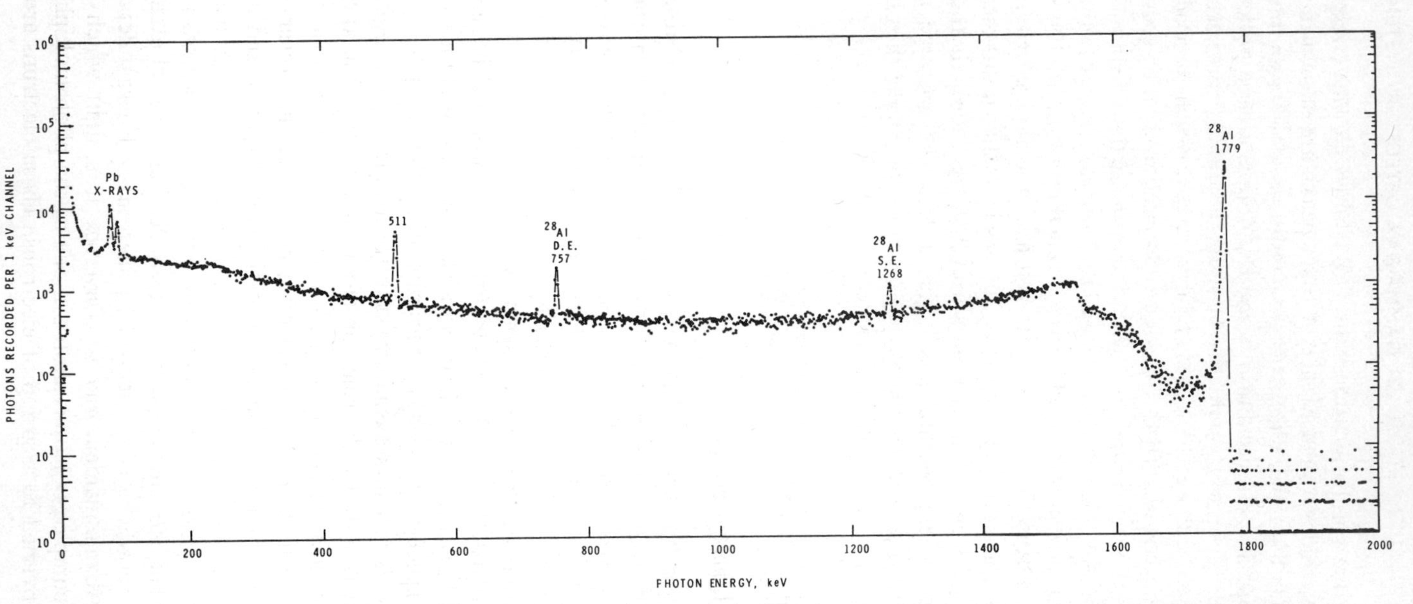

Fig. 10.3 The ^{28}Al spectrum taken with a Ge(Li) detector, showing the single escape, double escape, and annihilation peaks.

peaks result in the spectrum of energies 0.511 and 1.02 MeV, respectively, lower than the incident photon energy. Such peaks are shown in the ^{28}Al spectrum given in Fig. 10.3. Under certain conditions, such as an interference with the photopeak, or a condition where the double escape peak is larger than the photopeak, these peaks may be used in gamma ray spectrometry. Also shown in this figure is a peak of 0.511 MeV which is produced by annihilation photons scattered from pair production events occurring in materials near but not part of the sensitive volume of the detector.

While on the subject of escape peaks it is important to note that in the measurement of gamma rays in the energy range of about 30–100 keV with NaI(Tl) detectors an iodine escape peak is observed in the spectrum which is 28 keV lower in energy than the photopeak. The intensity of this escape peak relative to the photopeak ranges from about 25% at 30 keV to about 4.7% at 100 keV [5]. Where "windowless" Ge(Li) detectors are used a similar germanium escape peak which is 10.0 keV lower in energy than the photopeak has been observed at photon energies as high as 279 keV [7, 8].

3 GAMMA-RAY DETECTION SYSTEMS

Scintillation Detectors

Scintillation spectrometry became possible when high gain photomultiplier tubes were developed which gave a pulse of electrons proportional to the incident photon energy. The first counters using phototubes were not true spectrometers but charged particle counters. The first charged particle scintillation spectrometer was made by Kallman [9] who used a napthalene crystal. The gamma ray scintillation spectrometer followed shortly thereafter as a result of Hofstadter's [10] discovery of the thallium activated sodium iodide [NaI(Tl)] crystal. This technique quickly became the standard method for gamma ray spectrometry within the limits of its relatively poor resolution. It still remains the most used gamma ray spectrometric technique, but is being displaced rapidly by the semiconductor detector for the applications where the additional selectivity (higher resolution) characteristics of this detector outweigh its somewhat lower efficiency. Since this lower efficiency is largely due to the limited size of the semiconductor, increased use of the semiconductor detector will quickly follow each increase in size and efficiency.

Figure 10.4 shows a typical NaI(Tl) detector. The scintillation material (phosphor) of the detector is surrounded by reflecting material except where it is optically coupled to the photomultiplier tube. Energy deposited in the scintillator by a gamma ray is converted into light which is transmitted almost completely to the photocathode of the photomultiplier where this light is converted to a group of electrons. These electrons are focused

Fig. 10.4 A typical 3 × 3 Na(Tl) detector with photomultiplier.

electro-statistically onto the first dynode of the photomultiplier where interaction results in a number of ejected electrons about a factor of four larger than the number incident on that dynode. This group of electrons is then focused onto the next dynode where the multiplication process is repeated. After the usual ten dynodes the pulse of electrons is larger by about 0.5 to 5 × 10^6. This pulse of electrons induces a voltage drop at the anode in the range of a few millivolts to a few volts. This signal is routed to a multichannel analyzer for spectrum analysis.

Thallium-activated sodium iodide has the largest light output of the scintillation phosphors, which property together with its relatively high atomic number, and density (3.607 g/cc), its transparency to its own fluorescent light which is emitted in a band 3700–4500 Å wide, its relatively short fluorescence decay time of 0.25 μsec, and its ability to be grown as a large single crystal has caused it to be the most used scintillator for gamma spectrometry. Crystals of the same size and shape will have exactly the same efficiencies for photons and even though the pulse heights and resolutions may differ, the counts in a given peak will be the same. This property make the voluminous data presented by Heath [11] for the 3 × 3 = in. cylindrical crystal directly useful so that this size of detector has become almost a standard for gamma-ray spectrometry. These advantages far outweigh its disadvantages, some of which are hygroscopicity, relatively poor resolution [the best resolution was reported by Bell [5] to be 6.0% for the ^{137}Cs (^{137}Ba) 0.662 MeV gamma ray], and the fact that significant deviations occur in plots of energy versus pulse height for these detectors [12, 13]. The hygroscopicity requires careful canning and handling. The nonlinearity is generally readily overcome by careful calibration for gamma emitters and requires special care when the detector is used for highly ionizing particles.

Semiconductor Detectors

Germanium was first demonstrated to be useful as a radiation detector in 1949 by McKay [14], but it was not until 1962 that Freck and Wakefield [15] made the first use of a germanium detector in gamma-ray spectrometry. Since that time progress has been rapid, mainly because of the greatly improved resolution obtainable with this detector over the scintillation detector. This advantage outweighed the main disadvantage of the high thermal noise at room temperature which requires that the detector be operated at low-temperature, generally that of liquid nitrogen. These detectors also undergo a change at room temperature in which the lithium precipitates so that even when not in use the detector must be kept cold to keep them from being damaged.

About 2.9 eV of energy are required to produce an electron-hole pair in a germanium detector. This is far less than the 500 eV required per photoelectron in NaI(Tl) scintillation detectors. Since the relative resolution is essentially proportional to the negative square root of the signal one can calculate that the resolution of the germanium detector should be on the order of a factor 13 better than the NaI(Tl) detector. This is about the difference actually observed by experiment.

Ge(Li) semiconductor detectors are essentially the solid state equivalents of gaseous ionization chambers. Figure 10.5 shows two of the common types, the one on the left being a parallel plate type and the one on the right being a cylindrical (coaxial) type. These detectors are made from single crystals of pure germanium of the *p*-type which has an excess of positive holes resulting from electron acceptor impurities. A thin layer of lithium metal is plated on one flat surface of the planar detector crystal or on the outer cylindrical surface of the coaxial detector creating a layer of lithium rich *n*-type material with an excess of electrons easily raised from a valence band to a conduction band. By means of an impressed high temperature and an electric field some positive lithium ions are caused to drift partially through the volume of the material compensating for the electron acceptor material and resulting in "intrinsic" germanium which acts like ultrapure germanium.

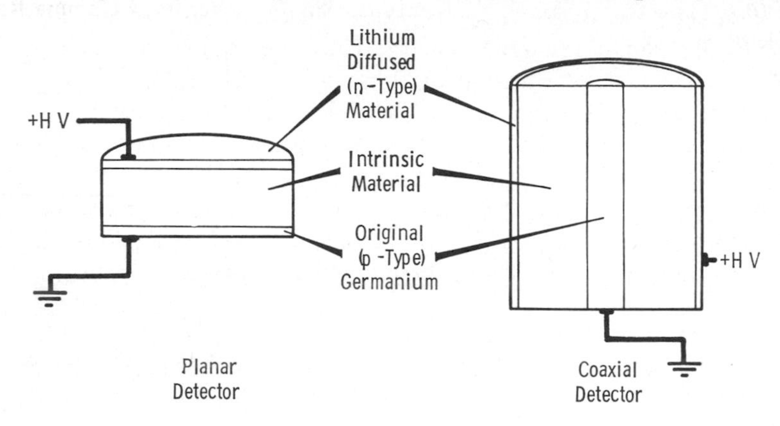

Fig. 10.5 Schematic diagrams of two common types of lithium-drifted germanium detectors.

The drifting process is discontinued while a layer of *p*-type germanium still remains.

During use as a detector an electric field is imposed with the positive electrode attached to the lithium layer and the negative electrode attached to the *p*-type germanium layer. This is called "reverse biasing" since its effect is to oppose the normal flow of electrons in the detector. Interactions by gamma rays cause ionization in the intrinsic material creating electron-hole pairs which separate and move to the electrodes under the influence of the impressed electric field. The resulting signal is amplified and routed to a multichannel analyzer for spectrum analysis. For additional detail with respect to the characteristics and use of Ge(Li) detectors see reviews by Hollander and Perlman [4] and Camp [16].

Efficiency, Background, and Resolution Considerations

Detector Efficiencies

In those cases where a high counting efficiency is required the NaI(Tl) scintillation detector is the detector of choice. Thallium-activated sodium iodide crystals are obtainable in diameters at least as large as 13 in. and in thicknesses over 10 in., while semiconductor detectors are presently obtainable only in sizes as large as about 80 cc. Table 10.1 gives the counting efficiencies of NaI(Tl) crystals of various sizes for complete absorption of gamma rays of energies representative of common counting requirements.

Table 10.1. Effect of Sodium Iodide Crystal Size on Percent of Gamma Rays Completely Absorbed

	Total Absorption Counting Efficiencies at Various Gamma-Ray Energies		
(in.)	0.662 MeV	1.38 MeV	2.76 MeV
Photoelectric process only*	10%	3%	1%
1.5 × 1	16%	—	3%
3 × 3	28%	17%	10%
8 × 8	38%	30%	26%
11 × 6	37%	29%	25%

* Peak-to-total value if no multiple processes occurred.

For comparison the efficiency of large (40 cc) Ge(Li) diodes for ^{137}Cs is less than 10% of that for a 3 × 3-in. NaI(Tl) crystal. The efficiencies given in the table are for a point source on the center of the end of the crystal (essentially a 2π geometry). If a duplicate crystal were placed on top of the sample or if a well crystal of similar dimensions to this double crystal were used the efficiencies would be double. Thus for the ^{137}Cs 0.662 gamma ray a 75% efficiency would be obtained with two 8 × 8-in. crystals and for the ^{24}Na 2.76 gamma rays a 52% efficiency would be obtained. With these large crystals (and for the energy range 0–3 MeV) the efficiencies are within a factor of 2 of 100% and although improvement could be gained by larger crystals, other factors, such as increased backgrounds, would partially nullify the advantage gained.

In order to assess the relative importance of counter efficiency versus background counting rate it is useful to examine their relationship to the statistical error of a given count as given in the equation

$$E = \frac{\sigma}{R} = \frac{\sqrt{B} + \sqrt{B + R}}{R\sqrt{T}} \tag{10.2}$$

where E = statistical error as defined by standard deviation, σ, divided by the sample counting rate, R

σ = standard deviation

R = sample counting rate

B = background counting rate

T = summed counting times of sample count plus background count

From this equation it can be seen that efficiency may be increased (statistical error decreased) for a given counting time by either decreasing the background rate, B, or increasing the sample counting rate, R. Decreasing the background rate will have a smaller proportionate effect because it appears only as the square root. In fact, if the sample and background rates are initially equal the equation indicates that doubling the sample counting rate is equivalent to reducing the background counting rate by an order of magnitude. Even for the case where the sample counting rate is much smaller than the background counting rate, doubling the sample counting rate is as beneficial as reducing the background counting rate by a factor of 4.

From statistical considerations the available counting time should be apportioned so that the background counting time is

$$T_B = \frac{1}{1 + \sqrt{(R + B)/B}} \tag{10.3}$$

It is interesting to note that the optimum background counting time is always less than the sample counting time. However, for practical convenience backgrounds are often taken for a time period equal to the sample counting time and stored in the analyzer in such a fashion as to result in a sample count with the background subtracted.

The equation used in the calculations above is also useful for comparing the relative efficiencies of two counting methods. The background under the 0.662 MeV peak for ^{137}Cs is about 20 cpm for a 3 × 3-in. crystal and 95 cpm for the 11 × 6-in. crystal. Use of the larger crystal would give an increase in efficiency of a factor of 1.32 (from 28 to 37%), but accompanied by a factor of 4.7 increase in background. The figures for the higher energy 2.76 MeV gamma rays are efficiency factor 2.6, background factor 12 (2.76 MeV bkg. is 1 cpm per 10 keV for the large crystal; for the 3 × 3-in. crystal the background is 0.085 cpm per 10 keV). If the equation above is applied to a ^{137}Cs sample with an activity of 100 dpm, the 3 × 3-in. crystal error value would be lower than the 11 × 6-in. crystal by about 40%. With an AC shield around the 11 × 6-in. crystal the background counting rate at the ^{137}Cs channels is 25 cpm rather than 95 cpm. Comparing this system with the non-AC shielded 3 × 3-in. crystal indicates that the larger crystal has an error figure lower by about 15% and is the more efficient detector.

Such a simple illustration does not tell the whole story, however. The larger detector has a greater relative efficiency for large samples and a special advantage when used to measure mixtures of radionuclides. For example, the point source counting efficiency for ^{60}Co on a 11 × 6-in. crystal is 31%. The efficiency for an extended source 10 in. in diameter is still a high 22%. For a 5-in. diameter crystal the point source efficiency is 17%. This has already fallen to 8% for a 5-in. source and would be much lower for a

10-in. diameter source. Thus for a 1-in.-thick source 10 in. in diameter on the large crystal compared to a 1-in.-thick source 5 in. in diameter you have a larger volume (and hence activity) by a factor of 4 and an increased efficiency factor of 2.5 for about an order of magnitude improvement (neglecting any background differences).

So far in the discussion we have considered the radionuclide being measured as being present alone except for background. This is seldom the case so we must consider the interference (generally from Compton interactions) that other, higher energy radionuclides introduce. Those gamma rays which interact with the detector and do not impart all of their energy in single or multiple collisions, result in the lower energy Compton continuum of the spectrum which is not identifiable to a specific radionuclide and hence is not useful for spectrometry. Furthermore, the measurement of a photopeak from another radionuclide in this area has that Compton portion as an additional background as can be seen in Fig. 10.6 taken from an early paper [17] on scintillation spectrometry. The Compton contribution from ^{95}Zr-^{95}Nb underlying the ^{103}Ru photopeak is 56% of the ^{95}Zr-^{95}Nb photopeak value and can be subtracted, but this operation reduces the accuracy and sensitivity obtainable. Thus it is desirable that the Compton contribution be as low as possible. For ^{137}Cs on a 3 × 3-in. crystal the Compton rate per channel is about 25% of the peak counting rate. For an 11 × 6-in. crystal it is about 8%. Since the events leading to counts in the Compton region result in a degraded gamma ray escaping the detector, the AC shield reduces the recorded Compton events whenever the AC shield is activated by the degraded gamma. For this reason the anticoincidence shields are sometimes called "Compton-canceling" shields, because this effect is more valuable in mixtures of radionuclides than the reduction of background also achieved. Anticoincidence shields have been used which reduce the Compton region of NaI(Tl) crystal spectra by as much as a factor of 7 for ^{137}Cs. Thus for the 11 × 6-in. crystal the ^{137}Cs Compton correction for photopeaks in the ^{137}C Compton continuum region would only be about 1% of the ^{137}Cs peak counting rate.

Resolution

Resolution is a measure of the ability of the detector to differentiate photopeaks of near identical energies. For NaI(Tl) detectors it is usually expressed as the percentage of the full width of the photopeak at one-half the maximum height (FWHM) of the energy of the ^{137}Cs (^{137}Ba) gamma ray. For example, a typical crystal would have a FWHM of about 45 keV and would be quoted as having a resolution of (45/661) × 100, or 7%. The resolution of Ge(Li) detectors is usually expressed as the FWHM in keV for several gamma rays of different energies. Camp [16] has suggested that the

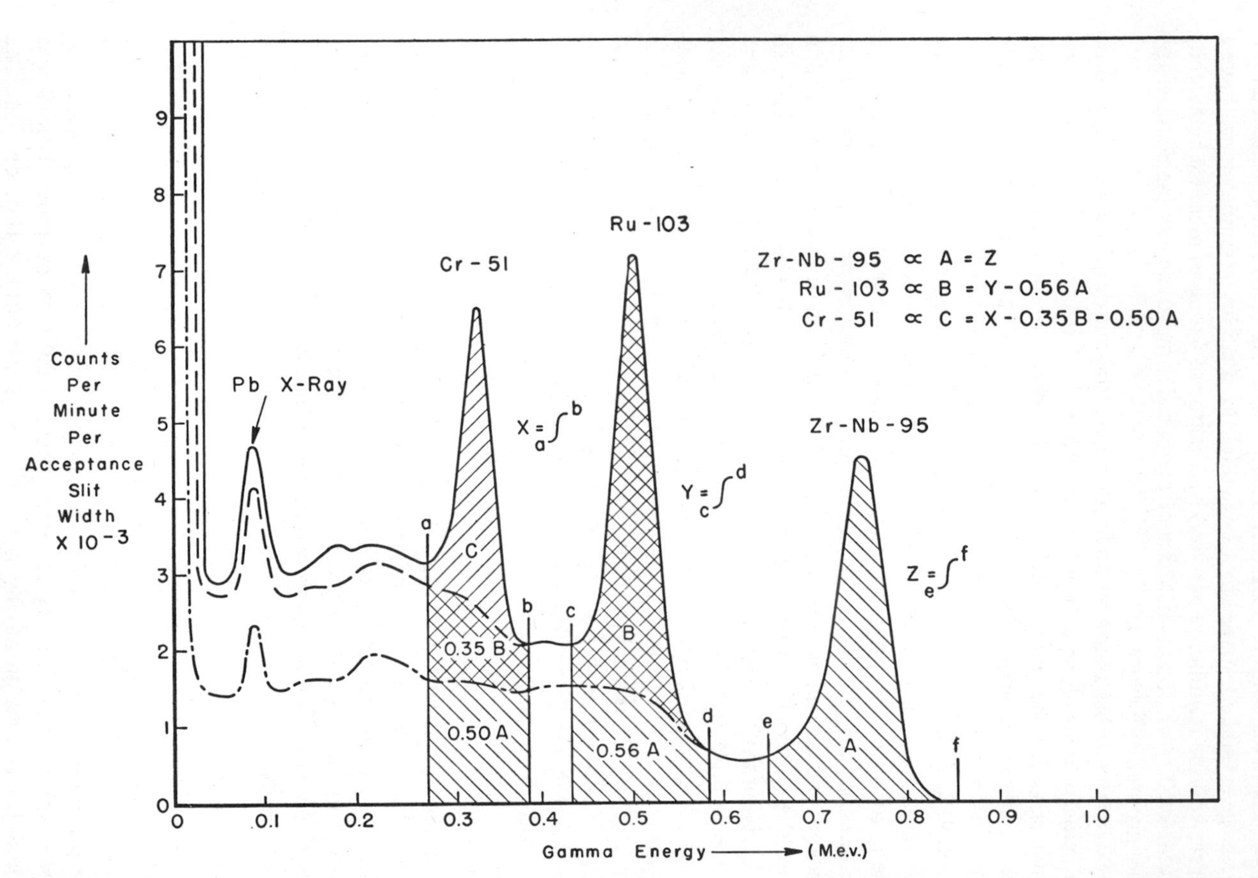

Fig. 10.6 A spectrum of three radionuclides showing the effect of the Compton continuum on the lower energy peaks and a matrix for determining the applicable corrections.

resolutions (FWHM) of the 123 keV line of ^{57}Co and the 1332 keV line of ^{60}Co have become standards to express the resolution performance of the Ge(Li) detector. As examples, for one detector he gives values of 1.115 keV FWHM at 123 keV and 2.36 keV FWHM at 1332 keV. Even better resolution was reported by Palms et al. [18] who obtained values of 450 eV at 14 keV and 1.7 keV at 1332 keV. Although these values may still not be the ultimate expected, they are factors of 10 to 40 better than the NaI(Tl) detector depending on the energy compared. This great difference in resolution can be seen graphically in Fig. 10.7 which compares spectra obtained with a 5 in. dia. × 6-in.-thick NaI(Tl) crystal and a 20 cc Ge(Li) diode on a neutron

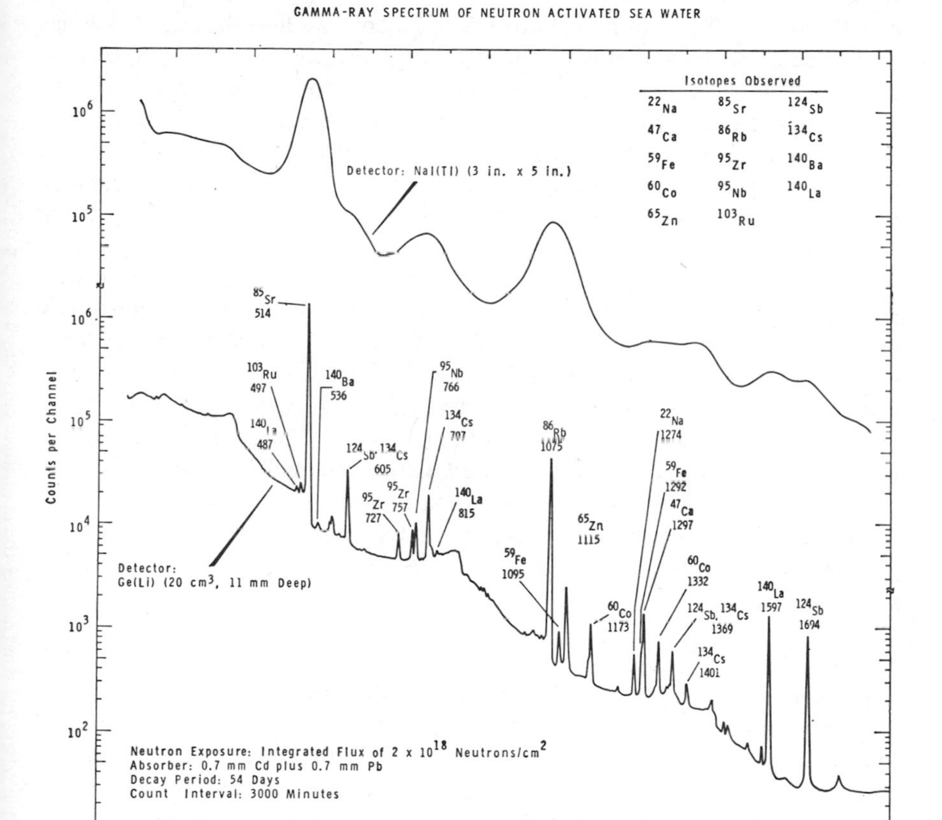

Fig. 10.7 Spectra of neutron activated seawater taken with NaI(Tl) and Ge(Li) detectors, showing selectivity obtained by greater resolution.

activated seawater salt sample [19]. It can be readily seen that the better resolution of the diode improves the ability to differentiate photopeaks of near identical energies. Not only does better resolution give greater selectivity but it also has an important effect on sensitivity. Other factors being equal if one detector had a better resolution than another, greater sensitivity would result. New Ge(Li) diodes have better resolutions and good ones can be obtained which have resolutions 19 times better for ^{137}Cs than a 7% resolution NaI(Tl) crystal. Such a crystal was described by Cooper et al. [19]. It was a 5-sided coaxial diode with an active volume of 20 cc and a resolution (FWHM) of 2.38 keV at the 662 keV ^{137}Cs line. The efficiency for the ^{137}Cs 662 keV line was 1% with a background for that section of the spectrum of 0.033 cpm. If we compare this with a 7% resolution 3 × 3-in. NaI(Tl) crystal which has an efficiency of 28% and a background of 20 cpm we find that the diode has an efficiency for ^{137}Cs of only about 4% that of the 3 × 3-in. NaI(Tl) crystal, but with a much lower background. The background is lower partly due to the smaller size and partly due to the improved resolution. With improved resolution the peaks are much narrower so that fewer channels need be taken to define the peak and since the background per channel is about the same in a given energy region, a lower background results. Comparison made using the error equation given above shows that the error for the diode count is only three times that for the NaI(Tl) crystal. Because of the much greater resolution of the diode it is still within about a factor of three in sensitivity even though it is a factor of 70 smaller and made of lower Z material. When the diode was surrounded by an AC shield the background was reduced to 0.008 cpm giving a sensitivity factor difference of 2.7. Although this difference is small a spectacular reduction was obtained in the Compton region to about 0.4% of the ^{137}Cs peak value. The best value obtained for a large NaI(Tl) crystal in an AC shield gives a Compton region of 1% of the peak value. Thus even this small diode has a special advantage in determining a radionuclide in a mixture of radionuclides because the Compton correction required is small.

Background and Shielding

Ionizing radiation produced directly or indirectly from cosmic radiation is the principal source of background with natural radioactivity from various sources next. Background reduction is generally obtained by one or both of two methods. Most of the soft component of the cosmic radiation and gamma rays from the natural radionuclides in the environment are absorbed by massive shields of higher atomic number metals. Shields are commonly 4 in. or more of lead, 6 in. or more of iron, several inches of mercury or combinations of these three materials. Inside a massive shield the charged particle component may be about 20%, the photon component about 80%, with the neutron component only a few percent. The effect of penetrating charged particles and photons of the cosmic flux can be reduced by a surrounding detector (or detectors) which inactivates the main detector whenever an event is recorded by this surrounding counter. This is called an anticoincidence (AC) shield. Cosmic rays at ground level generally consist of air showers with thousands to millions of components (neutrons, protons, μ-mesons, electrons, photons, etc.). Some of these components are delayed by having different speeds and different path lengths, but they all arrive within about a microsecond. This is the order of the time constant for counter electronics. Thus if any one of the particles or photons of the shower is detected, then the anticoincidence counter will remove all of the contribution of the shower.

Photons are the main uncharged particles in cosmic ray showers, but there are some neutrons present which produce interactions not removed by association with a time-coincident charged particle. These neutrons can be removed by capture after being thermalized. Paraffin or polyethylene is generally used for thermalizing, with boron added as an efficient capture nucleus. This process of thermalization and capture takes about 200 μsec, setting a new time constant during which the main detector must be inoperable following an event in the anticoincidence shield. Such a "dead time" restricts the use of this technique to the lower counting rates. The most effective neutron shield would be made from low Z material and there must be no high Z material between the neutron shield and the sensitive volume of the counter, since otherwise it would provide a site for cosmic ray component interaction and subsequent increase in background. Gamma-ray shielding is best done by high Z materials but this is incompatible with neutron shield requirements. The optimum solutions to the conflicting requirements of the neutron and the gamma shield are best resolved by experiment, although in most cases the gamma ray detection efficiency of counters approaches an order of magnitude greater than that for neutron detection so the main emphasis should be placed on selection of the gamma ray shield.

Shields have an adverse effect also on gamma ray spectra in the nature of backscattered radiation. To minimize this effect the shield is made as large as practical to remove the walls some distance from the detector [11]. X-Rays backscattered to the detector from interactions in the lead shield are an interference and can be reduced by lining the shield with several materials of successively lower Z to take advantage of critical absorption techniques. A typical graded shield would consist of a 30 mil Cd sheet with an inner 15 mil Cu sheet [11]. If an anticoincidence shield is used backscatter is effectively shielded against except for that originating in the source itself, or the materials of construction of the source holders and detectors themselves.

Massive high atomic number material shielding is especially necessary when an anticoincidence shield is used, since otherwise the anticoincidence shield counting rate would be so high as to result in excessive "dead time" of the main detector. Eight inches of lead are sufficient to remove virtually all the photon cascade, although in practice four inches of lead provide a nearly equivalent reduction.

Table 10.2 shows the reductions in background counting rates which are

Table 10.2. The Effect of Various Shields on the Background Counting Rate of a 3 × 3 in. NaI(Tl) Scintillation Detector [20]

	Background (cpm per 30 keV Channel)		
Shielding	0.5 MeV	1.0 MeV	2.0 MeV
None	325	76	5.8
4 in. Pb	4.0	1.7	0.36
4 in. Hg, AC Shield*	1.9	0.3	0.04
8 in. Pb, 300 ft concrete†	1.0	0.2	0.02
4 in. Pb, 4 in. Hg, 300 ft concrete, AC	0.16	0.05	0.01

* The anticoincidence (AC) shield was a 9-in. diameter NaI(Tl) detector with a center hole for the 3 × 3-in. primary detector.

† Counting room near base level inside Grand Coulee Dam, Washington.

possible for several energy regions by several types of shields. Although backgrounds are functions of material contamination, geographic latitude and altitude, as well as other factors, estimates can be made from this table which should prove useful in planning shielding requirements. Massive shielding of 4 in. of lead is almost standard and accomplishes a background reduction of factors of 80 to 16 in the energy region 0.5 to 2.0 MeV depending on the energy region. An increase in shielding of 4 in. of mercury, a 300 ft overburden of concrete and an anticoincidence shield gave a further decrease in background of 24 to 36 depending on the energy region. Further significant reduction will be more difficult to achieve and will require considerable effort to obtain uncontaminated materials of construction.

4 INSTRUMENTATION

Use of gamma-ray spectrometry is sufficiently great that commercial interests provide state-of-the-art instruments almost immediately following improved developments. Services are also generally available to modify and update the instrumentation so that continued high quality performance is possible for a laboratory without extensive instrument development effort. In addition to the detector, a gamma-ray spectrometer system requires a detector power supply, a preamplifier, an amplifier, a pulse height analyzer and storage unit, and a data readout device or devices.

Stable, noise-free high voltage power supplies are required capable of delivering 1000 to 2000 V to the photomultiplier for scintillation counting or to the Ge(Li) detector for semiconductor counting. A preamplifier is used to shape and prepare the pulse for transmission to the amplifier. This component is especially important for Ge(Li) detector counting and in order to obtain the best resolution the field-effect transistor (FET) used in the first stage of the preamplifier is operated at 140°K. In order to obtain this temperature the FET is placed near the detector which is cooled to liquid nitrogen temperature and a heat leak provided to keep the FET at 140°K. Camp reported [16] a 30 to 40% improvement in noise reduction over room temperature performance by operating this FET at 140°K.

Linear amplifiers using transistors have the best gain stability with respect to temperature variation. They can be obtained with stabilities such that a 10°C change in temperature produces less than a 0.01% change in gain. As with all of the instruments used for high resolution Ge(Li) spectrometry, the amplifier should be as noise-free and as stable as possible. Where extended time counts are to be made a gain stabilizer may be required. This device is a variable amplifier that eliminates the effects of small gain changes by electronically stabilizing the zero of the analog-to-digital converter (ADC) of the analyzer and maintaining the conversion gain constant. Using this device a long term stability may be obtained which allows a change of no more than one channel in 10,000 during a count.

In the pulse height analyzer the pulse delivered from the detector is converted to a train of pulses proportional to the pulse height. This part of the analyzer is called the analog-to-digital converter (ADC). The pulses in this train are counted by a scaler and the information stored in a ferrite core memory unit in a "channel" corresponding to the number of pulses in the chain, and hence to the energy of the absorbed gamma ray. The number of channels required to make full use of the data obtained is a function of the resolution of the detector. At least 5 channels per photopeak are desirable so that approximately 200 to 400 channels are commonly used for NaI(Tl) scintillation counting and 2000 to 4000 channels are used for Ge(Li) detector counting when the energy range to be covered is on the order of 0 to 3 MeV. For best operation the ADC should be linear to within about 0.01%. This performance is not available over the whole range of channels in present ADCs. The operation of digitizing a pulse requires approximately 100–300 μsec depending on the channel. Thus for a counting rate of 1000 counts/sec the analyzer is busy (dead) and cannot accept another pulse for 10 to 30% of the time. This is called the "dead time" and measurements are commonly made on "live time" which reports only the actual time for which the analyzer was capable of accepting pulses for analysis. The multichannel analyzer has the capability of subtracting as well as adding so that background spectra or other spectra such as a standard spectra can be subtracted by the analyzer.

A great variety of data readout equipment is available for use with the multichannel analyzers. Oscilloscopes, electric typewriters, digital printers, strip chart and *X-Y* recorders, and paper tape and magnetic tape recorders are commonly used. Because much of the advanced gamma ray spectrometry data are subjected to computer analysis, magnetic tape recording is desirable. A visual display, such as an *X-Y* plot, is also usually obtained to be used in the selection of computer programs, or in the preparation of reports.

5 GAMMA-RAY SPECTRA MEASUREMENTS

Calibration

Since neither scintillation nor semiconductor spectrometers are absolute instruments, they require energy and intensity calibration. Energy calibration is generally accomplished by establishing energy-channel relationships through the use of a series of radionuclides providing a set of calibration lines covering the regions of interest. Although there are not sufficient long-lived radionuclides for which precise gamma ray energies are known, Gunnink et al. [21] report some new measurements on seven radionuclides which provide a useful set of self-consistent and precise standard energies for calibrating gamma ray spectrometers. Brahmavar and Hamilton [22], give a set of eight radionuclides providing 27 calibration lines covering the energy range of 500 to 2600 keV. In each of these sets, one of the radionuclides is ^{24}Na, however, with the inconveniently short half-life of 15 hours. A review of standard gamma ray energies available in 1967 was published by Marion [23]. Because of nonlinearities calibration points closely spaced in energy are needed for high accuracy work. Determination of the nonlinearity of the system may be accomplished through use of a precision pulse generator.

Intensity calibration of a gamma-ray spectrometer requires the consideration of many factors. Through the use of radionuclides standardized by other methods, relative efficiency curves can be established for a given detector-configuration arrangement. Measurements can be made on radionuclides emitting gamma rays with energies in the calibration range and with the same source-detector configuration which have accuracies of the order of 1–5%. These accuracies will be somewhat reduced in mixtures of radionuclides where corrections must be made for the other radionuclides present. Examples of efficiency curves are given for the 3 × 3-in. NaI(Tl) detector by Heath [11] and for Ge(Li) detectors by Mowatt [24] and by Cooper [25]. For the best accuracy and for cases where the source is neither small nor weightless, calibration should be accomplished through the use of sources of the same size and material containing the standardized radionuclide. The presence of a thick source also alters the background due to shielding by the sample itself. This distortion can be corrected by using a blank material of the same shape and composition as the sample for the background measurement.

Analysis of Spectra

The simplest method of determining the counts in a given photopeak is to smooth the Compton continuum under the photopeak and subtract it from the Compton plus photopeak values. This method gives only an approximation when applied to NaI(Tl) spectra but gives surprisingly good results for Ge(Li) spectra since the peaks are so narrow that the Compton smoothing can be accomplished with good accuracy. Although some work has been done on computer analysis of Ge(Li) spectra, this Compton subtraction method is almost universally used. By relating the values obtained to the efficiencies determined by calibration with a standard source, a quantitative measure can be made of the radionuclide in the sample.

More rigorous methods used for NaI(Tl) spectra are the "spectrum stripping" and "matrix" methods. In the spectrum stripping method the contribution of one radionuclide at a time is subtracted starting at the highest energy photopeak which is assumed to represent only one radionuclide. This can even be accomplished visually using the oscilloscope display provided on most multichannel analyzers and using the standard spectra in the subtract mode. In the matrix method, those portions of a spectrum in areas where other radionuclides have their photopeaks are determined as a function of the photopeak for each given radionuclide. When done for each radionuclide in a mixture a set of simultaneous equations (a matrix) can be set up by which the count in the photopeak area for any of these radionuclides can be determined and corrections made to each photopeak for contributions from the other radionuclides present. This type of operation is indicated in Fig. 10.6.

When many radionuclides are present in the mixture more complex means are required to process the data because of the extensive overlapping of peaks in NaI(Tl) spectra. The problem is essentially one of curve fitting and this is generally accomplished by a computer using least squares analysis. The sample spectrum is compared with a library of spectra of known gamma emitters or to mathematical or empirical curves calculated for the detector. Reports by Nicholson, Schlosser and Brauer [26], Salmon [27], Heath [11], and Parr and Lucas [28] are representative of the many publications on this subject.

6 COINCIDENCE TECHNIQUES

In some samples, such as mixed fission or neutron activation products, a multitide of peaks are present, some masked by other peaks, or lost from view in the large Compton continuum. Although separation into groups or individual elements by chemical means is possible, many advantages are to be found in instrumental analysis. Several such methods are available using multiple detectors and coincidence techniques. More than half of the known gamma-emitting radionuclides decay with the emission of two or more gamma rays in cascade, or simultaneously within the time scale of gamma detector operation. In addition, all nuclides emitting gamma rays with energies in excess of 1.02 MeV produce electron-positron pairs in the detector with subsequent production of two 0.511 photons upon annihilation of the positron. These annihilation photons and the gamma ray producing them are also simultaneous within the time scale of gamma counting. By using two or more detectors, and requiring simultaneous (coincident) events for recording, a separation is accomplished into two groups, one containing cascade or pair producing gamma emitters and the other including only single or noncoincident gamma emitters. Hoogenboom used two detectors and recorded sum pulses which corresponded to the absorption of the full energies of the two cascade gamma rays [29]. He named this technique the sum coincidence method.

Another useful technique for simplifying gamma ray spectra is pair spectrometry. This technique was first used independently by Johansson [30] and Maienschein and Bair [31]. This is a three crystal triple-coincidence system wherein only events are recorded in the main crystal which are in coincidence with the detection of each of the annihilation photons in two lateral crystals. The main photopeak observed in the spectrum is the double escape peak. These peaks are recorded only for gamma rays with energies in excess of 1.022 MeV so that a degree of separation useful for certain mixtures is obtained.

A most useful general technique for improving selectivity and yet obtaining all of the information in a single count is the technique of multidimensional gamma ray spectrometry [32]. A schematic representation of this technique is shown in Fig. 10.8. The sample is placed between two crystals so that not only are single gamma ray events recorded but also coincident events in which both crystals simultaneously detect a gamma ray. If a 4096 channel analyzer is used in a 64 by 64 channel rectangular array, the single (noncoincident) events are stored on the X and Y axes depending on which crystal detected them, and the coincident events are stored on the plane of channels bounded by the X and Y axes and whose coordinates are determined by the energies of the events in the two detector crystals. This technique results in a great improvement in selectivity. A reduction in background is obtained in the areas used to determine a radionuclide decaying by cascade gamma emission which amounts to two to four orders of magnitude. A reduction in efficiency also occurs, but if large crystals and good sample geometries are used this reduction is usually less than one order of magnitude. Because of the added

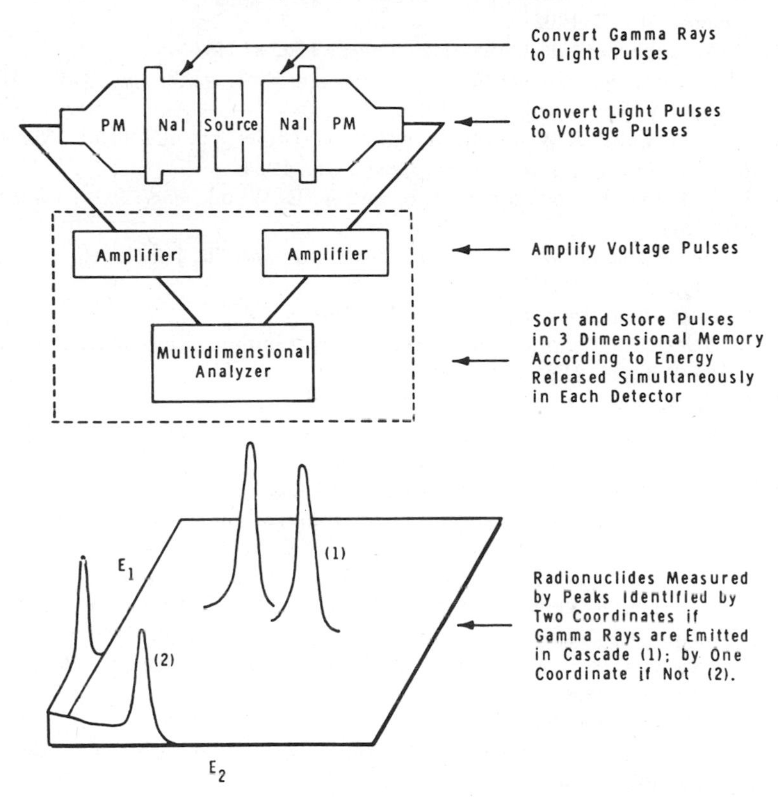

Fig. 10.8 Schematic diagram illustrating the principle of multidimensional gamma-ray spectrometry.

dimension the Compton events occur mainly in bands a few channels wide extending from each axis to each photopeak. Photopeaks from other radionuclides have a good chance of falling in areas little affected by these Compton areas.

Gamma-gamma coincidence techniques using two germanium diode detectors also have practical applications as is indicated by Thomas [33] and Cooper [34].

References

1. E. Rutherford and E. N. da C. Andrade, *Phil. Mag.* **27,** 854 (1914).
2. J. W. M. DuMond, *Rev. Sci. Instr.* **18,** 626 (1947).
3. E. J. Seppi, H. Henrickson, F. Boehm, and J. W. M. DuMond, *Nucl. Instr. Methods* **16,** 17 (1902).
4. J. M. Hollander and I. Perlman, *Science* **154,** 84 (1966).
5. *Alpha, Beta- and Gamma-Ray Spectrometry*, K. Siegbahn, ed., North-Holland Publishing Co., Amsterdam, 1965.
6. G. Friedlander, J. W. Kennedy, and J. M. Miller, pp. 109–110 in *Nuclear and Radiochemistry*, 2nd ed., Wiley, New York, 1964.
7. J. M. Palms, P. Venugopala Rao, and R. E. Wood, *Nucl. Instr. Methods* **64,** 310 (1968).
8. J. Ungrin and M. W. Johns, *Nucl. Instr. Methods* **70,** 112 (1969).
9. H. Kallmann, *Natur Technik*, July 1947.
10. R. Hofstadter, *Phys. Rev.* **74,** 100 (1948).
11. R. L. Heath, Scintillation Spectrometry Gamma Ray Spectrum Catalog, USAEC Report IDO-16408 (1964).
12. D. Engelkemeir, *Rev. Sci. Instr.* **27** 589, (1950).
13. R. L. Heath, *IRE Trans. Nucl. Sci. NS-9* [3] 294, (1962).
14. K. G. McKay, *Phys. Rev.* **76,** 1537 (1949).
15. D. V. Freck and J. Wakefield, *Nature* **193,** 669 (1962).
16. D. C. Camp, "Applications and Optimization of the Lithium-Drifted Germanium Detector System," UCRL-50156, March 1, 1967.
17. R. E. Connally, *Anal. Chem.* **28,** 1847–1853 (1956).
18. J. M. Palms, P. Venugopala Rao, and R. E. Wood, *Nucl. Instr. Methods* **64,** 310–316 (1968).
19. J. A. Cooper, N. A. Wogman, H. E. Palmer, and R. W. Perkins, *Health Physics* **15,** 419–433 (1968).
20. J. H. Kaye, unpublished data.
21. R. Gunnink, R. A. Meyer, J. B. Niday, and R. P. Anderson, *Nucl. Instr. Methods* **65,** 26–34 (1968).
22. S. M. Brahmavar and J. H. Hamilton, *Nucl. Instr. Methods* **69,** 353 (1969).
23. J. B. Marion, University of Maryland Technical Report 656 (Rev) (ORO-2098-58), August 1967.
24. R. S. Mowatt, *Nucl. Instr. Methods* **70,** 237–244 (1969).
25. J. A. Cooper, USAEC Report BNWL-1051, Part 2, p. 120, June, 1969.
26. W. L. Nicholson, J. E. Schlosser, and F. P. Brauer, USAEC Report HW-75806, 1962.
27. L. Salmon, Report NAS-NS-3107 165–183, 1963.
28. R. M. Parr and H. P. Lucas, Jr., IEEE Transactions on Nuclear Science NS-11, No. 3, 349–357, June 1964.
29. A. M. Hoogenboom, *Nucl. Instr.* **3,** 57–68 (1958).

30. S. A. E. Johansson, *Nature* **166**, 794 (1950).
31. F. C. Maienschein and J. K. Bair, *Phys. Rev.* **82A**, 317 (1951).
32. R. W. Perkins, *Nucl. Instr. Methods* **33**, 71–76 (1965).
33. C. W. Thomas, USAEC Report, BNWL-481-2, p. 52, December 1967.
34. J. A. Cooper, USAEC Report, BNWL-SA-3575, October 1970.

INDEX